SURFACTANTS AND INTERFACIAL PHENOMENA

SURFACTANTS AND INTERFACIAL PHENOMENA

FOURTH EDITION

Milton J. Rosen
Joy T. Kunjappu

A JOHN WILEY & SONS, INC., PUBLICATION

Published by John Wiley & Sons, Inc., Hoboken, New Jersey
Published simultaneously in Canada

Library of Congress Cataloging-in-Publication Data:
Rosen, Milton J.
Surfactants and interfacial phenomena / Milton J. Rosen and Joy T. Kunjappu. – 4th ed.
p. cm.
Includes index.
ISBN 978-0-470-54194-4 (hardback)
1. Surface active agents. 2. Surface chemistry. I. Kunjappu, Joy T. II. Title.
TP994.R67 2012
668'.1–dc23

2011046735

Printed in the United States of America

10 9 8 7 6 5 4 3 2 1

CONTENTS

PREFACE

Ever since surfactant science detached its umbilical cord from the body of colloid science and established its unique identity as an independent entity in the 1950s, great leaps in its theoretical and applied fronts have occurred. The real thrust of surfactant science and technology is centered on applications, although its importance in understanding frontline areas like the origin of life and soft matter (a subfield of condensed matter studies) technology is built upon the self-organizing power of surfactants into structures such as micelles or bilayers.

This book, when originally conceived in the 1970s, forecast the impending revolution that surfactant science was to witness in the future, and was developed to bridge the gap between fundamental knowledge and industrial applications. Later editions of the book incorporated advances in theory with a special link to end uses.

The importance of surfactants continues to emerge, as evidenced by the use of polymeric surfactants during the Gulf oil spill of 2010 to disperse the floating oil film in the ocean, in processing materials such as the silicon chip, and in the still emerging areas of *in vivo* biotechnology and *in vitro* nanotechnology.

The present edition has been updated to embrace these cutting-edge and state-of-the-art topics in surfactant application by the addition of three new chapters: Chapter 13, "Surfactants in Biology"; Chapter 14, "Surfactants in Nanotechnology"; and Chapter 15, "Surfactants and Molecular Modeling."

Most of the previously existing chapters have been revised with some new materials in the form of expanded, rewritten or new sections, and/or additional references and problems. Specifically, the following new sections (in **bold**) are added or existing ones revised (in *italics*): *environmental effects of surfactants*, **electronic searching of surfactant literature**, *zwitterioncs* (Chapter 1); *mechanism of adsorption and aggregation* (Chapter 2); **rheology of surfactant solutions** (Chapter 3); *solubilization* (Chapter 4); *accurate depiction of equations in film elasticity*, **foaming and antifoaming in organic media** (Chapter 7); *microemulsions*, *demulsification* (Chapter 8); *limitations of the DLVO theory*, **design**

of new dispersants (Chapter 9); **biosurfactants and enzymes in detergent formulations** (Chapter 10); and *problems* (Chapters 1, 2, 5–10, and 12).

ACKNOWLEDGMENTS

We thank a group of Brooklyn College students: Rameez Shoukat, Danielle Nadav, Meryem Choudhry, Ariana Gluck, Alex John, Abdelrahim Abdel, Khubaib Gondal, and Yara Adam, whose participation in a workshop conducted by one of the authors (JTK) on "Science Writing" helped accelerate the pace of writing of this edition. Thanks are also due to Drs. Viraht Sahni (Brooklyn College), Teresa Antony (New York City College of Technology) and Richard Magliazzo (Brooklyn College), all from the City University of New York, and Dr. John F. Scamerhorn of the University of Oklahoma for either providing useful leads in developing the topics or for commenting on sections of the manuscript.

We would also like to thank Ms. Anita Lekhwani and Ms. Catherine Odal at John Wiley & Sons and Ms. Stephanie Sakson at Toppan Best-set Premedia for offering their editorial skills and help from acquisition to production of the book.

Milton J. Rosen
Joy T. Kunjappu
(ProfessorKunjappu@gmail.com)

1 Characteristic Features of Surfactants

Surfactants are among the most versatile products of the chemical industry, appearing in such diverse products as the motor oils we use in our automobiles, the pharmaceuticals we take when we are ill, the detergents we use in cleaning our laundry and our homes, the drilling muds used in prospecting for petroleum, and the flotation agents used in beneficiation of ores. The last decades have seen the extension of surfactant applications to such high-technology areas as electronic printing, magnetic recording, biotechnology, microelectronics, and viral research.

A surfactant (a contraction of the term ***surf**ace-**act**ive **agent***) is a substance that, when present at low concentration in a system, has the property of adsorbing onto the surfaces or interfaces of the system and of altering to a marked degree the surface or interfacial free energies of those surfaces (or interfaces). The term *interface* indicates a boundary between any two immiscible phases; the term *surface* denotes an interface where one phase is a gas, usually air.

The interfacial free energy is the minimum amount of work required to create that interface. The interfacial free energy per unit area is what we measure when we determine the interfacial tension between two phases. It is the minimum amount of work required to create unit area of the interface or to expand it by unit area. The interfacial (or surface) tension is also a measure of the difference in nature of the two phases meeting at the interface (or surface). The greater the dissimilarity in their natures, the greater the interfacial (or surface) tension between them.

When we measure the surface tension of a liquid, we are measuring the interfacial free energy per unit area of the boundary between the liquid and the air above it. When we expand an interface, therefore, the minimum work required to create the additional amount of that interface is the product of the interfacial tension γ_I and the increase in area of the interface; $W_{min} = \gamma_I \times \Delta_{\text{interfacial area}}$. A surfactant is therefore a substance that at low concentrations adsorbs at some or all of the interfaces in the system and significantly changes the amount of work required to expand those interfaces.

Surfactants and Interfacial Phenomena, Fourth Edition. Milton J. Rosen and Joy T. Kunjappu.

Surfactants usually act to reduce interfacial free energy rather than to increase it, although there are occasions when they are used to increase it.

The questions that immediately arise are the following: Under what conditions can surfactants play a significant role in a process? How does one know when to expect surfactants to be a significant factor in some system under investigation? How and why do they work as they do?

I. CONDITIONS UNDER WHICH INTERFACIAL PHENOMENA AND SURFACTANTS BECOME SIGNIFICANT

The physical, chemical, and electrical properties of matter confined to phase boundaries are often profoundly different from those of the same matter in bulk. For many systems, even those containing a number of phases, the fraction of the total mass that is localized at phase boundaries (interfaces, surfaces) is so small that the contribution of these "abnormal" properties to the general properties and behavior of the system is negligible. There are, however, many important circumstances under which these "different" properties play a significant, if not a major, role.

One such circumstance is when the phase boundary area is so large relative to the volume of the system that a substantial fraction of the total mass of the system is present at boundaries (e.g., in emulsions, foams, and dispersions of solids). In this circumstance, surfactants can always be expected to play a major role in the system.

Another such circumstance is when the phenomena occurring at phase boundaries are so unusual relative to the expected bulk phase interactions that the entire behavior of the system is determined by interfacial processes (e.g., heterogeneous catalysis, corrosion, detergency, or flotation). In this circumstance also, surfactants can play an important role in the process. It is obviously necessary to understand the causes of this abnormal behavior of matter at the interfaces and the variables that affect this behavior in order to predict and control the properties of these systems.

II. GENERAL STRUCTURAL FEATURES AND BEHAVIOR OF SURFACTANTS

The molecules at a surface have higher potential energies than those in the interior. This is because they interact more strongly with the molecules in the interior of the substance than they do with the widely spaced gas molecules above it. Work is therefore required to bring a molecule from the interior to the surface.

Surfactants have a characteristic molecular structure consisting of a structural group that has very little attraction for the solvent, known as a *lyophobic group*, together with a group that has strong attraction for the solvent, called the *lyophilic group*. This is known as an *amphipathic* structure. When a molecule with an amphipathic structure is dissolved in a solvent, the lyophobic group may distort the structure of the solvent, increasing the free energy of the system. When that occurs, the system responds in some fashion in order to minimize contact between the lyophobic group and the solvent. In the case of a surfactant dissolved in aqueous medium, the lyophobic (hydrophobic) group distorts the structure of the water (by breaking hydrogen bonds between the water molecules and by structuring the water in the vicinity of the hydrophobic group). As a result of this distortion, some of the surfactant molecules are expelled to the interfaces of the system, with their hydrophobic groups oriented so as to minimize contact with the water molecules. The surface of the water becomes covered with a single layer of surfactant molecules with their hydrophobic groups oriented predominantly toward the air. Since air molecules are essentially nonpolar in nature, as are the hydrophobic groups, this decrease in the dissimilarity of the two phases contacting each other at the surface results in a decrease in the surface tension of the water. On the other hand, the presence of the lyophilic (hydrophilic) group prevents the surfactant from being expelled completely from the solvent as a separate phase, since that would require dehydration of the hydrophilic group. The amphipathic structure of the surfactant therefore causes not only concentration of the surfactant at the surface and reduction of the surface tension of the water, but also orientation of the molecule at the surface with its hydrophilic group in the aqueous phase and its hydrophobic group oriented away from it.

The chemical structures of groupings suitable as the lyophobic and lyophilic portions of the surfactant molecule vary with the nature of the solvent and the conditions of use. In a highly polar solvent such as water, the lyophobic group may be a hydrocarbon or fluorocarbon or siloxane chain of proper length, whereas in a less polar solvent, only some of these may be suitable (e.g., fluorocarbon or siloxane chains in polypropylene glycol). In a polar solvent such as water, ionic or highly polar groups may act as lyophilic groups, whereas in a nonpolar solvent such as heptane, they may act as lyophobic groups. As the temperature and use conditions (e.g., presence of electrolyte or organic additives) vary, modifications in the structure of the lyophobic and lyophilic groups may become necessary to maintain surface activity at a suitable level. Thus, for surface activity in a particular system, the surfactant molecule must have a chemical structure that is amphipathic *in that solvent under the conditions of use*.

The hydrophobic group is usually a long-chain hydrocarbon residue, and less often a halogenated or oxygenated hydrocarbon or siloxane chain; the hydrophilic group is an ionic or highly polar group. Depending on the nature of the hydrophilic group, surfactants are classified as

1. *Anionic*. The surface-active portion of the molecule bears a negative charge, for example, $RCOO^-Na^+$ (soap), $RC_6H_4SO_3^-Na^+$ (alkylbenzene sulfonate).

2. *Cationic.* The surface-active portion bears a positive charge, for example, $RNH_3^+Cl^-$ (salt of a long-chain amine), $RN(CH_3)_3^+Cl^-$ (quaternary ammonium chloride).
3. *Zwitterionic.* Both positive and negative charges may be present in the surface-active portion, for example, $RN^+H_2CH_2COO^-$ (long-chain amino acid), $RN^+(CH_3)_2CH_2CH_2SO_3^-$ (sulfobetaine).
4. *Nonionic.* The surface-active portion bears no apparent ionic charge, for example, $RCOOCH_2CHOHCH_2OH$ (monoglyceride of long-chain fatty acid), $RC_6H_4(OC_2H_4)_xOH$ (polyoxyethylenated alkylphenol), $R(OC_2H_4)_x$ OH(polyoxyethylenated alcohol).

A. General Use of Charge Types

Most natural surfaces are negatively charged. Therefore, if the surface is to be made hydrophobic (water-repellent) by use of a surfactant, then the best type of surfactant to use is a cationic. This typc of surfactant will adsorb onto the surface with its positively charged hydrophilic head group oriented toward the negatively charged surface (because of electrostatic attraction) and its hydrophobic group oriented away from the surface, making the surface water-repellent. On the other hand, if the surface is to be made hydrophilic (water-wettable), then cationic surfactants should be avoided. If the surface should happen to be positively charged, however, then anionics will make it hydrophobic and should be avoided if the surface is to be made hydrophilic.

Nonionics adsorb onto surfaces with either the hydrophilic or the hydrophobic group oriented toward the surface, depending upon the nature of the surface. If polar groups capable of H bonding with the hydrophilic group of the surfactant are present on the surface, then the surfactant will probably be adsorbed with its hydrophilic group oriented toward the surface, making the surface more hydrophobic; if such groups are absent from the surface, then the surfactant will probably be oriented with its hydrophobic group toward the surface, making it more hydrophilic.

Zwitterionics, since they carry both positive and negative charges, can adsorb onto both negatively charged and positively charged surfaces without changing the charge of the surface significantly. On the other hand, the adsorption of a cationic onto a negatively charged surface reduces the charge on the surface and may even reverse it to a positive charge (if sufficient cationic is adsorbed). In similar fashion, the adsorption of an anionic surfactant onto a positively charged surface reduces its charge and may reverse it to a negative charge. The adsorption of a nonionic onto a surface generally does not affect its charge significantly, although the effective charge density may be reduced if the adsorbed layer is thick.

Differences in the nature of the hydrophobic groups are usually less pronounced than those in the nature of the hydrophilic group. Generally, they are long-chain hydrocarbon residues. However, they include such different structures as

1. Straight-chain, long alkyl groups (C_8–C_{20})
2. Branched-chain, long alkyl groups (C_8–C_{20})
3. Long-chain (C_8–C_{15}) alkylbenzene residues
4. Alkylnaphthalene residues (C_3 and greater-length alkyl groups)
5. Rosin derivatives (rosin is obtained from plant resins)
6. High-molecular-weight propylene oxide polymers (polyoxypropylene glycol derivatives)
7. Long-chain perfluoroalkyl groups
8. Polysiloxane groups
9. Lignin derivatives

B. General Effects of the Nature of the Hydrophobic Group

1. Length of the Hydrophobic Group Increase in the length of the hydrophobic group (1) decreases the solubility of the surfactant in water and increases its solubility in organic solvents, (2) causes closer packing of the surfactant molecules at the interface (provided that the area occupied by the hydrophilic group at the interface permits it), (3) increases the tendency of the surfactant to adsorb at an interface or to form aggregates, called *micelles,* (4) increases the melting point of the surfactant and of the adsorbed film and the tendency to form liquid crystal phases in the solution, and (5) increases the sensitivity of the surfactant, if it is ionic, to precipitation from water by counterions.

2. Branching, Unsaturation The introduction of branching or unsaturation into the hydrophobic group (1) increases the solubility of the surfactant in water or in organic solvents (compared to the straight-chain, saturated isomer), (2) decreases the melting point of the surfactant and of the adsorbed film, (3) causes looser packing of the surfactant molecules at the interface (the cis isomer is particularly loosely packed; the trans isomer is packed almost as closely as the saturated isomer) and inhibits liquid crystal phase formation in solution, (4) may cause oxidation and color formation in unsaturated compounds, (5) may decrease biodegradability in branched-chain compounds, and (6) may increase thermal instability.

3. Aromatic Nucleus The presence of an aromatic nucleus in the hydrophobic group may (1) increase the adsorption of the surfactant onto polar surfaces, (2) decrease its biodegradability, and (3) cause looser packing of the surfactant molecules at the interface. Cycloaliphatic nuclei, such as those in rosin derivatives, are even more loosely packed.

4. Polyoxypropylene or Polyoxyethylene (POE) Units Polyoxypropylene units increase the hydrophobic nature of the surfactant, its adsorption onto polar surfaces, and its solubility in organic solvents. POE units decrease the hydrophobic character or increase the hydrophilicity of the surfactant.

5. Perfluoroalkyl or Polysiloxane Group The presence of either of these groups as the hydrophobic group in the surfactant permits reduction of the surface tension of water to lower values than those attainable with a hydrocarbon-based hydrophobic group. Perfluoroalkyl surfaces are both water- and hydrocarbon-repellent.

With such a variety of available structures, how does one choose the proper surfactant for a particular purpose? Alternatively, why are only certain surfactants used for a particular purpose and not other surfactants? Economic factors are often of major importance—unless the cost of using the surfactant is trivial compared to other costs, one usually chooses the most inexpensive surfactant that will do the job. In addition, such considerations as environmental effects (biodegradability, toxicity to and bioconcentration in aquatic organisms; Section IIIA) and, for personal care products, skin irritation (Section IIIB) are important considerations. The selection of the best surfactants or combination of surfactants for a particular purpose in a rational manner, without resorting to time-consuming and expensive trial-and-error experimentation, requires a knowledge of (1) the characteristic features of currently available surfactants (general physical and chemical properties and uses), (2) the interfacial phenomena involved in the job to be done and the role of the surfactant in these phenomena, (3) the surface chemical properties of various structural types of surfactants and the relation of the structure of a surfactant to its behavior in various interfacial phenomena. The following chapters attempt to cover these areas.

III. ENVIRONMENTAL EFFECTS OF SURFACTANTS

A. Surfactant Biodegradability

Surfactants are "performance" chemicals; that is, they are used to perform a particular function in some process or product, in contrast to other organic chemicals that may be used to produce another chemical or product. Since they are used in products or processes that impact on the environment, there are concerns regarding their effect, particularly their biodegradability in the environment and their toxicity, and of their biodegradation products to marine organisms and human beings.

Of late, these concerns in the public mind have become so serious that, to many people, the term "chemical" has become synonymous with "toxic chemical."* As a result, many manufacturers and users of chemicals, including

* One of the authors actually heard a college student, upon seeing "organic chemistry" written on a door, exclaim, "if it is organic, how can it be chemistry?"

surfactants, have paid serious attention to the biodegradability and toxicity of surfactants. In addition, they have sought new surfactants based upon renewable resources, so-called "green" surfactants (Section IVE, below).

An excellent review of surfactant biodegradability (Swisher, 1987) points out that biodegradability increases with increased linearity of the hydrophobic group and is reduced, for isomeric materials, by branching in that group, particularly by terminal quaternary carbon branching. A single methyl branch in the hydrophobic group does not change the biodegradation rate, but additional ones do.

In isomeric alkylbenzene and alkylphenol derivatives, degradation decreases as the phenyl group is moved from a position near the terminal end of a linear alkyl group to a more central position.

In POE nonionics, biodegradation is retarded by an increase in the number of oxyethylene groups. The inclusion of oxypropylene or oxybutylene groups in the molecule tends to retard biodegradation. Secondary ethoxylates degrade more slowly than primary ethoxylates even when both have linear hydrophobic groups.

In cationic quaternary ammonium surfactants, compounds with one linear alkyl chain attached to the nitrogen degrade faster than those with two, and these degrade faster than those with three. The replacement of a methyl group attached to the nitrogen by a benzyl group retards the rate of degradation slightly. Pyridinium compounds biodegrade significantly more slowly than the corresponding trimethylammonium compounds, while imidazolinium compounds biodegrade rapidly. Carboxylic acids have been identified as the metabolic end products of linear alcohol ethoxylates (AEs) and alkyl aryl sulfonates.

B. Surfactant Toxicity; Skin Irritation

Since surfactants are used in many products and formulations, such as cleaning solutions, cutting fluids, inks, and paints (Kunjappu, 2001), their skin irritability is important, and they can end up in aquifers and other water sources. $\mathbf{LD_{50}}$ (the *median lethal dose*—the dose required to kill half the members of a tested population) and $\mathbf{IC_{50}}$ (the *half maximal inhibitory concentration*—a measure of the effectiveness of a compound in inhibiting biological or biochemical function) data are used to represent the toxicity.

The toxicity of surfactants to marine organisms and their concentration in them depends upon their tendency to adsorb onto them and their ability to penetrate their cell membranes (Rosen et al., 1999). The parameter $\Delta G^{\circ}_{ad}/a^{s}_{m}$, where ΔG°_{ad} is the standard free energy of adsorption of the surfactant at the aqueous solution–air interface (Chapter 2, Section IIIF) and a^{s}_{m} is the minimum cross-sectional area of the surfactant at that interface (Chapter 2, Section IIIB), was found to correlate well for several anionic and nonionic surfactants with rotifer toxicity. The same parameter was found to correlate well for a series of cationic surfactants with rotifer and green algae toxicity and, for a series of linear alkylbenzenesulfonates (LASs), with bioconcentration in fish (Rosen et al., 2001).

Thus, toxicity increases with an increase in the length of the hydrophobic group and, for isomeric materials, decreases with branching or movement of the phenyl group to a more central position in the linear alkyl chain; in linear POE alcohols, toxicity increases with decrease in the number of oxyethylene units in the molecule, all due to the expected changes in the values of both ΔG°_{ad} and of a^{s}_{m}. Consequently, from the data in this section and in Section IVA above, it appears that some chemical structures in the surfactant molecule that promote biodegradability (such as increased length and linearity of the hydrophobic group or decreased oxyethylene content) increase its toxicity to or bioconcentration in marine organisms.

Cationic surfactants are found to be more toxic than anionics, and the anionics are more toxic than the nonionics. Although anionic surfactants are more irritable to skin than nonionics, sodium dodecyl sulfate (SDS) is used in many personal care products. Sodium alkyl ether sulfates are much milder than alkyl sulfates, and are used in many hand dishwashing formulations. The widely distributed, negatively charged groups in lipids, proteins, and nucleic acids may be responsible for the higher toxicity of ionic surfactants because of possible electrostatic interaction, which may explain the acute toxicity and genotoxicity of some of these surfactants.

Even in small doses, some surfactants produce dermatological problems. EC_{50}, *half maximal effective concentration*, refers to the concentration of a drug, antibody, or toxicant that induces a response halfway between the baseline and maximum after some specified exposure time (for SDS, the $EC_{50} = 0.071\%$ w/v for the human epidermis (Cannon et al., 1994). Polyol surfactants like alkyl glucosides, and zwitterionics like betaines and amidobetaines are known to be mild toward skin. The biocidal effects are studied by the effect on mucous membrane and on the bacterial surface. Biological toxicity has also been evaluated from the partition of the surfactant between oil and water (Salager et al., 2000).

IV. CHARACTERISTIC FEATURES AND USES OF COMMERCIALLY AVAILABLE SURFACTANTS

Surfactants are major industrial products with millions of metric tons produced annually throughout the world. Table 1.1 lists surfactant consumption in the United States and Canada for the year 2000. Table 1.1 shows consumption of the various surfactant charge types by percentage (A) and the consumption of the five major types of surfactant by tonnage (B). The projected average increase in surfactant consumption is 2.4% annually, although exact updated numbers are not available at this point (see table source line).

TABLE 1.1 Surfactant Consumption—United States and Canada (Excluding Soap), 2000

A. Surfactant, by Charge Type	
Type	%
Anionics	59
Cationics	10
Nonionics	24
Zwitterionics and others	7
Total	100
B. Major Surfactants, by Tonnage	
Surfactant	Thousand Metric Tons
Linear alkylbenzenesulfonates	420
Alcohol ethoxysulfates	380
Alcohol sulfates	140
Alcohol ethoxylates	275
Alkylphenol ethoxylates	225
Other	1625
Total	3065

Source: Colin A. Houston and Associates, Inc.

A. Anionics

1. Carboxylic Acid Salts

Sodium and Potassium Salts of Straight-Chain Fatty Acids, $RCOO^-M^+$ (Soaps)

PROPERTIES Below 10 carbons, too soluble for surface activity; above 20 carbons (straight chain), too insoluble for use in aqueous medium but usable for nonaqueous systems (e.g., detergents in lubricating oils or dry-cleaning solvents).

ADVANTAGES Easily prepared by neutralization of free fatty acids or saponification of triglycerides in simple equipment. Can be made in situ (e.g., for use as an emulsifying agent) (1) by adding fatty acid to oil phase and alkaline material to aqueous phase or (2) by partial saponification of triglyceride oil. Excellent physical properties for use in toilet soap bars.

DISADVANTAGES (1) Form water-insoluble soaps with divalent and trivalent metallic ions; (2) insolubilized readily by electrolytes, such as NaCl; (3) unstable at pH below 7, yielding water-insoluble free fatty acid.

MAJOR TYPES AND THEIR USES Sodium salts of tallow (animal fat) acids. (Tallow acids are oleic, 40–45%; palmitic, 25–30%; stearic, 15–20%.) Used in toilet soap bars and for degumming of silk, where alkaline solution is required. For industrial use in hard water, lime soap dispersing agents (sulfonates and sulfates) are added to prevent precipitation of insoluble lime soaps.

Sodium and Potassium Salts of Coconut Oil Fatty Acids (Coconut fatty acids are C_{12}, 45–50%; C_{14}, 16–20%; C_{16}, 8–10%; oleic, 5–6%; $<C_{12}$, 10–15%). Used as electrolyte-resistant soaps (seawater washing) and in liquid soaps, especially as the potassium soaps.

Sodium and Potassium Salts of Tall Oil Acids (Tall oil, a by-product of paper manufacture, is a mixture of fatty acids and rosin acids from wood; 50–70% fatty acid, mainly oleic and linoleic, 30–50% rosin acids related to abietic acid, the main constituent of rosin.) Mainly "captive" use or in situ preparation for various industrial cleaning operations. Used as foaming agents for concrete.

ADVANTAGES Inexpensive. More water-soluble and hard-water-resistant than tallow soaps. Lower viscosity solutions than tallow soaps at high concentrations, better wetting.

Soaps of synthetic long-chain fatty acids are produced in Europe but not in the United States at present.

Amine Salts Triethanolamine salts are used in nonaqueous solvents and in situ preparation as an emulsifying agent (free fatty acid in oil phase, triethanolamine in aqueous phase). Ammonia, morpholine, and other volatile amine salts are used in polishes, where evaporation of the amine following hydrolysis of the salt leaves only water-resistant material in film.

Other Types

ACYLATED AMINOACIDS (See Section IVE).

Acylated Polypeptides (From partially hydrolyzed protein from scrap leather and other waste protein.) Used in hair preparations and shampoos, alkaline cleaning preparations, wax strippers. Good detergency and resistance to hard water.

ADVANTAGES Soluble in concentrated aqueous solutions of alkaline salts. Nonirritating to skin; reduces skin irritation produced by other surfactants (e.g., SDS). Substantive to hair. Imparts soft "hand" to textiles.

DISADVANTAGES Precipitated by high concentrations of Ca^{2+} or Mg^{2+}, acids (below pH 5). Lower foaming than lauryl sulfates. Requires foam booster (e.g., alkanolamides) when foaming is important.

Polyoxyethylenated Fatty Alcohol Carboxylates (Alkyl Ether Carboxylates), $RO(CH_2CH_2O)_xCH_2COO^-M^+$ (x = 4, Usually) Products of the reaction of the terminal OH group of an AE with sodium monochloroacetate. Less basic than soaps of comparable chain length, ascribed to the ether oxygen atom adjacent to the carboxylate group in the molecule.

USES Hair care and skin care detergents, for the product based on C_{12-14} alcohol with low EO content. Emulsifying agent, solubilizing agent, dispersion agent. Textile and metal detergent. Industrial detergent for products having a short alkyl chain (C_{4-8}) because of low foaming power.

ADVANTAGES Low skin irritancy. Good resistance to hard water. Good stability in alkaline medium.

2. *Sulfonic Acid Salts*

LAS, $RC_6H_4SO_3^-M^+$ Three processes for the production of alkylbenzenes *(alkylate)* are used commercially. All are based on linear alkenes. They include alkylation with HF, $AlCl_3$, and solid acid alkylation catalysts. The product from all alkylation technologies is a mixture of linear alkyl benzene with the phenyl group at all positions in the alkyl chain with the exception of the 1-phenyl position. Alkylation by $AlCl_3$ and the current commercial solid acid alkylation catalysts favors the same higher 2- and 3- positions, and these are called *high 2-phenyl alkylates*. The HF alkylation process gives a more uniform or statistical distribution of phenyl groups along the hydrocarbon chain and is considered a low 2-phenyl alkylate. There are some differences as well as many similarities between the two types of alkylate. Alkylate produced from the older HF alkylation technology (low 2-phenyl) is still a large percentage of the production; however, all new plants as well as improved $AlCl_3$ alkylation plants are all high 2-phenyl alkylate. The high 2-phenyl alkylate has advantages for the growing production of liquid detergents, while the low 2-phenyl alkylate is used mainly in powder detergent applications. The sulfonation product is sold mainly as the sodium salt, but calcium salt (which may be oil-soluble or dispersible) and amine salts, which are also organic solvent-soluble or dispersible, are also sold. The chain length of the alkyl portions is about 12 carbons in most cases. LAS is relatively cheap, but requires acid-resistant equipment for manufacturing and sophisticated SO_3 sulfonation equipment for large-scale production. This applies also to alcohol sulfates (ASs) and ether sulfates (see "Sulfuric Acid Ester Salts"), which may be manufactured in the same or similar sulfonation equipment. Major amounts are sold as free sulfonic acid for neutralization (by processors) with amines. The sodium salt is the most widely used surfactant in industrial and high-foaming household detergents. The triethanolamine salt is in liquid detergents and cosmetics; the isopropylamine salt is in dry cleaning, since it is hydrocarbon-soluble; and the dimethylamine salt is in agricultural emulsions and dry-cleaning solvents (to solubilize the water used to remove water-soluble stains).

ADVANTAGES Completely ionized, water-soluble, free sulfonic acid; therefore solubility is not affected by low pH. Calcium and magnesium salts are water-soluble and therefore not affected by hard water. Sodium salt is sufficiently soluble in the presence of electrolyte (NaCl, Na_2SO_4) for most uses. Resistant to hydrolysis in hot acid or alkali.

DISADVANTAGES Sodium alkylbenzenesulfonate (LAS) is not soluble in organic solvents except alcohols. LAS is readily, rapidly, and completely biodegradable under aerobic conditions, which is the critical property for removal in the environment. However, LAS undergoes only primary biodegradation under anaerobic conditions. No evidence of complete biodegradation of LAS under anaerobic conditions has been reported. May cause skin irritation.

The introduction of a methyl group at an internal position in the linear alkyl chain of the hydrophobic group increases the water solubility and the performance properties of LAS.

Higher Alkylbenzenesulfonates C_{13}–C_{15} homologs are more oil-soluble, and are useful as lubricating oil additives.

Benzene-, Toluene-, Xylene-, and Cumenesulfonates Are used as hydrotropes, for example, for increasing the solubility of LAS and other ingredients in aqueous formulations, for thinning soap gels and detergent slurries.

Ligninsulfonates These are by-products of paper manufacture, prepared mainly as sodium and calcium salts, also as ammonium salts. They are used as dispersing agents for solids and as *O/W* (oil-in-water) emulsion stabilizers. They are sulfonated polymers of molecular weight 1000–20,000 of complex structure containing free phenolic, primary and secondary alcoholic, and carboxylate groupings. The sulfonate groups are at the α- and β-positions of C_3 alkyl groups joining the phenolic structures. They reduce the viscosity of and stabilize aqueous slurries of dyestuffs, pesticides, and cement.

ADVANTAGES They are among the most inexpensive surfactants and are available in very large quantities. They produce very little foam during use.

DISADVANTAGES Very dark color, soluble in water but insoluble in organic solvents, including alcohol. They produce no significant surface tension lowering.

Petroleum Sulfonates Products of the refining of selected petroleum fractions with concentrated sulfuric acid or oleum in the production of white oils. Metal or ammonium salts of sulfonated complex cycloaliphatic and aromatic hydrocarbons.

USES Tertiary oil recovery. Sodium salts of lower molecular weight (~435–450) are used as *O/W* emulsifying agents in soluble metal cutting oils, frothing agents in ore flotation, components of dry-cleaning soaps; sodium salts of higher molecular weight (465–500) are used as rust preventatives and pigment dispersants in organic solvents. Ammonium salts are used as ashless rust inhibitors and soluble dispersants in fuel oils and gasoline. Magnesium, calcium, and barium salts are used as sludge dispersants for fuel oils and as corrosion inhibitors for diesel lubricating oils.

ADVANTAGES Inexpensive.

DISADVANTAGES Dark in color. Contain unsulfonated hydrocarbon.

N-Acyl-N-Alkyltaurates, $RCON(R')CH_2CH_2SO_3^-M^+$ (Taurin is 2-Aminoethylsulfonic Acid) The solubility, foaming, detergency, and dispersing powers of the *N*-methyl derivatives are similar to those of the corresponding fatty acid soaps in soft water, but these materials are effective both in hard and soft water, are not sensitive to low pH, and are better wetting agents. They show good stability to hydrolysis by acids and alkali, good skin compatibility, and good lime soap dispersing power.

USES In bubble baths (together with soap) and in toilet bars together with soap, since they show no decrease in foaming or lathering in combination with the latter (in contrast with other anionics). In alkaline bottle washing compounds and for seawater laundering, since their salts are soluble, even in water containing high electrolyte concentrations. Impart soft feel ("hand") to fibers and fabrics (similar to soaps and fatty ASs, in contrast with nonionics and alkyl aryl sulfonates). Used as wetting and dispersing agents in wettable pesticide powders.

Paraffin Sulfonates, Secondary n-Alkanesulfonates (SASs) Produced in Europe by sulfoxidation of C_{14}–C_{17} n-paraffins with SO_2 and O_2. The n-paraffin hydrocarbons are separated from kerosene by molecular sieves.

USES Performance similar to that of LAS. Used in liquid household detergents, primarily light duty liquids (LDLs). Used as an emulsifier for the polymerization of vinyl polymers. Also used in various polymers (polyvinyl chloride [PVC] and polystyrene) as an antistatic agent. Unpurified paraffin sulfonates containing about 50% paraffin are used in fat liquoring of leather.

ADVANTAGES Solubility in water is reported to be somewhat better, viscosity of aqueous solutions somewhat lower, skin compatibility somewhat better, and biodegradability at low temperature somewhat better than those of LAS.

α-Olefin Sulfonates (AOSs) Produced by reaction of SO_3 with linear α-olefins. The product is a mixture of alkenesulfonates and hydroxyalkanesulfonates (mainly S- and 4-hydroxy).

ADVANTAGES Reported to be somewhat more biodegradable than LAS; less irritating to the skin. Show excellent foaming and detergency in hard water. High solubility in water allows products with high concentrations of actives.

Arylalkanesulfonates, $R(CH_2)_mCH(\Phi R^1)(CH_2)_nSO_3^-M^+$ Prepared by sulfonating an olefin (alkene) and then treating it with an aromatic compound. Used in agriculture, asphalt, detergents, enhanced oil recovery from petroleum reservoirs, lubricants.

ADVANTAGES Relatively inexpensive. A large variety of structures are possible by varying the nature of the olefin and the aromatic compound, including gemini (Chapter 12) disulfonates.

Sulfosuccinate Esters, $ROOCCH_2CH(SO_3^-M^+)COOR$ Used as wetting agents for paints, printing inks, textiles, agricultural emulsions. The dioctyl (2-ethylhexyl) ester is soluble in both water and organic solvents, including hydrocarbons, and is therefore used in dry-cleaning solvents. Monoesters used in cosmetics; in combination with other anionic surfactants, they reduce the eye and skin irritation of the latter.

ADVANTAGES Can be produced electrolyte-free, and is thus completely soluble in organic solvents and usable where electrolyte must be avoided. Amide monoesters are among least eye-irritating of anionic surfactants.

DISADVANTAGES Hydrolyzed by hot alkaline and acidic solutions. Dialkyl esters are irritating to skin (monoesters are not).

Alkyldiphenylether(di)sulfonates (DPESs), $RC_6H_3(SO_3^-Na^+)OC_6H_4SO_3^-Na^+$ Prepared by alkylating diphenyl ether and then sulfonating the reaction product. The C_{16} homolog is used as a detergent in cleaning products, the C_{16} and C_{12} homologs as emulsion stabilizers in emulsion polymerization, the C_{10} homolog in formulations containing high electrolyte content, the C_6 homolog as hydrotrope.

ADVANTAGES NaOCl shows good stability in solutions of DPES.

DISADVANTAGE The commercial product is a mixture of mono- and disulfonated mono-, di-, and trialkyldiphenylethers, each showing different performance properties.

Alkylnaphthalenesulfonates Mainly butyl- and isopropyl naphthalene sulfonates, for use as wetting agents for powders (agricultural wettables, powdered pesticides). Also used as wetting agents in paint formulations.

ADVANTAGES Available in nonhygroscopic powder form for mixing into formulated powders.

Naphthalenesulfonic Acid—Formaldehyde Condensates

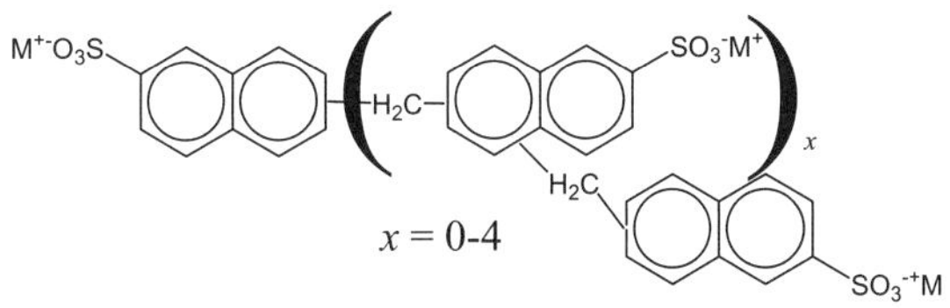

$x = 0\text{-}4$

USES Similar to those for ligninsulfonates (dispersing agents for solids in aqueous media, grinding aids for solids). Advantages over the usual ligninsulfonates are lighter color, even less foam.

Isethionates, $RCOOCH_2CH_2SO_3^-M^+$ (Isethionic Acid Is 2-Hydroxyethylsulfonic Acid) Used in cosmetic preparations, synthetic toilet soap bars, shampoos, bubble baths.

ADVANTAGES Excellent detergency and wetting power, good lime soap dispersing power, good forming power. Less irritating to skin than AS (below).

DISADVANTAGE Hydrolyzed by hot alkali.

3. Sulfuric Acid Ester Salts

Sulfated Primary Alcohols (ASs), $ROSO_3^-M^+$ Primary ASs are one of the "workhorse" surfactants and are formed by the direct sulfation of an alcohol.

The alcohol may be derived either from oleochemical or from petrochemical sources. Oleochemical ASs contain a highly linear hydrophobe, whereas the hydrophobe in petrochemical ASs may range from highly linear to highly branched, depending on the method of manufacture. For performance reasons, a mixture of alcohol chain lengths ranging from dodecyl to hexadecyl is preferred for ASs.

The most common commercial method of sulfation is "thin film" sulfation in which SO_3 vapor reacts with a thin film of alcohol. An alternative route, using chlorosulfonic acid, is convenient for laboratory sulfation and is sometimes practiced commercially. Both methods are capable of producing ASs with excellent color.

ADVANTAGES ASs have excellent foaming properties, especially if some unsulfated alcohol is retained in the product. ASs are also good detergents in the absence of high water hardness. Food-grade-quality ASs are also used in food and pharmaceutical applications.

DISADVANTAGES ASs readily hydrolyze in hot acid medium. They may cause skin and eye irritation. In the absence of builders, ASs readily form calcium and magnesium salts in the presence of high water hardness, reducing their effectiveness as cleaners.

TYPES AVAILABLE AND THEIR USE Sodium salts are most common. Sodium AS can be used in laundry powders, as a dyeing "retarder" when amino groups are present on the fiber, as a toothpaste foaming agent, as an emulsifier in food and cosmetic products, and as a dyestuff dispersion agent in aqueous solution. Magnesium "lauryl" sulfate is used where a less hygroscopic powder is needed and has greater solubility in hard water and higher alkali tolerance than the corresponding sodium salt.

Diethanol, triethanol, and ammonium salts are used in hand dishwashing liquids and in hair shampoos and cosmetics, where their higher water solubility and slightly acidic pH make them desirable.

Sulfated alcohols that are produced from alcohols that have a methyl branch in the hydrophobic group are more water-soluble than AS made from primary linear alcohols with the same number of carbon atoms in the hydrophobic group and are considerably more tolerant than the latter to calcium ion in the water. Their biodegradability is comparable to that of AS. They have been introduced into some laundry detergents.

Sulfated Polyoxyethylenated Straight-Chain Alcohols, $R(OC_2H_4)_xSO_4^-M^+$ *(AES)* R usually contains 12 carbon atoms; x usually has an average of 3, but with a broad range of distribution in polyoxyethylenated chain length; and the product usually contains about 14% of unreacted alcohol. Commercial materials having a narrow range of POE chain length have been developed by the use of new catalysts. These new materials contain less nonoxyethylenated hydrophobe (about 4%). The surface and bulk properties of these new materials are almost the same as those of conventional AES. The hardness tolerance of these new materials is better than that of conventional AES and less irritating to the skin because of the less unreacted hydrophobe.

ADVANTAGES OVER AS More water-soluble, more electrolyte resistant, much better lime soap dispersing agents, foam more resistant to water hardness and protein soil. NH_4^+ salt is less irritating to skin and eyes, produces higher viscosity solutions (advantages in shampoos).

USES In light-duty liquid detergents to improve foaming characteristics; together with nonionic in heavy-duty liquids free of phosphates; in shampoos.

Sulfated Triglyceride Oils (Sulf[on]ated Oils) Produced by sulfation of the hydroxy group and/or a double bond in the fatty acid portion of the triglyceride. (Iodine values of triglycerides used range from 40 to 140.) Mainly castor oil used (fatty acid present is mainly 12-hydroxyoleic acid), but also fish oils, tallow, sperm oil (25% oleyl, 50% C_{16} saturated fatty acid, remainder saturated C_{18} and C_{16} unsaturated). First synthetic surfactant (1850). Mainly used as textile wetting, cleaning, and finishing agent. Also used as emulsifying agent in textile finishing, in metal cutting oils, and in liquoring compositions for leather.

ADVANTAGES Cheap, easy to produce near room temperature by mixing oil and concentrated H_2SO_4. Product is a complex mixture since hydrolysis to sulfated di- and monoglycerides, and even free fatty acid, occurs during manufacture, and sulfonation occurs to a slight extent (in the α-position of fatty acid), yielding a wide range of properties. Adsorbs onto fibers to yield a soft "hand." Produces very little foam and decreases foaming of other surfactants.

DISADVANTAGES Readily hydrolyzed in hot acidic or hot alkaline solutions.

Fatty Acid Monoethanolamide Sulfates, $RCONHCH_2CH_2OSO_3^-Na^+$ RCO is usually derived from coconut oil. Produced by amidation of fatty acid with monoethanolamine, followed by sulfation.

USES Shampoos, dishwashing detergents, light-duty liquid detergents, industrial detergents, wetting agents, emulsifying agents.

ADVANTAGES OVER AS Less irritating to skin, more electrolyte-resistant, much better lime soap dispersing agent, foam more resistant to water hardness. Better cleansing power for oily soil.

DISADVANTAGES Hydrolyzed readily in hot acidic medium.

Polyoxyethylenated Fatty Acid Monoethanolamide Sulfates, $RCONHCH_2CH_2O(CH_2CH_2O)_3SO_3^-Na^+$ RCO is usually derived from coconut oil. Produced by amidation of fatty acid or fatty acid methyl ester with monoethanolamine, followed by polyoxyethylenation and sulfation.

USES Shampoos, body shampoos, dishwashing detergent.

ADVANTAGES Better-stabilized foam, less irritating to skin than AES. Produces higher viscosity water solutions. Skin irritation with this type of material is lower than with that of the corresponding fatty acid monoethanolamido sulfates.

DISADVANTAGES Hydrolyzed readily in hot acidic medium.

4. Phosphoric and Polyphosphoric Acid Esters, $R(OC_2H_4)_xOP(O)(O^-M^+)_2$ ***and*** $[R(OC_2H_4)_xO]_2P(O)O^-M^+$ Mainly phosphated POE alcohols and phenols, some sodium alkyl phosphates (not oxyethylenated). These materials are available in free acid form or as sodium or amine salts. Products are mixtures of monobasic and dibasic phosphates.

ADVANTAGES The free acids have good solubility in both water and organic solvents, including some hydrocarbon solvents, and can be used in free acid form since acidity is comparable to that of phosphoric acid. Low foaming. Not hydrolyzed by hot alkali; color unaffected. These materials show good resistance to hard water and concentrated electrolyte.

DISADVANTAGES Only moderate surface activity as wetting, foaming, or washing agents. Somewhat more expensive than sulfonates. Sodium salts usually not soluble in hydrocarbon solvents.

USES The polyoxyethylenated materials are used as emulsifying agents in agricultural emulsions (pesticides, herbicides), especially those blended with concentrated liquid fertilizer solutions, where emulsion stability in the presence of high electrolyte concentration is required; dry-cleaning detergents; metal cleaning and processing; hydrotropes (short-chain products).

The nonoxyethylenated monoalkyl phosphates cause little skin irritation and are used in personal care products. The sodium salt of monododecyl phosphate, unlike soap, works in a weakly acidic medium and can therefore be used as a detergent in face cleansers and in body shampoos. The potassium or alkanolammonium salt of monohexadecyl phosphate is used as an emulsifying agent in skin care products. The dialkyl phosphate must be avoided in the synthesis of these products, since it reduces foaming and water solubility.

5. Fluorinated Anionics Perfluorocarboxylic acids are much more completely ionized than fatty acids, hence are unaffected in aqueous solution by acids or polyvalent cations. They show good resistance to strong acids and bases, reducing and oxidizing agents, and heat (in excess of 600°F in some cases). They are much more surface active than the corresponding carboxylic acids and can reduce the surface tension of water to much lower values than are obtainable with surfactants containing hydrocarbon groups. They are also surface active in organic solvents. Perfluoroalkyl sulfonates, too, have outstanding chemical and thermal stability.

USES Emulsifiers for aqueous lattices of fluorinated monomers. Suppression of chromic acid mist and spray from chromium plating baths. "Light water" control of oil and gasoline fires. Formation of surfaces that are both hydrophobic and oleophobic on textiles, paper, and leather. Inhibition of evaporation of volatile organic solvents.

DISADVANTAGES Much more expensive than other types of surfactants; resistant to biodegradation even when straight chain.

Fluorinated Polyoxetanes

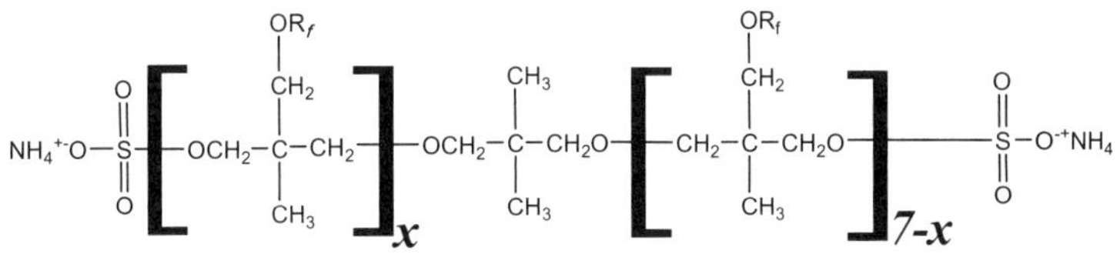

$$NH_4^{+-}O-S(=O)_2-\left[OCH_2-C(CH_2OR_f)(CH_3)-CH_2\right]_x-OCH_2-C(CH_3)_2-CH_2O-\left[CH_2-C(CH_2OR_f)(CH_3)-CH_2O\right]_{7-x}-S(=O)_2-O^{-+}NH_4$$

$R_f = CH_2CF_3,\ CH_2CF_2CF_3,\ CH_2CH_2(CF_2)_4F$

Ring-opening cationic polymerization of a perfluoroalkyl-substituted oxetane monomer using a Lewis acid catalyst and a diol initiator leads to an amphiphilic α, ω-diol. Sulfation of the terminal hydroxyl groups leads to an anionic bolaamphiphile.

USES Are effective and efficient wetting, flow, and leveling aids in aqueous and some solvent-borne coatings. Produce little foam when agitated.

ADVANTAGES Designed to have less environmental impact than traditional, smaller fluorosurfactants with longer ($\sim C_8F_{17}$) perfluoroalkyl chains.

B. Cationics

ADVANTAGES They are compatible with nonionics and zwitterionics. Surface-active moiety has a positive charge, thus adsorbs strongly onto most solid surfaces (which are usually negatively charged), and can impart special characteristics to the substrate. Some examples are given in Table 1.2. This adsorption also makes possible the formation of emulsions that "break" in

TABLE 1.2 Some Uses of Cationics Resulting from Their Adsorption onto Solid Substrates

Substrate	Use
Natural and synthetic fibers	Fabric softeners, antistatics, textile auxiliaries
Fertilizers	Anticaking agents
Weeds	Herbicides
Aggregates	Adhesion promoters in asphalt
Metals	Corrosion inhibitors
Pigments	Dispersants
Plastics	Antistatics
Skin, keratin	Toiletries, hair conditioners
Ores	Flotation agents
Microorganisms	Germicides

Source: M. K. Schwitzer, Chemistry and Industry, 822 (1972).

contact with negatively charged substrates, allowing deposition of active phase on substrate.

DISADVANTAGES Most types are not compatible with anionics (amine oxides are an exception). Generally more expensive than anionics or nonionics. Show poor detergency, poor suspending power for carbon.

1. Long-Chain Amines and Their Salts, $RNH_3^+X^-$ Primary amines derived from animal and vegetable fatty acids and tall oil; synthetic C_{12}–C_{18} primary, secondary, or tertiary amines adsorb strongly onto most surfaces, which are usually negatively charged. Very soluble and stable in strongly acidic solutions. Sensitive to pH changes—become uncharged and insoluble in water at pH above 7.

USES Cationic emulsifying agents at pH below 7. Corrosion inhibitors for metal surfaces, to protect them from water, salts, acids. Saturated, very long-chain amines best for this purpose, since these give close-packed hydrophobic surface films. Used in fuel and lubricating oils to prevent corrosion of metal containers. Anticaking agents for fertilizers, adhesion promotors for painting damp surfaces. Ore flotation collectors, forming nonwetting films on specific minerals, allowing them to be separated from other ores.

DISADVANTAGES Poor leveling is characteristic of cationic wax or wax–resin emulsions.

2. Acylated Diamines and Polyamines and Their Salts Uses and properties are similar to those above. Products of the type $(RCONHCH_2\text{-}CH_2)_2NH$ are used as adhesion promoters for asphalt coating of wet or damp road surfaces.

OTHER USES Ore flotation, to produce hydrophobic surface on ore or impurities; pigment coating, to make hydrophilic pigment lipophilic (adsorbed diamine salt yields positively charged surface, which then adsorbs fatty acid anion to give strongly chemisorbed lipophilic monolayer).

3. Quaternary Ammonium Salts

ADVANTAGES Electrical charge on the molecule is unaffected by pH changes—positive charge remains in acidic, neutral, and alkaline media.

DISADVANTAGES Since water solubility is retained at all pHs, they are more easily removed from surfaces onto which they may be adsorbed (insolubility of nonquaternary amines in water at pH above 7 is often an advantage). The long-chain dialkyl dimethylammonium chlorides are resistant to biodegradation. Alkyl pyridinium salts in alkaline aqueous solution are unstable and darken; alkyl trimethylammonium halides are stable even in hot aqueous alkaline solution.

USES *N*-alkyltrimethylammonium chlorides, $RN^+(CH_3)_3Cl^-$, are used as dye transfer inhibitors in rinse cycle fabric softeners. They are also used as emulsifying agents for acidic emulsions or where adsorption of emulsifying agent onto substrate is desirable (e.g., in insecticidal emulsions, adsorption of emulsifying agent onto substrate breaks emulsion and releases active ingredient as water-insoluble material). Highly effective germicides for industrial use. (Bis [long-chain alkyl] derivatives are less effective than monoalkyls; oxyethylenation drastically reduces germicidal effect; chlorinated aromatic ring increases it.)

N-benzyl-*N*-alkyldimethylammonium halides, $RN^+(CH_2C_6H_5)(CH_3)_2Cl^-$, are used as germicides, disinfectants, sanitizers. They are compatible with alkaline inorganic salts and nonionics and are used together with them in detergent sanitizers for public dishwashing (restaurants, bars). They are also used as hair conditioners (after shampoo rinses), since they adsorb onto hair, imparting softness and antistatic properties. The cetyl derivative is used in oral antiseptics. Cetylpyridinium bromide is used in mouthwashes. Behenyl (C_{22}) trimethylammonium chloride is used in hair rinses and hair conditioners, since it adsorbs more strongly onto hair than shorter-chain cationics, showing softening and antistatic properties.

Dialkyldimethylammonium salts of the type $R_2N^+(CH_3)_2Cl^-$ and imidazolinium salts of structure

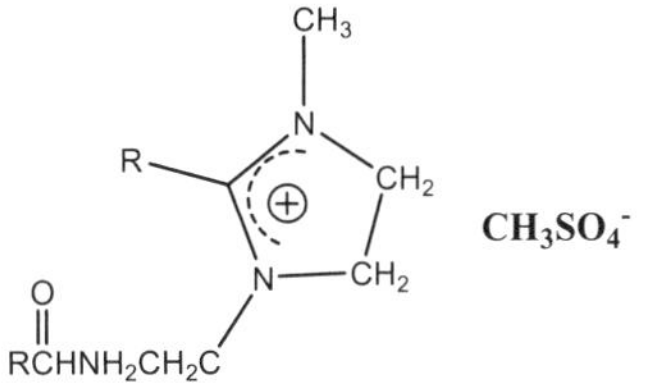

(R from tallow or hydrogenated tallow) are used as textile softeners industrially and for home use in the rinse cycle of washing machines. They impart fluffy, soft "hand" to fabrics by adsorbing onto them with hydrophobic groups oriented away from fiber.

At present, triethanolamine esterquats (TEAEQs), with a formal structure of $(RCO_2CH_2CH_2)_2N^+(CH_3)CH_2CH_2OH \bullet CH_3SO_4^-$ are the fabric softeners of choice in Europe and elsewhere, replacing the imidazolinium and dialkyldimethylammonium types.

ADVANTAGES OF TEAEQ Ease of biodegradation and environmentally friendly profile.

DISADVANTAGES OF TEAEQ Although the diester quat is the desired ingredient, with the best performance characteristics, the commercial TEAEQ is a mixture containing major amounts of monoester quat, the triester quat, and the triester amine. It therefore gives medium performance compared to the other mentioned types of fabric softeners.

4. *Polyoxyethylenated Long-Chain Amines, $RN\,[(CH_2CH_2O)_x\,H]_2$* Combine increased water solubility imparted by POE chains with cationic characteristics of the amino group. As the oxyethylene content increases, cationic properties decrease, and materials become more like nonionics in nature (e.g., solubility in water does not change much with pH change; incompatibility with anionics diminishes). If oxyethylene content is high enough, materials do not require acidic solution for water solubility.

USES In production of xanthate rayon to improve tensile strength of regenerated cellulose filaments and to keep spinnerets free of incrustations. Emulsifying agents for herbicides, insecticides, polishes, and wax emulsions, which "break" on contact with the substrate and deposit the oil phase on it.

ADVANTAGES Salts with inorganic or low-molecular-weight organic acids are water-soluble, those with high-molecular-weight organic acids are oil-soluble, even when the free POE amines are oil-insoluble. Show inverse solubility in water on heating, like other POE derivatives.

5. *Quaternized POE Long-Chain Amines $RN(CH_3)[(C_2H_4O)_xH]_2^+Cl^-$* Used as textile antistatic agent (ionic charge dissipates static charge; polyethylene group adsorbs water, which also dissipates charge). Also used as dyeing leveler (retarder) by competing transiently for dye sites on fabrics during the dyeing process, thereby decreasing the rate of dyeing at its most active sites—where it is most rapidly adsorbed—to that of the less active sites. This causes more uniform dyeing. Used as corrosion inhibitors for metallic surfaces. $(RCONHCH_2CH_2)_2N^+(CH_3)(CH_2–CH_2O)_xH \bullet CH_3SO_4^-$ (RCO from tallow) is used as fabric softener in rinse cycle of laundry washing. Promotes adhesion in asphalt (by adsorption to form hydrophobic, oleophilic surface film on substrate). Dispersing agent for clay in greases, emulsifying agent for polar compounds (e.g., fatty acids and amines) in *O/W* emulsions. Trifluoroacetate salts are used to produce foam that reduces chromic acid spray and mist in chromium plating. $[RCONH(CH_2)_3N(CH_3)_2CH_2CH_2OH]^+\ NO_3^-$ is used as a surface or internal antistatic for plastics.

6. *Amine Oxides, $RN^+(CH_3)_2O^-$* Usually *N*-alkyldimethylamine oxides. These are usually classified as cationics, although they are actually zwitterionics, and will be so classified in the following chapters (including the tables). They are compatible with anionics, cationics and nonionics, and other zwitterionics. Show excellent wetting in concentrated electrolyte solutions. The molecule adds a proton under the proper conditions, for example, at low pH or in the presence of anionic surfactants, to form the cationic conjugate acid.

The conjugate acid forms 1:1 salts with anionics that are much more surface active than either the anionic or the amine oxide. Used as foam stabilizer for anionics in detergents, liquid dishwashing compounds, and shampoos. Also increase the viscosity of the shampoo and manageability of hair. Cetyl dimethylamine oxide is used in electroplating baths. The stearyl derivative imparts a smooth "hand" to fabrics and hair.

ADVANTAGE OVER ALKANOLAMIDE FROM STABILIZERS Effective at lower concentrations.

C. Nonionics

ADVANTAGES They are compatible with all other types of surfactants. Generally available as 100% active material free of electrolyte. Can be made resistant to hard water, polyvalent metallic cations, electrolyte at high concentration; soluble in water and organic solvents, including hydrocarbons. POE nonionics are generally excellent dispersing agents for carbon.

DISADVANTAGES The products are liquids or pastes, rarely nontacky solids. They are poor foamers (may be an advantage sometimes); have no electrical effects (e.g., no strong adsorption onto charged surfaces). Ethylene oxide (EO) derivatives show inverse temperature effect on solubility in water, may become insoluble in water on heating. Commercial material is a mixture of products with a wide distribution of POE chain lengths. POE chains with terminal hydroxyl show yellowing (due to oxidation) in strong alkali that can be prevented by etherifying *(capping)* the hydroxyl.

1. Polyoxyethylenated Alkylphenols, Alkylphenol "Ethoxylates" (APEs), $RC_6H_4(OC_2H_4)_xOH$ Mainly polyoxyethylenated *p*-nonylphenol, *p*-octylphenol, or *p*-dodecylphenol (sometimes, dinonylphenol), derived from disobutylene, propylene trimer, or propylene tetramer.

ADVANTAGES Length of alkyl group on phenol or POE chain can be varied to give range of products varying in solubility from water-insoluble, aliphatic hydrocarbon-soluble products (1–5 mol of EO) to water-miscible, aliphatic hydrocarbon-insoluble ones. POE linkages are stable to hot dilute acid, alkali (except for some yellowing in the latter), and oxidizing agents result from hydratable multiple ether linkages. Advantage over polyoxyethylenated alcohols in that there is never any free alkylphenol in APE, since phenolic OH is more reactive than alcohol OH. Thus, no toxicity or dermatology problems associated with free phenol or other problems associated with presence of free hydrophobe.

DISADVANTAGES Even though APEs will completely biodegrade under aerobic conditions, the rates are slower than with other nonionic surfactants such as linear AEs. The aerobic biodegradation intermediates are more toxic to fish and other aquatic organisms than the parent APE. Also, there are reports that APEs may show endocrine disruptive activity in model systems in laboratory tests, although no unequivocal demonstration of APE endocrine disruptive activity in actual environmental systems has been found from human epidemiological data (Falconer et al., 2006).

USES Mainly industrial because of low degradability. Water-insoluble types used for *W/O* emulsifying agents, foam control agents, cosolvents; water-soluble types for *O/W* emulsifying agents for paints, agricultural emulsions, miscellaneous industrial and cosmetic emulsions. Materials with high EO content (>15 mol EO) are used as detergents and emulsifiers in strong electrolyte systems and as foam entrainment agents in concrete. Also used in liquid detergents and as dyeing retarders for cellulose (surfactant forms complex with dye molecules). Excellent dispersing agents for carbon.

2. *Polyoxyethylenated Straight-Chain Alcohols, AE, $R(OC_2H_4)_xOH$* AEs, like ASs and alcohol ethoxysulfates, can be made from either oleochemical or petrochemical alcohols. Consequently, the linearity of the hydrophobe can vary from highly linear when the alcohol is derived from oleochemical sources and some petrochemical sources to highly branched from other petrochemical sources. Often a blend of several carbon chain length alcohols is used to produce commercial products. To make these surfactants, EO is added to a blend of alcohols in the presence of a catalyst, often NaOH or KOH, until the average degree of ethoxylation is achieved. The result is a mixture that varies in both the carbon chain length and the distribution of ethoxymers. "Peaking" catalysts can be used to narrow the distribution of ethoxymers. Oleyl derivatives are more fluid than saturated alcohol derivatives; lubricating properties are more pronounced in the saturated alcohol derivatives than in the unsaturated ones. Used for industrial purposes similar to those of APE. In low- and controlled-foam laundry detergents.

ADVANTAGES The AE structure can be optimized for performance since the average hydrophobe, hydrophile, and distribution of the ethoxymers can be varied. AES biodegrade more readily than APEs. AES are more tolerant of high ionic strength and hard water than anionic surfactants and exhibit better stability in hot alkaline solutions than ethoxylated fatty acids. They also have excellent compatibility with enzymes in laundry formulations, are more water-soluble, and have better wetting powers than corresponding fatty acid ethoxylates. They are somewhat better than the corresponding APE for emulsification. More water-soluble than LAS, for use in high active, heavy-duty liquid detergents free of phosphates. More effective detergency than LAS under cool washing conditions and on synthetic fabrics.

DISADVANTAGES High concentrations of AES in laundry powders often "bleed" from the powder, giving poor powder properties. Because AEs are composed of a distribution of ethoxymers, some unethoxylated alcohol remains in commercial products. If present in sufficient quantity, this can impart an objectionable odor to the ethoxylate. This can be ameliorated to some extent by using a "peaking" ethoxylation catalyst.

Aqueous solutions of these "peaked" materials show lower toxicities, lower viscosities, lower gel temperatures, and remain fluid over a wider concentration range. In spray-drying operations, there is less evolution of volatile material, since they contain less unreacted hydrophobe than conventional materials. They wet cotton more efficiently, show higher initial foam heights (but lower foam stability), reduce interfacial tension against mineral oil more efficiently and effectively than the corresponding conventional types. When sulfated to produce AES, the product has less non-polyoxyethylenated alkyl sulfate and, consequently, less skin irritation and a greater tendency to thicken upon salt addition.

USES AEs are excellent detergents for removal of oily soil and are often used in laundry products, especially liquids. They are also excellent emulsifiers and suspending agents in numerous industrial applications, where they compete with APEs.

3. Polyoxyethylenated Polyoxypropylene Glycols Block copolymers of EO and propylene oxide. Materials with low EO content have very little foam; materials of high molecular weight with low EO content are wetting agents. Materials with high EO content are dispersing agents. Products range in molecular weight from 1000 to 30,000. Can form aqueous gels when hydrophobe (polypropylene oxide) molecular weight is greater than 1750.

USES High-molecular-weight materials with high EO content are used as dispersants for pigments in latex paints or for scale removal in boilers; low-molecular-weight materials with low EO content are used as foam control agents in laundry detergents and in rinsing aids for dishwashing. Petroleum demulsifiers.

ADVANTAGES Both hydrophobic group $-(CH_2CH(CH_3)O)-_x$ and hydrophilic group $(CH_2CH_2O)_y$ can be varied at will to "tailor-make" products with specific properties. Products with high-molecular-weight hydrophobes and high EO contents are nontacky solids (in contrast to other POE nonionics). Better wetting agents than ester-type nonionics.

DISADVANTAGES Polyoxypropylene group is less biodegradable than POE group.

4. Polyoxyethylenated Mercaptans, $RS(C_2H_4O)_xH$ Unstable to oxidizing agents, such as chlorine, hypochlorites, peroxides, and strong acids. (This may be an advantage when inactivation of surfactant after use is desired.) Stable in hot, strong alkali. Good lime soap dispersants.

USES Textile detergents (cleaning and scouring of wool), metal cleaning, shampoos.

ADVANTAGES Some evidence that quaternary amonium compounds are more effective as detergent sanitizers when formulated with polyoxyethylenated mercaptans rather than with other polyoxyethylenated nonionics.

DISADVANTAGES Have slight, unpleasant odor that is difficult to mask.

5. *Long-Chain Carboxylic Acid Esters*

ADVANTAGES In some cases, very easily made in simple equipment. Outstanding emulsifying properties compared to other nonionic types.

DISADVANTAGES Readily hydrolyzed by hot acids or hot alkalis. Lower foam than other nonionic types (may be an advantage for some uses).

Glyceryl and Polyglyceryl Esters of Natural Fatty Acids

ADVANTAGES Glyceryl esters are easily made by glycerolysis of triglycerides or, somewhat more expensively, by esterification of fatty acids with glycerol in simple equipment. Edible, hence usable in food and pharmaceutical products. May be liquid, soft plastic, or hard wax, depending on fatty acid composition. Can be modified by reaction with acetic, lactic, or tartaric acids. Polyglycerol esters of fatty acids are made by esterification of polymerized glycerol.

DISADVANTAGES Mixture of mono- and diglycerides (glycerides of, ~90% monoester content must be made by distillation of usual reaction product). Monoglyceride is a better emulsifier than diglyceride.

USES Cosmetic emulsifiers, food emulsifiers for bread, ice cream, margarine, synthetic cream, and other dairy products.

Propylene Glycol, Sorbitol, and Polyoxyethylenated Sorbitol Esters Propylene glycol esters are more lipophilic than the corresponding glycerol esters; sorbitol esters are more hydrophilic (unless dehydrated in course of manufacture). Polyoxyethylenation of sorbitol (and anhydrosorbitol produced during manufacture) gives wide range of solubilities and hydrophilic–lipophilic balances to products.

ADVANTAGES Edible, thus useful for food and drug use (e.g., soluble vitamins).

USES Food and pharmaceutical emulsifiers.

Polyoxyethylenated Glycol Esters and Polyoxyethylenated Fatty Acids (Including Tall Oil) Prepared either by esterification of polyoxyethylenated glycol with fatty acid or by addition of ethylene oxide to fatty acid. Tall oil derivatives have lower foaming properties than corresponding fatty acid derivatives. Advantage over glyceryl esters is that length of hydrophilic group, and hence solubility and hydrophilic–lipophilic balance of product, can be varied as desired. Generally better emulsifying agents than AE or APE.

DISADVANTAGES Generally poor wetting properties; hydrolyzed by hot alkaline solutions.

USES Emulsification of all sorts, especially in cosmetics and for textile use, except where hot alkaline solutions are encountered. Textile antistats.

6. Alkanolamine "Condensates," Alkanolamides Mainly of diethanolamine or monoisopropanolamine. Good stability to hydrolysis by hot alkali, poor to fair stability to hot acids.

1:1 Alkanolamine–Fatty Acid "Condensates" Made by reaction of methyl or triglyceride ester of fatty acid with equimolar amount of alkanolamine (about 90% alkanolamide content in product from methyl ester, 80% from triglyceride). Mainly based on coconut or purified coconut (lauric) esters.

Diethanolamides are insoluble but dispersible in water, soluble in organic solvents except some aliphatic hydrocarbons. Compatible with both anionics and cationics over wide pH range. Poor wetting and detergent properties, but synergistic to surfactants showing these properties. Show corrosion-inhibiting properties for steel. Easily prepared.

USES Foam stabilizers for LAS in laundry and dishwashing detergents (alternative to amine oxides). Thickeners for liquid detergents and shampoos (containing sodium lauryl sulfate).

2:1 Alkanolamine–Fatty Acid "Condensates" Made by reaction of 2 mol alkanolamine with 1 mol free fatty acid. Contains about 60–70% alkanolamide, 25–30% alkanolamine, 3–5% fatty acid (as soap of alkanolamine). Mainly based on coconut fatty acid.

ADVANTAGES OVER 1:1 CONDENSATE Diethanolamine–coconut fatty acid "condensate" is soluble in both water and organic solvents except aliphatic hydrocarbons. Excellent detergent, emulsifier, and viscosity thickener in aqueous medium at low concentrations.

DISADVANTAGES Complex mixture; foam stabilization depends only on amide content (60–70%). Fatty acid content makes it incompatible with cationics.

USES Textile detergent, shampoo ingredient, emulsifying agent, rust inhibitor, dry-cleaning soap, fuel oil additive.

7. Tertiary Acetylenic Glycols, $R_1R_2C(OH)C{\equiv}CC(OH)R_1R_2$, and Their "Ethoxylates," $R_1R_2C[(OC_2H_4)_xOH]C{\equiv}CC[(OC_2H_4)_xOH]R_1R_2$

ADVANTAGES OF THE GLYCOLS Excellent wetting agents at low concentrations and nonfoaming; nonwaxy solids (rare among nonionics); volatile with steam when not ethoxylated, thus readily removed from system after use.

DISADVANTAGES OF THE GLYCOLS Very low solubility in water; decompose in acidic medium; relatively expensive.

Polyoxyethylenation of the hydroxyl groups with a few oxyethylene units increases solubility in water without significant change in surface properties, but resulting products are liquid and nonvolatile with steam.

USES Wetting agents for use in powdered solids (dyestuffs, wettable pesticide powders); synergistic with anionics and nonionics to decrease foam, reduce viscosity, and increase wetting in aqueous solution; rinse aids in dishwashing; wetting agents in emulsion paints.

8. Polyoxyethylenated Silicones These are the reaction products of a reactive silicone intermediate, such as

$$(H_3C)_3SiO{-}[Si(CH_3)_2O]_x{-}[Si(CH_3)(H)O]_y{-}Si(CH_3)_3$$

with a capped allyl polyalkylene oxide, such as $CH_2{=}CH{-}CH_2{-}(OC_2H_4)_z{-}OR^1$, to yield

$$(H_3C)_3SiO{-}[Si(CH_3)_2O]_x{-}[Si(CH_3)((CH_2)_3(OC_2H_4)_zOR^1)O]_y{-}Si(CH_3)_3$$

The capped allyl polyalkylene oxide can also be based upon propylene oxide or a mixed ethylene oxide–propylene oxide copolymer. The resulting structure is a "comb" polymer, with pendant capped hydrophilic groups. In aqueous solution, the hydrophilic groups may form a sheath around the hydrophobic silicone backbone to minimize its contact with the water.

Minimum surface tensions for products of this type in aqueous solution fall in the 20- to 25-dyn/cm range at 25°C. They are excellent wetting agents at concentrations of a few hundredths of a percent for cotton and show good lubricating properties of textile fibers. They are also excellent wetting agents for polyester and polyethylene. They are low to moderate foamers in aqueous solution. They can also be used to lower the surface tension of nonaqueous solvents such as polyalkylene glycols.

9. N-Alkylpyrrolid(in)ones

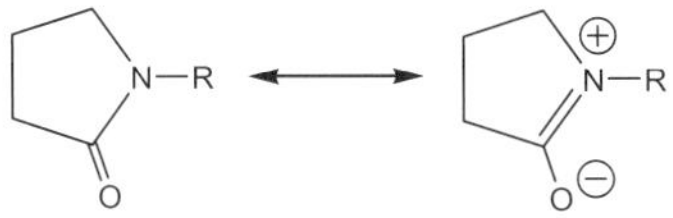

These are nonionic surfactants that, because of their dipolar resonance form, also show some of the properties of zwitterionics. They have limited solubility in water and do not form micelles by themselves in it at room temperature, but do form mixed micelles with other surfactants (e.g., LAS).

Their surface activity is high, the *n*-dodecyl compound depressing it to about 26 dyn/cm at a concentration of 0.002%. The *n*-octyl compound is an excellent low-foaming wetting agent. It also interacts synergistically with anionic surfactants, for example, LAS, to increase their foaming and wetting properties. *N*-alkylpyrrolid(in)ones, like polyvinyl pyrrolidone, act as complexing agents, particularly for phenols and other organic compounds that are capable of forming hydrogen bonds with the pyrrolidone ring.

10. Alkylpolyglycosides These are long-chain acetals of polysaccharides. A representative type is shown in Figure 1.1. Commercial products currently available have relatively short alkyl chains (averaging 10 and 12.5 carbon atoms). They show wetting, foaming, detergency, and biodegradation properties similar to those of corresponding AEs, but have higher solubility in water and in solutions of electrolytes. They are also soluble and stable in sodium hydroxide solutions, in contrast to AE. Although effective fatty soil removers,

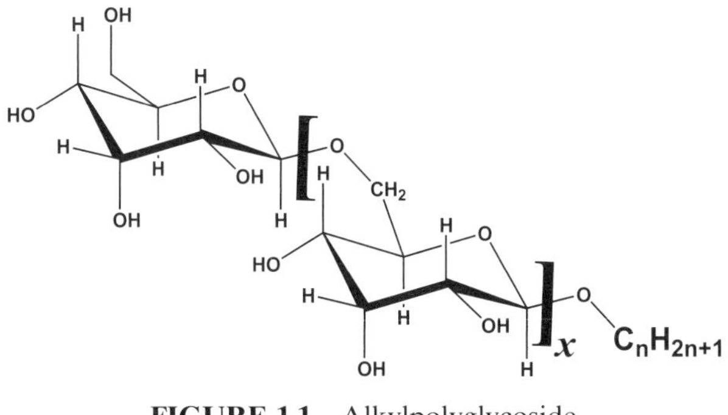

FIGURE 1.1 Alkylpolyglycoside.

they show very low skin irritation and are recommended for hand liquid dishwashing and hard surface cleaners.

D. Zwitterionics

ADVANTAGES They are compatible with all other types of surfactants. They are less irritating to skin and eyes than other types, and they may be adsorbed onto negatively or positively charged surfaces without forming hydrophobic film.

DISADVANTAGES They are often insoluble in most organic solvents, including ethanol.

1. pH-Sensitive Zwitterionics These are ampholytic materials, which may show the properties of anionics at high pHs and those of cationics at low pHs. In the vicinity of their isoelectric points they exist mainly as zwitterionics and show minimum solubility in water, and minimum foaming, wetting, and detergency.

β-N-Alkylaminopropionic Acids, $RN^{+}H_2CH_2CH_2COO^{-}$ Isoelectric point at pH~4. Very soluble in aqueous solutions of strong acids and alkalies, even in the presence of electrolytes like NaCl. Solubility is low in most organic solvents, including ethanol and isopropyl alcohol. Adsorb from aqueous solution onto skin, textiles, fibers, and metals. On hair and textile fibers they confer lubricity, softness, and antistatic properties; on metals they act as corrosion inhibitors. They solubilize many organic and inorganic compounds (e.g., quaternary ammonium salts, phenols, polyphosphates) in aqueous solutions. They are effective emulsifying agents for long-chain alcohols and slightly polar compounds, not good for paraffinic oils. Emulsions can be converted from anionic to cationic by pH adjustment. Emulsions more easily prepared at alkaline than at acidic pHs. *N*-dodecyl derivative is an excellent wetting agent and foam producer at alkaline pHs, less of a foamer at acid pHs.

USES Bactericides, corrosion inhibitors, pigment dispersion aids, cosmetics, alkaline cleaners with high alkali and electrolyte content.

N-Alkyl-β-Iminodipropionic Acids

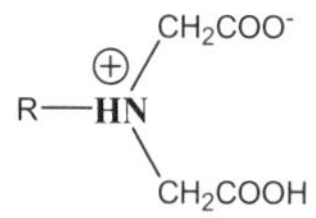

$$R-\overset{\oplus}{H}N\begin{matrix} CH_2COO^{-} \\ CH_2COOH \end{matrix}$$

Isoelectric point, pH 1.7–3.5. More soluble in water than corresponding monopropionic acid derivatives. Show very little skin and eye irritation. May be removed from substrates onto which they have adsorbed at pHs below their isoelectric points by raising the pH.

USES Fabric softeners (removed by increase in the pH to the alkaline side).

Imidazoline Carboxylates

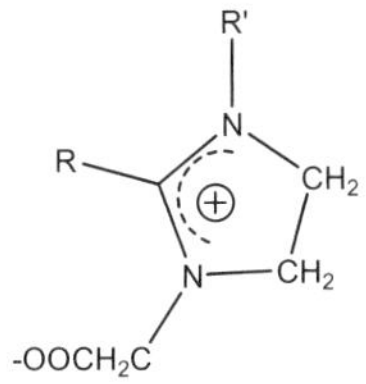

R'
N
R CH_2
⊕
N CH_2
$^{-}OOCH_2C$

R from RCOOH of commercially available fatty acids. When R′ is H, they are ampholytic and show cationic properties at low pHs, anionic properties at high pHs. When R′ = CH_2Z, pH sensitivity is more closely related to that of *N*-alkylbetaines (below). Compatible with anionics, cationics, and nonionics, soluble in water in the presence of high concentrations of electrolytes, acids, and alkalies. When R′ contains a second carboxylic acid group, products show very little skin and eye irritation. A new class of salts with large organic cations, like 1-alkyl-3-methyl imidazolium cations, that have melting temperatures close to room temperature (<100°C), are known as ionic liquids, and serve both as good ionic solvents and surfactants.

USES Cosmetic and toilet preparations, fabric softener (which can be removed from substrate by increase in pH to the alkaline side). Ionic liquids are used as "green" solvents in catalytic reactions, in capillary electrophoresis, and in various chromatogrpahies, including supercritical (sc) fluid applications with $scCO_2$.

N-Alkylbetaines, $RN^+(CH_3)_2CH_2COO^-$ These materials are zwitterionic at pHs at and above their isoelectric points (neutral and alkaline pHs) and cationic below their isoelectric points (acid pHs). They show no anionic properties. Compatible with all classes of surfactants at all pHs, except that at low pHs, they yield precipitates with anionics. Acid and neutral aqueous solutions are compatible with alkaline earth and other metallic ions ($A1^{3+}$, Cr^{3+}, Cu^{2+}, Ni^{2+}, Zn^{2+}). They show minimum skin irritation at pH 7. Show constant adsorption onto negatively charged surfaces (as cationics), irrespective of pH. Slightly better wetting and foaming properties at acidic than at alkaline pHs. Hard water has no effect on foaming properties in aqueous solution. Emulsification properties are similar to those of β-*N*-alkylaminopropionic acids (not good for paraffinic oils).

USES Similar to those of β-*N*-alkylaminopropionic acids.

Amidoamines and Amidobetaines These are products, related to the above, of typical structures: $RCONHCH_2CH_2N^+H(CH_2CH_2OH)CH_2COO^-$, $RCONHCH_2CH_2N^+H(CH_2CH_2OH)CH_2CH_2COO^-$, and $RCONHCH_2CH_2CH_2N^+(CH_3)_2COO^-$, that are used in cosmetics and personal care products (shampoos, liquid soaps, facial cleaners) because of their mildness on the skin and compatibility with anionic, cationic, and nonionic surfactants. The RCO group is usually ~C_{12}

Amine Oxides, $RN^+(CH_3)_2O^-$ See "Cationics," Section IVB.

2. pH-Insensitive Zwitterionics These materials are zwitterionics at all pHs (at no pH do they act merely like anionics or cationics).

Sulfobetaines, Sultaines, $RN^+(CH_3)_2(CH_2)_xSO_3^-$ Adsorb onto charged surfaces at all pHs without forming hydrophobic films. Good lime soap dispersants. Show little skin irritation.

USES Similar to other zwitterionics. Lime soap dispersants in soap–detergent formulations. Dispersants for textile finishing agents.

E. Newer Surfactants Based Upon Renewable Raw Materials

There has been intense interest in recent years in using renewable, readily biodegradable resources for both the hydrophilic and hydrophobic groups of commercial surfactants in order to provide them with a favorable environmental ("green") image. The search has centered upon natural fats as the source for hydrophobic groups and upon naturally occurring carbohydrates and amino acids (from proteins) for hydrophilic groups.

Soaps, of course, are based upon renewable fats and lignin sulfonates upon wood, while sulfated alcohols and sulfated triglycerides among the anionics, and glyceryl, polyglyceryl, sucrose, and sorbitol fatty acid esters and alkylpolyglycosides among the nonionics can be based upon renewable resources and thus considered "green."

1. α-Sulfofatty Acid Methyl Esters (SME), $RCH(SO_3^-Na^+)COOCH_3$ Produced by the reaction of SO_3 with fatty acid methyl esters (derived from triglycerides by transesterification with methanol). Generally from C_{12} to C_{18} fatty acid methyl esters.

Advantages Derived from relatively low-cost, renewable raw materials. Good biodegradability. The tallow methyl ester sulfonate has somewhat better detergency than LAS in both hard and soft water, while the palm kernel derivative is somewhat poorer than LAS in soft water but better than it in hard water. Excellent lime soap dispersion properties, which enables effective formulation with soaps. Larger solubilizing capacity for unsaturated oily soil than LAS. Can be produced electrolyte-free.

Disadvantages Production of low-color SME generally requires complex manufacturing process. Process must be tuned to minimize sulfonated free fatty acid, which has reduced detergency and solubility in water relative to SME. Methyl ester group is prone to hydrolysis at low and high pH; consequently, SME is difficult to incorporate in spray-dried detergents.

Uses Primary or auxiliary anionic surfactant in heavy-duty laundry detergents or light-duty liquid detergents.

2. Acylated Aminoacids These materials have good foaming properties, are less sensitive to hard water than soap, are nonirritating to the skin, and have antibacterial activity and good biodegradability. They are relatively expensive but are used in cosmetic, skin cleaning, and food formulations. The *N*-lauroyl (or cocoyl) derivatives generally show optimal properties.

a. N-Acyl L-Glutamates (AG), $RCONHCH(COO^-M^+)CH_2CH_2COO^-M^+$, $M{=}H^+$ *or Cation* Produced by N-acylation of L-glutamic acid with fatty acid chloride in a mixed solvent of water and water-miscible organic solvent. RCO is usually from coconut and tallow acids. AG is a dibasic acid, so both mono- and di-neutralized materials are possible. The carboxyl group at the α-position is neutralized prior to that at the γ-position. The water solubility of monosodium AG is low, so organic amines, that is, triethanolamine or diethanolamine, or K^+ are used as counterions.

ADVANTAGES The mono-neutralized AG works in aqueous solutions of weak acids, which is a favorable characteristic for cosmetic products. Mild to the skin. Decreases the skin irritancy of AS or AES.

USES Mono-neutralized products based upon C_{12} fatty acid are used as detergents in face cleaners (to remove soil) and face cleansers (to remove makeup); those based on C_{18} fatty acids, as emulsifying agents in skin care products.

b. N-Acyl Glycinates, $RCONHCH_2COO^-M^+$ Produced by reaction similar to that of AG above. RCO is usually derived from coconut oil for detergent use.

ADVANTAGES *N*-acylglycinates have better foaming power, especially in the vicinity of pH 9, than sulfated linear primary alcohols (AS), sulfated POE straight-chain alcohols (AES), or alkyl ether carboxylates.

USES Potassium *N*-cocoyl glycinate is often used in face cleaners (to remove soil) and face cleansers (to remove makeup). Mild to skin. For baby care products. Creamy foam.

c. N-Acyl DL-Alaninates, $RCONHCH(CH_3)COO^-M^+$ RCO is usually derived from coconut oil.

USES AND ADVANTAGES Better foaming power for triethanolammonium *N*-dodecanoyl alaninate than AS, AES, and alkyl ether carboxylates in the pH region between weakly acidic and neutral. Good foaming power even in the presence of silicone oil. Used as a detergent in face cleaners and face cleansers. Mild to skin. For baby care products. Fine, creamy foam.

d. Other Acylated Aminoacids *N*-lauroyl sarcosinate, $C_{11}H_{23}CON(CH_3)CH_2COO^-Na^+$, used in toothpaste, is strongly foaming, enzyme-inhibiting, with good detergency (like soap). *N*-oleylsarcosinate is a polyester fiber lubricant. *N*-lauryl-arginylphenylalanine shows strong antimicrobial activity against gram-positive and some gram-negative bacteria.

3. Nopol Alkoxylates

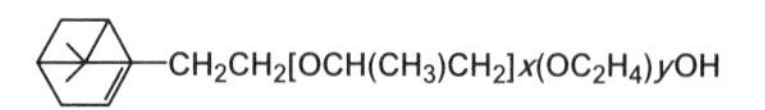

These are surfactants based upon nopol, an alcohol made by the reaction of β-pinene with formaldehyde. The nopol is reacted first with propylene oxide and then with ethylene oxide.

Advantages Based upon renewable pine oil. Show good dynamic surface tension reduction, good wetting, extremely low foam, and good rinsing properties. Very low ecotoxicity profile compared to linear C_{12}, branched C_{13}, or nonylphenolethoxylates.

Uses Spray cleaning and other wetting applications.
For additional information on the utilization of surfactants for specific applications, see the following resources:

1. *Industrial Utilization of Surfactants: Principles and Practice*, M. J. Rosen and M. Dahanayake, AOCS, 2000.
2. *Surfactants in Agrochemicals*, T. F. Tadros, Marcel Dekker, 1994.
3. *Surfactants in Chemical/Process Engineering,* D. T. Wasan, M. E. Ginn, and D. O. Shah, editors, Marcel Dekker, 1988.
4. *Surfactants in Cosmetics*, M. M. Rieger and L. D. Rhein, 2nd edition, Marcel Dekker, 2002.
5. *Surfactants in Emerging Technologies*, M. J. Rosen, editor, Marcel Dekker, 1987.

V. SOME USEFUL GENERALIZATIONS

Anionics are generally not compatible with cationics; that is, they precipitate each other from aqueous solution unless they have water-solubilizing groups in addition to their charges in their hydrophilic heads.

Carboxylic acid salts are more sensitive to low pH, polyvalent cations, and inert electrolyte in the aqueous phase than salts of organic phosphoric acids, and these in turn are more sensitive than organic sulfates or sulfonates.

Branched-chain unsaturated, or ring-containing surfactants are generally more soluble in both water and hydrocarbons and show lower viscosity in aqueous media than straight-chain materials with the same number of carbon atoms; the latter are much more biodegradable but more toxic to marine organisms than the former. Fluorocarbon chains, even when straight, are resistant to biodegradation.

Organic sulfates are readily hydrolyzed by hot acids; esters are readily hydrolyzed by hot alkali (or hot acids). Amides are more resistant to hydrolysis by hot acids or alkali than organic sulfates or esters, respectively.

Nonquaternary cationics are generally sensitive to high pH, polyvalent anions, and inert electrolyte in the aqueous phase; quaternary ammonium salts, on the other hand, are generally insensitive to these additives.

Oxyethylenation of any type of surfactant usually results in an increase in its solubility in water and a decrease in its sensitivity to pH change or electrolyte. Oxypropylenation increases its solubility in organic solvents but decreases its solubility in water.

Oxyethylenation of hydrophobes that are acidic (carboxylic acids, phenols) or basic (amines) leaves essentially no nonoxyethylenated hydrophobe, whereas oxyethylenation of alcohols generally leaves an appreciable amount of unreacted hydrophobe.

Edible ester-type surfactants can be based on glycerol, sorbitol, or propylene glycol. The foam stabilization and viscosity-thickening properties of diethanolamine–fatty acid condensates are related directly to their diethanolamide content; on the other hand, solubility in water is shown only by the 2:1 condensate.

Mercaptan-based nonionics are prone to develop a somewhat unpleasant odor and are unstable to oxidizing agents.

N-alkyl amino acids are sensitive to pH change, developing the characteristics of cationics at low pHs and those of anionics at high pHs. Zwitterionics containing a quaternized *N* and one carboxylate group (alkyl betaines) show the characteristics of cationics at low pHs, but show no anionic characteristics at high pHs. Sulfobetaines are insensitive to pH change.

Biodegradability increases with linearity of the hydrophobic group, and decreases, for isomeric materials, with branching in that group.

The toxicity of surfactants depends upon their tendency to absorb onto cell membranes and their ability to penetrate them.

VI. ELECTRONIC SEARCHING OF THE SURFACTANT LITERATURE

Scientific literature searching strategies have undergone tremendous transformation from scanning the contents and indexes of text books, browsing directories, encyclopedias, and technical dictionaries, searching abstracts, and direct reading of printed research journals, to modern electronic database searching methods.

Dissemination of information through electronic media and their searching capability through different search engines that allow data accessibility, coupled with the speed and accuracy of computers, have greatly facilitated search of the scientific literature. Some of the electronic information is available in public domains, especially those published with aid from governmental funding agencies.

Currently, free and reasonably reliable access to scholarly information is possible through www.googlescholar.com that provides an immediate survey of many of the recently published works through scientific key words (in some cases full text of articles are freely available); similarly, www.wikipedia.org is also an inexpensive means of collecting *preliminary* information on many scientific and nonscientific topics, though its credibility may need to be verified from more established sources before being cited.

A sure way to dig deep up into scientific literature is to utilize the paid services of individual E-journals that give full access to all the back issues through *archiving* facilities. Most of the established publications like *Nature*, *Science*, and various Chemical Society journals have easy-to-download *apps* for smart phones and handheld tablet computers that make fast tracking of cutting-edge information almost in real time. A searcher develops her/his search *key words* improvised for a given search session in contrast to the limited key words cited in the article. But the best way to gather comprehensive scientific data electronically is to gain paid access to hypermedia database services such as Web of Science (ISI), Scopus (Elsevier), JSTOR, Academic Search (Ebsco), and PubMed (NIH; some free access also). Some of the above services also quote citations to a particular work, which help estimate the impact of that work on science. A derived useful strategy to traverse into the latest in an area is to follow the citations of a work in recent literature.

The above considerations discussed generically can be applied to the specific instance of surfactant literature by the use of appropriate key words. The selection of key words such as micelles, adsorption, microemulsion and surface tension are typical. In fact, almost any term from the index of this book can serve the purpose. A somewhat exhaustive index of most science and technology journals in the world may be obtained from the site, http://www.ch.cam.ac.uk/resources-index, that contains the names of the journals of interest related directly or indirectly to surfactants. Moreover, there are many dedicated online journals that do not possess print editions. But the information registered in the general virtual electronic media is subject to continuous upgrading with the result that some of the cross-references may become obsolete with the passage of time, in contrast to the dedicated print journals, whose electronic versions also will remain invariant.

REFERENCES

Cannon, C. L., P. J. Neal, J. A. Southee, J. Kubilus, and M. Klausner (1994) *Toxicol. In Vitro I* **8**, 889.

Falconer, I. R., H. F. Chapman, M. R. Moore, and G. Ranmuthugala (2006) *Environ. Toxicol.* **21**, 181–191.

Kunjappu, J. T. *Essays in Ink Chemistry (For Paints and Coatings Too)*, Nova Science Publishers, Inc., New York, 2001.

Rosen, M. J., L. Fei, and S. W. Morrall (1999) *J. Surfactants Deterg.* **2**, 343.

Rosen, M. J., E. Li, S. W. Morrall, and D. J. Versteeg (2001) *Environ. Sci. Technol.* **35**, 54.

Salager, J. L., N. Marquez, A. Graciaa, and J. Lachaise (2000) *Langmuir* **16**, 5534.

Swisher, R. D. *Surfactant Biodegradation*, 2nd ed., Marcel Dekker, New York, 1987.

PROBLEMS

Write the structural formula for a surfactant type in current use that fits the general description in each case (for Problems 1.1–1.11):

1.1 Suitable for use in warm alkaline aqueous solution, but decomposes in warm acidic solution

1.2 An edible nonionic surfactant

1.3 Suitable, at neutral pH, for making most solid surfaces hydrophobic

1.4 An anionic surfactant unsuitable for use in a detergent bar for washing hands

1.5 A surfactant based entirely upon synthetic polymers

1.6 A zwitterionic surfactant whose structure does not change with change in pH

1.7 A surfactant that has the same chemical elements in both its hydrophilic and hydrophobic groups

1.8 An anionic surfactant unsuitable for use in hot alkaline solution.

1.9 An anionic surfactant based upon renewable resources that is nonirritating to the skin.

1.10 A surfactant used as a germicide.

1.11 A nonionic surfactant that is volatile with steam.

1.12 When you gradually dissolve a surfactant in water, what are the changes that you observe in the solution and at the interface?

1.13 Solely from molecular structural considerations, comment on the dependence of the hydrophilic and hydrophobic nature of a typical surfactant in the following cases: (1) hydrocarbon chain length is varied, (2) number of EO groups and the length of alkyl group are varied, (3) hydrocarbon chain is fluorinated partially and fully, (4) siloxyl groups are introduced into the structure, (5) multi-ionic heads and multiple hydrocarbon chains are introduced.

1.14 Can you provide a molecular argument based on electronic properties to explain why the surface tensions of a perfluorinated and siloxylated surfactant are lower than those of regular surfactants?

1.15 Why do you think linear alkylbenzenesulfonates have a high-volume consumption (see Table 1.1)?

1.16 Can you come up with a reasonable explanation for the observation that cationic surfactants are more toxic than anionics and nonionics?

1.17 Identify a "totally new" class of surfactant that is not discussed here that may have potential industrial applications.

2 Adsorption of Surface-Active Agents at Interfaces: The Electrical Double Layer

A fundamental characteristic of surfactants is their tendency to adsorb at interfaces in an oriented fashion. This adsorption has been studied to determine (1) the concentration of surfactant at the interface, since this is a measure of how much of the interface has been covered (and thus changed) by the surfactant; the performance of the surfactant in many interfacial processes (e.g., foaming, detergency, emulsification) depends on its concentration at the interface; (2) the orientation and packing of the surfactant at the interface, since this determines how the interface will be affected by the adsorption, that is, whether it will become more hydrophilic or more hydrophobic; (3) the rate at which this adsorption occurs, since this determines the performance in phenomena such as high-speed wetting or spreading; and (4) the energy changes, ΔG, ΔH, and ΔS, in the system, resulting from the adsorption, since these quantities provide information on the type and mechanism of any interactions involving the surfactant at the interface and the efficiency and effectiveness of its operation as a surface-active material.

In comparing the performance of different surfactants in interfacial phenomena, as in most phenomena, it is usually necessary to distinguish between the *amount* of surfactant required to produce a given amount of change in the phenomenon under investigation and the *maximum change* in the phenomenon that the surfactant can produce, regardless of the amount used. The former parameter is the *efficiency* of the surfactant, the latter its *effectiveness*. These two parameters do not necessarily run parallel to each other—in fact, in many cases they run counter to each other.

Throughout this text, *efficiency* is used as a measure of the equilibrium concentration of surfactant in the liquid phase necessary to produce a given amount of effect, and *effectiveness* is used as a measure of the maximum effect the surfactant can produce in that interfacial process irrespective of concentration.

In dilute solutions of surface-active agents, the amount of change in any interfacial phenomenon produced by the adsorption of surfactant at the

Surfactants and Interfacial Phenomena, Fourth Edition. Milton J. Rosen and Joy T. Kunjappu.

interface is a function of the concentration of surfactant absorbed at the interface. Thus, *efficiency* is determined by the ratio of surfactant concentration at the interface to that in the bulk (liquid) phase, $C_{\text{interface}}/C_{\text{bulk}}$. This ratio is determined by the free energy change ΔG involved in the transfer of a surfactant molecule from the interior of the bulk phase to the interface by the equation $C_{\text{interface}}/C_{\text{bulk}} = \exp(-\Delta G/RT)$, where $R = 1.99$ cal (or 8.31 J)/deg/mol and T = absolute temperature; therefore, efficiency is related to the free energy change associated with that transfer.

The advantage of measuring the effect of a surfactant on some interfacial phenomenon by some parameter that is related to the free energy change associated with the action of the surfactant in that phenomenon is that the total free energy change can be broken into the individual free energy changes associated with the action of the various structural groupings in the molecule; that is, $\Delta G_{\text{total}} = \Sigma_i \, \Delta G_i$, where ΔG_i is the free energy change associated with the action of any structural group in the molecule. This enables correlations to be made between the various structural groupings in the surfactant and its interfacial properties. In this fashion, the efficiency with which a surfactant is adsorbed at an interface can be related to the various structural groups in the molecule.

Since the effect of a surfactant on an interfacial phenomenon is a function of the concentration of surfactant at the interface, we can define the *effectiveness* of a surfactant in adsorbing at an interface as the maximum concentration that the surfactant can attain at that interface, that is, the surface concentration of surfactant at surface saturation. The effectiveness of adsorption is related to the interfacial area occupied by the surfactant molecule; the smaller the effective cross-sectional area of the surfactant at the interface, the greater its effectiveness of adsorption. Effectiveness of adsorption, therefore, depends on the structural groupings in the surfactant molecule and its orientation at the interface. Another parameter characterizing the performance of surfactants, important in high-speed interfacial phenomena such as wetting and spreading, is the *rate* of adsorption of the surfactant at the relevant interface(s). This will be discussed in Section IV of Chapter 5.

Before going further into the adsorption of surfactants at interfaces, it is advisable to discuss the so-called electrical double layer at interfaces, since this is necessary for an understanding of the electrical aspects of adsorption.

I. THE ELECTRICAL DOUBLE LAYER

At any interface, there is always an unequal distribution of electrical charges between the two phases. This unequal distribution causes one side of the interface to acquire a net charge of a particular sign and the other side to acquire a net charge of the opposite sign, giving rise to a potential across the interface and the so-called *electrical double layer.* Of course, since overall electrical neutrality must be maintained, the net charge on one side of the

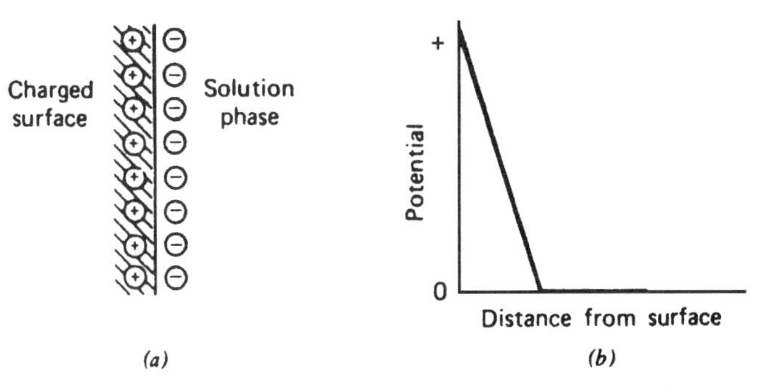

FIGURE 2.1 Helmholtz model of the electrical double layer: (a) distribution of counterions in the vicinity of the charged surface; (b) variation of electrical potential with distance from the charged surface.

interface must be balanced by an exactly equal net charge of opposite sign on the other side of the interface.

A major problem for investigation has been the determination of the exact distribution of the neutralizing charges *(counterions* or *gegenions)* in the solution surrounding a charged surface, since this distribution determines the rate at which the electrical potential will change with distance from the charged surface. An early theory concerning the distribution of these neutralizing charges in the solution surrounding a plane charged surface was that of Helmholtz (von Helmholtz, 1879), who envisioned all the counterions as being lined up parallel to the charged surface at a distance of about one molecular diameter (Figure 2.la). According to this model, the electrical potential should fall rapidly to zero within a very short distance from the charged surface (Figure 2.lb).

This model allowed Helmholtz to treat the electrical double layer mathematically as a parallel plate condenser. However, this model was untenable, since thermal agitation tends to diffuse some of the counterions throughout the solution. Accordingly, it was superseded by a model proposed by Gouy (1910, 1917) and Chapman (1913), who envisioned a diffuse distribution of the counterions, with the concentration of the counterions (and the potential) falling off rapidly at first with distance from the charged surface, because of a screening effect (Figure 2.2a), and then more and more gradually with distance (Figure 2.2b). This model, useful for planar charged surfaces with low charge densities, or for distances not too close to the surface, was inadequate for surfaces with high charge densities, especially at small distances from the charged surface, since it neglected the ionic diameters of the charges in solution and treated them as point charges. It was therefore modified by Stern (1924), who divided the solution side of the double layer into two parts: (1) a layer of strongly held counterions, adsorbed close to the charged surface on fixed sites (to correct the basic defect of the Gouy–Chapman model), and (2) a diffuse layer of counterions similar to that of their model (Figures 2.3 and

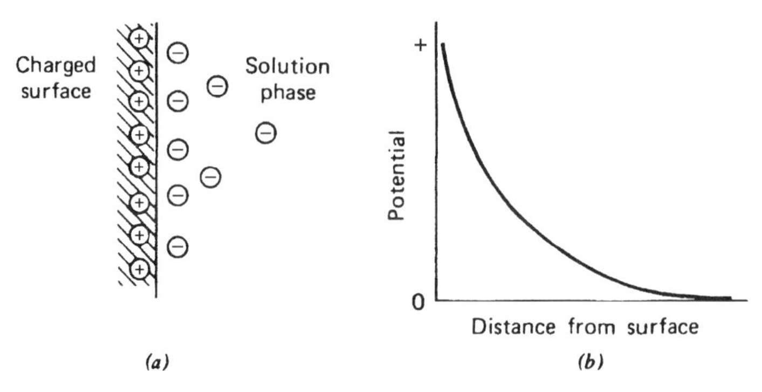

FIGURE 2.2 Gouy–Chapman model of the electrical double layer: (a) distribution of counterions in the vicinity of the charged surface; (b) variation of electrical potential with distance from the charged surface.

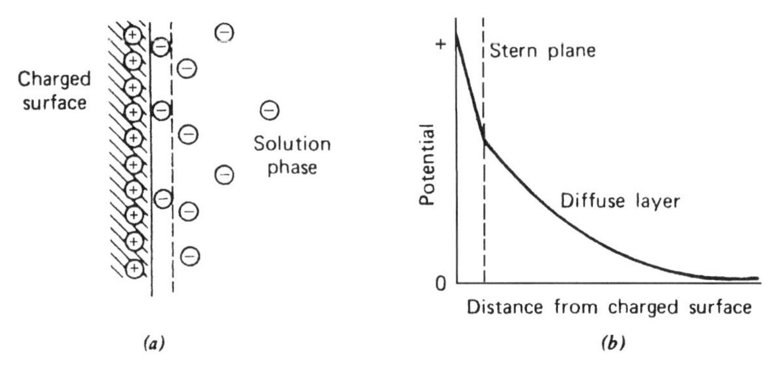

FIGURE 2.3 Stern model of the electrical double layer: (a) distributioin of counterions in the vicinity of the charged surface; (b) variation of electrical potential with distance from the charged surface.

2.4). According to this model, the electrical potential drops rapidly in the fixed portion (Stern layer) of the double layer and more gradually in the diffuse portion. The fixed counterions in the Stern layer may even change the sign of the potential resulting from the charged surface (Figure 2.4).

Mathematical treatment of the diffuse portion of the electrical double layer (Adamson, 1976) yields the very useful concept of an effective thickness $1/\kappa$ of that layer. This is the distance from the charged surface into the solution within which the major portion of electrical interactions with the surface can be considered to occur. The effective thickness, often called the *Debye length*, is given by

$$\frac{1}{\kappa} = \left(\frac{\varepsilon_r \varepsilon_0 RT}{4\pi F^2 \sum_i C_i Z_i^2} \right)^{1/2} \tag{2.1}$$

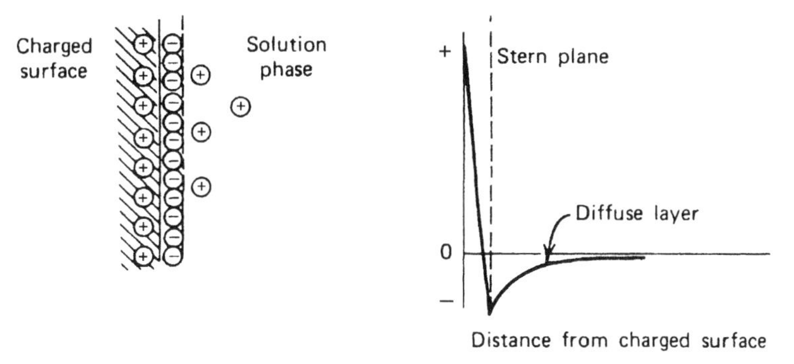

FIGURE 2.4 Stern model of the electrical double layer, showing reversal of the sign of the charged surface caused by adsorption of counterions in the Stern layer.

where $\varepsilon_r = \varepsilon/\varepsilon_0$ = the relative static permittivity or dielectric constant of the solution (ε = the static permittivity of the solution and ε_0 = the permittivity of a vaccum),

R = the gas constant,

T = the absolute temperature,

F = the Faraday constant,

C_i = the molar concentration of any ion in the solution phase.

From the preceding relation, it is apparent that $1/\kappa$ is inversely proportional to the valence Z of the ions in the solution phase and to the square root of their concentrations. It is also directly proportional to the square roots of the absolute temperature and the relative static permittivity (or dielectric constant) of the medium. It is therefore to be expected that in a solvent of high dielectric constant, such as water, electrical effects extend much further into the solution phase than in a solvent of low dielectric constant, such as a hydrocarbon. Also, in the presence of an electrolyte, electrical effects have shorter ranges than in its absence that is, the electrical double layer is *compressed*.

For 1:1 electrolytes at room temperature in aqueous solution, $1/\kappa \approx 3$ Å for 1 M, 10 Å for 0.1 M, 30 Å for 0.01 M, 100 Å for 1×10^{-3} M, and 300 Å for 1×10^{-4} M solutions.

A term often associated with the electrical double layer, and one that is often misused, is the *zeta potential*, or electrokinetic potential. This is the potential of a charged particle as calculated from electrokinetic phenomena (electroosmosis, electrophoresis, streaming potential, or sedimentation potential). It is the potential of the charged surface at the plane of shear between the particle and the surrounding solution as the particle and the solution move with respect to each other. Zeta potentials are conveniently measured (Adamson, 1976), and it is very tempting to place this plane of shear at the solution side of the Stern layer, since this is the boundary of the fixed ion layer and would give us an experimentally calculable value for the potential at that boundary. Unfortunately, although some authors do identify the zeta potential with the potential at the solution edge of the Stern layer, the plane of shear is *not* necessarily at the solution edge of the Stern layer and is at some undetermined point somewhere farther out in the diffuse layer due to bound water moving with the charged particle or being held to it when the solution moves. The zeta potential is consequently smaller in magnitude than the Stern potential, but, unfortunately, exactly how much smaller is not known definitely.

II. ADSORPTION AT THE SOLID–LIQUID INTERFACE

The adsorption of surfactants at the solid–liquid interface is strongly influenced by a number of factors: (1) the nature of the structural groups on the solid surface—whether the surface contains highly charged sites or essentially nonpolar groupings and the nature of the atoms of which these sites or groupings are constituted; (2) the molecular structure of the surfactant being adsorbed (the adsorbate)—whether it is ionic or nonionic, and whether the hydrophobic group is long or short, straight-chain or branched, aliphatic, or aromatic; and (3) the environment of the aqueous phase—its pH, its electrolyte content, the presence of any additives such as short-chain polar solutes (alcohol, urea, etc.), and its temperature. Together these factors determine the mechanism by which adsorption occurs, and the efficiency and effectiveness of adsorption. For a detailed review of cationic adsorption at the solid–liquid interface, see Atkins et al. (2003).

A. Mechanisms of Adsorption and Aggregation

There are a number of mechanisms by which surface-active solutes may adsorb onto solid substrates from aqueous solution. In general, adsorption of surfactants involves single ions (Kölbel and Hörig, 1959; Griffith and Alexander, 1967) rather than micelles (Chapter 3):

1. *Ion Exchange (Figure 2.5).** Involves replacement of counterions adsorbed onto the substrate from the solution by similarly charged surfactant ions (Somasundaran et al., 1964; Law and Kunze, 1966; Wakamatsu and Fuerstenau, 1968; Rupprecht and Liebl, 1972).

* It should be clearly understood that the rigid arrangement of the hydrophobic groups depicted in Figures 2.5 to 2.10 is only for convenience in illustrating the mechanisms of adsorption. In reality, the hydrophobic groups may assume all conformations, including the interweaving of hydrophobic chains of adjacent molecules, consistent with their surface areas per molecule and the relative orientation of their hydrophobic and hydrophilic groups with respect to the interface.

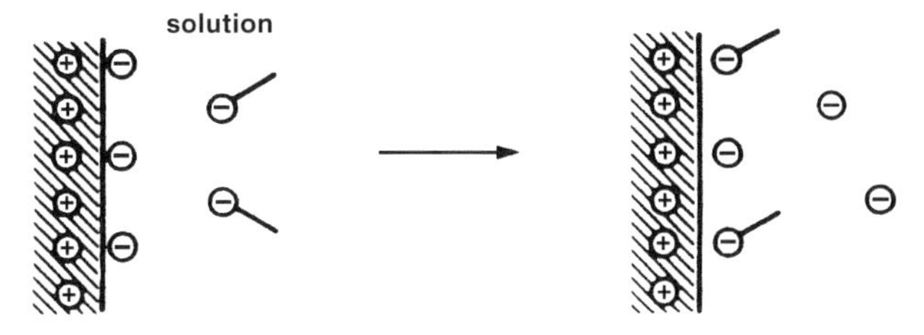

FIGURE 2.5 Ion exchange. Reprinted with permission from Rosen (1975).

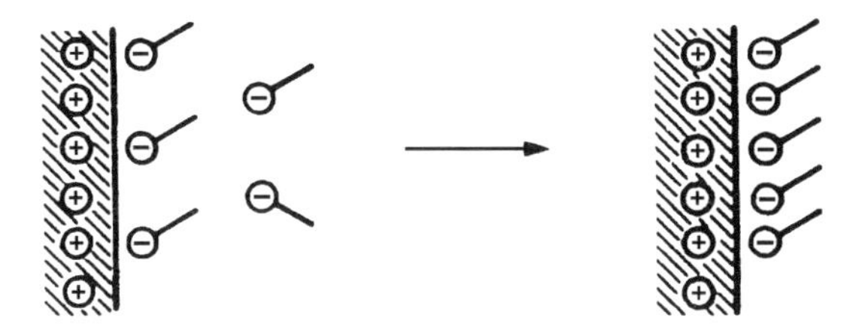

FIGURE 2.6 Ion pairing. Reprinted with permission from Rosen (1975).

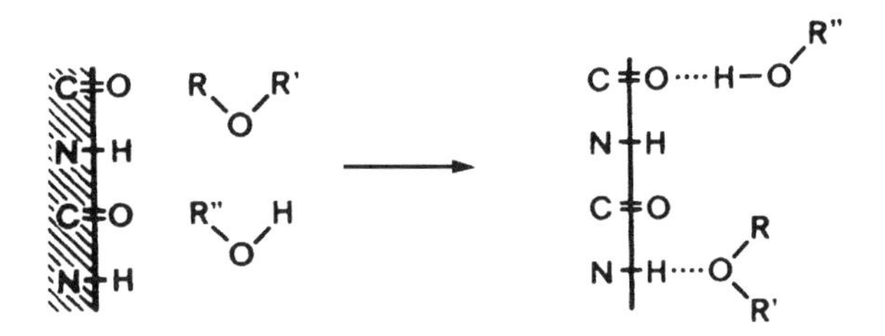

FIGURE 2.7 Hydrogen bonding. Reprinted with permission from Rosen (1975).

2. *Ion Pairing (Figure 2.6).* Adsorption of surfactant ions from solution onto oppositely charged sites unoccupied by counterions (Law and Kunze, 1966; Rupprecht and Liebl, 1972).
3. *Acid–Base Interaction (Fowkes, 1987).* Via either hydrogen bond formation (Figure 2.7) between substrate and adsorbate (Law and Kunze, 1966; Snyder, 1968; Rupprecht and Liebl, 1972) or Lewis acid–Lewis base reaction (Figure 2.8).
4. *Adsorption by Polarization of π Electrons.* Occurs when the adsorbate contains electron-rich aromatic nuclei and the solid adsorbent has strongly positive sites. Attraction between electron-rich aromatic nuclei of the adsorbate and positive sites on the substrate results in adsorption (Snyder, 1968).
5. *Adsorption by Dispersion Forces.* Occurs via London–van der Waals dispersion forces acting between adsorbent and adsorbate molecules

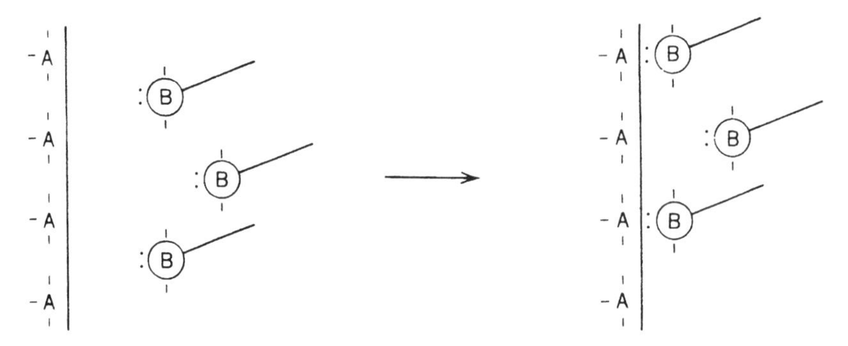

FIGURE 2.8 Adsorption via Lewis acid–Lewis base interaction.

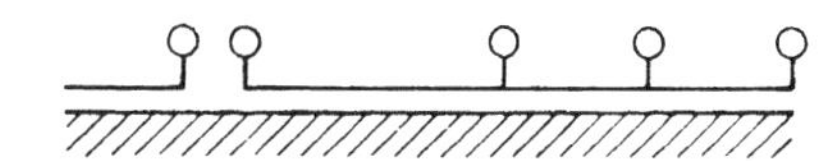

FIGURE 2.9 Adsorption via disperison forces on nonpolar surface.

(Figure 2.9). Adsorption by this mechanism generally increases with an increase in the molecular weight of the adsorbate and is important not only as an independent mechanism, but also as a supplementary mechanism in all other types. For example, it accounts in part for the pronounced ability of surfactant ions to displace equally charged simple inorganic ions from solid substrates by an ion exchange mechanism (Kölbel and Hörig, 1959; Law and Kunze, 1966).

6. *Hydrophobic Bonding.* Occurs when the combination of mutual attraction between hydrophobic groups of the surfactant molecules and their tendency to escape from an aqueous environment becomes large enough to permit them to adsorb onto the solid adsorbent by aggregating their chains (Wakamatsu and Fuerstenau, 1968; Dick et al., 1971; Giles et al., 1974; Gao et al., 1987).

These surfactant aggregates, termed *hemimicelles* (hemimicellar aggregates—Gaudin et al., 1955; Fuerstenau, 1957; Somasundaran et al., 1964, 1966; Somasundaran and Fuerstenau, 1966), or *admicelles* (adsorbed micelles—Harwell et al., 1985), or *solloids* (adsorbed aggregates on solids—Somasundaran and Kunjappu, 1989; Kunjappu, 1994a, 1994b; Kunjappu and Somasundaran, 1995) were assumed to be more or less flat. Later investigations (Manne and Gaub, 1995; Grosse and Estel, 2000; Wolgemuth et al., 2000) indicate that these aggregates, when in the form of monolayers (Figure 2.10a), may also be hemispherical and, when in the form of bilayers (Figure 2.10b), may also be in the form of cylinders (Figure 2.10b). They will be designated *surface aggregates* to distinguish them from micelles (Chapter 3) in the solution phase. Flat sheets of a large number of surfactants have been found (Grant et al., 1998) on

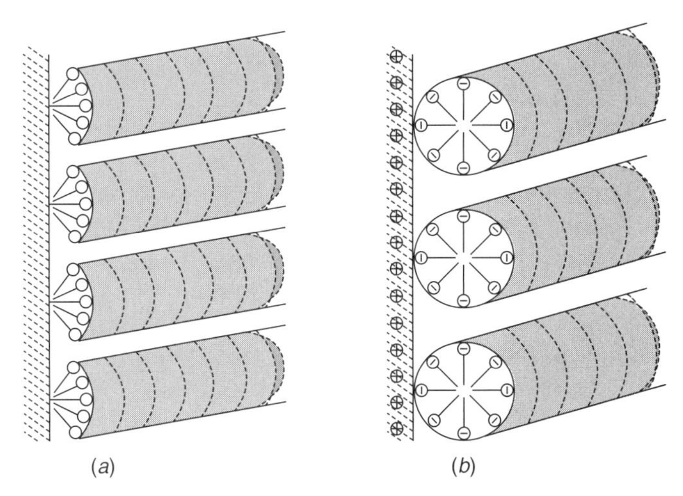

FIGURE 2.10 Adsorption from aqueous solution via hydrophobic bonding on (a) an uncharged surface and (b) via electrostatic interaction on an oppositely charged surface.

amorphous silica made hydrophobic by reaction with diethyloctylchlorosilane, hemispherical structures have been observed (Manne et al., 1994; Wanless and Ducker, 1996; Grant and Ducker, 1997; Jaschke et al., 1997) on hydrophobic graphite and gold, and spherical or cylindrical structures have been seen on hydrophilic silica (Subramanian and Ducker, 2000). Critical analysis of the adsorption of ionic surfactants from aqueous solution onto a polar solid show that the first layer adsorbs with its hydrophilic group oriented toward the solid forming a hemimicelle with its hydrophobic groups oriented toward the aqueous phase. With increasing concentration of the surfactant, additional surfactant adsorbs onto the adsorbed initial layer of hemimicelle with hydrophobic group oriented toward the adsorbed hemimicelle and its hydrophilic group oriented toward the aqueous phase (*reverse orientation model*—Kunjappu and Somasundaran, 1989; Kunjappu, 1994a, 1994b). In aqueous systems, the structures formed depend upon the interaction of the surfactant molecules with the solid surface in such fashion as to minimize exposure of the hydrophobic groups to the aqueous phase.

Orientation of the adsorbed surfactant molecules with their hydrophobic groups predominantly away from the solid substrate will make the surface more hydrophobic than before surfactant adsorption; orientation with their hydrophilic groups predominantly away from the surface will make it more hydrophilic than before adsorption of the surfactant. There are a number of simple ways of determining the predominant orientation of the surfactant molecules adsorbed on the solid: (1) If the solid can be obtained in the form of a smooth, nonporous planar film or plate, then measurement of the contact angle (Chapter 6, Section IA2) of a drop of water placed on the solid surface before and after surfactant adsorption can be used. The greater the contact angle, the greater the hydrophobicity of the surface. If the solid can be obtained in the form of finely divided particles, then if surfactant adsorption has made it more hydrophilic, it will disperse better in water after surfactant adsorption than before; if it has made it more hydrophobic, then the particles will either float to the surface when shaken with water or settle out of the water more rapidly than before surfactant adsorption. Alternatively, the particles may be shaken with a mixture of equal volumes of water and a nonpolar solvent (e.g., hexane). If the particles have become more hydrophobic by surfactant adsorption, they will disperse better in the nonpolar phase than before surfactant adsorption; if they have become more hydrophilic, they will disperse better in the aqueous phase than before surfactant adsorption.

Direct observation of the surface aggregates with atomic and molecular resolution is possible now with scanning probe microscopies like atomic force microscopy (AFM; see Chapter 14—Section IIIB) and scanning tunneling microscopy (STM) (Burgess et al., 1999; Gross, 2011). Developed in the 1980s, these techniques rely on the information gathered by a tungsten tip (in STM) or a diamond tip (in AFM) with atomic dimension that moves very close on the surface. STM probes the tunneling current between the scanning tip and the surface; AFM translates the interatomic forces between the tip and the surface into contour maps of the surface with atomic resolution. STM requires an electrically conducting surface whereas AFM can be applied to nonconducting surfaces also. They can provide magnifications up to 100 million times.

B. Adsorption Isotherms

At the solid–liquid interface, we are interested in determining (1) the amount of surfactant adsorbed per unit mass or unit area of the solid adsorbent, that is, the surface concentration of the surfactant (*adsorbate*) at a given temperature, since this is a measure of how much of the surface of the adsorbent has been covered, and hence changed, by the adsorption; (2) the equilibrium concentration of surfactant in the liquid phase required to produce a given surface concentration of surfactant at a given temperature, since this measures the *efficiency* with which the surfactant is adsorbed; (3) the concentration of surfactant on the adsorbent at surface saturation at a given temperature, since this determines the *effectiveness* with which the surfactant is adsorbed; (4) the orientation of the adsorbed surfactant and any other parameters that may shed light on the mechanism by which the surfactant is adsorbed, since a knowledge of the mechanism allows us to predict how a surfactant with a given molecular structure will adsorb at the interface; and (5) the effect of adsorption on other properties of the adsorbent. An adsorption isotherm is a mathematical expression that relates the concentration of adsorbate at the interface to its equilibrium concentration in the liquid phase. Since most of the information that we desire can be obtained from the adsorption isotherm, the isotherm is the usual method of describing adsorption at the liquid–solid interface.

The fundamental equation for calculating the amount of one component (component 1) of a binary solution adsorbed onto a solid adsorbent is (Aveyard and Haydon, 1973, p. 201)

$$\frac{n_0 \Delta \chi_1}{m} = n_1^s \chi_2 - n_2^s \chi_1 \tag{2.2}$$

where n_0 = the total number of moles of solution before adsorption,

$\Delta\chi_1 = \chi_{1,0} - \chi_1$,

$\chi_{1,0}$ = the mole fraction of component 1 before adsorption,

χ_1, χ_2 = the mole fractions of components 1 and 2 at adsorption equilibrium,

m = the mass of the adsorbent, in grams,

n_1^s, n_2^s = the number of moles of components 1 and 2 adsorbed per gram of adsorbent at adsorption equilibrium.

When the liquid phase is a dilute solution of a surfactant (component 1) that is much more strongly adsorbed onto the solid adsorbent than the solvent (component 2), then $n_0\Delta\chi_1 \approx \Delta n_1$ where Δn_1 = the change in the number of moles of component 1 in solution, $n_2^s \approx 0$, and $x_2 \approx 1$. Thus,

$$n_1^s = \frac{\Delta n_1}{m} = \frac{\Delta C_1 V}{m} \tag{2.3}$$

where $\Delta C_1 = C_{1,0} - C_{1,}$

$C_{1,0}$ = the molar concentration (in moles/liter) of component 1 before adsorption, in the liquid phase,

C_1 = the molar concentration of component 1 (the surfactant) at adsorption equilibrium, in the liquid phase,

V = the volume of the liquid phase, in liters.

For n_1^s to be determined with suitable accuracy, the value of ΔC_1, the change in the molar concentration of the surfactant solution upon adsorption, must be appreciable when compared to $C_{1,0}$, its initial concentration. For this to be so, the solid adsorbent must have a large surface area/gram (i.e., be finely divided).

For dilute solutions of surfactants then, the number of moles of surface-active solute adsorbed per unit mass of the solid substrate can be calculated from the concentrations of the solute in the liquid phase before and after the solution is mixed with the finely divided solid adsorbent and the mixture is shaken until equilibrium has been reached. Then n_1^s is plotted against C_1 to yield the adsorption isotherm. A variety of analytical techniques are available for determining the change in concentration of the surfactant (Rosen and Goldsmith, 1972).

The surface concentration Γ_1, in mol/cm^2, of the surfactant may be calculated when a_s, the surface area per unit mass of the solid adsorbent, in cm^2/g (the specific surface area), is known:

$$\Gamma_1 = \frac{\Delta C_1 V}{a_s \times m} \tag{2.4}$$

For solid substrates that cannot be obtained in finely divided form, but can be obtained as a planar, smooth, nonporous surface on film, surface concentrations can sometimes be calculated from contact angles (Chapter 6, Section IA1).

The adsorption isotherm can then be plotted in terms of Γ_1 as a function of C_1. The surface area per adsorbate molecule on the adsorbent a_1^s in square angstroms* is

$$a_1^s = \frac{10^{16}}{N\Gamma_1} \tag{2.5}$$

where N is Avogadro's number.

1. The Langmuir Adsorption Isotherm A type of adsorption isotherm commonly observed in adsorption from solutions of surfactants is the Langmuir-type isotherm (Langmuir, 1918), expressed by

$$\Gamma_1 = \frac{\Gamma_m C_1}{C_1 + a} \tag{2.6}$$

where Γ_m = the surface concentration of the surfactant, in mol/cm^2, at monolayer adsorption,

C_1 = the concentration of the surfactant in the liquid phase at adsorption equilibrium, in mol/L,

a = a constant [= 55.3 exp($\Delta G^0/RT$)], in mol/L, at absolute temperature T, in the vicinity of room temperature and where ΔG^0 is free energy of adsorption at infinite dilution.

This type of adsorption is valid in theory only under the following conditions (Betts, 1960):

1. The adsorbent is homogeneous.
2. Both solute and solvent have equal molar surface areas.
3. Both surface and bulk phases exhibit ideal behavior (e.g., no solute–solute or solute–solvent interactions in either phase).
4. The adsorption film is monomolecular.

* For the value in nm^2, divide by 10^2.

Many surfactant solutions show Langmuir-type behavior even when these restrictions are not met.

When adsorption follows the Langmuir equation, determination of the values of Γ_m and a permits calculation of the area per adsorbed molecule at surface saturation and the free energy of adsorption at infinite dilution. To determine whether adsorption is following the Langmuir equation and to permit calculation of the values of Γ_m and a, the equation is usually transformed into linear form by inverting it. Thus,

$$\frac{C_1}{\Gamma_1} = \frac{C_1}{\Gamma_m} + \frac{a}{\Gamma_m} \tag{2.7}$$

or

$$\frac{1}{\Gamma_1} = \frac{a}{\Gamma_m C_1} + \frac{1}{\Gamma_m} \tag{2.8}$$

A plot of C_1/Γ_1 versus C_1 (Equation 2.7) should be a straight line whose slope is $1/\Gamma_m$ and whose intercept with the ordinate is $1/\Gamma_m$. Alternatively, a plot of $1/\Gamma_1$ versus $1/C_1$ should be a straight line with slope = a/Γ_m and intercept = $1/\Gamma_m$ (Equation 2.8).

When the specific surface area of the solid adsorbent a_s is not known, n_1^s may be plotted against C_1, and the Langmuir equation takes the form

$$n_1^s = \frac{n_m^s C_1}{C_1 + a} \tag{2.9}$$

The linear forms are

$$\frac{C_1}{n_1^s} = \frac{C_1}{n_m^s} + \frac{a}{n_m^s} \tag{2.10}$$

and

$$\frac{1}{n_1^s} = \frac{a}{n_m^s C_1} + \frac{1}{n_m^s} \tag{2.11}$$

From Equation 2.6, $a = C_1$ when $\Gamma_1 = \Gamma_m/2$; from Equation 2.9, when $n_1^s = n_m^s/2$. Therefore, a may also be determined from a plot of C_1^s versus C_1 (or n_1^s vs. C_1) at the point where $\Gamma_1 = \Gamma_m/2$ (or where $n_1^m = n_m^s/2$, that is, it equals the equilibrium surfactant concentration in the liquid phase required for one-half monolayer coverage of the adsorbent surface. In mol/L, $a = 55.3 \exp(\Delta G^0/RT)$ in the vicinity of room temperature, and

$$-\log a = -\Delta G^0 / 2.3RT - 1.74 \tag{2.12}$$

Since $-\log a$ is therefore a function of the free energy change involved in the transfer of the surfactant molecule from the liquid phase to the solid substrate, it is a suitable measure of the efficiency of adsorption of the surfactant when adsorption follows the Langmuir equation.

The fact that experimental adsorption data fit the Langmuir equation does *not* mean that the assumptions on which the Langmuir model is based are fulfilled. In the case of surfactants, these assumptions, particularly the absence of lateral interactions, are almost never valid. In spite of this, many surfactants show Langmuir-type adsorption from solution because of the mutual compensation of several factors that affect the shape of the Langmuir isotherm. Some of these factors and the manner in which they modify the shape of the isotherm are as follows (Kitchener, 1965):

1. *Micellization of the Surfactant.* This causes a flattening of the curve, possibly below the level for close packing, because of the almost insignificant increase in activity of the surfactant with increase in its concentration in the liquid phase, once micellation occurs (see Chapter 3).
2. *Surface Potential.* If of the same sign as the surfactant ion, this reduces adsorption and thus reduces the slope of the isotherm; if of opposite sign, it increases adsorption and the slope of the isotherm.
3. *Heterogeneity of the Solid Adsorbent.* Adsorption onto high-energy sites on the substrate yields isotherms with higher slopes than adsorption onto low-energy sites. The summation of adsorptions onto sites of varying energy may yield an isotherm resembling a BET (multilayer) isotherm or a Freundlich isotherm ($n_1^s = kC_1^{1/n}$, where k and n are constants, with n generally greater than unity).
4. *Lateral Interaction.* Where lateral interactions are attractive, a common situation with surfactants, the slope of the isotherm becomes steeper and may become S-shaped or stepped (Kitchener, 1965; Giles et al., 1974).

A two-step adsorption mechanism has been proposed (Gu and Zhu, 1990; Gu et al., 1992) for the various types of S-shaped adsorption isotherms (non-Langmuir) that are sometimes obtained. In the first step, the surfactant molecules are adsorbed as individual molecules or ions. In the second step, the adsorption increases dramatically as surface aggregates form through interaction of the hydrophobic chains of the surfactant molecules with each other.

The authors have suggested the equation

$$\Gamma_1 = \Gamma_\infty K C_1^n / (1 + K C_1^n) \tag{2.12a}$$

where Γ_∞ is the limiting surfactant adsorption at high C_1 concentration, K is the equilibrium constant of the surface aggregation process, and n is the average aggregation number of the surface aggregate as a general adsorption isotherm.

Equation 2.12a can be transformed to the logarithmic form:

$$\log[\Gamma_1/(\Gamma_\infty - \Gamma_1)] = \log K + n\log C \tag{2.12b}$$

A plot of $\log[\Gamma_1/(\Gamma_\infty - \Gamma_1)]$ versus $\log C$ permits evaluation of K and n when the data give a straight line. When $n = 1$, Equation 2.12a becomes the Langmuir adsorption isotherm in the form $\Gamma_1 = \Gamma_\infty KC_1/(1 + KC_1)$ where $K + 1/a$. If surface aggregation occurs, then n should be greater than 1.

Adsorption isotherms of poorly purified solutes on heterogeneous or impure adsorbents often pass through a maximum in adsorption. Although such phenomena are possible in adsorption from concentrated solutions or from the gas phase, it is difficult to justify on theoretical grounds the existence of these phenomena in adsorption from dilute solutions of surfactants. They often disappear upon purification of the adsorbent and the solute, and are believed to be due to the presence of impurities (Kitchener, 1965).

C. Adsorption from Aqueous Solution onto Adsorbents with Strongly Charged Sites

Adsorbents with strongly charged sites include such substrates as wool and other polyamides at pH above and below their isoelectric points, oxides such as alumina above and below their points of zero charge, and cellulosic and silicate surfaces at high pH. Adsorption onto these surfaces is a complex process during which adsorption of the solute may occur successively by ion exchange, ion pairing, and hydrophobic bonding mechanisms.

1. Ionic Surfactants The adsorption isotherm for an ionic surfactant onto an oppositely charged substrate, for example, sodium alkanesulfonates (Somasundaran et al., 1966) and alkylbenzenesulfonates (Dick et al., 1971; Scamehorn et al., 1982) on positively charged Al_2O_3, is typically S-shaped. The shape of the isotherm (Figure 2.11) is believed to reflect three distinct modes of adsorption. In region 1, the surfactant adsorbs mainly by ion exchange, possibly with the hydrophobic group more or less prone on the substrate (Scamehorn et al., 1982). The charge density, or potential at the Stern layer of the solid, remains almost constant. In region 2, there is a marked increase in adsorption resulting from interaction of the hydrophobic chains of oncoming surfactant ions with those of previously adsorbed surfactant. This aggregation of the hydrophobic groups, which may occur at concentrations well below the critical micelle concentration (CMC) (Chapter 3, Section I) of the surfactant, has been called *hemimicelle formation* (Wakamatsu and Fuerstenau, 1968) or *cooperative adsorption* (Giles et al., 1974). In this adsorption region, the original charge of the solid is neutralized by the adsorption of oppositely charged surfactant ions and eventually reversed, so that at the end of region 2, the solid has acquired a charge of the same sign as the surfactant ion. The processes in regions 1 and 2 are diagrammed in Figure 2.12. In region 3, the slope of the

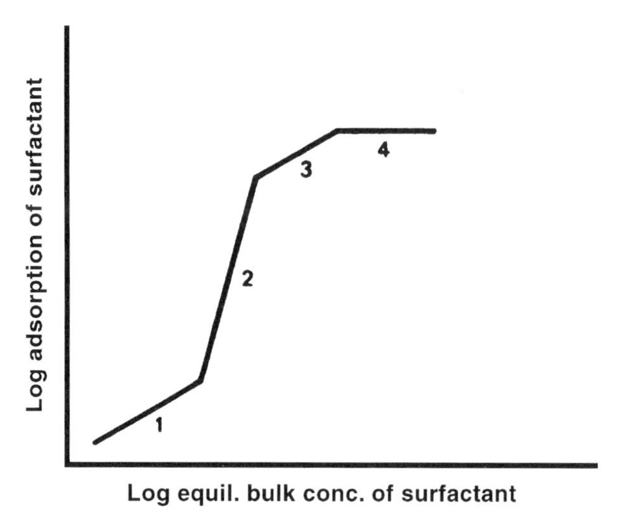

FIGURE 2.11 S-shaped adsorption isotherm for an ionic surfactant on an oppositely charged substrate.

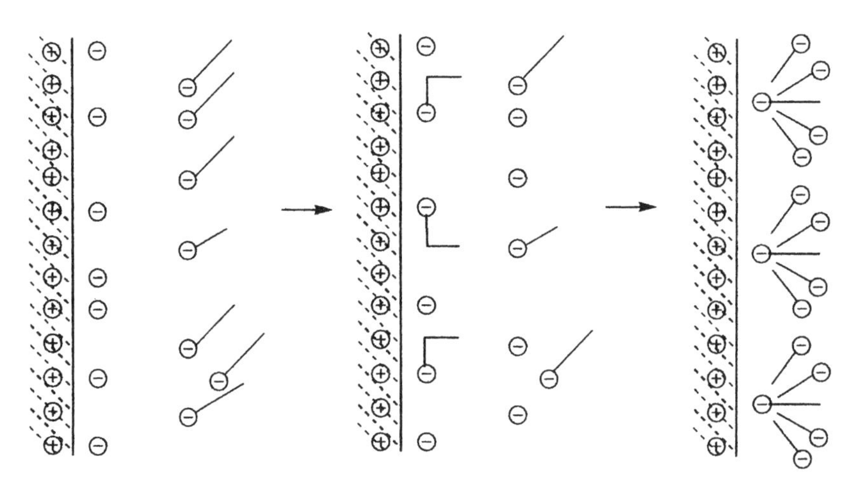

FIGURE 2.12 Adsorption of anionic surfactant onto an oppositely charged substrate by ion exchange (region 1) and by aggregation of oncoming hydrophobic groups with those of previously adsorbed surfactants (region 2).

isotherm is reduced because adsorption now must overcome electrostatic repulsion between the oncoming ions and the similarly charged solid. Adsorption in this fashion is usually complete (region 4) in the neighborhood of the CMC (Griffith and Alexander, 1967; Greenwood et al., 1968; Groot, 1968; Somasundaran et al., 1983) since adsorption appears to involve single ions rather than micelles.

When the mutual attraction of the hydrophobic groups is insufficient to overcome the mutual repulsion of the ionic hydrophilic groups (e.g., when the hydrophobic groups are short or when there are two or more similarly charged ionized groups in the surfactant molecule and the ionic strength of the aqueous solution is low), then aggregation of the hydrophobic chains may not occur and region 2 may be absent. In these cases the isotherm may be inverted L-shaped when the ionic strength of the solution phase is low, with adsorption in region 1 continuing by ion exchange and ion pairing until the original charge of the substrate has been neutralized and the substrate acquires a charge of the same sign as the surfactant ion. At this point the slope of the isotherm will be reduced, and adsorption will continue in the same manner as for region 3 of the S-shaped isotherm. When the ionic strength of the aqueous solution is high, electrical interactions in both regions 1 and 3 are weak, and the slopes of the isotherm in these regions will tend to become equal. In this case, if aggregation of the hydrophobic groups does not occur to a significant extent, the shape of the isotherm may be pseudolinear in regions 1–3 (Rosen and Nakamura, 1977).

The *efficiency* of adsorption due to nonelectrical interaction of an ionic surfactant onto an oppositely charged solid adsorbent can be measured by the log of the reciprocal, or negative log, of the equilibrium concentration of the surfactant in the liquid phase ($\log 1/C_0$ or $-\log C_0$) when the potential at the Stern layer becomes zero (point of zero charge, [pzc]). This follows from the Stern–Grahame equation for adsorption in the Stern layer (Grahame, 1947) at low surface coverage with the free energy change for the adsorption broken into electrical and nonelectrical parts:

$$\Gamma_\delta = 2 \times 10^{-3} r C_1 \exp \frac{-ZF\psi_\delta - \phi}{RT} \tag{2.13}$$

where Γ_δ = the concentration of surfactant ions adsorbed in the Stern layer of a positively charged surface, in mol/cm^2,

r = the effective radius of the adsorbed surfactant ion, in cm,

Z = the valence of the surfactant ion, including the sign,

F = the Faraday constant,

ψ_δ = the potential in the Stern plane,

ϕ = the nonelectrical free energy change upon adsorption.

When ψ_δ is zero,

$$C_{1(0)} = \frac{(\Gamma_\delta)_0}{2 \times 10^{-3} r} \exp \frac{\phi}{RT} \tag{2.14}$$

where $C_{1(0)}$ and $(\Gamma_\delta)_0$ represent the concentrations at zero potential in the Stern layer, and

$$-\log C_{1(0)} = -\log\frac{(\Gamma_\delta)_0}{2r} - \frac{\phi}{2.3RT} - 3 \tag{2.15}$$

It has been shown (Wakamatsu and Fuerstenau, 1968) that the variation in $\log[(\Gamma_\delta)_0/2r]$ is very small compared to the variation in ϕ with changes in the structure of the hydrophobic group, and therefore $-\log C_{1(0)}$ is essentially a function of ϕ and a suitable measure for the efficiency of adsorption due to nonelectrical interactions.

Since region 3 commences at the pzc (when the potential in the Stern layer reaches zero), in those adsorption isotherms where the distinction between regions 2 and 3 is clearly indicated by a change in slope, $C_{1(0)}$ can be taken as the concentration at which the slope decreases and region 3 commences.

An increase in the length of the hydrophobic group increases the efficiency of adsorption, because the free energy decrease associated both with the removal of the hydrophobic chain from contact with the water and with the tendency to aggregate or adsorb via dispersion forces, which increases with the increase in the length of the chain. Efficiencies calculated from data in the literature are listed in Table 2.1. For this purpose, the phenyl ring may be considered to have an effective length of about three and one-half carbon atoms in a straight carbon chain. Carbon atoms on short branches of an alkyl hydrophobic group, on the shorter portion of a hydrophobic group when the hydrophilic group is not terminally located, or between two hydrophilic groups, appear to have about one-half the effective length of a carbon atom on a straight alkyl chain with terminal hydrophilic groups. An increase in the size of the hydrophilic group also appears to increase the efficiency of adsorption by ion exchange or ion pairing.

The *effectiveness* of adsorption (i.e., the amount adsorbed at surface saturation), however, may increase, decrease, or show no change with increase in the length of the hydrophobic group, depending on the orientation of the adsorbate at the adsorbent–solution interface. If adsorption is perpendicular to the substrate surface in a close-packed arrangement, an increase in the length of a straight-chain hydrophobic group appears to cause no significant change in the number of moles of surfactant adsorbed per unit area of surface at surface saturation (Tamamushi and Tamaki, 1957), presumably because the cross-sectional area occupied by the chain oriented perpendicular to the interface does not change with increase in the number of units in the chain. In perpendicular orientation, moreover, the effectiveness of adsorption may be determined by the size of the hydrophilic group when the cross-sectional area of that group is greater than that of the hydrophobic chain; the larger the hydrophilic group, the smaller the amount adsorbed at surface saturation. If the arrangement is predominantly perpendicular but not close-packed, or if it is somewhat tilted away from the perpendicular, there may be some increase in effectiveness of adsorption with increase in the length of the hydrophobic group, resulting from greater van der Waals attraction and consequent closer packing of longer chains (Connor and Ottewill, 1971). The positioning of a

TABLE 2.1 Efficiency of Adsorption, $-\log C_{1(0)}$, of Surfactants on Solid Substrates

Surfactant	Solid Substrate	pH	Temperature (°C)	$-\log C_{1(0)}$	Reference
Decylammonium acetate	Quartz	6.5–6.9	22–25	1.7_5	Somasundaran et al. (1964)
Dodecylammonium acetate	Quartz	6.5–6.9	22–25	2.6_0	Somasundaran et al. (1964)
Tetradecylammonium acetate	Quartz	6.5–6.9	22–25	3.4_5	Somasundaran et al. (1964)
Hexadecylammonium acetate	Quartz	6.5–6.9	22–25	4.3_0	Somasundaran et al. (1964)
Octadecylammonium acetate	Quartz	6.5–6.9	22–25	5.1_5	Somasundaran et al. (1964)
Sodium decanesulfonate	α-Alumina	7.2 (I.S. = 2×10^{-3} *M*)	25	2.7_5	Wakamatsu and Fuerstenau (1968)
Sodium dodecanesulfonate	α-Alumina	4.2 (I.S. = 2×10^{-3} *M*)	25	4.4_0	Wakamatsu and Fuerstenau (1973)
Sodium dodecanesulfonate	α-Alumina	5.2 (I.S. = 2×10^{-3} *M*)	25	4.0_0	Wakamatsu and Fuerstenau (1973)
Sodium dodecanesulfonate	α-Alumina	6.2 (I.S. = 2×10^{-3} *M*)	25	3.8_5	Wakamatsu and Fuerstenau (1973)
Sodium dodecanesulfonate	α-Alumina	7.2 (I.S. = 2×10^{-3} *M*)	25	3.5_5	Wakamatsu and Fuerstenau (1968, 1973)
Sodium dodecanesulfonate	α-Alumina	8.2 (I.S. = 2×10^{-3} *M*)	25	3.3_5	Wakamatsu and Fuerstenau (1968, 1973)
Sodium dodecanesulfonate	α-Alumina	8.6 (I.S. = 2×10^{-3} *M*)	25	3.2_0	Wakamatsu and Fuerstenau (1968, 1973)
Sodium dodecanesulfonate	α-Alumina	6.9 (I.S. = 2×10^{-3} *M*)	45	3.6_2	Somasundaran et al. (1966)
Sodium tetradecanesulfonate	α-Alumina	7.2 (I.S. = 2×10^{-3} *M*)	25	4.2_5	Wakamatsu and Fuerstenau (1968)
Sodium hexadecanesulfonate	α-Alumina	7.2 (I.S. = 2×10^{-3} *M*)	25	5.0_0	Wakamatsu and Fuerstenau (1968)
$Na^{+-}O_3SC_6H_4(CH_2)_6C_6H_4SO_3^-Na^+$	α-Alumina	7.2 (I.S. = 2×10^{-3} *M*)	25	2.7_8	Rosen and Nakamura (1977)
$Na^{+-}O_3C_6H_4(CH_2)_8C_6H_4SO_3^-Na^+$	α-Alumina	7.2 (I.S. = 2×10^{-3} *M*)	25	3.3_2	Rosen and Nakamura (1977)
$Na^{+-}O_3C_6H_4(CH_2)_{10}C_6H_4SO_3^-Na^+$	α-Alumina	7.2 (I.S. = 2×10^{-3} *M*)	25	3.7_0	Rosen and Nakamura (1977)

(Continued)

TABLE 2.1 (*Continued*)

Surfactant	Solid Substrate	pH	Temperature (°C)	$-\log C_{1(0)}$	Reference
$Na^{+-}O_3C_6H_4(CH_2)_{12}C_6H_4\ SO_3^-Na^+$	α-Alumina	7.2 (I.S. = 2×10^{-3} *M*)	25	4.2_4	Rosen and Nakamura (1977)
Dodecylammonium chloride	Silver iodide			4.0	Ottewill and Rastogi (1960)
Octylpyridinium bromide	Silver iodide			3.8	Ottewill and Rastogi (1960)
Dodecylpyridinium bromide	Silver iodide			5.5_4	Ottewill and Rastogi (1960)
Cetylpyridinium bromide	Silver iodide			6.2_5	Ottewill and Rastogi (1960)
Dodecyltrimethylammonium bromide	Silver iodide			5.0	Ottewill and Rastogi (1960)
Dodecylquinolinium bromide	Silver iodide			6.1_3	Ottewill and Rastogi (1960)
Sodium decyl sulfate	Silver iodide sol	pAg = 3 (I.S. = 1.5×10^{-3} *M*)	20 ± 1	3.4_0	Watanabe (1960)
Sodium dodecyl sulfate	Silver iodide sol	pAg = 3 (I.S. = 1.5×10^{-3} *M*)	20 ± 1	4.3_8	Watanabe (1960)
Sodium dodecanesulfonate	Silver iodide sol	pAg = 3 (I.S. = 1.5×10^{-3} *M*)	20 ± 1	3.9_2	Watanabe (1960)
Sodium tetradecyl sulfate	Silver iodide sol	pAg = 3 (I.S. = 1.5×10^{-3} *M*)	20 ± 1	4.7_8	Watanabe (1960)
Sodium dodecylbenzenesulfonate	Silver iodide sol	pAg = 3 (I.S. = 1.5×10^{-3} *M*)	20 ± 1	4.8_4	Watanabe (1960))
Sodium pentanesulfonate	Silver iodide	3 (I.S. = 1×10^{-3} *M*)	20 ± 2	2.4_6	Osseo-Asare et al. (1975)
		4 (I.S. = 1×10^{-3} *M*)	20 ± 2	2.6_6	Osseo-Asare et al. (1975)
		5 (I.S. = 1×10^{-3} *M*)	20 ± 2	3.1_0	Osseo-Asare et al. (1975)
Sodium octanesulfonate	Silver iodide	3 (I.S. = 1×10^{-3} *M*)	20 ± 2	2.6_0	Osseo-Asare et al. (1975)
		4 (I.S. = 1×10^{-3} *M*)	20 ± 2	2.8_2	Osseo-Asare et al. (1975)
		5 (I.S. = 1×10^{-3} *M*)	20 ± 2	3.8_2	Osseo-Asare et al. (1975)
Sodium decanesulfonate	Silver iodide	3 (I.S. = 1×10^{-3} *M*)	20 ± 2	3.8_9	Osseo-Asare et al. (1975)
		4 (I.S. = 1×10^{-3} *M*)	20 ± 2	4.1_2	Osseo-Asare et al. (1975)
		5 (I.S. = 1×10^{-3} *M*)	20 ± 2	4.6_4	Osseo-Asare et al. (1975)
Sodium dodecanesulfonate	Silver iodide	3 (I.S. = 1×10^{-3} *M*)	20 ± 2	4.5_0	Osseo-Asare et al. (1975)
		4 (I.S. = 1×10^{-3} *M*)	20 ± 2	4.7_0	Osseo-Asare et al. (1975)
		5 (I.S. = 1×10^{-3} *M*)	20 ± 2	4.9_6	Osseo-Asare et al. (1975)
Sodium tetradecanesulfonate	Silver iodide	3 (I.S. = 1×10^{-3} *M*)	20 ± 2	5.1_5	Osseo-Asare et al. (1975)
		4 (I.S. = 1×10^{-3} *M*)	20 ± 2	5.2_5	Osseo-Asare et al. (1975)
		5 (I.S. = 1×10^{-3} *M*)	20 ± 2	5.4_7	Osseo-Asare et al. (1975)

I.S., total ionic strength.

benzenesulfonate group in a more central position in a linear alkylbenzene-sulfonate resulted in a decrease in the effectiveness of adsorption (Somasundaran et al., 1983).

However, if the orientation of the adsorbate is parallel to the interface, as may occur when the surfactant has two ionic groups of charge opposite to that of the substrate at opposite ends of the surfactant molecule, or when the hydrophobic chain interacts strongly with the surface (e.g., electron-rich aromatic nuclei in the adsorbate and positively charged sites on the adsorbent [Snyder, 1968]), then effectiveness of adsorption may decrease with increase in chain length, because this may increase the cross-sectional area of the molecule on the surface, and thus saturation of the surface will be accomplished by a smaller number of molecules (Kölbel and Kuhn, 1959).

2. Nonionic Surfactants Polyoxyethylene (POE) surfactants may adsorb onto silica surfaces via hydrogen bonding between SiOH groups on the surface and the oxygens of the oxyethylene groups (Rupprecht and Liebl, 1972; Aston et al., 1982; Nevskaia et al., 1996). On negatively charged silica surfaces, the oxygens of the POE chain may interact electrostatically with negative sites on the surface by picking up protons from the water and acquiring positive charges. Evidence for this is the increase in the pH of the solution with adsorption (M. J. Rosen and Z. H. Zhu, unpublished data). Adsorption isotherms are of the Langmuir type, with both efficiency and effectiveness decreasing with increase in the length of the POE group. The latter effect is due to the larger area occupied by the surfactant molecule at the interface as the length of the POE group is increased. At low coverage, the surfactant molecule may lie prone on the surface; at higher coverages, the hydrophobic group may be displaced from the surface by the hydrophilic group, and lateral interactions between adjacent hydrophobic groups (hemimicelle formation) may occur (Partyka et al., 1984). Maximum adsorption, which occurs near the CMC of the surfactant (Chapter 3), has been ascribed to both monolayer (Aston et al., 1982) and bilayer formation (Partyka et al., 1984).

A POE nonionic, by itself, showed very weak adsorption onto positively charged alumina (Somasundaran et al., 1983), while dodecyl-β-D-maltoside adsorbed strongly (Zhang et al., 1997), presumably because of the negative charge on the latter.

3. pH Change This usually causes marked changes in the adsorption of ionic surfactants onto charged solid substrates. As the pH of the aqueous phase is lowered, a solid surface will usually become more positive, or less negative, because of adsorption onto charged sites of protons from the solution, with consequent increase in the adsorption of anionic surfactants and decrease in the adsorption of cationics (van Senden and Koning, 1968; Connor and Ottewill, 1971). The reverse is true when the pH of the aqueous phase is raised. These effects are shown markedly by mineral oxides, such as silica and alumina, and by wool and other polyamides.

Change in the pH also may affect surfactant molecules, notably those containing carboxylate groups (soaps) or nonquaternary ammonium groups. In these cases, change in the pH may convert the surfactant from one containing an ionic group capable of strong adsorption onto oppositely charged sites on the adsorbent to a neutral molecule capable of adsorption only through hydrogen bonding or dispersion forces. Changes in pH also may affect nonionic surfactants, notably those having POE chains, because the ether linkages in these chains can be protonated at low pHs, yielding positively charged groupings that may adsorb onto negatively charged substrates.

4. Ionic Strength Addition of neutral electrolyte, such as NaCl or KBr, causes a decrease in the adsorption of ionic surfactants onto an oppositely charged adsorbent and an increase in their adsorption onto a similarly charged adsorbent.

These effects are presumably due to the decreased attraction between oppositely charged species and the decreased repulsion between similarly charged species at higher ionic strength. Both the efficiency and effectiveness of adsorption of ionic surfactants onto similarly charged substrates are increased by an increase in the ionic strength of the aqueous phase (Sexsmith and White, 1959; Groot, 1968; Connor and Ottewill, 1971).

The presence of polyvalent cations, especially Ca^{2+}, in the solution causes an increase in the adsorption of anionics. This may be due to the adsorption of Ca^{2+} onto the adsorbent, yielding ⊕-charged sites onto which negatively charged surfactant can adsorb (van Senden and Koning, 1968).

5. Temperature Temperature increase generally causes a decrease in the efficiency and effectiveness of adsorption of ionic surfactants, the change being relatively small compared to that caused by pH change. However, a rise in temperature usually results in an increase in the adsorption of nonionic surfactants containing a POE chain as the hydrophilic group (Corkill et al., 1964; Cases et al., 1982). This has been attributed to the decreased solute–solvent interaction (i.e., dehydration of the POE group) as the temperature is raised (Corkill et al., 1966; Partyka et al., 1984).

D. Adsorption from Aqueous Solution onto Nonpolar, Hydrophobic Adsorbents

Common substrates in this class are carbon and polyethylene or polypropylene. Adsorption isotherms for well-purified monofunctional anionic and cationic surfactants are similar on these adsorbents and are of the Langmuir type (Figures 2.13 and 2.14). They appear to show surface saturation in the vicinity of the CMC of the adsorbate, with an orientation of the adsorbate perpendicular to the substrate. Adsorption onto these substrates is mainly by dispersion forces. The orientation of the adsorbate initially may be parallel to the surface

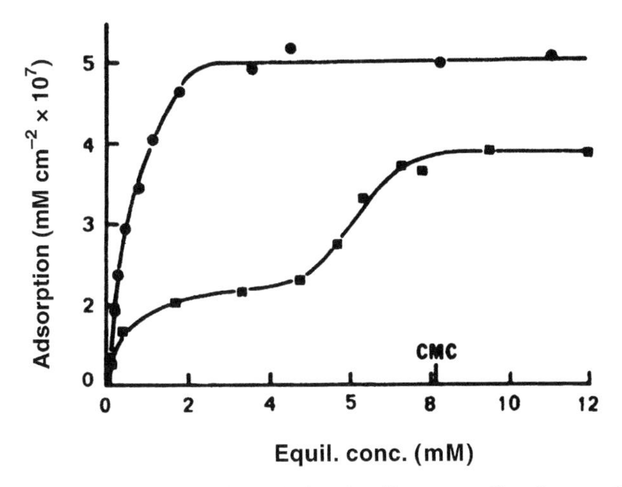

FIGURE 2.13 Adsorption of sodium dodecyl sulfate onto Graphon at 25°C: ■, from pure water; •, from aqueous 0.1 *M* NaCl. Reprinted with permission from Greenwood et al. (1968).

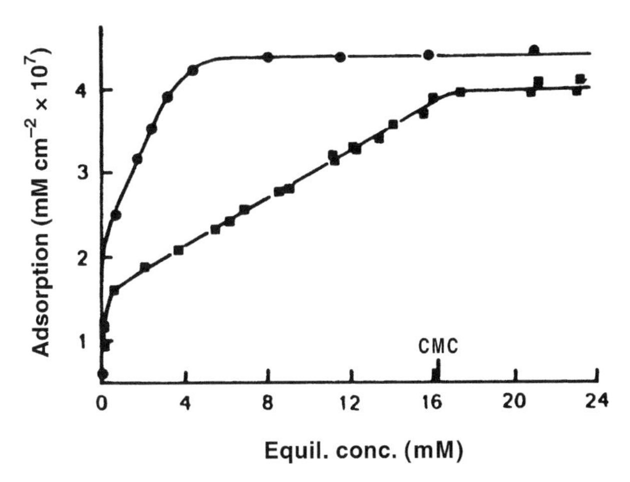

FIGURE 2.14 Adsorption of dodecyltrimethylammonium bromide onto Graphon at 25°C: ■, from pure water; •, from aqueous 0.1 *M* KBr. Reprinted with permission from Greenwood et al. (1968).

of the solid or slightly tilted or L-shaped, with the hydrophobic group close to the surface and the hydrophilic group oriented toward the aqueous phase. As adsorption continues, the adsorbed molecules may become oriented more and more perpendicular to the surface with hydrophilic heads oriented toward the water (Corkill et al., 1967). In some cases, the adsorption isotherm shows an inflection point (Figure 2.13) that has been ascribed to a change in orientation of the surfactant from parallel to perpendicular. An increase in the length of the hydrophobic group increases efficiency and slightly increases effectiveness of adsorption. The increase in efficiency is due to the increase in the magnitude of the $-\Delta G$ of adsorption with increase in the number of units in the hydrophobic chain; the increase in effectiveness may be due to tighter packing of the hydrophobic chains (Weber, 1964; Zettlemoyer et al., 1968). Here, as in the case of adsorption onto surfaces having strongly charged sites, the phenyl ring in a *p*-benzenesulfonate may be considered to have an effective length of about three and one-half carbon atoms in a straight alkyl chain. POE nonionics appear to adsorb more efficiently onto hydrophobic surfaces than onto hydrophilic ones (Aston et al., 1982). An increase in the length of the POE chain appears to decrease both efficiency of adsorption, presumably because the $-\Delta G$ of adsorption is decreased in magnitude as the number of oxyethylene units is increased, and effectiveness, because the cross-sectional area of the molecule at the interface increases with increase in the number of oxyethylene (OE) units (Abe and Kuno, 1962; Kronberg, 1984). Increase in the length of the hydrophobic group, however, appears to increase the efficiency of adsorption (Corkill et al., 1966).

The rate of adsorption has been shown to be a function of the position of the hydrophilic group in the molecule, with surfactants containing the hydrophilic group in a central location in the molecule adsorbing faster than those in which the hydrophilic group is terminally located (Zettlemoyer et al., 1968). The effect here may be due to either the more compact structure in aqueous solution, and hence the greater diffusion coefficient of surfactants with a hydrophilic group in a central position, or their greater CMC (see p. 123) and the consequent higher activity of their monomeric form in solution (Mukerjee, 1968). The rate of adsorption on carbon also has been shown to be dependent on the presence in the aqueous phase of additives that affect the structure of water. Additives that are structure breakers, such as urea and *N*-methylacetamide, appear to increase the rate of adsorption, whereas those that promote structure, such as xylose and fructose, decrease the rate of adsorption (Schwuger, 1971a).

Neutral electrolyte addition increases both the efficiency of adsorption of ionic surfactants, by decreasing the electrical repulsion between the similarly charged adsorbed ions and oncoming ions, and the effectiveness of adsorption, probably by decreasing the electrical repulsion between the similarly charged adsorbed ions, permitting closer packing (Figures 2.13 and 2.14). The addition of small amounts of cationics to aqueous solutions of anionics (Schwuger, 1971b), or small amounts of metal carboxylates to cationic solutions (Suzuki, 1967), has also been shown to increase the adsorption of the predominant ionic surfactant.

For solid, nonpolar, hydrophobic substrates that are not available in finely divided form but can be made in the form of nonporous planar sheets or films, surfactant concentrations at the solid–liquid interface can be determined from contact angles (Chapter 6, Section IA1).

E. Adsorption from Aqueous Solution onto Polar Adsorbents without Strongly Charged Sites

Adsorption of surfactants onto substrates such as cotton, polyesters, and polyamides in neutral solution is mainly by a combination of hydrogen bonding and adsorption via dispersion forces. For example, free fatty acids from the hydrolysis of soaps are adsorbed, probably by H bonding, onto polyester and nylon 66 (Gavet et al., 1973). Adsorption isotherms for POE nonionics (alcohol ethoxylate [AE] and alkylphenol ethoxylate [APE] types) from aqueous solution onto polyester fiber are of the Langmuir type, with areas occupied by the surfactant molecules approximating those occupied at the water–air interface (Gum and Goddard, 1982). Where the substrate has groups such as –OH or –NH, surfactants containing a POE chain will probably be adsorbed via H bonding. Thus, under laundering conditions, adsorption on nylon and cotton has been reported (Gordon and Shebs, 1968) to be much greater for nonionics than anionics by a factor of 2:1. POE *n*-dodecanol adsorbs onto cotton from aqueous solutions at 25°C to form a close-packed monolayer with the molecules lying flat on the substrate (Schott, 1967). An increase in the number of units in the POE chains causes a decrease in the efficiency, effectiveness, and rate (Gordon and Shebs, 1968) of the adsorption. Increase in the length of the hydrophobic chain, on the other hand, increases the efficiency of the adsorption.

A study of the adsorption of POE 1° linear alcohols on desized cotton at 60°C from a heavy-duty built formulation and its desorption by rinsing at 90°C under laundering conditions (Waag, 1968) indicates that at constant ethylene oxide percentage, compounds with short hydrophobic groups are adsorbed more than those with longer hydrophobic groups and that a higher percentage of these shorter-chain compounds than of the longer-chain compounds is retained on the fabric after four rinses. Furthermore, for the *n*-dodecanol-ethylene oxide adducts, only after the percentage of ethylene oxide in the molecule is above 60% does the percentage of the original surfactant remaining on the fabric after rinsing decrease. The fact that shorter-chain compounds are more strongly adsorbed (almost irreversibly) than longer-chain compounds appears to indicate adsorption with the hydrophilic group oriented toward the surface and with the hydrophobic groups toward the aqueous phase.

When the substrate is not capable of donating a hydrogen for bonding of the adsorbate (polyesters, polyacrylonitrile), adsorption is often mainly by dispersion forces; the character of the adsorption will then be similar to that on nonpolar, hydrophobic surfaces.

F. Effects of Adsorption from Aqueous Solution on the Surface Properties of the Solid Adsorbent

1. Substrates with Strongly Charged Sites As mentioned earlier, the adsorption of surface-active counterions by an ion-exchange mechanism causes no change in the electrical potential of the adsorbent. However, if adsorption of surface-active counterions continues by an ion-pairing mechanism, then the potential at the Stern layer of the adsorbent decreases and eventually is completely neutralized. During this process the tendency of the surface to repel other similarly charged surfaces diminishes and ceases when the charge on the adsorbent has been neutralized. Thus, solid adsorbents in the form of finely divided particles, dispersed in the aqueous phase in part because of their mutual electrical repulsion, usually flocculate at some point as their charge is neutralized by the adsorption of oppositely charged surfactant ions.

Furthermore, since adsorption by an ion-exchange or ion-pairing mechanism results in the orientation of the adsorbed surfactant with its hydrophobic group toward the aqueous phase (Figure 2.5), such adsorption causes the surface to become increasingly hydrophobic (Law and Kunze, 1966; Robb and Alexander, 1967). This is shown by an increase in the contact angle at the solid–water–air interface as adsorption increases up to the point where the zeta potential is reduced to zero. Negatively charged mineral surfaces such as quartz, when treated with cationic surfactants (e.g., cetyltrimethylammonium bromide), show this effect, becoming more difficult to wet by water and more easily wet by nonpolar compounds (McCaffery and Mungan, 1970). Adsorption in this manner may account for the reduced swelling of wool fibers in aqueous solution after adsorption of anionic surfactant onto the positively charged sites (Machinson, 1967) and the elimination of shrinkage resistance from oxidized wool by cationic softeners (Stigter, 1971). In both cases, adsorption of oppositely charged surfactant ions makes the wool surface more hydrophobic. If adsorption of surfactant ions onto the adsorbent is continued beyond the pzc, however, then the charge at the Stern layer is reversed, and the substrate acquires a charge whose sign is that of the adsorbate ion. Orientation of the adsorbed surfactant ion during this process is with the hydrophilic head toward the aqueous phase, imparting increasing hydrophilic character to the substrate as adsorption continues. The contact angle decreases again, and the tendency to disperse in water increases (Parfitt and Wharton, 1972).

Adsorption in this manner may also account for the increased reactivity of wool cystine disulfide bonds to attack by alkali in the presence of cationic surfactants and their decreased reactivity in the presence of anionics (Meichelbeck and Knittel, 1971). The adsorption of cationic surfactants onto the wool surface, which is negatively charged in an alkaline medium, can impart a positive charge to the surface, thus increasing its attraction for hydroxide and sulfite ions, with consequent increase in its rate of reaction with these ions. In analogous fashion, the acid hydrolysis of peptide bonds in the wool is increased by the presence of anionic surfactants (which adsorb onto the wool surface, positively charged in an acid medium, and impart to it a negative charge). The presence of cationic surfactants, on the other hand, decreases the acid hydrolysis of these bonds, whereas nonionic surfactants have no effect.

The adsorption of surfactant ions onto solid surfaces is one of the major factors governing detergency. The greater retention of carbon black in the presence of anionic surfactants by polyester than by wool, for example, has been explained by the greater attraction of the wool (with charged sites) for the surfactant than for the nonpolar carbon, and the reverse in the case of the hydrophobic polyester (Von Hornuff and Mauer, 1972). The action of surfactants in retarding and leveling the dyeing of fabrics also involves competitive adsorption onto charged sites, with surfactant ions of charge similar to that of the dyes adsorbing competitively onto oppositely charged sites on the fiber, thus reducing the effective rate of adsorption of the dyestuff. In all cases, the more strongly adsorbed the surfactant, the greater its retarding action.

2. Nonpolar Adsorbents Adsorption of surfactants at any concentration onto a well-purified substrate of this type (i.e., free of impurities with polar groups on the surface) occurs with the adsorbate oriented with its hydrophilic group toward the aqueous phase. Thus, adsorption increases the hydrophilicity of the adsorbent and, in the case of ionic surfactants, increases its surface charge density, making it more wettable by the aqueous phase (Elton, 1957; Ginn, 1970, p. 372) and more dispersible (if in finely dividing form). These factors account, for example, for the greater dispersibility of carbon black in aqueous medium in the presence of nonionic or ionic surfactants (Corkill et al., 1967; Greenwood et al., 1968). In the case of POE nonionics, adsorption may produce a steric barrier to the close approach of another similarly covered particle, since such approach would result in the restriction of the movement of the randomly coiled POE chains, with consequent decrease in the entropy of the system. Adsorption of a nonionic surfactant can thereby also produce an energy barrier to flocculation of a solid if the latter is in finely divided form. These effects, in part, account for the greater desorption of carbon and other hydrophobic pigments from cotton in the presence of surfactants.

G. Adsorption from Nonaqueous Solution

The adsorption of surfactants in fuel oil onto pulverized coal has been studied in connection with the development of coal–oil mixtures (COM), that is, stable dispersions of finely pulverized coal in fuel oil. The stabilization of such dispersions by a cationic surfactant has been shown (Kosman and Rowell, 1982) to involve adsorption of the cationic via its positively charged head group onto nucleophilic sites on the coal, with its hydrocarbon group oriented toward the oil phase. The adsorption of alkylaromatics on carbon black from *n*-heptane indicates adsorption in an orientation parallel to the interface, with the alkyl chains remaining mobile on the surface (van der Waarden, 1951). Increased length of the alkyl chains increases the degree of dispersion of the carbon.

Adsorption of sodium bis(2-ethylhexyl)sulfosuccinate from benzene solution onto carbon blacks follows the Langmuir equation and depends on the amount of oxygen on the surface. No adsorption onto heat-treated hydrophobic carbon (Graphon) could be detected (Abram and Parfitt, 1962).

H. Determination of the Specific Surface Areas of Solids

The most commonly used and most reliable method for determining the specific surface area of finely divided solids is by adsorption of nitrogen or argon at liquid air temperature. However, for use in measuring the adsorption and orientation of surfactants at the solid–liquid interface, this determination is sometimes better done by adsorption from solution rather than adsorption from the gas phase. The use of an adsorbate of size comparable to that of the surfactant molecule gives a value that may be more indicative of the surface area available for adsorption of surfactant than the use of much smaller (gaseous) adsorbates, which can enter pores and crevices in the surface not accessible to (larger) surfactant molecules. Moreover, adsorption from solution is experimentally much easier than adsorption from the gas phase, which requires vacuum apparatus. However, the orientation of the adsorbate molecule on the adsorbent being studied and the point of monolayer formation must be known with assurance for the method to have validity. Some of the caveats for using adsorption from solution for the determination of specific areas of solids are discussed by Kipling (1965) and Gregg and Sing (1967). Using adsorbates of known cross-sectional area at the particular solid–liquid interface under investigation, the saturation adsorption for monolayer formation is determined (e.g., by use of the Langmuir equation in linear form when adsorption fits the equation), and from the results, the specific area, a_s, in cm^2/g, is obtained:

$$a_s = \frac{n_m^s \times a_m^s \times N}{10^{16}} \tag{2.16}$$

where n_m^s = the number of moles of solute adsorbed per gram of solid at monolayer saturation,

a_m^s = the surface area occupied per molecule of adsorbate at monolayer adsorption, in square angstroms.

Solutes used for this determination by solution adsorption include stearic acid from benzene solution (Daniel, 1951; Kipling and Wright, 1962) and *p*-nitrophenol from aqueous or xylene solution (Giles and Nakhwa, 1962). Giles and Nakhwa (1962) has discussed the requirements for a suitable adsorbate for the purpose.

III. ADSORPTION AT THE LIQUID–GAS (L/G) AND LIQUID–LIQUID (L/L) INTERFACES

The direct determination of the amount of surfactant adsorbed per unit area *L*/*G* or *L*/*L* interface, although possible, is not generally undertaken because of the difficulty of isolating the interfacial region from the bulk phase(s) for purposes of analysis when the interfacial region is small, and of measuring the interfacial area when it is large. Instead, the amount of material adsorbed per unit area of interface is calculated indirectly from surface or interfacial tension measurements. As a result, a plot of surface (or interfacial) tension as a function of (equilibrium) concentration of surfactant in one of the liquid phases, rather than an adsorption isotherm, is generally used to describe adsorption at these interfaces. From such a plot, the amount of surfactant adsorbed per unit area of interface can readily be calculated by use of the Gibbs adsorption equation.

A. The Gibbs Adsorption Equation

The Gibbs adsorption equation, in its most general form (Gibbs, 1928),

$$d\gamma = -\sum_i \Gamma_i d\mu_i \tag{2.17}$$

where $d\gamma$ = the change in surface or interfacial tension of the solvent,
Γ_i = the surface excess concentration* of any component of the system,
$d\mu_i$ = the change in chemical potential of any component of the system,

is fundamental to all adsorption processes where monolayers are formed. At equilibrium between the interfacial and bulk phase concentrations, $d\mu_i = RT d\ln a_i$, where a_i = the activity of any component in the bulk (liquid) phase, R = the gas constant, and T = the absolute temperature. Thus,

$$\begin{aligned} d\gamma &= -RT\sum_i \Gamma_i d\ln a_i \\ &= -RT\sum_i \Gamma_i d\ln x_i f_i \\ &= -RT\sum_i \Gamma_i d(\ln x_i + \ln f_i) \end{aligned} \tag{2.18}$$

where x_i is the mole fraction of any component in the bulk phase and f_i its activity coefficient.

For solutions consisting of the solvent and only one solute, $d\gamma = -RT(\Gamma_0 d\ln a_0 + \Gamma_1 d\ln a_1)$, where subscripts 0 and 1 refer to the solvent and the solute, respectively. For dilute solutions (10^{-2} M or less) containing only one nondissociating surface-active solute, the activity of the solvent and the activity coefficient of the solute can both be considered to be constant, and the mole fraction of the solute x_1 may be replaced by its molar concentration C_1. Thus,

* The surface excess concentration is defined here as the excess, per unit area of interface, of the amount of any component actually present in the system over that present in a reference system of the same volume in which the bulk concentrations in the two phases remain uniform up to a hypothetical (Gibbs) dividing surface.

$$d\gamma = -RT\Gamma_1 d\ln C_1 \\ = -2.303\, RT\Gamma_1 d\log C_1 \tag{2.19}$$

which is the form in which the Gibbs equation is commonly used for solutions of nonionic surfactants containing no other materials. When γ is in dyn/cm (= ergs/cm^2) and $R = 8.31 \times 10^7$ ergs mol^{-1} K^{-1}, then Γ_1 is in mol/cm^2; when γ is in mN/m (= mJ/m^2) and $R = 8.31$ J mol^{-1} K^{-1}, then Γ_1, is in mol/1000 m^2.

For ionic surfactants,

$$d\gamma = -nRT\Gamma_1 d\ln C_1 = -2.303\, nRT\Gamma_1 d\log C_1 \tag{2.19a}$$

where n is the number of solute species whose concentration at the interface changes with change in the value of C_1. Thus, for solutions of a completely dissociated surfactant of the 1:1 electrolyte type, A^+B^-, as the only solute,

$$d\gamma = RT(\Gamma_{A^+} d\ln a_{A^+} + \Gamma_{B^-} d\ln a_{B^-})$$

Since $\Gamma_{A^+} = \Gamma_{B^-} = \Gamma_1$, to maintain electroneutrality and $a_{A^+} = a_{B^-} = C_1 \times f_\pm$ without significant error,

$$d\gamma = -2RT\Gamma_1 d(\ln C_1 + \ln f_\pm) \tag{2.20}$$

where f_+ is the mean activity coefficient of the surfactant. For dilute solutions (10^{-2} M or less),

$$d\gamma = -2RT\Gamma d\ln C_1 \\ = -4.606\, RT\Gamma d\log C_1 \tag{2.21}$$

without significant error.

For mixtures of two different surfactants, the value of n for the mixture, $n_{mix} = n_1X_1 + n_2X_2$, where n_1, and n_2 are the n values for individual surfactants 1 and 2 of the mixture and X_1 and X_2 their respective mole fractions at the interface (Equation 2.46).

For mixtures of ionic and nonionic surfactants in aqueous solution in the absence of added electrolyte, the coefficient decreases from 4.606 to 2.303 with a decrease in the concentration of the ionic surfactant at the interface (Hua and Rosen, 1982).

For dilute solutions of a completely dissociated surface-active 1:1 electrolyte in the presence of a swamping, constant amount of electrolyte containing a common nonsurfactant counterion,

$$d\gamma = -RT\Gamma_1 d\ln C_1 \\ = -2.303\, RT\Gamma_1 d\log C_1 \tag{2.22}$$

since under these conditions the change in the concentration of the nonsurfactant common ion at the interface with adsorption is essentially zero. This is the same form as for a nonionic surfactant in dilute solution (Equation 2.19). For less than swamping concentrations of a 1:1 non-surface-active electrolyte, for example, NaCl (Matijevic and Pethica, 1958),

$$\begin{aligned} d\gamma &= -\mathrm{y}RT\Gamma d\ln C_1 \\ &= -2.303\,\mathrm{y}RT\Gamma d\log C_1 \end{aligned} \tag{2.23}$$

where $\mathrm{y} = 1 + C_1/(C_1 + C_{NaCl})$.

Where activity coefficients are expected to deviate significantly from unity, for example, when divalent or multivalent ions are present in the solution or concentrations of the surfactant exceed 10^{-2} M, $d\log C_1$ in the appropriate equation may be replaced by $d(\log C_1 + \log f_{\pm})$, and $\log f_{\pm}$ calculated in water at 25°C, from the Debye–Hückel equation,

$$\log f_{\pm} = \frac{-0.509|Z_+Z_-|\sqrt{I}}{1+0.33\alpha\sqrt{I}} \tag{2.24}$$

where the total ionic strength of the solution

$$I = \frac{1}{2}\sum_i C_i Z_i^2$$

and α is the mean distance of approach of the ions, in Å (Boucher et al., 1968). $\mathrm{Log}f_{\pm}$ can be assumed to equal $(\log f_+ + \log f_-)/2$, and α is taken as 0.3 for small counterions (Na^+, K^+, Br^-, Cl^-) and 0.6 for the surfactantion.

B. Calculation of Surface Concentrations and Area Per Molecule at the Interface by Use of the Gibbs Equation

For surface-active solutes, the surface excess concentration, Γ_1, can be considered to be equal to the actual surface concentration without significant error. The concentration of surfactant at the interface may therefore be calculated from surface or interfacial tension data by use of the appropriate Gibbs equation. Thus, for dilute solutions of a nonionic surfactant, or for a 1:1 ionic surfactant in the presence of a swamping amount of electrolyte containing a common nonsurfactant ion, from Equation 2.19,

$$\Gamma_1 = -\frac{1}{2.303\,RT}\left(\frac{\partial\gamma}{\partial\log C_1}\right)_T \tag{2.25}$$

and the surface concentration can be obtained from the slope of a plot of γ versus $\log C_1$ at constant temperature (when γ is in dyn/cm or ergs/cm^2 and

$R = 8.31 \times 10^7$ ergs/mol/K, then Γ_1 is in mol/cm^2; when γ is in mNm^{-1} or mJm^{-2} and $R = 8.31$ Jmol$^{-1}K^{-1}$, Γ_1 is in mol/1000 m^2).

For solutions of 1:1 ionic surfactant in the absence of any other solutes, in similar fashion,

$$\Gamma_1 = -\frac{1}{4.606\,RT}\left(\frac{\partial\gamma}{\partial \log C_1}\right)_T \tag{2.26}$$

When activity coefficients are used, γ is plotted versus $(\log C_1 + \log f_\pm)$ to obtain Γ_1.

The area per molecule at the interface provides information on the degree of packing and the orientation of the adsorbed surfactant molecule when compared with the dimensions of the molecule as obtained by use of molecular models. From the surface excess concentration, the area per molecule at the interface, a_1^s, in square angstroms is calculated from the relation

$$a_1^s = \frac{10^{16}}{N\Gamma_1} \tag{2.27}$$

where N = Avogardo's number and Γ_1 is in mol/cm^2.*

A typical $\gamma - \log C_1$ plot for a dilute solution of an individual surfactant (surfactants are often used at concentrations of less than 1×10^{-2} M) is shown in Figure 2.15. The break in the curve occurs at the CMC, the concentration at which the monomeric form, in which the surfactant exists in very dilute solution, aggregates to form a surfactant cluster known as a *micelle* (Chapter 3). Above this concentration, the surface tension of the solution remains essentially constant since only the monomeric form contributes to the reduction of the surface or interfacial tension. For concentrations below but near the CMC,

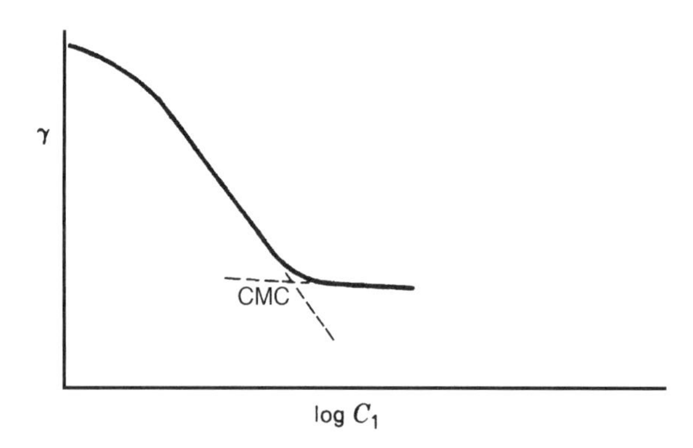

FIGURE 2.15 Plot of surface tension versus log of the bulk phase concentration for an aqueous solution of a surfactant.

* For the value in nm^2, divide by 100. When Γ is in mol/1000 m^2, a_1^s, in square angstroms, equals $10^{23}/N\Gamma$.

the slope of the curve is essentially constant, indicating that the surface concentration has reached a constant maximum value. In this range, the interface is considered to be saturated with surfactant (van Voorst Vader, 1960a), and the continued reduction in the surface tension (Elworthy and Mysels, 1966) is due mainly to the increased activity (Miles and Shedlovsky, 1945; Mulley and Metcalf, 1962) of the surfactant in the bulk phase rather than at the interface (Equation 2.17). For ionic surfactants in the presence of a constant concentration of counterion, this region of saturated adsorption may extend down to one-third of the CMC.

C. Effectiveness of Adsorption at the *L/G* and *L/L* Interfaces

The surface excess concentration (≈surface concentration) at surface saturation Γ_m is a useful measure of the *effectiveness* of adsorption of the surfactant at the *L/G* or *L/L* interface, since it is the maximum value that adsorption can attain. The effectiveness of adsorption is an important factor in determining such properties of the surfactant as foaming, wetting, and emulsification, since tightly packed, coherent interfacial films have very different interfacial properties than loosely packed, noncoherent films. Table 2.2 lists values for the effectiveness of adsorption Γ_m, in mol/cm^2, and the area per molecule at the interface at surface saturation a_m^s, in square angstroms (which is inversely proportional to the effectiveness of adsorption) for a wide variety of anionic, cationic, nonionic, and zwitterionic surfactants at various interfaces.

Since the cross-sectional area of an aliphatic chain oriented perpendicular to the interface is about 20 Å^2 and that of a benzene ring oriented in the same fashion is about 25 Å^2, it is apparent that the hydrophobic chains of surfactants adsorbed at the aqueous solution–air (A/A) or aqueous solution–hydrocarbon interfaces are generally not in the close-packed arrangement normal to the interface at saturation adsorption. On the other hand, since the cross-sectional area of a $-CH_2-$ group lying flat in the interface is about 7 Å^2, the chains in the usual ionic surfactant with a hydrophilic group at one end of the molecule are not lying flat in the interface either, but are somewhat tilted with respect to the interface.

For surfactants with a single hydrophilic group, either ionic or nonionic, the area occupied by a surfactant molecule at the surface appears to be determined by the area occupied by the hydrated hydrophilic group rather than by the hydrophobic group. On the other hand, if a second hydratable, hydrophilic group is introduced into the molecule, that portion of the molecule between the two hydrophilic groups tends to lie flat in the interface, and the area occupied by the molecule in the interface is increased.

The small areas/molecule obtained for polymeric surfactants such as POE polyoxypropylene block copolymers suggest that there is considerable folding of the surfactant molecule at the A/A interface with the hydrophobic polyoxypropylene group looping away from the aqueous phase into the air and the POE groups extending into the aqueous phase (Alexandridis et al., 1994).

TABLE 2.2 Effectiveness of Adsorption Γ_m, Area/Molecule at Surface Saturation a_m^s, and Efficiency of Adsorption pC_{20} of Surfactants from Aqueous Solution at Various Interfaces

Compound	Interface	Temp. (°C)	Γ_m, (mol/cm^2 × 10^{10})	a_m^s, Å^2	pC_{20}	Reference
Anionics						
$C_{11}H_{23}COO^-Na^+$	0.11 *M* NaCl(aq.)–air	20	3.5	47	—	van Voorst Vader (1960a)
$C_{11}H_{23}COO^-Na^+$	0.11 *M* NaCl(aq.)–heptane	20	3.7	45	—	van Voorst Vader (1960a)
$C_{11}H_{23}COO^-K^+$	0.11 *M* NaCl(aq.)–air	20	3.8_5	43	—	van Voorst Vader (1960a)
$C_{11}H_{23}COO^-K^+$	0.11 *M* NaCl(aq.)–heptane	20	3.8_5	44	—	van Voorst Vader (1960a)
$C_{15}H_{31}COO^-Na^+$	0.1 *M* NaCl(aq.)–air	25	—	—	4.7	Rosen and Zhu (1989)
$C_{15}H_{31}COO^-Na^+$	0.1 *M* NaCl(aq.)–air	60	5.4	31	4.7	Rosen and Zhu (1989)
$C_{10}H_{21}OCH_2COO^-Na^+$, pH 10.5	0.1 *M* NaCl(aq.)	30	5.4	31	3.2	Tsubone and Rosen (2001)
$C_{11}H_{23}CONHCH_2COO^-Na^+$	0.1 *M* NaOH(aq.)	45	3.4_5	48	—	Miyagishi et al. (1989)
$C_{11}H_{23}CONHCH(CH_3)COO^-Na^+$	0.1 *M* NaOH(aq.)	45	2.9	57	—	Miyagishi et al. (1989)
$C_{11}H_{23}CONHCH(C_2H_5)COO^-Na^+$	0.1 *M* NaOH(aq.)	45	2.8	60	—	Miyagishi et al. (1989)
$C_{11}H_{23}CONHCH[CH(CH_3)_2]COO^-Na^+$	0.1 *M* NaOH(aq.)	45	2.7	62	—	Miyagishi et al. (1989)
$C_{11}H_{23}CONHCH[CH_2CH(CH_3)_2]–COO^-Na^+$	0.1 *M* NaOH(aq.)	45	2.7	61	—	Miyagishi et al. (1989)
$C_{11}H_{23}CON(CH_3)CH_2COO^-Na^+$, pH 10.5	H_2O	30	2.1	81	2.5	Tsubone and Rosen (2001)
$C_{11}H_{23}CON(CH_3)CH_2COO^-Na^+$, pH 10.5	0.1 *M* NaOH(aq.)–air	30	2.9	58	3.3	Tsubone and Rosen (2001)

$C_{11}H_{23}CON(C_2H_5)CH_2COO^-Na^+$	"Hard river" water (I.S. = 6.6×10^{-3} *M*)	25	2.7_7	$59._9$	3.8_4	Zhu et al. (1998a)
$C_{11}H_{23}CON(C_4H_9)CH_2COO^-Na^+$	H_2O–air	25	1.5	107	3.6_2	Zhu et al. (1998a)
$C_{11}H_{23}CON(C_4H_9)CH_2COO^-Na^+$	"Hard river" water (I.S. = 6.6×10^{-3} *M*)[a]	25	2.9	$57._3$	4.7_6	Zhu et al. (1998a)
$C_{11}H_{23}CON(CH_3)CH_2CH_2COO^-Na^+$, pH 10.5	H_2O	30	1.6	104	2.7	Tsubone and Rosen (2001)
$C_{11}H_{23}CON(CH_3)CH_2CH_2COO^-Na^+$, pH 10.5	0.1 *M* NaCl(aq.)	30	2.5	66	3.4	Tsubone and Rosen (2001)
$C_{13}H_{27}CON(C_3H_7)CH_2COO^-Na^+$	H_2O	25	1.5_8	105	4.3_0	Zhu et al. (1998a)
$C_{13}H_{27}CON(C_3H_7)CH_2\text{-}COO^-Na^+$	"Hard river" water (I.S. = 6.6×10^{-3} *M*)	25	3.5_0	$47._4$	5.2_8	Zhu et al. (1998a)
$C_{17}H_{35}CON[(CH_2)_3OMe]CH_2\text{-}COO^-Na^+$	H_2O	25	1.0_5	158	5.3_8	Zhu et al. (1998a)
$C_{17}H_{33}CON[(CH_2)_3OMe]CH_2\text{-}COO^-Na^+$	"Hard river" water (I.S. = 6.6×10^{-3} *M*)[a]	25	3.3_3	$49._9$	5.8_6	Zhu et al. (1998a)
$C_{10}H_{21}SO_3^-Na^+$	H_2O–air	10	3.3_7	49	1.7_0	Dahanayake et al. (1986)
$C_{10}H_{21}SO_3^-Na^+$	H_2O–air	25	3.2_2	52	1.6_9	Dahanayake et al. (1986)
$C_{10}H_{21}SO_3^-Na^+$	H_2O–air	40	3.0_5	54	1.6_6	Dahanayake et al. (1986)
$C_{10}H_{21}SO_3^-Na^+$	0.1 *M* NaCl(aq.)–air	10	4.0_6	41	2.2_9	Dahanayake et al. (1986)
$C_{10}H_{21}SO_3^-Na^+$	0.1 *M* NaCl(aq.)–air	25	3.8_5	43	2.2_9	Dahanayake et al. (1986)
$C_{10}H_{21}SO_3^-Na^+$	0.1 *M* NaCl(aq.)–air	40	3.6_7	45	2.2_7	Dahanayake et al. (1986)
$C_{10}H_{21}SO_3^-Na^+$	0.5 *M* NaCl(aq.)–air	10	4.4_6	37	2.8_9	Dahanayake et al. (1986)
$C_{10}H_{21}SO_3^-Na^+$	0.5 *M* NaCl(aq.)–air	25	4.2_4	41	2.8_7	Dahanayake et al. (1986)

(Continued)

TABLE 2.2 (*Continued*)

Compound	Interface	Temp. (°C)	Γ_m, (mol/cm^2 × 10^{10})	a_m^s, Å^2	pC_{20}	Reference
$C_{10}H_{21}SO_3^-Na^+$	0.5 *M* NaCl(aq.)–air	40	4.0_4	41	2.8_4	Dahanayake et al. (1986)
$C_{11}H_{23}SO_3^-Na^+$	0.1 *M* NaCl(aq.)–air	20	3.2	52	—	van Voorst Vader (1960a)
$C_{12}H_{25}SO_3^-Na^+$	H_2O–air	10	3.0_2	55	2.3_8	Dahanayake et al. (1986)
$C_{12}H_{25}SO_3^-Na^+$	H_2O–air	25	2.9_3	57	2.3_6	Dahanayake et al. (1986)
$C_{12}H_{25}SO_3^-Na^+$	H_2O–air	40	2.7_3	60	2.3_3	Bujake and Goodard (1965); Dahanayake et al. (1986)
$C_{12}H_{25}SO_3^-Na^+$	H_2O–air	60	2.5	65	2.1_4	Rosen (1976)
$C_{12}H_{25}SO_3^-Na^+$	0.1 *M* NaCl(aq.)–air	10	3.9_2	42	3.4_1	Dahanayake et al. (1986)
$C_{12}H_{25}SO_3^-Na^+$	0.1 *M* NaCl(aq.)–air	25	3.7_6	44	3.3_8	Dahanayake et al. (1986)
$C_{12}H_{25}SO_3^-Na^+$	0.1 *M* NaCl(aq.)–air	40	3.5_5	47	3.3_0	Dahanayake et al. (1986)
$C_{12}H_{25}SO_3^-Na^+$	0.5 *M* NaCl(aq.)–air	10	3.9_8	42	4.1_1	Dahanayake et al. (1986)
$C_{12}H_{25}SO_3^-Na^+$	0.5 *M* NaCl(aq.)–air	25	3.8_5	42	4.0_6	Dahanayake et al. (1986)
$C_{12}H_{25}SO_3^-Na^+$	0.5 *M* NaCl(aq.)–air	40	3.6_0	44	3.9_3	Dahanayake et al. (1986)
$C_{12}H_{25}SO_3^-Na^+$	0.1 *M* NaCl(aq.)–PTFE	25	3.0	56	—	Gu and Rosen (1989)
$C_{12}H_{25}SO_3^-K^+$	H_2O–air	25	3.4	49	2.4_3	Rosen (1974)

$C_{16}H_{33}SO_3^-K^+$	H_2O–air	60	2.8	58	3.3_5	M. J. Rosen and J. Solash, unpublished data; Rosen (1974)
$C_8H_{17}SO_4^-Na^+$	H_2O–heptane	50	2.3	72	1.6_1	Kling and Lange (1957)
$C_9H_{19}SO_4^-Na^+$	H_2O–heptane	20	3.0	56 ± 2	—	van Voorst Vader (1960a)
$C_{10}H_{21}SO_4^-Na^+$	H_2O–air	27	2.9	57	1.8_9	Dreger et al. (1944); Kling and Lange (1957)
$C_{10}H_{21}SO_4^-Na^+$	0.1 *M* NaCl(aq.)	22	3.7	45	—	Betts and Pethica (1957)
$C_{10}H_{21}SO_4^-Na^+$	H_2O–heptane	50	3.0_5	54	2.1_1	Kling and Lange (1957); van Voorst Vader (1960a)
$C_{10}H_{21}SO_4^-Na^+$	0.032 *M* NaCl(aq.)–heptane	50	3.2	52	—	Lange (1957)
$C_{12}H_{25}SO_4^-Na^+$	H_2O–air	25	3.1_6	53	2.5_1	Dahanayake et al. (1986)
$C_{12}H_{25}SO_4^-Na^+$	H_2O–air	60	2.6_5	63	2.2_4	Rosen and Solash (1969)
$C_{12}H_{25}SO_4^-Na^+$	0.1 *M* NaCl(aq.)	25	4.0_3	41	3.6_7	Dahanayake et al. (1986)
$C_{12}H_{25}SO_4^-Na^+$	H_2O–heptane	20	3.1	53	—	van Voorst Vader (1960a)
$C_{12}H_{25}SO_4^-Na^+$	H_2O–heptane	50	2.9_5	56	2.7_2	Kling and Lange (1957); van Voorst Vader (1960a)

(*Continued*)

TABLE 2.2 (*Continued*)

Compound	Interface	Temp. (°C)	Γ_m, (mol/cm^2 × 10^{10})	a_m^s, Å^2	pC_{20}	Reference
$C_{12}H_{25}SO_4^-Na^+$	0.008 *M* NaCl(aq.)–heptane	50	3.2	52	—	Lange (1957)
$C_{12}H_{25}SO_4^-Na^+$	0.1 *M* NaCl(aq.)–heptane	20	3.3_2	50	—	Vijayendran and Bursh (1979)
$C_{12}H_{25}SO_4^-Na^+$	H_2O–octane	25	3.3_2	50	2.7_6	Rehfeld (1967)
$C_{12}H_{25}SO_4^-Na^+$	H_2O–decane	25	3.5	48	2.7_5	Rehfeld (1967)
$C_{12}H_{25}SO_4^-Na^+$	H_2O–heptadecane	25	3.3_2	50	2.7_5	Rehfeld (1967)
$C_{12}H_{25}SO_4^-Na^+$	H_2O–cyclohexane	25	3.1_0	54	2.8_2	Rehfeld (1967)
$C_{12}H_{25}SO_4^-Na^+$	H_2O–benzene	25	2.3_3	71	2.5_7	Rehfeld (1967)
$C_{12}H_{25}SO_4^-Na^+$	H_2O–1-hexene	25	2.5_1	66	2.4_1	Rehfeld (1967)
$C_{12}H_{25}SO_4^-Na^+$	0.1 *M* NaCl(aq.)–ethylbenzene	20	3.0_0	55	—	Vijayendran and Bursh (1979)
$C_{12}H_{25}SO_4^-Na^+$	0.1 *M* NaCl(aq.)–ethylpropionate	20	1.2_7	131	—	Vijayendran and Bursh (1979)
Branched $C_{12}H_{25}SO_4^-Na^{+a}$	H_2O–air	25	1.7	$95._1$	2.9	Varadaraj et al. (1992)
Branched $C_{12}H_{25}SO_4^-Na^{+a}$	0.1 *M* NaCl(aq.)–air	25	3.3	$49._9$	3.6	Varadaraj et al. (1992)
$(C_{11}H_{23})(CH_3)CHSO_4^-Na^+$	H_2O–air	25	2.9_5	56	—	Dreger et al. (1944); van Voorst Vader (1960a)
$C_{14}H_{29}SO_4^-Na^+$	H_2O–air	25	3.0	56	3.1_1	Huber (1991); Rosen et al. (1996)
$C_{14}H_{29}SO_4^-Na^+$	H_2O–heptane	50	3.2	52	3.3_1	Kling and Lange (1957); van Voorst Vader (1960a)

$C_{14}H_{29}SO_4^-Na^+$	0.002 *M* NaCl(aq.)–heptane	50	3.2_5	51	—	Lange (1957)
$(C_7H_{15})_2CHSO_4^-Na^+$	H_2O–air	25	3.2_5	51	—	Dreger et al. (1944); van Voorst Vader (1960a)
$C_{16}H_{33}SO_4^-Na^+$	H_2O–air	25	—	—	3.7_0	Livingston and Drogin (1965)
$C_{16}H_{33}SO_4^-Na^+$	0.1 *M* NaCl(aq.)–air	25	—	—	5.2_4	Caskey and Barlage (1971)
$C_{16}H_{33}SO_4^-Na^+$	H_2O–heptane	50	3.0_5	54	3.8_9	Kling and Lange (1957); van Voorst Vader (1960a)
$C_4H_9OC_{12}H_{25}SO_4^-Na^+$	H_2O–air	25	1.1_3	147	2.7_7	Livingston and Drogin (1965)
$C_{12}H_{25}OC_4H_9SO_4^-Na^+$	0.01 *M* NaCl(aq.)–air	20	3.1_5	$52._5$	—	van Voorst Vader (1960b)
$C_{14}H_{29}OC_2H_4SO_4^-Na^+$	H_2O–air	25	2.1	66	3.9_2	Livingston and Drogin (1965)
$(C_{10}H_{21})(C_7H_{15})CHSO_4^-Na^+$	H_2O–air	20	3.3	50	—	van Voorst Vader (1960a); Livingston and Drogin (1965)
$(C_{10}H_{21})(C_7H_{15})CHSO_4^-Na^+$	H_2O–heptane	20	2.8_5	58	—	van Voorst Vader (1960a); Livingston and Drogin (1965)
$C_{18}H_{37}SO_4^-Na^+$	H_2O–heptane	50	2.3	72	4.4_2	Kling and Lange (1957)

(*Continued*)

TABLE 2.2 (*Continued*)

Compound	Interface	Temp. (°C)	Γ_m, (mol/cm^2 × 10^{10})	a_m^s, Å^2	pC_{20}	Reference
$C_{10}H_{21}OC_2H_4SO_3^-Na^+$	H_2O–air	25	3.2_2	52	2.1_0	Dahanayake et al. (1986)
$C_{10}H_{21}OC_2H_4SO_3^-Na^+$	0.1 *M* NaCl(aq.)–air	25	3.8_5	43	2.9_3	Dahanayake et al. (1986)
$C_{12}H_{25}OC_2H_4SO_3^-Na^+$	H_2O–air	25	2.9_2	57	2.7_5	Dahanayake et al. (1986)
$C_{12}H_{25}OC_2H_4SO_3^-Na^+$	0.1 *M* NaCl(aq.)–air	25	3.7_3	44	4.0_7	Dahanayake et al. (1986)
$C_{10}H_{21}(OC_2H_4)_2SO_4^-Na^+$	0.01 *M* NaCl(aq.)–air	20	2.2	74	—	van Voorst Vader (1960b)
$C_{10}H_{21}(OC_2H_4)_2SO_4^-Na^+$	0.01 *M* NaCl(aq.)–heptane	20	2.3	73	—	van Voorst Vader (1960b)
$C_{10}H_{21}(OC_2H_4)_2SO_4^-Na^+$	0.03 *M* NaCl(aq.)–air	20	2.8	59	—	van Voorst Vader (1960b)
$C_{12}H_{25}(OC_2H_4)_4SO_4^-Na^+$	H_2O–air	25	—	—	3.0_2	Rosen et al. (1996)
$C_{12}H_{25}OC_2H_4SO_4^-Na^+$	0.1 *M* NaCl(aq.)–air	10	4.0_3	41	4.2_9	Dahanayake et al. (1986)
$C_{12}H_{25}OC_2H_4SO_4^-Na^+$	0.1 *M* NaCl(aq.)–air	25	3.8_1	44	4.2_3	Dahanayake et al. (1986)
$C_{12}H_{25}OC_2H_4SO_4^-Na^+$	0.1 *M* NaCl(aq.)–air	40	3.6_0	46	4.0_9	Dahanayake et al. (1986)
$C_{12}H_{25}(OC_2H_4)_2SO_4^-Na^+$	H_2O–air	10	2.7_6	60	2.9_6	Dahanayake et al. (1986)
$C_{12}H_{25}(OC_2H_4)_2SO_4^-Na^+$	H_2O–air	25	2.6_2	63	2.9_2	Dahanayake et al. (1986)
$C_{12}H_{25}(OC_2H_4)_2SO_4^-Na^+$	H_2O–air	40	2.5_0	66	2.8_6	Dahanayake et al. (1986)
$C_{12}H_{25}(OC_2H_4)_2SO_4^-Na^+$	0.1 *M* NaCl(aq.)–air	10	3.6_5	46	4.4_0	Dahanayake et al. (1986)

$C_{12}H_{25}(OC_2H_4)_2SO_4^-Na^+$	0.1 *M* NaCl(aq.)–air	25	3.4_6	48	4.3_6	Dahanayake et al. (1986)
$C_{12}H_{25}(OC_2H_4)_2SO_4^-Na^+$	0.1 *M* NaCl(aq.)–air	40	3.3_0	50	4.2_3	Dahanayake et al. (1986)
o-$C_8H_{17}C_6H_4SO_3^-Na^+$	H_2O–air	25	2.5	66	—	Gray et al. (1955)
p-$C_8H_{17}C_6H_4SO_3^-Na^+$	H_2O–air	25	3.0	55	—	Gray et al. (1955)
p-$C_8H_{17}C_6H_4SO_3^-Na^+$	H_2O–air	70	3.4	49	1.9_6	Lange (1964)
$(C_5H_{11})(C_3H_7)CHCH_2C_6H_4SO_3^-Na^+$	H_2O–air	75	2.7_5	60	—	Greiss (1955); van Voorst Vader (1960a)
p-$C_{10}H_{21}C_6H_4SO_3^-Na^+$	H_2O–air	70	3.9	43	2.5_3	Lange (1964)
p-$C_{10}H_{21}C_6H_4SO_3^-Na^+$	H_2O–air	75	2.1	78	2.5_2	Greiss (1955)
$C_{10}H_{21}$-2-$C_6H_4SO_3^-Na^+$	"Hard river" water (I.S. = 6.6×10^{-3} *M*)	30	3.4_5	$48._1$	4.1	Zhu et al. (1998b)
$C_{11}H_{23}$-2-$C_6H_4SO_3^-Na^+$	"Hard river" water (I.S. = 6.6×10^{-3} *M*)	30	3.6_9	$45._0$	4.6	Zhu et al. (1998b)
$C_{11}H_{23}$-5-$C_6H_4SO_3^-Na^+$	"Hard river" water (I.S. = 6.6×10^{-3} *M*)	30	3.2_4	$51._2$	4.5	Zhu et al. (1998b)
p-$C_{12}H_{25}C_6H_4SO_3^-Na^+$	H_2O–air	70	3.7	45	3.1_0	Lange (1964)
p-$C_{12}H_{25}C_6H_4SO_3^-Na^+$	H_2O–air	75	3.2	52	3.1_4	Greiss (1955); van Voorst Vader (1960a)
$C_{12}H_{25}C_6H_4SO_3^-Na^{+b}$	0.1 *M* NaCl(aq.)–air	25	3.6	46	4.9	Zhu and Rosen (1987)
$C_{12}H_{25}C_6H_4SO_3^-Na^{+b}$	0.1 *M* NaCl(aq.)–air	60	2.8	59	4.9	Rosen and Zhu (1989)
$C_{12}H_{25}C_6H_4SO_3^-Na^{+b}$	0.1 *M* NaCl(aq.)–Parafilm	25	4.5_6	$36._4$	4.7	Murphy et al. (1990)
$C_{12}H_{25}C_6H_4SO_3^-Na^{+b}$	0.1 *M* NaCl(aq.)–Teflon	25	4.2_3	$38._4$	4.5	Murphy et al. (1990)

(*Continued*)

TABLE 2.2 (*Continued*)

Compound	Interface	Temp. (°C)	Γ_m, (mol/cm^2 × 10^{10})	a_m^s, Å^2	pC_{20}	Reference
$C_{12}H_{25}$-2-$C_6H_4SO_3^-Na^+$	"Hard river" water (I.S. = 6.6 × 10^{-3} *M*)	30	4.1_6	$39._9$	4.9	Zhu et al. (1998b)
$C_{12}H_{25}$-3-$C_6H_4SO_3^-Na^+$	"Hard river" water (I.S. = 6.6 × 10^{-3} *M*)	30	3.9_8	$41._7$	4.7	Zhu et al. (1998b)
$C_{12}H_{25}$-4-$C_6H_4SO_3^-Na^+$	"Hard river" water (I.S. = 6.6 × 10^{-3} *M*)	30	3.4_4	$48._3$	4.9	Zhu et al. (1998b)
$C_{12}H_{25}$-5-$C_6H_4SO_3^-Na^+$	H_2O–air	75	2.3	71	2.8_9	Greiss (1955)
$C_{12}H_{25}$-5-$C_6H_4SO_3^-Na^+$	"Hard river" water (I.S. = 6.6 × 10^{-3} *M*)	30	3.3_8	$49._1$	4.7	Zhu et al. (1998b)
$C_{12}H_{25}$-6-$C_6H_4SO_3^-Na^+$	H_2O–air	75	2.2	74	2.5_2	Greiss (1955)
$C_{12}H_{25}$-6-$C_6H_4SO_3^-Na^+$	"Hard river" water (I.S. = 6.6 × 10^{-3} *M*)	30	3.1_5	$52._7$	4.9	Zhu et al. (1998b)
$C_{13}H_{27}$-2-$C_6H_4SO_3^-Na^+$	"Hard river" water (I.S. = 6.6 × 10^{-3} *M*)	30	4.0_5	$41._0$	5.5	Zhu et al. (1998b)
$C_{13}H_{27}$-5-$C_6H_4SO_3^-Na^+$	H_2O–air	30	2.1_5	$77._2$	4.0	Zhu et al. (1998b)
$C_{13}H_{27}$-5-$C_6H_4SO_3^-Na^+$	"Hard river" water (I.S. = 6.6 × 10^{-3} *M*)	30	3.5_8	$46._4$	5.3	Zhu et al. (1998b)
p-$C_{14}H_{29}C_6H_4SO_3^-Na^+$	H_2O–air	70	2.7	61	3.6_4	Lange (1964)
$C_{14}H_{29}$-6-$C_6H_4SO_3^-Na^+$	H_2O–air	75	3.0_0	57	—	Greiss (1955); van Voorst Vader (1960a)
p-$C_{16}H_{33}C_6H_4SO_3^-Na^+$	H_2O–air	70	1.9	87	4.2_1	Lange (1964)
$C_{16}H_{33}$-8-$C_6H_4SO_3^-Na^+$	H_2O–air	45	1.6_1	103	5.4_5	Lascaux et al. (1983)
$C_{16}H_{33}$-8-$C_6H_4SO_3^-Na^+$	0.05 *M* NaCl(aq.)–air	45	3.2_7	51	6.6_4	Lascaux et al. (1983)
$C_{10}H_{21}C_6H_3(SO_3^-Na^+)OC_6H_5$	0.1 *M* NaCl(aq.)–air	25	3.4_5	48	5.5	Rosen et al. (1992)
$C_{10}H_{21}C_6H_3(SO_3^-Na^+)OC_6H_4SO_3^-Na^+$	1 *N* NaCl(aq.)–air	25	2.2	75	3.6	Rosen et al. (1992)

$(C_{10}H_{21})_2C_6H_2(SO_3^-Na^+)OC_6H_4SO_3^-Na^+$	0.1 *M* NaCl(aq.)–air	25	1.6	101	—	Rosen et al. (1992)
$C_{11}H_{23}CON(CH_3)CH_2CH_2SO_3^-Na^+$, pH 10.5	H_2O	30	2.2	77	2.3	Tsubone and Rosen (2001)
$C_{11}H_{23}CON(CH_3)CH_2CH_2SO_3^-Na^+$, pH 10.5	0.1 *M* NaCl(aq.)	30	3.0	56	3.6	Tsubone and Rosen (2001)
$C_4H_9OOCCH_2CH(SO_3^-Na^+)COOC_4H_9$	H_2O–air	25	2.1_3	78	1.4_4	Williams et al. (1957)
$C_6H_{13}OOCCH_2CH(SO_3^-Na^+)COOC_6H_{13}$	H_2O–air	25	1.8_0	92	2.9_4	Williams et al. (1957)
$C_6H_{13}OOCCH_2CH(SO_3^-Na^+)COOC_6H_{13}$	H_2O–benzene	23	1.8_5	89	—	Lange (1957)
$C_6H_{13}OOCCH_2CH(SO_3^-Na^+)COOC_6H_{13}$	0.01 *M* NaCl(aq.)–benzene	23	2.0	84	—	Lange (1957)
$C_4H_9CH(C_2H_5)CH_2OOCCH_2CH(SO_3^-Na^+)COOCH_2CH(C_2H_5)C_4H_9$	H_2O–air	25	1.5_6	106	4.0_5	Williams et al. (1957)
$C_4H_9CH(C_2H_5)CH_2OOCCH_2CH(SO_3^-Na^+)COOCH_2CH(C_2H_5)C_4H_9$	0.003 *M* NaCl(aq.)–air	20	1.4_5	115	—	van Voorst Vader (1960b)
$C_8H_{17}COO(CH_2)_2SO_3^-Na^+$	H_2O–air	30	3.2	52	—	Hikota et al. (1970)
$C_{12}H_{25}COO(CH_2)_2SO_3^-Na^+$	H_2O–air	30	2.8_5	58	—	Hikota et al. (1970)
$C_8H_{17}OOC(CH_2)_2SO_3^-Na^+$	H_2O–air	30	2.9	57	—	Hikota et al. (1970)
$C_{10}H_{21}OOC(CH_2)_2SO_3^-Na^+$	H_2O–air	30	2.8	59	—	Hikota et al. (1970)
$C_{12}H_{25}OOC(CH_2)_2SO_3^-Na^+$	H_2O–air	30	2.6	65	—	Hikota et al. (1970)
$C_6H_{13}OOCCH(C_7H_{15})SO_3^-Na^+$	0.01 *M* NaCl(aq.)–air	25	2.8	59	—	Boucher et al. (1968)

(*Continued*)

TABLE 2.2 (*Continued*)

Compound	Interface	Temp. (°C)	Γ_m, (mol/cm^2 × 10^{10})	a_m^s, Å^2	pC_{20}	Reference
$C_6H_{13}OOCCH(C_7H_{15})SO_3^-Na^+$	0.04 *M* NaCl(aq.)–air	25	2.9	57	—	Boucher et al. (1968)
$C_7H_{13}OOCCH(C_7H_{15})SO_3^-Na^+$	0.01 *M* NaCl(aq.)–air	25	2.9	57	—	Boucher et al. (1968)
$C_7H_{15}OOCCH(C_7H_{15})SO_3^-Na^+$	0.04 *M* NaCl(aq.)–air	25	3.0	56	—	Boucher et al. (1968)
$C_4H_9OOCCH(C_{10}H_{21})SO_3^-Na^+$	0.01 *M* NaCl(aq.)–air	20	2.4	70	—	van Voorst Vader (1960b)
$CH_3OOCCH(C_{12}H_{25})SO_3^-Na^+$	0.01 *M* NaCl(aq.)–air	25	3.0	55	—	Boucher et al. (1968)
$CH_3OOCCH(C_{12}H_{25})SO_3^-Na^+$	0.04 *M* NaCl(aq.)–air	25	3.3	51	—	Boucher et al. (1968)
$CH_3OOCCH(C_{14}H_{29})SO_3^-Na^+$	0.01 *M* NaCl(aq.)–air	25	3.8	44	—	Boucher et al. (1968)
$CH_3OOCCH(C_{14}H_{29})SO_3^-Na^+$	0.04 *M* NaCl(aq.)–air	25	3.5	47	—	Boucher et al. (1968)
$C_9H_{19}C_6H_4(OC_2H_4)_{9.5}OP(O)(OH)_2$	H_2O–air (pH 2.5)	25	1.9	86	—	Groves et al. (1972)
$C_9H_{19}C_6H_4(OC_2H_4)_{8.5}OP(O)(OH)_2$	H_2O–air (0.05 *M* phosphate buffer, pH 6.86)	25	2.8_5	58	—	Groves et al. (1972)
$C_9H_{19}C_6H_4(OC_2H_4)_{8.5}OP(O)(OH)_2$	H_2O–hexane (pH 2.5)	20	2.1_5	77	—	Groves et al. (1972)
$C_9H_{19}C_6H_4(OC_2H_4)_{8.5}OP(O)(OH)_2$	H_2O–hexane (0.05 *M* phosphate buffer, pH 6.88)	20	3.0	56	—	Groves et al. (1972)

$Na^+ \cdot {}^-O_3S-C_6H_4-O(CH_2)_6O-C_6H_4-SO_3^-Na^+$	H_2O–air	25	0.3_6	460	—	Rosen et al. (1976)
$Na^+ \cdot {}^-O_3S-C_6H_4-O(CH_2)_{10}O-C_6H_4-SO_3^-Na^+$	H_2O–air	40	0.6_4	260	—	Rosen et al. (1976)
$Na^+ \cdot {}^-O_3S-C_6H_4-O(CH_2)_{10}O-C_6H_4-SO_3^-Na^+$	H_2O–air	60	0.2_2	750	—	Rosen et al. (1976)
$Na^+ \cdot {}^-O_3S-C_6H_4-O(CH_2)_{12}O-C_6H_4-SO_3^-Na^+$	H_2O–air	70	0.2_2	760	—	Rosen et al. (1976)
$Na^{+-}O_4S(CH_2)_{16}SO_4^-Na^+$	0.001 *M* NaCl(aq.)–air	25	1.7_5	95	—	Elworthy (1959)
$Na^{+-}O_4S(CH_2)_{16}SO_4^-Na^+$	0.2 *M* NaCl(aq.)–air	25	1.9	88	—	Elworthy (1959)
$Na^{+-}O_4S(CH_2)_{16}SO_4^-Na^+$	1 *M* NaCl(aq.)–air	25	1.9	86	—	Elworthy (1959)
$C_7F_{15}SO_3^-Na^+$	H_2O–air	25	3.1	53	2.7_6	Shinoda et al. (1972)
$C_8F_{17}SO_3^-Li^+$	H_2O–air	25	3.0	55	3.2_0	Shinoda et al. (1972)
$C_8F_{17}SO_3^-Na^+$	H_2O–air	25	3.1	53	3.2_3	Shinoda et al. (1972)
$C_8F_{17}SO_3^-K^+$	H_2O–air	25	3.7	45	3.5_6	Shinoda et al. (1972)
$C_8F_{17}SO_3^-NH_4^+$	H_2O–air	25	4.1	41	3.4_0	Shinoda et al. (1972)
$C_8F_{17}SO_3^-NH_3C_2H_4OH^+$	H_2O–air	25	3.9	43	3.4_4	Shinoda et al. (1972)
$C_7F_{15}COO^-Na^+$	H_2O–air	25	4.0	42	2.5_0	Shinoda et al. (1972)
$C_7F_{15}COO^-K^+$	H_2O–air	25	3.9	43	2.5_7	Shinoda et al. (1972)

(*Continued*)

TABLE 2.2 (*Continued*)

Compound	Interface	Temp. (°C)	Γ_m, (mol/cm^2 × 10^{10})	a_m^s, Å^2	pC_{20}	Reference
$(CF_3)_2CF(CF_2)_4COO^-Na^+$	H_2O–air	25	3.8	44	2.5_7	Shinoda et al. (1972)
Cationics						
$C_{10}H_{21}N(CH_3)_3^+Br^-$	0.1 *M* NaCl(aq.)–air	25	3.3_9	49	1.8_0	Li et al. (2001)
$C_{12}H_{25}N(CH_3)_3^+Cl^-$	0.1 *M* NaCl(aq.)–air	25	4.3_9	38	2.7_1	Li et al. (2001)
$C_{14}H_{29}N(CH_3)_3^+Br^-$	H_2O–air	30	2.7	61	—	Venable and Nauman (1964)
$C_{14}H_{29}N(CH_3)_3^+Br^-$	0.1 *M* NaCl(aq.)–air	25	2.3	59	$3._8$	Rosen and Sulthana (2001)
$C_{14}H_{29}N(C_3H_7)_3^+Br^-$	H_2O–air	30	1.9	89	—	Venable and Nauman (1964)
$C_{14}H_{29}N(C_3H_7)_3^+Br^-$	0.05 *M* KBr(aq.)–air	30	2.6	64	—	Venable and Nauman (1964)
$C_{16}H_{33}N(CH_3)_3^+Cl^-$	0.1 *M* NaCl(aq.)–air	25	3.6	46	5.0_0	Caskey and Barlage (1971)
$C_{16}H_{33}N(C_3H_7)_3^+Br^-$	H_2O–air	30	1.8	91	—	Venable and Nauman (1964)
$C_{18}H_{37}N(CH_3)_3^+Br^-$	H_2O–air	25	2.6	64	—	Brashier and Thornhill (1968)
$C_8H_{17}Pyr^+Br^-$	H_2O–air	20	2.3	73	1.2_8	Bury and Browning (1953)
$C_{10}H_{21}Pyr^+Br^-$	H_2O–air	25	—	—	1.82	Venable and Nauman (1964)
$C_{12}H_{25}Pyr^+Br^-$	H_2O–air	25	3.3	50	2.3_3	Rosen et al. (1982b)
$C_{12}H_{25}Pyr^+Br^-$	0.1 *M* NaCl(aq.)–air	10	3.7	45	3.4_8	Rosen et al. (1982b)
$C_{12}H_{25}Pyr^+Br^-$	0.1 *M* NaCl(aq.)–air	25	3.5	48	3.4_0	Rosen et al. (1982b)

$C_{12}H_{25}Pyr^+Cl^-$	0.1 *M* NaCl(aq.)–air	40	3.3	51	3.3_0	Rosen et al. (1982b)
$C_{12}H_{25}Pyr^+Cl^-$	H_2O–air	10	2.7	61	2.1_2	Rosen et al. (1982b)
$C_{12}H_{25}Pyr^+Cl^-$	H_2O–air	25	2.7	62	2.1_0	Rosen et al. (1982b)
$C_{12}H_{25}Pyr^+Cl^-$	H_2O–air	40	2.6	63	2.0_7	Rosen et al. (1982b)
$C_{12}H_{25}Pyr^+Cl^-$	0.1 *M* NaCl(aq.)–air	25	3.0	55	2.9_8	Rosen et al. (1982b)
$C_{14}H_{29}Pyr^+Br^-$	H_2O–air	30	2.7_5	60	2.9_4	Venable and Nauman (1964)
$C_{14}H_{29}Pyr^+Br^-$	0.05 *M* KBr(aq.)–air	30	3.4_5	48	—	Venable and Nauman (1964)
$C_{14}Pyr^+Cl^-$	0.1 *M* KCl(aq.)	25	3.4_6	46	—	Semmler and Kohler (1999)
$C_{16}Pyr^+Cl^-$	H_2O–air	25	3.3_7	49	—	Semmler and Kohler (1999)
$C_{16}Pyr^+Cl^-$	0.1 *M* KCl(aq.)	25	5.0_4	33	—	Semmler and Kohler (1999)
$C_{12}N^+H_2CH_2CH_2OHCl^-$	H_2O–air	25	1.9_3	86	2.1_9	Omar and Abdel-Khalek (1997)
$C_{12}N^+H(CH_2CH_2OH)_2Cl^-$	H_2O–air	25	2.4_9	67	2.3_1	Omar and Abdel-Khalek (1997)
$C_{12}N^+(CH_2CH_2OH)_3Cl^-$	H_2O–air	25	2.9_1	57	2.3_4	Omar and Abdel-Khalek (1997)

(*Continued*)

TABLE 2.2 (*Continued*)

Compound	Interface	Temp. (°C)	Γ_m, (mol/cm^2 × 10^{10})	a_m^s, Å^2	pC_{20}	Reference
Nonionics						
Nonionics (Homogeneous Head Group)						
$C_8H_{17}OCH_2CH_2OH$	H_2O–air	25	5.2	32	3.1_7	Shinoda et al. (1959)
$C_8H_{17}CHOHCH_2OH$	H_2O–air	25	5.1	33	3.6_3	Kwan and Rosen (1980)
$C_8H_{17}CHOHCH_2CH_2OH$	H_2O–air	25	5.3	32	3.5_9	Kwan and Rosen (1980)
$C_{12}H_{25}CHOHCH_2CH_2OH$	H_2O–air	25	5.1	33	5.7_7	Kwan and Rosen (1980)
Octyl-β-D-glucoside	H_2O–air	25	4.0	41	—	Shinoda et al. (1959)
Decyl-α-glucoside	H_2O–air	25	3.7_7	44	—	Aveyard et al. (1998)
Decyl-β-glucoside	H_2O–air	25	4.0_5	41	—	Aveyard et al. (1998)
Decyl-β-glucoside	0.1 *M* NaCl(aq.)–air	25	4.1_8	40	3.7_6	Li et al. (2001)
Dodecyl-β-glucoside	H_2O–air	25	4.6_1	36	—	Aveyard et al. (1998)
Decyl-β-maltoside	H_2O–air	25	2.9_6	56	—	Aveyard et al. (1998)
Decyl-β-maltoside	0.01 *M* NaCl(aq.)–air	22	—	—	3.5_8	Liljekvist and Kronberg (2000)
Decyl-β-maltoside	0.1 *M* NaCl(aq.)–air	25	3.3_7	49	3.5_2	Li et al. (2001)
Dodecyl-β-maltoside	H_2O–air	25	3.3_2	50	—	Aveyard et al. (1998)
Dodecyl-β-maltoside	0.1 *M* NaCl(aq.)–air	25	3.6_7	45	4.6_4	Li et al. (2001)

N-(2-ethyl hexyl)2-pyrrolid(in)one	H_2O–air	25	3.5_7	$46._5$	3.0_0	Rosen et al. (1988)
N-(2-ethyl hexyl)2-pyrrolid(in)one	H_2O, pH 7.0–polyethylene	25	3.2_6	$50._9$	—	Rosen and Wu (2001)
N-octyl-2-pyrrolid(in)one	H_2O–air	25	4.3_8	$37._9$	3.1_4	Rosen et al. (1988)
N-octyl-2-pyrrolid(in)one	H_2O, pH 7.0–polyethylene	25	4.2_5	$39._0$	—	Rosen and Wu (2001)
N-octyl-2-pyrrolid(in)one	"Hard river" water (I.S. = 6.6×10^{-3} *M*)	25	4.0_1	$41._4$	3.3_4	Rosen et al. (1996)
N-octyl-2-pyrrolid(in)one	H_2O–0.1 *M* NaCl(aq.)	25	4.2_7	$38._9$	3.2_1	Rosen et al. (1988)
N-octyl-2-pyrrolid(in)one	0.1 *M* NaCl(aq.)–Parafilm	25	4.1_4	$40._3$	3.2_8	Rosen et al. (1988)
N-octyl-2-pyrrolid(in)one	0.1 *M* NaCl(aq.)–Teflon	25	3.7_9	$43._8$	3.0_4	Rosen et al. (1988)
N-decyl-2-pyrrolid(in)one	H_2O–air	25	4.6_1	$36._0$	4.1_9	Rosen et al. (1988)
N-decyl-2-pyrrolid(in)one	H_2O–Parafilm	25	4.5_4	$36._6$	4.2_4	Rosen et al. (1988)
N-decyl-2-pyrrolid(in)one	H_2O–Teflon	25	4.2_4	$39._2$	4.0_4	Rosen et al. (1988)
N-decyl-2-pyrrolid(in)one	"Hard river" water (I.S. = 6.6×10^{-3} *M*)	25	4.1_7	$39._8$	4.3_8	Rosen et al. (1996)
N-dodecyl-2-pyrrolid(in)one	H_2O–air	25	5.0_8	$32._7$	5.3_0	Rosen et al. (1988)
N-dodecy1-2-pyrrolid(in)one	"Hard river" water (I.S. = 6.6×10^{-3} *M*)	25	5.1_1	$32._5$	5.3_7	Rosen et al. (1996)

(*Continued*)

TABLE 2.2 (*Continued*)

Compound	Interface	Temp. (°C)	Γ_m, (mol/cm^2 × 10^{10})	a_m^s, Å^2	pC_{20}	Reference
N-dodecyl-2-pyrrolid(in)one	0.1 *M* NaCl(aq.)–air	25	5.1_5	$32._2$	5.3_4	Rosen et al. (1988)
$C_{11}H_{23}CON(C_2H_4OH)_2$	H_2O–air	25	3.7_5	44	4.3_8	Zhu and Rosen (1984)
$C_{11}H_{23}CON(C_2H_4O)_4OH$	H_2O–air	23	3.4	49	—	Kjellin et al. (2002)
$C_{10}H_{21}CON(CH_3)CH_2(CHOH)_4CH_2OH$	0.1 *M* NaCl(aq.)–air	25	3.8_0	44	3.8_0	Zhu et al. (1999)
$C_{11}H_{23}CON(CH_3)CH_2CHOHCH_2OH$	0.1 *M* NaCl(aq.)–air	25	4.3_4	38	4.6_4	Zhu et al. (1999)
$C_{11}H_{23}CON(CH_3)CH_2(CHOH)_3CH_2OH$	0.1 *M* NaCl(aq.)–air	25	4.2_9	39	4.4_7	Zhu et al. (1999)
$C_{11}H_{23}CON(CH_3)CH_2(CHOH)_4CH_2OH$	0.1 *M* NaCl(aq.)–air	25	4.1_0	$40._5$	4.4_0	Zhu et al. (1999)
$C_{12}H_{25}CON(CH_3)CH_2(CHOH)_4CH_2OH$	0.1 *M* NaCl(aq.)–air	25	4.6_0	36	5.0_2	Zhu et al. (1999)
$C_{13}H_{27}CON(CH_3)CH_2(CHOH)_4CH_2OH$	0.1 *M* NaCl(aq.)–air	25	4.6_8	$35._5$	5.4_3	Zhu et al. (1999)
$C_{10}H_{21}N(CH_3)CO(CHOH)_4CH_2OH$	H_2O–air	20	3.9_6	42	3.6_0	Burczyk et al. (2001)
$C_{12}H_{25}N(CH_3)CO(CHOH)_4CH_2OH$	H_2O–air	20	3.9_9	42	4.7_8	Burczyk et al. (2001)
$C_{14}H_{29}N(CH_3)CO(CHOH)_4CH_2OH$	H_2O–air	20	3.9_7	42	5.5_5	Burczyk et al. (2001)
$C_{16}H_{33}N(CH_3)CO(CHOH)_4CH_2OH$	H_2O–air	20	3.6_5	45	6.1_1	Burczyk et al. (2001)
$C_{18}H_{37}N(CH_3)CO(CHOH)_4CH_2OH$	H_2O–air	20	3.9_7	42	6.4_6	Burczyk et al. (2001)
$(C_2H_5)_2CHCH_2(OC_2H_4)_6OH$	H_2O–air	20	2.1_5	77	—	Elworthy and Florence (1964)
$C_6H_{13}(OC_2H_4)_6OH$	H_2O–air	25	2.7	62	2.4_8	Elworthy and Florence (1964)
$C_8H_{17}(OC_2H_4)_6OH$	H_2O–air	25	1.5_0	111	3.1_4	Varadaraj et al. (1991)
$C_8H_{17}(OC_2H_4)_5OH$	0.1 *M* NaCl(aq.)–air	25	3.4_6	48	3.1_6	Varadaraj et al. (1991)

$(C_4H_9)_2CHCH_2(OC_2H_4)_6OH$	H_2O–air	20	2.8	61	—	Elworthy and Florence (1964)
$C_{10}H_{21}(OC_2H_4)_4OH$	H_2O–air	25	4.0_7	41	—	Eastoe et al. (1997)
$C_{10}H_{21}(OC_2H_4)_5OH$	H_2O–air	25	3.1_1	53	—	Eastoe et al. (1997)
$C_{10}H_{21}(OC_2H_4)_6OH$	H_2O–air	23.5	3.0	55	4.2_7	Carless et al. (1964)
$C_{10}H_{21}(OC_2H_4)_6OH$	"Hard river" water (I.S. = 6.6×10^{-3} *M*)	25	2.8_3	$58._7$	4.2_7	Rosen et al. (1996)
$C_{10}H_{21}(OC_2H_4)_8OH$	H_2O–air	25	2.3_8	70	4.2_0	Meguro et al. (1981)
$C_{10}H_{21}(OC_2H_4)_8OH$	0.01 *M* NaCl(aq.)–air	22	—	—	4.2_4	Liljekvist and Kronberg (2000)
$C_{12}H_{25}(OC_2H_4)_3OH$	H_2O–air	25	3.9_8	42	5.3_4	Rosen et al. (1982a)
$C_{12}H_{25}(OC_2H_4)_4OH$	H_2O–air	25	3.6_3	46	5.3_4	Rosen et al. (1982a)
$C_{12}H_{25}(OC_2H_4)_4OH$	H_2O–hexadecane	25	3.1_6	$52._6$	—	Rosen and Murphy (1991)
$C_{12}H_{25}(OC_2H_4)_5OH$	"Hard river" water (I.S. = 6.6×10^{-3} *M*)	25	3.8_8	$42._8$	5.3_8	Rosen et al. (1996)
$C_{12}H_{25}(OC_2H_4)_5OH$	H_2O–air	25	3.3_1	50	5.3_7	Rosen et al. (1982a)
$C_{12}H_{25}(OC_2H_4)_5OH$	0.1 *M* NaCl(aq.)–air	25	3.3_1	50	5.4_6	Rosen and Murphy (1991)
$C_{12}H_{25}(OC_2H_4)_6OH$	H_2O–air	25	3.2_1	52	—	Eastoe et al. (1997)

(*Continued*)

TABLE 2.2 (*Continued*)

Compound	Interface	Temp. (°C)	Γ_m, (mol/cm^2 × 10^{10})	a_m^s, Å^2	pC_{20}	Reference
$C_{12}H_{25}(OC_2H_4)_6OH$	"Hard river" water (I.S. = 6.6×10^{-3} *M*)	25	3.1_9	$52._0$	5.2_7	Rosen et al. (1996)
$C_{12}H_{25}(OC_2H_4)_7OH$	H_2O–air	10	2.8_5	58	5.1_5	Rosen et al. (1982a)
$C_{12}H_{25}(OC_2H_4)_7OH$	H_2O–air	25	2.9_0	57	5.2_6	Rosen et al. (1982a)
$C_{12}H_{25}(OC_2H_4)_7OH$	H_2O–air	40	2.7_7	60	5.2_8	Rosen et al. (1982a)
$C_{12}H_{25}(OC_2H_4)_7OH$	0.1 *M* NaCl(aq.)–air	25	3.6_5	$45._5$	5.2	Rosen and Sulthana (2001)
$C_{12}H_{25}(OC_2H_4)_8OH$	H_2O–air	10	2.5_6	65	5.0_5	Rosen et al. (1982a)
$C_{12}H_{25}(OC_2H_4)_8OH$	H_2O–air	25	2.5_2	66	5.2_0	Rosen et al. (1982a)
$C_{12}H_{25}(OC_2H_4)_8OH$	H_2O–air	40	2.4_6	67	5.2_2	Rosen et al. (1982a)
$C_{12}H_{25}(OC_2H_4)_8OH$	H_2O–heptane	25	2.6_2	$63._6$	5.2_7^c	Rosen and Murphy (1991)
$C_{12}H_{25}(OC_2H_4)_8OH$	H_2O–hexadecane	25	2.6_4	63	5.2_4^c	Rosen and Murphy (1991)
6-branched $C_{13}H_{27}(OC_2H_4)_5OH$	0.1 *M* NaCl(aq.)–air	25	2.8_7	58	5.1_6	Varadaraj et al. (1991)
$C_{13}H_{27}(OC_2H_4)_5OH$	H_2O–air	25	1.9_6	85	5.3_4	Varadaraj et al. (1991)
$C_{13}H_{27}(OC_2H_4)_5OH$	0.1 *M* NaCl(aq.)–air	25	3.8_9	43	5.6_2	Varadaraj et al. (1991)
$C_{13}H_{27}(OC_2H_4)_8OH$	H_2O–air	25	2.7_8	60	5.6_2	Meguro et al. (1981)
$C_{14}H_{29}(OC_2H_4)_8OH$	H_2O–air	25	3.4_3	48	6.0_2	Meguro et al. (1981)

$C_{14}H_{29}(OC_2H_4)_8OH$	"Hard river" water (I.S. = 6.6×10^{-3} *M*)	25	2.6_7	$62._2$	6.1_4	Rosen et al. (1996)
$C_{15}H_{31}(OC_2H_4)_8OH$	H_2O–air	25	3.5_9	46	6.3_1	Meguro et al. (1981)
$C_{16}H_{33}(OC_2H_4)_6OH$	H_2O–air	25	4.4	38	6.8_0	Elworthy and MacFarlane (1962)
$C_{16}H_{33}(OC_2H_4)_6OH$	"Hard river" water (I.S. = 6.6×10^{-3} *M*)	25	3.2_3	$51._4$	6.7_8	Rosen et al. (1996)
$C_{16}H_{33}(OC_2H_4)_7OH$	H_2O–air	25	3.8	44	—	Elworthy and MacFarlane (1962)
$C_{16}H_{33}(OC_2H_4)_9OH$	H_2O–air	25	3.1	53	—	Elworthy and MacFarlane (1962)
n-$C_{16}H_{33}(OC_2H_4)_{12}OH$	H_2O–air	25	2.3	72	—	Elworthy and MacFarlane (1962)
n-$C_{16}H_{33}(OC_2H_4)_{15}OH$	H_2O–air	25	2.0_5	81	—	Elworthy and MacFarlane (1962)
n-$C_{16}H_{33}(OC_2H_4)_{21}OH$	H_2O–air	25	1.4	120	—	Elworthy and MacFarlane (1962)
p-*t*-$C_8H_{17}C_6H_4(OC_2H_4)_3OH$	H_2O–air	25	3.7	45	—	Crook et al. (1964)
p-*t*-$C_8H_{17}C_6H_4(OC_2H_4)_3OH$	H_2O–air	85	3.2	52	—	Crook et al. (1964)
p- *t*-$C_8H_{17}C_6H_4(OC_2H_4)_4OH$	H_2O–air	25	3.3_5	50	—	Crook et al. (1964)

(*Continued*)

TABLE 2.2 (*Continued*)

Compound	Interface	Temp. (°C)	Γ_m, (mol/cm^2 × 10^{10})	a_m^s, Å^2	pC_{20}	Reference
p-t-$C_8H_{17}C_6H_4(OC_2H_4)_5OH$	H_2O–air	25	3.1	53	—	Crook et al. (1964)
p-t-$C_8H_{17}C_6H_4(OC_2H_4)_6OH$	H_2O–air	25	3.0	56	—	Crook et al. (1964)
p-t-$C_8H_{17}C_6H_4(OC_2H_4)_6OH$	H_2O–air	55	2.9	58	—	Crook et al. (1964)
p-t-$C_8H_{17}C_6H_4(OC_2H_4)_6OH$	H_2O–air	85	2.7	61	—	Crook et al. (1964)
p-t-$C_8H_{17}C_6H_4(OC_2H_4)_7OH$	H_2O–air	25	2.9	58	4.9_3	Crook et al. (1963, 1964)
p-t-$C_8H_{17}C_6H_4(OC_2H_4)_8OH$	H_2O–air	25	2.6	64	4.8_9	Crook et al. (1963, 1964)
p-t-$C_8H_{17}C_6H_4(OC_2H_4)_9OH$	H_2O–air	25	2.5	66	4.8_0	Crook et al. (1963, 1964)
p-t-$C_8H_{17}C_6H_4(OC_2H_4)_{10}OH$	H_2O–air	25	2.2	$74._5$	4.7_2	Crook et al. (1963, 1964)
p-t-$C_8H_{17}C_6H_4(OC_2H_4)_{10}OH$	H_2O–air	55	2.1	79	—	Crook et al. (1964)
p-t-$C_8H_{17}C_6H_4(OC_2H_4)_{10}OH$	H_2O–air	85	2.1	80	—	Crook et al. (1964)
$(CH_3)_3SiOSi(CH_3)[CH_2(C_2H_4O)_5H]$-$OSi(CH_3)_3$	H_2O–air	23 ± 2	5.0	$33._5$	—	Gentle and Snow (1995)
$(CH_3)_3SiOSi(CH_3)[CH_2(C_2H_4O)_9H]$-$OSi(CH_3)_3$	H_2O–air	23 ± 2	5.1	$32._6$	—	Gentle and Snow (1995)
$(CH_3)_3SiOSi(CH_3)[CH_2(C_2H_4O)_{13}H]$-$OSi(CH_3)_3$	H_2O–air	23 ± 2	4.2	$39._2$	—	Gentle and Snow (1995)
$(CH_3)_3SiOSi(CH_3)[CH_2(C_2H_4O)_{8.5}CH_3]$-$OSi(CH_3)_3$	H_2O pH 7.0	25	2.5_2	66	5.9_5	Gentle and Snow (1995)

$(CH_3)_3SiOSi(CH_3)[CH_2(C_2H_4O)_{8.5}CH_3]$-$OSi(CH_3)_3$	H_2O, pH 7.0–polyethylene	25	2.7_2	61	—	Rosen and Wu (2001)
$C_6F_{13}C_2H_4SC_2H_4(OC_2H_4)_2OH$	H_2O–air	25	4.7_4	35	—	Matos et al. (1989)
$C_6F_{13}C_2H_4SC_2H_4(OC_2H_4)_3OH$	H_2O–air	25	4.4_6	37	—	Matos et al. (1989)
$C_6F_{13}C_2H_4SC_2H_4(OC_2H_4)_5OH$	H_2O–air	25	3.5_6	$46._5$	—	Matos et al. (1989)
$C_6F_{13}C_2H_4SC_2H_4(OC_2H_4)_7OH$	H_2O–air	25	3.1_9	52	—	Matos et al. (1989)
Zwitterionics						
$C_{12}H_{25}N(CH_3)_2O$	H_2O–air	25	3.5	47	3.6_2	Rosen (1974)
$C_8H_{17}CH(COO^-)N^+(CH_3)_3$	H_2O–air	27	2.8	60	—	Tori et al. (1963)
$C_{10}H_{21}CH(COO^-)N^+(CH_3)_3$	H_2O–air	10	3.0	55	—	Tori and Nakagawa (1963)
$C_{10}H_{21}CH(COO^-)N^+(CH_3)_3$	H_2O–air	27	2.8	60	—	Tori et al. (1963)
$C_{10}H_{21}CH(COO^-)N^+(CH_3)_3$	H_2O–air	60	2.5	66	—	Tori and Nakagawa (1963)
$C_{12}H_{25}CH(COO^-)N^+(CH_3)_3$	H_2O–air	27	3.1	54	—	Tori et al. (1963)
$C_{10}H_{21}N^+(CH_3)_2CH_2COO^-$	H_2O–air	23	4.1_5	40	2.5_9	Beckett and Woodward (1963)
$C_{12}H_{25}N^+(CH_3)_2CH_2COO^-$	H_2O–air	25	3.2	52	—	Chevalier et al. (1991)
$C_{14}H_{29}N^+(CH_3)_2CH_2COO^-$	H_2O–air	23	3.5_3	47	4.6_2	Beckett and Woodward (1963)

(Continued)

TABLE 2.2 (*Continued*)

Compound	Interface	Temp. (°C)	Γ_m, (mol/cm^2 × 10^{10})	a_m^s, Å^2	pC_{20}	Reference
$C_{16}H_{33}N^+(CH_3)_2CH_2COO^-$	H_2O–air	23	4.1_3	40	5.5_4	Beckett and Woodward (1963)
$C_{12}H_{25}N^+(CH_3)_2(CH_2)_3COO^-$	H_2O–air	25	2.5	67	—	Chevalier et al. (1991)
$C_{12}H_{25}N^+(CH_3)_2(CH_2)_5COO^-$	H_2O–air	25	2.4	68	—	Chevalier et al. (1991)
$C_{12}H_{25}N^+(CH_3)_2(CH_2)_7COO^-$	H_2O–air	25	2.1_5	77	—	Chevalier et al. (1991)
$C_{10}H_{21}CH(Pyr^+)COO^-$	H_2O–air	25	3.5_9	46	2.8_7	Zhao and Rosen (1984)
$C_{12}H_{25}CH(Pyr^+)COO^-$	H_2O–air	25	3.5_7	46	3.9_8	Zhao and Rosen (1984)
$C_{14}H_{29}CH(Pyr^+)COO^-$	H_2O–air	40	3.4_0	49	4.9_2	Zhao and Rosen (1984)
$C_{10}H_{21}N^+(CH_2C_6H_5)(CH_3)CH_2COO^-$	H_2O–air	25	2.9_1	57	3.3_6	Dahanayake and Rosen (1984)
$C_{12}H_{25}N^+(CH_2C_6H_5)(CH_3)CH_2COO^-$	H_2O–air	10	2.9_6	56	4.4_2	Dahanayake and Rosen (1984)
$C_{12}H_{25}N^+(CH_2C_6H_5)(CH_3)CH_2COO^-$	H_2O–air	25	2.8_6	58	4.4_2	Dahanayake and Rosen (1984)
$C_{12}H_{25}N^+(CH_2C_6H_5)(CH_3)CH_2COO^-$	H_2O–air	40	2.7_6	60	4.3_2	Dahanayake and Rosen (1984)
$C_{12}H_{25}N^+(CH_2C_6H_5)(CH_3)CH_2COO^-$	0.1 *M* NaCl(aq.), pH 5.7	25	3.1_3	$53._0$	4.6	Rosen and Sulthana (2001)

$C_{12}H_{25}N^+(CH_2C_6H_5)(CH_3)CH_2COO^-$	H_2O–heptane	25	2.7_6	60	—	Murphy and Rosen (1988)
$C_{12}H_{25}N^+(CH_2C_6H_5)(CH_3)CH_2COO^-$	H_2O–isooctane	25	2.7_7	60	—	Murphy and Rosen (1988)
$C_{12}H_{25}N^+(CH_2C_6H_5)(CH_3)CH_2COO^-$	H_2O–heptamethylnonane	25	2.7_8	60	—	Murphy and Rosen (1988)
$C_{12}H_{25}N^+(CH_2C_6H_5)(CH_3)CH_2COO^-$	H_2O–dodecane	25	2.8_3	59	—	Murphy and Rosen (1988)
$C_{12}H_{25}N^+(CH_2C_6H_5)(CH_3)CH_2COO^-$	H_2O–hexadecane	25	2.9_0	57	—	Murphy and Rosen (1988)
$C_{12}H_{25}N^+(CH_2C_6H_5)(CH_3)CH_2COO^-$	H_2O–cyclohexane	25	2.6_4	63	—	Murphy and Rosen (1988)
$C_{12}H_{25}N^+(CH_2C_6H_5)(CH_3)CH_2COO^-$	H_2O–toluene	25	2.5_1	66	—	Murphy and Rosen (1988)
$C_8H_{17}N^+(CH_2C_6H_5)(CH_3)CH_2CH_2SO_3^-$	H_2O–air	25	2.7_2	61	2.2_3	Dahanayake and Rosen (1984)
$C_{10}H_{21}N^+(CH_2C_6H_5)(CH_3)CH_2CH_2SO_3^-$	H_2O–air	25	2.7_2	61	3.3_4	Dahanayake and Rosen (1984)
$C_{12}H_{25}N^+(CH_2C_6H_5)(CH_3)CH_2CH_2SO_3^-$	H_2O–air	10	2.8_1	59	4.5_2	Dahanayake and Rosen (1984)
$C_{12}H_{25}N^+(CH_2C_6H_5)(CH_3)CH_2CH_2SO_3^-$	H_2O–air	25	2.7_2	61	4.4_4	Dahanayake and Rosen (1984)
$C_{12}H_{25}N^+(CH_2C_6H_5)(CH_3)CH_2CH_2SO_3^-$	H_2O–air	40	2.5_9	64	4.3_2	Dahanayake and Rosen (1984)
$C_{12}H_{25}CHOHCH_2N^+(CH_3)_2CH_2CH_2OP\text{-}(O)(OH)O^-$	H_2O–air	25	3.8	43.8	—	Tsubone and Uchida (1990)

(Continued)

TABLE 2.2 (***Continued***)

Compound	Interface	Temp. (°C)	Γ_m, (mol/cm^2 × 10^{10})	a_m^s, Å^2	pC_{20}	Reference
Anionic—Cationic Salts						
$C_2H_5N^+(CH_3)_3 \cdot C_{12}H_{25}SO_4^-$	H_2O–air	25	2.6_3	63	3.0_4	Lange and Schwuger (1971)
$C_4H_9N^+(CH_3)_3 \cdot C_{10}H_{21}SO_4^-$	H_2O–air	25	2.8_5	58	2.5_7	Lange and Schwuger (1971)
$C_6H_{13}N^+(CH_3)_3 \cdot C_8H_{17}SO_4^-$	H_2O–air	25	2.5_0	67	2.5_7	Lange and Schwuger (1971)
$C_8H_{17}N^+(CH_3)_3 \cdot C_6H_{13}SO_4^-$	H_2O–air	25	2.5_3	66	2.5_7	Lange and Schwuger (1971)
$C_{10}H_{21}N^+(CH_3)_3 \cdot C_4H_9SO_4^-$	H_2O–air	25	2.5_0	67	2.5_7	Lange and Schwuger (1971)
$C_{12}H_{25}N^+(CH_3)_3 \cdot CH_3SO_4^-$	H_2O–air	25	2.7_0	61	2.3_2	Lange and Schwuger (1971)
$C_{12}H_{25}N^+(CH_3)_3 \cdot C_2H_5SO_4^-$	H_2O–air	25	2.8_5	58	2.5_7	Lange and Schwuger (1971)

$C_{12}H_{25}N^+(CH_3)_3 . C_4H_9SO_4^-$	H_2O–air	25	2.6_7	62	3.0_2	Lange and Schwuger (1971)
$C_{12}H_{25}N^+(CH_3)_3 . C_6H_{13}SO_4^-$	H_2O–air	25	2.5_8	64	3.7_0	Lange and Schwuger (1971)
$C_{12}H_{25}N^+(CH_3)_3 . C_8H_{17}SO_4^-$	H_2O–air	25	2.7_2	61	4.2_7	Lange and Schwuger (1971)
$C_{12}H_{25}N^+(CH_3)_3 . C_{12}H_{25}SO_4^-$	H_2O–air	25	2.7_4	61	5.3_2	Lange and Schwuger (1971)
$C_{10}H_{21}N^+(CH_3)_3 . C_{10}H_{21}SO_4^-$	H_2O–air	25	3.3_5	58	—	Corkill et al. (1963)
$C_{12}H_{25}N^+(CH_3)_2OH . C_{12}H_{25}SO_3^-$	H_2O–air	25	2.1_4	78	5.6_6	Rosen et al. (1964)
$C_{16}H_{33}N^+(CH_3)_3 . C_{12}H_{25}SO_4^-$	H_2O–air	30	2.8_0	59	—	Tomasic et al. (1999)

[a] From branched dodecyl alcohol with 4.4 methyl branches in the molecule.
[b] Commercial material.
[c] pC^{30}.
I.S., total ionic strength; PTFE, polytetrafluoroethylene; Pyr^+, pyridinium.

The data in Table 2.2 indicate the following relations between the structure of the surfactant and its effectiveness of adsorption at the A/A and aqueous solution–hydrocarbon interfaces.

Change in the length of the hydrophobic group of straight-chain ionic surfactants beyond 10 carbon atoms appears to have almost no effect on the effectiveness of adsorption at the aqueous solution–heptane (A/H) interface and very little effect on the effectiveness at the A/A interface.

A phenyl group that is part of a hydrophobic group has the effect of about three and one-half $-CH_2-$ groups in a straight hydrophobic chain. When the number of carbon atoms in a straight-chain hydrophobic group exceeds 16 at either the A/A or aqueous solution–hydrocarbon interface, there is a significant decrease in the effectiveness of adsorption, which has been attributed (Mukerjee, 1967) to coiling of the long chain, with a consequent increase in the cross-sectional area of the molecule at the interface.

The positioning of the hydrophilic group in a central, rather than in a terminal, position in a straight alkyl chain or branching of the alkyl chain results in an increase in the area per molecule at the liquid–air interface.

Replacement of the usual hydrocarbon-based hydrophobic group by a fluorocarbon-based hydrophobic group appears to cause only a small increase in the effectiveness of adsorption at the A/A interface, in contrast to its large effect on most other interfacial properties.

In ionic surfactants, those with more tightly bound counterions (ions with small hydrated radii, e.g., Cs^+, K^+, NH_4^+) appear to be more effectively adsorbed than those with less tightly bound ones (Na^+, Li^+, F^-), although the effect, except for tetraalkylammonium salts (Tamaki, 1967) is rather small. In quaternary ammonium salts of structure $R(CH_2)_mN(R')_3^+X^-$ (e.g., $C_{14}H_{29}N(CH_3)_3^+Br^-$ and $C_{14}H_{29}N\ (C_3H_7)_3^+Br^-$), an increase in the size of R′ results in an increase in a_m^s and a consequent decrease in Γ_m. Salt formation between an ionic surfactant and an oppositely charged surfactant of approximately equal hydrophobic chain length (e.g., $C_{10}H_{21}N(CH_3)_3^+.C_{10}H_{21}SO_4^-$ or $C_{12}H_{25}N(CH_3)_3^+.C_{12}H_{25}SO_3^-$) produces a large increase in the effectiveness of adsorption, with the area per molecule at the interface approaching that of a close-packed film with the hydrophobic chains oriented perpendicular to the interface. This is probably the result of the combined effects of mutual attraction of ionic groups and mutual attraction of hydrophobic chains.

In POE materials in which the POE group constitutes either the entire hydrophilic group, as in POE nonionics (Weil et al., 1958), or a portion of the hydrophilic group, as in $C_{16}H_{33}(OC_2H_4)_xSO_4^-Na^+$ or $C_9H_{19}C_6H_4(OC_2H_4)_xOPO(OH)_2$, the POE chain, immersed in the aqueous phase in the form of a coil whose cross-sectional area increases with the number of OE units (Schick, 1962), determines a_m^s and therefore Γ_m. As the number of EO units increases, a_m^s increases and Γ_m decreases. In nonionic POE materials containing the same mole ratio of OE, effectiveness increases with increase in the length of the hydrophobic group, due to the larger lateral interaction.

Other factors that produce significant changes in Γ_m are

1. *The Addition of Neutral Electrolyte (NaCl, KBr) to an Aqueous Solution of an Ionic Surfactant Containing No Electrolyte.* This results in increased adsorption at the A/A interface because of the decrease in repulsion between the oriented ionic heads at the interface when the ionic strength of the solution is increased (see Section I of this chapter). For nonionics, there appears to be only a small increase in the saturation adsorption upon addition of neutral electrolyte (Shinoda et al., 1961; Schick, 1962) and little change on the addition of either water structure-breaking (urea, *N*-methylacetamide) or structure-promoting (fructose, xylose) additives (Schwuger, 1969, 1971b).
2. *The Nature of the Nonaqueous Phase in Adsorption at the L/L Interface.* It has been found that saturation adsorption increases with increase in the interfacial tension between the two phases (van Voorst Vader, 1960a).

   When the nonaqueous phase is a straight-chain saturated hydrocarbon, the value of Γ_m is close to that at the A/A interface, with possibly a slight increase in the effectiveness of adsorption as the length of the alkane is increased. When the nonaqueous phase is a short-chain unsaturated or aromatic hydrocarbon, however, there is a significant decrease in the effectiveness of adsorption at the aqueous solution–hydrocarbon interface (Rehfeld, 1967; Murphy and Rosen, 1988).
3. *Temperature.* An increase in temperature results in an increase in the area per molecule, presumably due to increased thermal motion, with a consequent decrease in the effectiveness of adsorption.

D. The Szyszkowski, Langmuir, and Frumkin Equations

In addition to the Gibbs equation, three other equations have been suggested that relate concentration of the surface-active agent at the interface, surface or interfacial tension, and equilibrium concentration of the surfactant in a liquid phase. The Langmuir equation (Langmuir, 1917),

$$\Gamma_1 = \frac{\Gamma_m C_1}{C_1 + a} \qquad (2.6)$$

discussed previously, relates surface concentrations with bulk concentration. The Szyszkowski equation (Szyszkowski, 1908),

$$\gamma_0 - \gamma = \pi = 2.303RT\Gamma_m \log\left(\frac{C_1}{a} + 1\right) \qquad (2.28)$$

where γ_0 is the surface tension of the solvent and π is the surface pressure (the reduction in surface tension), and relates surface tension with bulk concentration. The Frumkin equation (Frumkin, 1925),

$$\gamma_0 - \gamma = \pi = -2.303RT\Gamma_m \log\left(1 - \frac{\Gamma_1}{\Gamma_m}\right) \tag{2.29}$$

relates surface tension and surface (excess) concentration. These equations, first formulated as empirical relations, can be obtained from a general surface equation of state (Lucassen-Reynders and van den Tempel, 1967) if one assumes ideal surface behavior (i.e., surface activity coefficients close to unity). This assumption has been found to be generally valid for ionic surfactants at the A/A and aqueous solution–hydrocarbon interfaces (Lucassen-Reynders, 1966), with the exception of C_{18} or longer compounds at the A/A interface.

E. Efficiency of Adsorption at the *L/G* and *L/L* Interfaces

Significance of C_{20} Value: As in the case of adsorption at the solid–liquid interface, in comparing the performance of surfactants at the *L/G* or *L/L* interfaces, it is useful to have a parameter that measures the concentration of surfactant in the liquid phase required to produce a given amount of adsorption at the interface, the efficiency of adsorption of the surfactant, especially when it can be related to the free energy change involved in the adsorption. A convenient measure of the efficiency of adsorption is the negative logarithm of the concentration of surfactant in the bulk phase required to produce a 20 mN/m(dyn/cm) reduction in the surface or interfacial tension of the solvent, $-\log C_{(-\Delta\gamma=20)} \equiv pC_{20}$. This is based on the following considerations: The ideal measure of efficiency of adsorption would be some function of the minimum concentration of surfactant in the bulk phase necessary to produce maximum (saturation) adsorption at the interface. However, determining this concentration would require a complete $\gamma - \log C_1$ plot for each surfactant being investigated. Observation of $\gamma - \log C_1$ plots in the literature reveals that when the surface (or interfacial) tension of the pure solvent has been decreased about 20 dyn/cm by adsorption of the surfactant, the surface (excess) concentration Γ_1 of the surfactant is close to its saturation value. This is confirmed by use of the Frumkin Equation 2.29. From Table 2.2, Γ_m is $1 - 4.4 \times 10^{-10}$ mol/cm^2. Solving for Γ_1 in the Frumkin equation, when $\gamma_0 - \gamma = \pi = 20$ dyn/cm and $\Gamma_1 = 1 - 4.4 \times 10^{-10}$ mol/cm^2, $\Gamma_1 = 0.84$ to $0.999\ \Gamma_m$ at 25°C, indicating that when the surface (or interfacial) tension has been reduced by 20 dyn/cm, the surface concentration is 84–99.9% saturated.

Thus, the bulk liquid phase concentration of surfactant required to depress the surface (or interfacial) tension of the solvent by 20 dyn/cm (m N m^{-1}) is a good measure of the efficiency of adsorption of the surfactant; that is, it is close to the minimum concentration needed to produce saturation adsorption at the interface. The negative logarithm of the bulk phase concentration of surfactant in mol dm^{-3}, pC_{20}, rather than the concentration C_{20} itself, is used because the negative logarithm can be related to standard free energy change $\Delta G°$ involved in the transfer of the surfactant molecule from the interior of the bulk liquid phase to the interface (see below).

The advantage of measuring the effect of a surfactant in an interfacial phenomenon by some parameter that is related to the standard free energy change associated with the action of the surfactant in that phenomenon is that the total standard free energy change can be broken into the individual standard free energy changes associated with the action of the various structural groupings in the molecule. This enables correlations to be made between the various structural groupings in the surfactant and its interfacial properties. In this fashion, the efficiency with which a surfactant is adsorbed at an interface can be related to the various structural groups in the molecule.

The relation of pC_{20} to the free energy change on adsorption at infinite dilution $\Delta G°$ can be seen by use of the Langmuir and Szyszkowski Equations 2.6 and 2.28, respectively. Since at $\pi = 20$ mN/m(dyn/cm), $\Gamma_1 = 0.84$ to $0.999\ \Gamma_m$, from the Langmuir equation $C_1 = 5.2$ to $999 \times a$; thus, the quantity $\log([C_1/a] + 1) \approx \log(C_1/a)$ and the Szyszkowski equation becomes $\gamma_0 - \gamma = \pi = -2.303RT\Gamma_m \log(C_1/a)$. In this case, then,

$$\log(1/C_1)_{\pi=20} = -\left(\log a + \frac{\gamma_0 - \gamma}{2.303RT\Gamma_m}\right)$$

$$\text{Since } a = 55\exp(\Delta G°/RT) \text{ and } \log a = 1.74 + \Delta G°/2.303\,RT, \tag{2.30}$$

$$\log\left(\frac{1}{C_1}\right)_{\pi=20} \equiv pC_{20} = -\left(\frac{\Delta G°}{2.303RT} + 1.74 + \frac{20}{2.303RT\Gamma_m}\right)$$

For a straight-chain surfactant of structure, $CH_3(CH_2)_nW$, where W is the hydrophilic portion of the molecule, the standard free energy of adsorption $\Delta G°$ can be broken into the standard free energy changes associated with the transfer of the terminal methyl group, the $-CH_2-$ groups of the hydrocarbon chain, and the hydrophilic group, from the interior of the liquid phase to the interface at $\pi = 20$, that is, under conditions where the surface (or interfacial) tension has been reduced by 20 mN/m(dyn/cm):

$$\Delta G° = m.\Delta G°(-CH_2-) + \Delta G°(W) + \text{constant}$$

where m = the total number of carbon atoms ($n + 1$) in the hydrocarbon chain, and the constant equals $\Delta G°(CH_3-) - \Delta G°(-CH_2-)$.

When, for a homologous series of surfactants (with the same hydrophilic group), the value of Γ_m (or a_m^s) does not change much (Table 2.2) with increase in the number of carbon atoms in the molecule at constant microenvironmental conditions (temperature, ionic strength of the solution), and $\Delta G°$ (W) can be considered to be a constant, the relation between pC_{20} and $\Delta G°$ ($-CH_2-$), under these conditions, is

$$pC_{20} = \left[\frac{-\Delta G^0(-CH_2-)}{2.3RT}\right] m + \text{constant} \tag{2.31}$$

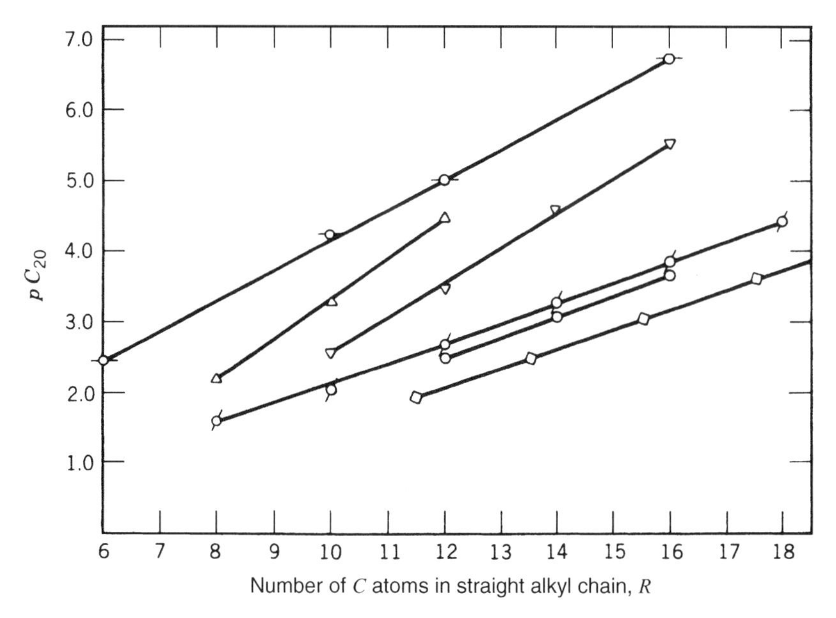

FIGURE 2.16 Effect of length of the hydrophobic group R on the efficiency of adsorption at the aqueous solution–air (*A*/*A*) and aqueous solution–heptane (*A*/*H*) interfaces: –⊙–$R(OC_2H_4)_6OH$ at 25°C (*A*/*A*); △$RN^+(CH_2C_6H_5)(CH_3)CH_2CH_2SO_3^-$ at 25°C (*A*/*A*); ▽$RN^+ (CH_3)_2CH_2COO^-$ at 23°C (*A*/*A*); ○$RSO_4^-Na^+$ at 50°C (*A*/*H*); ⊙$RSO_4^-Na^+$ at 25°C (*A*/*A*); ◇p-$R'C_6H_4SO_4^-Na^+$ (R = R′ + 3.5) at 70°C (*A*/*A*). Data from Table 2.2.

This equation indicates that the efficiency factor pC_{20} is a linear function of the number of carbon atoms in a straight-chain hydrophobic group, increasing as the number of carbon atoms increases. Figure 2.16 shows this linear relation for several homologous series of surfactants of different charge type.

The larger the value of pC_{20}, the more efficiently the surfactant is adsorbed at the interface and the more efficiently it reduces surface or interfacial tension, that is, the smaller the bulk liquid phase concentrations required either to attain saturation adsorption or to reduce the surface or interfacial tension by 20 mN/m(dyn/cm). Since this is a logarithmic relation, a value of pC_{20} one unit greater means 10 times the efficiency, that is, 1/10 the bulk phase concentration required to produce surface saturation.

Table 2.2 lists the efficiency of adsorption pC_{20} for a number of surfactants of different structural type at the A/A and aqueous solution–hydrocarbon interfaces. The data indicate the following relations between the structure of the surfactant and its efficiency of adsorption at A/A and aqueous solution–hydrocarbon interfaces.

The efficiency of adsorption at these interfaces increases linearly with an increase in the number of carbon atoms in a straight-chain hydrophobic group (Figure 2.16), reflecting the negative free energy of adsorption of a methylene group at these interfaces. If the hydrophobic group of an ionic surfactant is increased by two $-CH_2-$ groups, pC_{20} appears to be increased by 0.56–0.6 for adsorption at the A/A or aqueous solution–hydrocarbon interfaces, meaning that a surface concentration close to the saturation value can be obtained with only 25–30% of the bulk phase surfactant concentration previously required.

For POE nonionic surfactants at the A/A interface, pC_{20} appears to be increased by about 0.9 when the chain length is increased by two methylene groups, meaning that a bulk phase surfactant concentration of only one-seventh is required. The larger slope of the POE nonionic curve is due to the fact that, in this case, Γ_m increases with increase in the number of C atoms in the alkyl chain (Table 2.2).

In contrast to the situation with effectiveness of adsorption, where an increase in the length of the hydrophobic group beyond 16 carbon atoms appears to cause a decrease in effectiveness, efficiency of adsorption appears to increase steadily with increase in the length of the hydrophobic group up to at least 20 carbon atoms.

As in effectiveness of adsorption, a phenyl group in the hydrophobic portion appears to be equivalent to about three and one-half carbon atoms in a straight carbon chain. Methylene groups between two hydrophilic groups appear to be equivalent to about one-half a $-CH_2-$ group in a straight carbon chain with a single, terminal hydrophilic group.

When the hydrophobic group has side chains, the carbon atoms on the side chains seem to have about two-thirds the effect of carbon atoms in a straight alkyl chain with a single terminal hydrophilic group. When the hydrophilic group is at a nonterminal position in the hydrophobic group, the latter appears to act as if it were branched at the position of the hydrophilic group with the carbon atoms on the shorter portion of the hydrophobic group having about two-thirds the effect of the carbon atoms in the longer portion.

In POE compounds of structure $R(OC_2H_4)_xSO_4^-Na^+$, where $x = 1, 2,$ or 3, and in $RCONH(C_2H_4OH)_2$, the first oxyethylene group appears to be equivalent to about 2.5 $-CH_2-$ groups in a straight alkyl chain, with the additional oxyethylene groups having little or no effect.

Short alkyl groups (totaling four carbon atoms or less), including the pyridine nucleus, surrounding the N in quaternary ammonium salts or amine oxides, appear to have little effect. In these cases, efficiency of adsorption seems to be determined almost exclusively by the length of the long carbon chain attached to the N.

In POE nonionics, an increase in the number of oxyethylene units in the hydrophilic group above six units, in contrast to its large effect in decreasing the effectiveness of adsorption, seems to cause only a small decrease in the efficiency of adsorption. This appears to indicate a very small change in the free energy of transfer of the molecule from bulk phase interior to the interface with change in the number of oxyethylene units above six in the hydrophilic head.

In nonionic surfactants generally, the efficiency of adsorption is much greater than in ionic surfactants with the same number of carbon atoms in the hydrophobic group. This is because in the adsorption of ionic surfactants, electrical repulsion between the ionic heads of surfactant ions already at the interface and the similarly charged oncoming surfactant ions increases the positive free energy of transfer of the hydrophilic head from the interior of the bulk phase to the interface.

A change in the sign of the charge of a univalent ionic hydrophilic group produces little if any effect on the efficiency of adsorption. However, the replacement of the counterion by one that is more tightly bound increases the efficiency.

This increased efficiency is probably the result of the greater neutralization of the charge on the surfactant ion by the more tightly bound counterion. This would result in a smaller electrical repulsion between already adsorbed surfactant ions and oncoming surfactant ions at the interface.

The addition of inert electrolyte containing a common, nonsurfactant ion to a solution of ionic surfactant in an aqueous medium causes a large increase in the efficiency of its adsorption at the liquid–air interface (Boucher et al., 1968). In sharp contrast to their lack of significant effect on the effectiveness of adsorption, the addition of water structure breakers, such as urea and *N*-methylacetamide, to aqueous solutions of nonionic surfactants results in a decrease in efficiency of adsorption at the interface (Schwuger, 1969), whereas the addition of structure formers, such as fructose and xylose, increases the efficiency of adsorption (Schwuger, 1971b).

In summary, the efficiency of adsorption of a surfactant at the A/A interface, as measured by the pC_{20} value, is increased by the following factors:

1. Increase in the number of carbon atoms in the hydrophobic chain.
2. A straight alkyl chain as the hydrophobic group, rather than a branched alkyl chain containing the same number of carbon atoms.
3. A single hydrophilic group situated at the end of the hydrophobic group, rather than one (or more) at a central position.
4. A nonionic or zwitterionic hydrophilic group, rather than an ionic one.
5. For ionic surfactants, decrease in the effective charge of the hydrophilic group by
   a. use of a more tightly bound (less hydrated) counterion,
   b. increase in the ionic strength of the aqueous phase.

Temperature increase in the 10–40°C range causes an increase in the efficiency of adsorption for POE nonionics but a decrease for ionics and zwitterionics.

F. Calculation of Thermodynamic Parameters of Adsorption at the *L*/*G* and *L*/*L* Interfaces

Standard thermodynamic parameters, ΔG°, ΔH°, and ΔS°, tell us what is happening in a process. The standard free energy change upon adsorption ΔG°_{ad}

tells us whether adsorption (in the standard states) is spontaneous (ΔG°_{ad} negative) or not and the magnitude of the driving force. The standard enthalpy change upon adsorption ΔH°_{ad} indicates whether bond making (ΔH°_{ad} negative) or bond breaking (ΔH°_{ad} positive) predominates in the adsorption process. The standard entropy change ΔS°_{ad} indicates whether the system becomes more structured (ΔS°_{ad} negative) or more random (ΔS°_{ad} positive).

To calculate standard parameters of adsorption, ΔG°, ΔH°, and ΔS°, it is necessary to define standard states of the surface and bulk phases. If one uses the usual convention of unit concentrations in the bulk and interface as the standard states (Adam, 1940), then it is necessary to choose an appropriate thickness for the interface, which is not readily accomplished. Another convention has consequently been proposed (Betts and Pethica, 1957, 1960) in which the standard states for the surface and bulk phases are unit surface pressure (unit fugacity) and unit activity, respectively, and $-\Delta G^{\circ} = RT\ln \pi^{*}/a$, where π^{*} is the surface fugacity and a is the bulk phase activity of the solute. For very dilute solutions of surfactants (π = 0–3 dyn/cm), the reduction in surface tension (or surface pressure π) varies linearly with the molar bulk phase concentration C_1 of the surfactant:

$$\left(\frac{\partial \pi}{\partial C_1}\right)_{C_1 \to 0} = \alpha \quad \text{(Traube's constant)}$$

Fugacity and activity coefficients are assumed to approach unity under the above conditions, and a standard free energy of adsorption ΔG°_{Tr} can be obtained

$$\Delta G^{\circ}_{Tr} = -2.303\, RT \log\left(\frac{\partial \pi}{\partial C_1}\right)_{C_1 \to 0}$$

and a standard molar free energy of adsorption can be calculated from the linear, low concentration region of the γ (or π)-C_1 curve (Tamaki, 1967; Gillap et al., 1968; Naifu and Gu, 1979; Spitzer and Heerze, 1983).

Unfortunately, good surface tension data in this region are difficult to obtain, since traces of impurities adsorbed from the air or present in the solvent or in the surfactant can markedly affect the results. Second, there are only a few studies in the literature on this region of the surface tension–concentration curves, since investigators of the effect of surfactants on the surface tension of solvents generally are interested in the region where surfactants show the maximum effect, rather than the region where they show little effect.

Standard free energies at the A/A interface can be calculated from surface tension data in the vicinity of the CMC, where such data are commonly and conveniently taken, by use of Equation 2.32 (Rosen and Aronson, 1981):

$$\Delta G^{\circ}_{ad} = RT \ln a_{\pi}/\omega - \pi A^{s}_{m} \tag{2.32}$$

where a_π = activity of the surfactant in the aqueous phase at a surface pressure of π (= $\gamma_0 - \gamma$) in the region of surface saturation (i.e., where $\Gamma = \Gamma_m$ and the molar area of the surfactant $A^s = A^s_m$).

The standard state for the surface phase is a hypothetical monolayer of the surfactant at its closest packing (minimum surface area/molecule) but at a surface pressure of zero. For nonionic surfactants at dilute concentrations ($<1 \times 10^{-2}$ M) in the solution phase, we can substitute mole fractions for activities and the relation becomes

$$\Delta G^\circ_{\text{ad}} = RT \ln C_\pi / \omega - \pi A^s_m \tag{2.33}$$

where C_π, = molar concentration of surfactant in the aqueous phase at a surface pressure of π and ω is the number of moles of water per liter of water. When C_π is in mol/L, π in dyn/cm (mJm^{-2}), a^s_m, in Å^2 per molecule, and $R = 8.314$ J mol^{-1} K^{-1}, this becomes

$$\Delta G^\circ_{\text{ad}} (\text{in } J/\text{mol}) = 2.3RT \log C_\pi / \omega - 6.023\pi \cdot a^s_m \tag{2.33a}$$

For an ionic surfactant of the type AB,

$$\Delta G^\circ_{\text{ad}} (\text{in } J/\text{mol}) = 2.3RT[\log C_{\text{A}} / \omega + \log f_{\text{A}} + \log C_{\text{B}} / \omega + \log f_{\text{B}}] - 6.023\pi \cdot a^s_m \tag{2.33b}$$

The activity coefficients f_{A} and f_{B} can be evaluated by Equation 2.24. The standard free energies of adsorption calculated by use of Equation 2.33b are independent of the ionic strength of the solution.

When the surface tension of the solvent has been reduced by 20 mN/m(dyn/cm), that is, $\pi = 20$ mN/m(dyn/cm), then the relation (Equation 2.33a) for a nonionic surfactant becomes

$$\Delta G^\circ_{\text{ad}} = -(2.303RT)pC_{20} - 6.023 \times 20a^s_m - 2.303\ RT \log \omega \tag{2.34}$$

and

$$pC_{20} = -\left(\frac{\Delta G^\circ_{\text{ad}}}{2.303\ RT} + \frac{6.023 \times 20a^s_m}{2.303\ RT} \right) - \log \omega \tag{2.35}$$

Since $a^s_m = 10^{16}/\Gamma_m N$, this relation is similar to the one (Equation 2.30) obtained previously from the Langmuir and Szyszkowski equations. For surfactants whose a^s_m values do not vary much,

$$pC_{20} = \frac{-\Delta G^\circ_{\text{ad}}}{2.303\ RT} - K \tag{2.36}$$

and the pC_{20} value is a measure of the standard free energy of adsorption. Since

$$\Delta G^{\circ}_{ad} = \Delta H^{\circ}_{ad} - T\Delta S^{\circ}_{ad} \tag{2.37}$$

standard entropies and enthalpies of adsorption, ΔS°_{ad} and ΔH°_{ad}, respectively, can be calculated from the relation

$$d\Delta G^{\circ}_{ad} / dT = -\Delta S^{\circ}_{ad} \tag{2.38}$$

if ΔH°_{ad} is constant over the temperature range investigated. Alternatively,

$$T^2 d(\Delta G^{\circ}_{ad} / T)\mathrm{d}T = -\Delta H^{\circ}_{ad} \tag{2.38a}$$

can be used, if ΔS°_{ad} is constant over that temperature range.

Standard thermodynamic parameters of adsorption calculated by use of Equations 2.33, 2.34, 2.37, and 2.38 or 2.38a are listed in Table 2.3. All the ΔG°_{ad} values are negative, indicating that adsorption of these compounds at the A/A and aqueous solution–air hydrocarbon interfaces is spontaneous.

It is evident that the positive entropy change upon adsorption is by far the major contributor to the negative values of the free energy change and thus the main driving force for adsorption at the interface in these compounds. The $-\Delta G^{\circ}_{ad}$ per $-CH_2-$ group at 25°C is 3.0–3.5 kJ; increase in the length of the alkyl chain therefore increases the tendency of the compound to adsorb.

It is noteworthy that, for all the compounds listed, ΔG°_{ad} and ΔH°_{ad} become more negative with increase in temperature, which appears to indicate that some dehydration of the hydrophilic group is required for adsorption. At higher temperatures, the surfactant is less hydrated, requires less dehydration to adsorb, and adsorbs more readily.

In the POE nonionics listed, the ΔG°_{ad} becomes slightly more negative with increase in the EO content of the molecule, reflecting the increasing value of ΔS°_{ad} with this change. This increase and the concomitant increase in ΔH°_{ad} seem to indicate that adsorption at the A/A interface is accompanied by partial dehydration of the POE chain, with the amount of dehydration per molecule increasing with increase in the number of EO units.

This increase in $-\Delta G^{\circ}_{ad}$ with increase in the EO content of the molecule is seen also in the EO alkyl sulfates (Zoeller and Blankschtein, 1998), where the addition of the first EO group to the alkyl sulfate molecule increases the $-\Delta G^{\circ}_{ad}$ value by about 3 kJ/mol and the addition of the second by about half that value.

The presence of a second liquid (hydrocarbon) phase increases the $-\Delta G^{\circ}_{ad}$ value by a few kJ/mol, with the increase being largest for cyclohexane (of the hydrocarbons investigated) and becoming smaller with increase in the chain length of the hydrocarbon.

Relationships have been found between the adsorption properties described above of surfactants and their environmental effects (toxicity, bioconcentration) on aquatic organisms (algae, fish, rotifers). The log of the EC_{50} (the surfactant molar concentration in the water at which the organism population is reduced by 50% relative to a no-dose control) and the log of the BCF (the

TABLE 2.3 Standard Thermodynamic Parameters of Adsorption for Surfactants at the Aqueous Solution–Air Interface or Aqueous Solution–Hydrocarbon Interface[a]

Compound	Tem. (°C)	ΔG°_{ad} (kJ/mol)[b]	ΔH°_{ad}(kJ/mol)	$T\Delta S^{\circ}_{ad}$ (kJ/mol)	Reference[c]
$C_{10}H_{21}SO_3^-Na^+$	10	$-43._{3}$			Dahanayake et al. (1986)
			+2	$+4_{7}$	
$C_{10}H_{21}SO_3^-Na^+$	25	$-45._{7}$			Dahanayake et al. (1986)
			−4	$+4_{3}$	
$C_{10}H_{21}SO_3^-Na^+$	40	$-47._{9}$			Dahanayake et al. (1986)
$C_{12}H_{25}SO_3^-Na^+$	10	$-50._{7}$			Dahanayake et al. (1986)
			−7	$+4_{5}$	
$C_{12}H_{25}SO_3^-Na^+$	25	$-53._{0}$			Dahanayake et al. (1986)
			-1_{0}	$+4_{4}$	
$C_{12}H_{25}SO_3^-Na^+$	40	$-55._{3}$			Dahanayake et al. (1986)
$C_{12}H_{25}SO_4^-Na^+$	25	$-54._{4}$	—	—	Rehfeld (1967); Dahanayake et al. (1986)
$C_{12}H_{25}SO_4^-Na^+$ (aq. soln.–octane interface)	25	$-56._{9}$	—	—	Rehfeld (1967)
$C_{12}H_{25}SO_4^-Na^+$ (aq. soln.–heptadecane interface)	25	$-56._{5}$	—	—	Rehfeld (1967)
$C_{12}H_{25}SO_4^-Na^+$ (aq. soln.–cyclohexane interface)	25	$-58._{0}$	—	—	Rehfeld (1967)
$C_{12}H_{25}SO_4^-Na^+$ (aq. soln.–benzene interface)	25	$-57._{9}$	—	—	Rehfeld (1967)
$C_{12}H_{25}SO_4^-Na^+$ (aq. soln.–butyl benzene interface)	25	$-55._{8}$			Rehfeld (1967)
$C_{10}H_{21}OC_2H_4SO_3^-Na^+$	10	$-47._{2}$			Dahanayake et al. (1986)
			+3	$+5_{1}$	
$C_{10}H_{21}OC_2H_4SO_3^-Na^+$	25	$-49._{7}$			Dahanayake et al. (1986)
			−6	$+4_{5}$	
$C_{10}H_{21}OC_2H_4SO_3^-Na^+$	40	$-51._{9}$			Dahanayake et al. (1986)
$C_{12}H_{25}OC_2H_4SO_3^-Na^+$	10	$-54._{5}$			Dahanayake et al. (1986)
			−6	$+4_{9}$	

$C_{12}H_{25}OC_2H_4SO_3^-Na^+$	25	$-57._{0}$			Dahanayake et al. (1986)
			-1_1	$+4_7$	
$C_{12}H_{25}OC_2H_4SO_3^-Na^+$	40	$-59._{3}$			Dahanayake et al. (1986)
$C_{12}H_{25}OC_2H_4SO_4^-Na^+$	10	$-54._{7}$			Dahanayake et al. (1986)
			-3	$+5_3$	
$C_{12}H_{25}OC_2H_4SO_4^-Na^+$	25	$-57._{5}$			Dahanayake et al. (1986)
			-1_7	$+4_1$	
$C_{12}H_{25}OC_2H_4SO_4^-Na^+$	40	$-59._{5}$			Dahanayake et al. (1986)
$C_{12}H_{25}(OC_2H_4)SO_4^-Na^+$	10	$-56._{4}$			Dahanayake et al. (1986)
			-5	$+5_3$	
$C_{12}H_{25}(OC_2H_4)SO_4^-Na^+$	25	$-59._{1}$			Dahanayake et al. (1986)
			-8	$+5_2$	
$C_{12}H_{25}(OC_2H_4)_2SO_4^-Na^+$	40	$-61._{7}$			Dahanayake et al. (1986)
$C_{12}H_{25}Pyr^+Br^{-d}$	10	$-50._{0}$			Rosen et al. (1982b)
			-7	$+4_4$	
$C_{12}H_{25}Pyr^+Br^-$	25	$-52._{3}$			Rosen et al. (1982b)
			-9	$+4_4$	
$C_{12}H_{25}Pyr^+Br^-$	40	$-54._{5}$			Rosen et al. (1982b)
$C_{12}H_{25}Pyr^+Cl^-$	10	$-49._{0}$			Rosen et al. (1982b)
			-1_1	$+3_9$	
$C_{12}H_{25}Pyr^+Cl^-$	25	$-51._{1}$			Rosen et al. (1982b)
			-1_1	$+4_1$	
$C_{12}H_{25}Pyr^+Cl^-$	40	$-53._{1}$			Rosen et al. (1982b)
$C_8H_{17}OCH_2CH_2OH$	25	$-31._{8}$	—	—	Shinoda et al. (1959)
$C_8H_{17}CHOHCH_2OH$	25	$-34._{7}$	—	—	Kwan and Rosen (1980)
$C_8H_{17}CHOHCH_2CH_2OH$	25	$-34._{3}$	—	—	Kwan and Rosen (1980)
$C_{10}H_{21}CHOHCH_2CH_2OH$	25	$-40._{4}$	—	—	Kwan and Rosen (1980)
$C_{12}H_{25}CHOHCH_2CH_2OH$	25	$-46._{9}$	—	—	Kwan and Rosen (1980)

(*Continued*)

TABLE 2.3 (*Continued*)

Compound	Tem. (°C)	ΔG°_{ad} (kJ/mol)[b]	ΔH°_{ad}(kJ/mol)	$T\Delta S^{\circ}_{ad}$ (kJ/mol)	Reference[c]
N-hexyl-2 pyrrolid(in)one, pH 7.0	25	$-28._8$			Rosen and Wu (2001)
N-(2-ethylhexyl)-2-pyrrolid(in)one, pH 7.0	25	$-33._0$			Rosen and Wu (2001)
N-octyl-2-pyrrolid(in)one, pH 7.0	25	$-33._1$			Rosen and Wu (2001)
N-decyl-pyrrolid(in)one, pH 7.0	25	$-38._7$			Rosen and Wu (2001)
$C_{12}H_{25}(OC_2H_4)_3OH$	10	$-43._0$			Rosen et al. (1982a)
			+4	$+4_8$	
$C_{12}H_{25}$ $(OC_2H_4)_3OH$	25	$-45._5$			Rosen et al. (1982a)
			-1_2	$+3_5$	
$C_{12}H_{25}$ $(OC_2H_4)_3OH$	40	$-47._2$			Rosen et al. (1982a)
$C_{12}H_{25}$ $(OC_2H_4)_5OH$	10	$-43._7$			Rosen et al. (1982a)
			+3	$+4_8$	
$C_{12}H_{25}$ $(OC_2H_4)_5OH$	25	$-46._2$			Rosen et al. (1982a)
			−5	$+4_3$	
$C_{12}H_{25}$ $(OC_2H_4)_5OH$	40	$-48._3$			Rosen et al. (1982a)
$C_{12}H_{25}$ $(OC_2H_4)_7OH$	10	$-44._2$			Rosen et al. (1982a)
			+7	$+5_2$	
$C_{12}H_{25}$ $(OC_2H_4)_7OH$	25	$-46._9$			Rosen et al. (1982a)
			−3	$+4_5$	
$C_{12}H_{25}$ $(OC_2H_4)_7OH$	40	$-49._1$			Rosen et al. (1982a)
$C_{12}H_{25}$ $(OC_2H_4)_8OH$	10	$-44._7$			Rosen et al. (1982a)
			+6	$+5_2$	
$C_{12}H_{25}$ $(OC_2H_4)_8OH$	25	$-47._7$			Rosen et al. (1982a)
			−2	$+4_7$	
$C_{12}H_{25}(OC_2H_4)_8OH$	40	$-49._7$			Rosen et al. (1982a)
$C_{12}H_{25}(OC_2H_4)_8OH$ (aq. soln.—cyclohexane)	25	$-52._8$			Rosen and Murphy (1991)
$C_{12}H_{25}(OC_2H_4)_8OH$ (aq. soln.—heptane)	25	$-51._5$			Rosen and Murphy (1991)
$C_{12}H_{25}(OC_2H_4)_8OH$ (aq. soln.—hexadecane)	25	$-51._5$			Rosen and Murphy (1991)

t-$C_8H_{17}(OC_2H_4)_3OH$	25	$-44._8$	—	—	Crook et al. (1964)
t-$C_8H_{17}(OC_2H_4)_5OH$	25	$-45._6$	—	—	Crook et al. (1964)
t-$C_8H_{17}(OC_2H_4)_7OH$	25	$-45._1$	—	—	Crook et al. (1964)
t-$C_8H_{17}(OC_2H_4)_9OH$	25	$-45._4$	—	—	Crook et al. (1964)
$(CH_3)_3SiOSi(CH_3)[CH_2(CH_2CH_2O)_{8.5}\text{–}CH_3]$ $OSi(CH_3)_3$, pH 7.0	25	$-51._9$			Rosen and Wu (2001)
$C_{10}H_{21}N^+(CH_3)(CH_2C_6H_5)\text{-}CH_2COO^-$	10	$-34._2$			Dahanayake and Rosen (1984)
			$-0._2$	$+3_5$	
$C_{10}H_{21}N^+(CH_3)(CH_2C_6H_5)\text{-}CH_2COO^-$	25	$-36._0$			Dahanayake and Rosen (1984)
			-4	$+3_3$	
$C_{10}H_{21}N^+(CH_3)(CH_2C_6H_5)\text{-}CH_2COO^-$	40	$-37._6$			Dahanayake and Rosen (1984)
$C_{12}H_{25}N^+(CH_3)(CH_2C_6H_5)\text{-}CH_2COO^-$	10	$-40._3$			Dahanayake and Rosen (1984)
			-6	$+3_5$	
$C_{12}H_{25}N^+(CH_3)(CH_2C_6H_5)\text{-}CH_2COO^-$	25	$-42._1$			Dahanayake and Rosen (1984)
			-1_2	$+3_1$	
$C_{12}H_{25}N^+(CH_3)(CH_2C_6H_5)\text{-}CH_2COO^-$	40	$-43._6$			Dahanayake and Rosen (1984)
$C_{12}H_{25}N^+(CH_3)(CH_2C_6H_5)\text{-}CH_2COO^-$ (aq. soln.–heptane interface)	25	$-46._7$	-2_1 (35°)	$+2_6$ (35°)	Murphy and Rosen (1988)
$C_{12}H_{25}N^+(CH_3)(CH_2C_6H_5)\text{-}CH_2COO^-$ (aq. soln.–isooctane interface)	25	$-46._6$	-1_9 (35°)	$+2_9$ (35°)	Murphy and Rosen (1988)
$C_{12}H_{25}N^+(CH_3)(CH_2C_6H_5)\text{-}CH_2COO^-$ (aq. soln.–hexadecane interface)	25	$-45._5$	-1 (35°)	$+4_6$ (35°)	Murphy and Rosen (1988)
$C_{12}H_{25}N^+(CH_3)(CH_2C_6H_5)\text{-}CH_2COO^-$ (aq. soln.–cyclohexane interface)	25	$-48._0$	-1_4 (35°)	$+3_4$ (35°)	Murphy and Rosen (1988)

(*Continued*)

TABLE 2.3 (*Continued*)

Compound	Tem. (°C)	ΔG°_{ad} (kJ/mol)[b]	ΔH°_{ad}(kJ/mol)	$T\Delta S^\circ_{ad}$ (kJ/mol)	Reference[c]
$C_{12}H_{25}N^+(CH_3)(CH_2C_6H_5)$-$CH_2COO^-$ (aq. soln.–toluene interface)	25	$-46._9$	-2_4 (35°)	$+2_4$ (35°)	Murphy and Rosen (1988)
$C_8H_{17}N^+(CH_3)(CH_2C_6H_5)$-$CH_2CH_2SO_3^-$	10	$-28._2$			Dahanayake and Rosen (1984)
			+6	$+3_5$	
$C_8H_{17}N^+(CH_3)(CH_2C_6H_5)$-$CH_2CH_2SO_3^-$	25	$-30._0$			Dahanayake and Rosen (1984)
			−8	$+2_2$	
$C_8H_{17}N^+(CH_3)(CH_2C_6H_5)$-$CH_2CH_2SO_3^-$	40	$-31._1$			Dahanayake and Rosen (1984)
$C_{10}H_{21}N^+(CH_3)(CH_2C_6H_5)$-$CH_2CH_2SO_3^-$	10	$-34._6$			Dahanayake and Rosen (1984)
	10				
			−1	$+3_5$	
$C_{10}H_{21}N^+(CH_3)(CH_2C_6H_5)$-$CH_2CH_2SO_3^-$	25	$-36._4$			Dahanayake and Rosen (1984)
			-1_5	$+2_2$	
$C_{10}H_{21}N^+(CH_3)(CH_2C_6H_5)$-$CH_2CH_2SO_3^-$	40	$-37._5$			Dahanayake and Rosen (1984)
$C_{12}H_{25}N^+(CH_3)(CH_2C_6H_5)$-$CH_2CH_2SO_3^-$	10	$-41._0$			Dahanayake and Rosen (1984)
			−9	$+3_3$	
$C_{12}H_{25}N^+(CH_3)(CH_2C_6H_5)$-$CH_2CH_2SO_3^-$	25	$-42._7$			Dahanayake and Rosen (1984)
			-1_5	$+2_8$	
$C_{12}H_{25}N^+(CH_3)(CH_2C_6H_5)$-$CH_2CH_2SO_3^-$	40	$-44._0$			Dahanayake and Rosen (1984)
$C_{10}H_{21}CH(Pyr^+)COO^-$[d]	25	$-31._9$	—	—	Zhao and Rosen (1984)
$C_{12}H_{25}CH(Pyr^+)COO^-$[d]	25	$-38._2$	—	—	Zhao and Rosen (1984)
$C_{14}H_{29}CH(Pyr^+)COO^-$[d]	40	$-45._8$	—	—	Zhao and Rosen (1984)

[a] Values are for the aqueous solution–air interface unless otherwise indicated.

[b] Values are independent of total ionic strength and are averages of values at different electrolyte contents.

[c] Parameters calculated from data in listed reference.

[d] Pyr^+, pyridinium.

ratio of surfactant concentration in the fish relative to that in the water) have both been shown (Rosen et al., 1999, 2001) to be linearly related to the parameter $\Delta G^{\circ}_{\mathrm{ad}}/a^{s}_{m}$ for a series of anionic, cationic, and nonionic surfactants. The values of a^{s}_{m} and $\Delta G^{\circ}_{\mathrm{ad}}$ were obtained by the methods described above in Sections IIIB and IIIF, respectively.

G. Adsorption from Mixtures of Two Surfactants

Mixtures of two or more different types of surfactants (De Lisi et al., 1997; Nakano et al., 2002) often show a "synergistic" interaction; that is, the interfacial properties of the mixture are more pronounced than those of the individual components by themselves. As a result, in many industrial products and processes, mixtures of different types of surfactants, rather than individual materials, are used. A study of the adsorption of the individual surface-active components in the mixture and of the interaction between them affords an understanding of the role of each and makes possible the selection in a rational, systematic manner of components for optimal properties.

The Gibbs adsorption Equation 2.17 for two surface-active solutes in dilute solution can be written as

$$d\gamma = RT(\Gamma_1 d\ln a_1 + \Gamma_2 d\ln a_2) \tag{2.39}$$

where Γ_1, Γ_1 are the surface (excess) concentrations of the two solutes at the interface and a_1, a_2 their respective activities in the solution phase. From this expression, since molar concentrations can be substituted for activities in dilute solution,

$$\Gamma_1 = \frac{1}{RT}\left(\frac{-\partial\gamma}{\partial \ln C_1}\right)_{C_2} = \frac{1}{2.303\,RT}\left(\frac{-\partial\gamma}{\partial \log C_1}\right)_{C_2} \tag{2.40}$$

and

$$\Gamma_2 = \frac{1}{RT}\left(\frac{-\partial\gamma}{\partial \ln C_2}\right)_{C_1} = \frac{1}{2.303\,RT}\left(\frac{-\partial\gamma}{\partial \log C_2}\right)_{C_1} \tag{2.41}$$

Therefore, the concentration of each surfactant at the interface can be calculated from the slope of a $\gamma - \ln C$ (or $\log C$) plot of each surfactant, holding the solution concentration of the other surfactant constant.

When the absolute concentrations of the surfactants at the interface are not required, but only their relative concentrations, that is, their relative effectiveness of adsorption, then these can be obtained in convenient fashion by use of nonideal solution theory.

From the thermodynamics of the system, it has been shown (Rosen and Hua, 1982) that the molar concentrations of the two surfactants in the solution phase are given by the expressions

$$C_1 = C_1^0 f_1 X_1 \tag{2.42}$$

and

$$C_2 = C_2^0 f_2 X_2 \tag{2.43}$$

where f_1 and f_2 are the activity coefficients of the surfactants (1 and 2, respectively) in the interface; X_1 is the mole fraction of surfactant 1 in the total surfactant at the interface (i.e., $X_1 = 1 - X_2$), C_1^0 is the molar concentration required to attain a given surface tension in a solution of pure surfactant 1, and C_2^0 is the molar concentration required to attain the same surface tension in a solution of pure surfactant 2.

From nonideal solution theory, the activity coefficients at the interface can be approximated by the expressions

$$\ln f_1 = \beta^\sigma (1 - X_1)^2 \tag{2.44}$$

$$\ln f_2 = \beta^\sigma (X_1)^2 \tag{2.45}$$

where β^σ is a parameter related to the interaction between the two surfactants at the interface. From Equations 2.42–2.45,

$$\frac{(X_1)^2 \ln(C_1 / C_1^0 X_1)}{(1 - X_1)^2 \ln\left[\dfrac{C_2}{C_2^0(1 - X_1)}\right]} = 1 \tag{2.46}$$

Surface tension–total surfactant concentration (C_t) curves for the two pure surfactants and a mixture of them at a fixed value of α, the mole fraction of surfactant 1 in the total surfactant in the solution phase, are used (Figure 2.17)

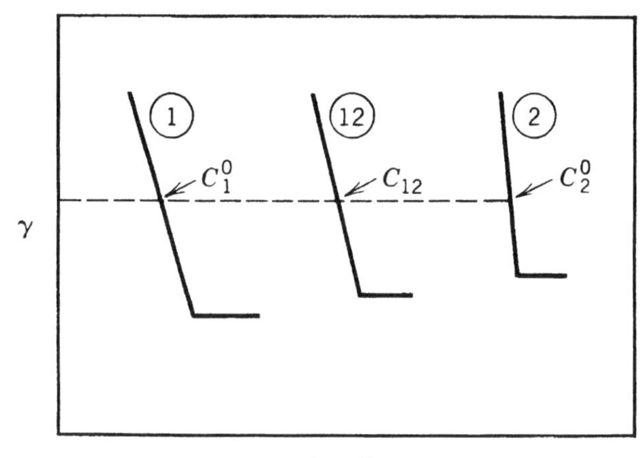

FIGURE 2.17 Evaluation of X_1 and X_2. (1) Pure surfactant 1. (2) Pure surfactant 2. (12) Mixture of 1 and 2 at a fixed value of α.

to determine $C_1(= \alpha_1 C_{12})$, C_1^0, $C_2[= (1 - \alpha_1)C_{12}]$ and C_2^0, the molar concentrations at the same surface tension. Substitution of these values into Equation 2.46 permits it to be solved iteratively for X_1 and X_2 $(= 1 - X_1)$. The ratio of surfactant 1 : surfactant 2 at the interface at that particular value of α is then X_1/X_2.

The conditions for synergistic interaction between the two surfactants are discussed in Chapter 11.

REFERENCES

Abe, R. and H. Kuno (1962) *Kolloid Z.* **181**, 70.

Abram, J. C. and G. D. Parfitt, *Proc. 5th Conf. Carbon*, London, 1962, pp. 97–102.

Adam, N. K., *The Physics and Chemistry of Surfaces*, Oxford University Press, Oxford, 1940.

Adamson, A. W., *Physical Chemistry of Surfaces*, 3rd ed., Interscience, New York, 1976.

Alexandridis, P., V. Athanassiou, S. Fukuda, and T. A. Hatton (1994) *Langmuir* **10**, 2604.

Aston, J. R., D. N. Furlong, F. Grieser, P. J. Scales, and G. G. Warr, in *Adsorption at the Gas/Solid and Liquid/Solid Interface*, J. Rouquerol and K. S. W. Sing (ed.), Elsevier, Amsterdam, 1982, pp. 97–102.

Atkins, R., V. S. J. Craig, E. J. Wanless, and S. Briggs (2003) *Adv. Colloid Interface Sci.* **103**, 219.

Aveyard, R. and D. A. Haydon, *An Introduction to the Principles of Surface Chemistry*, Cambridge University Press, Cambridge, 1973, pp. 201–212.

Aveyard, R., B. P. Binks, J. Chen, J. Equena, D. I. Fletcher, R. Buscall, and S. Davies (1998) *Langmuir* **14**, 4699.

Beckett, A. H. and R. J. Woodward (1963) *J. Pharm. Pharmacol.* **15**, 422.

Betts, J. J. and B. A. Pethica, 2nd Int. Congr. Surface Activity, London, 1957, I, p. 152.

Betts, J. J. and B. A. Pethica (1960) *Trans. Faraday Soc.* **56**, 1515.

Boucher, E. S., T. M. Grinchuk, and A. C. Zettlemoyer (1968) *J. Am. Oil Chem. Soc.* **45**, 49.

Brashier, G. K. and C. K. Thornhill (1968) *Proc. La. Acad. Sci.* **31**, 101.

Bujake, J. E. and E. D. Goodard (1965) *Trans. Faraday Soc.* **61**, 190.

Burczyk, B., K. A. Wilk, A. Sokolowski, and L. Syper (2001) *J. Colloid Interface Sci.* **240**, 552.

Burgess, I., C. A. Jeffrey, X. Cai, G. Szymanski, Z. Galus, and J. Lipkowski (1999) *Langmuir* **15**, 2607.

Bury, C. R. and J. Browning (1953) *Trans. Faraday Soc.* **49**, 209.

Carless, J. E., R. A. Challis, and B. A. Mulley (1964) *J. Colloid Sci.* **19**, 201.

Cases, J. M., D. Canet, N. Doerler, and J. E. Poirier, *Adsorption at the Gas-Solid and Liquid Solid Interface*, Elsevier, Amsterdam, 1982.

Caskey, J. A. and W. B., Jr. Barlage (1971) *J. Colloid Interface Sci.* **35**, 46.

Chapman, D. L. (1913) *Philos. Mag.* **25**, 475.

Chevalier, Y., Y. Storets, S. Pourchet, and P. LePerchec (1991) *Langmuir* **7**, 848.

Connor, P. and R. H. Ottewill (1971) *J. Colloid Interface Sci.* **37**, 642.

Corkill, J. M., J. F. Goodman, C. P. Ogden, and J. R. Tate (1963) *Proc. R. Soc.* **273**, 84.

Corkill, J. M., J. T. Goodman, and S. P. Harrold (1964) *Trans. Faraday Soc.* **60**, 202.

Corkill, J. M., J. F. Goodman, and J. R. Tate (1966) *Trans. Faraday Soc.* **62**, 979.

Corkill, J. M., J. F. Goodman, and J. R. Tate (1967) *Soc. Chem. Ind. (London)*, 363–369.

Crook, E. H., D. B. Fordyce, and G. F. Trebbi (1963) *J. Phys. Chem.* **67**, 1987.

Crook, E. H., G. F. Trebbi, and D. B. Fordyce (1964) *J. Phys. Chem.* **68**, 3592.

Dahanayake, M. and M. J. Rosen, in *Structure/Performance Relationships in Surfactants*, M. J. Rosen (ed.), ACS Symp. Series 253, American Chemical Society, Washington, DC, 1984, p. 49.

Dahanayake, M., A. W. Cohen, and M. J. Rosen (1986) *J. Phys. Chem.* **90**, 2413.

Daniel, S. G. (1951) *Trans. Faraday Soc.* **47**, 1345.

De Lisi, R., A. Inglese, S. Milioto, and A. Pellento (1997) *Langmuir* **13**, 192.

Dick, S. G., D. W. Fuerstenau, and T. W. Healy (1971) *J. Colloid Interface Sci.* **37**, 595.

Dreger, E. E., G. I. Keim, G. D. Miles, L. Shedlovsky, and J. Ross (1944) *Ind. Eng. Chem.* **36**, 610.

Eastoe, J., J. S. Dalton, P. G. A. Rogueda, E. R. Crooks, A. R. Pitt, and E. A. Simister (1997) *J. Colloid Interface Sci.* **188**, 423.

Elton, G. A., 2nd Int. Congr. Surface Activity, London, England, September 1957, III, p. 161.

Elworthy, P. H. (1959) *J. Pharm. Pharmacol.* **11**, 624.

Elworthy, P. H. and A. T. Florence (1964) *Kolloid Z. Z. Polym.* **195**, 23.

Elworthy, P. H. and C. B. MacFarlane (1962) *J. Pharm. Pharmacol.* **14**, 100.

Elworthy, P. H. and K. J. Mysels (1966) *J. Colloid Interface Sci.* **21**, 331.

Fowkes, F. M. (1987) *J. Adhes. Sci. Tech.* **1**, 7.

Frumkin, A. (1925) *Z. Phys. Chem.* **116**, 466.

Fuerstenau, D. W. (1957) *Trans. AIME* **208**, 1365.

Gao, Y., J. Du, and T. Gu (1987) *J. Chem. Soc., Faraday Trans. I* **83**, 2671.

Gaudin, A. M. and D. W. Fuerstenau (1955) *Trans. AIME* **202**, 958.

Gavet, L., A. Couval, H. Bourdiau, and P. Rochas (1973) *Bull. Sci. Inst. Text. Fr.* **2**, 275.

Gentle, T. E. and S. A. Snow (1995) *Langmuir* **11**, 2905.

Gibbs, J. W., *The Collected Works of J. W Gibbs*, Vol. I, Longmans, Green, London, 1928.

Giles, C. H. and S. N. Nakhwa (1962) *J. Appl. Chem.* **12**, 266.

Giles, C. H., A. P. D'Silva, and I. A. Easton (1974) *J. Colloid Interface Sci.* **47**, 766.

Gillap, W. R., N. D. Weiner, and M. Gibaldi (1968) *J. Phys. Chem.* **72**, 2218.

Ginn, M. E., in *Cationic Surfactants*, E. Jungermann (ed.), Dekker, New York, 1970, p. 352ff and 372.

Gordon, B. and W. T. Shebs, 5th Int. Congr. Surface-Active Substances, Barcelona, Spain, September 1968, III, p. 155.

Gouy, G. (1910) *J. Phys.* **9**, 457.

Gouy, G. (1917) *Ann. Phys.* **7**, 129.

Grahame, D. C. (1947) *Chem. Rev.* **41**, 441.

Grant, L. M. and W. A. Ducker (1997) *J. Phys. Chem. B* **101**, 5337.

Grant, L. M., F. Tiberg, and W. A. Ducker (1998) *J. Phys. Chem. B* **102**, 4288.

Gray, F. W., J. F. Gerecht, and I. J. Krems (1955) *J. Org. Chem.* **20**, 511.

Greenwood, F. G., G. D. Parfitt, N. H. Picton, and D. G. Wharton, in *Adsorption from Aqueous Solution*, W. J., Jr. Weber and E. Matijevic (eds.), Adv. Chem. Series 79, American Chemical Society, Washington, DC, 1968, pp. 135–144.

Gregg, S. J. and K. S. W. Sing, *Adsorption, Surface Area, and Porosity*, Academic, London, 1967, Chap. 7.

Greiss, W. (1955) *Fette, Seifen, Anstrichmi* **57**, 24, 168, 236.

Griffith, J. C. and A. E. Alexander (1967) *J. Colloid Interface Sci.* **25**, 311.

Groot, R. C., 5th Int. Cong. Surface-Active Substances, Barcelona, Spain, September 1968, II, p. 581.

Gross, L. (2011) *Nat. Chem.* **3**, 273.

Grosse, I. and K. Estel (2000) *Colloid Polym. Sci.* **278**, 1000.

Groves, M. J., R. M. A. Mustafa, and J. E. Carless (1972) *J. Pharm. Pharmacol.* **24** (Suppl.), 104.

Gu, B. and M. J. Rosen (1989) *J. Colloid Interface Sci.* **129**, 537.

Gu, T. and B.-Y. Zhu (1990) *Colloids Surf.* **44**, 81.

Gu, T., B.-Y. Zhu, and H. Rupprecht (1992) *Prog. Colloid Polym. Sci.* **88**, 74.

Gum, M. L. and E. D. Goddard (1982) *J. Am. Oil Chem. Soc.* **59**, 142.

Harwell, J. H., J. C. Hoskins, R. S. Schechter, and W. H. Wade (1985) *Langmuir* **1**, 251.

Hikota, T., K. Morohara, and K. Meguro (1970) *Bull. Chem. Soc. Jpn.* **43**, 3913.

Hua, X. Y. and M. J. Rosen (1982) *J. Colloid Interface Sci.* **87**, 469.

Huber, K. (1991) *J. Colloid Interface Sci.* **147**, 321.

Jaschke, M., H.-J. Butt, H. E. Gaub, and S. Manne (1997) *Langmuir* **13**, 1381.

Kipling, J. J., *Adsorption from Solutions of Non-Electrolytes*, Academic, New York, 1965, Chap. 17.

Kipling, J. J. and E. H. M. Wright (1962) *J. Chem. Soc.* 855.

Kitchener, J. A. (1965) *J. Photogr. Sci.* **13**, 152.

Kjellin, U. R. M., P. M. Claesson, and P. Linse (2002) *Langmuir* **18**, 6745.

Kling, W. and H. Lange, 2nd Int. Congr. Surface Activity, London, 1957, I, p. 295.

Kölbel, H. and K. Hörig (1959) *Angew. Chem.* **71**, 691.

Kölbel, H. and P. Kuhn (1959) *Angew. Chem.* **71**, 211.

Kosman, J. J. and R. L. Rowell (1982) *Colloids Surf.* **4**, 245.

Kronberg, B., P. Stenius, and Y. Thorssell (1984) *Colloids Surf.* **12**, 113.

Kunjappu, J. T., D.Sc. Thesis (sequel to Ph.D. thesis, 1985), University of Mumbai, 1994a.

Kunjappu, J. T. (1994b) *J. Colloid Interface Sci.* **162**, 261.

Kunjappu, J. T. and P. Somasundaran (1989) *J. Phys. Chem.* **93**, 7744.

Kunjappu, J. T. and P. Somasundaran (1995) *J. Colloid Interface Sci.* **175**, 520.

Kwan, C.-C. and M. J. Rosen (1980) *J. Phys. Chem.* **84**, 547.

Lange, H. (1957) *Kolloid-Z.* **152**, 155.

Lange, H., 4th Int. Congr. Surface-Active Substances, Brussels, 1964, II, p. 497.

Lange, H. and M. J. Schwuger (1971) *Kolloid Z. Z. Polym.* **243**, 120.

Langmuir, I. (1917) *J. Am. Chem. Soc.* **39**, 1848.

Langmuir, I. (1918) *J. Am. Chem. Soc.* **40**, 1361.

Lascaux, M. P., O. Dusart, R. Granet, and S. Piekarski (1983) *J. Chim. Phys.* **80**, 615.

Law, J. P., Jr. and G. W. Kunze (1966) *Soil Sci. Soc. Am. Proc.* **30**, 321.

Li, F., M. J. Rosen, and S. B. Sulthana (2001) *Langmuir* **17**, 1037.

Liljekvist, P. and B. Kronberg (2000) *J. Colloid Interface Sci.* **222**, 159.

Livingston, J. R. and R. Drogin (1965) *J. Am. Oil Chem. Soc.* **42**, 720.

Lucassen-Reynders, E. H. (1966) *J. Phys. Chem.* **70**, 1777.

Lucassen-Reynders, E. H. and M. van den Tempel, 4th Int. Congr. Surface Active Substances, Brussels, 1967, p. 779.

Machinson, K. R. (1967) *J. Text. Inst. Trans.* **58**, 1.

Manne, S. and H. E. Gaub (1995) *Science* **270**, 1480.

Manne, S., J. P. Cleveland, H. E. Gaub, G. D. Stucky, and P. K. Hansma (1994) *Langmuir* **10**, 4409.

Matijevic, E. and B. A. Pethica (1958) *Trans. Faraday Soc.* **54**, 1382, 1390, 1400.

Matos, S. L., J.-C. Ravey, and G. Serratrice (1989) *J. Colloid Interface Sci.* **128**, 341.

McCaffery, F. G. and N. Mungan (1970) *J. Can. Petrol. Technol.* **9**, 185.

Meguro, K., Y. Takasawa, N. Kawahasi, Y. Tabata, and M. Ueno (1981) *J. Colloid Interface Sci.* **83**, 50.

Meichelbeck, H. and H. Knittel (1971) *Fette, Seifen, Anstrichmi* **73**, 25.

Miles, G. D. and L. Shedlovsky (1945) *J. Phys. Chem.* **49**, 71.

Miyagishi, S., T. Asakawa, and M. Nishida (1989) *J. Colloid Interface Sci.* **131**, 68.

Mukerjee, P. (1967) *Adv. Colloid Interface Sci.* **1**, 264.

Mukerjee, P. (1968) *Nature* **217**, 1046.

Mulley, B. A. and A. D. Metcalf (1962) *J. Colloid Sci.* **17**, 523.

Murphy, D. S. and M. J. Rosen (1988) *J. Phys. Chem.* **92**, 2870.

Murphy, D. S., Z. H. Zhu, X. Y. Hua, and M. J. Rosen (1990) *J. Am. Oil Chem. Soc.* **67**, 197.

Naifu, Z. and T. Gu (1979) *Sci. Sinica* **22**, 1033.

Nakano, T.-Y., G. Sugihara, T. Nakashima, and S.-C. Yu (2002) *Langmuir* **18**, 8777.

Nevskaia, D. M., A. Guerrera-Ruiz, and J. de Lopez-Gonzalez (1996) *J. Colloid Interface Sci.* **181**, 571.

Omar, A. M. A. and N. A. Abdel-Khalek (1997) *Tenside Surf. Det.* **34**, 178.

Osseo-Asare, K., D. W. Fuerstenau, and R. H. Ottewill, in *Adsorption at Interfaces*, K. L. Mittal (ed.), Symposium Series No. 8, American Chemical Society, Washington, DC, 1975, pp. 63–78.

Ottewill, R. H. and M. C. Rastogi (1960) *Trans. Faraday Soc.* **56**, 866.

Parfitt, G. D. and D. G. Wharton (1972) *J. Colloid Interface Sci.* **38**, 431.
Partyka, S., S. Zaini, M. Lindheimer, and B. Brun (1984) *Colloids Surf.* **12**, 255.
Rehfeld, S. J. (1967) *J. Phys. Chem.* **71**, 738.
Robb, D. J. M. and A. E. Alexander (1967) *Soc. Chem. Ind. (London)* Monograph No. 25, 292.
Rosen, M. J. (1974) *J. Am. Oil Chem. Soc.* **51**, 461.
Rosen, M. J. (1975) *J. Am. Oil Chem. Soc.* **52**, 431.
Rosen, M. J. (1976) *J. Colloid Interface Sci.* **56**, 320.
Rosen, M. J. and S. Aronson (1981) *Colloids Surf.* **3**, 201.
Rosen, M. J. and H. A. Goldsmith, *Systematic Analysis of Surface-Active Agents*, 2nd ed., Wiley-Interscience, New York, 1972.
Rosen, M. J. and X. Y. Hua (1982) *J. Colloid Interface Sci.* **86**, 164.
Rosen, M. J. and D. S. Murphy (1991) *Langmuir* **7**, 2630.
Rosen, M. J. and Y. Nakamura (1977) *J. Phys. Chem.* **80**, 873.
Rosen, M. J. and J. Solash (1969) *J. Am. Oil Chem. Soc.* **46**, 399.
Rosen, M. J. and S. B. Sulthana (2001) *J. Colloid Interface Sci.* **239**, 528.
Rosen, M. J. and V. Wu (2001) *Langmuir* **17**, 7296.
Rosen, M. J. and Z. H. Zhu (1989) *J. Colloid Interface Sci.* **133**, 473.
Rosen, M. J., D. Friedman, and M. Gross (1964) *J. Phys. Chem.* **68**, 3219.
Rosen, M. J., M. Baum, and F. Kasher (1976) *J. Am. Oil Chem. Soc.* **53**, 742.
Rosen, M. J., A. W. Cohen, M. Dahanayake, and X.-Y. Hua (1982a) *J. Phys. Chem.* **86**, 541.
Rosen, M. J., M. Dahanayake, and A. W. Cohen (1982b) *Colloids Surf.* **5**, 159.
Rosen, M. J., Z. H. Zhu, B. Gu, and D. S. Murphy (1988) *Langmuir* **4**, 1273.
Rosen, M. J., Z. H. Zhu, and X. Y. Hua (1992) *J. Am. Oil Chem. Soc.* **64**, 30.
Rosen, M. J., Y.-P. Zhu, and S. W. Morrall (1996) *J. Chem. Eng. Data* **41**, 1160.
Rosen, M. J., L. Fei, Y.-P. Zhu, and S. W. Morrall (1999) *J. Surfactants Deterg.* **2**, 343.
Rosen, M. J., E. Li, S. W. Morrall, and D. J. Versteeg (2001) *Environ. Sci. Sechnol.* **35**, 954.
Rupprecht, H. and H. Liebl (1972) *Kolloid Z. Z. Polym.* **250**, 719.
Scamehorn, J. F., R. S. Schechter, and W. H. Wade (1982) *J. Colloid Interface Sci.* **85**, 463.
Schick, M. J. (1962) *J. Colloid Sci.* **17**, 801.
Schott, H. (1967) *J. Colloid Interface Sci.* **23**, 46.
Schwuger, M. J. (1969) *Kolloid-Z. Z. Polym.* **232**, 775.
Schwuger, M. J. (1971a) *Bee Bunsenes. Ges. Phys. Chem.* **75**, 167.
Schwuger, M. J. (1971b) *Kolloid-Z. Z. Polym.* **243**, 129.
Semmler, A. and H.-H. Kohler (1999) *J. Colloid Interface Sci.* **218**, 137.
Sexsmith, F. H. and H. J. White (1959) *J. Colloid Sci.* **14**, 598.
Shinoda, K., T. Yamanaka, and K. Kinoshita (1959) *J. Phys. Chem.* **63**, 648.
Shinoda, K., T. Yamaguchi, and R. Hori (1961) *Bull. Chem. Soc. Jpn.* **34**, 237.
Shinoda, K., M. Hato, and T. Hayashi (1972) *J. Phys. Chem.* **76**, 909.

Snyder, L. R. (1968) *J. Phys. Chem.* **72**, 489.

Somasundaran, P. and D. W. Fuerstenau (1966) *J. Phys. Chem.* **70**, 90.

Somasundaran, P. and J. T. Kunjappu (1989) *Colloids Surf.* **37**, 245.

Somasundaran, P., T. W. Healy, and D. W. Fuerstenau (1964) *J. Phys. Chem.* **68**, 3562.

Somasundaran, P., T. W. Healy, and D. W. Fuerstenau (1966) *J. Colloid Interface Sci.* **22**, 599.

Somasundaran, P., R. Middleton, and K. V. Viswanathan, in *Structure/Performance Relationships in Surfactants*, M. J. Rosen (ed.), ACS Symp. Series, 253, American Chemical Society, Washington, DC, 1983, p. 269.

Spitzer, J. J. and L. D. Heerze (1983) *Can. J. Chem.* **61**, 1067.

Stern, O. (1924) *Z. Electrochem.* **30**, 508.

Stigter, D. (1971) *J. Am. Oil Chem. Soc.* **48**, 340.

Subramanian, V. and W. A. Ducker (2000) *Langmuir* **16**, 4447.

Suzuki, H. (1967) *Yukagaku* **16**, 667 (*C.* A. **68**, 41326h [1968]).

Szyszkowski, B. (1908) *Z. Phys. Chem.* **64**, 385.

Tamaki, K. (1967) *Bull. Chem. Soc. Jpn.* **40**, 38.

Tamamushi, B. and K. Tamaki, 2nd Int. Congr. Surface Activity, London, England, September 1957, III, p. 449.

Tomasic, V., I. Stefanic, and N. Filipovic-Vincekovic (1999) *Coll. Polym. Sci.* **277**, 153.

Tori, K. and T. Nakagawa (1963) *Kolloid-Z. Z. Polym.* **189**, 50.

Tori, K., K. Kuriyama, and T. Nakagawa (1963) *Kolloid-Z. Z. Polym.* **191**, 48.

Tsubone, K. and M. J. Rosen (2001) *J. Colloid Interface Sci.* **244**, 394.

Tsubone, K. and N. Uchida (1990) *J. Am. Oil Chem. Soc.* **67**, 394.

van der Waarden, M. (1951) *J. Colloid Sci.* **6**, 443.

van Senden, K. G. and J. Koning (1968) *Fette, Seifen, Anstrichmi* **70**, 36.

van Voorst Vader, F. (1960a) *Trans. Faraday Soc.* **56**, 1067.

van Voorst Vader, F. (1960b) *Trans. Faraday Soc.* **56**, 1078.

Varadaraj, R., J. Bock, S. Zushma, N. Brons, and T. Colletti (1991) *J. Colloid Interface Sci.* **147**, 387.

Varadaraj, R., J. Bock, S. Zushma, and N. Brons (1992) *Langmuir* **8**, 14.

Venable, R. L. and R. V. Nauman (1964) *J. Phys. Chem.* **68**, 3498.

Vijayendran, B. R. and T. P. Bursh (1979) *J. Colloid Interface Sci.* **68**, 383.

von Helmholtz, H. (1879) *Wied. Ann. Phys.* **7**, 337.

Von Hornuff, G. and W. Mauer (1972) *Deut. Text. Tech.* **22**, 290.

Waag, A., Chim. Phys. Appl. Prat. Agents de Surface. 5th C.R. Int. Congr. Detergence, Barcelona, 1968 (Publ. 1969) 3, 143.

Wakamatsu, T. and D. W. Fuerstenau, in *Adsorption from Aqueous Solution*, W. J., Jr. Weber and E. Matijevic (eds.), American Chemical Society, Washington, DC, 1968, pp. 161–172.

Wakamatsu, T. and D. W. Fuerstenau (1973) *Trans. Soc. Min. Eng. AIME* **254**, 123.

Wanless, E. J. and W. A. Ducker (1996) *J. Phys. Chem.* **100**, 3207.

Watanabe, A. (1960) *Bull. Inst. Chem. Res. Kyoto Univ.* **38**, 179.

Weber, W. J., Jr. (1964) *J. Appl. Chem.* **14**, 565.

Weil, J. K., R. G. Bistline, and A. J. Stirton (1958) *J. Phys. Chem.* **62**, 1083.

Williams, E. F., N. T. Woodbury, and J. K. Dixon (1957) *J. Colloid Sci.* **12**, 452.

Wolgemuth, J. L., R. K. Workman, and S. Manne (2000) *Langmuir* **16**, 3077.

Zettlemoyer, A. C., V. S. Rao, E. Boucher, and R. Fix, 5th Int. Congr. Surface-Active Substances, Barcelona, Spain, September 1968, III, p. 613.

Zhang, L., P. Somasundaran, and C. Maltesh (1997) *J. Colloid Interface Sci.* **191**, 202.

Zhao, F. and M. J. Rosen (1984) *J. Phys. Chem.* **88**, 6041.

Zhu, B. Y. and M. J. Rosen (1984) *J. Colloid Interface Sci.* **99**, 435.

Zhu, Y.-P., M. J. Rosen, and S. W. Morrall (1998a) *J. Surfactants Deterg.* **1**, 1.

Zhu, Y.-P., M. J. Rosen, S. W. Morrall, and J. Tolls (1998b) *J. Surfactants Deterg.* **1**, 187.

Zhu, Y.-P., M. J. Rosen, P. K. Vinson, and S. W. Morrall (1999) *J. Surfactants Deterg.* **2**, 357.

Zoeller, N. and D. Blankschtein (1998) *Langmuir* **14**, 7155.

PROBLEMS

2.1 A nonionic surface-active solute in aqueous solution at 30°C gives the following γ-log C data (C in moles dm^{-3}):

γ(mJm^{-2}):	71.4	60.0	52.0	40.6	29.2	29.2	29.2
log C:	−6.217	−5.992	−5.688	−5.255	−4.822	−4.691	−4.552

The slope of the γ–log C plot is linear at $\gamma < 60$ mJm^{-2} down to the c.m.c.

(a) Calculate the surface excess concentration, Γ, in moles cm^{-2} at $\gamma < 60$ mJm^{-2}.

(b) Calculate the minimum surface area/molecule, in Å^2.

(c) Calculate ΔG^0_{ad}, in kJ mol^{-1}.

2.2 Without looking at the tables, place the following compounds in order of increasing efficiency of adsorption (increasing pC_{20} value at the aqueous solution–air interface):

(a) $CH_3(CH_2)_{10}CH_2SO_4^-Na^+$

(b) $CH_3(CH_2)_9$–C_6H_4–$SO_3^-Na^+$

(c) $CH_3(CH_2)_8CH_2N^+(CH_3)_3Cl^-$

(d) $CH_3(CH_2)_4CH(C_3H_7)CH_2SO_4^-Na^+$

(e) $CH_3(CH_2)_{10}CH_2(OC_2H_4)_6OH$

2.3 Without looking at the tables, place the following compounds in order of increasing effectiveness of adsorption (increasing Γ_m) at the aqueous solution–air interface. Use ≃ if two or more compounds have approximately equal Γ_m values:

(a) $C_{10}H_{21}SO_4^-Na^+$ (in H_2O)
(b) $C_{12}H_{25}SO_4^-Na^+$ (in H_2O)
(c) $C_{16}H_{33}SO_4^-Na^+$ (in H_2O)
(d) $C_{16}H_{33}SO_4^-Na^+$ (in 0.1 *M* NaCl)
(e) $C_{18}H_{37}SO_4^-Na^+$ (in H_2O)

2.4 If $1/k \simeq 10$ Å for 0.1 M NaCl in aqueous solution at room temperature, calculate $1/k$ for 0.1 *M* $CaCl_2$ under the same conditions.

2.5 2.0 g of a solid, whose specific surface area is 50 m^2/g, is shaken with 100 mL of a 1×10^{-2} *M* solution of a surfactant. After equilibrium is reached, the concentration of the surfactant solution is 7.22×10^{-3} *M*. Calculate the average area occupied per surfactant molecule on the solid surface in Å^2.

2.6 The molar concentrations of two individual surfactants required to yield a surface tension value of 36 dyn/cm in aqueous solution are 2.6×10^{-3} and 1.15×10^{-3}, respectively. The total molar surfactant concentration required to yield a surface tension of 36 dyn/cm is 6.2×10^{-4} for a mixture of the two surfactants in which the mole fraction of the first surfactant (on a surfactant-only basis) is 0.41. Calculate the value of X_1, the mole fraction of surfactant 1 in the total surfactant at the aqueous solution–air interface for this mixture.

2.7 **(a)** 50 mL of a cationic surfactant solution is placed in a uniform glass beaker of 5 cm radius for the determination of its surface tension. Since glass carries a negative charge, the cationic surfactant adsorbs onto it. Assuming that it forms at least an adsorbed monolayer on the glass, with an area/molecule at that interface of 50 Å^2, and at the liquid–air interface of 60 Å^2, calculate the surfactant concentration at which adsorption at the glass interface and at the surface will result in about a 10% reduction in the surfactant bulk phase concentration.

(b) What procedures could be taken to avoid this error when surfactant solutions must be used at that concentration?

2.8 Prepare a list of all the terms like *efficiency*, *effectiveness*, and others that you come across in this chapter and provide their definition/meaning.

2.9 Calculate Γ_1 for the adsorption of a surfactant at water–air interface at 25°C, assuming a value for $\Gamma_m = 4.4 \times 10^{-10}$ mol/cm^2 using Equation 2.29 when the surface tension of the solution is reduced by 15 mN/m (dyn/cm).

2.10 Use SI units for all the dimensions of variables in Equation 2.1 and show that $1/\kappa$ has the unit of length in meter (m).

3 Micelle Formation by Surfactants

We now turn our attention to a property of surfactants that may be as fundamental, and certainly is as important, as their property of being adsorbed at interfaces. This property is micelle formation—the property that surface-active solutes have of forming colloidal-sized clusters in solution.* Micelle formation, or micellization, is an important phenomenon not only because a number of important interfacial phenomena, such as detergency and solubilization, depend on the existence of micelles in solution, but also because it affects other interfacial phenomena, such as surface or interfacial tension reduction, that do not directly involve micelles. Micelles have become a subject of great interest to the organic chemist and the biochemist—to the former because of their unusual catalysis of organic reactions (Fendler and Fendler, 1975) and to the latter because of their similarity to biological membranes and globular proteins.

I. THE CRITICAL MICELLE CONCENTRATION (CMC)

Almost from the very beginning of the study of the properties of surfactant solutions (actually, soap solutions), it was recognized that their bulk properties were unusual and indicated the presence of colloidal particles in the solution.

When the equivalent conductivity (specific conductance per gram equivalent of solute) of an anionic surfactant of the type $Na^+ R^-$ in water is plotted against the square root of the normality of the solution, the curve obtained, instead of being the smoothly decreasing curve characteristic of ionic electrolytes of this type, has a sharp break in it at low concentrations (Figure 3.1). This break in the curve, with its sharp reduction in the conductivity of the

* Only those polar solvents that have two or more potential hydrogen-bonding centers and thus are capable of forming three-dimensional hydrogen-bonded networks appear capable of showing micelle formation (Ray, 1971). In nonpolar solvents, clusters of surfactants may form, but they are generally not of colloidal size, and their behavior is not analogous to that of micelles in aqueous media.

Surfactants and Interfacial Phenomena, Fourth Edition. Milton J. Rosen and Joy T. Kunjappu.

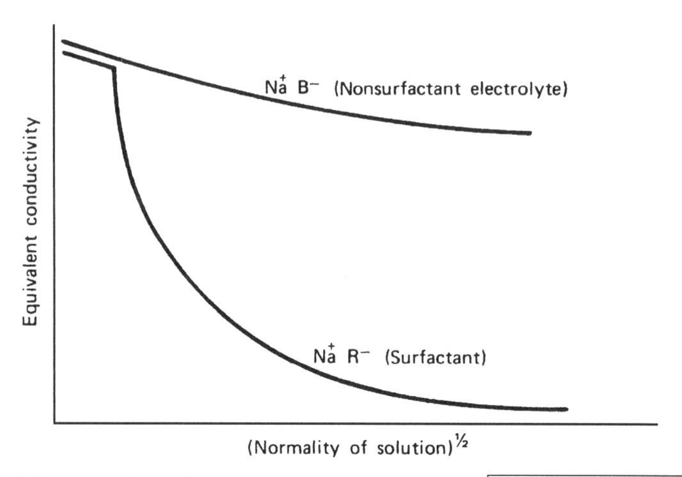

FIGURE 3.1 Plot of equivalent conductivity versus $\sqrt{\text{normality of solution}}$ for an aqueous solution of surfactant of type Na^+ R^- (*normality* is a unit of concentration used in old literature, which is related to *molarity*).

solution, indicating a sharp increase in the mass per unit charge of the material in solution, is interpreted as evidence of the formation at that point of micelles from the unassociated molecules of surfactant, with part of the charge of the micelle neutralized by associated counterions.

The concentration at which this phenomenon occurs is called the CMC. Similar breaks in almost every measurable physical property that depends on size or number of particles in solution, including micellar solubilization of solvent-insoluble material (Chapter 4) and reduction of surface or interfacial tension (Chapter 5), are shown by all types of surfactants—nonionic, anionic, cationic, and zwitterionic in aqueous media.

In some cases, particularly where the hydrophobic group is long (e.g., $>C_{16}$), a second break in the conductivity–surfactant concentration curve has been observed. It has been suggested (Treiner and Makayssi, 1992) that this indicates change in the micellar structure (Section II).

The determination of the value of the CMC can be made by use of any of the(se) physical properties, but most commonly, the breaks in the electrical conductivity,* surface tension, light scattering (Ford et al., 1966), or fluorescence spectroscopy–concentration curves have been used for this purpose. CMCs have also very frequently been determined from the change in the spectral characteristics of some dyestuff added to the surfactant solution when

* When electrolyte (other than the ionic surfactant) is present in the aqueous solution, it is preferable to use a plot of $\Delta k/\Delta c$ vs. $(c)^{1/2}$, where Δk is the change in the specific conductivity of the surfactant-containing solution, Δc the change in the surfactant concentration, and c the average value of c over the Δc range (Fujiwara et al., 1997).

the CMC of the latter is reached. However, this method is open to the serious objection that the presence of the dyestuff may affect the value of the CMC. An excellent critical evaluation of the methods for determining CMCs is included in the comprehensive compilation of CMCs in aqueous solution by Mukerjee and Mysels (1971).

Based on a vast amount of data concerning the phenomenon, a picture of the process of micellization and the structure of the micelles formed has slowly emerged. It has previously been mentioned (Chapter 1) that when they are dissolved in water, materials that contain a hydrophobic group distort the structure of the water and therefore increase the free energy of the system. They therefore concentrate at the surface, where, by orienting so that their hydrophobic groups are directed away from the solvent, the free energy of the solution is minimized. However, there is another means of minimizing the free energy in these systems. The distortion of the solvent structure can also be decreased (and the free energy of the solution reduced) by the aggregation of the surface-active molecules into clusters (micelles) with their hydrophobic groups directed toward the interior of the cluster and their hydrophilic groups directed toward the solvent. Micellization is therefore an alternative mechanism to adsorption at the interfaces for removing hydrophobic groups from contact with the water, thereby reducing the free energy of the system. When there is little distortion of the structure of the solvent by the lyophobic group (e.g., in water, when the hydrophobic group of the surfactant is short), then there is little tendency for micellization to occur. This is often the case in nonaqueous solvents, and therefore micelles of size comparable to those formed in aqueous media are seldom found in other solvents.

Although removal of the lyophobic group from contact with the solvent may result in a decrease in the free energy of the system, the surfactant molecule, in transferring from solution in the solvent to the micelle, may experience some loss of freedom in being confined to the micelle and, in the case of ionic surfactants, from electrostatic repulsion from other similarly charged surfactant molecules in the micelle. These forces increase the free energy of the system and thus oppose micellization. Whether micellization occurs in a particular case and, if so, at what concentration of monomeric surfactant, therefore depends on the balance between the factors promoting micellization and those opposing it.

As will be seen in the following chapters, micelles have a vast number of uses. An interesting use of micelles of anionic surfactants that involves both their adsorption and solubilization (Chapter 4) properties is for the removal of pollutants such as metallic ions and organic material from water. Metallic ions bind to the negatively charged surface of micelles of anionic surfactants, and organic material is solubilized in the interior of the micelles. The micellar solution is forced through an ultrafiltration membrane with pores small enough to block the passage of the micelles with their associated metallic ions and organic material (Fillipi et al., 1999).

II. MICELLAR STRUCTURE AND SHAPE

A. The Packing Parameter

The shape of the micelle produced in aqueous media is of importance in determining various properties of the surfactant solution, such as its viscosity, its capacity to solubilize water-insoluble material (Chapter 4), and its cloud point (Chapter 4, Section IIIB).

At the present time, the major types of micelles appear to be (1) relatively small, spherical structures (aggregation number <100), (2) elongated cylindrical, rodlike micelles with hemispherical ends (prolate ellipsoids), (3) large, flat lamellar micelles (disklike extended oblate spheroids), and (4) vesicles—more or less spherical structures consisting of bilayer lamellar micelles arranged in one or more concentric spheres.

In aqueous media, the surfactant molecules are oriented, in all these structures, with their polar heads predominantly toward the aqueous phase and their hydrophobic groups away from it. In vesicles, there will also be an aqueous phase in the interior of the structure. In ionic micelles, the aqueous solution–micelle interfacial region contains the ionic head groups, the Stern layer of the electrical double layer with the bound counterions, and water. The remaining counterions are contained in the Gouy–Chapman portion of the double layer that extends further into the aqueous phase. For polyoxyethylenated nonionics, the structure is essentially the same, except that the outer region contains no counterions, but includes coils of hydrated POE chains.

The interior region of the micelle, containing the hydrophobic groups, has a radius approximately equal to the length of the fully extended hydrophobic chain. The aqueous phase is believed to penetrate into the micelle beyond the hydrophobic head groups, and the first few methylene groups of the hydrophobic chain adjacent to the hydrophobic head are often considered in the hydration sphere. It is therefore useful to divide the interior region into an outer core that may be penetrated by water and an inner core from which water is excluded (Muller et al., 1972).

In nonpolar media, the structure of the micelle is similar but reversed, with the hydrophilic heads comprising the interior region surrounded by an outer region containing the hydrophobic groups and nonpolar solvent (Hirschhorn, 1960). Dipole–dipole interactions hold the hydrophilic heads together in the core (Singleterry, 1955).

Changes in temperature, concentration of surfactant, additives in the liquid phase, and structural groups in the surfactant may all cause change in the size, shape, and aggregation number of the micelle, with the structure varying from spherical through rod- or disklike to lamellar in shape (Winsor, 1968).

A theory of micellar structure, based upon the geometry of various micellar shapes and the space occupied by the hydrophilic and hydrophobic groups of the surfactant molecules, has been developed by Israelachvili et al. (1976, 1977) and Mitchell and Ninham (1981). The volume V_H occupied by the hydrophobic groups in the micellar core, the length of the hydrophobic group in the core l_c, and the cross-sectional area $a_{()}$ occupied by the hydrophilic group at the micelle–solution interface are used to calculate a "packing parameter," $V_H/l_c a_{()}$, which determines the shape of the micelle.

Value of $V_H/l_c a_{()}$	Structure of the Micelle
$0-\frac{1}{3}$	Spheroidal in aqueous media
$\frac{1}{3}-\frac{1}{2}$	Cylindrical in aqueous media
$\frac{1}{2}-1$	Lamellar in aqueous media
>1	Inverse (reversed) micelles in nonpolar media

B. Surfactant Structure and Micellar Shape

From Tanford (1980), $V_H = 27.4 + 26.9n$ Å^3, where n is the number of carbon atoms of the chain embedded in the micellar core (the total number of carbon atoms in the chain, or one less); $l_c \leq 1.5 + 1.265n$ Å, depending upon the extension of the chain. For saturated, straight chains, l_c may be 80% of the fully extended chain.

The solubilization of hydrocarbons in the interior of the micelle (Chapter 4, Section I) increases the value of V_H.

The value of $a_{()}$varies not only with the structure of the hydrophilic head group, but also with changes in the electrolyte content, temperature, pH, and the presence of additives in the solution. Additives, such as medium-chain alcohols that are solubilized in the vicinity of the head groups (Chapter 4, Section IIIA), increase the value of $a_{()}$. With ionic surfactants, $a_{()}$ decreases with increase in the electrolyte content of the solution, due to compression of the electrical double layer, and also with increase in the concentration of the ionic surfactant, since that increases the concentration of counterions in the solution. This decrease in the value of $a_{()}$ promotes change in the shape of the micelle from spherical to cylindrical. For POE nonionic surfactants, an increase in temperature may cause a change in shape if temperature increase results in increased dehydration of the POE chain.

Some ionic surfactants form long, wormlike micelles in aqueous media, especially in the presence of electrolyte or other additives that decrease the repulsion between the ionic head groups (Raghavan and Kaler, 2001). These giant, wormlike micelles give rise to unusually strong viscoelasticity because of the entanglement of these structures.

When the value of the parameter $V_H/l_c a_{()}$ reaches a value of approximately 1, the surfactant can form either normal lamellar micelles in aqueous media or reversed micelles in nonpolar media. As the value of the parameter gets larger and larger than 1, the reverse micelles in nonpolar media tend to become less asymmetrical and more spherical in shape.

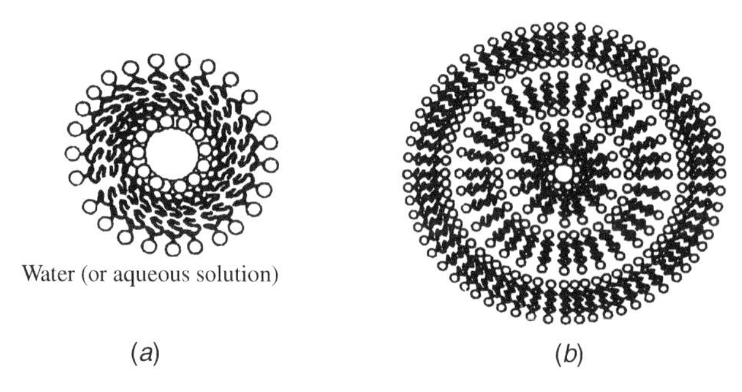

FIGURE 3.2 Vesicles: (a) unilamellar; (b) multilamellar.

In aqueous media, surfactants with bulky or loosely packed hydrophilic groups and long, thin hydrophobic groups tend to form spherical micelles, while those with short, bulky hydrophobic groups and small, close-packed hydrophilic groups tend to form lamellar or cylindrical micelles.

Surfactants having two long alkyl chains may, upon sonification in aqueous media, form vesicles (Figure 3.2). Thus, fatty acid esters of sucrose, especially the diesters, form vesicles upon sonification (Ishigami and Machida, 1989). Since vesicles are curved, closed lamellar bilayers, there are critical geometric and flexibility requirements for their formation. The packing parameter, $V_H/l_c a_0$, must be close to 1. However, some structure must be present in the molecule to keep the hydrophobic groups from becoming closely packed; otherwise, the flexibility requirement will not be met. And, since the hydrophobic groups cannot be closely packed, the hydrophilic head groups must also not pack closely to retain the packing parameter value close to 1. Vesicles have been formed from short-chain POE alcohols and perfluoroalcohols with short POE groups (Ravey and Stebe, 1994) and from cetyl trimethylammonium *p*-toluenesulfonate plus sodium dodecyl benzene sulfonate, but not plus sodium dodecyl sulfate. The sulfate head group in the latter compound is stated to pack too closely with the trimethylammonium group to form vesicles; the benzenesulfonate group packs more loosely (Salkar et al., 1998). Mixtures of dodecyl dimethylammonium bromide with dodecyl trimethylammonium chloride spontaneously form vesicles. The spontaneous formation is attributed to differences in the packing parameter of the two surfactants (Viseu et al., 2000). Interest in vesicles stems from their possible medical use as carriers for toxic drug delivery, although there are other applications.

The shape of the micelle may change when material is solubilized by the micelle (Chapter 4, Section IIIA) and by change in molecular environmental factors (Figure 4.4).

C. Liquid Crystals

When there is a sufficient number of micelles in the solution phase, they start to pack together in a number of geometric arrangements, depending upon the

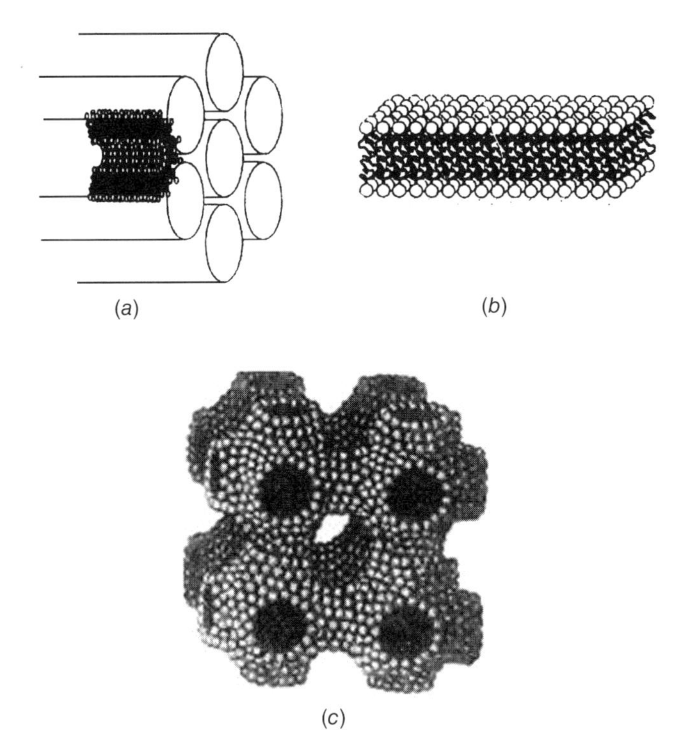

FIGURE 3.3 Hexagonal (a), lamellar (b), and bicontinuous (c) liquid crystal structures.

shape of the individual micelles. These packing arrangements are known as *liquid crystals*. Liquid crystals have the ordered molecular arrangement of solid crystals but the mobility of liquids. Because of this ordered arrangement of the molecules, they increase the viscosity of the solution phase, sometimes very considerably. Spherical micelles pack together into cubic liquid crystals, cylindrical micelles pack to form hexagonal liquid crystals, and lamellar micelles form lamellar liquid crystals (Figure 3.3). It is easier to pack surfactant molecules having a bulky head group into hexagonal phases, while surfactants having two alkyl groups pack better into a lamellar phase. Both normal cylindrical micelles in aqueous media and reverse cylindrical micelles in nonpolar media can form hexagonal liquid crystals. Because some types of micelles change their structure from spherical to cylindrical to lamellar with increase in surfactant concentration, hexagonal phases are usually encountered at lower surfactant concentrations than lamellar phases. With increase in surfactant concentration, some cylindrical micelles become branched and interconnected, leading to a bicontinuous liquid crystalline phase (Figure 3.3c) in which there are no distinct micelles. Hexagonal and lamellar phases are anisotropic and can be detected by their radiance under the polarizing microscope. Hexagonal liquid crystals appear as fanlike structures or with a variety of nongeometrical structures; lamellar liquid crystals appear as Maltese crosses or as oil

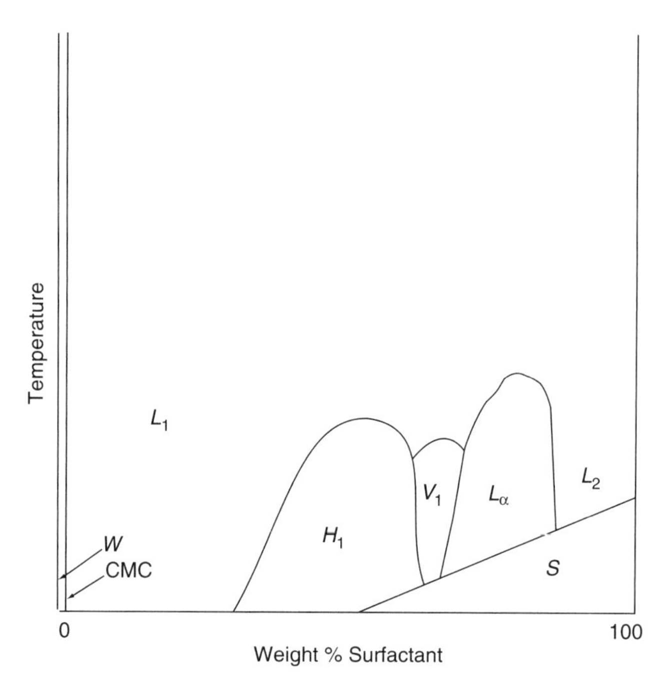

FIGURE 3.4 Phase diagram showing location of hexagonal (H_1), normal bicontinuous cubic (V_1), and lamellar ($L\alpha$) liquid phases, aqueous nonmicellar solution (W), micellar solution (L_1), liquid surfactant containing water (L_2), and solid surfactant (S).

streaks. Hexagonal phases are more viscous than lamellar phases, which in turn are more viscous than ordinary solutions. Spherical micelles pack together at high surfactant concentrations to form cubic liquid crystals that are very high-viscosity gels. Bicontinuous structures also form cubic phases. Cubic phases may therefore be formed from normal spherical or reverse spherical micelles, or from normal bicontinuous or reverse bicontinuous structures. These are all isotropic structures, as are spherical micelles, and cannot be observed under the polarizing microscope. They can be identified by use of water-soluble and oil-soluble dyes (Kunieda et al., 2003).

Plots that show the conditions (temperature, composition) at which various phases exist in a system are known as *phase diagrams*. Figure 3.4 is one type of phase diagram showing the effects of temperature and surfactant concentration on the various solution phases of an aqueous surfactant system. The order of the various liquid crystal phases with increase in surfactant concentration—micellar ⇒ hexagonal ⇒ bicontinuous cubic ⇒ lamellar—is found in many surfactant systems.

The effect of temperature increase is typical for surfactants whose solubility increases with temperature increase, converting all liquid crystal phases to micellar solutions when the temperature is high enough. At high surfactant concentration and low temperature, solid surfactant may precipitate.

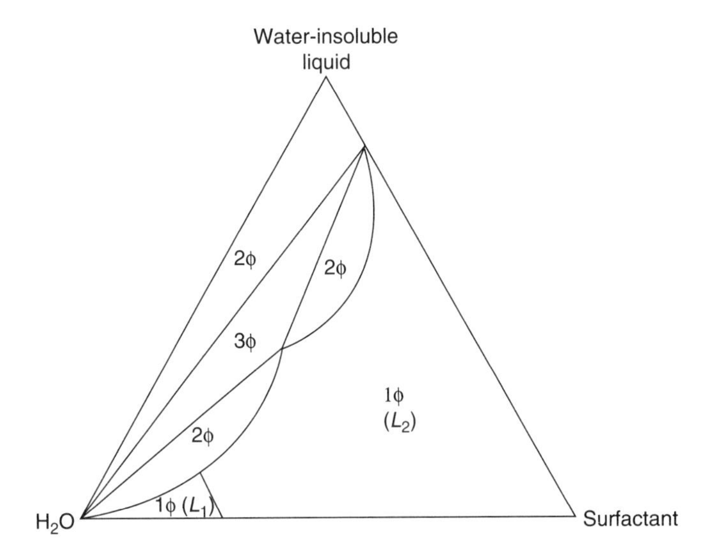

FIGURE 3.5 Simplified isothermal ternary phase diagram (liquid crystalline phases omitted) showing the number of phases (Φ) in each location. L_1 = aqueous micellar phase, L_2 = reverse micellar phase, with microemulsion phase adjacent to the 3Φ region.

Liquid crystal structures are important not only in the viscosity modification of surfactant solutions, but also in the stabilization of foams and emulsions, in detergency, in lubrication (Boschkova et al., 2002), and in other applications.

Another type of phase diagram, when two or more components in addition to water are present, shows the effect of the composition of the system at constant temperature on the number and location of the different phases in the system. These are known as (*isothermal*) *ternary phase diagrams* (Friberg, 1969). Each vertex of the triangle represents the point of 100% of the solvent, the surfactant, and any other component (or combination of components at a constant ratio of the two). Phase diagrams of this type are often used to show the location of microemulsion phases (Chapter 8, Section II), when hydrocarbon and water phases are present, in addition to the surfactant. A highly simplified phase diagram of this type is shown in Figure 3.5. The locations and numbers of phases change with temperature and with the nature of the surfactant and the water-insoluble liquid.

D. Rheology of Surfactant Solutions

Rheology refers to behavior that connects applied stress with the resulting deformation. The rheology of surfactant solutions generally pertains to their flow behavior and in that sense is closely related to solution viscosity. Surfactant solutions at low concentration ranges, where they form uniform spherical micelles, behave as Newtonian fluids, which means that the applied shearing stress is directly proportional to the rate of sheer. But at higher surfactant concentrations, the aggregate shapes acquire asymmetry in the form of large micelles or cylinders or bilayers, and the solutions show non-Newtonian behavior and the viscosity increases sharply. Polymeric surfactants also form nonspherical aggregates in solution and show Newtonian behavior.

Surfactant solutions can also become *viscoelastic*, that is, they can be viscous and elastic simultaneously in response to deformation, as in gels (see below). Cationic surfactant–salicylate systems and many polymer–surfactant mixtures show viscoelasticity effect. This is due to an increase in the micellar size and intermicellar interactions.

Solid dispersions stabilized by surfactants (see Chapter 9), as in inks and paints, behave as non-Newtonian complex fluids (Kunjappu, 2001). Their deviation from Newtonian behavior is mainly of two types: *pseudoplasticity* in which viscosity decreases with increase in shear stress (shear thinning) and *dilatancy* in which viscosity increases with increase in shear stress (shear thickening). Most of the inks show pseudoplasticity. In dilatancy, the particles cannot move fast enough past one another and the system rigidifies.

Another rheology-related property of surfactants is *thixotropy*, the flow behavior in which viscosity is reduced by agitation or stirring or as a consequence of time. Thixotropic systems can form gels on standing. A gel is a colloidal solid having a network structure such that both solid and liquid components are highly interdispersed. The opposite phenomenon of thixotropy is *rheopexy*, where under a steady shear rate the viscosity goes up before reaching a maximum value.

III. MICELLAR AGGREGATION NUMBERS

Nuclear magnetic resonance (NMR) self-diffusion coefficients (Lindman, 1983), small-angle neutron scattering (SANS) (Cebula and Ottewill, 1982; Triolo et al., 1983; Corti et al., 1984), freezing point and vapor pressure methods (Herrington and Sahi, 1986), and fluorescent probes (Atik et al., 1979) have been used to calculate aggregation numbers of several different types of surfactants (Lianos and Zana, 1980, 1981, 1982; Zana, 1980; Lianos et al., 1983). Some aggregation numbers of surfactants are listed in Table 3.1.

From geometric considerations, the aggregation numbers n of micelles in aqueous media should increase rapidly with increase in the length of the hydrophobic group l_c of the surfactant molecule and decrease with increase in the cross-sectional area of the hydrophilic group a_0 or the volume of the hydrophobic group V_H. For example, in a spherical micelle in aqueous media, the surface area, $n \times a_0 = 4\pi(l_c + \Delta)^2$, or $n = 4\pi(l_c + \Delta)^2/a_0$, where Δ is the added length of the radius of the sphere due to the hydrophilic group (Lianos and Zana, 1982). Similarly, the volume of the hydrophobic core $n \times V_H = \frac{4}{3}\pi(l_c)^3$ or $n = \frac{4}{3}\pi(l_c)^3/V_H$.

TABLE 3.1 Aggregation Numbers of Some Surfactant Micelles

Compound	Solvent	Temp. (°C)	Aggregation Number	Reference
Anionics				
$C_8H_{17}SO_3^-Na^+$	H_2O	23	25	Tartar and Lelong (1955)
$(C_8H_{17}SO_3^-)_2Mg^{2+}$	H_2O	23	51	Tartar and Lelong (1955)
$C_{10}H_{21}SO_3^-Na^+$	H_2O	30	40	Tartar and Lelong (1955)
$(C_{10}H_{21}SO_3^-)_2Mg^{2+}$	H_2O	60	103	Tartar and Lelong (1955)
$C_{12}H_{25}SO_3^-Na^+$	H_2O	40	54	Tartar and Lelong (1955)
$(C_{12}H_{25}SO_3^-)_2Mg^{2+}$	H_2O	60	107	Tartar and Lelong (1955)
$C_{14}H_{29}SO_3^-Na^+$	H_2O	60	80	Tartar and Lelong (1955)
$C_{14}H_{29}SO_3^-Na^+$	0.01 *M* NaCl	23	138	Tartar and Lelong (1955)
$C_{10}H_{21}SO_4^-Na^+$	H_2O	23	50	Tartar and Lelong (1955)
$C_{12}H_{25}SO_4^-Na^+$	H_2O	25	80	Sowada (1994)
$C_{12}H_{25}SO_4^-Na^+$	0.1 *M* NaCl	25	112	Sowada (1994)
$C_{12}H_{25}SO_4^-Na^+$	0.2 *M* NaCl	25	118	Sowada (1994)
$C_{12}H_{25}SO_4^-Na^+$	0.4 *M* NaCl	25	126	Sowada (1994)
$C_6H_{13}OOCCH_2SO_3^-Na^+$	H_2O	25	16	Jobe and Reinsborough (1984)
$C_8H_{17}OOCCH_2SO_3^-Na^+$	H_2O	25	37, 42	Jobe and Reinsborough (1984)
$C_{10}H_{21}OOCCH_2SO_3^-Na^+$	H_2O	25	69, 71	Jobe and Reinsborough (1984)
$C_6H_{13}OOCCH_2CH(SO_3^-Na^+)COOC_6H_{13}$	H_2O	25	30, 36	Jobe and Reinsborough (1984)
$C_8H_{17}OOCCH_2CH(SO_3^-Na^+)COOC_8H_{17}$	H_2O	25	59, 56	Jobe and Reinsborough (1984)
$C_{10}H_{21}$-1-$\Phi SO_3^-Na^+$	H_2O (0.05 *M* conc.)	25	60	Binana-Limbele et al. (1991a)
$C_{10}H_{21}$-1-$\Phi SO_3^-Na^+$	0.1 *M* NaCl (0.05 *M* conc.)	25	78	Binana-Limbele et al. (1991a)
p-C_{10}-5-$\Phi SO_3^-Na^+$	H_2O (0.05 *M* conc.)	25	47	Binana-Limbele et al. (1991a)
p-C_{10}-5-$\Phi SO_3^-Na^+$	H_2O (0.1 *M* conc.)	25	76	Binana-Limbele et al. (1991a)
p-C_{10}-5-$\Phi SO_3^-Na^+$	0.1 *M* NaCl (0.1 *M* conc.)	25	81	Binana-Limbele et al. (1991a)
p-C_{12}-3-$\Phi SO_3^-Na^+$	H_2O (0.05 *M* conc.)	25	77	Binana-Limbele et al. (1991a)

(*Continued*)

TABLE 3.1 (*Continued*)

Compound	Solvent	Temp. (°C)	Aggregation Number	Reference
Cationics				
$C_{10}H_{21}N^+(CH_3)_3Br^-$	H_2O	20	39	Lianos and Zana (1981)
$C_{10}H_{21}N^+(CH_3)_3Cl^-$	H_2O	25	36	Sowada (1994)
$C_{12}H_{25}N^+(CH_3)_3Br^-$	H_2O (0.04 *M* conc.)	25	42	Rodenas et al. (1994)
$C_{12}H_{25}N^+(CH_3)_3Br^-$	H_2O (0.10 *M* conc.)	25	69	Rodenas et al. (1994)
$C_{12}H_{25}N^+(CH_3)_3Br^-$	0.02 *M* KBr (0.04 *M* conc.)	25	49	Rodenas et al. (1994)
$C_{12}H_{25}N^+(CH_3)_3Br^-$	0.08 *M* KBr (0.04 *M* conc.)	25	59	Rodenas et al. (1994)
$C_{12}H_{25}N^+(CH_3)_3Cl^-$	H_2O	25	50	Sowada (1994)
$[C_{12}H_{25}N^+(CH_3)_3]_2SO_4^{2-}$	H_2O	23	65	Tartar and Lelong (1955)
$C_{14}H_{29}N^+(CH_3)_3Br^-$	H_2O (1.05×10^{-1} *M* conc.)	5	131	Gorski and Kalus (2001)
$C_{14}H_{29}N^+(CH_3)_3Br^-$	H_2O (1.05×10^{-1} *M* conc.)	10	122	Gorski and Kalus (2001)
$C_{14}H_{29}N^+(CH_3)_3Br^-$	H_2O (1.05×10^{-1} *M* conc.)	20	106	Gorski and Kalus (2001)
$C_{14}H_{29}N^+(CH_3)_3Br^-$	H_2O (1.05×10^{-1} *M* conc.)	40	88	Gorski and Kalus (2001)
$C_{14}H_{29}N^+(CH_3)_3Br^-$	H_2O (1.05×10^{-1} *M* conc.)	60	74	Gorski and Kalus (2001)
$C_{14}H_{29}N^+(CH_3)_3Br^-$	H_2O (1.05×10^{-1} *M* conc.)	80	73	Gorski and Kalus (2001)
$C_{14}H_{29}N^+(C_2H_5)_3Br^-$	H_2O	20	55	Lianos and Zana (1982)
$C_{14}H_{29}N^+(C_4H_9)_3Br^-$	H_2O	20	35	Lianos and Zana (1982)
$C_{16}H_{33}N^+(CH_3)_3Br^-$	H_2O (0.005 *M* conc.)	25	44	Rodenas et al. (1994)
$C_{16}H_{33}N^+(CH_3)_3Br^-$	H_2O (0.021 *M* conc.)	25	75	Rodenas et al. (1994)
$C_{16}H_{33}N^+(CH_3)_3Br^-$	0.1 *M* KBr (0.005 *M* conc.)	25	57	Rodenas et al. (1994)
$C_{16}H_{33}N^+(CH_3)_3Br^-$	0.1 *M* KBr (0.021 *M* conc.)	25	71	Rodenas et al. (1994)
Zwitterionics				
$C_8H_{17}N^+(CH_3)_2CH_2COO^-$	H_2O	21	24	Tori and Nakagawa (1963a)
$C_8H_{17}CH(COO^-)N^+(CH_3)_3$	H_2O	21	31	Tori and Nakagawa (1963a)
$C_{12}H_{25}N^+(CH_3)_2CH_2COO^-$	H_2O	25	80–85	Chorro et al. (1996)

$C_{12}H_{25}N^+(CH_3)_2(CH_2)_3COO^-$	H_2O	25	55–56	Kamenka et al. (1995a)
$C_{12}H_{25}N^+(CH_3)_2(CH_2)_5COO^-$	H_2O	25	39–43	Kamenka et al. (1995a)
$C_{12}H_{25}N^+(CH_3)_2(CH_2)_3SO_3^-$	H_2O	25	59–67	Kamenka et al. (1995a)
Anionic–Cationic Salts				
$C_8H_{17}NH_3^+.C_2H_5COO^-$	C_6H_6	30	5 ± 1	Fendler et al. (1973a)
$C_8H_{17}NH_3^+.C_2H_5COO^-$	CCl_4	30	3 ± 1	Fendler et al. (1973a)
$C_8H_{17}NH_3^+.C_2H_5COO^-$	C_6H_6	30	3 ± 1	Fendler et al. (1973a)
$C_8H_{17}NH_3^+.C_3H_7COO^-$	CCl_4	30	4 ± 1	Fendler et al. (1973a)
$C_8H_{17}NH_3^+.C_5H_{11}COO^-$	C_6H_6	30	3 ± 1	Fendler et al. (1973a)
$C_8H_{17}NH_3^+.C_5H_{11}COO^-$	CCl_4	30	5 ± 1	Fendler et al. (1973a)
$C_8H_{17}NH_3^+.C_8H_{17}COO^-$	C_6H_6	30	3 ± 1	Fendler et al. (1973a)
$C_8H_{17}NH_3^+.C_8H_{17}COO^-$	CCl_4	30	5 ± 1	Fendler et al. (1973a)
$C_8H_{17}NH_3^+.C_{11}H_{23}COO^-$	C_6H_6	30	7 ± 1	Fendler et al. (1973a)
$C_8H_{17}NH_3^+.C_{13}H_{27}COO^-$	C_6H_6	30	3 ± 1	Fendler et al. (1973a)
$C_8H_{17}NH_3^+.C_{13}H_{27}COO^-$	CCl_4	30	3 ± 1	Fendler et al. (1973a)
$C_4H_9NH_3^+.C_2H_5COO^-$	C_6H_6	—	4	Fendler et al. (1973b)
$C_4H_9NH_3^+.C_2H_5COO^-$	CCl_4	—	3	Fendler et al. (1973b)
$C_6H_{13}NH_3^+.C_2H_5COO^-$	C_6H_6	—	7	Fendler et al. (1973b)
$C_6H_{13}NH_3^+.C_2H_5COO^-$	CCl_4	—	7	Fendler et al. (1973b)
$C_8H_{17}NH_3^+.C_2H_5COO^-$	C_6H_6	—	5	Fendler et al. (1973b)
$C_8H_{17}NH_3^+.C_2H_5COO^-$	CCl_4	—	5	Fendler et al. (1973b)
$C_{10}H_{21}NH_3^+.C_2H_5COO^-$	C_6H_6	—	5	Fendler et al. (1973b)
$C_{10}H_{21}NH_3^+.C_2H_5COO^-$	CCl_4	—	4	Fendler et al. (1973b)
Nonionics				
$C_8H_{17}O(C_2H_4O)_6H$	H_2O	18	30	Balmbra et al. (1964)
$C_8H_{17}O(C_2H_4O)_6H$	H_2O	30	41	Balmbra et al. (1964)
$C_8H_{17}O(C_2H_4O)_6H$	H_2O	40	51	Balmbra et al. (1964)
$C_8H_{17}O(C_2H_4O)_6H$	H_2O	60	210	Balmbra et al. (1964)
$C_{10}H_{21}O(C_2H_4O)_6H$	H_2O	35	260	Balmbra et al. (1964)

(*Continued*)

TABLE 3.1 ***(Continued)***

Compound	Solvent	Temp. (°C)	Aggregation Number	Reference
$C_{12}H_{25}O(C_2H_4O)_2H$	C_6H_6	—	34	Becher (1960)
$C_{12}H_{25}O(C_2H_4O)_6H$	H_2O	15	140	Balmbra et al. (1962)
$C_{12}H_{25}(OC_2H_4)_6OH$	H_2O	20	254–345	Lianos and Zana (1981)
$C_{12}H_{25}O(C_2H_4O)_6H$	H_2O	25	400	Balmbra et al. (1962)
$C_{12}H_{25}O(C_2H_4O)_6H$	H_2O	35	1,400	Balmbra et al. (1962)
$C_{12}H_{25}O(C_2H_4O)_6H$	H_2O	45	4,000	Balmbra et al. (1962)
$C_{12}H_{25}O(C_2H_4O)_8H$[a]	H_2O	25	123	Becher (1961)
$C_{12}H_{25}O(C_2H_4O)_{12}H$[a]	H_2O	25	81	Becher (1961)
$C_{12}H_{25}O(C_2H_4O)_{18}H$[a]	H_2O	25	51	Becher (1961)
$C_{12}H_{25}O(C_2H_4O)_{23}H$[a]	H_2O	25	40	Becher (1961)
$C_{13}H_{27}O(C_2H_4O)_6H$	C_6H_6	—	99	Becher (1960)
$C_{14}H_{29}O(C_2H_4O)_6H$	H_2O	35	7,500	Balmbra et al. (1964)
$C_{16}H_{33}O(C_2H_4O)_6H$	H_2O	34	16,600	Balmbra et al. (1964)
$C_{16}H_{33}O(C_2H_4O)_6H$	H_2O	25	2,430	Elworthy and MacFarlane (1963)
$C_{16}H_{33}O(C_2H_4O)_7H$	H_2O	25	594	Elworthy and MacFarlane (1963); Elworthy and McDonald (1964)
$C_{16}H_{33}O(C_2H_4O)_9H$	H_2O	25	219	Elworthy and MacFarlane (1963)
$C_{16}H_{33}O(C_2H_4O)_{12}H$	H_2O	25	152	Elworthy and MacFarlane (1963)
$C_{16}H_{33}O(C_2H_4O)_{21}H$	H_2O	25	70	Elworthy and MacFarlane (1963)
$C_9H_{19}C_6H_4O(C_2H_4O)_{10}H$[b]	H_2O	25	276	Schick et al. (1962)
$C_9H_{19}C_6H_4O(C_2H_4O)_{15}H$[b]	H_2O	25	80	Schick et al. (1962)

$C_9H_{19}C_6H_4O(C_2H_4O)_{15}H$[b]	0.5 *M* urea	25	82	Schick et al. (1962)
$C_9H_{19}C_6H_4O(C_2H_4O)_{15}H$[b]	0.86 *M* urea	25	83	Schick et al. (1962)
$C_9H_{19}C_6H_4O(C_2H_4O)_{20}H$[b]	H_2O	25	62	Schick et al. (1962)
$C_9H_{19}C_6H_4O(C_2H_4O)_{30}H$[b]	H_2O	25	44	Schick et al. (1962)
$C_9H_{19}C_6H_4O(C_2H_4O)_{50}H$[b]	H_2O	25	20	Schick et al. (1962)
$C_{10}H_{21}O(C_2H_4O)_8CH_3$	H_2O	30	83	Nakagawa et al. (1960)
$C_{10}H_{21}O(C_2H_4O)_8CH_3$	H_2O +2.3% *n*-decane	30	90	Nakagawa et al. (1960)
$C_{10}H_{21}O(C_2H_4O)_8CH_3$	H_2O +4.9% *n*-decane	30	105	Nakagawa et al. (1960)
$C_{10}H_{21}O(C_2H_4O)_8CH_3$	H_2O +3.4% *n*-decanol	30	89	Nakagawa et al. (1960)
$C_{10}H_{21}O(C_2H_4O)_8CH_3$	H_2O +8.5% *n*-decanol	30	109	Nakagawa et al. (1960)
$C_{10}H_{21}O(C_2H_4O)_{11}CH_3$	H_2O	30	65	Nakagawa et al. (1960)
α-Monocaprin	C_6H_6	—	42	Debye and Prins (1958)
α-Monolaurin	C_6H_6	—	73	Debye and Prins (1958)
α-Monomyristin	C_6H_6	—	86	Debye and Prins (1958)
α-Monopalmitin	C_6H_6	—	15	Debye and Prins (1958)
α-Monostearin	C_6H_6	—	11	Debye and Prins (1958)
Sucrose monolaurate	H_2O	0–60	52	Herrington and Sahi (1986)
Sucrose monooleate	H_2O	0–60	99	Herrington and Sahi (1986)

[a] Commercial material.
[b] Molecularly distilled commercial material.

In agreement with the geometric considerations mentioned above, aggregation numbers in aqueous solution increase with increase in the length of the hydrophobic group (greater l_c), decrease in the number of oxyethylene OE units in polyoxyethylenated nonionics (smaller a_0), and increase in the binding of the counterions to the micelle in ionics (smaller a_0). They decrease with increase in the size of the hydrophilic group (larger a_0). Surfactants in which the hydrophobic group is based on a dimethylsiloxane rather than on a hydrocarbon chain appear to have aggregation numbers less than 5 in aqueous solution (Schwarz and Reid, 1963), possibly because of the bulky dimethylsiloxane chain (large V_H).

Ionic surfactants containing a single long alkyl chain and zwitterionics (containing a single long alkyl chain) in which the electrical charges are not on adjacent atoms show aggregation numbers of less than 100 in aqueous solutions containing low or moderate concentrations of NaCl (<0.1 *M*), and these vary only slightly with the surfactant concentration up to about 0.1–0.3 *M* (Lianos and Zana, 1981). This is indicative of spherical micelle formation. At high salt content, however, n increases sharply with surfactant concentration (Mazer et al., 1976), with formation of rodlike cylindrical or disklike lamellar micelles. The formation of rodlike micelles and the sharp increase in aggregation number result in an increase in the viscosity of the aqueous solution (Kumar et al., 2002).

Ionic surfactants with two long (six or more carbons) alkyl chains have high V_H values relative to l_c, and probably do not form spherical micelles. They have values of n that increase with surfactant concentration, the increase becoming more pronounced with increase in the length of the chains. Some of these micellar solutions are in equilibrium with lamellar liquid crystal structures (Lianos et al., 1983).

The addition of neutral electrolyte to solutions of ionic surfactants in aqueous solution causes an increase in the aggregation number, presumably because of compression of the electrical double layer surrounding the ionic heads. The resulting reduction of their mutual repulsion in the micelle permits closer packing of the head groups (a_0 is reduced), with a consequent increase in n.

The addition of certain large anions, such as sodium salicylate (2-hydroxybenzoate), sodium *p*-toluenesulfonate, or sodium 3-hydroxynaphthalene-2-carboxylate, to aqueous solutions of quaternary cationic surfactants such as $C_{16}H_{33}N^+(CH_3)_3Br^-$ produces long, threadlike micelles (Shikata et al., 1987; Imae, 1990; Hassan and Yakhmi, 2000). Above a certain critical surfactant concentration, these threadlike micelles entangle to form highly viscous solutions. These long, threadlike micelles can act as drag reducers for aqueous solutions; that is, they reduce the turbulence of the solution flowing in tubes or pipes (Harwigsson and Hellsten, 1996; Zakin et al., 1998).

For zwitterionics of the betaine and sulfobetaine types, $C_{12}H_{25}N^+(CH_3)_2(CH_2)_mCOO^-$ and $C_{12}H_{25}N^+(CH_3)_2(CH_2)_3SO_3^-$, respectively, the micellar aggregation number varies very little with change in surfactant concentration or electrolyte content (Kamenka et al., 1995a).

For POE nonionics, n increases considerably with surfactant concentration in the range below 0.1 M, even in pure water, with aggregation numbers of several hundred or more, indicating that the micelles are not spherical in shape.

The effect of neutral electrolyte on the aggregation number of micelles of POE nonionics in aqueous solution is somewhat unclear, with both increases and decreases being observed on the addition of electrolyte (Binana-Limbele et al., 1991b). In either case, however, the effect appears to be small.

An increase in the temperature appears to cause a small decrease in the aggregation number in aqueous medium of ionics, presumably because a_0 is increased due to thermal agitation. For POE nonionics, there is a slow increase in n until the temperature reaches 40°C below the "cloud point" (Chapter 4, Section IIIA), the temperature at which the solution, on being heated, begins to show turbidity because of dehydration of the POE chains, and then starts rapidly to increase (Binana-Limbele et al., 1991a). Dehydration of the POE chains causes a decrease in a_0. It also appears to produce a sharp increase in the asymmetry of the nonionic micelle. Nonionic surfactants with cloud points above 100°C show no change in aggregation number at temperatures below 60°C.

If small amounts of hydrocarbons or long-chain polar compounds are added to an aqueous solution of a surfactant above its CMC, these normally water-insoluble materials may be solubilized in the micelles (Chapter 4). This solubilization generally causes an increase in the aggregation number of the micelle, and as the amount of material solubilized by the micelle increases, the aggregation continues to increase until the solubilization limit is reached.

There is much less information on aggregation numbers of micelles in nonaqueous solvents, and some of it is controversial. From the data available, the average aggregation number in nonpolar media increases with increase in dipole–dipole attraction or intermolecular bonding between the polar head groups and decreases with increase in the number of alkyl chains per surfactant molecule, the length of the chains, the steric requirements of the chain close to the polar head group, and the temperature (Ruckenstein and Nagarajan, 1980). The addition of water that is solubilized in the interior of a micelle in hydrocarbon medium has been shown to cause an increase in the aggregation number (Mathews and Hirschhorn, 1953), similar to the effect of hydrocarbon addition to micelles in aqueous medium. Temperature change from 25°C to 90°C had almost no effect on the aggregation numbers of some dialkylnaphthalenesulfonates in n-decane (Heilweil, 1964).

In polar solvents, such as chloroform or ethanol, either micellization does not occur or, if it does, the aggregation number is very small, presumably because the polar surfactant molecules can dissolve in the solvent without distorting its liquid structure significantly. As might be expected, in these solvents, surfactants have also almost no tendency to adsorb at the interfaces.

In addition to its effect on the value of l_c, mentioned above, the number of carbon atoms in the hydrophobic group may also affect the aggregation number in other ways (Nagarajan, 2002). For example, increase in the number of carbon atoms in the hydrophobic group decreases the value of the CMC (see Section IV). For ionic surfactants, this means a decrease in the ionic strength of the solution and a resulting increase in the value of a_0, producing a smaller aggregation number than expected from the increase in the value of l_c.

IV. FACTORS AFFECTING THE VALUE OF THE CMC IN AQUEOUS MEDIA

Since the properties of solutions of surface-active agents change markedly when micelle formation commences, many investigations have been concerned with determining values of the CMC in various systems, and a great deal of work has been done on elucidating the various factors that determine the CMC at which micelle formation becomes significant, especially in aqueous media. An extensive compilation of the CMCs of surfactants in aqueous media has been published (Mukerjee and Mysels, 1971). Some typical CMC values are listed in Table 3.2.

Among the factors known to affect the CMC in aqueous solution are (1) the structure of the surfactant, (2) the presence of added electrolyte in the solution, (3) the presence in the solution of various organic compounds, (4) the presence of a second liquid phase, and (5) temperature of the solution. Some examples of the effects of these factors are apparent from the data in Table 3.2.

A. Structure of the Surfactant

In general, the CMC in aqueous media decreases as the hydrophobic character of the surfactant increases.

1. The Hydrophobic Group In aqueous medium, the CMC decreases as the number of carbon atoms in the hydrophobic group increases to about 16, and a general rule for ionic surfactants is that the CMC is halved by the addition of one methylene group to a straight-chain hydrophobic group attached to a single terminal hydrophilic group. For nonionics and zwitterionics, the decrease with increase in the hydrophobic group is somewhat larger, an increase by two methylene units reducing the CMC to about one-tenth its previous value (compared to one-quarter in ionics). A phenyl group that is part of a hydrophobic group with a terminal hydrophilic group is equivalent to about three and one-half methylene groups. When the number of carbon atoms in a

TABLE 3.2 Critical Micelle Concentrations of Some Surfactants in Aqueous Media

Compound	Solvent	Temp. (°C)	CMC (*M*)	Reference
Anionics				
$C_{10}H_{21}OCH_2COO^-Na^+$	0.1 *M* NaCl, pH 10.5	30	2.8×10^{-3}	Tsubone and Rosen (2001)
$C_{12}H_{25}COO^-K^+$	H_2O, pH 10.5	30	1.2×10^{-2}	Tsubone and Rosen (2001)
$C_9H_{19}CONHCH_2COO^-Na^+$	H_2O	40	3.8×10^{-2}	Desai and Bahadur (1992)
$C_{11}H_{23}CONHCH_2COO^-Na^+$	H_2O	40	1.0×10^{-2}	Desai and Bahadur (1992)
$C_{11}H_{23}CONHCH_2COO^-Na^+$	0.1 *M* NaOH (aq.)	45	3.7×10^{-3}	Miyagishi et al. (1989)
$C_{11}H_{23}CON(CH_3)CH_2COO^-Na^+$	H_2O, pH 10.5	30	1.0×10^{-2}	Tsubone and Rosen (2001)
$C_{11}H_{23}CON(CH_3)CH_2COO^-Na^+$	0.1 *M* NaCl, pH 10.5	30	3.5×10^{-3}	Tsubone and Rosen (2001)
$C_{11}H_{23}CON(CH_3)CH_2CH_2COO^-Na^+$	H_2O, pH 10.5	30	7.6×10^{-3}	Tsubone and Rosen (2001)
$C_{11}H_{23}CON(CH_3)CH_2CH_2COO^-Na^+$	0.1 *M* NaCl, pH 10.5	30	2.7×10^{-3}	Tsubone and Rosen (2001)
$C_{11}H_{23}CONHCH(CH_3)COO^-Na^+$	0.1 *M* NaOH (aq.)	45	3.3×10^{-3}	Miyagishi et al. (1989)
$C_{11}H_{23}CONHCH(C_2H_5)COO^-Na^+$	0.1 *M* NaOH (aq.)	45	2.1×10^{-3}	Miyagishi et al. (1989)
$C_{11}H_{23}CONHCH[CH(CH_3)_2]COO^-Na^+$	0.1 *M* NaOH (aq.)	45	1.4×10^{-3}	Miyagishi et al. (1989)
$C_{11}H_{23}CONHCH[CH_2CH(CH_3)_2]COO^-Na^+$	0.1 *M* NaOH (aq.)	45	5.8×10^{-4}	Miyagishi et al. (1989)
$C_{13}H_{27}CONHCH_2COO^-Na^+$	H_2O	40	4.2×10^{-3}	Desai and Bahadur (1992)
$C_{15}H_{31}CONHCH[CH(CH_3)_2]COO^-Na^+$	H_2O	25	1.9×10^{-3}	Ohta et al. (2003)
$C_{15}H_{31}CONHCH[CH_2CH(CH_3)_2]COO^-Na^+$	H_2O	25	1.5×10^{-3}	Ohta et al. (2003)
$C_8H_{17}SO_3^-Na^+$	H_2O	40	1.6×10^{-1}	Klevens (1948)
$C_{10}H_{21}SO_3^-Na^+$	H_2O	10	4.8×10^{-2}	Dahanayake et al. (1986)
$C_{10}H_{21}SO_3^-Na^+$	H_2O	25	4.3×10^{-2}	Dahanayake et al. (1986)
$C_{10}H_{21}SO_3^-Na^+$	H_2O	40	4.0×10^{-2}	Dahanayake et al. (1986)
$C_{10}H_{21}SO_3^-Na^+$	0.1 *M* NaC1	10	2.6×10^{-2}	Dahanayake et al. (1986)
$C_{10}H_{21}SO_3^-Na^+$	0.1 *M* NaC1	25	2.1×10^{-2}	Dahanayake et al. (1986)
$C_{10}H_{21}SO_3^-Na^+$	0.1 *M* NaC1	40	1.8×10^{-2}	Dahanayake et al. (1986)
$C_{10}H_{21}SO_3^-Na^+$	0.5 M NaC1	10	7.9×10^{-3}	Dahanayake et al. (1986)

(*Continued*)

TABLE 3.2 (*Continued*)

Compound	Solvent	Temp. (°C)	CMC (*M*)	Reference
$C_{10}H_{21}SO_3^-Na^+$	0.5 *M* NaC1	25	7.3×10^{-3}	Dahanayake et al. (1986)
$C_{10}H_{21}SO_3^-Na^+$	0.5 *M* NaC1	40	6.5×10^{-3}	Dahanayake et al. (1986)
$C_{12}H_{25}SO_3^-Na^+$	H_2O	25	$1.2_4 \times 10^{-2}$	Dahanayake et al. (1986)
$C_{12}H_{25}SO_3^-Na^+$	H_2O	40	$1.1_4 \times 10^{-2}$	Dahanayake et al. (1986)
$C_{12}H_{25}SO_3^-Na^+$	0.1 *M* NaCI	25	2.5×10^{-3}	Dahanayake et al. (1986)
$C_{12}H_{25}SO_3^-Na^+$	0.1 *M* NaC1	40	2.4×10^{-3}	Dahanayake et al. (1986)
$C_{12}H_{25}SO_3^-Na^+$	0.5 *M* NaCI	40	7.9×10^{-3}	Dahanayake et al. (1986)
$C_{12}H_{25}SO_3^-Li^+$	H_2O	25	1.1×10^{-2}	Mohle et al. (1993)
$C_{12}H_{25}SO_3^-NH_4^+$	H_2O	25	8.9×10^{-3}	Mohle et al. (1993)
$C_{12}H_{25}SO_3^-K^+$	H_2O	25	9.3×10^{-3}	Mohle et al. (1993)
$C_{14}H_{29}SO_3^-Na^+$	H_2O	40	2.5×10^{-3}	Klevens (1948)
$C_{16}H_{33}SO_3^-Na^+$	H_2O	50	7.0×10^{-4}	Klevens (1948)
$C_8H_{17}SO_4^-Na^+$	H_2O	40	1.4×10^{-1}	Evans (1956)
$C_{10}H_{21}SO_4^-Na^+$	H_2O	40	3.3×10^{-2}	Evans (1956)
$C_{11}H_{23}SO_4^-Na^+$	H_2O	21	1.6×10^{-2}	Huisman (1964)
Branched $C_{12}H_{25}SO_4^-Na^+$	H_2O	25	1.42×10^{-2}	Varadaraj et al. (1992)
Branched $C_{12}H_{25}SO_4^-Na^+$	0.1 *M* NaC1	25	3.8×10^{-3}	Varadaraj et al. (1992)
$C_{12}H_{25}SO_4^-Na^+$	H_2O	25	8.2×10^{-3}	Elworthy and Mysels (1966)
$C_{12}H_{25}SO_4^-Na^+$	H_2O	40	8.6×10^{-3}	Flockhart (1961)
$C_{12}H_{25}SO_4^-Na^+$	"Hard river" water (I.S. = 6.6×10^{-3} *M*)[c]	25	$>1.58 \times 10^{-3}$	Rosen et al. (1996)
$C_{12}H_{25}SO_4^-Na^+$	0.1 *M* NaCl	21	5.6×10^{-3}	Huisman (1964)
$C_{12}H_{25}SO_4^-Na^+$	0.3 *M* NaCl	21	3.2×10^{-3}	Huisman (1964)
$C_{12}H_{25}SO_4^-Na^+$	0.1 *M* NaCl	25	1.62×10^{-3}	Huisman (1964)
$C_{12}H_{25}SO_4^-Na^+$	0.2 *M* NaCl (aq.)	25	8.3×10^{-4}	Sowada (1994)
$C_{12}H_{25}SO_4^-Na^+$	0.4 *M* NaCl (aq.)	25	5.2×10^{-4}	Sowada (1994)

$C_{12}H_{25}SO_4^-Na^+$	0.3 *M* urea	25	9.0×10^{-3}	Schick (1964)
$C_{12}H_{25}SO_4^-Na^+$	H_2O-cyclohexane	25	7.4×10^{-3}	Rehfeld (1967)
$C_{12}H_{25}SO_4^-Na^+$	H_2O-octane	25	8.1×10^{-3}	Rehfeld (1967)
$C_{12}H_{25}SO_4^-Na^+$	H_2O-decane	25	8.5×10^{-3}	Rehfeld (1967)
$C_{12}H_{25}SO_4^-Na^+$	H_2O-heptadecane	25	8.5×10^{-3}	Rehfeld (1967)
$C_{12}H_{25}SO_4^-Na^+$	H_2O-cyclohexane	25	7.9×10^{-3}	Rehfeld (1967)
$C_{12}H_{25}SO_4^-Na^+$	H_2O-carbon tetrachloride	25	6.8×10^{-3}	Rehfeld (1967)
$C_{12}H_{25}SO_4^-Na^+$	H_2O-benzene	25	6.0×10^{-3}	Rehfeld (1967)
$C_{12}H_{25}SO_4^-Na^+$	0.1 *M* NaCl (aq.)-heptane	20	1.4×10^{-3}	Vijayendran and Bursh (1979)
$C_{12}H_{25}SO_4^-Na^+$	0.1 *M* NaCl (aq.)-ethylbenzene	20	1.1×10^{-3}	Vijayendran and Bursh (1979)
$C_{12}H_{25}SO_4^-Na^+$	0.1 *M* NaCl (aq.)-ethylacetate	20	1.8×10^{-3}	Vijayendran and Bursh (1979)
$C_{12}H_{25}SO_4^-Li^+$	H_2O	25	8.9×10^{-3}	Mysels and Princen (1959)
$C_{12}H_{25}SO_4^-K^+$	H_2O	40	7.8×10^{-3}	Meguro and Kondo (1956)
$(C_{12}H_{25}SO_4^-)_2Ca^{2+}$	H_2O	70	3.4×10^{-3}	Corkill and Goodman (1962)
$C_{12}H_{25}SO_4^-N(CH_3)_4^+$	H_2O	25	5.5×10^{-3}	Mysels and Princen (1959)
$C_{12}H_{25}SO_4^-N(C_2H_5)_4^+$	H_2O	30	4.5×10^{-3}	Meguro and Kondo (1959)
$C_{12}H_{25}SO_4^-N(C_3H_7)_4^+$	H_2O	25	2.2×10^{-3}	Mukerjee (1967)
$C_{12}H_{25}SO_4^-N(C_4H_9)_4^+$	H_2O	30	1.3×10^{-3}	Meguro and Kondo (1959)
$C_{13}H_{27}SO_4^-Na^+$	H_2O	40	4.3×10^{-3}	Götte and Schwuger (1969)
$C_{14}H_{29}SO_4^-Na^+$	H_2O	25	2.1×10^{-3}	Lange and Schwuger (1968)
$C_{14}H_{29}SO_4^-Na^+$	H_2O	40	2.2×10^{-3}	Flockhart (1961)
$C_{15}H_{31}SO_4^-Na^+$	H_2O	40	1.2×10^{-3}	Götte and Schwuger (1969)
$C_{16}H_{33}SO_4^-Na^+$	H_2O	40	5.8×10^{-4}	Evans (1956)
$C_{13}H_{27}CH(CH_3)CH_2SO_4^-Na^+$	H_2O	40	8.0×10^{-4}	Götte and Schwuger (1969)
$C_{12}H_{25}CH(C_2H_5)CH_2SO_4^-Na^+$	H_2O	40	9.0×10^{-4}	Götte and Schwuger (1969)

(*Continued*)

TABLE 3.2 (*Continued*)

Compound	Solvent	Temp. (°C)	CMC (*M*)	Reference
$C_{11}H_{23}CH(C_3H_7)CH_2SO_4^-Na^+$	H_2O	40	1.1×10^{-3}	Götte and Schwuger (1969)
$C_{10}H_{21}CH(C_4H_9)CH_2SO_4^-Na^+$	H_2O	40	1.5×10^{-3}	Götte and Schwuger (1969)
$C_9H_{19}CH(C_5H_{11})CH_2SO_4^-Na^+$	H_2O	40	2×10^{-3}	Götte and Schwuger (1969)
$C_8H_{17}CH(C_6H_{13})CH_2SO_4^-Na^+$	H_2O	40	2.3×10^{-3}	Götte and Schwuger (1969)
$C_7H_{15}CH(C_7H_{15})CH_2SO_4^-Na^+$	H_2O	40	3×10^{-3}	Götte and Schwuger (1969)
$C_{12}H_{25}CH(SO_4^-Na^+)C_3H_7$	H_2O	40	1.7×10^{-3}	Evans (1956)
$C_{10}H_{21}CH(SO_4^-Na^+)C_5H_{11}$	H_2O	40	2.4×10^{-3}	Evans (1956)
$C_8H_{17}CH(SO_4^-Na^+)C_7H_{15}$	H_2O	40	4.3×10^{-3}	Evans (1956)
$C_{18}H_{37}SO_4^-Na^+$	H_2O	50	2.3×10^{-4}	Götte (1960)
$C_{10}H_{21}OC_2H_4SO_3^-Na^+$	H_2O	25	1.59×10^{-2}	Dahanayake et al. (1986)
$C_{10}H_{21}OC_2H_4SO_3^-Na^+$	0.1 *M* NaCl	25	5.5×10^{-3}	Dahanayake et al. (1986)
$C_{10}H_{21}OC_2H_4SO_3^-Na^+$	0.5 *M* NaCl	25	2.0×10^{-3}	Dahanayake et al. (1986)
$C_{12}H_{25}OC_2H_4SO_4^-Na^+$	H_2O	25	3.9×10^{-3}	Dahanayake et al. (1986)
$C_{12}H_{25}OC_2H_4SO_4^-Na^+$	"Hard river" water (I.S. = 6.6×10^{-3} *M*)[c]	25	8.1×10^{-4}	Rosen et al. (1996)
$C_{12}H_{25}OC_2H_4SO_4^-Na^+$	0.1 *M* NaCl	25	4.3×10^{-4}	Dahanayake et al. (1986)
$C_{12}H_{25}OC_2H_4SO_4^-Na^+$	0.5 *M* NaCl	25	1.3×10^{-4}	Dahanayake et al. (1986)
$C_{12}H_{25}(OC_2H_4)_2SO_4^-Na^+$	H_2O	10	3.1×10^{-3}	Dahanayake et al. (1986)
$C_{12}H_{25}(OC_2H_4)_2SO_4^-Na^+$	H_2O	25	2.9×10^{-3}	Dahanayake et al. (1986)
$C_{12}H_{25}(OC_2H_4)_2SO_4^-Na^+$	H_2O	40	2.8×10^{-3}	Dahanayake et al. (1986)
$C_{12}H_{25}(OC_2H_4)_2SO_4^-Na^+$	"Hard river" water (I.S. = 6.6×10^{-3} *M*)[c]	25	5.5×10^{-4}	Rosen et al. (1996)
$C_{12}H_{25}(OC_2H_4)_2SO_4^-Na^+$	0.1 *M* NaCl	10	3.2×10^{-4}	Dahanayake et al. (1986)
$C_{12}H_{25}(OC_2H_4)_2SO_4^-Na^+$	0.1 *M* NaCl	25	2.9×10^{-4}	Dahanayake et al. (1986)
$C_{12}H_{25}(OC_2H_4)_2SO_4^-Na^+$	0.1 *M* NaCl	40	2.8×10^{-4}	Dahanayake et al. (1986)
$C_{12}H_{25}(OC_2H_4)_2SO_4^-Na^+$	0.5 *M* NaCl	10	1.1×10^{-4}	Dahanayake et al. (1986)

$C_{12}H_{25}(OC_2H_4)_2SO_4^-Na^+$	0.5 *M* NaCl	25	1.0×10^{-4}	Dahanayake et al. (1986)
$C_{12}H_{25}(OC_2H_4)_2SO_4^-Na^+$	0.5 *M* NaCl	40	1.0×10^{-4}	Dahanayake et al. (1986)
$C_{12}H_{25}(OC_2H_4)_3SO_4^-Na^+$	H_2O	50	2.0×10^{-3}	Götte (1960)
$C_{12}H_{25}(OC_2H_4)_4SO_4^-Na^+$	H_2O	50	1.3×10^{-3}	Götte (1960)
$C_{16}H_{33}(OC_2H_4)_5SO_4^-Na^+$	H_2O	25	2.5×10^{-5}	Varadaraj et al. (1991a)
$C_8H_{17}CH(C_6H_{13})CH_2(OC_2H_4)_5SO_4^-Na^+$	H_2O	25	8.6×10^{-5}	Varadaraj et al. (1991a)
$C_6H_{13}OOCCH_2SO_3^-Na^+$	H_2O	25	1.7×10^{-1}	Jobe and Reinsborough (1984)
$C_8H_{17}OOCCH_2SO_3^-Na^+$	H_2O	25	6.6×10^{-2}	Jobe and Reinsborough (1984)
$C_{10}H_{21}OOCCH_2SO_3^-Na^+$	H_2O	25	2.2×10^{-2}	Jobe and Reinsborough (1984)
$C_8H_{17}OOC(CH_2)_2SO_3^-Na^+$	H_2O	30	4.6×10^{-2}	Hikota et al. (1970)
$C_{10}H_{21}OOC(CH_2)_2SO_3^-Na^+$	H_2O	30	1.1×10^{-2}	Hikota et al. (1970)
$C_{12}H_{25}OOC(CH_2)_2SO_3^-Na^+$	H_2O	30	2.2×10^{-3}	Hikota et al. (1970)
$C_{14}H_{29}OOC(CH_2)_2SO_3^-Na^+$	H_2O	40	9×10^{-4}	Hikota et al. (1970)
$C_4H_9OOCCH_2CH(SO_3^-Na^+)COOC_4H_9$	H_2O	25	2.0×10^{-1}	Williams et al. (1957)
$C_5H_{11}OOCH_2(SO_3^-Na^+)COOC_5H_{11}$	H_2O	25	5.3×10^{-2}	Williams et al. (1957)
$C_6H_{13}OOCH_2CH(SO_3^-Na^+)COOC_6H_{13}$	H_2O	25	1.4×10^{-2}	Jobe and Reinsborough (1984)
$C_4H_9CH(C_2H_5)CH_2OOCCH_2CH(SO_3^-Na^+)$ $COOCH_2CH(C_2H_5)C_4H_9$	H_2O	25	2.5×10^{-3}	Williams et al. (1957)
$C_8H_{17}OOCCH_2CH(SO_3^-Na^+)COOC_8H_{17}$	H_2O	25	9.1×10^{-4}	Nave et al. (2000)
$C_{12}H_{25}CH(SO_3^-Na^+)COOCH_3$	H_2O	13	2.8×10^{-3}	Ohbu (1998)
$C_{12}H_{25}CH(SO_3^-Na^+)COOC_2H_5$	H_2O	25	2.25×10^{-3}	Ohbu (1998)
$C_{12}H_{25}CH(SO_3^-Na^+)COOC_4H_9$	H_2O	25	1.35×10^{-3}	Ohbu (1998)
$C_{14}H_{29}CH(SO_3^-Na^+)COOCH_3$	H_2O	23	7.3×10^{-4}	Ohbu (1998)
$C_{16}H_{33}CH(SO_3^-Na^+)COOCH_3$	H_2O	33	1.8×10^{-4}	Ohbu (1998)
$C_{11}H_{23}CON(CH_3)CH_2CH_2SO_4^-Na^+$	H_2O pH 10.5	30	8.9×10^{-3}	Tsubone and Rosen (2001)
$C_{11}H_{23}CON(CH_3)CH_2CH_2SO_4^-Na^+$	0.1 *M* NaCl, pH 10.5	30	1.6×10^{-3}	Tsubone and Rosen (2001)
$C_{12}H_{25}NHCOCH_2SO_4^-Na^+$	H_2O	35	5.2×10^{-3}	Mizushima et al. (1999)
$C_{12}H_{25}NHCO(CH_2)_3SO_4^-Na^+$	H_2O	35	4.4×10^{-3}	Mizushima et al. (1999)
p-$C_8H_{17}C_6H_4SO_3^-Na^+$	H_2O	35	1.5×10^{-2}	Gershman (1957)

(*Continued*)

TABLE 3.2 (*Continued*)

Compound	Solvent	Temp. (°C)	CMC (*M*)	Reference
$p\text{-}C_{10}H_{21}C_6H_4SO_3^-Na^+$	H_2O	50	3.1×10^{-3}	Gershman (1957)
$C_{10}H_{21}\text{-}2\text{-}C_6H_4SO_3^-Na^+$	H_2O	30	4.6×10^{-3}	Van Os et al. (1991)
$C_{10}H_{21}\text{-}3\text{-}C_6H_4SO_3^-Na^+$	H_2O	30	6.1×10^{-3}	Van Os et al. (1991)
$C_{10}H_{21}\text{-}5\text{-}C_6H_4SO_3^-Na^+$	H_2O	30	8.2×10^{-3}	Van Os et al. (1991)
$C_{11}H_{23}\text{-}2\text{-}C_6H_4SO_3^-Na^+$	H_2O	35	2.5×10^{-3}	Zhu et al. (1998)
$C_{11}H_{23}\text{-}2\text{-}C_6H_4SO_3^-Na^+$	"Hard river" water (I.S. = 6.6×10^{-3} *M*)	30	2.5×10^{-4}	Zhu et al. (1998)
$p\text{-}C_{12}H_{25}C_6H_4SO_3^-Na^+$	H_2O	60	1.2×10^{-3}	Gershman (1957)
$C_{12}H_{25}C_6H_4SO_3^-Na^{+d}$	0.1 *M* NaCl	25	1.6×10^{-4}	Murphy et al. (1990)
$C_{12}H_{25}\text{-}2\text{-}C_6H_4SO_3^-Na^+$	H_2O	30	1.2×10^{-3}	Zhu et al. (1998)
$C_{12}H_{25}\text{-}2\text{-}C_6H_4SO_3^-Na^+$	"Hard river" water (I.S. = 6.6×10^{-3} *M*)	30	6.3×10^{-5}	Zhu et al. (1998)
$C_{12}H_{25}\text{-}3\text{-}C_6H_4SO_3^-Na^+$	H_2O	30	2.4×10^{-3}	Van Os et al. (1991)
$C_{12}H_{25}\text{-}5\text{-}C_6H_4SO_3^-Na^+$	H_2O	30	3.2×10^{-3}	Zhu et al. (1998)
$C_{12}H_{25}\text{-}5\text{-}C_6H_4SO_3^-Na^+$	"Hard river" water (I.S. = 6.6×10^{-3} *M*)	30	4.6×10^{-4}	Zhu et al. (1998)
$C_{13}H_{27}\text{-}2\text{-}C_6H_4SO_3^-Na^+$	H_2O	35	7.2×10^{-4}	Zhu et al. (1998)
$C_{13}H_{27}\text{-}2\text{-}C_6H_4SO_3^-Na^+$	"Hard river" water (I.S. = 6.6×10^{-3} *M*)	30	1.1×10^{-5}	Zhu et al. (1998)
$C_{13}H_{27}\text{-}5\text{-}C_6H_4SO_3^-Na^+$	H_2O	30	7.6×10^{-4}	Zhu et al. (1998)
$C_{13}H_{27}\text{-}5\text{-}C_6H_4SO_3^-Na^+$	"Hard river" water (I.S. = 6.6×10^{-3} *M*)	30	8.3×10^{-5}	Zhu et al. (1998)
$C_{16}H_{33}\text{-}7\text{-}C_6H_4SO_3^-Na^+$	H_2O	45	5.1×10^{-5}	Lascaux et al. (1983)
$C_{16}H_{33}\text{-}7\text{-}C_6H_4SO_3^-Na^+$	0.051 *M* NaCl	45	3.2×10^{-6}	Lascaux et al. (1983)

Fluorinated Anionics				
$C_7F_{15}COO^-K^+$	H_2O	25	2.9×10^{-2}	Shinoda and Katsura (1964)
$C_7F_{15}COO^-Na^+$	H_2O	25	3.0×10^{-2}	Shinoda and Hirai (1977)
$C_7F_{15}COO^-Li^+$	H_2O	25	3.3×10^{-2}	Muzzalupo et al. (1995)
$(CF_3)_2CF(CF_2)_4COO^-Na^+$	H_2O	25	3.0×10^{-2}	Shinoda and Hirai (1977)
$C_8F_{17}COO^-Na^+$	H_2O	35	1.1×10^{-2}	Nakano et al. (2002)
$C_8F_{17}COO^-Li^+$	H_2O	25	4.9×10^{-3}	Muzzalupo et al. (1995)
$C_8F_{17}SO_3^-Li^+$	H_2O	25	6.3×10^{-3}	Shinoda and Hirai (1977)
$C_4F_9CH_2OOCCH(SO_3^-Na^+)CH_2COOCH_2C_4F_9$	H_2O	30	1.6×10^{-3}	Downer et al. (1999)
Cationics				
$C_8H_{17}N^+(CH_3)_3Br^-$	H_2O	25	1.4×10^{-1}	Klevens (1948)
$C_{10}H_{21}N^+(CH_3)_3Br^-$	H_2O	25	6.8×10^{-2}	Klevens (1948)
$C_{10}H_{21}N^+(CH_3)_3Br^-$	0.1 *M* NaCl	25	$4.2_7 \times 10^{-2}$	Li et al. (2001)
$C_{10}H_{21}N^+(CH_3)_3Cl^-$	H_2O	25	6.8×10^{-2}	Sowada (1994)
$C_{12}H_{25}N^+(CH_3)_3Br^-$	H_2O	25	1.6×10^{-2}	Klevens (1948)
$C_{12}H_{25}N^+(CH_3)_3Br^-$	"Hard river" water (I.S. = 6.6×10^{-3} *M*)	25	$1.2_6 \times 10^{-2}$	Rosen et al. (1996)
$C_{12}H_{25}N^+(CH_3)_3Br^-$	0.01 *M* NaBr	25	1.2×10^{-2}	Tanaka and Ikeda (1991)
$C_{12}H_{25}N^+(CH_3)_3Br^-$	0.1 *M* NaBr	25	4.2×10^{-3}	Tanaka and Ikeda (1991)
$C_{12}H_{25}N^+(CH_3)_3Br^-$	0.5 *M* NaBr	31.5	1.9×10^{-3}	Anacker and Ghose (1963)
$C_{12}H_{25}N^+(CH_3)_3Cl^-$	H_2O	25	2.0×10^{-2}	Osugi et al. (1965)
$C_{12}H_{25}N^+(CH_3)_3Cl^-$	0.1 *M* NaCl	25	$5.7_6 \times 10^{-3}$	Li et al. (2001)
$C_{12}H_{25}N^+(CH_3)_3Cl^-$	0.5 *M* NaCl	31.5	3.8×10^{-3}	Anacker and Ghose (1963)
$C_{12}H_{25}N^+(CH_3)_3F^-$	0.5 *M* NaF	31.5	8.4×10^{-3}	Anacker and Ghose (1963)
$C_{12}H_{25}N^+(CH_3)_3NO_3^-$	0.5 *M* $NaNO_3$	31.5	8×10^{-4}	Anacker and Ghose (1963)
$C_{14}H_{29}N^+(CH_3)_3Br^-$	H_2O	25	3.6×10^{-3}	Lianos and Zana (1982)
$C_{14}H_{29}N^+(CH_3)_3Br^-$	"Hard river" water (I.S. = 6.6×10^{-3} *M*)	25	$2.4_5 \times 10^{-3}$	Rosen et al. (1996)
$C_{14}H_{29}N^+(CH_3)_3Br^-$	H_2O	40	4.2×10^{-3}	Gorski and Kalus (2001)

(*Continued*)

TABLE 3.2 (*Continued*)

Compound	Solvent	Temp. (°C)	CMC (*M*)	Reference
$C_{14}H_{29}N^+(CH_3)_3Br^-$	H_2O	60	5.5×10^{-3}	Gorski and Kalus (2001)
$C_{14}H_{29}N^+(CH_3)_3Cl^-$	H_2O	25	4.5×10^{-3}	Hover and Marmo (1961)
$C_{16}H_{33}N^+(CH_3)_3Br^-$	H_2O	25	9.8×10^{-4}	Okuda et al. (1987)
$C_{16}H_{33}N^+(CH_3)_3Br^-$	0.001 *M* KCl	30	5×10^{-4}	Varjara and Dixit (1996)
$C_{16}H_{33}N^+(CH_3)_3Cl^-$	H_2O	30	1.3×10^{-3}	Raston (1947)
$C_{18}H_{37}N^+(CH_3)_3Br^-$	H_2O	40	3.4×10^{-4}	Swanson-Vethamuthu et al. (1998)
$C_{10}H_{21}Pyr^+Br^{-b}$	H_2O	25	4.4×10^{-2}	Skerjanc et al. (1999)
$C_{10}H_{21}Pyr^+Br^{-b}$	H_2O	25	6.3×10^{-2}	Mehrian et al. (1993)
$C_{11}H_{23}Pyr^+Br^{-b}$	H_2O	25	2.1×10^{-2}	Skerjanc et al. (1999)
$C_{12}H_{25}Pyr^+Br^{-b}$	H_2O	10	$1.1_7 \times 10^{-2}$	Rosen et al. (1982b)
$C_{12}H_{25}Pyr^+Br^{-b}$	H_2O	25	$1.1_4 \times 10^{-2}$	Rosen et al. (1982b)
$C_{12}H_{25}Pyr^+Br^{-b}$	H_2O	40	$1.1_2 \times 10^{-2}$	Rosen et al. (1982b)
$C_{12}H_{25}Pyr^+Br^{-b}$	0.1 *M* NaBr	10	$2.7_5 \times 10^{-3}$	Rosen et al. (1982b)
$C_{12}H_{25}Pyr^+Br^{-b}$	0.1 *M* NaBr	25	$2.7_5 \times 10^{-3}$	Rosen et al. (1982b)
$C_{12}H_{25}Pyr^+Br^{-b}$	0.1 *M* NaBr	40	$2.8_5 \times 10^{-3}$	Rosen et al. (1982b)
$C_{12}H_{25}Pyr^+Br^{-b}$	0.5 *M* NaBr	10	$1.0_7 \times 10^{-3}$	Rosen et al. (1982b)
$C_{12}H_{25}Pyr^+Br^{-b}$	0.5 *M* NaBr	25	$1.0_8 \times 10^{-3}$	Rosen et al. (1982b)
$C_{12}H_{25}Pyr^+Br^{-b}$	0.5 *M* NaBr	40	$1.1_6 \times 10^{-3}$	Rosen et al. (1982b)
$C_{12}H_{25}Pyr^+Cl^{-b}$	H_2O	10	$1.7_5 \times 10^{-2}$	Rosen et al. (1982b)
$C_{12}H_{25}Pyr^+Cl^{-b}$	H_2O	25	1.7×10^{-2}	Rosen et al. (1982b)
$C_{12}H_{25}Pyr^+Cl^{-b}$	H_2O	40	1.7×10^{-2}	Rosen et al. (1982b)
$C_{12}H_{25}Pyr^+Cl^{-b}$	0.1 *M* NaCl	10	5.5×10^{-3}	Rosen et al. (1982b)
$C_{12}H_{25}Pyr^+Cl^{-b}$	0.1 *M* NaCl	25	4.8×10^{-3}	Rosen et al. (1982b)
$C_{12}H_{25}Pyr^+Cl^{-b}$	0.1 *M* NaCl	40	4.5×10^{-3}	Rosen et al. (1982b)
$C_{12}H_{25}Pyr^+Cl^{-b}$	0.5 *M* NaCl	10	1.9×10^{-3}	Rosen et al. (1982b)

$C_{12}H_{25}Pyr^+Cl^{-b}$	0.5 *M* NaCl	25	$1.7_8 \times 10^{-3}$	Rosen et al. (1982b)
$C_{12}H_{25}Pyr^+Cl^{-b}$	0.5 *M* NaCl	40	$1.7_8 \times 10^{-3}$	Rosen et al. (1982b)
$C_{12}H_{25}Pyr^+I^{-b}$	H_2O	25	5.3×10^{-3}	Mandru (1972)
$C_{13}H_{27}Pyr^+Br^{-b}$	H_2O	25	5.3×10^{-3}	Skerjanc et al. (1999)
$C_{14}H_{29}Pyr^+Br^{-b}$	H_2O	25	2.7×10^{-3}	Skerjanc et al. (1999)
$C_{14}H_{29}Pyr^+Cl^{-b}$	H_2O	25	3.5×10^{-3}	Mehrian et al. (1993)
$C_{14}H_{29}Pyr^+Cl^{-b}$	0.1 *M* NaCl	25	4×10^{-4}	Mehrian et al. (1993)
$C_{15}H_{31}Pyr^+Br^{-b}$	H_2O	25	1.3×10^{-3}	Skerjanc et al. (1999)
$C_{16}H_{33}Pyr^+Br^{-b}$	H_2O	25	6.4×10^{-4}	Skerjanc et al. (1999)
$C_{16}H_{33}Pyr^+Cl^-$	H_2O	25	9.0×10^{-4}	Hartley (1938)
$C_{18}H_{37}Pyr^+Cl^-$	H_2O	25	2.4×10^{-4}	Evers and Kraus (1948)
$C_{12}H_{25}N^+(C_2H_5)(CH_3)_2Br^-$	H_2O	25	1.4×10^{-2}	Lianos et al. (1983)
$C_{12}H_{25}N^+(C_4H_9)(CH_3)_2Br^-$	H_2O	25	7.5×10^{-3}	Lianos et al. (1983)
$C_{12}H_{25}N^+(C_6H_{13})(CH_3)_2Br^-$	H_2O	25	3.1×10^{-3}	Lianos et al. (1983)
$C_{12}H_{25}N^+(C_8H_{17})(CH_3)_3Br^-$	H_2O	25	1.1×10^{-3}	Lianos et al. (1983)
$C_{14}H_{29}N^+(C_2H_5)_3Br^-$	H_2O	25	3.1×10^{-3}	Lianos and Zana (1982)
$C_{14}H_{29}N^+(C_3H_7)_3Br^-$	H_2O	25	2.1×10^{-3}	Venable and Nauman (1964); Lianos and Zana (1982)
$C_{14}H_{29}N^+(C_4H_9)_3Br^-$	H_2O	25	1.2×10^{-3}	Lianos and Zana (1982)
$C_{10}H_{21}N^+(CH_2C_6H_5)(CH_3)_2Cl^-$	H_2O	25	3.9×10^{-2}	de Castillo et al. (2000)
$C_{12}H_{25}N^+(CH_2C_6H_5)(CH_3)_2Cl^-$	H_2O	25	8.8×10^{-3}	Rodriguez and Czapkiewicz (1995)
$C_{14}H_{29}N^+(CH_2C_6H_5)(CH_3)_2Cl^-$	H_2O	25	2.0×10^{-3}	Rodriguez and Czapkiewicz (1995)
$C_{12}H_{25}NH_2^+CH_2CH_2OH^-Cl^-$	H_2O	25	4.5×10^{-2}	Omar and Abdel-Khalek (1997)
$C_{12}H_{25}N^+H(CH_2CH_2OH)_2Cl^-$	H_2O	25	3.6×10^{-2}	Omar and Abdel-Khalek (1997)
$C_{12}H_{25}N^+H(CH_2CH_2OH)_3Cl^-$	H_2O	25	2.5×10^{-2}	Omar and Abdel-Khalek (1997)
$(C_{10}H_{21})_2N^+(CH_3)_2Br^-$	H_2O	25	$1.8_5 \times 10^{-3}$	Lianos et al. (1983)
$(C_{12}H_{25})_2N^+(CH_3)_2Br^-$	H_2O	25	$1.7_6 \times 10^{-4}$	Lianos et al. (1983)

(*Continued*)

TABLE 3.2 (*Continued*)

Compound	Solvent	Temp. (°C)	CMC (*M*)	Reference
Anionic–Cationic Salts				
$C_6H_{13}SO_4^- \bullet {}^+N(CH_3)_3C_6H_{13}$	H_2O	25	1.1×10^{-1}	Corkill et al. (1966)
$C_6H_{13}SO_4^- \bullet {}^+N(CH_3)_3C_8H_{17}$	H_2O	25	2.9×10^{-2}	Lange and Schwuger (1971)
$C_8H_{17}SO_4^- \bullet {}^+N(CH_3)_3C_6H_{13}$	H_2O	25	1.9×10^{-2}	Lange and Schwuger (1971)
$C_4H_9SO_4^- \bullet {}^+N(CH_3)_3C_{10}H_{21}$	H_2O	25	1.9×10^{-2}	Lange and Schwuger (1971)
$CH_3SO_4^- \bullet {}^+N(CH_3)_3C_{12}H_{25}$	H_2O	25	1.3×10^{-2}	Lange and Schwuger (1971)
$C_2H_5SO_4^- \bullet {}^+N(CH_3)_3C_{12}H_{25}$	H_2O	25	9.3×10^{-3}	Lange and Schwuger (1971)
$C_{10}H_{21}SO_4^- \bullet {}^+N(CH_3)_3C_4H_9$	H_2O	25	9.3×10^{-3}	Lange and Schwuger (1971)
$C_8H_{17}SO_4^- \bullet {}^+N(CH_3)_3C_8H_{17}$	H_2O	25	7.5×10^{-3}	Corkill et al. (1965)
$C_4H_9SO_4^- \bullet {}^+N(CH_3)_3C_{12}H_{25}$	H_2O	25	5.0×10^{-3}	Lange and Schwuger (1971)
$C_6H_{13}SO_4^- \bullet {}^+N(CH_3)_3C_{12}H_{25}$	H_2O	25	2.0×10^{-3}	Lange and Schwuger (1971)
$C_{10}H_{21}SO_4^- \bullet {}^+N(CH_3)_3C_{12}H_{21}$	H_2O	25	4.6×10^{-4}	Corkill et al. (1963a)
$C_8H_{17}SO_4^- \bullet {}^+N(CH_3)_3C_{12}H_{25}$	H_2O	25	5.2×10^{-4}	Lange and Schwuger (1971)
$C_{12}H_{25}SO_4^- \bullet {}^+N(CH_3)_3C_{12}H_{25}$	H_2O	25	4.6×10^{-5}	Lange and Schwuger (1971)
Zwitterionics				
$C_8H_{17}N^+(CH_3)_2CH_2COO^-$	H_2O	27	2.5×10^{-1}	Tori and Nakagawa (1963a)
$C_{10}H_{21}N^+(CH_3)_2CH_2COO^-$	H_2O	23	1.8×10^{-2}	Beckett and Woodward (1963)
$C_{12}H_{25}N^+(CH_3)_2CH_2COO^-$	H_2O	25	2.0×10^{-3}	Chevalier et al. (1991)
$C_{12}H_{25}N^+(CH_3)_2CH_2COO^-$	0.1 *M* NaCl	25	1.6×10^{-3}	Zajac et al. (1997)
$C_{14}H_{29}N^+(CH_3)_2CH_2COO^-$	H_2O	25	2.2×10^{-4}	Zajac et al. (1997)
$C_{16}H_{33}N^+(CH_3)_2CH_2COO^-$	H_2O	23	2.0×10^{-5}	Beckett and Woodward (1963)
$C_{12}H_{25}N^+(CH_3)_2(CH_2)_3COO^-$	H_2O	25	4.6×10^{-3}	Zajac et al. (1997)
$C_{12}H_{25}N^+(CH_3)_2(CH_2)_5COO^-$	H_2O	25	2.6×10^{-3}	Chevalier et al. (1991)
$C_{12}H_{25}N^+(CH_3)_2(CH_2)_7COO^-$	H_2O	25	1.5×10^{-3}	Chevalier et al. (1991)
$C_8H_{17}CH(COO^-)N^+(CH_3)_3$	H_2O	27	9.7×10^{-2}	Tori and Nakagawa (1963a)
$C_8H_{17}CH(COO^-)N^+(CH_3)_3$	H_2O	60	8.6×10^{-2}	Tori and Nakagawa (1963b)

$C_{10}H_{21}CH(COO^-)N^+(CH_3)_3$	H_2O	27	1.3×10^{-2}	Tori and Nakagawa (1963b)
$C_{12}H_{25}CH(COO^-)N^+(CH_3)_3$	H_2O	27	1.3×10^{-3}	Tori and Nakagawa (1963b)
p-$C_{12}H_{25}Pyr^+COO^{-b}$	H_2O	50	1.9×10^{-3}	Amrhar et al. (1994)
m-$C_{12}H_{25}Pyr^+COO^{-b}$	H_2O	50	1.5×10^{-3}	Amrhar et al. (1994)
$C_{10}H_{21}CH(Pyr^+)COO^{-b}$	H_2O	25	5.2×10^{-3}	Zhao and Rosen (1984)
$C_{12}H_{25}CH(Pyr^+)COO^{-b}$	H_2O	25	6.0×10^{-4}	Zhao and Rosen (1984)
$C_{14}H_{29}CH(Pyr^+)COO^{-b}$	H_2O	40	7.4×10^{-5}	Zhao and Rosen (1984)
$C_{10}H_{21}N^+(CH_3)(CH_2C_6H_5)\ CH_2COO^-$	H_2O, pH 5.5–5.9	25	5.3×10^{-3}	Dahanayake and Rosen (1984)
$C_{10}H_{21}N^+(CH_3)(CH_2C_6H_5)CH_2COO^-$	H_2O, pH 5.5–5.9	40	4.4×10^{-3}	Dahanayake and Rosen (1984)
$C_{12}H_{25}N^+(CH_3)(CH_2C_6H_5)CH_2COO^-$	H_2O, pH 5.5–5.9	25	5.5×10^{-4}	Dahanayake and Rosen (1984)
$C_{12}H_{25}N^+(CH_3)(CH_2C_6H_5)CH_2COO^-$	0.1 *M* NaCl, pH 5.7	25	4.2×10^{-4}	Rosen and Sulthana (2001)
$C_{12}H_{25}N^+(CH_3)(CH_2C_6H_5)CH_2COO^-$	H_2O-cyclohexane	25	3.7×10^{-4}	Murphy and Rosen (1988)
$C_{12}H_{25}N^+(CH_3)(CH_2C_6H_5)CH_2COO^-$	H_2O-isooctane	25	4.2×10^{-4}	Murphy and Rosen (1988)
$C_{12}H_{25}N^+(CH_3)(CH_2C_6H_5)CH_2COO^-$	H_2O-heptane	25	4.4×10^{-4}	Murphy and Rosen (1988)
$C_{12}H_{25}N^+(CH_3)(CH_2C_6H_5)CH_2COO^-$	H_2O-dodecane	25	4.9×10^{-4}	Murphy and Rosen (1988)
$C_{12}H_{25}N^+(CH_3)(CH_2C_6H_5)CH_2COO^-$	H_2O-heptamethylnonane	25	5.0×10^{-4}	Murphy and Rosen (1988)
$C_{12}H_{25}N^+(CH_3)(CH_2C_6H_5)CH_2COO^-$	H_2O-hexadecane	25	5.3×10^{-4}	Murphy and Rosen (1988)
$C_{12}H_{25}N^+(CH_3)(CH_2C_6H_5)CH_2COO^-$	H_2O-toluene	25	1.9×10^{-4}	Murphy and Rosen (1988)
$C_{12}H_{25}N^+(CH_3)(CH_2C_6H_5)CH_2COO^-$	0.1 *M* NaBr, pH 5.9	25	3.8×10^{-4}	Zhu and Rosen (1985)
$C_{10}H_{21}N^+(CH_3)(CH_2C_6H_5)CH_2CH_2SO_3^-$	H_2O, pH 5.5–5.9	40	4.6×10^{-3}	Dahanayake and Rosen (1984)
$C_{12}H_{25}N^+(CH_3)_2(CH_2)_3SO_3^-$	H_2O	25	3.0×10^{-3}	Zajac et al. (1997)
$C_{12}H_{25}N^+(CH_3)_2(CH_2)_3SO_3^-$	0.1 *M* NaCl	25	2.6×10^{-3}	Zajac et al. (1997)
$C_{14}H_{29}N^+(CH_3)_2(CH_2)_3SO_3^-$	H_2O	25	3.2×10^{-4}	Zajac et al. (1997)
$C_{12}H_{25}N(CH_3)_2O$	H_2O	27	2.1×10^{-3}	Hermann (1962)
Nonionics				
$C_8H_{17}CHOHCH_2OH$	H_2O	25	2.3×10^{-3}	Kwan and Rosen (1980)
$C_8H_{17}CHOHCH_2CH_2OH$	H_2O	25	2.3×10^{-3}	Kwan and Rosen (1980)
$C_{10}H_{21}CHOHCH_2OH$	H_2O	25	1.8×10^{-4c}	Kwan and Rosen (1980)

(*Continued*)

TABLE 3.2 (*Continued*)

Compound	Solvent	Temp. (°C)	CMC (*M*)	Reference
$C_{12}H_{25}CHOHCH_2CH_2OH$	H_2O	25	1.3×10^{-5}	Kwan and Rosen (1980)
n-Octy1-β-D-glucoside	H_2O	25	2.5×10^{-2}	Shinoda et al. (1961)
n-Decy1-α-D-glucoside	H_2O	25	8.5×10^{-4}	Aveyard et al. (1998)
n-Decy1-β-D-glucoside	H_2O	25	2.2×10^{-3}	Shinoda et al. (1961)
n-Decy1-β-D-glucoside	0.1 *M* NaCl (aq.), pH = 9	25	1.9×10^{-3}	Li et al. (2001)
n-Dodecy1-α-D-glucoside	H_2O	60	7.2×10^{-5}	Bocker and Thiem (1989)
Dodecy1-β-D-glucoside	H_2O	25	1.9×10^{-4}	Shinoda et al. (1961)
Decy1-β-D-maltoside	H_2O	25	2.0×10^{-3}	Aveyard et al. (1998)
Decy1-β-D-maltoside	0.1 *M* NaCl (aq.), pH = 9	25	$1.9_5 \times 10^{-3}$	Li et al. (2001)
Dodecy1-α-D-maltoside	H_2O	20	1.5×10^{-4}	Bocker and Thiem (1989)
Dodecy1-β-D-maltoside	H_2O	25	1.5×10^{-4}	Aveyard et al. (1998)
Dodecy1-β-D-maltoside	0.1 *M* NaCl (aq.), pH = 9	25	1.6×10^{-4}	Li et al. (2001)
$C_{12.5}H_{26}$alkylpolyglucoside (degree of polym., 1.3)[d]	H_2O	25	1.9×10^{-4}	Balzer (1993)
Tetradecy1-α-D-maltoside	H_2O	20	2.2×10^{-5}	Bocker and Thiem (1989)
Tetradecy1-β-D-maltoside	H_2O	20	1.5×10^{-5}	Bocker and Thiem (1989)
n-$C_4H_9(OC_2H_4)_6OH$	H_2O	20	8.0×10^{-1}	Elworthy and Florence (1964)
n-$C_4H_9(OC_2H_4)_6OH$	H_2O	40	7.1×10^{-1}	Elworthy and Florence (1964)
$(CH_3)_2CHCH_2(OC_2H_4)_6OH$	H_2O	20	9.1×10^{-1}	Elworthy and Florence (1964)
$(CH_3)_2CHCH_2(OC_2H_4)_6OH$	H_2O	40	8.5×10^{-1}	Elworthy and Florence (1964)
n-$C_6H_{13}(OC_2H_4)_6OH$	H_2O	20	7.4×10^{-2}	Elworthy and Florence (1964)
n-$C_6H_{13}(OC_2H_4)_6OH$	H_2O	40	5.2×10^{-2}	Elworthy and Florence (1964)
$(C_2H_5)_2CHCH_2(OC_2H_4)_6OH$	H_2O	20	1.0×10^{-1}	Elworthy and Florence (1964)

$(C_2H_5)_2CHCH_2(OC_2H_4)_6OH$	H_2O	40	8.7×10^{-2}	Elworthy and Florence (1964)
$C_8H_{17}OC_2H_4OH$	H_2O	25	4.9×10^{-3}	Shinoda et al. (1959)
$C_8H_{17}(OC_2H_4)_3OH$	H_2O	25	7.5×10^{-3}	Corkill et al. (1964)
$C_8H_{17}(OC_2H_4)_5OH$	H_2O	25	9.2×10^{-3}	Varadaraj et al. (1991b)
$C_8H_{17}(OC_2H_4)_5OH$	0.1 *M* NaCl	25	5.8×10^{-3}	Varadaraj et al. (1991b)
$C_8H_{17}(OC_2H_4)_6OH$	H_2O	25	9.9×10^{-3}	Corkill et al. (1964)
$(C_3H_7)_2CHCH_2(OC_2H_4)_6OH$	H_2O	20	2.3×10^{-2}	Elworthy and Florence (1964)
$C_{10}H_{21}(OC_2H_4)_4OH$	H_2O	25	6.8×10^{-4}	Hudson and Pethica (1964)
$C_{10}H_{21}(OC_2H_4)_5OH$	H_2O	25	7.6×10^{-4}	Eastoe et al. (1997)
$C_{10}H_{21}(OC_2H_4)_6OH$	H_2O	25	9.0×10^{-4}	Corkill et al. (1964)
$C_{10}H_{21}(OC_2H_4)_6OH$	"Hard river" water (I.S. = 6.6×10^{-3} M)	25	8.7×10^{-4}	Rosen et al. (1996)
$C_{10}H_{21}(OC_2H_4)_8OH$	H_2O	15	1.4×10^{-3}	Meguro et al. (1981)
$C_{10}H_{21}(OC_2H_4)_8OH$	H_2O	25	1.0×10^{-3}	Meguro et al. (1981)
$C_{10}H_{21}(OC_2H_4)_8OH$	H_2O	40	7.6×10^{-4}	Meguro et al. (1981)
$(C_4H_9)_2CHCH_2(OC_2H_4)_6OH$	H_2O	20	3.1×10^{-3}	Elworthy and Florence (1964)
$(C_4H_9)_2CHCH_2(OC_2H_4)_9OH$	H_2O	20	3.2×10^{-3}	Elworthy and Florence (1964)
$C_{11}H_{23}(OC_2H_4)_8OH$	H_2O	15	4.0×10^{-4}	Meguro et al. (1981)
$C_{11}H_{23}(OC_2H_4)_8OH$	H_2O	25	3.0×10^{-4}	Meguro et al. (1981)
$C_{11}H_{23}(OC_2H_4)_8OH$	H_2O	40	2.3×10^{-4}	Meguro et al. (1981)
$C_{12}H_{25}(OC_2H_4)_2OH$	H_2O	10	3.8×10^{-5}	Rosen et al. (1982a)
$C_{12}H_{25}(OC_2H_4)_2OH$	H_2O	25	3.3×10^{-5}	Rosen et al. (1982a)
$C_{12}H_{25}(OC_2H_4)_2OH$	H_2O	40	3.2×10^{-5}	Rosen et al. (1982a)
$C_{12}H_{25}(OC_2H_4)_3OH$	H_2O	10	6.3×10^{-5}	Rosen et al. (1982a)
$C_{12}H_{25}(OC_2H_4)_3OH$	H_2O	25	5.2×10^{-5}	Rosen et al. (1982a)
$C_{12}H_{25}(OC_2H_4)_3OH$	H_2O	40	5.6×10^{-5}	Rosen et al. (1982a)
$C_{12}H_{25}(OC_2H_4)_4OH$	H_2O	10	8.2×10^{-5}	Rosen et al. (1982a)
$C_{12}H_{25}(OC_2H_4)_4OH$	H_2O	25	6.4×10^{-5}	Rosen et al. (1982a)
$C_{12}H_{25}(OC_2H_4)_4OH$	H_2O	40	5.9×10^{-5}	Rosen et al. (1982a)

(*Continued*)

TABLE 3.2 (*Continued*)

Compound	Solvent	Temp. (°C)	CMC (*M*)	Reference
$C_{12}H_{25}(OC_2H_4)_4OH$	"Hard river" water (I.S. = 6.6×10^{-3} *M*)	25	4.8×10^{-5}	Rosen et al. (1996)
$C_{12}H_{25}(OC_2H_4)_5OH$	H_2O	10	9.0×10^{-5}	Rosen et al. (1982a)
$C_{12}H_{25}(OC_2H_4)_5OH$	H_2O	25	6.4×10^{-5}	Rosen et al. (1982a)
$C_{12}H_{25}(OC_2H_4)_5OH$	H_2O	40	5.9×10^{-5}	Rosen et al. (1982a)
$C_{12}H_{25}(OC_2H_4)_5OH$	0.1 *M* NaCl	25	6.4×10^{-5}	Varadaraj et al. (1991b)
$C_{12}H_{25}(OC_2H_4)_5OH$	0.1 *M* NaCl	40	5.9×10^{-5}	Varadaraj et al. (1991b)
$C_{12}H_{25}(OC_2H_4)_6OH$	H_2O	20	8.7×10^{-5}	Corkill et al. (1961)
$C_{12}H_{25}(OC_2H_4)_6OH$	"Hard river" water (I.S. = 6.6×10^{-3} *M*)	25	6.9×10^{-5}	Rosen et al. (1996)
$C_{12}H_{25}(OC_2H_4)_7OH$	H_2O	10	12.1×10^{-5}	Rosen et al. (1982a)
$C_{12}H_{25}(OC_2H_4)_7OH$	H_2O	25	8.2×10^{-5}	Rosen et al. (1982a)
$C_{12}H_{25}(OC_2H_4)_7OH$	H_2O	40	7.3×10^{-5}	Rosen et al. (1982a)
$C_{12}H_{25}(OC_2H_4)_7OH$	0.1 M NaCl (aq.)	25	7.9×10^{-5}	Rosen and Sulthana (2001)
$C_{12}H_{25}(OC_2H_4)_8OH$	H_2O	10	$1.5_6 \times 10^{-4}$	Rosen et al. (1982a)
$C_{12}H_{25}(OC_2H_4)_8OH$	H_2O	25	$1.0_9 \times 10^{-4}$	Rosen et al. (1982a)
$C_{12}H_{25}(OC_2H_4)_8OH$	H_2O	40	9.3×10^{-5}	Rosen et al. (1982a)
$C_{12}H_{25}(OC_2H_4)_8OH$	H_2O-cyclohexane	25	$1.0_1 \times 10^{-4}$	Rosen and Murphy (1991)
$C_{12}H_{25}(OC_2H_4)_8OH$	H_2O-heptane	25	$0.9_9 \times 10^{-4}$	Rosen and Murphy (1991)
$C_{12}H_{25}(OC_2H_4)_8OH$	H_2O-hexadecane	25	$1.0_2 \times 10^{-4}$	Rosen and Murphy (1991)
$C_{12}H_{25}(OC_2H_4)_9OH$	H_2O	23	10.0×10^{-5}	Lange (1965)
$C_{12}H_{25}(OC_2H_4)_{12}OH$	H_2O	23	14.0×10^{-5}	Lange (1965)
6-branched $C_{13}H_{27}(OC_2H_4)_5OH$	H_2O	25	2.8×10^{-4}	Varadaraj et al. (1991b)
6-branched $C_{13}H_{27}(OC_2H_4)_5OH$	H_2O	40	2.1×10^{-4}	Varadaraj et al. (1991b)
$C_{13}H_{27}(OC_2H_4)_5OH$	H_2O	25	4.9×10^{-5}	Varadaraj et al. (1991b)
$C_{13}H_{27}(OC_2H_4)_5OH$	0.1 *M* NaCl	25	2.1×10^{-5}	Varadaraj et al. (1991b)

$C_{13}H_{27}(OC_2H_4)_8OH$	H_2O	15	3.2×10^{-5}	Meguro et al. (1981)
$C_{13}H_{27}(OC_2H_4)_8OH$	H_2O	25	2.7×10^{-5}	Meguro et al. (1981)
$C_{13}H_{27}(OC_2H_4)_8OH$	H_2O	40	2.0×10^{-5}	Meguro et al. (1981)
$C_{14}H_{29}(OC_2H_4)_6OH$	H_2O	25	1.0×10^{-5}	Corkill et al. (1964)
$C_{14}H_{29}(OC_2H_4)_6OH$	"Hard river" water (I.S. = 6.6×10^{-3} *M*)	25	6.9×10^{-5}	Rosen et al. (1996)
$C_{14}H_{29}(OC_2H_4)_8OH$	H_2O	15	1.1×10^{-5}	Meguro et al. (1981)
$C_{14}H_{29}(OC_2H_4)_8OH$	H_2O	25	9.0×10^{-6}	Meguro et al. (1981)
$C_{14}H_{29}(OC_2H_4)_8OH$	H_2O	40	7.2×10^{-6}	Meguro et al. (1981)
$C_{14}H_{29}(OC_2H_4)_8OH$	"Hard river" water (I.S. = 6.6×10^{-3} *M*)	25	1.0×10^{-5}	Rosen et al. (1996)
$C_{15}H_{31}(OC_2H_4)_8OH$	H_2O	15	4.1×10^{-6}	Meguro et al. (1981)
$C_{15}H_{31}(OC_2H_4)_8OH$	H_2O	25	3.5×10^{-6}	Meguro et al. (1981)
$C_{15}H_{31}(OC_2H_4)_8OH$	H_2O	40	3.0×10^{-6}	Meguro et al. (1981)
$C_{16}H_{33}(OC_2H_4)_6OH$	H_2O	25	$1.6_6 \times 10^{-6}$	Rosen et al. (1996)
$C_{16}H_{33}(OC_2H_4)_6OH$	"Hard river" water (I.S. = 6.6×10^{-3} *M*)	25	2.1×10^{-6}	Rosen et al. (1996)
$C_{16}H_{33}(OC_2H_4)_7OH$	H_2O	25	1.7×10^{-6}	Elworthy and MacFarlane (1962)
$C_{16}H_{33}(OC_2H_4)_9OH$	H_2O	25	2.1×10^{-6}	Elworthy and MacFarlane (1962)
$C_{16}H_{33}(OC_2H_4)_{12}OH$	H_2O	25	2.3×10^{-6}	Elworthy and MacFarlane (1962)
$C_{16}H_{33}O(C_2H_4O)_{15}H$	H_2O	25	3.1×10^{-6}	Elworthy and MacFarlane (1962)
$C_{16}H_{33}O(C_2H_4O)_{21}H$	H_2O	25	3.9×10^{-6}	Elworthy and MacFarlane (1962)
p-t-$C_8H_{17}C_6H_4O(C_2H_4O)_2H$	H_2O	25	1.3×10^{-4}	Crook et al. (1963)
p-t-$C_8H_{17}C_6H_4O(C_2H_4O)_3H$	H_2O	25	9.7×10^{-5}	Crook et al. (1963)
p-t-$C_8H_{17}C_6H_4O(C_2H_4O)_4H$	H_2O	25	1.3×10^{-4}	Crook et al. (1963)
p-t-$C_8H_{17}C_6H_4O(C_2H_4O)_5H$	H_2O	25	1.5×10^{-4}	Crook et al. (1963)
p-t-$C_8H_{17}C_6H_4O(C_2H_4O)_6H$	H_2O	25	2.1×10^{-4}	Crook et al. (1963)
p-t-$C_8H_{17}C_6H_4O(C_2H_4O)_7H$	H_2O	25	2.5×10^{-4}	Crook et al. (1963)
p-t-$C_8H_{17}C_6H_4O(C_2H_4O)_8H$	H_2O	25	2.8×10^{-4}	Crook et al. (1964)

(*Continued*)

TABLE 3.2 (***Continued***)

Compound	Solvent	Temp. (°C)	CMC (*M*)	Reference
p-t-$C_8H_{17}C_6H_4O(C_2H_4O)_9H$	H_2O	25	3.0×10^{-4}	Crook et al. (1964)
p-t-$C_8H_{17}C_6H_4O(C_2H_4O)_{10}H$	H_2O	25	3.3×10^{-4}	Crook et al. (1964)
p-$C_9H_{19}C_6H_4(OC_2H_4)_8OH$	H_2O	—	1.3×10^{-4}	Voicu et al. (1994)
$C_9H_{19}C_6H_4(OC_2H_4)_{10}OH^f$	H_2O	25	7.5×10^{-5}	Schick and Gilbert (1965)
$C_9H_{19}C_6H_4(OC_2H_4)_{10}OH^f$	3 *M* urea	25	10×10^{-5}	Schick and Gilbert (1965)
$C_9H_{19}C_6H_4(OC_2H_4)_{10}OH^f$	6 *M* urea	25	24×10^{-5}	Schick and Gilbert (1965)
$C_9H_{19}C_6H_4(OC_2H_4)_{10}OH^f$	3 *M* guanidinium Cl	25	14×10^{-5}	Schick and Gilbert (1965)
$C_9H_{19}C_6H_4(OC_2H_4)_{10}OH^f$	1.5 *M* dioxane	25	10×10^{-5}	Schick and Gilbert (1965)
$C_9H_{19}C_6H_4(OC_2H_4)_{10}OH^f$	3 *M* dioxane	25	18×10^{-5}	Schick and Gilbert (1965)
$C_9H_{19}C_6H_4(OC_2H_4)_{31}OH^f$	H_2O	25	1.8×10^{-4}	Schick and Gilbert (1965)
$C_9H_{19}C_6H_4(OC_2H_4)_{31}OH^f$	3 *M* urea	25	3.5×10^{-4}	Schick and Gilbert (1965)
$C_9H_{19}C_6H_4(OC_2H_4)_{31}OH^f$	3 *M* urea	25	7.4×10^{-4}	Schick and Gilbert (1965)
$C_9H_{19}C_6H_4(OC_2H_4)_{31}OH^f$	3 *M* guanidinium Cl	25	4.3×10^{-4}	Schick and Gilbert (1965)
$C_9H_{19}C_6H_4(OC_2H_4)_{31}OH^f$	3 *M* dioxane	25	5.7×10^{-4}	Schick and Gilbert (1965)
$C_6H_{13}[OCH_2CH(CH_3)]_2(OC_2H_4)_{9.9}OH$	H_2O	20	4.7×10^{-2}	Kucharski and Chlebicki (1974)
$C_6H_{13}[OCH_2CH(CH_3)]_3(OC_2H_4)_{9.7}OH$	H_2O	20	3.2×10^{-2}	Kucharski and Chlebicki (1974)
$C_6H_{13}[OCH_2CH(CH_3)]_4(OC_2H_4)_{9.9}OH$	H_2O	20	1.9×10^{-2}	Kucharski and Chlebicki (1974)
$C_7H_{15}[OCH_2CH(CH_3)]_3(OC_2H_4)_{9.7}OH$	H_2O	20	1.1×10^{-2}	Kucharski and Chlebicki (1974)
Sucrose monolaurate	H_2O	25	3.4×10^{-4}	Herrington and Sahi (1986)
Sucrose monooleate	H_2O	25	5.1×10^{-6}	Herrington and Sahi (1986)
$C_{11}H_{23}CON(C_2H_4OH)_2$	H_2O	25	$2.6_4 \times 10^{-4}$	Rosen et al. (1964)
$C_{15}H_{31}CON(C_2H_4OH)_2$	H_2O	35	11.5×10^{-6}	Hayes et al. (1980)
$C_{11}H_{23}CONH(C_2H_4O)_4H$	H_2O	23	5.0×10^{-4}	Kjellin et al. (2002)
$C_{10}H_{21}CON(CH_3)(CHOH)_4CH_2OH$	0.1 *M* NaCl	25	$1.5_8 \times 10^{-3}$	Zhu et al. (1999)
$C_{11}H_{23}CON(CH_3)CH_2CHOHCH_2OH$	0.1 *M* NaCl	25	$2.3_4 \times 10^{-4}$	Zhu et al. (1999)
$C_{11}H_{23}CON(CH_3)CH_2(CHOH)_3CH_2OH$	0.1 *M* NaCl	25	$3.3_1 \times 10^{-4}$	Zhu et al. (1999)

$C_{11}H_{23}CON(CH_3)CH_2(CHOH)_4CH_2OH$	0.1 M NaCl	25	$3.4_7 \times 10^{-4}$	Zhu et al. (1999)
$C_{12}H_{25}CON(CH_3)CH_2(CHOH)_4CH_2OH$	0.1 M NaCl	25	$7.7_6 \times 10^{-5}$	Zhu et al. (1999)
$C_{13}H_{27}CON(CH_3)CH_2(CHOH)_4CH_2OH$	0.1 M NaCl	25	$1.4_8 \times 10^{-5}$	Zhu et al. (1999)
$C_{10}H_{21}N(CH_3)CO(CHOH)_4CH_2OH$	H_2O	20	$1.2_9 \times 10^{-3}$	Burczyk et al. (2001)
$C_{12}H_{25}N(CH_3)CO(CHOH)_4CH_2OH$	H_2O	20	$1.4_6 \times 10^{-4}$	Burczyk et al. (2001)
$C_{14}H_{29}N(CH_3)CO(CHOH)_4CH_2OH$	H_2O	20	$2.3_6 \times 10^{-5}$	Burczyk et al. (2001)
$C_{16}H_{33}N(CH_3)CO(CHOH)_4CH_2OH$	H_2O	20	$7.7_4 \times 10^{-6}$	Burczyk et al. (2001)
$C_{18}H_{37}N(CH_3)CO(CHOH)_4CH_2OH$	H_2O	20	$2.8_5 \times 10^{-6}$	Burczyk et al. (2001)
Fluorinated Nonionics				
$C_6F_{13}CH_2CH_2(OC_2H_4)_{11.5}OH$	H_2O	20	4.5×10^{-4}	Mathis et al. (1982)
$C_6F_{13}CH_2CH_2(OC_2H_4)_{14}OH$	H_2O	20	6.1×10^{-4}	Mathis et al. (1982)
$C_8F_{17}CH_2CH_2N(C_2H_4OH)_2$	H_2O	20	1.6×10^{-4}	Mathis et al. (1982)
$C_6F_{13}C_2H_4SC_2H_4(OC_2H_4)_2OH$	H_2O	25	2.5×10^{-3}	Matos et al. (1989)
$C_6F_{13}C_2H_4SC_2H_4(OC_2H_4)_3OH$	H_2O	25	2.8×10^{-3}	Matos et al. (1989)
$C_6F_{13}C_2H_4SC_2H_4(OC_2H_4)_5OH$	H_2O	25	3.7×10^{-3}	Matos et al. (1989)
$C_6F_{13}C_2H_4SC_2H_4(OC_2H_4)_7OH$	H_2O	25	4.8×10^{-3}	Matos et al. (1989)
Siloxane-Based Nonionics				
$(CH_3)_3SiOSi(CH_3)[CH_2(C_2H_4O)_5H]OSi(CH_3)_3$	H_2O	23 ± 2	7.9×10^{-5}	Gentle and Snow (1995)
$(CH_3)_3SiOSi(CH_3)[CH_2(C_2H_4O)_9H]OSi(CH_3)_3$	H_2O	23 ± 2	1.0×10^{-4}	Gentle and Snow (1995)
$(CH_3)_3SiOSi(CH_3)[CH_2(C_2H_4O)_{13}H]OSi(CH_3)_3$	H_2O	23 ± 2	6.3×10^{-4}	Gentle and Snow (1995)

[a] From branched dodecyl alcohol with 4.4 branches in the molecule.

[b] Pyr^+, pyridinium.

[c] Below Krafft point (p. 241) supersaturated solution.

[d] Commercial product.

[e] Solubility too low to reach CMC.

[f] Hydrophilic group not homogeneous, but distribution of POE chains reduced by molecular distillation. Hydrophobic group equivalent to 10.5 C atoms in the straight chain.

I.S. = ionic strength.

straight-chain hydrophobic group exceeds 16, however, the CMC no longer decreases so rapidly with increase in the length of the chain, and when the chain exceeds 18 carbons, it may remain substantially unchanged with further increase in the chain length (Greiss, 1955). This may be due to the coiling of these long chains in water (Mukerjee, 1967).

When the hydrophobic group is branched, the carbon atoms on the branches appear to have about one-half the effect of carbon atoms on a straight chain (Götte and Schwuger, 1969). When carbon–carbon double bonds are present in the hydrophobic chain, the CMC is generally higher than that of the corresponding saturated compound, with the *cis* isomer generally having a higher CMC than the *trans* isomer. This may be the result of a steric factor in micelle formation. Surfactants with either bulky hydrophobic or bulky hydrophilic groups have larger CMC values than those with similar, but less bulky, groups. The increase in the CMC upon introduction of a bulky hydrophobic group in the molecule is presumably due to the difficulty of incorporating the bulky hydrophobic group in the interior of a spherical or cylindrical micelle.

The introduction of a polar group such as -O- or -OH into the hydrophobic chain generally causes a significant increase in the CMC in aqueous medium at room temperature, the carbon atoms between the polar group and the hydrophilic head appearing to have about one-half the effect on the CMC that they would have, were the polar group absent. When the polar group and the hydrophilic group are both attached to the same carbon atom, that carbon atom seems to have no effect on the value of the CMC.

In POE polyoxypropylene block copolymers with the same number of OE units in the molecule, the CMC decreases significantly with increase in the number of oxypropylene units (Alexandridis et al., 1994).

The replacement of a hydrocarbon-based hydrophobic group by a fluorocarbon-based one with the same number of carbon atoms appears to cause a decrease in the CMC (Shinoda and Hirai, 1977). By contrast, the replacement of the terminal methyl group of a hydrocarbon-based hydrophobic group by a trifluoromethyl group has been shown to cause the CMC to increase. For 12,12,12-trifluorododecyltrimethylammonium bromide and 10,10,10-trifluorodecyltrimethylammonium bromide, the CMCs are twice those of the corresponding nonfluorinated compounds (Muller et al., 1972).

2. The Hydrophilic Group In aqueous medium, ionic surfactants have much higher CMCs than nonionic surfactants containing equivalent hydrophobic groups; 12-carbon straight-chain ionics have CMCs of approximately 1×10^{-2} *M*, whereas nonionics with the same hydrophobic group have CMCs of approximately 1×10^{-4} *M*. Zwitterionics appear to have slightly smaller CMCs than ionics with the same number of carbon atoms in the hydrophobic group.

As the hydrophilic group is moved from a terminal position to a more central position, the CMC increases. The hydrophobic group seems to act as if it had become branched at the position of the hydrophilic group, with the carbon atoms on the shorter section of the chain having about half their usual effect on the CMC (Evans, 1956). This may be another example of the steric effect in micelle formation noted above.

It has been found (Stigter, 1974) that the CMC is higher when the charge on an ionic hydrophilic group is closer to the α-carbon atom of the (alkyl) hydrophobic group. This is explained as being due to an increase in electrostatic self-potential of the surfactant ion when the ionic head group moves from the bulk water to the vicinity of the nonpolar micellar core during the process of micellization; work is required to move an electric charge closer to a medium of lower dielectric constant. The order of decreasing CMC in some *n*-alkyl ionics was aminium salts > carboxylates (with one more carbon atom in the molecule) > sulfonates > sulfates. This same order had been noted earlier (Klevens, 1953).

As expected, surfactants containing more than one hydrophilic group in the molecule show larger CMCs than those with one hydrophilic group and the equivalent hydrophobic group.

In quaternary cationics, pyridinium compounds have smaller CMCs than the corresponding trimethylammonium compounds. This may be due to the greater ease of packing the planar pyridinium, compared to the tetrahedral trimethylammonium group, into the micelle. In the series $C_{12}H_{25}N^+(R)_3Br^-$, the CMC decreases with increase in the length of R, presumably due to the increased hydrophobic character of the molecule.

For the usual type of polyoxyethylenated nonionic (in which the hydrophobic group is a hydrocarbon residue), the CMC in aqueous medium increases with increase in the number of OE units in the polyoxyethylene chain. However, the change per OE unit is much smaller than that per methylene unit in the hydrophobic chain. The greatest increase per OE unit seems to be obtained when the POE chain is short and the hydrophobic group is long. Since commercial POE nonionics are mixtures containing POE chains with different numbers of OE units clustered about some mean value, their CMCs are slightly lower than those of single species materials containing the same hydrophobic group and with OE content corresponding to that mean value, probably because the components with low OE content in the commercial material reduce the CMC more than it is raised by those with high OE content (Crook et al., 1963). Polyoxyethylenated fatty amides have lower CMC values than their corresponding polyoxyethylenated fatty alcohols, presumably due to hydrogen bonding between the head groups, in spite of their increased hydrophilicity (Folmer et al., 2001).

When the hydrophobic group of the POE nonionic is oleyl, or 9,10-dibromo-, 9,10-dichloro-, or 9,10-dihydroxystearyl, the CMC decreases with increase in the number of EO units in the molecule (Garti and Aserin, 1985). The effect here may be due to the bulky nature of the hydrophobic group in these molecules, which produces an almost parallel arrangement of the surfactant molecules in the micelle, similar to that at the planar liquid–air interface. At that interface, the introduction of an EO group causes a slight increase in the *hydrophobic* nature of the molecule, as evidenced by an increase in the value of $-\Delta G^0_{ad}$ (p. 104). Such an increase in the hydrophobic character of the molecule when the surfactant molecules are arranged in the micelle in a similar, more or less parallel fashion should produce a decrease in the CMC. For silicone-based nonionics of the type $(CH_3)_3SiO[Si\text{-}(CH_3)_2O]_x$-$Si(CH_3)_2CH_2(C_2H_4O)_yCH_3$, too, the CMC appears to decrease with increase in the OE content of the molecule (Kanner et al., 1967). Here, too, the hydrophobic group is bulky. However, only a few compounds have been studied.

In POE polyoxypropylene block copolymers with a constant number of oxypropylene units in the molecule, the CMC increases with increase in the number of OE units. At a constant POE/polyoxypropylene ratio, increase in the molecular weight of the surfactant decreases the CMC (Alexandridis et al., 1994).

3. The Counterion in Ionic Surfactants; Degree of Binding to the Micelle A plot of the specific conductivity, κ, of an ionic surfactant versus its concentration, *C*, in the aqueous phase is linear, with a break at the CMC, above which the (decreased) slope of the plot again becomes linear (Figure 3.6). The break in the plot is due to the binding of some of the counterions of the ionic surfactant to the micelle. The degree of ionization, α, of the micelle near its CMC can be obtained from the ratio, S_2/S_1, of the slopes above and below the break indicative of the CMC (Yiv and Zana, 1980). The degree of binding of the counterion to the micelle, for a surfactant with a single ionic head group in the molecule, is $(1 - \alpha)$.

The larger the hydrated radius of the counterion, the weaker the degree of binding; thus, $NH_4^+ > K^+ > Na^+ > Li^+$ and $I^- > Br^- > Cl^-$.*

In a number of series of cationic surfactants, Zana (1980) has shown (Table 3.3) that the degree of binding (or ionization) is related to the surface area per head group, a_m^s, in the ionic micelle, with the degree of binding increasing as the surface area per head group decreases (i.e., as the surface charge density increases). This is also apparent from the data of Granet and Piekarski (1988) and of Binana-Limbele et al. (1988) in Table 3.3, where, with increase in the length of the 2-alkyl side chain in the decanesulfonates or of the POE group in the carboxylates, respectively, and the presumable resulting increase in the surface area per head group, the degree of binding decreases.

* For anionic surfactants of structure $RC(O)N(R^1)CH_2CH_2COO^-Na^+$, it has been found (Tsubone and Rosen, 2001; Tsubone et al., 2003a, 2003b) that this break in the conductance–surfactant concentration may be smaller than expected or absent, yielding binding $(1 - \alpha)$ values much smaller than those of comparable surfactants without the amide group (Table 3.3). This may be due to protonation of the carboxylate group and hydrogen-bonded ring formation with the amido group, with simultaneous release of the Na^+, upon micellization. This absence of a break in the conductance–surfactant concentration plot at the CMC is even more prone to occur in gemini surfactants (Chapter 12) with the above structure in the molecule.

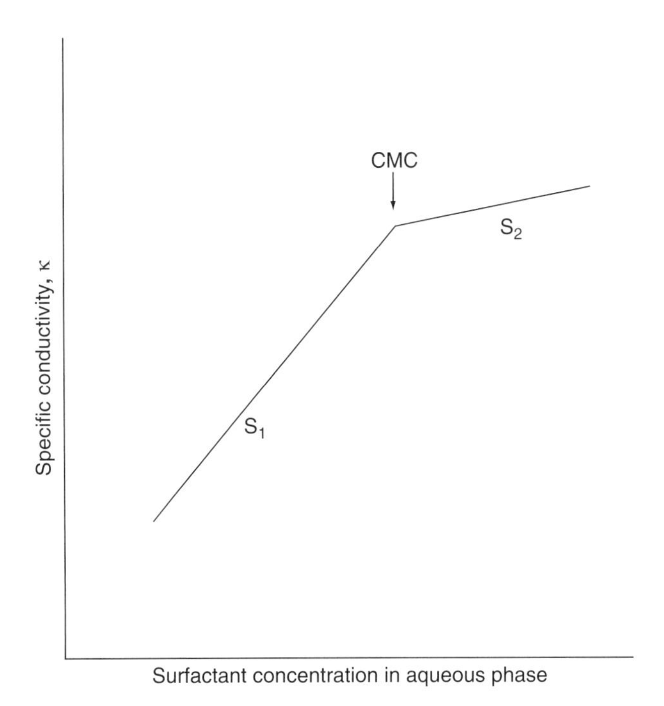

FIGURE 3.6 Plot of specific conductivity, κ, versus surfactant concentration in the aqueous phase, showing change in slopes before (S_1) and after (S_2) the CMC.

The degree of counterion binding also decreases with solubilization of short (C_2–C_6) alcohols in the palisade layer of the micelle, whereas the solubilization of octane, which occurs in the interior of the micelle (Chapter 4, Section IIA), does not affect the degree of counterion binding (Bostrom et al., 1989). This is presumably because solubilization in the palisade layer increases the surface area per ionic head group, whereas solubilization in the interior of the micelle does not. It is also decreased by the addition of urea, replacing water from the interface (Souza et al., 1995). Counterion binding also increases with increase in the electrolyte content of the solution (Asakawa et al., 2001) and may also increase when surfactant concentration increase produces micellar growth (Quirion and Magid, 1986; Iijima et al., 2000), presumably because both of these are accompanied by a decrease in head group area. Ionic micelles that have more tightly bound counterions are more nonionic in character than those with a greater degree of ionization, have lower water solubility, and are more likely to have nonspherical micelles and to show viscoelasticity in aqueous solution.

The binding of the counterions Na^+ and Cl^- to zwitterionics of the betaine and sulfobetaine types starts well above the CMC and hence does not affect its value. The degree of binding of the Cl^- is always larger than that of the Na^+ (Kamenka et al., 1995b).

TABLE 3.3 Degree of Binding (1 − α), Surface Area at Surface Saturation (a_m^s) per Head Group, and CMC of Some Ionic Micelles in H_2O

Compound	Temp. (°C)	Degree of Binding	(a_m^s) (Å²)	CMC (*M*)	Reference
Anionics					
$C_{11}H_{23}COO^-Na^+$	25	0.54	—	2.8×10^{-3}	De Lisi et al. (1997)
$C_{12}H_{25}(OC_2H_4)_5OCH_2COO^-Na^+$ (pH 8.2)	25	0.19	—	3.5×10^{-3}	Binana-Limbele et al. (1988)
$C_{12}H_{25}(OC_2H_4)_9OCH_2COO^-Na^+$ (pH 8.2)	25	0.14	—	5.9×10^{-3}	Binana-Limbele et al. (1988)
$C_{11}H_{23}CON(CH_3)CH_2COO^-Na^+$ (0.1 *M* NaCl, pH 10.5)	30	0.41	58	3.5×10^{-3}	Tsubone and Rosen (2001)
$C_{11}H_{23}CON(CH_3)CH_2CH_2COO^-Na^+$ (0.1 *M* NaCl, pH 10.5)	30	0.36	66	2.7×10^{-3}	Tsubone and Rosen (2001)
$C_7F_{15}COO^-Na^+$	25	0.44	—	3.0×10^{-2}	De Lisi et al. (1997)
$C_8F_{17}COO^-Na^+$	40	0.47	—	1.0×10^{-2}	Nakano et al. (2002)
$C_9H_{19}CH_2SO_3^-Na^+$	45	0.88	—	4.0×10^{-2}	Granet and Piekarski (1988)
$C_9H_{19}CH(C_2H_5)SO_3^-Na^+$	45	0.86	—	1.8×10^{-2}	Granet and Piekarski (1988)
$C_9H_{19}CH(C_4H_9)SO_3^-Na^+$	45	0.83	—	6.8×10^{-3}	Granet and Piekarski (1988)
$C_9H_{19}CH(C_6H_{13})SO_3^-Na^+$	45	0.77	—	1.7×10^{-3}	Granet and Piekarski (1988)
$C_9H_{19}CH(C_8H_{17})SO_3^-Na^+$	45	0.67	—	2.4×10^{-4}	Granet and Piekarski (1988)
$C_9H_{19}CH(C_9H_{19})SO_3^-Na^+$	45	0.59	—	8.9×10^{-5}	Granet and Piekarski (1988)
$C_{11}H_{23}CON(CH_3)CH_2CH_2SO_3^-Na^+$ (0.1 *M* NaCl, pH 10.5)	30	0.42	56	1.6×10^{-3}	Tsubone and Rosen (2001)
$C_{12}H_{25}SO_4^-Na^+$	25	0.82	—	8.1×10^{-3}	Sowada (1994)
$C_{12}H_{25}SO_4^-Na^+$ (0.1 *M* NaCl)	25	0.88	—	1.6×10^{-3}	Sowada (1994)
$C_{12}H_{25}SO_4^-Na^+$ (0.2 *M* NaCl)	25	0.86	—	8.3×10^{-4}	Sowada (1994)
$C_{12}H_{25}SO_4^-Na^+$ (0.4 *M* NaCl)	25	0.87	—	5.2×10^{-4}	Sowada (1994)
$C_{14}H_{29}SO_4^-Na^+$	40	0.85	—	2.23×10^{-3}	Nakano et al. (2002)

$C_{12}H_{25}OC_2H_4SO_4^-Na^+$	25	0.84	—	—	Zoeller and Blankschtein (1998)
$C_{12}H_{25}(OC_2H_4)_2SO_4^-Na^+$	25	0.87	—	—	Zoeller and Blankschtein (1998)
$C_{12}H_{25}(OC_2H_4)_4SO_4^-Na^+$	25	0.84	—	—	Zoeller and Blankschtein (1998)
$C_{12}H_{25}(OC_2H_4)_6SO_4^-Na^+$	25	0.81	—	—	Zoeller and Blankschtein (1998)
Cationics					
$C_8H_{17}N^+(CH_3)_3Br^-$	25	0.64	83.3	2.9×10^{-1}	Zana (1980); Zana et al. (1981)
$C_{10}H_{21}N^+(CH_3)_3Br^-$	25	0.73	79.7	6.4×10^{-2}	Zana (1980); Zana et al. (1981)
$C_{10}H_{21}N^+(CH_3)_3Cl^-$	25	0.75	—	6.8×10^{-2}	Sowada (1994)
$C_{10}H_{21}N^+(CH_2C_6H_5)(CH_3)_2Cl^-$	25	0.46	—	3.9×10^{-2}	de Castillo et al. (2000)
$C_{12}H_{25}N^+(CH_3)_3Br^-$	25	0.78	77.2	1.5×10^{-2}	Zana (1980); Zana et al. (1981)
$C_{12}H_{25}N^+(CH_3)_2(C_2H_5)Br^-$	25	0.72	89	1.4×10^{-2}	Zana (1980); Zana et al. (1981)
$C_{12}H_{25}N^+(CH_3)_2(C_2H_5)Br^-$	25	0.68	96	1.1×10^{-2}	Zana (1980); Zana et al. (1981)
$C_{12}H_{25}N^+(CH_3)_3Cl^-$	25	0.66	—	2.2×10^{-2}	Zana et al. (1995)
$C_{14}H_{29}N^+(CH_3)_3Br^-$	25	0.81	75.3	3.2×10^{-3}	Zana (1980); Zana et al. (1981)
$C_{14}H_{29}N^+(CH_3)_3Br^-$	40	0.75	—	4.2×10^{-3}	Gorski and Kalus (2001)
$C_{16}H_{33}N^+(CH_3)_3Br^-$	25	0.84	74	8.5×10^{-3}	Zana (1980); Zana et al. (1981)
$C_{16}H_{33}N^+(CH_3)_3Cl^-$	25	0.63	—	1.4×10^{-3}	Sepulveda and Cortes (1985)
$C_{10}H_{21}Pyr^+Br^-$	25	0.62	—	4.4×10^{-2}	Skerjanc et al. (1999)
$C_{11}H_{23}Pyr^+Br^-$	25	0.64	—	2.1×10^{-2}	Skerjanc et al. (1999)
$C_{12}H_{25}Pyr^+Br^-$	25	0.66	—	1.0×10^{-2}	Skerjanc et al. (1999)
$C_{13}H_{27}Pyr^+Br^-$	25	0.66	—	5.3×10^{-3}	Skerjanc et al. (1999)
$C_{14}H_{29}Pyr^+Br^-$	25	0.69	—	2.7×10^{-3}	Skerjanc et al. (1999)
$C_{15}H_{31}Pyr^+Br^-$	25	0.69	—	1.3×10^{-3}	Skerjanc et al. (1999)
$C_{16}H_{33}Pyr^+Br^-$	25	0.69	—	6.4×10^{-4}	Skerjanc et al. (1999)

The CMC in aqueous solution for a particular surfactant reflects the degree of binding of the counterion to the micelle. Increased binding of the counterion, in aqueous systems, causes a decrease in the CMC of the surfactant. The extent of binding of the counterion increases also with increase in the polarizability and charge of the counterion and decreases with increase in its hydrated radius. Thus, in aqueous medium, for the anionic lauryl sulfates, the CMC decreases in the order $Li^+ > Na^+ > K^+ > Cs^+ > N(CH_3)_4^+ > N(C_2H_5)_4^+ > Ca^{2+}$, Mg^{2+}, which is the same order as the increase in the degree of binding of the cation (Robb and Smith, 1974). The depression of the CMC from Li^+ to K^+ is small, but for the other counterions it is quite substantial. When the counterion is a cation of a primary amine, RNH_3^+, the CMC decreases with increase in the chain length of the amine (Packter and Donbrow, 1963). For the cationic dodecyltrimethylammonium and dodecylpyridinium salts, the order of decreasing CMC in aqueous medium is $F^- > Cl^- > Br^- > I^-$ (Mukerjee, 1967), which is the same order as the increase in degree of binding of the anion (Ottewill and Parreira, 1962).

On the other hand, when comparing surfactants of different structural types, the value of the CMC does not always increase with decrease in the degree of binding of the counterion. Thus, although in the series $RN^+(CH_3)_3Br^-$ the degree of binding increases and the CMC decreases with increase in the length of R, the decrease in the CMC is due mainly to the increased hydrophobicity of the surfactant as a result of the increase in the alkyl chain length, and only to a minor extent due to the smaller area per head group, a_m^s. This is seen also in the series, $C_{12}H_{25}N^+(CH_3)_2(R^1)Br^-$ and $C_9H_{19}CH(R^1)SO_3^-Na^+$ (Table 3.3), where, although the degree of binding decreases with increase in the length of the alkyl chain, R^1, the CMC decreases due to the increase in the hydrophobicity of the surfactant with increase in R.

4. Empirical Equations Many investigators have developed empirical equations relating the CMC to the various structural units in surface-active agents. Thus, for homologous straight-chain ionic surfactants (soaps, alkanesulfonates, alkyl sulfates, alkylammonium chlorides, alkyltrimethylammonium bromides) in aqueous medium, a relation between the CMC and the number of carbon atoms N in the hydrophobic chain was found (Klevens, 1953) in the form

$$\log \mathrm{CMC} = \mathrm{A} - BN, \tag{3.1}$$

where A is a constant for a particular ionic head at a given temperature, and B is a constant ≈ 0.3 ($= \log 2$) at 35°C for the ionic types cited previously. The basis for the rule mentioned earlier about the CMC being halved for each increase in the hydrophobic chain by one carbon atom is apparent from this relation. Nonionics and zwitterionics also show this relation, but the value of B is ≈ 0.5, which is the basis for the rule that the CMC decreases by a factor of about 10 for each two methylene groups added to

TABLE 3.4 Constants for the Relation log CMC = $A - BN$

Surfactant Series	Temp. (°C)	A	B	Reference
Na carboxylates (soaps)	20	1.8_5	0.30	Markina (1964)
K carboxylates (soaps)	25	1.9_2	0.29	Klevens (1953)
Na (K) alkyl 1-sulfates or -sulfonates	25	1.5_1	0.30	Rosen (1976)
Na alkane-l-sulfonates	40	1.5_9	0.29	Klevens (1953)
Na alkane-1-sulfonates	55	1.1_5	0.26	Schick and Fowkes (1957)
Na alkyl-l-sulfates	45	1.4_2	0.30	Klevens (1953)
Na alkyl-l-sulfates	60	1.3_5	0.28	Rosen (1976)
Na alkyl-2-sulfates	55	1.2_8	0.27	Schick and Fowkes (1957)
Na *p*-alkylbenzenesulfonates	55	1.6_8	0.29	Schick and Fowkes (1957)
Na *p*-alkylbenzenesulfonates	70	1.3_3	0.27	Lange (1964)
Alkyltrimethylammonium bromides	25	2.0_1	0.32	Zana (1980)
Alkyltrimethylammonium chlorides (0.1 *M* NaCl)	25	1.2_3	0.33	Caskey and Barlage (1971)
Alkyltrimethylammonium bromides	60	1.7_7	0.29	Klevens (1953)
Alkylpyridinium bromides	30	1.7_2	0.31	Venable and Nauman (1964)
$C_nH_{2n+1}(OC_2H_4)_6OH$	25	1.8_2	0.49	Rosen (1976)
$C_nH_{2n+1}(OC_2H_4)_8OH$	15	2.1_8	0.51	Meguro et al. (1981)
$C_nH_{2n+1}(OC_2H_4)_8OH$	25	1.8_9	0.50	Meguro et al. (1981)
$C_nH_{2n+1}(OC_2H_4)_8OH$	40	1.6_6	0.48	Meguro et al. (1981)
$C_nH_{2n+1}N^+(CH_3)_2CH_2COO^-$	23	3.1_7	0.49	Beckett and Woodward (1963)

the hydrophobic chain. Table 3.4 lists some values of A and B determined from experimental data.

B. Electrolyte

In aqueous solution, the presence of electrolyte causes a change in the CMC, the effect being more pronounced for anionic and cationic than for zwitterionic surfactants and more pronounced for zwitterionics than for nonionics. Experimental data indicate that for the first two classes of surfactants, the effect of the concentration of electrolyte is given by (Corrin and Harkins, 1947; Barry et al., 1970)

$$\log \text{CMC} = -a \log C_i + b, \tag{3.2}$$

where a and b are constants for a given ionic head at a particular temperature and C_i is the total counterion concentration in equivalents per liter. The depression of the CMC in these cases is due mainly to the decrease in the thickness of the ionic atmosphere surrounding the ionic head groups in the presence of the additional electrolyte and the consequent decreased electrical repulsion between them in the micelle. For sodium laurate and sodium naphthenate, the order of decreasing effectiveness of the anion in depressing the CMC is $PO_4^{3-} > B_4O_7^{2-} > OH^- > CO_3^{2-} > HCO_3^- > SO_4^{2-} > NO_3^-$ $> Cl^-$ (Demchenko et al., 1962).

For nonionics and zwitterionics, the preceding relation does not hold. Instead, the effect is given better by (Shinoda et al., 1961; Tori and Nakagawa, 1963b; Ray and Nemethy, 1971)

$$\log \text{CMC} = -KC_s + \text{constant}\ (C_s < 1), \tag{3.3}$$

where K is a constant for a particular surfactant, electrolyte, and temperature, and C_s is the concentration of electrolyte in moles per liter. For alkylbetaines, the value of K increases with increase in the length of the hydrophobic group and the charge on the anion of the electrolyte (Tori and Nakagawa, 1963b).

The change in the CMC of nonionics and zwitterionics on the addition of electrolyte has been attributed (Mukerjee, 1967; Ray, 1971) mainly to the "salting out" or "salting in" of the hydrophobic groups in the aqueous solvent by the electrolyte, rather than to the effect of the latter on the hydrophilic groups of the surfactant. Salting in or salting out by an ion depends upon whether the ion is a water structure breaker or a water structure maker. Ions with a large ionic charge/radius ratio, such as F^-, are highly hydrated and are water structure makers. They salt out the hydrophobic groups of the monomeric form of the surfactant and decrease the CMC. Ions with a small ionic charge/radius ratio, such as CNS^-, are water structure breakers; they salt in the hydrophobic groups of the monomeric form of the surfactant and increase the CMC. The total effect of an electrolyte appears to approximate the sum of its effects on the various parts of the solute molecule in contact with the aqueous phase. Since the hydrophilic groups of the surfactant molecules are in contact with the aqueous phase in both the monomeric and micellar forms of the surfactant, while the hydrophobic groups are in contact with the aqueous phase only in the monomeric form, the effect of the electrolyte on the hydrophilic groups in the monomeric and micellar forms may cancel each other, leaving the hydrophobic groups in the monomers as the moiety most likely to be affected by the addition of electrolyte to the aqueous phase.

The effects of the anion and the cation in the electrolyte are additive. For anions, the effect on the CMC of POE nonionics appears to depend on the charge/radius (water structure) effect. Thus, the order of effectiveness in decreasing the CMC is $\frac{1}{2}SO_4^{2-} > F^- > BrO_3^- > Cl^- > Br^- > NO_3^- > I^- > CNS^-$. For cations, the order is $NH_4^+ > K^+ > Na^+ > Li^+ > \frac{1}{2}Ca^{2+}$ (Schick, 1962; Ray and Nemethy, 1971). Here, the reason for the order is not clear. Similar anion and cation effects have been observed on the CMC value of *n*-dodecy1-β-maltoside solutions (Zhang et al., 1996).

Investigation of the effect of electrolyte on the CMC of a high-molecular-weight nonionic of the POE polyoxypropylene type (Pandit et al., 2000) found that the CMC decreased in the order $Na_3PO_4 > Na_2SO_4 > NaCl$. The addition of NaSCN increased the CMC, consistent with its action as a water structure breaker.

Tetraalkylammonium cations increase the CMC values of POE nonionics, the order of their effectiveness in increasing them being $(C_3H_7)_4N^+ > (C_2H_5)_4N^+ > (CH_3)_4N^+$. This is the order of their effectiveness in water structure breaking.

C. Organic Additives

Small amounts of organic materials may produce marked changes in the CMC in aqueous media. Since some of these materials may be present as impurities or by-products of the manufacture of surfactants, their presence may cause significant differences in supposedly similar commercial surfactants. A knowledge of the effects of organic materials on the CMC of surfactants is therefore of great importance for both theoretical and practical purposes.

To understand the effects produced, it is necessary to distinguish between two classes of organic materials that markedly affect the CMCs of aqueous solutions of surfactants: class I, materials that affect the CMC by being incorporated into the micelle; and class II, materials that change the CMC by modifying solvent–micelle or solvent–surfactant interactions.

1. Class I Materials Materials in the first class are generally polar organic compounds, such as alcohols and amides. They affect the CMC at much lower liquid phase concentrations than those in the second class. Water-soluble compounds in this class may operate as members of the first class at low bulk phase concentrations (Miyagishi, 1976) and, at high bulk phase concentrations, as members of the second class.

Members of class I reduce the CMC. Shorter-chain members of the class are probably adsorbed mainly in the outer portion of the micelle close to the water–micelle "interface." The longer-chain members are probably adsorbed mainly in the outer portion of the core, between the surfactant molecules. Adsorption of the additives in these fashions decreases the work required for micellization, in the case of ionic surfactants probably by decreasing the mutual repulsion of the ionic heads in the micelle.

Depression of the CMC appears to be greater for straight-chain compounds than for branched ones and increases with chain length to a maximum when the length of the hydrophobic group of the additive approximates that of the surfactant. An explanation for these observations (Schick and Fowkes, 1957) is that those molecules that are most effective at reducing the CMC are solubilized in the outer portion of the micelle core and are there under lateral pressure tending to force them into the inner portion of the core. This pressure increases with the cross-sectional area of the molecule. Thus, straight-chain molecules, with smaller cross-sectional areas than branched-chain ones, have a greater tendency to remain in this outer portion and consequently reduce the CMC more than do the latter, which are forced into the interior. Another factor may be the greater degree of interaction between the hydrophobic group of the surfactant and the hydrophobic chain of the additive when the latter is straight rather than branched. This also would tend to keep straight-chain molecules, compared to branched-chain ones, in the outer portions of the micelle. It also explains the greater effect on the CMC of additives containing hydrophobic groups of approximately the same length as those of the surfactant molecules comprising the micelle, since in this condition maximum interaction between hydrophobic groups in additive and surfactant is obtained.

Additives that have more than one group capable of forming hydrogen bonds with water in a terminal polar grouping appear to produce greater depressions of the CMC than those with only one group capable of hydrogen bonding to water. Here the explanation offered (Schick and Fowkes, 1957) is that hydrogen bonding between the polar groups of the additive and water molecules helps counterbalance the lateral pressure tending to push the additive into the interior of the micelle. Therefore, a higher proportion of those additives with more than one group capable of forming hydrogen bonds with water will remain in the outer core between the surfactant molecules than will be the case with those with only one group of this type; consequently, the CMC will be reduced more by the former type of additive.

Just as polar compounds that are believed to penetrate into the inner portion of the core produce only small depressions of the CMC, so, too, hydrocarbons, which are solubilized in the inner portion of the core, decrease the CMC only slightly. Very short-chain polar compounds (e.g., dioxane and ethanol) at low bulk phase concentrations also depress the CMC, but the effect here, too, is small (Shirahama and Matuura, 1965). In these compounds, adsorption probably occurs on the surface of the micelle, close to the hydrophilic head.

2. Class II Materials Members of class II change the CMC, but at bulk phase concentrations usually considerably higher than those at which class I members are effective. The members of this class change the CMC by modifying the interaction of water with the surfactant molecule or with the micelle, doing this by modifying the structure of the water, its dielectric constant, or its solubility parameter (cohesive energy density). Members of this class include urea, formamide, *N*-methylacetamide, guanidinium salts, short-chain alcohols, water-soluble esters, dioxane, ethylene glycol, and other polyhydric alcohols such as fructose and xylose.

Urea, formamide, and guanidinium salts are believed to increase the CMC of surfactants in aqueous solution, especially polyoxyethylenated nonionics, because of their disruption of the water structure (Schick and Gilbert, 1965). This may increase the degree of hydration of the hydrophilic group, and since hydration of the hydrophilic group opposes micellization, may cause an increase in the CMC. These water structure breakers may also increase the CMC by decreasing the entropy effect accompanying micellization (Section VIII). The hydrophobic hydrocarbon chain of the surfactant is believed to create structure in the liquid water phase when it is dissolved in it, and its removal from it via micellization consequently produces an increase in the entropy of the system that favors micellization. The presence of structure breakers in the aqueous phase may disrupt the organization of the water produced by the dissolved hydrophobic group, thereby decreasing the entropy increase on micellization. Since the entropy increase favoring micellization is decreased, a higher bulk concentration of surfactant is needed for micelle formation; that is, the CMC is increased.

Materials that promote water structure, such as xylose or fructose (Schwuger, 1971), for similar reasons decrease the CMC of the surfactant.

The effect of urea on the CMC of ionic surfactants is smaller and complex. Although urea addition was found to increase the CMCs of $C_{12}H_{25}SO_4^-Li^+$ and $C_{14}H_{29}SO_4^-.{}^+NH_2(C_2H_5)_2$, it decreased slightly the CMCs or $C_8F_{17}SO_4^-Li^+$ and $C_8F_{17}COO^-Li^+$. It is suggested that the effect here may be due to direct action of the urea, replacing the water surrounding the hydrophilic head group (Asakawa et al., 1995).

Dioxane, ethylene glycol, water-soluble esters, and short-chain alcohols at high bulk phase concentrations may increase the CMC because they decrease the cohesive energy density, or solubility parameter, of the water, thus increasing the solubility of the monomeric form of the surfactant and hence the CMC (Schick and Gilbert, 1965). An alternative explanation for the action of these compounds in the case of ionic surfactants is based on the reduction of the dielectric constant of the aqueous phase that they produce (Herzfeld et al., 1950). This would cause increased mutual repulsion of the ionic heads in the micelle, thus opposing micellization and increasing the CMC.

D. The Presence of a Second Liquid Phase

The CMC of the surfactant in the aqueous phase is changed very little by the presence of a second liquid phase in which the surfactant does not dissolve appreciably and which, in turn, either does not dissolve appreciably in the aqueous phase or is solubilized only in the inner core of the micelles (e.g., saturated aliphatic hydrocarbons). When the hydrocarbon is a short-chain unsaturated, or aromatic hydrocarbon, however, the value of the CMC is significantly less than that in air, with the more polar hydrocarbon causing a larger decrease (Rehfeld, 1967; Vijayendran and Bursh, 1979; Murphy and Rosen, 1988). This is presumably because some of this second liquid phase adsorbs in the outer portion of the surfactant micelle and acts as a class I material (Section C). On the other hand, the more polar ethyl acetate increases the CMC of sodium dodecyl sulfate slightly, presumably either because it has appreciable solubility in water and thus increases its solubility parameter, with consequent increase in the CMC of the surfactant, or because the surfactant has appreciable solubility in the ethyl acetate phase, thus decreasing its concentration in the aqueous phase with consequent increase in the CMC.

E. Temperature

The effect of temperature on the CMC of surfactants in aqueous medium is complex, the value appearing first to decrease with temperature to some minimum and then to increase with further increase in temperature. Temperature increase causes decreased hydration of the hydrophilic group, which favors micellization. However, temperature increase also causes disruption of the structured water surrounding the hydrophobic group, an effect that disfavors micellization. The relative magnitude of these two opposing effects, therefore, determines whether the CMC increases or decreases over a particular temperature range. From the data available, the minimum in the CMC–temperature curve appears to be around 25°C for ionics (Flockhart, 1961) and around 50°C for nonionics (Crook et al., 1963; Chen et al., 1998). For bivalent metal alkyl sulfates, the CMC appears to be practically independent of the temperature (Mujamoto, 1960). Data on the effect of temperature on zwitterionics are limited. They appear to indicate a steady decrease in the CMC of alkylbetaines with increase in the temperature in the range 6–60°C (Tori and Nakagawa, 1963b; Dahanayake and Rosen, 1984). Whether further increase in temperature will cause an increase in the CMC is not evident from the data.

V. MICELLIZATION IN AQUEOUS SOLUTION AND ADSORPTION AT THE AQUEOUS SOLUTION–AIR OR AQUEOUS SOLUTION–HYDROCARBON INTERFACE

Although similar factors, such as the structure of the surfactant molecule and the microenvironmental conditions surrounding it, affect both micellization and adsorption, the effect of these factors on these two phenomena are generally not equal.

Steric factors in the surfactant molecule, such as the presence of a bulky hydrophilic or hydrophobic group, inhibit micellization more than they do adsorption at the aqueous solution–air interface. On the other hand, electrical factors such as the presence of an ionic, rather than a nonionic, hydrophilic group in the surfactant molecule, appear to inhibit adsorption at the aqueous solution–air interface more than they do micellization.

A. The CMC/C_{20} Ratio

A convenient way of measuring the relative effects of some structural or microenvironmental factor on micellization and on adsorption is to determine its effect on the CMC/C_{20} ratio, where C_{20} (Chapter 2, Section IIIE) is the concentration of surfactant in the bulk phase that produces a reduction of 20 dyn/cm in the surface tension of the solvent. An increase in the CMC/C_{20} ratio as a result of the introduction of some factor indicates that micellization is inhibited more than adsorption or adsorption facilitated more than micellization; a decrease in the CMC/C_{20} ratio indicates that adsorption is inhibited more than micellization or micellization facilitated more than adsorption. The CMC/C_{20} ratio, therefore, affords insights into the adsorption and micellization processes. The CMC/C_{20} ratio is also an important factor in determining the value to which the surface tension of the solvent can be reduced by the presence in its solution of the surfactant (Chapter 5, Section IIB).

Some CMC/C_{20} ratios are listed in Table 3.5. The data show that for single-chain compounds of all types listed the CMC/C_{20} ratio

1. Is not increased substantially by increase in the length of the alkyl chain of the hydrophobic group (from C_{10} to C_{16}) in ionic surfactants.
2. Is increased by the introduction of branching in the hydrophobic group or positioning of the hydrophilic group in a central position in the molecule.
3. Is increased by the introduction of a larger hydrophilic group.
4. Is increased greatly for ionic surfactants by increase in the ionic strength of the solution or by the use of a more tightly bound counterion, especially one containing an alkyl chain of six or more carbon atoms. For a nonionic surfactant, the effect of the addition of electrolyte is more complex, depending upon the nature of the electrolyte added, its salting-in or salting-out effect, and its possible complex formation with the nonionic. In some cases, the CMC/C_{20} ratio is increased, in other cases it is decreased by the addition of electrolyte, and in still others there is little effect.
5. Is decreased by an increase in temperature in the range 10–40°C.
6. Is increased considerably by the replacement of a hydrocarbon chain by a fluorocarbon- or silicone-based chain.

TABLE 3.5 CMC/C_{20} Ratios

Compound	Solvent	Temp. (°C)	CMC/C_{20} Ratio	Reference
Anionics				
$C_{10}H_{21}OCH_2COO^-Na^+$	0.1 *M* NaCl (aq.), pH 10.5	30	4.9	Tsubone and Rosen (2001)
$C_{11}H_{23}CON(CH_3)CH_2COO^-Na^+$	H_2O, pH 10.5	30	3.5	Tsubone and Rosen (2001)
$C_{11}H_{23}CON(CH_3)CH_2COO^-Na^+$	0.1 *M* NaCl (aq.), pH 10.5	30	6.5	Tsubone and Rosen (2001)
$C_{11}H_{23}CON(CH_3)CH_2CH_2COO^-Na^+$	H_2O, pH 10.5	30	3.7	Tsubone and Rosen (2001)
$C_{11}H_{23}CON(CH_3)CH_2CH_2COO^-Na^+$	0.1 *M* NaCl (aq.), pH 10.5	30	6.9	Tsubone and Rosen (2001)
$(CF_3)_2CF(CF_2)COO^-Na^+$	H_2O	25	11.1	Shinoda and Hirai (1977)
$C_7F_{15}COO^-Na^+$	H_2O	25	9.5	Shinoda and Hirai (1977)
$C_7F_{15}COO^-K^+$	H_2O	25	10.8	Shinoda and Hirai (1977)
$C_{10}H_{21}SO_3^-Na^+$	H_2O	10	2.4	Dahanayake et al. (1986)
$C_{10}H_{21}SO_3^-Na^+$	H_2O	25	2.1	Dahanayake et al. (1986)
$C_{10}H_{21}SO_3^-Na^+$	H_2O	40	1.8	Dahanayake et al. (1986)
$C_{10}H_{21}SO_3^-Na^+$	0.1 *M* NaCl (aq.)	25	4.1	Dahanayake et al. (1986)
$C_{10}H_{21}SO_3^-Na^+$	0.5 *M* NaCl (aq.)	25	5.4	Dahanayake et al. (1986)
$C_{10}H_{21}OC_2H_4SO_3^-Na^+$	H_2O	25	2.0	Dahanayake et al. (1986)
$C_{10}H_{21}OC_2H_4SO_3^-Na^+$	0.1 *M* NaCl (aq.)	25	4.5	Dahanayake et al. (1986)
$C_{10}H_{21}OC_2H_4SO_3^-Na^+$	0.5 *M* NaCl (aq.)	25	7.1	Dahanayake et al. (1986)
$C_{12}H_{25}SO_3^-Na^+$	H_2O	25	2.8	Dahanayake et al. (1986)
$C_{12}H_{25}SO_3^-Na^+$	0.1 *M* NaCl (aq.)	25	5.9	Dahanayake et al. (1986)
$C_{16}H_{33}SO_3^-K^+$	H_2O	60	1.9	M. J. Rosen and J. Solash, unpublished data
Branched $C_{12}H_{25}SO_4^-Na^{+b}$	H_2O	25	11.3	Varadaraj et al. (1992)
Branched $C_{12}H_{25}SO_4^-Na^{+b}$	0.1 *M* NaCl	25	15.2	Varadaraj et al. (1992)

$C_{12}H_{25}SO_4^-Na^+$	H_2O	25	2.6	Dahanayake et al. (1986)
$C_{12}H_{25}SO_4^-Na^+$	0.1 *M* NaCl (aq.)	25	6.0	Dahanayake et al. (1986)
$C_{12}H_{25}SO_4^-Na^+$	H_2O (1-hexene satd.)	25	1.5	Rehfeld (1967)
$C_{12}H_{25}SO_4^-Na^+$	H_2O (benezene satd.)	25	2.2	Rehfeld (1967)
$C_{12}H_{25}SO_4^-Na^+$	H_2O (cyclohexane satd.)	25	4.9	Rehfeld (1967)
$C_{12}H_{25}SO_4^-Na^+$	H_2O (octane satd.)	25	4.7	Rehfeld (1967)
$C_{12}H_{25}SO_4^-Na^+$	H_2O (heptadecane satd.)	25	4.8	Rehfeld (1967)
$C_{14}H_{29}SO_4^-Na^+$	H_2O	25	2.6	Lange and Schwuger (1968)
$C_{12}H_{25}OC_2H_4SO_4^-Na^+$	H_2O	25	2.6	Dahanayake et al. (1986)
$C_{12}H_{25}OC_2H_4SO_4^-Na^+$	0.1 *M* NaCl	25	7.3	Dahanayake et al. (1986)
$C_{12}H_{25}OC_2H_4SO_4^-Na^+$	0.5 *M* NaCl	25	8.3	Dahanayake et al. (1986)
$C_{12}H_{25}O(C_2H_4)_2SO_4^-Na^+$	H_2O	10	2.8	Dahanayake et al. (1986)
$C_{12}H_{25}O(C_2H_4)_2SO_4^-Na^+$	H_2O	25	2.5	Dahanayake et al. (1986)
$C_{12}H_{25}O(C_2H_4)_2SO_4^-Na^+$	H_2O	40	2.0	Dahanayake et al. (1986)
$C_{12}H_{25}O(C_2H_4)_2SO_4^-Na^+$	0.1 *M* NaCl (aq.)	25	6.7	Dahanayake et al. (1986)
$C_{12}H_{25}O(C_2H_4)_2SO_4^-Na^+$	0.5 *M* NaCl (aq.)	25	10.0	Dahanayake et al. (1986)
p-$C_8H_{17}C_6H_4SO_3^-Na^+$	H_2O	70	1.4	Lange (1964)
p-$C_{10}H_{21}C_6H_4SO_3^-Na^+$	H_2O	70	1.4	Lange (1964)
$C_{11}H_{23}$-2-$C_6H_4SO_3^-Na^+$	"Hard river" water (I.S. = 6.6×10^{-3} *M*)	30	9.7	Zhu et al. (1998)
p-(1,3,5,7-tetra Me) $C_8H_{17}C_6H_4SO_3^-Na^+$	H_2O	75	2.8	Greiss (1955)
$C_{12}H_{25}$-2-$C_6H_4SO_3^-Na^+$	"Hard river" water (I.S. = 6.6×10^{-3} *M*)	30	5.0	Zhu et al. (1998)
$C_{12}H_{25}$-4-$C_6H_4SO_3^-Na^+$	"Hard river" water (I.S. = 6.6×10^{-3} *M*)	30	17.4	Zhu et al. (1998)
$C_{12}H_{25}$-6-$C_6H_4SO_3^-Na^+$	"Hard river" water (I.S. = 6.6×10^{-3} *M*)	30	21.5	Zhu et al. (1998)
p-$C_{12}H_{25}C_6H_4SO_3^-Na^+$	H_2O	70	1.3	Lange (1964)

(*Continued*)

TABLE 3.5 (*Continued*)

Compound	Solvent	Temp. (°C)	CMC/C_{20} Ratio	Reference
$p\text{-}C_{12}H_{25}C_6H_4SO_3^-Na^+$	H_2O	75	$1._5$	Greiss (1955)
$C_{12}H_{25}C_6H_4SO_3^-Na^{+c}$	0.1 *M* NaCl	25	$11._6$	Murphy et al. (1990)
$C_{13}H_{27}\text{-2-}C_6H_4SO_3^-Na^+$	"Hard river" water (I.S. = 6.6×10^{-3} *M*)	30	$3._1$	Zhu et al. (1998)
$C_{13}H_{27}\text{-5-}C_6H_4SO_3^-Na^+$	H_2O	30	$7._6$	Zhu et al. (1998)
$C_{13}H_{27}\text{-5-}C_6H_4SO_3^-Na^+$	"Hard river" water (I.S. = 6.6×10^{-3} *M*)	30	$15._8$	Zhu et al. (1998)
$p\text{-}C_{14}H_{29}C_6H_4SO_3^-Na^+$	H_2O	70	$1._5$	Lange (1964)
$p\text{-}C_{16}H_{33}C_6H_4SO_3^-Na^+$	H_2O	70	$1._9$	Lange (1964)
$C_{16}H_{33}\text{-7-}C_6H_4SO_3^-Na^+$	H_2O	45	$14._4$	Lascaux et al. (1983)
$C_4H_9OOCCH_2CH(SO_3^-Na^+)COOC_4H_9$	H_2O	25	$5._6$	Williams et al. (1957)
$C_6H_{13}OOCCH_2CH(SO_3^-Na^+)COOC_6H_{13}$	H_2O	25	$11._0$	Williams et al. (1957)
$C_4H_9CH(C_2H_5)CH_2COOCH_2CH(SO_3^-Na^+)\text{-}COOCH_2CH(C_2H_5)C_4H_9$	H_2O	25	28.	Williams et al. (1957)
$C_{11}H_{23}CON(CH_3)CH_2CH_2SO_3^-Na^+$	H_2O, pH 10.5	30	$2._0$	Tsubone and Rosen (2001)
$C_{11}H_{23}CON(CH_3)CH_2CH_2SO_3^-Na^+$	0.1 *M* NaCl (aq.), pH 10.5	30	$5._5$	Tsubone and Rosen (2001)
Cationics				
$C_{10}H_{21}N^+(CH_3)_3Br^-$	0.1 *M* NaCl (aq.)	25	$2._7$	Li et al. (2001)
$C_{12}H_{25}N^+(CH_3)_3Br^-$	H_2O	25	$2._4$	Zhu and Rosen (1985)
$C_{12}H_{25}N^+(CH_3)_3Br^-$	0.1 *M* NaBr (aq.)	25	$6._9$	Zhu and Rosen (1985)
$C_{12}H_{25}N^+(CH_3)_3Cl^-$	0.1 *M* NaCl (aq.)	25	$3._0$	Li et al. (2001)
$C_{14}H_{29}N^+(CH_3)_3Br^-$	H_2O	30	$2._4$	Venable and Nauman (1964)

$C_{12}H_{25}Pyr^+Br^-$[a]	H_2O	10	$2._7$	Rosen et al. (1982b)
$C_{12}H_{25}Pyr^+Br^-$[a]	H_2O	25	$2._5$	Rosen et al. (1982b)
$C_{12}H_{25}Pyr^+Br^-$[a]	H_2O	40	$2._1$	Rosen et al. (1982b)
$C_{12}H_{25}Pyr^+Br^-$[a]	0.1 *M* NaBr (aq.)	25	$6._9$	Rosen et al. (1982b)
$C_{12}H_{25}Pyr^+Br^-$[a]	0.5 *M* NaBr (aq.)	25	$8._9$	Rosen et al. (1982b)
$C_{12}H_{25}Pyr^+Cl^-$	H_2O	10	$2._3$	Rosen et al. (1982b)
$C_{12}H_{25}Pyr^+Cl^-$	H_2O	25	$2._0$	Rosen et al. (1982b)
$C_{12}H_{25}Pyr^+Cl^-$	H_2O	40	$1._8$	Rosen et al. (1982b)
$C_{12}H_{25}Pyr^+Cl^-$	0.1 *M* NaCl (aq.)	25	$4._6$	Rosen et al. (1982b)
$C_{12}H_{25}Pyr^+Cl^-$	0.5 *M* NaCl (aq.)	25	$5._5$	Rosen et al. (1982b)
$C_{12}H_{25}NPyr^+I^-$[a]	H_2O	25	$2._4$	Mandru (1972)
Anionic–Cationic Salts				
$C_{10}H_{21}N^+(CH_3)_3.C_{10}H_{21}SO_4^-$	H_2O	25	$9._1$	Corkill (1963a, b)
$C_{12}H_{25}SO_3^-.HON(CH_3)_2C_{12}H_{25}$	H_2O	25	$13._6$	M. J. Rosen and M. Gross, unpublished data
Zwitterionics				
$C_{10}H_{21}N^+(CH_3)_2CH_2COO^-$	H_2O	23	$7._0$	Beckett and Woodward (1963)
$C_{10}H_{21}CH(Pyr^+)COO^-$	H_2O	25	$3._9$	Zhao and Rosen (1984)
$C_{10}H_{21}N^+(CH_3)(CH_2C_6H_5)CH_2COO^-$	H_2O	10	$13._8$	Dahanayake and Rosen (1984)
$C_{10}H_{21}N^+(CH_3)(CH_2C_6H_5)CH_2COO^-$	H_2O	25	$12._0$	Dahanayake and Rosen (1984)
$C_{10}H_{21}N^+(CH_3)(CH_2C_6H_5)CH_2COO^-$	H_2O	40	$8._7$	Dahanayake and Rosen (1984)
$C_{10}H_{21}CH(COO^-).N^+(CH_3)_3$	H_2O	27	$5._7$	Tori and Nakagawa (1963b)
$C_{10}H_{21}N^+(CH_3)(CH_2C_6H_5)CH_2CH_2SO_3^-$	H_2O	40	$7._6$	Dahanayake and Rosen (1984)

(*Continued*)

TABLE 3.5 (*Continued*)

Compound	Solvent	Temp. (°C)	CMC/C_{20} Ratio	Reference
$C_{12}H_{25}N^+.(CH_3)_2CH_2COO^-$	H_2O	23	$6_{.5}$	Beckett and Woodward (1963)
$C_{12}H_{25}CH(Pyr^+)COO^-$	H_2O	25	$5_{.7}$	Zhao and Rosen (1984)
$C_{12}H_{25}N^+(CH_3)(CH_2C_6H_5)CH_2COO^-$	H_2O	10	$15_{.8}$	Dahanayake and Rosen (1984)
$C_{12}H_{25}N^+(CH_3)(CH_2C_6H_5)CH_2COO^-$	H_2O	25	$14_{.4}$	Dahanayake and Rosen (1984)
$C_{12}H_{25}N^+(CH_3)(CH_2C_6H_5)CH_2COO^-$	H_2O	40	$11_{.0}$	Dahanayake and Rosen (1984)
$C_{12}H_{25}N^+(CH_3)(CH_2C_6H_5)CH_2COO^-$	0.1 *M* NaCl (aq.), pH 5.7	25	$15_{.1}$	Rosen and Sulthana (2001)
$C_{12}H_{25}CH(COO^-).N^+(CH_3)_3$	H_2O	27	$7_{.8}$	Tori and Nakagawa (1963b)
$C_{14}H_{29}N^+.(CH_3)_2CH_2COO^-$	H_2O	25	$7_{.5}$	Beckett and Woodward (1963)
$C_{14}H_{29}CH(Pyr^+)COO^-$	H_2O	40	$6_{.2}$	Zhao and Rosen (1984)
$C_{16}H_{33}N^+.(CH_3)_2CH_2COO^-$	H_2O	23	$6_{.9}$	Beckett and Woodward (1963)
Nonionics				
$C_8H_{17}OCH_2CH_2OH$	H_2O	25	$7_{.2}$	Shinoda et al. (1959)
$C_8H_{17}CHOHCH_2OH$	H_2O	25	$9_{.6}$	Kwan and Rosen (1980)
$C_8H_{17}CHOHCH_2CH_2OH$	H_2O	25	$8_{.9}$	Kwan and Rosen (1980)
$C_{12}H_{25}CHOHCH_2CH_2OH$	H_2O	25	$7_{.7}$	Kwan and Rosen (1980)

β-Decyl glucoside	0.1 *M* NaCl, pH = 9	25	$11._{1}$	Li et al. (2001)
β-Decyl maltoside	0.1 *M* NaCl, pH = 9	25	$6._{5}$	Li et al. (2001)
β-Decyl maltoside	0.1 *M* NaCl, pH = 9	25	$7._{1}$	Li et al. (2001)
$C_{11}H_{23}CON(CH_2CH_2OH)_2$	H_2O	25	$6._{3}$	M. J. Rosen and M. Gross, unpublished data
$C_{11}H_{23}CON(CH_3)CH_2CHOHCH_2OH$	0.1 *M* NaCl	25	$10._{9}$	Zhu et al. (1999)
$C_{10}H_{21}CON(CH_3)CH_2(CHOH)_4CH_2OH$	0.1 *M* NaCl	25	$10._{5}$	Zhu et al. (1999)
$C_{11}H_{23}CON(CH_3)CH_2(CHOH)_4CH_2OH$	0.1 *M* NaCl	25	$8._{7}$	Zhu et al. (1999)
$C_{12}H_{25}CON(CH_3)CH_2(CHOH)_4CH_2OH$	0.1 *M* NaCl	25	$7._{8}$	Zhu et al. (1999)
$C_{13}H_{27}CON(CH_3)CH_2(CHOH)_4CH_2OH$	0.1 *M* NaCl	25	$4._{0}$	Zhu et al. (1999)
$C_{10}H_{21}N(CH_3)CO(CHOH)_4CH_2OH$	H_2O	20	$5._{2}$	Burczyk et al. (2001)
$C_{12}H_{25}N(CH_3)CO(CHOH)_4CH_2OH$	H_2O	20	$8._{8}$	Burczyk et al. (2001)
$C_{14}H_{29}N(CH_3)CO(CHOH)_4CH_2OH$	H_2O	20	$8._{5}$	Burczyk et al. (2001)
$C_{16}H_{33}N(CH_3)CO(CHOH)_4CH_2OH$	H_2O	20	$10._{1}$	Burczyk et al. (2001)
$C_{18}H_{37}N(CH_3)CO(CHOH)_4CH_2OH$	H_2O	20	$8._{1}$	Burczyk et al. (2001)
$C_8H_{17}(OC_2H_4)_5OH$	H_2O	25	$12._{7}$	Varadaraj et al. (1991b)
$C_8H_{17}(OC_2H_4)_5OH$	H_2O	40	$15._{1}$	Varadaraj et al. (1991b)
$C_8H_{17}(OC_2H_4)_5OH$	0.1 *M* NaCl	25	$8._{4}$	Varadaraj et al. (1991b)
$C_{10}H_{21}(OC_2H_4)_8OH$	H_2O	25	$16._{7}$	Meguro et al. (1981)
4-branched $C_{12}H_{25}(OC_2H_4)_5OH$	H_2O	25	$23._{0}$	Varadaraj et al. (1991b)
4-branched $C_{12}H_{25}(OC_2H_4)_5OH$	H_2O	40	$37._{6}$	Varadaraj et al. (1991b)
4-branched $C_{12}H_{25}(OC_2H_4)_5OH$	0.1 *M* NaCl	40	$19._{2}$	Varadaraj et al. (1991b)
$C_{12}H_{25}(OC_2H_4)_3OH$	H_2O	25	$11._{4}$	Rosen et al. (1982a)
$C_{12}H_{25}(OC_2H_4)_4OH$	H_2O	10	$17._{9}$	Rosen et al. (1982a)
$C_{12}H_{25}(OC_2H_4)_4OH$	H_2O	25	$13._{7}$	Rosen et al. (1982a)
$C_{12}H_{25}(OC_2H_4)_4OH$	H_2O	40	$11._{8}$	Rosen et al. (1982a)
$C_{12}H_{25}(OC_2H_4)_5OH$	H_2O	25	$15._{0}$	Rosen et al. (1982a)
$C_{12}H_{25}(OC_2H_4)_5OH$	0.1 *M* NaCl	25	$18._{5}$	Varadaraj et al. (1991a, 1991b)
$C_{12}H_{25}(OC_2H_4)_7OH$	H_2O	10	$17._{1}$	Rosen et al. (1982a)

(*Continued*)

TABLE 3.5 (*Continued*)

Compound	Solvent	Temp. (°C)	CMC/C_{20} Ratio	Reference
$C_{12}H_{25}(OC_2H_4)_7OH$	H_2O	25	14.9	Rosen et al. (1982a)
$C_{12}H_{25}(OC_2H_4)_7OH$	H_2O	40	13.9	Rosen et al. (1982a)
$C_{12}H_{25}(OC_2H_4)_8OH$	H_2O	10	17.5	Rosen et al. (1982a)
$C_{12}H_{25}(OC_2H_4)_8OH$	H_2O	25	17.3	Rosen et al. (1982a)
$C_{12}H_{25}(OC_2H_4)_8OH$	H_2O	40	15.4	Rosen et al. (1982a)
6-branched $C_{13}H_{27}(OC_2H_4)_5OH$	H_2O	25	43.0	Varadaraj et al. (1991b)
6-branched $C_{13}H_{27}(OC_2H_4)_5OH$	0.1 *M* NaCl	25	35.7	Varadaraj et al. (1991b)
$C_{13}H_{27}(OC_2H_4)_5OH$	H_2O	25	10.7	Varadaraj et al. (1991b)
$C_{13}H_{27}(OC_2H_4)_5OH$	H_2O	40	19.0	Varadaraj et al. (1991b)
$C_{13}H_{27}(OC_2H_4)_5OH$	0.1 *M* NaCl	25	8.8	Varadaraj et al. (1991b)
$C_{13}H_{27}(OC_2H_4)_8OH$	H_2O	25	11.3	Meguro et al. (1981)
$C_{14}H_{29}(OC_2H_4)_8OH$	H_2O	25	8.4	Meguro et al. (1981)
$C_{15}H_{31}(OC_2H_4)_8OH$	H_2O	25	7.1	Meguro et al. (1981)
p-t-$C_8H_{17}C_6H_4(OC_2H_4)_3OH$	H_2O	25	11.1	Crook et al. (1964)
p-t-$C_8H_{17}C_6H_4(OC_2H_4)_3OH$	H_2O	55	8.9	Crook et al. (1964)
p-t-$C_8H_{17}C_6H_4(OC_2H_4)_4OH$	H_2O	25	17.3	Crook et al. (1964)
p-t-$C_8H_{17}C_6H_4(OC_2H_4)_4OH$	H_2O	55	10.7	Crook et al. (1964)
p-t-$C_8H_{17}C_6H_4(OC_2H_4)_6OH$	H_2O	25	18.2	Crook et al. (1964)
p-t-$C_8H_{17}C_6H_4(OC_2H_4)_6OH$	H_2O	55	10.9	Crook et al. (1964)
p-t-$C_8H_{17}C_6H_4(OC_2H_4)_8OH$	H_2O	25	21.5	Crook et al. (1964)
p-t-$C_8H_{17}C_6H_4(OC_2H_4)_{10}OH$	H_2O	25	17.4	Crook et al. (1964)

[a] Pyr^+, pyridinium.
[b] From dodecyl alcohol with 4.4 methyl branches in the molecule.
[c] Commercial material.
I.S. = ionic strength.

7. Is increased considerably by the replacement of air as the second phase at the interface by a saturated aliphatic hydrocarbon and decreased slightly when the second liquid phase is a short-chain aromatic or unsaturated hydrocarbon.

The greater steric effect on micellization than on adsorption at the aqueous solution–air interface is illustrated by (2), (3), (5), and (6); the greater effect of the electrical factor on adsorption than on micellization is illustrated by (4). The greater difficulty of accommodating a bulky hydrophobic group in the interior of a spherical or cylindrical micelle rather than at a planar interface (e.g., air–water) is presumably the reason for observations (2) and (6) above. The increase in the CMC/C_{20} ratio with replacement of air by a saturated aliphatic hydrocarbon is due to an increased tendency to adsorb at the latter interface (as evidenced by larger pC_{20} values, Table 2.2), while the micellization tendency is not changed significantly. The small decrease in the ratio when the second phase is an aromatic or unsaturated hydrocarbon is due to the increased tendency to form micelles, which is almost equaled by the increased tendency to adsorb.

For POE nonionics, (1) the ratio increases with increase in the number of OE units in the POE chain at constant hydrophobic chain length, the effect becoming less pronounced as the number of EO units increases, and (2) the ratio decreases with increase in the length of the alkyl chain, at constant number of EO units in the POE chain. The first effect is due to the increase in the size of the hydrophilic head with this change; the second effect may reflect the larger diameter of the micelle as the alkyl chain is increased, with a resulting larger surface area to accommodate the hydrophilic head groups.

The decrease in the CMC/C_{20} ratio with increase in temperature (10–40°C) presumably occurs either because the size of the hydrophilic group decreases as a result of dehydration with this change or because the surface area of the micelle increases with this change.

In general, then, ionic surfactants (both anionic and cationic) with a single straight-chain hydrophobic group, in distilled water against air at room temperature, show low CMC/C_{20} ratios of 3 or less, while POE nonionics under the same conditions show ratios of about 7 or more. Increase in the electrolyte content of the solution causes the CMC/C_{20} ratios of ionics to approach those of nonionics. Zwitterionics have CMC/C_{20} ratios intermediate between those of ionics and POE nonionics.

VI. CMCs IN NONAQUEOUS MEDIA

When the structure of the solvent is not distorted significantly by the presence in it of a surfactant, a CMC of the type observed in aqueous media is not present (Ruckenstein and Nagarajan, 1980). There is no sharp change in aggregation number over a narrow concentration range and consequently no marked change in the surface or bulk properties of the solution in that region. In nonpolar solvents, the surfactant molecules may aggregate due to dipole–dipole interactions between the hydrophilic head groups, producing structures that have been called *reverse micelles*, with the head groups oriented toward each other in the interior of the structure and the hydrophobic groups oriented toward the nonpolar solvent. However, in the absence of additives such as water, the aggregation numbers are generally so small (seldom exceeding 10) that analogies with micelles in aqueous media are misleading. When the polarity of the solvent is large, solvent–surfactant interaction is not very different from that between surfactant molecules themselves and the latter consequently remain essentially individually dissolved. In ethylene glycol, glycerol, and similar solvents having multiple hydrogen-bonding capacity, surfactant aggregates are assumed to have the normal structure.

Some investigators have assigned a CMC value to the range where a discontinuity appears in the plot of some property of the surfactant solution in nonaqueous media, even when the change is not sharp. Some values are listed in Table 3.6.

VII. EQUATIONS FOR THE CMC BASED ON THEORETICAL CONSIDERATIONS

Equations relating the CMC to the various factors that determine it have been derived from theoretical considerations by Hobbs (1951), Shinoda (1953), and Molyneux et al. (1965). These equations are based on the fact that for nonionics, the CMC is related to the free energy change ΔG_{mic} associated with the aggregation of the individual surfactant molecules to form micelles by the expression

$$\Delta G_{mic} = 2.3RT \log x_{CMC}. \tag{3.4}$$

x_{CMC} is the mole fraction of the surfactant in the liquid phase at the CMC. In aqueous solutions where the CMC is generally $<10^{-1}$, $x_{CMC} = CMC/\omega$ without significant error, and

$$\Delta G_{mic} = 2.3RT(\log CMC - \log \omega) \tag{3.5}$$

from which

$$\log CMC = \frac{\Delta G_{mic}}{2.3RT} + \log \omega, \tag{3.6}$$

where ω is the molar concentration of water (55.3 at 25°C).

ΔG_{mic} can be broken into contributions from the component parts of the surfactant molecule, $CH_3(CH_2)_mW$, where W = the hydrophilic group, in the following fashion:

TABLE 3.6 CMCs of Surfactants in Nonaqueous Media

Surfactant	Temp. (°C)	Solvent	CMC (*M*)	Reference
$C_4H_9NH_3^+.C_2H_5COO^-$	30	Benzene	$(4.5–5.5) \times 10^{-2}$	Fendler et al. (1973b)
$C_4H_9NH_3^+.C_2H_5COO^-$	30	CCl_4	$(2.3–2.6) \times 10^{-2}$	Fendler et al. (1973b)
$C_8H_{17}NH_3^+.C_2H_5COO^-$	30	Benzene	$(1.5–1.7) \times 10^{-2}$	Fendler et al. (1973b)
$C_8H_{17}NH_3^+.C_2H_5COO^-$	30	CCl_4	$(2.6–3.1) \times 10^{-2}$	Fendler et al. (1973b)
$C_{12}H_{25}NH_3^+.C_2H_5COO^-$	30	Benzene	$(3–7) \times 10^{-3}$	Fendler et al. (1973b)
$C_{12}H_{25}NH_3^+.C_2H_5COO^-$	30	CCl_4	$(2.1–2.5) \times 10^{-2}$	Fendler et al. (1973b)
$C_8H_{17}NH_3^+.C_5H_{11}COO^-$	30	Benzene	$(4.1–4.5) \times 10^{-2}$	Fendler et al. (1973b)
$C_8H_{17}NH_3^+.C_5H_{11}COO^-$	30	CCl_4	$(4.2–4.5) \times 10^{-2}$	Fendler et al. (1973b)
$C_8H_{17}NH_3^+.C_{13}H_{27}COO^-$	30	Benzene	$(1.9–2.2) \times 10^{-2}$	Fendler et al. (1973b)
$C_8H_{17}NH_3^+.C_{13}H_{27}COO^-$	30	CCl_4	$(2.8–4.0) \times 10^{-2}$	Fendler et al. (1973b)
$C_{12}H_{25}NH_3^+.C_3H_7COO^-$	10	Benzene	3×10^{-3}	Kitahara (1956)
$C_{12}H_{25}NH_3^+.C_5H_{11}COO^-$	10	Benzene	18×10^{-3}	Kitahara (1956)
$C_{12}H_{25}NH_3^+.C_7H_{15}COO^-$	10	Benzene	20×10^{-3}	Kitahara (1956)
$C_{18}H_{37}NH_3^+.C_3H_7COO^-$	10	Benzene	5×10^{-3}	Kitahara (1956)
Na bis(2-ethylhexyl)-sulfosuccinate	20	Benzene	3×10^{-3}	Kon-no and Kitahara (1965)
Na bis(2-ethylhexyl)-sulfosuccinate	25	Pentane	4.9×10^{-4}	Eicke and Rehak (1976)
$C_9H_{19}\ C_6H_4(OC_2H_4)_9OH$	27.5	Glycerol	8.0×10^{-6}	Ray (1971)
$C_9H_{19}\ C_6H_4(OC_2H_4)_9OH$	27.5	Ethylene glycol	7.1×10^{-4}	Ray (1971)
$C_9H_{19}\ C_6H_4(OC_2H_4)_9OH$	27.5	Propylene glycol	5.0×10^{-2}	Ray (1971)
$C_{12}H_{25}\ (OC_2H_4)_2OH$	—	Benzene	7.6×10^{-3}	Becher (1960)
$C_{13}H_{27}\ (OC_2H_4)_6OH$	—	Benzene	2.6×10^{-3}	Becher (1960)
$C_{12}H_{25}Pyr^+Br^-$ [a]	40	Benzene	5.5×10^{-3}	Miyagishi et al. (1977)
$C_{18}H_{37}Pyr^+Br^-$	40	Benzene	4.4×10^{-3}	Miyagishi et al. (1977)
$C_6F_{13}(CH_2)_3(OC_2H_4)_2OH$	—	C_6F_6	9.3×10^{-3}	Mathis (1982)
$C_8F_{17}C_2H_4N(C_2H_4OH)_2$	—	C_6F_6	3.65×10^{-4}	Mathis (1982)

[a] Pyr^+, pyridinium.

$$\Delta G_{\text{mic}} = \Delta G_{\text{mic}}\,(-\text{CH}_3) + m\,\Delta G_{\text{mic}}\,(-\text{CH}_2-) + \Delta G_{\text{mic}}\,(-W). \tag{3.7}$$

Studies on the solubility of alkanes in water indicate that $\Delta G_{\text{mic}}(\text{-CH}_3)$ does not change with increase in the length of the alkyl chain and can be represented by $\Delta G_{\text{mic}}\ (\text{-CH}_3) = \Delta G_{\text{mic}}\ (\text{-CH}_2\text{-}) + k$, where k is a constant. Thus,

$$\log \text{CMC} = \frac{\Delta G_{\text{mic}}(-W) + k}{2.3RT} + \log \omega + \left[\frac{\Delta G_{\text{mic}}(-\text{CH}_2-)}{2.3RT}\right] N, \tag{3.8}$$

where $N = m + 1$, the total number of carbon atoms in the hydrophobic group.

If we assume that the contribution of the hydrophilic head group $\Delta G(\text{-}W)$ and the fraction of counterions bound to the micelle, α, do not change with increase in the length of the hydrophobic group, then for any homologous series of surfactants, the relations between the CMC and the number of carbon atoms in the hydrophobic group can be put into the form

$$\log \text{CMC} = A - BN, \tag{3.1}$$

where

$$A = \frac{-\Delta G_{\text{mic}}(-W) + k}{2.3RT} + \log \omega \tag{3.9}$$

and

$$B = \left[\frac{-\Delta G_{\text{mic}}(-\text{CH}_2-)}{2.3RT}\right]. \tag{3.10}$$

Thus, A and B are constants reflecting the free energy changes involved in transferring the hydrophilic group and a methylene unit of the hydrophobic group, respectively, from an aqueous environment to the micelle. This accounts for both the form of the empirical relation between the CMC and the number of carbon atoms in the hydrophobic group that has been discussed previously, and the relatively small variation of B in different homologous series of ionic surfactants.

We can also see from Equations 3.1 and 3.10 and the experimental values of B given in Table 3.4 that the free energy change $\Delta G\ (\text{-CH}_2\text{-})$ involved in the transfer of a methylene unit of the hydrophobic group from an aqueous environment to the interior of the micelle is negative, thus favoring micellization, which accounts for the fact that the CMC decreases with increase in the length of the hydrophobic group. From Equation 3.9 and the values of A in Table 3.4, we can see that the free energy change involved in the transfer of the hydrophilic group from an aqueous environment to the exterior of the micelle is positive, and therefore oppose micellization.

If the value of the CMC is replaced by the *activity* of the surfactant at the CMC (CMA), and log CMA is plotted against N (Nakagaki, 1984), then the value of the slope B for ionics is close to that for nonionics and zwitterionics, indicating similar values for ΔG_{mic} (-CH_2-), the free energy change involved in the transfer of a methylene group from the aqueous solution to the micelle, for all types of surfactants. The value of CMA is obtained for univalent surfactants such as sodium alkyl sulfates or alkyltrimethylammonium halides from

$$\text{CMA} = f_{\pm}^{2}\text{CMC}\,(\text{CMC} + C_i) \tag{3.11}$$

and, in the case of divalent surfactants, such as disodium alkyl phosphates, from

$$\text{CMA} = f_{\pm}^{3}\text{CMC}(\text{CMC} + C_i)^2, \tag{3.12}$$

where C_i is the concentration of added electrolyte with a common counterion, and $f_{\pm}$ is the mean ionic activity coefficient of the surfactant, calculated by

$$\log f_{\pm} = \frac{-A|Z^{+}.Z^{-}|(I)^{1/2}}{1+(I)^{1/2}}. \tag{3.13}$$

Z^+, Z^- are the valences of the ions comprising the surfactant, I is the ionic strength of the solution, and $A = 1.825 \times 10^6 (DT)^{3/2}$, with D being the dielectric constant of the solvent.

The value of ΔG_{mic}(-CH_2-) obtained in this fashion falls in the range –(2.8–3.3) kJ[–(708–777)cal/mol] for all types of surfactants (nonionics, zwitterionics, uni- and divalent ionics), irrespective of the presence or amount of added electrolyte.

For ionic surfactants $\Delta G(\text{-}W)$, the electrical energy E_{el} involved in transferring the ionic hydrophilic group from an aqueous environment to the micelle is given, when the aggregation number is not too small, by (Shinoda and Hirai, 1977)

$$E_{\text{el}} = (K_g / Z_i)RT\left(\ln\frac{2000\pi\sigma^2}{\varepsilon_r RT} - \ln C_i\right), \tag{3.14}$$

where (K_g/Z_i) is the slope of the plot of CMC versus total concentration C_i, in equivalents per liter, of the counterions of charge Z_i in the solution, σ is the charge density on the micelle surface, ε_r is the dielectric constant of the solvent, and K_g is the effective coefficient of electrical energy of micellization. From this,

$$\log \text{CMC} = K_g / Z_i\left(\log\frac{2000\pi\sigma^2}{\varepsilon_r RT} - \log C_i\right) + \left[\frac{\Delta G(-\text{CH}_2-)}{2.3RT}\right]N + \text{constant}. \tag{3.15}$$

Equation 3.15 predicts the effect of electrolyte on the CMC of ionic surfactants, indicating that the log of the CMC will decrease linearly with $\log C_i$, which is in accordance with experimental findings (Equation 3.3). It also indicates that the CMC of ionic surfactants will decrease with increase in the extent of binding of the counterion to the micelle since that decreases the charge density on the micellar surface. Organic additives that decrease the dielectric constant of the solvent will increase the CMC of the surfactant, both of which are consistent with the experimental results discussed previously. The effect of temperature on the CMC of ionic surfactants is difficult to predict from Equation 3.15. An increase in the temperature should cause a direct decrease in the CMC, but since an increase in temperature causes a decrease in the dielectric constant ε_r of the solvent and may also affect σ, the overall effect of an increase in temperature is not readily determinable from the equation alone.

VIII. THERMODYNAMIC PARAMETERS OF MICELLIZATION

As is evident from the previous discussion, a clear understanding of the process of micellization is necessary for rational explanation of the effects of structural and environmental factors on the value of the CMC and for predicting the effects on it of new structural and environmental variations. The determination of thermodynamic parameters of micellization ΔG_{mic}, ΔH_{mic}, and ΔS_{mic} has played an important role in developing such an understanding.

A standard free energy of micellization ΔG^0_{mic} may be calculated by choosing (Molyneux et al., 1965) for the standard initial state of the nonmicellar surfactant species a hypothetical state at unit mole fraction x, but with the individual ions or molecules behaving as at infinite dilution, and for the standard final state, the micelle itself. For nonionic surfactants, the standard free energy of micellization is given by

$$\Delta G^0_{\text{mic}} = RT \ln x_{\text{CMC}}. \tag{3.16}$$

When the CMC is $10^{-2}\,M$ or less, this can be approximated without significant error by

$$\Delta G^0_{\text{mic}} = 2.3RT \log(\text{CMC}/\omega), \tag{3.16a}$$

where the CMC is expressed in molar units and ω is the number of moles of water per liter of water at that absolute temperature T. For ionic surfactants, a standard free energy change of micellization, ΔG^0_{mic}, can be calculated by taking into account the degree of binding of the counterion to the micelle, $1 - \alpha$. Thus, for ionic surfactants of the 1:1 electrolyte type (Nakagaki, 1984; Zana, 1996),

$$\Delta G^0_{\text{mic}} = RT[1+(1-\alpha)]\ln x_{\text{CMC}} = 2.3RT(2-\alpha)\ln x_{\text{CMC}}, \tag{3.16b}$$

where α is the degree of ionization of the surfactant, measured by the ratio of the slopes of the specific conductivity versus C plotted above and below the CMC (Figure 3.6, Section IVA3), and x_{CMC} is the mole fraction of the surfactant in the liquid phase at the CMC.

For ionic surfactants with divalent counterions (Zana, 1996),

$$\Delta G^0 = RT[1+(1-\alpha)/2]\ln(\text{CMC}/\omega) = 2.3RT[1+(1-\alpha)/2]\log(\text{CMC}/\omega). \tag{3.16c}$$

Since

$$\Delta G^0_{\text{mic}} = \Delta H^0_{\text{mic}} - T\Delta S^0_{\text{mic}}, \tag{3.17}$$

$$d(\Delta G^0_{\text{mic}})/dT = \Delta S^0_{\text{mic}} \tag{3.18}$$

if ΔH^0_{mic} is constant over the temperature range investigated. Alternatively,

$$T^2 d\left(\Delta G^0_{\text{mic}}/T\right)/dT = \Delta H^0_{\text{mic}} \tag{3.19}$$

if ΔS^0_{mic} is constant over the temperature range investigated. These relations are strictly correct only when the variation in aggregation number of the micelles with temperature is negligible (Birdi, 1974), which is often not true for polyoxyethylenated nonionics. This has usually been disregarded by most investigators.

Some values of ΔG^0_{mic}, ΔH^0_{mic}, and ΔS^0_{mic}, are listed in Table 3.7. Values of ΔH^0_{mic} can also be determined calorimetrically, thus avoiding some of the problems mentioned here.

The data available (mainly for aqueous systems) indicate that the negative values of ΔG^0_{mic} are due mainly to the large positive values of ΔS^0_{mic}. ΔH^0_{mic} is often positive and, even when negative, is much smaller than the value of $T\Delta S^0_{\text{mic}}$. Therefore, the micellization process is governed primarily by the entropy gain associated with it, and the driving force for the process is the tendency of the lyophobic group of the surfactant to transfer from the solvent environment to the interior of the micelle.

This large entropy increase on micellization in aqueous medium has been explained in two ways: (1) structuring of the water molecules surrounding the hydrocarbon chains in aqueous medium, resulting in an increase in the entropy of the system when the hydrocarbon chains are removed from the aqueous medium to the interior of the micelle–"hydrophobic bonding" (Nemethy and Scheraga, 1962); (2) increased freedom of the hydrophobic chain in the nonpolar interior of the micelle compared to the aqueous environment (Stainsby and Alexander, 1950; Aranow and Witten, 1960, 1961, 1965). Any structural or environmental factors that may affect solvent–lyophobic group interactions or interactions between the lyophobic groups in the interior of the micelle will therefore affect ΔG^0_{mic} and consequently the value of the CMC.

TABLE 3.7 Thermodynamic Parameters of Micellization

Compound	Solvent	Temp. (°C)	ΔG^0_{mic} (kJ/mol)[a]	ΔH^0_{mic} (kJ/mol)[a]	$T\Delta S^0_{mic}$ (kJ/mol)[a]	Reference[b]
$C_{10}H_{21}SO_3^-Na^+$	H_2O	10	$-33._3$			Dahanayake et al. (1986)
				-3	-3_1	
$C_{10}H_{21}SO_3^-Na^+$	H_2O	25	$-34._9$			Dahanayake et al. (1986)
				$+8$	$+4_4$	
$C_{10}H_{21}SO_3^-Na^+$	H_2O	40	$-37._0$			Dahanayake et al. (1986)
$C_{12}H_{25}SO_3^-Na^+$	H_2O	10	$39._7$			Dahanayake et al. (1986)
				$+5$	$+4_6$	
$C_{12}H_{25}SO_3^-Na^+$	H_2O	40	$-42._0$			Dahanayake et al. (1986)
$C_{12}H_{25}SO_4^-Na^+$	H_2O	21	$-42._4$			Mukerjee (1967)
$C_{10}H_{21}OC_2H_4SO_3^-Na^+$	H_2O	10	$-34._7$			Dahanayake et al. (1986)
				-2_0	$+1_5$	
$C_{10}H_{21}OC_2H_4SO_3^-Na^+$	H_2O	25	$-35._5$			Dahanayake et al. (1986)
				-7	$+3_0$	
$C_{10}H_{21}OC_2H_4SO_3^-Na^+$	H_2O	40	$-37._0$			Dahanayake et al. (1986)
$C_{12}H_{25}OC_2H_4SO_4^-Na^+$	H_2O	10	$-42._2$			Dahanayake et al. (1986)
				-5	$+3_8$	
$C_{12}H_{25}OC_2H_4SO_4^-Na^+$	H_2O	25	$-44._1$			Dahanayake et al. (1986)
				-1_0	$+3_5$	

$C_{12}H_{25}OC_2H_4SO_4^-Na^+$	H_2O	40	$-45._8$			Dahanayake et al. (1986)
$C_{12}H_{25}(OC_2H_4)_2SO_4^-Na^+$	H_2O	10	$-41._7$			Dahanayake et al. (1986)
				+2	$+4_5$	
$C_{12}H_{25}(OC_2H_4)_2SO_4^-Na^+$	H_2O	25	$-44._0$			Dahanayake et al. (1986)
				−2	$+4_4$	
$C_{12}H_{25}(OC_2H_4)_2SO_4^-Na^+$	H_2O	40	$-46._2$			Dahanayake et al. (1986)
$C_{12}H_{25}Pyr^+Br^{-c}$	H_2O	10	$-36._4$			Rosen et al. (1982b)
				−2	$+3_5$	
$C_{12}H_{25}Pyr^+Br^-$	H_2O	25	$-38._2$			Rosen et al. (1982b)
				-1_4	$+2_5$	
$C_{12}H_{25}Pyr^+Br^-$	H_2O	40	$-39._4$			Rosen et al. (1982b)
$C_{12}H_{25}Pyr^+Cl^-$	H_2O	10	$-35._2$			Rosen et al. (1982b)
				+2	$+3_8$	
$C_{12}H_{25}Pyr^+Cl^-$	H_2O	25	$-37._1$			Rosen et al. (1982b)
				−4	$+3_4$	
$C_{12}H_{25}Pyr^+Cl^-$	H_2O	40	$-38._8$			Rosen et al. (1982b)
$C_{10}H_{21}N^+(CH_3)_2CH_2COO^-$	H_2O	23	$-19._8$			Beckett and Woodward (1963)
$C_{12}H_{25}N^+(CH_3)_2CH_2COO^-$	H_2O	23	$-25._4$			Beckett and Woodward (1963)
$C_{10}H_{21}N^+(CH_3)(CH_2C_6H_5)CH_2COO^-$	H_2O	10	$-21._4$			Dahanayake and Rosen (1984)
				+8	$+3_1$	

(*Continued*)

TABLE 3.7 (*Continued*)

Compound	Solvent	Temp. (°C)	ΔG^0_{mic} (kJ/mol)[a]	ΔH^0_{mic} (kJ/mol)[a]	$T\Delta S^0_{mic}$ (kJ/mol)[a]	Reference[b]
$C_{10}H_{21}N^+(CH_3)(CH_2C_6H_5)CH_2COO^-$	H_2O	25	$-23._0$			Dahanayake and Rosen (1984)
				−9	$+3_3$	
$C_{10}H_{21}N^+(CH_3)(CH_2C_6H_5)CH_2COO^-$	H_2O	40	$-24._6$			Dahanayake and Rosen (1984)
$C_{12}H_{25}N^+(CH_3)(CH_2C_6H_5)CH_2COO^-$	H_2O	10	$-26._9$			Dahanayake and Rosen (1984)
				+4	$+3_2$	
$C_{12}H_{25}N^+(CH_3)(CH_2C_6H_5)CH_2COO^-$	H_2O	25	$-28._6$			Dahanayake and Rosen (1984)
				−2	$+3_1$	
$C_{12}H_{25}N^+(CH_3)(CH_2C_6H_5)CH_2COO^-$	H_2O	40	$-30._1$			Dahanayake and Rosen (1984)
$C_{12}H_{25}N^+(CH_3)(CH_2C_6H_5)CH_2COO^-$	H_2O (dodecane satd.)	25	$-28._9$	−3 (35°)	$+2_7$ (35°)	Murphy and Rosen (1988)
$C_{12}H_{25}N^+(CH_3)(CH_2C_6H_5)CH_2COO^-$	H_2O (isooctane satd.)	25	$-29._2$	−3 (35°)	$+2_7$ (35°)	Murphy and Rosen (1988)
$C_{12}H_{25}N^+(CH_3)(CH_2C_6H_5)CH_2COO^-$	H_2O (toluene satd.)	25	$-30._8$	−2 (35°)	$+3_4$ (35°)	Murphy and Rosen (1988)

$C_{12}H_{25}N^+(CH_3)_2O^-$	H_2O	30	$-25._9$	+7	$+3_3$	Hermann (1962)
$C_{10}H_{21}(OC_2H_4)_8OH$	H_2O	25	$-27._0$	$+1_8$	$+4_5$	Meguro et al. (1981)
$C_{11}H_{23}(OC_2H_4)_8OH$	H_2O	25	$-30._0$	$+1_7$	$+4_7$	Meguro et al. (1981)
$C_{12}H_{25}(OC_2H_4)_2OH$	H_2O	25	$-35._5$	+3	$+3_9$	Rosen et al. (1982a)
$C_{12}H_{25}(OC_2H_4)_3OH$	H_2O	25	$-34._3$	+5	$+3_9$	Rosen et al. (1982a)
$C_{12}H_{25}(OC_2H_4)_4OH$	H_2O	25	$-33._8$	+8	$+4_2$	Rosen et al. (1982a)
$C_{12}H_{25}(OC_2H_4)_4OH$	55% w/w $HCONH_2$-H_2O	21	$-26._1$	—	—	McDonald (1970)
$C_{12}H_{25}(OC_2H_4)_2OH$	$HCONH_2$	25	$-17._0$	−2	$+1_5$	McDonald (1970)
$C_{10}H_{25}(OC_2H_4)_5OH$	H_2O	10	$-31._4$			Rosen et al. (1982a)
				$+1_6$	$+4_8$	
$C_{12}H_{25}(OC_2H_4)_5OH$	H_2O	25	$-33._9$			Rosen et al. (1982a)
				+4	$+3_9$	
$C_{12}H_{25}(OC_2H_4)_5OH$	H_2O	40	$-35._7$			Rosen et al. (1982a)
$C_{12}H_{25}(OC_2H_4)_6OH$	H_2O	25	$-33._0$	$+1_6$	$+4_9$	Corkill et al. (1964)
$C_{12}H_{25}(OC_2H_4)_6OH$	55% w/w $HCONH_2$-H_2O	25	$-25._2$	+2	$+2_7$	McDonald (1970)
$C_{12}H_{25}(OC_2H_4)_6OH$	$HCONH_2$	25	$-16._6$	−4	$+1_3$	McDonald (1970)
$C_{12}H_{25}(OC_2H_4)_7OH$	H_2O	25	$-33._2$	$+1_2$	$+4_5$	Rosen et al. (1982a)
$C_{12}H_{25}(OC_2H_4)_8OH$	H_2O	10	$-30._1$			Rosen et al. (1982a)
				$+1_7$	$+4_8$	

(Continued)

TABLE 3.7 (*Continued*)

Compound	Solvent	Temp. (°C)	ΔG^0_{mic} (kJ/mol)[a]	ΔH^0_{mic} (kJ/mol)[a]	$T\Delta S^0_{mic}$ (kJ/mol)[a]	Reference[b]
$C_{12}H_{25}(OC_2H_4)_8OH$	H_2O	25	$-32._6$			Rosen et al. (1982a)
				+9	$+4_3$	
$C_{12}H_{25}(OC_2H_4)_8OH$	H_2O	40	$-34._6$			Rosen et al. (1982a)
$C_{12}H_{25}(OC_2H_4)_8OH$	55% w/w $HCONH_2$ - H_2O	25	$-24._3$	+2	$+2_7$	McDonald (1970)
$C_{12}H_{25}(OC_2H_4)_8OH$	$HCONH_2$	25	$-16._2$	−3	$+1_3$	McDonald (1970)
$C_{12}H_{25}(OC_2H_4)_8OH$	H_2O - cyclohexane	25	$-32._8$			Rosen and Murphy (1991)
$C_{12}H_{25}(OC_2H_4)_8OH$	H_2O - heptane	25	$-33._0$			Rosen and Murphy (1991)
$C_{12}H_{25}(OC_2H_4)_8OH$	H_2O - hexadecane	25	$-32._7$			Rosen and Murphy (1991)
$C_{13}H_{27}(OC_2H_4)_8OH$	H_2O	25	$-35._9$	$+1_4$	+5o	Meguro et al. (1981)
$C_{14}H_{29}(OC_2H_4)_8OH$	H_2O	25	$-38._7$	$+1_3$	$+5_1$	Meguro et al. (1981)
$C_{15}H_{31}(OC_2H_4)_8OH$	H_2O	25	$-41._0$	$+1_1$	$+5_2$	Meguro et al. (1981)

[a] To convert to kcal/mol, divide by 4.18; values for ionic surfactants are independent of total ionic strength and are averages of values at different electrolyte contents.

[b] Parameters calculated from data in listed reference.

[c] Pyr^+, pyridinium.

In aqueous medium, an increase in the length of the hydrophobic group causes an increase in the value of ΔS^0_{mic}, and a usually smaller decrease in ΔH^0_{mic}, making ΔG^0_{mic} more negative by about 3 kJ per -CH_2- group. Variations in this value, ΔG^0_{mic} (-CH_2-), have been ascribed (Clint and Walker, 1975) to change in the degree of nonpolarity of the interior of the micelle with change in the polarity of the hydrophilic head, since penetration of water into the micelle, at least in the vicinity of the first five or six carbon atoms adjacent to the hydrophilic head, has been pointed out by several investigators (Clifford and Pethica, 1964; Benjamin, 1966; Walker, 1971).

In POE nonionics, both ΔH^0_{mic} and ΔS^0_{mic} appear to increase with increase in the number of OE units in the hydrophilic head, with the net result that ΔG^0_{mic} becomes slightly less negative. The increase in ΔH^0_{mic} is probably due to reduction in the degree of hydration of the OE groups on micellization. The change in ΔH^0_{mic} per OE unit above three units, ΔG^0_{mic} (-EO-), appears to be about one-tenth of the change in ΔG^0_{mic} per methylene group, ΔG^0_{mic} (-CH_2-), and is opposite in sign (Corkill et al., 1964), since EO groups oppose micellization, whereas methylene groups favor it. The terminal hydroxyl group appears to be the main structural unit opposing micellization (McDonald, 1970).

An increase in temperature seems to cause both ΔH^0_{mic} and ΔS^0_{mic} to become less positive (Hudson and Pethica, 1964) in POE nonionics, presumably because both the amount of water structured by the hydrophobic chain and the amount of water bound by the hydrophilic POE group in the nonmicellar species decrease with increase in temperature, resulting in a decrease in ΔS^0_{mic} and ΔH^0_{mic}, respectively. Since these two parameters have opposite effects on ΔG^0_{mic}, it may become more negative or less negative with temperature change, depending on the relative magnitude of the changes in ΔS^0_{mic} and ΔH^0_{mic}. From the available data, ΔG^0_{mic} appears to become more negative with increase in temperature up to about 50°C in most cases and then to become more positive with further increase in temperature (Crook et al., 1963).

In highly polar nonaqueous solvents, such as formamide, *N*-methylformamide, and *N,N*-dimethylformamide, from the limited data available, it appears that the driving force for micellization is again mainly entropic, that is, the tendency of the lyophobic group to transfer from the solvent environment to the interior of the micelle (McDonald, 1970).

IX. MIXED MICELLE FORMATION IN MIXTURES OF TWO SURFACTANTS

In many products or processes, two surfactants are used together to improve the properties of the system. In some cases, the two surfactants interact in such fashion that the CMC of the mixture (C^M_{12}) is always intermediate in value between those of the two components (C^M_1, C^M_2). In other cases, they interact in such fashion that C^M_{12} at some ratio of the two surfactants is less than either

C_1^M or C_2^M. When the latter case occurs, the system is said to exhibit *synergism* in mixed micelle formation. In still other cases, C_{12}^M at some ratio of the two surfactants may be larger than either C_1^M or C_2^M. Here the system is said to exhibit antagonism (*negative* synergism) in mixed micelle formation.

The CMC of the mixture is given by

$$\frac{1}{C_{12}^M} = \frac{\alpha}{f_1 C_1^M} + \frac{1-\alpha}{f_2 C_2^M}, \tag{3.20}$$

where α is the mole fraction of surfactant 1 in the solution phase on a surfactant-only basis (i.e., the mole fraction of surfactant 2 in the mixture is $1 - \alpha$), and f_1, f_2 are the activity coefficients of surfactants 1 and 2, respectively, in the mixed micelle. Using regular solution Equations (2.44 and 2.45) for the activity coefficients f_1 and f_2, Rubingh (1979) developed a convenient method (Equations 11.3 and 11.4) for predicting the CMC of any mixture of two surfactants from the CMC values (C_1^M, C_2^M) of the individual surfactants and one or more mixtures of them. When the values of the individual CMCs (Table 3.2) and of the interaction parameter for mixed micelle formation β^M are known (Table 11.1), the value of C_{12}^M can be calculated directly from these without any other experimental data. However, on commercial materials, the presence of surface-active impurities may cause serious deviations from values obtained without the use of some experimental data (Goloub et al., 2000).

When there is no interaction between the two surfactants, that is, the mixed is ideal, then $f_1 = f_2 = 1$ and Equation 3.20 becomes

$$\frac{1}{C_{12}^M} = \frac{\alpha}{C_1^M} + \frac{1-\alpha}{C_2^M} \tag{3.21}$$

or

$$C_{12}^M = \frac{C_1^M C_2^M}{C_1^M (1-\alpha) + C_2^M \alpha}. \tag{3.22}$$

The CMC value of any mixture can then be calculated at any value of α directly from the CMC values of the individual surfactants.

REFERENCES

Alexandridis, V., A. Athanassiou, S. Fukuda, and T. A. Hatton (1994) *Langmuir* **10**, 2604.

Amrhar, J., Y. Chevalier, B. Gallot, P. LePerchec, X. Auvray, and C. Petipas (1994) *Langmuir* **10**, 3435.

Anacker, E. W. and H. M. Ghose (1963) *J. Phys. Chem.* **67**, 1713.

Aranow, R. H. and L. Witten (1960) *J. Phys. Chem.* **64**, 1643.

Aranow, R. H. and L. Witten (1961) *J. Chem. Phys.* **35**, 1504.

Aranow, R. H. and L. Witten (1965) *J. Chem. Phys.* **43**, 1436.

Asakawa, T., M. Hashikawa, K. Amada, and S. Miyagishi (1995) *Langmuir* **11**, 2376.

Asakawa, T., H. Kitano, A. Ohta, and S. Miyagishi (2001) *J. Colloid Interface Sci.* **242**, 284.

Atik, S., M. Nam, and L. Singer (1979) *Chem. Phys. Lett.* **67**, 75.

Aveyard, R., B. P. Binks, J. Chen, J. Equena, P. D. I. Fletcher, R. Buscall, and S. Davies (1998) *Langmuir* **14**, 4699.

Balmbra, R. R., J. S. Clunie, J. M. Corkill, and J. F. Goodman (1962) *Trans. Faraday Soc.* **58**, 1661.

Balmbra, R. R., J. S. Clunie, J. M. Corkill, and J. F. Goodman (1964) *Trans. Faraday Soc.* **60**, 979.

Balzer, D. (1993) *Langmuir* **9**, 3375.

Barry, B. W., J. C. Morrison, and G. Russell (1970) *J. Colloid Interface Sci.* **33**, 554.

Becher, P. (1960) *J. Phys. Chem.* **64**, 1221.

Becher, P. (1961) *J. Colloid Sci.* **16**, 49.

Beckett, A. H. and R. J. Woodward (1963) *J. Pharm. Pharmacol.* **15**, 422.

Benjamin, L. (1966) *J. Phys. Chem.* **70**, 3790.

Binana-Limbele, W., R. Zana, and E. Platone (1988) *J. Colloid Interface Sci.* **124**, 647.

Binana-Limbele, W., N. M. Van Os, A. M. Rupert, and R. Zana (1991a) *J. Colloid Interface Sci.* **141**, 157.

Binana-Limbele, W., N. M. Van Os, A. M. Rupert, and R. Zana (1991b) *J. Colloid Interface Sci.* **144**, 458.

Birdi, K. S. Paper presented before 167th Am Chem. Soc. Meeting, Los Angeles, CA, April 1974.

Bocker, T. and J. Thiem (1989) *Tenside Surf. Det.* **26**, 318.

Boschkova, K., B. Kronberg, J. J. R. Stalgren, K. Persson, and M. R. Salageon (2002) *Langmuir* **18**, 1680.

Bostrom, G., S. Backlund, A. M. Blokhus, and H. Hoeiland (1989) *J. Colloid Interface Sci.* **128**, 169.

Burczyk, B., K. A. Wilk, A. Sokolowski, and L. Syper (2001) *J. Colloid Interface Sci.* **240**, 552.

Caskey, J. A. and W. B., Jr. Barlage (1971) *J. Colloid Interface Sci.* **35**, 46.

Cebula, D. J. and R. H. Ottewill (1982) *Coll. Polym. Sci.* **260**, 1118.

Chen, L.-J., S.-Y. Lin, C.-C. Huang, and E.-M. Chen (1998) *Coll. Surfs. A* **135**, 175.

Chevalier, Y., Y. Storet, S. Pourchet, and P. LePerchec (1991) *Langmuir* **7**, 848.

Chorro, M., N. Kamenka, B. Faucompre, S. Partyka, M. Lindheimer, and R. Zana (1996) *Coll. Surfs. A* **110**, 249.

Clifford, J. and B. A. Pethica (1964) *Trans. Faraday Soc.* **60**, 1483.

Clint, J. H. and T. Walker (1975) *J. Chem. Soc. Faraday Trans.* **171**, 946.

Corkill, J. M. and J. F. Goodman (1962) *Trans. Faraday Soc.* **58**, 206.

Corkill, J. M., J. F. Goodman, and R. H. Ottewill (1961) *Trans. Faraday. Soc.* **57**, 1627.

Corkill, J. M., J. F. Goodman, and C. P. Ogden (1963a) *Proc. R. Soc.* **273**, 84.

Corkill, J. M., J. F. Goodman, C. P. Ogden, and J. R. Tate (1963b) *Proc. R. Soc.* **273**, 84.

Corkill, J. M., J. F. Goodman, and S. P. Harrold (1964) *Trans. Faraday Soc.* **60**, 202.

Corkill, J. M., J. F. Goodman, and C. P. Ogden (1965) *Trans. Faraday Soc.* **61**, 583.

Corkill, J. M., J. F. Goodman, and S. P. Harrold (1966) *Trans. Faraday Soc.* **62**, 994.

Corrin, M. L. and W. D. Harkins (1947) *J. Am. Chem. Soc.* **69**, 684.

Corti, M., V. Degiorgio, J. Hayter, and M. Zulauf (1984) *Chem. Phys. Lett.* **109**, 579.

Crook, E. H., D. B. Fordyce, and G. F. Trebbi (1963) *J. Phys. Chem.* **67**, 1987.

Crook, E. H., G. F. Trebbi, and D. B. Fordyce (1964) *J. Phys. Chem.* **68**, 3592.

Dahanayake, M. and M. J. Rosen, in *Structure/Performance Relationships in Surfactants*, M. J. Rosen (ed.), ACS Symp. Series 253, American Chemical Society, Washington, DC, 1984, p. 49.

Dahanayake, M., A. W. Cohen, and M. J. Rosen (1986) *J. Phys. Chem.* **90**, 2413.

de Castillo, J. L., J. Czapkiewicz, A. Gonzalez Perez, and J. R. Rodriguez (2000) *Coll. Surfs. A* **166**, 161.

De Lisi, R., A. Inglese, S. Milioto, and A. Pellerito (1997) *Langmuir* **13**, 192.

Debye, P. and W. Prins (1958) *J. Colloid Sci.* **13**, 86.

Demchenko, P. A., N. N. Zakharova, and L. G. Demchenko (1962) *Ukr. Khlin. Zh.* **28**, 611 [C. A. **58**, 4745b (1963)].

Desai, A. and P. Bahadur (1992) *Tenside Surf. Det.* **29**, 425.

Downer, A., J. Eastoe, A. R. Pitt, E. A. Simiser, and J. Penfold (1999) *Langmuir* **15**, 7591.

Eastoe, J., J. S. Dalton, P. G. A. Rogueda, E. R. Crooks, A. R. Pitt, and E. A. Simister (1997) *J. Colloid Interface Sci.* **188**, 423.

Eicke, H. F. and J. Rehak (1976) *Hely. Chem. Acta* **59**, 2883.

Elworthy, P. H. and A. T. Florence (1964) *Kolloid-Z.* **195**, 23.

Elworthy, P. H. and C. B. MacFarlane (1962) *J. Pharm. Pharmacol. Suppl.* **14**, 100.

Elworthy, P. H. and C. B. MacFarlane (1963) *J. Chem. Soc.,* 907.

Elworthy, P. H. and C. McDonald (1964) *Kolloid-Z.* **195**, 16.

Elworthy, P. H. and K. J. Mysels (1966) *J. Colloid Sci.* **21**, 331.

Evans, H. C. (1956) *J. Chem. Soc.*, 579.

Evers, E. C. and C. A. Kraus (1948) *J. Am. Chem. Soc.* **70**, 3049.

Fendler, J. H. and E. J. Fendler, *Catalysis in Micellar and Macromolecular Systems*, Academic, New York, 1975.

Fendler, E. J., J. H. Fendler, R. T. Medary, and O. A. El Seoud (1973a) *J. Phys. Chem.* **77**, 1432.

Fendler, J. H., E. J. Fendler, R. T. Medary, and O. A. El Seoud (1973b) *J. Chem. Soc. Faraday Trans. I* **69**, 280.

Fillipi, B. R., L. W. Brandt, J. F. Scamehorn, and S. D. Christian (1999) *J Colloid Interface Sci.* **213**, 68.

Flockhart, B. D. (1961) *J. Colloid Sci.* **16**, 484.

Folmer, B. M., K. Holmberg, E. G. Klingskog, and K. Bergstrom (2001) *J. Surfactants Deterg.* **4**, 175.

Ford, W., R. H. Ottewill, and H. C. Parreira (1966) *J. Colloid Interface Sci.* **21**, 522.

Friberg, S. (1969) *J. Colloid Interface Sci.* **29**, 155.

Fujiwara, M., T. Okano, T. H. Nakashima, A. A. Nakamura, and G. Sugihara (1997) *Colloid Polym. Sci.* **275**, 474.

Garti, N. and A. Aserin (1985) *J. Disp. Sci. Tech.* **6**, 175.

Gentle, T. C. and S. A. Snow (1995) *Langmuir* **11**, 2905.

Gershman, J. W. (1957) *J. Phys. Chem.* **61**, 581.

Goloub, T. P., R. J. Pugh, and B. V. Zhmud (2000) *J. Colloid Interface Sci.* **229**, 72.

Gorski, N. and J. Kalus (2001) *Langmuir* **17**, 4211.

Götte, E., 3rd Intl. Congr. Surface Activity, Cologne, 1960, 1, p. 45.

Götte, E. and M. J. Schwuger (1969) *Tenside* **3**, 131.

Granet, R. and S. Piekarski (1988) *Colloids Surf.* **33**, 321.

Greiss, W. (1955) *Fette, Seife, Anstrichmi* **57**, 24, 168, 236.

Hartley, G. S. (1938) *J. Chem. Soc.*, 168.

Harwigsson, I. and M. Hellsten (1996) *J. Am. Oil Chem. Soc.* **73**, 921.

Hassan, P. A. and J. V. Yakhmi (2000) *Langmuir* **16**, 7187.

Hayes, M. E., M. El-Emary, R. S. Schechter, and W. H. Wade (1980) *J. Disp. Sci. Tech.* **1**, 297.

Heilweil, I. J. (1964) *J. Colloid Sci.* **19**, 105.

Hermann, K. W. (1962) *J. Phys. Chem.* **66**, 295.

Herrington, T. M. and S. S. Sahi (1986) *Colloids Surf.* **17**, 103.

Herzfeld, S. H., M. L. Cowin, and W. D. Harkins (1950) *J. Phys. Chem.* **54**, 271.

Hikota, T., K. Morohara, and K. Meguro (1970) *Bull. Chem. Soc. Jpn.* **43**, 3913.

Hirschhorn, E. (1960) *Soap Chem. Spec.* **36**, 51–54, 62–64, 105–109.

Hobbs, M. E. (1951) *J. Phys. Colloid Chem.* **55**, 675.

Hover, H. W. and A. Marmo (1961) *J. Phys. Chem.* **65**, 1807.

Hudson, R. A. and B. A. Pethica, in *Chem. Phys. Appl. Surface Active Substances*, Vol. 4, J. T. G. Overbeek (ed.), Proc. 4th Intl. Congr., 1964, Gordon & Breach, New York, 1964, p. 631.

Huisman, H. F. (1964) *K. Ned. Akad. Wet. Proc. Ser. B* **67**, 388.

Iijima, H., T. Kato, and O. Soderman (2000) *Langmuir* **16**, 318.

Imae, T. (1990) *J. Phys. Chem.* **94**, 5953.

Ishigami, Y. and H. Machida (1989) *J. Am. Oil Chem. Soc.* **66**, 599.

Israelachvili, J. N., D. J. Mitchell, and B. W. Ninham (1976) *J. Chem. Soc. Faraday Trans. 1* **72**, 1525.

Israelachvili, J. N., D. J. Mitchell, and B. W. Ninham (1977) *Biochm. Biophys. Acta* **470**, 185.

Jobe, D. J. and V. C. Reinsborough (1984) *Can. J. Chem.* **62**, 280.

Kamenka, N., Y. Chevalier, and R. Zana (1995a) *Langmuir* **11**, 3351.

Kamenka, N., M. Chorro, Y. Chevalier, H. Levy, and R. Zana (1995b) *Langmuir* **11**, 4234.

Kanner, B., W. G. Reid, and I. H. Petersen (1967) *Int. Eng. Chem. Prod. Res. Dev.* **6**, 88.

Kitahara, A. (1956) *Bull. Chem. Soc. Jpn.* **29**, 15.

Kjellin, U. R. M., P. M. Claesson, and P. Linse (2002) *Langmuir* **18**, 6745.

Klevens, H. B. (1948) *J. Phys. Colloid Chem.* **52**, 130.

Klevens, H. B. (1953) *J. Am. Oil Chem. Soc.* **30**, 74.

Kon-no, K. and A. Kitahara (1965) *Kogyo Kagaku Zasshi* **68**, 2058.

Kucharski, S. and J. Chlebicki (1974) *J. Colloid Interface Sci.* **46**, 518.

Kumar, S., Z. A. Khan, and K. ud-Din (2002) *J. Surfactants Deterg.* **5**, 55.

Kunieda, H., K. Aramaki, T. Izawa, M. H. Kabir, K. Sakamoto, and K. Watanabe (2003) *J. Oleo Sci.* **52**, 429.

Kunjappu, J. T., *Essays in Ink Chemistry (For Paints and Coatings Too)*, Nova Science Publishers, Inc., New York, 2001.

Kwan, C.-C. and M. J. Rosen (1980) *J. Phys. Chem.* **84**, 547.

Lange, H., *Proc. 4th Int. Congr. Surface Active Substances*, Brussels, Belgium, Vol. 2, p. 497, 1964.

Lange, H. (1965) *Kolloid-Z.* **201**, 131.

Lange, H. and M. J. Schwuger (1968) *Kolloid Z. Z. Polym.* **223**, 145.

Lange, H. and M. J. Schwuger (1971) *Kolloid Z. Z. Polym.* **243**, 120.

Lascaux, M. P., O. Dusart, R. Granet, and S. Piekarski (1983) *J. Chem. Phys.* **80**, 615.

Li, F., M. J. Rosen, and S. B. Sulthana (2001) *Langmuir* **17**, 1037.

Lianos, P. and R. Zana (1980) *J. Phys. Chem.* **84**, 3339.

Lianos, P. and R. Zana (1981) *J. Colloid Interface Sci.* **84**, 100.

Lianos, P. and R. Zana (1982) *J. Colloid Interface Sci.* **88**, 594.

Lianos, P., J. Lang, and R. Zana (1983) *J. Colloid Interface Sci.* **91**, 276.

Lindman, B. (1983) *J. Phys. Chem.* **87**, 1377, 4756.

Mandru, I. (1972) *J. Colloid Interface Sci.* **41**, 430.

Markina, Z. N. (1964) *Kolloid Z.* **26**, 76.

Mathews, M. B. and E. Hirschhorn (1953) *J. Colloid Sci.* **8**, 86.

Mathis, G., J. C. Ravey, and M. Buzier, in *Microemulsions (Proc. Conf. Phys. Chem. Microemulsions, 1980)*, I. D. Robb (ed.), Plenum, New York, 1982, pp. 85–102.

Matos, L., J.-C. Ravey, and G. Serratrice (1989) *J. Colloid Interface Sci.* **128**, 341.

Mazer, N., G. Benedek, and M. Carey (1976) *J. Phys. Chem.* **80**, 1075.

McDonald, C. (1970) *J. Pharm. Pharmacol.* **22**, 774.

Meguro, K. and T. Kondo (1956) *J. Chem. Soc. Jpn. Pure Chem. Sec.* **77**, 1236.

Meguro, K. and T. Kondo (1959) *J. Chem. Soc. Jpn. Pure Chem. Sec.* **80**, 823.

Meguro, K., Y. Takasawa, N. Kawahashi, Y. Tabata, and M. Ueno (1981) *J. Colloid Interface Sci.* **83**, 50.

Mehrian, T., A. de Keizer, A. J. Kortwegr, and J. Lyklema (1993) *Coll. Surf. A* **71**, 2551.

Mitchell, D. J. and B. W. Ninham (1981) *J. Chem. Soc. Faraday Trans. 2* **77**, 601.

Miyagishi, S. (1976) *Bull. Chem. Soc. Jpn.* **49**, 34.

Miyagishi, S., M. Nishida, M. Okano, and K. Fujita (1977) *Colloid Polym. Sci.* **255**, 585.

Miyagishi, S., T. Asakawa, and M. Nishida (1989) *J. Colloid Interface Sci.* **131**, 68.

Mizushima, H., T. Matsuo, N. Satah, H. Hoffman, and D. Grachner (1999) *Langmuir* **15**, 6664.

Mohle, L., S. Opitz, and U. Ohlench (1993) *Tenside Surf. Det.* **30**, 104.

Molyneux, P., C. T. Rhodes, and J. Swarbrick (1965) *Trans. Faraday Soc.* **61**, 1043.

Mujamoto, S. (1960) *Bull. Chem. Soc. Jpn.* **33**, 375.

Mukerjee, P. (1967) *Adv. Colloid Interface Sci.* **1**, 241.

Mukerjee, P., and K. J. Mysels, *Critical Micelle Concentrations of Aqueous Surfactant Systems*, NSRDS-NBS 36, U.S. Dept. of commerce, Washington, DC, 1971.

Muller, N., J. Pellerin, and W. Chem (1972) *J. Phys. Chem.* **76**, 3012.

Murphy, D. S. and M. J. Rosen (1988) *J. Phys. Chem.* **92**, 2870.

Murphy, D. S., Z. H. Zhu, X. Y. Hua, and M. J. Rosen (1990) *J. Am. Oil Chem. Soc.* **67**, 197.

Muzzalupo, R., G. A. Ranieri, and C. L. Mesa (1995) *Coll. Surfs. A* **104**, 327.

Mysels, K. J. and L. H. Princen (1959) *J. Phys. Chem.* **63**, 1696.

Nagarajan, R. (2002) *Langmuir* **18**, 31.

Nakagaki, M., in *Structure/Performance Relationships in Surfactants*, M. J. Rosen (ed.), ACS Symp. Series No. 253, Amer. Chem. Soc., Washington, DC, 1984, p. 73.

Nakagawa, T., K. Kuriyama, and H. Inoue (1960) *J. Colloid Sci.* **15**, 268.

Nakano, T.-Y., G. Sugihara, T. Nakashima, and S.-C. Yu (2002) *Langmuir* **18**, 8777.

Nave, S., J. Eastoe, and J. Penfold (2000) *Langmuir* **16**, 8733.

Nemethy, G. and H. A. Scheraga (1962) *J. Chem. Phys.* **36**, 3401

Ohbu, K. (1998) *Prog. Colloid Polym. Sci.* **109**, 85.

Ohta, A., N. Ozawa, S. Nakashima, T. Asakawa, and S. Miyagishi (2003) *Colloid Polym. Sci.* **281**, 363.

Okuda, H., T. Imac, and S. Ikeda (1987) *Colloids Surf.* **27**, 187.

Omar, A. M. A. and N. A. Abdel-Khalek (1997) *Tenside Surf. Det.* **34**, 178.

Osugi, J., M. Sato, and N. Ifuku (1965) *Rev. Phys. Chem. Jpn.* **35**, 32.

Ottewill, R. H. and H. C. Parreira Paper presented before Div. Colloid Surf. Chemistry, 142nd Natl. Meeting, Am. Chem. Soc., September 1962.

Packter, A. and M. Donbrow (1963) *J. Pharm. Pharmacol.* **15**, 317.

Pandit, N., T. Trygstad, S. Craig, M. Boharquez, and C. Koch (2000) *J. Colloid Interface Sci.* **222**, 213.

Quirion, F. and L. Magid (1986) *J. Phys. Chem.* **90**, 5435.

Raghavan, S. R. and E. W. Kaler (2001) *Langmuir* **17**, 300.

Ravey, J. C. and M. J. Stebe (1994) *Coll. Surf. A* **84**, 11.

Ray, A. (1971) *Nature (London)* **231**, 313.

Ray, A. and G. Nemethy (1971) *J. Am. Chem. Soc.* **93**, 6787.

Rehfeld, S. J. (1967) *J. Phys. Chem.* **71**, 738.

Robb, I. D. and R. Smith (1974) *J. Chem. Soc. Faraday Trans. 1* **70**, 187.

Rodenas, E., C. Doleet, M. Valiente, and E. C. Valeron (1994) *Langmuir* **10**, 2088.

Rodriguez, J. R. and J. Czapkiewicz (1995) *Coll. Surf. A* **101**, 107.

Rosen, M. J. (1976) *J. Colloid Interface Sci.* **56**, 320.

Rosen, M. J. and D. S. Murphy (1991) *Langmuir* **7**, 2630.

Rosen, M. J. and S. B. Sulthana (2001) *J. Colloid Interface Sci.* **239**, 528.

Rosen, M. J., D. Friedman, and M. Gross (1964) *J. Phys. Chem.* **68**, 3219.

Rosen, M. J., A. W. Cohen, M. Dahanayake, and X. Y. Hua (1982a) *J. Phys. Chem.* **86**, 541.

Rosen, M. J., M. Dahanayake, and A. W. Cohen (1982b) *Colloids Surf.* **5**, 159.

Rosen, M. J., Y.-P. Zhu, and S. W. Morrall (1996) *J. Chem. Eng. Data* **41**, 1160.

Rubingh, D., in *Solution Chemistry of Surfactants*, K. L. Mittal (ed.), Plenum, New York, 1979, p. 337ff.

Ruckenstein, E. and R. Nagarajan (1980) *J. Phys. Chem.* **84**, 1349.

Salkar, R. A., D. Mukesh, S. D. Samant, and C. Manohar (1998) *Langmuir* **14**, 3778.

Schick, M. J. (1962) *J. Colloid Sci.* **17**, 801.

Schick, M. J. (1964) *J. Phys. Chem.* **68**, 3585.

Schick, M. J. and F. M. Fowkes (1957) *J. Phys. Chem.* **61**, 1062.

Schick, M. J. and A. H. Gilbert (1965) *J. Colloid Sci.* **20**, 464.

Schick, M. J., S. M. Atlas, and F. R. Eirich (1962) *J. Phys. Chem.* **66**, 1326.

Schwarz, E. G. and W. G. Reid (1963) *Ind. Eng. Chem.* **56** (9), 26.

Schwuger, M. J. (1971) *Ber. Bunsenes. Ges. Phys. Chem.* **75**, 167.

Sepulveda, L. and J. Cortes (1985) *J. Phys. Chem.* **89**, 5322.

Shikata, T., Y. Sakaiguchi, H. Uragami, A. Tamura, and H. Hirata (1987) *J. Colloid Interface Sci.* **119**, 291.

Shinoda, K. (1953) *Bull. Chem. Soc. Jpn.* **26**, 101.

Shinoda, K. and T. Hirai (1977) *J. Phys. Chem.* **81**, 1842.

Shinoda, K. and K. Katsura (1964) *J. Phys. Chem.* **68**, 1568.

Shinoda, K., T. Yamanaka, and K. Kinoshita (1959) *J. Phys. Chem.* **63**, 648.

Shinoda, K., T. Yamaguchi, and R. Hori (1961) *Bull. Chem. Soc. Jpn.* **34**, 237.

Shirahama, K. and R. Matuura (1965) *Bull. Chem. Soc. Jpn.* **38**, 373.

Singleterry, C. R. (1955) *J. Am. Oil Chem. Soc.* **32**, 446.

Skerjanc, S., K. Kogej, and J. Cerar (1999) *Langmuir* **15**, 5023.

Souza, S. M. B., H. Chaimovich, and M. Politi (1995) *Langmuir* **11**, 1715.

Sowada, R. (1994) *Tenside Surf. Det.* **31**, 195.

Stainsby, G. and A. E. Alexander (1950) *Trans. Faraday Soc.* **46**, 587.

Stigter, D. (1974) *J. Phys. Chem.* **78**, 2480.

Swanson-Vethamuthu, M., E. Feitosa, and W. Brown (1998) *Langmuir* **14**, 1590.

Tanaka, A. and S. Ikeda (1991) *Colloids Surf.* **56**, 217.

Tanford, C., *The Hydrophobic Effect*, 2nd ed., Wiley, New York, 1980.

Tartar, H. V. and A. Lelong (1955) *J. Phys. Chem.* **59**, 1185.

Tori, K. and T. Nakagawa (1963a) *Kolloid-Z. Z. Polym.* **188**, 47.

Tori, K. and T. Nakagawa (1963b) *Kolloid-Z. Z. Polym.* **189**, 50.

Treiner, C. and A. Makayssi (1992) *Langmuir* **8**, 794.

Triolo, R., L. J. Magid, J. S. Johnson, and H. R. Child (1983) *J. Phys. Chem.* **87**, 4548.

Tsubone, K. and M. J. Rosen (2001) *J. Colloid Interface Sci.* **244**, 394.

Tsubone, K., Y. Arakawa, and M. J. Rosen (2003a) *J. Colloid Interface Sci.* **262**, 516.

Tsubone, K., T. Ogawa, and K. Mimura (2003b) *J. Surfactants Deterg.* **6**, 39.

Van Os, N. M., G. J. Daane, and G. Handrikman (1991) *J. Colloid Interface Sci.* **141**, 199.

Varadaraj, R., P. Valint, J. Bock, S. Zushma, and N. Brons (1991a) *J. Surfactants Deterg.* **144**, 340.

Varadaraj, R., J. Bock, P. Geissler, S. Zushma, N. Brons, and T. Colletti (1991b) *J. Colloid Interface Sci.* **147**, 396.

Varadaraj, R., J. Bock, S. Zushma, and N. Brons (1992) *Langmuir* **8**, 14.

Varjara, A. K. and S. G. Dixit (1996) *J. Colloid Interface Sci.* **177**, 359.

Venable, R. L. and R. V. Nauman (1964) *J. Phys. Chem.* **68**, 3498.

Vijayendran, B. R. and T. P. Bursh (1979) *J. Colloid Interface Sci.* **68**, 383.

Viseu, M. I., K. Edwards, C. S. Campos, and S. M. B. Costa (2000) *Langmuir* **16**, 2105.

Voicu, A., M. Elian, M. Balcan, and D. F. Anghel (1994) *Tenside Surf. Det.* **31**, 120.

Walker, T. (1971) *J. Colloid Interface Sci.* **45**, 372.

Williams, E. F., N. T. Woodbury, and J. K. Dixon (1957) *J. Colloid Sci.* **12**, 452.

Winsor, P. A. (1968) *Chem. Rev.* **68**, 1.

Yiv, S. and R. Zana (1980) *J. Colloid Interface Sci.* **77**, 449.

Zajac, J., C. Chorro, M. Lindheimer, and S. Partyka (1997) *Langmuir* **13**, 1486.

Zakin, J. L., B. Lu, and H.-W. Bewersdorf (1998) *Rev. Chem. Eng.* **14**, 253.

Zana, R. (1980) *J. Colloid Interface Sci.* **78**, 330.

Zana, R. (1996) *Langmuir* **12**, 1208.

Zana, R., S. Yiv, C. Strazielle, and P. Lianos (1981) *J. Colloid Interface Sci.* **80**, 208.

Zana, R., H. Levy, D. Papoutsi, and G. Beinert (1995) *Langmuir* **11**, 3694.

Zhang, L., P. Somasundaran, and C. Maltesh (1996) *Langmuir* **12**, 2371.

Zhao, F. and M. J. Rosen (1984) *J. Phys. Chem.* **88**, 6041.

Zhu, B. Y. and M. J. Rosen (1985) *J. Colloid Interface Sci.* **108**, 423.

Zhu, Y.-P., M. J. Rosen, S. W. Morrall, and J. Tolls (1998) *J. Surfactants Deterg.* **1**, 187.

Zhu, Y.-P., M. J. Rosen, P. K. Vinson, and S. W. Morrall (1999) *J. Surfactants Deterg.* **2**, 357.

Zoeller, N. and D. Blankschtein (1998) *Langmuir* **14**, 7155.

PROBLEMS

3.1 If we assume that the length of the alkyl chain of a surfactant in a micelle is 80% of its fully extended length, what would be the shape of the micelle of a surfactant whose hydrophobic group is a straight 12-carbon chain and whose hydrophilic group has cross-sectional area at the micellar surface of 60 Å^2?

3.2 Indicate in the table below the effect of each change on the aggregation number of a micelle. Use symbols: + = increase; – = decrease; 0 = little or no effect; ? = effect not clearly known.

For compounds $R(OC_2H_4)_xOH$ in water (R = straight chain):	Effect
(a) Increase in temperature	
(b) Increase in the number of carbon atoms in R	
(c) Increase in the value of x	

For compounds $RSO_4^-Na^+$ in water	Effect
(a) Addition of electrolyte to the solution	
(b) Replacement of Na^+ by Li^+	
(c) Replacement of water as a solvent by methyl alcohol	

3.3 Place in order of increasing CMC in aqueous solution (list answers by letters):

(a) $CH_3(CH_2)_{11}SO_3^-Na^+$

(b) $CH_3(CH_2)_{11}(OC_2H_4)_8OH$

(c) $CH_3(CH_2)_9SO_3^-Na^+$

(d) $CH_3(CH_2)_8CH(C_2H_5)SO_3^-Na^+$

(e) $H_3C(H_2C)_9$–C_6H_4–$SO_3^-Na^+$

(f) $CH_3(CH_2)_{11}(OC_2H_4)_4OH$

3.4 Calculate the ΔG^0_{mic}, in kJ/mo1, for a nonionic surfactant whose CMC is 4×10^{-4} mol/liter at 27°C.

3.5 Indicate in the table below the effect of each change on the CMC/C_{20} ratio of the surfactant in aqueous solution. Use symbols: + = increase; – = decrease; 0 = little or no effect; ? = effect not clearly known.

Change	Effect

(a) Increase in the length of the hydrophobic group

(b) Branched-, instead of straight-chain, isomeric hydrophobic group

(c) Addition of urea to aqueous solution

(d) Addition of NaCl to aqueous solution of ionic surfactant

(e) Decrease in the length of the POE chain (nonionic surfactant)

3.6 Derive the following relationships for mixed micelle formation in a mixture of the two surfactants in aqueous solution. Define all symbols.

(a) $f_1^M X_1^M = \dfrac{C_1^m}{C_1^M}$

(b) $C_{12}^M = \dfrac{C_1^M . C_2^M}{C_1^M(1-\alpha) + C_2^M \alpha}$ for ideal mixed micelle formation.

3.7 Without using the tables, place the following compounds in the order of decreasing CMC/C_{20} ratios. Use ≅ if values are approximately equal.

(a) $C_{12}H_{25}SO_4^-Na^+$, in H_2O, 25°C

(b) $C_{12}H_{25}SO_4^-$ Na^+, in H_2O, 40°C

(c) $C_{12}H_{25}SO_4^-Na^+$, in 0.1 *M* NaCl (aq.), 25°C

(d) $C_{12}H_{25}N(CH_3)_3^+Br^-$, in H_2O, 25°C

(e) $C_{12}H_{25}(OC_2H_4)_6OH$, in H_2O, 25°C

3.8 Why are ionic micelles that have more tightly bonded counterions more likely to have nonspherical micelles?

3.9 Explain why the data on micellar aggregation numbers in Table 3.1 for ionic surfactants often include the surfactant concentration at which the value was determined (the value in parentheses in the table), while the data for nonionics and zwitterionics do not include the concentration.

4 Solubilization by Solutions of Surfactants: Micellar Catalysis

One of the important properties of surfactants that is directly related to micelle formation is *solubilization*. Solubilization may be defined as the spontaneous dissolving of a substance (solid, liquid, or gas) by reversible interaction with the micelles of a surfactant in a solvent to form a thermodynamically stable isotropic solution with reduced thermodynamic activity of the solubilized material. Although both solvent-soluble and solvent-insoluble materials may be dissolved by the solubilization mechanism, the importance of the phenomenon from the practical point of view is that it makes possible the dissolving of substances in solvents in which they are normally insoluble. For example, although ethylbenzene is normally insoluble in water, almost 5 g of it may be dissolved in 100 mL of a 0.3 *M* aqueous solution of potassium hexadecanoate to yield a clear solution.

Solubilization into aqueous media is of major practical importance in such areas as the formulation of products containing water-insoluble ingredients, where it can replace the use of organic solvents or cosolvents; in detergency, where solubilization is believed to be one of the major mechanisms involved in the removal of oily soil; in micellar catalysis of organic reactions; in emulsion polymerization, where it appears to be an important factor in the initiation step; in the separation of materials for manufacturing or analytical purposes; and in enhanced oil recovery, where solubilization produces the ultralow interfacial tension required for mobilization of the oil. Solubilization into nonaqueous media is of major importance in dry cleaning. The solubilization of materials in biological systems (Florence et al., 1984) sheds light on the mechanisms of the interaction of drugs and other pharmaceutical materials with lipid bilayers and membranes.

Solubilization is distinguished from *emulsification* (the dispersion of one liquid phase in another) by the fact that in solubilization, the solubilized material (the "solubilizate") is in the same phase as the solubilizing solution and the system is consequently thermodynamically stable.

If the solubility of a normally solvent-insoluble material is plotted against the concentration of the surfactant solution that is solubilizing it, we find that

Surfactants and Interfacial Phenomena, Fourth Edition. Milton J. Rosen and Joy T. Kunjappu.

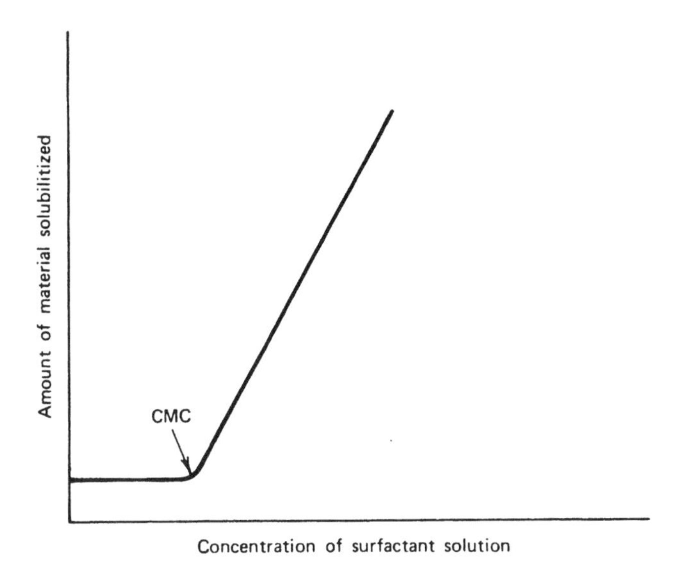

FIGURE 4.1 Plot of amount of material solubilized as a function of concentration of the surfactant in the bulk phase.

the solubility is very slight until a critical concentration is reached at which the solubility increases approximately linearly with the concentration of the surfactant. That critical concentration is the critical micelle concentration (CMC) of the surfactant in the presence of the solubilizate* (Figure 4.1). This indicates that solubilization is a micellar phenomenon, since it occurs only to a negligible extent at concentrations where micelles, if they exist at all, are found only in insignificant numbers.

I. SOLUBILIZATION IN AQUEOUS MEDIA

A. Locus of Solubilization

The exact location in the micelle at which solubilization occurs (i.e., the locus of solubilization) varies with the nature of the material solubilized and is of importance in that it reflects the type of interaction occurring between surfactant and solubilizate. Data on sites of solubilization are obtained from studies on the solubilizate before and after solubilization, using X-ray diffraction

* Since activity of the surfactant in the micelle is changed by the introduction of the solubilizate, the concentration of monomeric surfactant in the aqueous phase in equilibrium with it must change. Therefore, the presence of the solubilizate changes the CMC, in most cases reducing it. Methods for determining the CMC that use *probes* (solubilized materials) consequently give values that are generally less than the CMC of the surfactant in the absence of the probe.

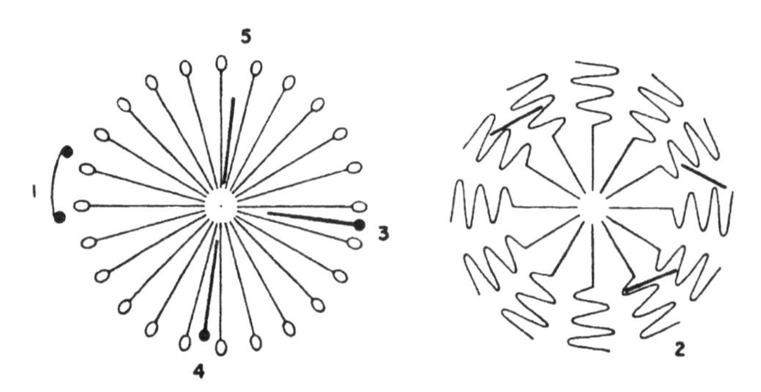

FIGURE 4.2 Loci of solubilization of material in a surfactant micelle. Reprinted with permission from Elworthy et al. (1968, p. 68).

(Hartley, 1949; Philipoff, 1950), ultraviolet spectroscopy (Riegelman et al., 1958), nuclear magnetic resonance (NMR) spectrometry (Eriksson, 1963; Eriksson and Gillberg, 1966), Raman spectroscopy (Kunjappu et al., 1989), and fluorescence spectroscopy (Kunjappu et al., 1990; Kunjappu, 1993, 1994a, 1994b; Saito et al., 1993; Paterson et al., 1999). Diffraction studies measure changes in micellar dimensions on solubilization, whereas UV, NMR, and fluorescence spectra indicate changes in the environment of the solubilizate on solubilization. Based on these studies, solubilization is believed to occur at a number of different sites in the micelle (Figure 4.2): (1) on the surface of the micelle, at the micelle–solvent interface; (2) between the hydrophilic head groups (e.g., in polyoxyethylene [POE] materials); (3) in the so-called palisade layer of the micelle between the hydrophilic groups and the first few carbon atoms of the hydrophobic groups that comprise the outer core of the micellar interior; (4) more deeply in the palisade layer; and (5) in the inner core of the micelle.

Based upon UV spectral studies and the interfacial activity of benzene in heptanes–water systems, Mukerjee (1979) postulated a two-state model for polar and polarizable solubilizates, involving a distribution between *adsorbed state* at the micellar–water interface and a *dissolved state* in the hydrocarbon core. Although a distribution favoring the adsorbed state is expected of solubilizates with high polarity, increased concentration of the solubilizate appears to produce a redistribution favoring the dissolved state. Heats of solution support the two-state model for the solubilization of molecules containing aromatic nuclei (Bury and Treiner, 1985). The distribution of benzene between the two loci depends also upon the hydrophilic group of the surfactant (Nagarajan et al., 1984).

Saturated aliphatic and alicyclic hydrocarbons and other types of molecules that are not polarized or not easily polarizable are solubilized in aqueous medium in the inner core of the micelle between the ends of the hydrophobic groups of the surfactant molecules. Their UV and NMR spectra indicate a completely nonpolar environment on solubilization.

Polarizable hydrocarbons, such as short-chain arenes (benzene, isopropylbenzene), have been shown to be solubilized in quaternary ammonium solutions initially by adsorption at the micelle–water interface, replacing water molecules that may have penetrated into the outer core of the micelle close to the polar heads, but solubilization of additional material is either deep in the palisade layer or located in the inner core of the micelle. The polarizability of the π-electron cloud of the aromatic nucleus and its consequent ability to interact with the positively charged quaternary ammonium groups at the micelle–water interface may account for the initial adsorption of these hydrocarbons in that location. In POE nonionics, benzene may be solubilized between the POE chains of the hydrophilic groups (Nakagawa, 1967).

Large polar molecules, such as long-chain alcohols or polar dyestuffs, are believed to be solubilized, in aqueous medium, mainly between the individual molecules of surfactant in the palisade layer (the layer immediately below the micellar surface) with the polar groups of the solubilizate oriented toward the polar groups of the surfactants and the nonpolar portions oriented toward the interior of the micelle. Interaction here is presumably by H bonding or dipole–dipole attraction between the polar groups of solubilizate and surfactant. The spectrum of the solubilizate in these cases indicates that at least part of the molecule is in a polar environment when solubilized. Depth of penetration in the palisade layer depends on the ratio of polar to nonpolar structures in the solubilizate molecule, longer-chain and less polar compounds penetrating more deeply than shorter-chain and more polar materials. In POE materials, the locus of solubilization for polar dyestuffs may change with change in the length of the POE chain, more of the solubilizate being solubilized in the vicinity of the oxyethylene groups as the length of the POE chain increases (Tokiwa, 1968; Schwuger, 1970).

Small polar molecules in aqueous medium are generally solubilized close to the surface in the palisade layer or by adsorption at the micelle–water interface. The spectra of these materials after solubilization indicate that they are in a completely, or almost completely, polar environment. Short-chain phenols, when solubilized in POE nonionics, appear to be located between the POE chains (Nakagawa, 1967). The localization of probes like ruthenium bipyridyl complexes has been addressed affirmatively in sodium dodecylsulfate (SDS) micelles, comicelled with nitroxide substituted surfactants of comparable chain length, in which the position of the nitroxide quencher group on the carbon chain was varied, by fluorescence spectroscopy (Kunjappu et al., 1990).

In concentrated aqueous surfactant solutions, although the shape of the micelles may be very different from that in dilute solution, the locus of solubilization for a particular type of solubilizate appears to be analogous to that in dilute solution; that is, polar molecules are solubilized mainly in the outer regions of the micellar structures, whereas nonpolar solubilizates are contained in the inner portions.

B. Factors Determining the Extent of Solubilization

Because of the importance of solubilization in the removal of oily soil by detergents and in the preparation of pharmaceutical, cosmetic, insecticide, and other types of formulations, a good deal of work has been done on elucidating the factors that determine the amount of solubilizate that can be solubilized by various types of surfactants. The situation is complicated by the existence of the different sites for the solubilization of different types of materials.

The solubilization capacity or solubilizing power of the micelle is defined (Stearns et al., 1947; Paterson et al., 1999) as the number of moles of solubilizate per mole of micellized surfactant, given by the ratio $(S_W - S_{CMC})/(C_{surf} - CMC)$, where S_W is the molar solubility of the solubilizate in the aqueous system, S_{CMC} its molar solubility at the CMC, and C_{surf} the molar concentration of the surfactant (Edwards et al., 1991). It often remains constant for a particular surfactant over a wide concentration range above the CMC, although some surfactants show increasing solubilizing power at higher concentrations. In general, solubilization capacity is greater for polar solubilizates than for nonpolar ones, especially for spherical micelles (because of the larger volume available at the surface of the micelle than in the interior), and decreases with increase in the molar volume of the solubilizate. Also, factors that promote micellization (e.g., electrolyte addition to ionic surfactants) increase solubilization capacity.

The extent to which a substance can be solubilized into a particular micelle depends upon the portion of the micelle that is the locus of the solubilization. The volume of that portion depends upon the shape of the micelle. As we have seen (Chapter 3, Section IIA), the shape of the micelle is determined by the value of the parameter $V_H/l_c a_o$. As that value increases, the micelle in aqueous medium becomes increasingly asymmetrical, with the result that the volume of the inner core increases relative to that of the outer portion. We can therefore expect that the solubilization of material in the core will increase relative to that in the outer region of the micelle with increase in asymmetry (increase in the value of $V_H/l_c a_o$). The amount solubilized in any location will also increase with increase in the volume of the micelle, for example, with increase in the diameter of a spherical micelle.

The effect of the curvation of the micelle on solubilization capacity has been pointed out by Mukerjee (1979, 1980). The convex surface produces a considerable Laplace pressure (Equation 7.1) inside the micelle. This may explain the lower solubilizing power of aqueous micellar solutions of hydrocarbon-chain surfactants for hydrocarbons, compared to that of bulk phase hydrocarbons, and the decrease in solubilization capacity with increase in molar volume of the solubilizate. On the other hand, reduction of the tension or the curvature at the micellar–aqueous solution interface should increase solubilization capacity through reduction in Laplace pressure. This may in part account for the increased solubilization of hydrocarbons by aqueous solutions of ionic surfactants upon the addition of polar solubilizates or upon the addition of electrolyte. The increase in the solubilization of hydrocarbons with decrease in interfacial tension has been pointed out by Bourrel and Chambu (1983).

1. Structure of the Surfactant For hydrocarbons and long-chain polar compounds that are solubilized in the interior of the micelle or deep in the palisade layer, the amount of material solubilized generally increases with increase in the size of the micelles. Therefore, any factor that causes an increase in either the diameter of the micelle or its aggregation number (Chapter 3, Section III) can be expected to produce an increase in the solubilization capacity for this type of material. Since aggregation numbers increase with increase in the degree of "dissimilarity" between solvent and surfactant, an increase in the chain length of the hydrophobic portion of the surfactant generally results in increased solubilization capacity for hydrocarbons in the interior of the micelle in aqueous media. Fluorocarbon chain surfactants appear to solubilize fluorocarbons better than do hydrocarbon chain surfactants (Asakawa et al., 1998).

Bivalent metal alkyl sulfates appear to show greater solubilizing power than the corresponding sodium salts for hydrocarbons, probably reflecting the greater micellar aggregation numbers, asymmetry, and volumes of the former compared to the latter (Satake and Matsuura, 1963).

In aqueous solutions of POE nonionics, the extent of solubilization of aliphatic hydrocarbons at a given temperature appears to increase as the length of the hydrophobic group increases and as the length of the POE chain decreases (Saito, 1967), reflecting the increase in the aggregation number of the micelles produced by these changes. Fluorescence studies of the solubilization of *n*-octane and *n*-octanol in aqueous solutions of POE nonionics indicate that it is the volumes of the hydrophilic and hydrophobic regions, rather than the size or aggregation number of the micelle, that determines the solubilization of polar and nonpolar solubilizates, respectively (Saito et al., 1993). In POE polyoxypropylene glycols, the solubilization of naphthalene increases with increase in the size of the polyoxypropylene group relative to that of the POE group (Paterson et al., 1999).

Nonionic surfactants, because of their lower critical micelle concentrations, are better solubilizing agents than ionics in very dilute solutions. In general, the order of solubilizing power for hydrocarbons and polar compounds that are solubilized in the inner core appears to be as follows: nonionics > cationics > anionics for surfactants with the same hydrophobic chain length (McBain and Richards, 1946; Saito, 1967; Tokiwa, 1968). The greater solubilizing power of cationics, compared to anionics of equivalent hydrophobic chain length, may be due to looser packing of the surfactant molecules in the micelles of the former (Klevens, 1950; Schott, 1967).

Polymeric quaternary ammonium surfactants, made from *n*-dodecyl bromide and poly(2-vinylpyridine), are better solubilizers for aliphatic and aromatic hydrocarbons than *N*-laurylpyridinium chloride, with the extent of solubilization increasing as the alkyl content of the polymeric quaternary is increased (Strauss and Jackson, 1951; Inoue, 1964).

For polar compounds, very few generalizations relating the degree of solubilization to the structure of the surfactant can be made from the available data, since solubilization can occur in both the inner and outer regions of the micelle. Thus, methyl isobutyl ketone and *n*-octyl alcohol show greater solubilization in 0.1 *M* sodium oleate than in potassium laurate of the same concentration at 25°C, whereas octylamine shows about equal solubilization in each (McBain and Richards, 1946). The solubilization of chloroform in soap micelles increases with increase in the number of carbon atoms in the soap and the solubilization of 1-heptanol increases with increase in the number of carbon atoms in sodium alkanesulfonates (Demchenko and Chernikov, 1973). Yellow OB (1-*o*-tolyl-azo-2-naphthylamine), which is solubilized in both the interior and the POE portion of the micelle in sodium dodecyl polyoxyethylenesulfates, $C_{12}H_{25}(OC_2H_4)_xSO_4^-Na^+$, where $x = 1–10$, shows increased solubilization with increase in the length of the POE chain in these compounds. On the other hand, this same solubilizate shows almost no change in extent of solubilization with increases in the length of the POE chain in the corresponding nonionic, unsulfated dodecyl POE glycols, $C_{12}H_{25}(OC_2H_4)_xOH$, where x = 6–20 (Tokiwa, 1968). This latter effect may be the result of two compensating factors: increased oxyethylene content and decreased aggregation number. Other oil-soluble azo dyes similarly show little change in the amount solubilized as the length of the POE chain in nonionics is increased Schwuger, 1970). In both nonionic and anionic POE surfactants, the extent of solubilization of Yellow OB is much greater than in sodium alkyl sulfates (C_8-C_{14}) without POE chains (Tokiwa, 1968).

Polymeric quaternary ammonium surfactants made from *n*-dodecyl bromide and poly(2-vinylpyridine) are better solubilizers for oil-soluble azo dyes and for *n*-decanol than monomeric quaternary cationics with similar (monomeric) structures (Tokiwa, 1963; Inoue, 1964). Solubilization of *n*-decanol in the polycationics increased as the alkyl chain content increased to a maximum at 24% alkyl content and resulted, at high decanol content, in intermolecular aggregation of the polycationic molecules (Inoue, 1964).

The introduction into the surfactant molecule of a second ionic head group affords some further insights into the solubilization of polar and nonpolar materials. A comparison of the two series of surfactants, the monosodium salts of the monoesters of maleic acid, $ROOCCH{=}CHCOO^-Na^+$, and the disodium salts of the corresponding monoesters of sulfosuccinic acid, $ROOCCH_2CH(SO_3^-Na^+)COO^-Na^+$, where $R = C_{12}$-C_{20}, shows that the introduction of the sulfonate group into the molecule decreases its solubilizing power for the nonpolar compound *n*-octane and increases its solubilizing power for the polar substance *n*-octyl alcohol (Reznikov and Bavika, 1966). This may be explained as follows: The introduction of the sulfonate groups increases the cross-sectional area a_0 of the hydrophilic portion of the surfactant molecule and consequently decreases the aggregation number of the micelles (Chapter 3, Section III). It also causes increased repulsion between the head groups in the micelles, with consequent increase in the space available for solubilization between the surfactant molecules in the palisade layer. The decreased aggregation number in the micelles causes reduced solubilization of nonpolar substances, whereas the increased repulsion between the head groups results in increased solubilization of polar molecules.

Consistent with the above, in aqueous solutions of two different surfactants that interact strongly with each other (Chapter 11, Table 11.1), mixed micelle formation is unfavorable for the solubilization of polar solubilizates that are solubilized in the palisade layer and favorable for the solubilization of nonpolar ones that are solubilized in the micellar inner core. This is due to the reduction of a_0 and the sphere-to-cylindrical micelle transition and the increase in aggregation number resulting from the interaction (Treiner et al., 1990).

2. Structure of the Solubilizate Crystalline solids generally show less solubility in micelles than do liquids of similar structure, the latent heat of fusion presumably opposing the change. For aliphatic and alkylaryl hydrocarbons, the extent of solubilization appears to decrease with increase in the chain length and to increase with unsaturation or cyclization if only one ring is formed (McBain and Richards, 1946). For condensed aromatic hydrocarbons, the extent of solubilization appears to decrease with increase in the molecular size (Schwuger, 1972). Branched-chain compounds appear to have approximately the same solubility as their normal chain isomers. Short-chain alkylaryl hydrocarbons may be solubilized both at the micelle–water interface and in the core, with the proportion in the core increasing with increase in the concentration of the solubilizate.

For polar solubilizates, the situation is complicated by the possibility of variation in the depth of penetration into the micelle as the structure of the solubilizate is changed. If the micelle is more or less spherical in shape, we can expect that space will become less available as the micelle is penetrated more deeply. Thus, polar compounds that are solubilized close to the micelle–water interface should be solubilized to a greater extent than nonpolar solubilizates that are located in the inner core. This is generally the case, if the surfactant concentration is not high (McBain and Richards, 1946; Nakagawa and Tori, 1960). We should also expect that polar compounds that are solubilized more deeply in the palisade layer would be less soluble than those whose locus of solubilization is closer to the micelle–water interface. Usually, the less polar the solubilizate (or the weaker its interaction with either the polar head of the surfactant molecules in the micelle or the water molecules at the micelle–water interface) and the longer its chain length, the smaller its degree of solubilization; this may reflect its deeper penetration into the palisade layer.

3. Effect of Electrolyte The addition of small amounts of neutral electrolyte to solutions of ionic surfactants appears to increase the extent of solubilization of hydrocarbons that are solubilized in the inner core of the micelle and to decrease that of polar compounds that are solubilized in the outer portion of the palisade layer (Klevens, 1950). The effect of neutral electrolyte addition on the ionic surfactant solution is to decrease the repulsion between the similarly charged ionic surfactant head groups, thereby decreasing the CMC (Chapter 3, Section IVB) and increasing the aggregation number (Chapter 3, Section III) and volume of the micelles. The increase in aggregation number of the micelles presumably results in an increase in hydrocarbon solubilization in the inner core of the micelle. The decrease in mutual repulsion of the ionic head groups causes closer packing of the surfactant molecules in the palisade layer and a resulting decrease in the volume available there for solubilization of polar compounds. This may account for the observed reduction in the extent of solubilization of some polar compounds. As the chain length of the polar compound increases, this reduction of solubility by electrolytes appears to decrease, and the solubility of *n*-dodecanol is increased slightly by the addition of neutral electrolyte. This is believed to be due to its location deep in the palisade layer close to the locus of solubilization of nonpolar materials (Klevens, 1950).

The addition of neutral electrolyte to solutions of nonionic POE surfactants increases the extent of solubilization of hydrocarbons at a given temperature in those cases where electrolyte addition causes an increase in the aggregation number of the micelles. The order of increase in solubilization appears to be the same as that for depression of the cloud point (CP) (Section IIIB) (Saito, 1967): $K^+ > Na^+ > Li^+$; $Ca^{2+} > Al^{3+}$; $SO_4^{2-} > Cl^-$. The effect of electrolyte addition on the solubilization of polar materials is not clear.

4. Effect of Monomeric Organic Additives The presence of solubilized hydrocarbon in the surfactant micelles generally increases the solubility of polar compounds in these micelles. The solubilized hydrocarbon causes the micelle to swell, and this may make it possible for the micelle to incorporate more polar material in the palisade layer. On the other hand, the solubilization of such polar material as long-chain alcohols, amines, mercaptans, and fatty acids into the micelles of a surfactant appears to increase their solubilization of hydrocarbons. The longer the chain length of the polar compound and the less capable it is of hydrogen bonding, the greater appears to be its power to increase the solubilization of hydrocarbons, that is, $RSH > RNH_2 > ROH$ (Klevens, 1949); Shinoda and Akamatu, 1958; Demchenko and Kudrya, 1970). One explanation for this is that the increased chain length and lower polarity result in a lower degree of order in the micelle, with a consequent increase in solubilizing power for hydrocarbons. Another is that the additives with longer chain length and lesser hydrogen bonding power are solubilized more deeply in the interior of the micelle, and hence expand this region, producing the same effect as a lengthening of the hydrocarbon chain of the micelle-producing molecule.

However, the addition of long-chain alcohols to aqueous solutions of sodium dodecyl sulfate decreased its solubilization of oleic acid. The extent of solubilization of the latter decreased as both the concentration and the chain length of the added alcohol increased. These effects are believed to be due to competition between oleic acid and added alcohol for sites in the palisade layer of the micelle (Matsuura et al., 1961).

*5. **Effect of Polymeric Organic Additives*** Macromolecular compounds, including synthetic polymers, proteins, starches, and cellulose derivatives, interact with surfactants to form complexes in which the surfactant molecules are absorbed onto the macromolecules, mainly by electrical and hydrophobic interactions. When the surfactant concentration in the complex is sufficiently high, the polymer–surfactant complex may show solubilization power, in some cases greater than that of the surfactant alone, and at concentrations below the CMC of the surfactant (Saito, 1957; Blei, 1959, 1960; Saito and Hirta, 1959; Breuer and Strauss, 1960). The addition of macromolecules of the proper structure to surfactant solutions can therefore increase the solubilizing power of the latter. Thus, sodium alkyl sulfates containing 10–16 carbon atoms, at concentrations below their CMCs, form complexes with serum albumin that solubilize oil-soluble azo dyes and isooctane. The moles of dye, solubilized per mole of surfactant, appear to increase with increase in the chain length of the surfactant, the number of surfactant molecules adsorbed per mole of protein, and the concentration of the protein (Blei, 1959, 1960; Breuer and Strauss, 1960). The amount of Yellow OB solubilized by sodium dodecyl sulfate–polymer complexes appears to increase with increase in the hydrophobic nature of the polymers (Arai and Horin, 1969) and on the addition of small amounts of NaCl (Horin and Arai, 1970).

The addition of POE glycols to aqueous solutions of sodium dodecyl sulfate and sodium *p*-octylbenzenesulfonate increased their solubilization power for the azo dye Yellow OB. As the degree of polymerization of the glycol increased, the extent of solubilization for the dye increased. The effect is believed to be due to the formation of two types of complexes between the surfactant micelles and the glycol. Low-molecular-weight POE glycols (degree of polymerization <10–15) are believed to form micelle–glycol complexes in which the glycol is adsorbed on the surface of the micelle in a manner similar to that of small polar compounds and the solubilized dye is located mainly in the inner core of the micelle. Higher-molecular-weight glycols are believed to form true polymer–surfactant complexes in which the glycol is in the form of a random coil bound to the surfactant with its hydrophilic groups oriented toward the aqueous phase. Here the dye is solubilized in the POE-rich region (Tokiwa and Tsujii, 1973b).

Generally, the more hydrophobic the polymer, the greater the adsorption of surfactant onto it from water, since hydrophilic groups on the macromolecule can interact with water and weaken surfactant–polymer interaction. The adsorption of anionic surfactants onto nonionic macromolecules appears to follow the approximate order polyvinylpyrrolidone ≈ polypropylene glycol > polyvinylacetate > methylcellulose > polyethylene glycol > polyvinyl alcohol. Long-chain alkyl ammonium chlorides appear to follow the same general order of interaction, except for much weaker interaction with polyvinylpyrrolidone (less than with polyethylene glycol). The very strong interaction of anionic surfactants, especially sulfated anionics, and very weak interaction of cationics with polyvinylpyrrolidone may be due to protonation of the latter in aqueous solution (Breuer and Robb, 1972; Roscigno et al., 2001). Nonionic surfactants interact only weakly with nonionic macromolecules (Saito, 1960).

The relation between the extent of solubilization and the structures of solubilizate and surfactant–polymer complex is not completely clear. Aromatic hydrocarbons appear to be more highly solubilized than aliphatic hydrocarbons by complexes of anionic surfactants and hydrophilic polymers with no proton-donating groups, such as polyvinylpyrrolidone, but the nature of the forces involved is not clear. Some cationic surfactant–polymer complexes are broken by the solubilization of aromatic hydrocarbons. It has been suggested that structural compatibility between solubilizate and polymer may be a factor and that the function of the surfactant is to increase the hydrophilic character of the polymer and to promote contact between polymer and solubilizate (Saito, 1967).

6. Mixed Anionic–Nonionic Micelles An investigation of the solubilization of Yellow OB by mixed micelles of anionics and a POE nonionic, $C_{12}H_{25}(OC_2H_4)_9OH$, indicated that increased solubilization of the dye occurs when there is interaction between the POE chain and the benzenesulfonate groups, $-C_6H_4-SO_3^-$, rather than the phenyl or sulfonate groups alone (Tokiwa and Tsujii, 1973a). The degree of interaction of the aromatic nucleus with the POE chain decreased with separation of the ring from the sulfonate groups, giving the following order of interaction:

$$C_8H_{17}-C_6H_4-SO_3^-Na^+ > C_4H_9-C_6H_4-C_4H_8SO_3^-Na^+ > C_6H_5-C_8H_{16}SO_3^-Na^+.$$

Only the first compound increased the extent of solubilization of Yellow OB by the nonionic. The addition of $C_{10}H_{21}SO_3^-Na^+$ to the nonionic decreased its solubilization of Yellow OB.

7. Effect of Temperature For ionic surfactants, an increase in temperature generally results in an increase in the extent of solubilization for both polar and nonpolar solubilizates, possibly because increased thermal agitation increases the space available for solubilization in the micelle. Thus, the solubilization of cyclohexane in an aqueous solution of sodium di(2-ethylhexyl) sulfosuccinate above 50°C increases with increase in the temperature (Kunieda and Shinoda, 1979).

For nonionic POE surfactants on the other hand, the effect of temperature increase appears to depend on the nature of the solubilizate. Nonpolar materials, such as aliphatic hydrocarbons and alkyl halides, which are solubilized in the inner core of the micelle, appear to show increased solubility as the

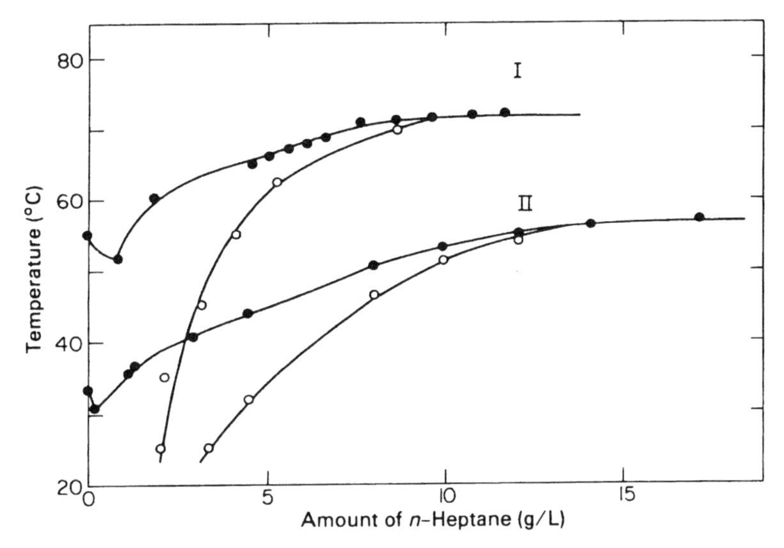

FIGURE 4.3 The effect of temperature on the solubilization of *n*-heptane in 1% aqueous solution of I, POE (9.2) nonylphenyl ether, and II, POE (9.0) dodecylphenyl ether. ●, Cloud point, ○, solubilization limit. Reprinted with permission from Shinoda, K. and S. Frieberg (1967), *J. Colloid Interface Sci.* **24**, 4.

temperature is raised, the increase becoming very rapid as the CP (Section IIIB) of the surfactant is approached (Saito, 1967).

Figure 4.3 illustrates this, and also the effect of increase in the length of the hydrophobic group. The upper curves in I and II, respectively, are for the CP in the presence of excess heptane; the lower curves, for the amount of heptane solubilized. This rapid increase just below the CP probably reflects the large increase in aggregation number of the micelles in this region (Chapter 3, Section III) and the transition from spherical to more asymmetric micelles. The solubility of the oil-soluble azo dye Sudan Red G also increases with increase in temperature (Schwuger, 1970). However, the solubility behavior of polar materials, whose locus of solubilization is the palisade layer of the micelle, appears to be very different, the amount of material solubilized generally going through a maximum as the temperature is increased to the CP (Nakagawa and Tori, 1960). When the temperature is raised above 10°C, there is at first a small or moderate increase in the extent of solubilization, presumably reflecting the increase in thermal agitation of the surfactant molecules in the micelles. This is followed by a decrease in the amount of material solubilized as further increase in temperature causes increased dehydration and tighter coiling of the POE chains, decreasing the space available in the palisade layer. As the CP is approached, this decrease in the amount solubilized may become marked, particularly for short-chain polar compounds that are solubilized close to the surface of the micelle.

8. Hydrotropy When there are strong chain–chain and head–head interactions between surfactant molecules (due to long, straight chains and close-packed heads), either insoluble crystal formation (low Krafft point, p. 241) or liquid crystal formation (Chapter 3, Section IIC) may occur. Since there is much less space available for solubilization in rigid liquid crystal structures than in the more flexible types of micelles, the onset of crystal formation usually limits the solubilization capacity of the solution. The tendency to form crystalline structures can be reduced by the addition of certain nonsurfactant organic additives called *hydrotropes*.

Hydrotropes have been known for decades as organic substances that increase the solubility and reduce the viscosity in water of other organic substances. They have a structure somewhat similar to those of surfactants in that they have a hydrophilic and a hydrophobic group in the molecule, but are different from surfactants in that the hydrophobic group is generally short, *cyclic,* and/or branched. Typical hydrotropes include sodium benzene-, toluene-, xylene-, cumene-, and *p*-cymenesulfonates, 1-hydroxy-2-naphthoate, 2-hydroxy-1-naphthalenesulfonate, and sodium 2-ethylhexyl sulfate.

The mechanism by means of which hydrotropes operate in surfactant solutions was elucidated by Friberg and Rydhag (1970, 1971) and Cox and Friberg (1981), who showed that it is due to the inhibition of the formation of surfactant liquid crystalline phases by the hydrotrope. Since they have structures similar to those of surfactants, hydrotropes can form mixed micellar structures with surfactants. However, since their hydrophilic heads are large and their hydrophobic groups are small (their $V_H/l_c a_o$ ratio [Chapter 3, Section II] is $\ll 1$), they tend to form spheroidal rather than lamellar or liquid crystalline structures and thus inhibit the formation of the latter. This destruction or inhibition of the liquid crystalline phase increases the solubility of the surfactant in the aqueous phase and the capacity of its micellar solution to solubilize material. Hydrotropic action occurs at concentrations at which the hydrotrope self-associates to form these mixed structures with the surfactant (Gonzalez et al., 2000).

Short-chain alkyl polyglycosides (C_4-C_{10}), because they have the large hydrophilic head group and short hydrophobic group needed to disrupt liquid-crystalline phases, are also effective hydrotropes. The C_8 and C_{10} homologs are very effective in raising the CPs (Section IIIB) of some alcohol ethoxylates (Matero et al., 1998).

C. Rate of Solubilization

For the solubilization of highly insoluble hydrocarbons by POE nonionics into water, the rate of solubilization has been found (Carroll, 1981; Carroll et al., 1982) to be directly proportional to the surfactant concentration above the CMC, and to increase with the polarity and decrease with the molecular weight of the oil. The rate is also strongly temperature dependent in the region of the CP (Section IIIB), increasing rapidly as that temperature is approached. The mechanism suggested involves diffusion of the micelles to the hydrocarbon–water interface, where they dissociate and adsorb as monomers. This adsorption produces concerted desorption from the interface of an equivalent amount of monomeric surfactant, but in the form of micelles containing a quantity of solubilizate.

A study of the rate of solubilization of *n*-hexadecane by mixtures of different sodium linear alkylarenesulfonates with a nonionic (C_{12}-C_{15} alcohol oxyethylenated with 9 mol of EO) in the presence of 3–10 m*M* Ca^{2+} showed that the rate of solubilization increases dramatically: (1) as the position of the phenylsulfonate group on the linear chain becomes more central and (2) as the position of the sulfonate group approaches that of the long alkyl group on the ring. Thus, the order of decreasing rate of solubilization was 2-alkyl-4,5-dimethylbenzene-sulfonate >> 3-alkyl-6-methylbenzenesulfonate > 4-alkyl-2,5-dimethylbenzenesulfonate ≈ 4-alkylbenzenesulfonate. For *n*-hexadecane solubilization, the rate increases with increase in the length of the alkyl chain of the alkylarenesulfonate to a maximum at 11–12 carbon atoms (Bolsman et al., 1988).

II. SOLUBILIZATION IN NONAQUEOUS SOLVENTS

Surfactants can also solubilize materials into solvents other than water. Even when surfactant aggregation does not occur or the aggregation number is small in a particular solvent in the absence of other material, the addition of solvent-insoluble material, such as water, may give rise to aggregation with consequent solubilization of the additive (Kitahara, 1980). The solubilization of water and aqueous solutions into organic solvents has been especially studied in connection with dry cleaning, and the solubilization of organic acids has been studied in connection with corrosion prevention in fuels and lubricants. Investigations have been confined almost exclusively to solubilizates that are small polar molecules, especially water, and to solvents that are hydrocarbons or chlorinated hydrocarbons. Since the surfactant molecules in these solvents have their polar or ionic heads buried in the inner core of the micelles and their hydrophobic groups oriented toward the solvent, solubilization of these small polar materials in these solvents occurs in the interior of the micelle. The surfactants used in these systems must necessarily be soluble in the solvent, and since many ionic surfactants are not soluble in hydrocarbons, only a few ionic surfactants have been used for this purpose. Anionics that have been most commonly used are the amine soaps of fatty acids and various metal dialkylsulfosuccinates (Mathews and Hirschhorn, 1953) and dinonylnaphthalenesulfonates (Honig and Singleterry, 1954); cationics used include dodecylammonium carboxylates (Palit and Venkateswarlu, 1954), didodecyldimethylammonium halides, and di(2-ethylhexyl)ammonium halides. Since many POE nonionics are soluble in aliphatic and aromatic hydrocarbons, structural limitations on the use of these materials for this purpose are not as restrictive as in the case of ionics. Polyanionic soaps made by copolymerization of maleic anhydride and dodecyl (or octadecyl) vinyl ether followed by treatment with morpholine have been used to solubilize water into nonaqueous solvents (Ito and Yamashita, 1964).

In micellar solutions of ionic surfactants the mechanism for solubilization of the small polar molecules appears to involve, initially at least, ion–dipole interaction between the solubilizate and the counterion of the surfactant present in the interior of the micelle, possibly followed by weaker interaction (e.g., via hydrogen bonding) between the solubilizate and the surfactant ion (Kaufman, 1964; Kitahara et al., 1969; Kon-no and Kitahara, 1971a). In solutions of nonionic POE surfactants, solubilization of polar molecules appears to be by interaction with the ether oxygens of the POE chain.

A classification of solubilization isotherms for small polar molecules into nonaqueous solvents by surfactants, based on the strength of interaction between solubilizate and surfactant, has been proposed by Kon-no and Kitahara (1972a). When the moles of material solubilized per mole of surfactant are plotted against the relative vapor pressure, p/p^0, of the system at constant temperature (where p is the vapor pressure of the water in the system and p^0 is the vapor pressure of pure water), isotherms are obtained whose shapes reflect the strength of the solubilizate–surfactant interaction. Systems having strong surfactant–solubilizate interaction are concave to the p/p^0 axis, whereas those showing weak interaction are convex to that axis. Systems with very weak solubilizate–surfactant interaction show almost linear isotherms.

The maximum amount of water solubilized into hydrocarbon solvents by ionic surfactants appears to increase with increase in the concentration of the surfactant, the valence of the counterion, and the length of the alkyl chain (Kon-no et al., 1971), and with the introduction of double bonds into the hydrophobic group (Demchenko et al., 1971). Straight-chain compounds appear to solubilize less water than branched-chain ones, possibly because the former form micelles that are more compact and rigid than the latter (Frank and Zografi, 1969; Kon-no and Kitahara, 1971b).

The addition of neutral electrolyte appears to decrease markedly the solubilization power of ionic surfactants for water, ions of charge opposite to that of the surfactant ion having a much greater effect than similarly charged ions (Kitahara and Kon-no, 1966). This is explained in the case of anionics (Kon-no and Kitahara, 1972a, 1972b) as caused by the decreased repulsion between the ionic heads of the surfactant molecules in the interior of the micelle resulting from the compression of the electrical double layer in the presence of the electrolyte. This decrease in repulsion permits the ionic heads to approach each other more closely, thereby decreasing the space available for solubilization of water. Temperature increase appears to cause an increase in the solubilization of water by ionic surfactants by increasing the distance between the ionic heads (Kon-no and Kitahara, 1972b).

Change in the nature and molecular weight of the solvent affects the extent of solubilization of water. The amount solubilized by sodium di(2-ethylhexyl) sulfosuccinate goes through a maximum with increase in the molecular weight of the solvent in the *n*-alkane series, with *n*-dodecane showing the greatest solubilizing power. Cyclohexane and toluene showed much lower solubilizing power (Frank and Zografi, 1969). In general, solubilizing power for water appears to decrease as the polarity of the solvent increases, presumably because of increased competition by the solvent for the surfactant molecules and the smaller aggregation number of the micelles in the more polar solvents.

The amount of water solubilized into hydrocarbon solvents by POE nonionics appears to increase with increase in the concentration of the surfactant and the length of the POE chain (Nakagaki and Sone, 1964; Saito, 1972). The amount of H_2O solubilized into aliphatic, aromatic, and chlorinated solvents showed little change with increase in the temperature from 15°C to 35°C for a series of POE nonionics and only a small increase for some ionic surfactants (Kitahara, 1980). The extent of solubilization of water into hydrocarbon solvents by POE nonionics is not affected as much as for ionic surfactants by the addition of electrolytes. Here the anion of the added electrolyte appears to have a much greater effect than the cation in decreasing the solubilizing power for water, the order being as follows: $Na_2SO_4 \gg NaCl > MgCl_2 > AlCl_3$. This order corresponds to that of increasing lyotropic numbers for anions and cations and is the same as that for their effect on the CMC of POE nonionics (Chapter 3, Section IVB). This indicates that their action must involve a salting out of the hydrogen bonds between the ether oxygen of the POE chains and the solubilized water molecules (Kitahara and Kon-no, 1966).

From the data available for surfactants with similar hydrophobic groups, the solubilizing power for water into hydrocarbon solvents appears to decrease in the following order: anionics > nonionics > cationics (Kon-no and Kitahara, 1971b).

In the presence of 15% pentanol, large amounts of water can be solubilized into heptane or toluene solutions of C_{12}-C_{16} alkylpyridinium or alkyltrimethylammonium bromides (Venable, 1985). In heptane/pentanol, the longer-chain surfactants appear to be more effective than the shorter ones, while in toluene/pentanol, the shorter ones appear to be more effective. In both solvent mixtures, the pyridinium salts are more effective solubilizers than the corresponding trimethylammonium salts. All the quaternaries investigated were more effective than sodium dodecyl sulfate.

These effects on the solubilization of water are in agreement with the prediction of Mitchell and Ninham that bringing the >1 value of the ratio $V_H/l_c a_o$ (Chapter 3, Section II) closer to 1 should increase the solubilization of water in inverted micelles. This is also consistent with increases in the solubilization of water observed upon the addition of benzene or nitrobenzene to solutions of sodium di(2-ethylhexyl)sulfosuccinate in isooctane (Maitra et al., 1983). The effects were explained as caused by desolvation of the surfactant by the additives, with consequent decrease in the value of V_H.

A. Secondary Solubilization

The secondary solubilization of such water-soluble materials as salts, sugars, and water-soluble dyes into nonaqueous solutions of surfactants containing solubilized water is of great importance in dry cleaning, since it is a major mechanism by means of which water-soluble stains can be removed. The data available indicate that the tightly bound, initially solubilized water is not available for this purpose and that only the subsequently solubilized, more loosely bound water is responsible for such secondary solubilization (Aebi and Weibush, 1959; Wentz et al., 1969; Kon-no and Kitahara, 1971a). The strength of the binding of water molecules to the ionic head group in sulfosuccinates appears to be a function of the size of the groups around the hydrophilic head, the heats of solubilization of water decreasing in the following order: Na di(*n*-octyl)sulfosuccinate > Na di(1-methylheptyl)sulfosuccinate > Na di (2-ethylhexyl)-sulfosuccinate. Water appears to be less strongly bound in the potassium di(2-ethylhexyl)sulfosuccinate than in the corresponding sodium salt, presumably because of the greater bulk of the K^+ compared to the Na^+ (Kon-no and Kitahara, 1971a).

III. SOME EFFECTS OF SOLUBILIZATION

A. Effect of Solubilization on Micellar Structure

The incorporation of solubilizate into a micelle may change the nature and shape of the micelle considerably. With the incorporation of increasing amounts of nonpolar material into its inner core, the value of V_H in the structure parameter $V_H/l_c a_o$ (Chapter 3, Section IIA) increases and a normal micelle in aqueous medium may become more and more asymmetric, eventually becoming lamellar in shape. Continued addition of nonpolar material may result in the conversion of the normal lamellar micelle to an inverted lamellar micelle and eventually to a spherical inverted micelle in the nonpolar medium. The reverse process, the conversion of an inverted micelle in the nonpolar medium to a normal micelle in aqueous medium upon the addition of increasing amounts of water, is also possible. These conversions are diagrammed in Figure 4.4. Liquid crystalline phases (Section IB8) may also appear, along with these micellar structures, depending upon the structure of the surfactant and the nonpolar solubilizate, at various ratios of surfactant to water and/or nonpolar material.

The addition of medium-chain alcohols that are solubilized close to the surface of the micelle in the palisade layer increases the value of a_o, resulting in a greater tendency to form spherical micelles. Increase in the ionic strength of the aqueous solution or increase in the concentration of an ionic surfactant in the aqueous phase, on the other hand, decreases the value of a_o and promotes the tendency to form cylindrical or lamellar micelles.

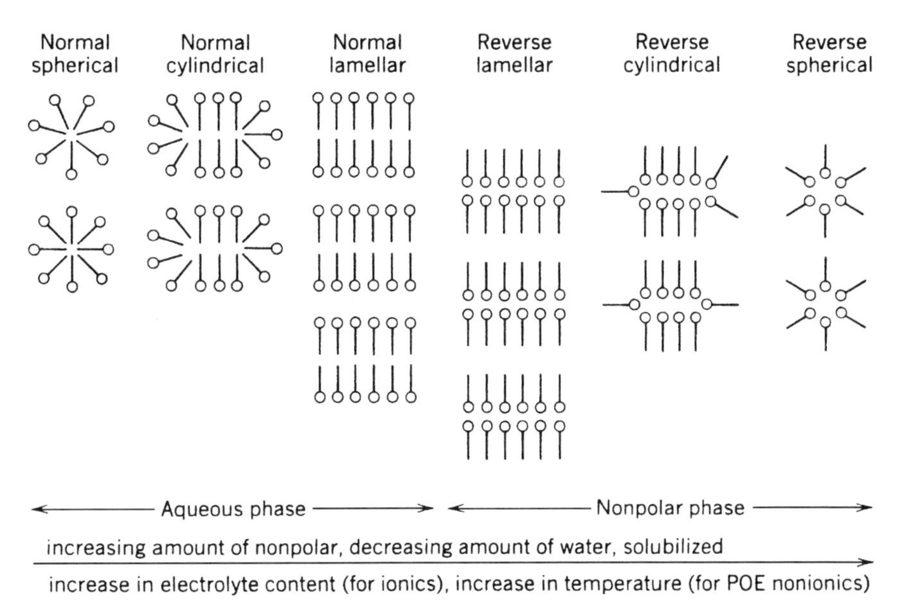

FIGURE 4.4 Effect of solubilization content and other molecular environmental factors on micellar structure. Note that interconversion of normal and reverse lamellar micelles involves only small changes in distances between hydrophilic and hydrophobic groups.

B. Change in the CPs of Aqueous Solutions of Nonionic Surfactants

Aqueous solutions of POE nonionics, if the oxyethylene content is below about 80%, become turbid on being heated to a temperature known as the *CP*, following which there is separation of the solution into two phases. This phase separation occurs within a narrow temperature range that is fairly constant for surfactant concentrations below a few percent (Nakagawa and Shinoda, 1963). The phases appear to consist of an almost micelle-free dilute solution of the nonionic surfactant at a concentration equal to its CMC at that temperature and a surfactant-rich micellar phase, which appears only when the solution is above its CP. The phase separation is reversible, and on cooling of the mixture to a temperature below the CP, the two phases merge to form once again a clear solution.

The separation is believed to be due to the sharp increase in the aggregation number of the micelles and the decrease in intermicellar repulsions (Staples and Tiddy, 1978; Tiddy, 1980) resulting from the decreased hydration of the oxyethylene oxygens in the POE hydrophilic group with increase in temperature (Chapter 3, Section III). As the temperature increases, micellar growth and increased intermicellar attraction cause the formation of particles, for example, rodlike micelles, that are so large that the solution becomes visibly turbid (Glatter et al., 2000). Phase separation occurs because of the difference in density of the micelle-rich and micelle-poor phases (Nakagawa and Shinoda, 1963). Some CPs are listed in Table 4.1.

The temperature at which clouding occurs depends on the structure of the POE nonionic surfactant. For a particular hydrophobic group, the larger the percentage of oxyethylene in the surfactant molecule, the higher the CP, although the relation between oxyethylene percentage and CP is not linear. A study of the effect of structural changes in the surfactant molecule on the CP (Schott, 1969) indicates that, at constant oxyethylene percentage, the CP is lowered by the following: decreased molecular weight of the surfactant, broader distribution of POE chain length in commercial materials, branching of the hydrophobic group, more central position of the POE hydrophilic group in the surfactant molecule, replacement of the terminal hydroxyl group of the hydrophilic group by a methoxyl, and replacement of the ether linkage between the hydrophilic and the hydrophobic group by an ester linkage. On the other hand, the replacement of the ether linkage by an amide linkage raised the CP (Kjcllin ct al., 2002).

Schott (2003) has found the following linear equation between the CP of water-soluble POE nonionics and the average number, p, of POE units in the molecule.

$$(p - p^0)/\mathrm{CP} = a + b(p - p^0), \tag{4.1}$$

where p^0 is the smallest value of p that confers solubility in cold water (in a homologous series of POE surfactant, it corresponds to CP = 0°C), and a and b are constants. This equation covers the entire range for the POE nonionics, including when $p > 100$.

The appearance of turbidity in the aqueous solution and its separation into two phases introduce certain disadvantages, esthetic as well as practical, in its utilization, and has resulted in investigations to determine the effect of solubilization on the temperature at which clouding appears. In general, long-chain nonpolar solubilizates, such as saturated aliphatic hydrocarbons, which are solubilized in the inner core of the micelle, appear to cause an increase in the CP of the solution, whereas polar and polarizable compounds, such as fatty acids and alcohols of moderate chain length, phenol, or benzene, which are solubilized in the outer regions of the micelle, depress it (Maclay, 1956; Nakagawa and Tori, 1960). Figure 4.3 shows the effect of solubilization of n-heptane on the CPs of two POE nonionics. The upper curves in I and II, respectively, are for the CP in the presence of excess heptane.

The increase in the CP on solubilization of long-chain nonpolar material in the interior of the micelle may be due to the resulting increase in the radius, and hence the surface of the micelle, leaving more room at the micelle–water interface for increased hydration of the POE chains. On the other hand, the decrease when polar compounds are solubilized in the outer regions may be due to decreased hydration of the POE chains as a result of competition for the hydratable sites by the polar solubilizate.

TABLE 4.1 Cloud Points of POE Nonionics

Substance	Solvent	Cloud Point (°C)	Reference
$n\text{-}C_6H_{13}(OC_2H_4)_3OH$[a]	H_2O	37	Mulley (1967)
$n\text{-}C_6H_{13}(OC_2H_4)_5OH$[a]	H_2O	75	Mulley (1967)
$n\text{-}C_6H_{13}(OC_2H_4)_6OH$[a]	H_2O	83	Mulley (1967)
$(C_2H_5)_2CHCH_2(OC_2H_4)_6OH$[a]	H_2O	78	Elworthy and Florence (1964)
$n\text{-}C_8H_{17}(OC_2H_4)_4OH$[a]	H_2O	35.5	Mulley (1967)
$n\text{-}C_8H_{17}(OC_2H_4)_6OH$[a]	H_2O	68	Shinoda (1967b)
$C_{10}H_{21}(OC_2H_4)_4OH$[a]	H_2O	21	Mitchell et al. (1983)
$C_{10}H_{21}(OC_2H_4)_5OH$[a]	H_2O	44	Mitchell et al. (1983)
$n\text{-}C_{10}H_{21}(OC_2H_4)_6OH$[a]	H_2O	60	Mulley (1967)
$(n\text{-}C_4H_9)_2CHCH_2(OC_2H_4)_6OH$[a]	H_2O	27	Elworthy and Florence (1964)
$C_{11}H_{23}CONH(CH_2CH_2)_4H$[a]	H_2O	52	Kjellin et al. (2002)
$n\text{-}C_{12}H_{25}(OC_2H_4)_3OH$[a]	H_2O	25	Cohen and Rosen (1981)
$C_{12}H_{25}(OC_2H_4)_4OH$[a]	H_2O	4	Mitchell et al. (1983)
$C_{12}H_{25}(OC_2H_4)_5OH$[a]	H_2O	27	Mitchell et al. (1983)
$n\text{-}C_{12}H_{25}(OC_2H_4)_6OH$[a]	H_2O	52	Cohen and Rosen (1981)
$n\text{-}C_{12}H_{25}(OC_2H_4)_7OH$[a]	H_2O	62	Cohen and Rosen (1981)
$n\text{-}C_{12}H_{25}(OC_2H_4)_7OH$[a]	H_2O	58.5	Schott (1969)
$n\text{-}C_{12}H_{25}(OC_2H_4)_8OH$[a]	H_2O	79	Mulley (1967)
$n\text{-}C_{12}H_{25}(OC_2H_4)_8OH$[b]	H_2O	73	Fineman et al. (1952)
$n\text{-}C_{12}H_{25}(OC_2H_4)_{9.4}OH$	H_2O	84	Kuwamura (1984)
$C_{12}H_{25}(OC_2H_4)_{9.2}OH$[a]	H_2O	75	Kuwamura (1984)
$n\text{-}C_{12}H_{25}(OC_2H_4)_{10}OH$[a]	H_2O	95	Mulley (1967)
$n\text{-}C_{12}H_{25}(OC_2H_4)_{10}OH$[b]	H_2O	88	Wrigley et al. (1957)
$n\text{-}C_{13}H_{27}(OC_2H_4)_{8.9}OH$[b]	H_2O	79	Kuwamura (1984)
$(n\text{-}C_6H_{13})_2CH(OC_2H_4)_{9.2}OH$[b]	H_2O	35	Kuwamura (1984)
$(n\text{-}C_4H_9)_3CH(OC_2H_4)_{9.2}OH$[b]	H_2O	34	Kuwamura (1984)
$n\text{-}C_{14}H_{29}(OC_2H_4)_6OH$[a]	H_2O	45	Mulley (1967)
$n\text{-}C_{16}H_{33}(OC_2H_4)_6OH$[a]	H_2O	32	Mulley (1967)
$n\text{-}C_{16}H_{33}(OC_2H_4)_{12.2}OH$	H_2O	97	Kuwamura (1984)
$(n\text{-}C_5H_{11})_3C(OC_2H_4)_{12.0}OH$	H_2O	48	Kuwamura (1984)
$C_{16}H_{33}(OC_2H_4)_{11.9}OH$	H_2O	80	Kuwamura (1984)
$C_8H_{17}C_6H_4(OC_2H_4)_7OH$[b]	H_2O	15	Mansfield and Locke (1964)
$C_8H_{17}C_6H_4(OC_2H_4)_{9\text{-}10}OH$[b]	H_2O	64.3	Schott and Han (1977)

(*Continued*)

TABLE 4.1 (*Continued*)

Substance	Solvent	Cloud Point (°C)	Reference
$C_8H_{17}C_6H_4(OC_2H_4)_{9\text{-}10}OH$[b]	0.2 *M* NH_4Cl	60.0	Schott and Han (1977)
$C_8H_{17}C_6H_4(OC_2H_4)_{9\text{-}10}OH$[b]	0.2 *M* NH_4Br	62.5	Schott and Han (1977)
$C_8H_{17}C_6H_4(OC_2H_4)_{9\text{-}10}OH$[b]	0.2 *M* NH_4NO_3	63.2	Schott and Han (1977)
$C_8H_{17}C_6H_4(OC_2H_4)_{9\text{-}10}OH$[b]	0.2 *M* $(CH_3)_4NCl$	59.6	Schott and Han (1977)
$C_8H_{17}C_6H_4(OC_2H_4)_{9\text{-}10}OH$[b]	0.2 *M* $(CH_3)_4NI$	67.0	Schott and Han (1977)
$C_8H_{17}C_6H_4(OC_2H_4)_{9\text{-}10}OH$[b]	0.2 *M* $(C_2H_5)_4NCl$	61.0	Schott and Han (1977)
$C_8H_{17}C_6H_4(OC_2H_4)_{9\text{-}10}OH$[b]	0.2 *M*$(C_3H_7)_4NI$	78.5	Schott and Han (1977)
$C_8H_{17}C_6H_4(OC_2H_4)_{10}OH$[b]	H_2O	75	Mansfield and Locke (1964)
$C_8H_{17}C_6H_4(OC_2H_4)_{13}OH$[b]	H_2O	89	Fineman et al. (1952)
$C_9H_{19}C_6H_4(OC_2H_4)_8OH$[b]	H_2O	34	Fineman et al. (1952)
$C_9H_{19}C_6H_4(OC_2H_4)_{9.2}OH$[b]	H_2O	56	Shinoda (1967b)
$C_9H_{19}C_6H_4(OC_2H_4)_{9.2}OH$[b]	*n*-$C_{16}H_{34}$-saturated H_2O	80	Shinoda (1967b)
$C_9H_{19}C_6H_4(OC_2H_4)_{9.2}OH$[b]	*n*-$C_{10}H_{22}$-saturated H_2O	79	Shinoda (1967b)
$C_9H_{19}C_6H_4(OC_2H_4)_{9.2}OH$[b]	*n*-C_7H_{16}-saturated H_2O	71.5	Shinoda (1967b)
$C_9H_{19}C_6H_4(OC_2H_4)_{9.2}OH$[b]	Cyclohexane-saturated H_2O	54	Shinoda (1967b)
$C_9H_{19}C_6H_4(OC_2H_4)_{9.2}OH$[b]	$C_2H_5C_6H_5$-saturated H_2O	30.5	Shinoda (1967b)
$C_9H_{19}C_6H_4(OC_2H_4)_{9.2}OH$[b]	Benzene-saturated H_2O	<0	Shinoda (1967b)
$C_9H_{19}C_6H_4(OC_2H_4)_{12.4}OH$[b]	H_2O	87	Fineman et al. (1952)
$C_{12}H_{25}C_6H_4(OC_2H_4)_9OH$[b]	H_2O	33	Shinoda (1967a)
$C_{12}H_{25}C_6H_4(OC_2H_4)_{11.1}OH$[b]	H_2O	50	Fineman et al. (1952)
$C_{12}H_{25}C_6H_4(OC_2H_4)_{15}OH$[b]	H2O	90	Fineman et al. (1952)

[a] Single compound.
[b] Distribution of POE chains.

Ions that are water structure formers, lower the CP of POE nonionics, $OH^- > F^- > Cl^- > Br^-$, by decreasing the availability of nonassociated water molecules to hydrate the ether oxygens of the POE chain. Ions that are water structure breakers (large, polarizable anions; soft bases: SCN^-, I^-) increase the CP by making more water molecules available to interact with the POE chain (Schott, 1984). Thus, chloride ions, which are water structure makers, lower the CP; iodide ions, which are structure breakers, raise it; bromide ions have no pronounced effect.

Ammonium ions and alkali metal cations, except lithium, tend to lower the CP of POE nonionics ($Na^+ > K^+ > Cs^+ > NH_4^+$) by salting them out, whereas polyvalent cations, H^+, and lithium ion form complexes with the ether oxygens of the POE chain, thereby increasing intermicellar repulsions and increasing the CP (Schott, 1973, 1996; Schott and Han, 1975). Hydrogen ions and polyvalent cations are particularly effective, with the result that the addition of hydrogen chloride raises the CP of $C_{12}H_{25}(OC_2H_4)_6OH$, LiCl has almost no effect, and NaCl lowers it (Nakanishi et al., 1984). All divalent cations and Ag^+ increase the CP (Schott, 1996) by complexing with the ether oxygens of the POE chain.

On the other hand, whereas tetraalkylammonium ions are water structure formers and this effect increases with increase in the length of the alkyl group from one to four carbon atoms, tetramethylammonium and tetraethylammonium chlorides decrease the CP (the former more than the latter), while tetrapropyl and tetrabutylammonium chlorides increase it (the latter more than the former). All these quaternary cations, as do the iodides, increase the CP of POE nonionics, with the tetrabutyl ammonium ion producing the largest increase. The CP raising in the case of the tetrapropyl and tetrabutylammonium cations is ascribed to mixed micelle formation, with the nonionic predominating over water structure formation (Schott and Han, 1977). The mixed micelles, with their cationic components, should presumably have greater intermicellar repulsions and stronger interaction with water, and consequently higher CPs than the original POE nonionic micelles.

The CPs of nonionics can also be increased by the addition of polyelectrolytes or ionic surfactants that interact with them and cause them to acquire a charge (Goddard, 1986; Saito, 1986).

Alkylpolyglycosides also show CPs, with the effect of electrolytes on their CPs being greater than for POE alcohols. Salts generally reduce their CPs, with cations having a greater effect than anions. On the other hand, NaOH sharply increases their CPs. Both effects are presumably due to the negative charge on the glucoside molecule in the pH range of 3–9 (Balzer, 1993).

C. Reduction of the CMC

See Chapter 3, Section IVC1.

D. Miscellaneous Effects of Solubilization

Other useful effects of solubilization are the binding of organic acids soluble in lubricating oils to the cations of metal sulfonate detergents in these oils, thereby decreasing the corrosion of metals by these acids (Bascom and Singleterry, 1958) and the solubilization of foamicidal oils by foam-producing surfactants, leading to increased foam life.

In some cases, solubilization of physiologically active materials enhances their potency; in other cases, it diminishes their potency. Moreover, the use of surfactants in preparations that are ingested by organisms may increase their solubilization of other physiologically active undesirable materials, such as bacterial toxins or carcinogens. Solubilization may also inactivate preservatives in pharmaceutical preparations by incorporating the former into the micelles of surfactants used in the formulation.

Cell membranes may also be affected by solubilizing surfactants, which may disrupt lysosomes, mitochondria, and erythrocytes. Triton X-100 is particularly effective in this respect. Details of these and other effects may be found in the monograph on solubilization by Elworthy et al. (1968).

IV. MICELLAR CATALYSIS

The reactions of organic compounds can be catalyzed markedly in micellar solution. Catalysis by both normal micelles in aqueous medium and by reversed micelles in nonpolar solvents is possible (Fendler and Fendler, 1975; Kitahara, 1980). In normal micelles in aqueous medium, enhanced reaction of the solubilized substrate generally, but not always, occurs at the micelle–aqueous solution interface; in reversed micelles in nonaqueous medium, this reaction occurs deep in the inner core of the micelle.

The effect of micelles on organic reactions can be attributed to both electrostatic and hydrophobic interactions. Electrostatic interaction may affect the rate of a reaction either by its effect on the transition state of the reaction or by its effect on the concentration of reactant in the vicinity of the reaction site. Thus, a cationic micelle with its multiplicity of positively charged hydrophilic heads may catalyze the reaction between a nucleophilic anion and a neutral substrate by delocalizing the negative charge developing in the transition state of this reaction, thereby decreasing the energy of activation of the reaction. It may also catalyze the reaction by increasing the concentration of nucleophilic anion at the micelle–water interface close to the reactive site of the substrate. For catalysis to occur, it is necessary (1) that the substrate be solubilized by the micelle and (2) that the locus of solubilization be such that the reactive site of the substrate is accessible to the attacking reagent. It is here that hydrophobic interactions become important, because they determine the extent and the locus of solubilization in the micelle (Section IA).

In the simplest case, where we assume that the surfactant does not complex with (i.e., solubilize) the substrate S, except when the former is in the form of micelles M and that complexing between the substrate and the micelle is in a 1:1 stoichiometric ratio, we can symbolize the formation of a reaction product P as (Fendler and Fendler, 1975)

$$\begin{array}{c} M + S \overset{K}{\rightleftarrows} MS \\ \downarrow k_0 \quad \downarrow k_m \\ P \quad P, \end{array}$$

where k_0 is the rate constant for the reaction of the substrate in the bulk phase, and k_m is the rate constant for the reaction of the substrate in the micelle. The overall rate constant for the reaction k_p is then given by the expression

$$k_p = k_0[F_0] + k_m[F_m], \tag{4.2}$$

where F_0 is the fraction of the uncomplexed substrate and F_m is the fraction of the complexed substrate. The equilibrium constant K for the interaction between substrate and micelle, usually called the *binding constant*, is then given by the relation

$$K = \frac{[F_m]}{[M][F_0]}, \tag{4.3}$$

from which

$$k_p = k_0[F_0] + k_m K[M][F_0] = (k_0 + k_m[M]K)F_0. \tag{4.4}$$

Since $F_0 + F_m = 1$,

$$K = \frac{[F_m]}{[M][F_0]} = \frac{[1 - F_0]}{[M][F_0]}$$

and

$$F_0 = \frac{1}{1 + K[M]}$$

from which

$$k_p = \frac{k_0 + k_m[M]K}{1 + K[M]}. \tag{4.5}$$

If we assume that $[M]$ is correctly given by the expression

$$[M] = \frac{(C - CMC)}{N}, \tag{4.6}$$

where C is the total concentration of surfactant, CMC its critical micelle concentration, and N the aggregation number in its micelles, then the expression for the overall rate constant can be put in the form

$$\frac{1}{k_0 - k_p} = \frac{1}{k_0 - k_m} + \left(\frac{1}{k_0 - k_m}\right)\left[\frac{N}{K(C - CMC)}\right]. \tag{4.7}$$

Since the overall rate constant for the reaction k_p and the rate constant for the reaction in the absence of micelles k_0 are readily obtained from kinetic data, a plot of $1/(k_0 - k_p)$ versus $[1/(C - CMC)]$, which should be a straight line with slope = $N/K(k_0 - k_m)$ and intercept = $1/(k_0 - k_m)$, allows the calculation of k_m, the rate constant for the substrate complexed with the micelle, and K, the binding constant of the substrate to the micelle. This treatment is also applicable to bimolecular micelle-inhibited reactions in which one reagent is excluded from the micelle, for example, by electrostatic repulsion between an ionic reagent and a similarly charged micelle (Menger and Portnoy, 1967). Quantitative treatment of more complex reactions and some of the problems involved has been discussed by Bunton (1979).

Since surfactant concentrations are usually below 10^{-1} M and often one or two orders of magnitude below that, there will generally be little enhancement of the rate of reaction in the presence of micelles unless the product k_mK is 10^2 or more. Since the binding constant K depends on the extent of hydrophobic bonding between surfactant and substrate, it can be expected that K will increase with increase in the chain length of both the surfactant and the substrate. However, if the hydrophobic group of the substrate is too long, it may be solubilized so deeply in the micelle that access to its reactive site by a reagent in an aqueous solution phase is hindered. In that case, solubilization will inhibit, rather than catalyze, the reaction.

In accordance with these principles, the alkaline hydrolysis in aqueous medium of *p*-nitrophenyl esters is catalyzed by cationic *n*-alkyltrimethylammonium bromide micelles and retarded by anionic sodium laurate micelles (Menger and Portnoy, 1967). Nonionic surfactants either decrease the rate or have no significant effect on the rates of hydrolysis of carboxylic acid esters. The ester is probably solubilized at the micelle–water interface. The transition state for alkaline hydrolysis of the ester linkage carries a negative charge due to the oncoming OH^-, and this charge can be stabilized by the adjacent positive charges of the hydrophilic heads of cationic micelles and destabilized by the adjacent negative charges in anionic micelles. In addition, the concentration of OH^- at the micelle–water interface is increased by the multiple positive charges on the cationic micelles and decreased by the multiple negative charges on the anionic micelles. Both of these effects may account for the enhancement and diminution of reaction rates in the respective cases. These effects also explain the observation that the rate of reaction with neutral nucleophiles, such as morpholine, is not accelerated by cationic micelles (Behme and Cordes, 1965). They also explain the inhibiting action of small concentrations of inorganic anions (F^-, Cl^-, Br^-, NO_3^-, $SO_4^=$) on micellar catalysis by cationic surfactants, since these anions compress the electrical double layer surrounding the positively charged hydrophilic head groups, thus weakening their interaction with negative charges. The extent of both rate enhancement by cationics and diminution by anionics increased as the chain length of the acyl group of the ester was increased, the order being *p*-nitrophenyl dodecanote > *p*-nitrophenylhexanoate >> *p*-nitrophenylacetate.

In the case of certain other esters, however (e.g., ethyl benzoate and acetyl salicylate [Nogami et al., 1962; Mitchell, 1964]), both anionic and cationic micelles retard the rate of hydrolysis. These effects are attributed either to small binding constants between substrate and micelle or to solubilization into micelles in such fashion as to remove the reaction site from the attacking reagent.

The increase in the rate of acid-catalyzed hydrolysis of esters in aqueous media by anionic micelles can be explained in similar fashion as being due to stabilization of the positively charged transition state or to concentration of H^+ at the micelle–water interface by the negatively charged adjacent hydrophilic head groups.

Plots of rate constant versus surfactant concentration often show a maximum at some surfactant concentration above the CMC. There are a number of explanations for this. First, the number of micelles increases with increase in the surfactant concentration. When the number of micelles exceeds that required to solubilize all of the substrate, there is a dilution of the concentration of substrate per micelle as the surfactant concentration is increased further. This causes a reduction in the rate constant. Second, the charged surface of an ionic micelle in aqueous media may cause not only the concentration at the micelle–solution interface of an oppositely charged reactant in the solution phase, but the adsorption of that reactant on it, or even the solubilization of the reactant into the micelle. Such adsorption or solubilization of the reactant will result in a decrease in its activity in the solution phase. An increase in the concentration of surfactant over that required to effect substantially complete solubilization of the substrate may therefore result in a decrease in the rate constant, even in those cases where rate enhancement by micelles occurs.

Aliphatic and aromatic nucleophilic substitution reactions are also subject to micellar effects, with results consistent with those in other reactions. In the reaction of alkyl halides with CN^- and $S_2O_3^{2-}$ in aqueous media, sodium dodecyl sulfate micelles decreased the second-order rate constants and dodecyltrimethylammonium bromide increased them (Winters and Grunwald, 1965; Bunton and Robinson, 1968). The reactivity of methyl bromide in the cationic micellar phase was 30–50 times that in the bulk phase and was negligible in the anionic micellar phase; a nonionic surfactant did not significantly affect the rate constant for *n*-pentyl bromide with $S_2O_3^{2-}$. Micellar effects on nucleophilic aromatic substitution reactions follow similar patterns. The reaction of 2,4-dinitrochlorobenzene or 2,4-dinitrofluorobenzene with hydroxide ion in aqueous media is catalyzed by cationic surfactants and retarded by sodium dodecyl sulfate (Bunton and Robinson, 1968, 1969). Cetyltrimethylammonium bromide micelles increased the reactivity of dinitrofluorobenzene 59 times, whereas sodium dodecyl sulfate decreased it by a factor of 2.5; for dinitrochlorobenzene, the figures are 82 and 13 times, respectively. A POE nonionic surfactant had no effect.

Diquaternary ammonium halides of the gemini type (Chapter 12) are particularly effective micellar catalysts for nucleophilic substitution and decarboxylation reactions (Bunton et al., 1971, 1972, 1973).

The hydrolysis of long-chain alkyl sulfates in aqueous solution is an example of a reaction where micellar effects can be observed without the complicating presence of a solubilizate. Here the rate of acid-catalyzed hydrolysis is increased about 50 times by micellization because of the high concentration of H^+ on the negatively charged micellar surface (Nakagaki, 1986). As the chain length of the alkyl group is increased, the rate constant increases, reflecting the lower CMC of the surfactant. On the other hand, alkaline hydrolysis of these compounds is retarded considerably by micelle formation. Micelle formation has a negligible effect on the neutral hydrolysis of these materials (Kurz, 1962; Nogami and Kanakubo, 1963).

A study of two-tailed (sodium dialkylsulfosuccinate) and two-headed (disodium monoalkylsulfosuccinate) surfactants revealed that these types have no advantage over similar single-headed, single-tailed (sodium alkylsulfoacetate) materials (Jobe and Reinsborough, 1984). The second tail does not increase substantially the binding of the substrate (pyridine-*z*-azo-*p*-dimethylaniline) to the micelles, and the second head decreased, rather than increased, the binding of the reagent (Ni^{2+}) to the micelle. The latter effect may be due to the competition of the addition $Na^{\pm}$ present.

The presence of micelles can also result in the formation of different reaction products. A diazonium salt, in an aqueous micellar solution of sodium dodecyl sulfate, yielded the corresponding phenol from reaction with OH^- in the bulk phase, but the corresponding hydrocarbon from material solubilized in the micelles (Abe et al., 1983).

Micellar effects are also apparent in reactions involving free radicals. Surfactants have been used extensively for the enhancement or inhibition of industrially and biologically important free radical processes, such as emulsion polymerization and the oxidation of hydrocarbons and unsaturated oils. An investigation of the free radical oxidation of benzaldehyde and *p*-methylbenzaldehyde by oxygen in aqueous nonionic surfactant solutions indicated that the rate of oxidation is increased when the aldehyde is solubilized in the interior region of the micelles. As the alkyl chain length of the surfactant was increased, the oxidation rate of *p*-methylbenzaldehyde increased because of the increased solubilization of the aldehyde in the interior region of the micelle. However, the oxidation rate for benzaldehyde was not increased by this change in the structure of the surfactant. Spectroscopic observations indicated that *p*-methylbenzaldehyde is solubilized in both the outer and inner regions of the micelles and increase in the length of the alkyl chain of the surfactant increases the proportion of aldehyde in the inner region, whereas benzaldehyde is solubilized only in the POE region of the micelle (Mitchell and Wan, 1965).

REFERENCES

Abe, M., N. Suzuki, and K. Ogino (1983) *J. Colloid Interface Sci.* **93**, 285.

Aebi, C. M. and J. R. Weibush (1959) *J. Colloid Sci.* **14**, 161.

Arai, H. and S. Horin (1969) *J. Colloid Interface Sci.* **30**, 372.

Asakawa, T., T. Kitaguchi, and S. Miyagishe (1998) *J. Surfactants Deterg.* **1**, 195.

Balzer, D. (1993) *Langmuir* **9**, 3375.

Bascom, W. D. and C. R. Singleterry (1958) *J. Colloid Sci.* **13**, 569.

Behme, M. T. A. and E. H. Cordes (1965) *J. Am. Chem. Soc.* **87**, 260.

Blei, I. (1959) *J. Colloid. Sci.* **14**, 358.

Blei, I. (1960) *J. Colloid Sci.* **15**, 370.

Bolsman, T. A. B. M., F. T. G. Veltmaat, and N. M. van Os (1988) *J. Am. Oil Chem. Soc.* **65**, 280.

Bourrel, M. and C. Chambu (April 1983) *Soc. Pet. Eng. J.* **23**, 327.

Breuer, M. M. and I. D. Robb (1972) *Chem. Ind. (London)* **13**, 530.

Breuer, M. M. and U. P. Strauss (1960) *J. Phys. Chem.* **64**, 228.

Bunton, C. A., in *Solution Chemistry of Surfactants*, Vol. 2, K. L. Mittal (ed.), Plenum, New York, 1979, p. 519.

Bunton, C. A., I. Robinson, J. Schaak, and M. F. Stam (1971) *J. Org. Chem.* **36**, 2346.

Bunton, C. A., A. Kamego, and M. J. Minch (1972) *J. Org. Chem.* **37**, 1388.

Bunton, C. A., M. J. Minch, J. Hidalgo, and L. Sepulveda (1973) *J. Am. Chem. Soc.* **95**, 3262.

Bunton, C. A. and L. Robinson (1968) *J. Am. Chem. Soc.* **90**, 5972.

Bunton, C. A. and L. Robinson (1969) *J. Org. Chem.* **34**, 780.

Bury, R. and C. Treiner (1985) *J. Colloid Interface Sci.* **103**, 1.

Carroll, B. J. (1981) *J. Colloid Interface Sci.* **79**, 126.

Carroll, B. J., B. G. C. O'Rourke, and A. J. I. Ward (1982) *J. Pharm. Pharmacol.* **34**, 287.

Cohen, A. W. and M. J. Rosen (1981) *J. Am. Oil Chem. Soc.* **58**, 1062.

Cox, J. M. and S. E. Friberg (1981) *J. Am. Oil Chem. Soc.* **58**, 743.

Demchenko, P. A. and O. Chernikov (1973) *Maslozhir Prom.* **7**, 18 [C. A. **79**, 106322C (1973)].

Demchenko, P. A. and T. P. Kudrya (1970) *Ukr. Khim. Zh.* **36**, 1147 [C. A. **74**, 88946Z (1971)].

Demchenko, P., L. Novitskaya, and B. Shapoval (1971) *Kolloid Zh.* **33**, 831 [C. A. **73**, 7672U (1972)].

Edwards, D. A., R. G. Luthy, and Z. Liu (1991) *Environ. Sci. Technol.* **25**, 127.

Elworthy, P. H. and A. T. Florence (1964) *Kolloid-Z. Z. Polym.* **195**, 23.

Elworthy, P. H., A. T. Florence, and C. B. MacFarlane, *Solubilization by Surface-Active Agents*, Chapman & Hall, London, 1968, p. 68 and 90.

Eriksson, J. C. (1963) *Acta Chem. Scand.* **17**, 1478.

Eriksson, J. C. and G. Gillberg (1966) *Acta Chem. Scand.* **20**, 2019.

Fendler, J. and E. Fendler, *Catalysis in Miceller and Macromolecular Systems*, Academic, New York, 1975.

Fineman, M. N., G. L. Brown, and R. J. Myers (1952) *J. Phys. Chem.* **56**, 963.

Florence, A. T., I. G. Tucker, and K. A. Walters, in *Structure/Performance Relationships in Surfactants*, M. J. Rosen (ed.), ACS Symp. Series 253, American Chemical Society, Washington, DC, 1984, p. 189.

Frank, S. G. and G. Zografi (1969) *J. Colloid Interface Sci.* **29**, 27.

Friberg, S. E. and L. Rydhag (1970) *Tenside* **7**, 2.

Friberg, S. E. and L. Rydhag (1971) *J. Am. Oil Chem. Soc.* **48**, 113.

Glatter, O., G. Fritz, H. Undner, J. Brunner-Popela, R. Mittebach, R. Strey, and S. U. Egdhaaf (2000) *Langmuir* **16**, 8692.

Goddard, E. D. (1986) *Colloids Surf.* **19**, 255.

Gonzalez, G., E. J. Nasser, and M. E. D. Zaniquelli (2000) *J. Colloid Interface Sci.* **230**, 223.

Hartley, G. S. (1949) *Nature* **163**, 767.

Honig, J. G. and C. R. Singleterry (1954) *J. Phys. Chem.* **58**, 201.

Horin, S. and H. Arai (1970) *J. Colloid Interface Sci.* **32**, 547.

Inoue, H. (1964) *Kolloid-Z. Z. Polym.* **196**, 1.

Ito, K. and Y. Yamashita (1964) *J. Colloid Sci.* **19**, 152.

Jobe, D. J. and V. C. Reinsborough (1984) *Aust. J. Chem.* **37**, 1593.

Kaufman, S. (1964) *J. Phys. Chem.* **68**, 2814.

Kitahara, A. (1980) *Adv. Colloid Interface Sci.* **12**, 109.

Kitahara, A. and K. Kon-no (1966) *J. Phys. Chem.* **70**, 3394.

Kitahara, A., K. Watanabe, K. Kon-no, and T. Ishikawa (1969) *J. Colloid Interface Sci.* **29**, 48.

Kjellin, U. R. K., P. M. Cloesson, and P. Linse (2002) *Langmuir* **18**, 6745.

Klevens, H. B. (1949) *J. Chem. Phys.* **17**, 1004.

Klevens, H. B. (1950) *J. Am. Chem. Soc.* **72**, 3780.

Kon-no, K. and A. Kitahara (1971a) *J. Colloid Interface Sci.* **35**, 409.

Kon-no, K. and A. Kitahara (1971b) *J. Colloid Interface Sci.* **37**, 469.

Kon-no, K. and A. Kitahara (1972a) *J. Colloid Interface Sci.* **41**, 86.

Kon-no, K. and A. Kitahara (1972b) *J. Colloid Interface Sci.* **41**, 47.

Kon-no, K., Y. Ueno, Y. Ishii, and A. Kitahara (1971) *Nippon Kagaku Zasshi* **92**, 381 [C. A. **75**, 80877c (1971)].

Kunieda, H. and K. Shinoda (1979) *J. Colloid Interface Sci.* **70**, 577.

Kunjappu, J. T. (1993) *J. Photochem. Photobiol. A Chem.* **71**, 269.

Kunjappu, J. T. (1994a) *J. Photochem. Photobiol. A Chem.* **78**, 237.

Kunjappu, J. T. (1994b) *Colloid Surf. A* **89**, 37.

Kunjappu, J. T., P. Somasundaran, and N. J. Turro (1989) *Chem. Phys. Lett.* **162**, 233.

Kunjappu, J. T., P. Somasundaran, and N. J. Turro (1990) *J. Phys. Chem.* **94**, 8464.

Kurz, J. L. (1962) *J. Phys. Chem.* **66**, 2239.

Kuwamura, T., in *Structure/Performance Relationships in Surfactants*, M. J. Rosen (ed.), ACS Symp. Series 253, American Chemical Society, Washington, DC, 1984, p. 32.

Maclay, W. N. (1956) *J. Colloid Sci.* **11**, 272.

Maitra, A., G. Vasta, and H.-F. Eicke (1983) *J. Colloid Interface Sci.* **93**, 383.

Mansfield, R. C. and J. E. Locke (1964) *J. Am. Oil Chem. Soc.* **41**, 267.

Matero, A., A. Mattson, and M. Svensson (1998) *J. Surfactants Deterg.* **1**, 485.

Mathews, M. B. and E. Hirschhorn (1953) *J. Colloid Sci.* **8**, 86.

Matsuura, R., K. Furudate, H. Tsutsumi, and S. Miida (1961) *Bull. Chem. Soc. Jpn.* **34**, 395.

McBain, J. W. and P. H. Richards (1946) *Ind. Eng. Chem.* **38**, 642.

Menger, F. M. and C. E. Portnoy (1967) *J. Am. Chem. Soc.* **89**, 4698.

Mitchell, A. G. (1964) *J. Pharm. Pharmacol.* **16**, 43.

Mitchell, A. G. and L. Wan (1965) *J. Pharm. Sci.* **54**, 699.

Mitchell, D. J., G. J. T. Waring, T. Bostock, and M. P. McDonald (1983) *J. Chem. Soc. Faraday Trans. 1* **79**, 975.

Mukerjee, P., in *Solution Chemistry of Surfactants*, K. L. Mittal (ed.), Plenum, New York, 1979, p. 153.

Mukerjee, P. (1980) *Pure Appl. Chem.* **52**, 1317.

Mulley, B. A., in *Nonionic Surfactants*, M. J. Schick (ed.), Dekker, New York, 1967, p. 372.

Nagarajan, R., M. A. Chaiko, and E. Ruckenstein (1984) *J. Phys. Chem.* **88**, 2916.

Nakagaki, M., presented at *77th Annual Meeting, American Oil Chemists Society, Honolulu, Hawaii*, May 1986.

Nakagaki, M. and S. Sone (1964) *Yukagaku Zasshi* **84**, 151 [C. A. **61**, 5911A (1964)].

Nakagawa, T., in *Nonionic Surfactants*, M. J. Schick (ed.), Dekker, New York, 1967, p. 297.

Nakagawa, T. and K. Shinoda, in *Colloidal Surfactants*, K. Shinoda, T. Nakagawa, B. Tamamushi, and T. Isemura (eds.), Academic, New York, 1963, p. 129.

Nakagawa, T. and K. Tori (1960) *Kolloid-Z.* **168**, 132.

Nakanishi, T., T. Seimiya, T. Sugawara, and H. Iwamura (1984) *Chem. Lett.* **12**, 2135.

Nogami, H., S. Awazu, and N. Nakajima (1962) *Chem. Pharm. Bull. (Tokyo)* **10**, 503.

Nogami, H. and Y. Kanakubo (1963) *Chem. Pharm. Bull. (Tokyo)* **11**, 943.

Palit, S. R. and V. Venkateswarlu (1954) *J. Chem. Soc.*, 2129.

Paterson, I. F., B. Z. Chowdhry, and A. S. Leharne (1999) *Langmuir* **15**, 6187.

Philipoff, W. (1950) *J. Colloid Sci.* **5**, 169.

Reznikov, I. G. and V. I. Bavika (1966) *Maslozhir Prom.* **32**, 27 [C. A. **65**, 10811G (1966)].

Riegelman, S., N. A. Allawala, M. K. Hrenoff, and L. A. Strait (1958) *J. Colloid Sci.* **13**, 208.

Roscigno, P., L. Paduano, G. D'Ernico, and V. Vitagliano (2001) *Langmuir* **17**, 4510.

Saito, S. (1957) *Kolloid-Z.* **154**, 49.

Saito, S. (1960) *J. Colloid Sci.* **15**, 283.

Saito, S. (1967) *J. Colloid Interface Sci.* **24**, 227.

Saito, S. (1972) *Nippon Kagaku Kaishi* **3**, 491 [C. A. **77**, 77053s (1972)].

Saito, S. (1986) *Colloid Surf.* **19**, 351.

Saito, S. and H. Hirta (1959) *Kolloid-Z.* **165**, 162.

Saito, Y., M. Abe, and T. Sato (1993) *J. Am. Oil Chem. Soc.* **70**, 717.

Satake, I. and R. Matsuura (1963) *Bull. Chem. Soc. Jpn.* **36**, 813.

Schott, H. (1967) *J. Phys. Chem.* **71**, 3611.

Schott, H. (1969) *J. Pharm. Sci.* **58**, 1443.

Schott, H. (1973) *J. Colloid Interface Sci.* **43**, 150.

Schott, H. (1984) *Colloids Surf.* **11**, 51.

Schott, H. (1996) *Tenside, Surf. Det.* **33**, 457.

Schott, H. (2003) *J. Colloid Interface Sci.* **260**, 219.

Schott, H. and S. K. Han (1975) *J. Pharm. Sci.* **64**, 658.

Schott, H. and S. K. Han (1977) *J. Pharm. Sci.* **66**, 165.

Schwuger, M. J. (1970) *Kolloid-Z. Z. Polym.* **240**, 872.

Schwuger, M. J. (1972) *Kolloid-Z. Z. Polym.* **250**, 703.

Shinoda, K. (1967a) *J. Colloid Interface Sci.* **24**, 4.

Shinoda, K., in *Solvent Properties of Surfactant Solutions*, K. Shinoda (ed.), Dekker, New York, 1967b, chapter 2.

Shinoda, K. and H. Akamatu (1958) *Bull. Chem. Soc. Jpn.* **31**, 497.

Staples, E. J. and G. J. T. Tiddy (1978) *J. Chem. Soc., Faraday Trans. 1* **74**, 2530.

Stearns, R. S., H. Oppenheimer, E. Simons, and W. D. Harkins (1947) *J. Chem. Phys.* **15**, 496.

Strauss, U. P. and E. G. Jackson (1951) *J. Polym. Sci.* **6**, 649.

Tiddy, G. J. T. (1980) *Phys. Rep.* **57**, 1.

Tokiwa, F. (1963) *Bull. Chem. Soc. Jpn.* **36**, 1589.

Tokiwa, F. (1968) *J. Phys. Chem.* **72**, 1214.

Tokiwa, F. and K. Tsujii (1973a) *Bull. Chem. Soc. Jpn.* **46**, 1338.

Tokiwa, F. and K. Tsujii (1973b) *Bull. Chem. Soc. Jpn.* **46**, 2684.

Treiner, C., M. Nortz, and C. Vaution (1990) *Langmuir* **6**, 1211.

Venable, R. L. (1985) *J. Am. Oil Chem. Soc.* **62**, 128.
Wentz, M., W. H. Smith, and A. Martin (1969) *J. Colloid Interface Sci.* **29**, 36.
Winters, L. J. and E. Grunwald (1965) *J. Am. Chem. Soc.* **87**, 4608.
Wrigley, A. N., F. D. Smith, and A. J. Stirton (1957) *J. Am. Oil Chem. Soc.* **34**, 39.

PROBLEMS

4.1 Predict the locations of the following solubilizates in a micelle of $C_{12}H_{25}SO_4^-Na^+$ in aqueous medium:

(a) Toluene
(b) Cyclohexane
(c) *n*-Hexyl alcohol
(d) *n*-Dodecyl alcohol

4.2 Predict the effect of the following changes on the solubilization capacity of a micelle of $R(OC_2H_4)xOH$ in aqueous medium for the two solubilizates given below. Use the symbols: + = increase; – = decrease; 0 ≅ little or no effect; ? = effect not clearly predictable.

Change	Effect for	
	n-Octane	*n*-Octylamine
(a) Increase in the value of x		
(b) Increase in the temperature to the cloud point		
(c) Addition of electrolyte		
(d) Addition of HCl		
(e) Increase in the chain length, R		

4.3 Predict the effect on the solubilization of water by micelles of $R(OC_2H_4)_xOH$ in heptane of:

(a) Increase in the value of x
(b) Increase in the temperature
(c) Addition of electrolyte
(d) Addition of HCl
(e) Increase in the chain length, R.

4.4 Explain why it is advisable to use a solution of $C_{11}H_{23}CO_2CH_2CH_2SO_3^-M^+$ at a concentration above its CMC in distilled water soon after it is prepared, if one wishes to obtain an accurate measurement of its surface tension. (The pH of distilled water is about 5.8.)

4.5 Predict and explain the effect of each of the following on the cloud point of a nonionic surfactant, $R(OC_2H_4)xOH$:

(a) Decrease in the pH of the solution below 7

(b) Saturation of the aqueous solution with *n*-hexane

(c) Addition of NaF to the solution

(d) Use of a commercial, rather than a pure grade, of $R(OC_2H_4)xOH$

(e) Addition of $CaCl_2$ to the solution

5 Reduction of Surface and Interfacial Tension by Surfactants

Reduction of surface or interfacial tension is one of the most commonly measured properties of surfactants in solution. Since it depends directly on the replacement of molecules of solvent at the interface by molecules of surfactant, and therefore on the surface (or interfacial) excess concentration of the surfactant, as shown by the Gibbs equation (Equation 2.17 in Chapter 2)

$$d\gamma = -\sum_i \Gamma_i d\mu_i,$$

it is also one of the most fundamental of interfacial phenomena.

The molecules at the surface of a liquid have potential energies greater than those of similar molecules in the interior of the liquid. This is because attractive interactions of molecules at the surface with those in the interior of the liquid are greater than those with the widely separated molecules in the gas phase. Because the potential energies of molecules at the surface are greater than those in the interior of the phase, an amount of work equal to this difference in potential energy must be expended to bring a molecule from the interior to the surface. The surface free energy per unit area, or surface tension, is a measure of this work; it is the minimum amount of work required to bring sufficient molecules to the surface from the interior to expand it by unit area. Although more correctly thought of as a surface free energy per unit area, surface tension is often conceptualized as a force per unit length *at a right angle to the force* required to pull apart the surface molecules in order to permit expansion of the surface by movement into it of molecules from the phase underneath it.

At the interface between two condensed phases, the dissimilar molecules in the adjacent layers facing each other across the interface (Figure 5.1) also have potential energies different from those in their respective phases. Each molecule at the interface has a potential energy greater than that of a similar molecule in the interior of its bulk phase by an amount equal to its interaction energy with the molecules in the interior of its bulk phase minus its interaction energy with the molecules in the bulk phase across the interface. For most

Surfactants and Interfacial Phenomena, Fourth Edition. Milton J. Rosen and Joy T. Kunjappu.

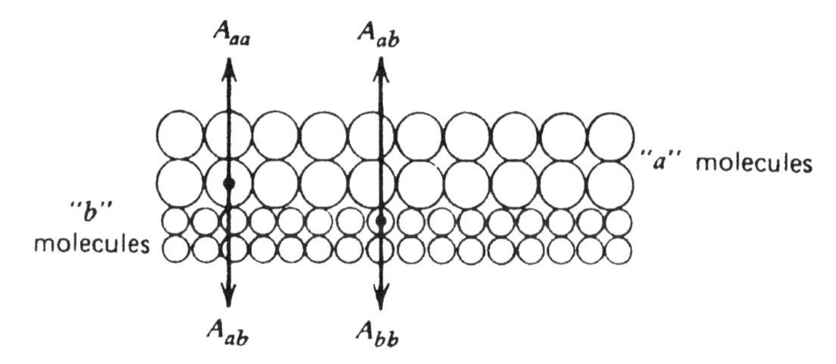

FIGURE 5.1 Simplified diagram of the interface between two condensed phases *a* and *b*.

purposes, however, only interactions with adjacent molecules need be taken into account. If we consider an interface between two pure liquid phases *a* and *b* (Figure 5.1), then the increased potential energy of the *a* molecules at the interface over those in the interior of that phase is $A_{aa} - A_{ab}$, where A_{aa} symbolizes the molecular interaction energy between *a* molecules at the interface and similar molecules in the interior of the bulk phase, and A_{ab} symbolizes the molecular interaction energy between *a* molecules at the interface and *b* molecules across the interface. Similarly, the increased potential of *b* molecules at the interface over those in the interior is $A_{bb} - A_{ab}$. The increased potential energy of all the molecules at the interface over those in the interior of the bulk phases, the interfacial free energy, is then $(A_{aa} - A_{ab}) + (A_{bb} - A_{ab})$ or $A_{aa} + A_{bb} - 2A_{ab}$, and this is the minimum work required to create the interface. The interfacial free energy per unit area of interface, the interfacial tension γ_I, is then given by the expression

$$\gamma_I = \gamma_a + \gamma_b - 2\gamma_{ab}, \tag{5.1}$$

where γ_a and γ_b are the surface free energies per unit area (the surface tensions) of the pure liquids *a* and *b*, respectively, and γ_{ab} is the $a - b$ interaction energy per unit area across the interface.

The value of the interaction energy per unit area across the interface γ_{ab} is large when molecules *a* and *b* are similar in nature to each other (e.g., water and short-chain alcohols). When γ_{ab} is large, we can see from Equation 5.1 that the interfacial tension γ_I will be small; when γ_{ab} is small, γ_I is large. The value of the interfacial tension is therefore a measure of the dissimilarity of the two types of molecules facing each other across the interface.

In the case where one of the phases is a gas (the interface is a surface), the molecules in that phase are so far apart relative to those in the condensed phase that tensions produced by molecular interaction in that phase can be disregarded. Thus, if phase *a* is a gas, γ_a and γ_{ab} can be disregarded and $\gamma_I \approx \gamma_b$, the surface tension of the condensed phase *b*.

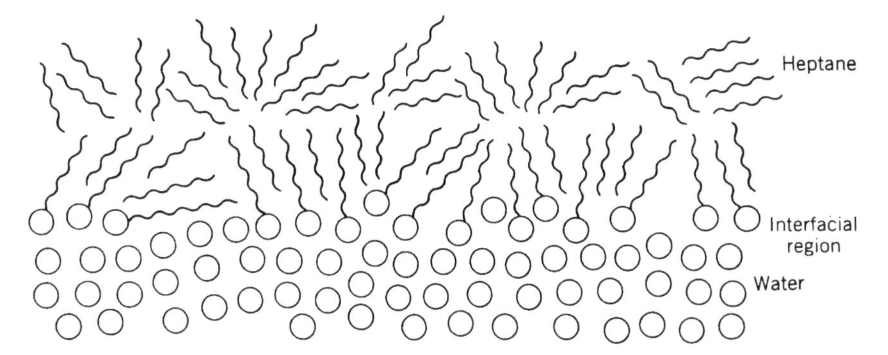

FIGURE 5.2 Diagrammatic representation of heptane–water interface with adsorbed surfactant.

When the two phases are immiscible liquids, γ_a and γ_b, their respective surface tensions, are experimentally determinable, permitting the evaluation of γ_{ab}, at least in some cases. If one of the phases is solid, on the other hand, experimental evaluation of γ_{ab} is difficult, if not impossible. However, here, too, the greater the similarity between a and b in structure or in the nature of their intramolecular forces, the greater the interaction between them (i.e., the greater the value of γ_{ab}) and the smaller the resulting interfacial tension between the two phases. When $2\gamma_{ab}$ becomes equal to $\gamma_{ab} + \gamma_{ab}$, the interfacial region disappears, and the two phases spontaneously merge to form a single one.

If we now add to a system of two immiscible phases (e.g., heptane and water), a surface-active agent that is adsorbed at the interface between them, it will orient itself there, mainly with the hydrophilic group toward the water and the hydrophobic group toward the heptane (Figure 5.2). When the surfactant molecules replace water and/or heptane molecules of the original interface, the interaction across the interface is now between the hydrophilic group of the surfactant and water molecules on one side of the interface and between the hydrophobic group of the surfactant and heptane on the other side of the interface. Since these interactions are now much stronger than the original interaction between the highly dissimilar heptane and water molecules, the tension across the interface is significantly reduced by the presence there of the surfactant. Since air consists of molecules that are mainly nonpolar, surface tension reduction by surfactants at the air–aqueous solution interface is similar in many respects to interfacial tension reduction at the heptane–aqueous solution interface.

We can see from this simple model why a necessary but not sufficient condition for surface or interfacial tension reduction is the presence in the surfactant molecule of both lyophobic and lyophilic portions. The lyophobic portion has two functions: (1) to produce spontaneous adsorption of the surfactant molecule at the interface and (2) to increase interaction across the interface between the adsorbed surfactant molecules there and the molecules in the adjacent phase. The function of the lyophilic group is to provide strong interaction between the molecules of surfactant at the interface and the molecules of solvent. If any of these functions is not performed, then the marked reduction of interfacial tension characteristic of surfactants will probably not occur. Thus, we would not expect ionic surfactants containing hydrocarbon chains to reduce the surface tensions of hydrocarbon solvents, in spite of the distortion of the solvent structure by the ionic groups in the surfactant molecules. Adsorption of such molecules at the air–hydrocarbon interface with the ionic groups oriented toward the predominantly nonpolar air molecules would result in *decreased* interaction across the interface, compared to that with their hydrophobic groups oriented toward the air.

For significant surface activity, a proper balance between lyophilic and lyophobic character in the surfactant is essential. Since the lyophilic (or lyophobic) character of a particular structural group in the molecule varies with the chemical nature of the solvent and such conditions of the system as temperature and the concentrations of electrolyte and/or organic additives, the lyophilic–lyophobic balance of a particular surfactant varies with the system and the conditions of use. In general, good surface or interfacial tension reduction is shown only by those surfactants that have an appreciable, but limited, solubility in the system under the conditions of use. Thus, surfactants which may show good surface tension reduction in aqueous systems may show no significant surface tension reduction in slightly polar solvents such as ethanol and polypropylene glycol in which they may have high solubility.

Measurement of the surface or interfacial tension of liquid systems is accomplished readily by a number of methods of which the most useful and precise for solutions of surfactants are probably the drop-weight and Wilhelmy plate methods. An excellent discussion of the various methods for determining surface and interfacial tension is included in the monograph on emulsions by Becher (1965).

For the purpose of comparing the performance of surfactants in reducing surface or interfacial tension, as in adsorption, it is necessary to distinguish between the efficiency of the surfactant (i.e., the bulk phase concentration of surfactant required to reduce the surface or interfacial tension by some significant amount) and its effectiveness, the maximum reduction in tension that can be obtained, regardless of bulk phase concentration of surfactant. These two parameters do not necessarily run parallel to each other and sometimes even run counter to each other.

The *efficiency* of a surfactant in reducing surface tension can be measured by the same quantity that is used to measure the efficiency of adsorption at the liquid–gas interface (Chapter 2, Section IIIE), pC_{20}, the negative log of the bulk phase concentration necessary to reduce the surface tension by 20 dyn/cm (mN/m). The *effectiveness* of a surfactant in reducing surface tension can be measured by the amount of reduction, or surface pressure, Π_{CMC} ($= \gamma_0 - \gamma_{CMC}$) attained at the critical micelle concentration (CMC), since reduction of the tension beyond the CMC is relatively insignificant (Figure 5.3).

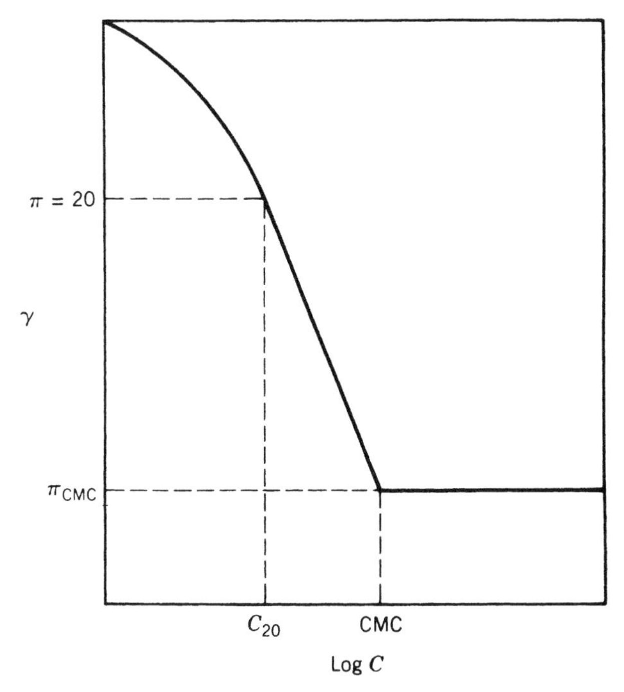

FIGURE 5.3 Surface tension -log C plot illustrating efficiency, -log C_{20}(pC_{20}), and effectiveness of surface tension reduction, Π_{CMC}.

I. EFFICIENCY IN SURFACE TENSION REDUCTION*

Since surface or interfacial tension reduction depends on the replacement of solvent molecules at the interface by surfactant molecules, the efficiency of a surfactant in reducing surface tension should reflect the concentration of the surfactant at the interface relative to that in the bulk liquid phase. A suitable measure for the efficiency with which a surfactant performs this function would therefore be the ratio of the concentration of surfactant at the surface to that in the bulk liquid phase at equilibrium, both concentrations being expressed in the same units, for example, $[C_1^s]/C_1$, where both concentrations are in moles/liter.

The surface concentration of surfactant $[C_1^s]$, in moles per liter, is related to its surface excess concentration Γ_1, in moles per cm^2, by the relation $[C_1^s] = (1000\Gamma_1/d) + C_1$, where d = the thickness of the interfacial region, in centimeters. For surfactants, Γ_1 is in the range $1\text{–}5 \times 10^{-10}$ mol/cm^2, while $d = 50 \times 10^{-8}$ cm or less and $C_1 = 0.01\ M$ or less. Thus, $[C_1^s] = 1000\Gamma_1/d$ without significant error, and $[C_1^s]/C_1 = 1000\Gamma_1/Cd$.

* Rosen (1974).

When the tension has been reduced by 20 mN/m (dyn/cm), the value of Γ is close to its maximum value, Γ_m, as shown in Chapter 2, and most surfactant molecules are lying slightly tilted to the interface. If we assume that the thickness of the interfacial region d is determined by the height of the surfactant normal to the interface, then d is inversely proportional to the minimum surface area per adsorbed molecule a_m^s; a larger value of a_m^s generally indicates a smaller angle of the surfactant with respect to the interface, a smaller value of a_m^s indicates an orientation of the surfactant more perpendicular to the interface. Since $a_m^s = (K/\Gamma_1) \propto (1/d)$, the quantity Γ_1/d may be considered to be a constant, and $[C_1^S]/C_1 = (K_1/C_1)_{\pi=20}$, where K and K_1 are constants. This indicates that the bulk concentration of surfactant necessary to produce a 20-mN/m (dyn/cm) reduction in tension, C_{20} (Chapter 2, Section IIIE), is not only a measure of the efficiency of adsorption at the liquid–gas interfaces, but also a measure of the efficiency of surface tension reduction by the surfactant (Figure 5.3).

Again, as discussed there, it is more useful and convenient to use the quantity C_{20} in the form of its negative logarithm pC_{20}, since the latter quantity is related (Equations 2.30, 2.31, 2.35, 2.36) to standard free energies of adsorption. The factors that determine the value of pC_{20} have been discussed in Chapter 2, Section IIIE.

For surfactants in aqueous solution, the efficiency increases with increase in the hydrophobic character of the surfactant. Equation 2.31 indicates that the efficiency factor pC_{20} is often a linear function of the number of carbon atoms in a straight-chain hydrophobic group, increasing as the number of carbon atoms increases. This has been shown (Rosen, 1974) to be valid for several homologous series of anionic, cationic, and nonionic surfactants. Some plots of pC_{20} as a function of the number of carbon atoms in the (straight hydrocarbon chain) hydrophobic group are given in Figure 2.16.

In its effect on the efficiency of surface tension reduction, as in both efficiency and effectiveness of adsorption, a phenyl group in the hydrophobic chain appears to be equivalent to about three and one-half $-CH_2-$ groups in a straight alkyl chain with a single terminal hydrophilic group. This same effect has been noted in measurements of relative adsorptivity at the aqueous solution–air interface (Shinoda and Masio, 1960) and of CMCs (Chapter 3, Section IVA). Replacement of a single straight-chain hydrophobic group by a branched or an unsaturated one containing the same number of carbon atoms, or by two or more groups with the same total number of carbon atoms, decreases the efficiency. When the hydrophobic group has side chains, the carbon atoms on the side chains appear to have about two-thirds the effect of carbon atoms in a straight alkyl chain with a single, terminal hydrophilic group. Thus, the efficiency of $C_6H_{13}CH(C_4H_9)CH_2C_6H_4SO_3^-Na^+$ at 75°C is between those of the sodium *p-n*-decyl- and *p-n*-dodecylbenzenesulfonates. When the hydrophilic group is at a nonterminal position in the hydrophobic group, the latter appears to act as if it were branched at the position of the hydrophilic group, with the carbon atoms on the shorter portion of the hydrophobic group having about two-thirds the effect of the carbon atoms in the longer portion. Thus, sodium *p-n*-dodecyl-6-benzenesulfonate, $C_6H_{13}CH(C_5H_{11})C_6H_4SO_3^-Na^+$, has a surface tension concentration curve at 75°C, which is practically identical with that of sodium *p-n*-decylbenzenesulfonate (Greiss, 1955).

For the series $RCOO(CH_2)_nSO_3^-Na^+$ where $n = 2, 3$, or 4, the -CH_2- groups between the two hydrophilic groups, -COO^- and -$SO_3^-Na^+$, appear to be equivalent to about one-half -CH_2- group in a straight alkyl chain with a single terminal hydrophilic group. In compounds of structure $R(OC_2H_4)_nSO_4^-Na^+$, where $n = 1$, 2, or 3, or in $RCONH(C_2H_4OH)_2$, at 25°C, the first oxyethylene (OE) group appears to be equivalent to about two and one-half -CH_2- groups in a straight chain, with the additional oxyethylene groups having little or no effect.

The replacement of the usual hydrocarbon-based hydrophobic group by a fluorocarbon-based hydrophobic group causes a very large increase in the efficiency of surface tension reduction (Shinoda et al., 1972), the C_7 perfluorosulfonate showing greater efficiency than the corresponding C_{12} hydrocarbon-based sulfonate.

A change in the sign of the charge of a univalent ionic hydrophilic group produces little, if any, effect on the efficiency. However, the replacement of the counterion by one that is more tightly bound increases the efficiency, presumably by decreasing the net electrical charge on the surfactant molecule. For similar reasons, the replacement of an ionic hydrophilic group by a nonionic one or the addition of neutral electrolyte to a solution of an ionic surfactant in pure water results in a large increase in the value of pC_{20}. Temperature increase in the 10–40°C range causes a small decrease in the pC_{20} value for ionics and zwitterionics, a somewhat larger *increase* for polyoxyethylene (POE) nonionics.

The addition of water structure promoters (fructose, xylose) or a water structure breaker (*N*-methylacetamide) to an aqueous solution of a POE nonionic has been shown to affect markedly the efficiency of the surfactant in reducing surface tension (Schwuger, 1971). Water structure promoters appear to increase the efficiency of the surfactant, whereas structure breakers decrease it. The reasons for these changes are probably the same as those that account for the effect of these additives on the CMCs of nonionics (Chapter 3).

II. EFFECTIVENESS IN SURFACE TENSION REDUCTION*

A. The Krafft Point

We have seen in Figure 2.15 that the surface tension of a solution of an individual surfactant decreases steadily as the bulk concentration of surfactant is increased until the concentration reaches a value known as the CMC, above which the tension remains virtually unchanged. The surface tension at the

* Rosen (1976).

CMC is therefore very close to the minimum tension (or maximum surface pressure) that the system can achieve. The surface pressure at this point, Π_{CMC}, is therefore a suitable measure of the "effectiveness" of a surfactant in reducing surface tension (Figure 5.3).

If the CMC exceeds the solubility of the surfactant at a particular temperature, then the minimum surface tension will be achieved at the point of maximum solubility, rather than at the CMC. The temperature at which the solubility of an ionic surfactant becomes equal to the CMC is known as the *Krafft point* (T_k). For surfactants being used below T_k, then, the maximum reduction in surface tension will be determined by the concentration of surfactant at solution saturation, and these materials may show lower effectiveness in reducing surface tension than similar materials that are being used above their Krafft points. Krafft points of some surfactants are given in Table 5.1. These are for purified compounds; mixtures of isomeric materials generally have T_k's that are considerably lower than those of individual compounds.

The Krafft point increases with increase in the number of carbon atoms in the hydrophobic group and decreases with branching or unsaturation in that group in a homologous series of ionic surfactants (Gu and Sjeblom, 1992). It also depends upon the nature of the counterion, increasing in the order $Li^+ < NH_4^+ < Na^+ < K^+$ for anionics. Oxyethylenation of alkyl sulfates decreases their Krafft points; oxypropylenation decreases them even further. Alkanesulfonates have higher Krafft points than their corresponding alkyl sulfates. The substitution of triethyl for trimethyl in the head groups of cationic alkyl trimethylammonium bromides leads to significant reduction in their Krafft points (Davey et al., 1998).

For surfactants that are being used above T_k, maximum reduction, for all practical purposes, is reached at the CMC. Since surfactants are normally used above their Krafft points, we restrict our discussion to that condition and consider maximum surface tension reduction to occur at the CMC.

B. Interfacial Parameter and Chemical Structural Effects

We have seen that at some point below, but near the CMC, the surface becomes essentially saturated with surfactant $\Gamma \approx \Gamma_m$. The relation between γ and $\log C_1$, the Gibbs equation, $d\gamma = -2.3nRT\Gamma_m \log C_1$ (Equation 2.19a), in that region therefore becomes essentially linear. This linear relation continues to the CMC (in fact, it is usually used to determine the CMC).

When the point, C_{20}, is on the linear portion of the curve (Figure 5.3), that is, if the surface is essentially saturated when the surface tension of the solvent has been reduced by 20 dyn/cm (20 mN/m), which is generally the case for most surfactants, then for the linear portion of the plot, the Gibbs adsorption equation becomes

$$\Delta\gamma = -\Delta\pi \approx -2.3nRT\Gamma_m\Delta \log C \quad (5.2)$$

TABLE 5.1 Krafft Points of Surfactants

Compound	Krafft Point (°C)	Reference
$C_{12}H_{25}SO_3^-Na^+$	38	Weil et al. (1963)
$C_{14}H_{29}SO_3^-Na^+$	48	Weil et al. (1963)
$C_{16}H_{33}SO_3^-Na^+$	57	Weil et al. (1963)
$C_{18}H_{37}SO_3^-Na^+$	70	Weil et al. (1963)
$C_{10}H_{21}SO_4^-Na^+$	8	Raison (1957)
$C_{12}H_{25}SO_4^-Na^+$	16	Weil et al. (1963)
$2\text{-}MeC_{11}H_{23}SO_4^-Na^+$	<0	Gotte (1969)
$C_{14}H_{29}SO_4^-Na^+$	30	Weil et al. (1963)
$2\text{-}MeC_{13}H_{27}SO_4^-Na^+$	11	Gotte (1969)
$C_{16}H_{33}SO_4^-Na^+$	45	Weil et al. (1963)
$2\text{-}MeC_{15}H_{31}SO_4^-Na^+$	25	Gotte (1969)
$C_{16}H_{33}SO_4^-.^+NH_2(C_2H_4OH)_2$	<0	Weil et al. (1959)
$C_{18}H_{37}SO_4^-Na^+$	56	Weil et al. (1963)
$2\text{-}MeC_{17}H_{35}SO_4^-Na^+$	30	Gotte (1969)
$Na^{+-}O_4S(CH_2)_{12}SO_4^-Na^+$	12	Ueno et al. (1974)
$Na^{+-}O_4S(CH_2)_{14}SO_4^-Na^+$	24.8	Ueno et al. (1974)
$Li^{+-}O_4S(CH_2)_{14}SO_4^-Li^+$	35	Ueno et al. (1974)
$Na^{+-}O_4S(CH_2)_{16}SO_4^-Na^+$	39.1	Ueno et al. (1974)
$K^{+-}O_4S(CH_2)_{16}SO_4^-K^+$	45.0	Ueno et al. (1974)
$Li^{+-}O_4S(CH_2)_{16}SO_4^-Li^+$	39.0	Ueno et al. (1974)
$Na^{+-}O_4S(CH_2)_{18}SO_4^-Na^+$	44.9	Ueno et al. (1974)
$K^{+-}O_4S(CH_2)_{18}SO_4^-K^+$	55.0	Ueno et al. (1974)
$C_8H_{17}COO(CH_2)_2SO_3^-Na^+$	0	Hikota et al. (1970)
$C_{10}H_{21}COO(CH_2)_2SO_3^-Na^+$	8.1	Hikota et al. (1970)
$C_{12}H_{25}COO(CH_2)_2SO_3^-Na^+$	24.2	Hikota et al. (1970)
$C_{14}H_{29}COO(CH_2)_2SO_3^-Na^+$	36.2	Hikota et al. (1970)
$C_8H_{17}OOC(CH_2)_2SO_3^-Na^+$	0	Hikota et al. (1970)
$C_{10}H_{21}OOC(CH_2)_2SO_3^-Na^+$	12.5	Hikota et al. (1970)
$C_{12}H_{25}OOC(CH_2)_2SO_3^-Na^+$	26.5	Hikota et al. (1970)
$C_{14}H_{29}OOC(CH_2)_2SO_3^-Na^+$	39.0	Hikota et al. (1970)
$C_{12}H_{25}CH(SO_3^-Na^+)COOCH_3$	6	Ohbu et al. (1998)
$C_{12}H_{25}CH(SO_3^-Na^+)COOC_2H_5$	1	Ohbu et al. (1998)
$C_{14}H_{29}CH(SO_3^-Na^+)COOCH_3$	17	Ohbu et al. (1998)
$C_{16}H_{33}CH(SO_4^-Na^+)COOCH_3$	30	Ohbu et al. (1998)
$C_{10}H_{21}CH(CH_3)C_6H_4SO_3^-Na^+$	31.5	Smith et al. (1966)
$C_{12}H_{25}CH(CH_3)C_6H_4SO_3^-Na$	46.0	Smith et al. (1966)
$C_{14}H_{29}CH(CH_3)C_6H_4SO_3^-Na^+$	54.2	Smith et al. (1966)
$C_{16}H_{33}CH(CH_3)C_6H_4SO_3^-Na^+$	60.8	Smith et al. (1966)
$C_{14}H_{29}OCH_2CH(SO_3^-Na^+)CH_3$	14	Weil et al. (1966)
$C_{14}H_{29}[OCH_2CH(CH_3)]_2SO_4^-Na^+$	<0	Weil et al. (1966)
$C_{16}H_{33}OCH_2CH_2SO_4^-Na^+$	36	Gotte (1969)
$C_{16}H_{33}(OCH_2CH_2)_2SO_4^-Na^+$	24	Gotte (1969)
$C_{16}H_{33}OCH_2CH(SO_4^-Na^+)CH_3$	27	Weil et al. (1966)
$C_{18}H_{37}OCH_2CH(SO_4^-Na^+)CH_3$	43	Weil et al. (1966)

(*Continued*)

TABLE 5.1 (*Continued*)

Compound	Krafft Point (°C)	Reference
$C_{16}H_{33}OCH_2CH_2SO_4^-Na^+$	36	Weil et al. (1959)
$C_{16}H_{33}(OC_2H_4)_2SO_4^-Na^+$	24	Weil et al. (1959)
$C_{16}H_{33}(OC_2H_4)_3SO_4^-Na^+$	19	Weil et al. (1959)
$C_{16}H_{33}[OCH_2CH(CH_3)]_2SO_4^-Na^+$	19	Gotte (1969)
$C_{18}H_{37}(OC_2H_4)_3SO_4^-Na^+$	32	Weil et al. (1959)
$C_{18}H_{37}(OC_2H_4)_4SO_4^-Na^+$	18	Weil et al. (1959)
$C_{18}H_{37}[OCH_2CH(CH_3)]_2SO_4^-Na^+$	31	Gotte (1969)
n-$C_7F_{15}COO^-Li^+$	<0	Shinoda et al. (1972)
n-$C_7F_{15}COO^-Na^+$	8.0	Shinoda et al. (1972)
n-$C_7F_{15}COO^-K^+$	25.6	Shinoda et al. (1972)
n-$C_7F_{15}COOH$	20	Shinoda et al. (1972)
n-$C_7F_{15}COO^-NH_4^+$	2.5	Shinoda et al. (1972)
$(CF_3)_2CF(CF_2)_4COO^-K^+$	<0	Shinoda et al. (1972)
$(CF_3)_2CF(CF_2)_4COO^-Na^+$	<0	Shinoda et al. (1972)
n-$C_7F_{15}SO_3^-Na^+$	56.5	Shinoda et al. (1972)
n-$C_8F_{17}SO_3^-Li^+$	<0	Shinoda et al. (1972)
n-$C_8F_{17}SO_3^-Na^+$	75	Shinoda et al. (1972)
n-$C_8F_{17}SO_3^-K^+$	80	Shinoda et al. (1972)
n-$C_8F_{17}SO_3^-NH_4^+$	41	Shinoda et al. (1972)
n-$C_8F_{17}SO_3^-.^+NH_3C_2H_4OH$	<0	Shinoda et al. (1972)
Cationics		
$C_{16}H_{33}N^+(CH_3)_3Br^-$	25	Davey et al. (1998)
$C_{16}H_{33}N^+(C_2H_5)_3Br^-$	<0	Davey et al. (1998)
$C_{18}H_{37}N^+(CH_3)_3Br^-$	36	Davey et al. (1998)
$C_{18}H_{37}N^+(C_2H_5)_3Br^-$	12	Davey et al. (1998)
$C_{16}H_{33}Pyr^+Br^-$	25	Davey et al. (1998)
Zwitterionics		
$C_{12}H_{25}N^+(CH_3)_2(CH_2)_{1\text{-}6}COO^-$	<1	Weers et al. (1991)
$C_{16}H_{33}N^+(CH_3)_2CH_2COO^-$	17	Weers et al. (1991)
$C_{16}H_{33}N^+(CH_3)_2(CH_2)_3COO^-$	13	Weers et al. (1991)
$C_{16}H_{33}N^+(CH_3)_2(CH_2)_5COO^-$	<0	Weers et al. (1991)
$C_{10}H_{21}(Pyr^+)COO^-$	<0	Zhao and Rosen (1984)
$C_{12}H_{23}CH(Pyr^+)COO^-$	23	Zhao and Rosen (1984)
$C_{14}H_{29}CH(Pyr^+)COO^-$	38	Zhao and Rosen (1984)
$C_{12}H_{25}N^+(CH_3)_2CH_2CH_2SO_3^-$	70	Weers et al. (1991)
$C_{12}H_{25}N^+(CH_3)_2(CH_2)_3SO_3^-$	<0	Weers et al. (1991)
$C_{16}H_{33}N^+(CH_3)_2CH_2CH_2SO_3^-$	90	Weers et al. (1991)
$C_{16}H_{33}N^+(CH_3)_2(CH_2)_3SO_3^-$	28	Weers et al. (1991)
$C_{16}H_{33}N^+(CH_3)_2(CH_2)_4SO_3^-$	30	Weers et al. (1991)

Pyr^+, pyridinium.

and

$$(\Pi_{CMC} - 20) \approx 2.3nRT\Gamma_m(\log \text{CMC} - \log C_{20})$$

or

$$\Pi_{CMC} \approx 20 + 2.3nRT\Gamma_m \log(\text{CMC}/C_{20}). \quad (5.3)$$

From this, we see that the effectiveness of a surfactant in reducing the surface tension of a solvent depends upon

1. the number of ions n whose surface concentration changes with change in the liquid-phase concentration of the surfactant;
2. the effectiveness of adsorption of the surfactant Γ_m; and
3. the CMC/C_{20} ratio.

The larger each of these quantities, the greater the reduction in surface tension attained at the CMC.

The factors that affect Γ_m, the effectiveness of adsorption, have been discussed previously (Chapter 2, Section IIIC). We can summarize their effects as follows:

1. Change in the length of the hydrophobic group (from 10 to 16 carbon atoms), or the introduction of some branching into the hydrophobic group, has very little effect on Γ_m in ionic surfactants.
2. As the size of the hydrophilic group, or its distance from a second hydratable group in the molecule, increases, Γ_m decreases.
3. For ionic surfactants, an increase in the ionic strength of the solution causes an increase in Γ_m.
4. For POE nonionics, an increase in the length of the POE chain at constant hydrophobic chain length causes a decrease in Γ_m; an increase in the hydrophobic chain length at constant POE chain length causes an increase in Γ_m.
5. Temperature increase causes a small decrease in Γ_m.

The factors that affect the CMC/C_{20} ratio have also been previously discussed (Chapter 3, Section VA). We have seen that

1. The ratio is increased only slightly, if at all, by an increase in the length of the hydrophobic group in ionic surfactants.
2. The ratio is increased by the introduction of branching into the hydrophobic group or the positioning of the hydrophilic group in a central position in the molecule.
3. The ratio is increased by the introduction of a larger hydrophilic group.
4. For ionic surfactants, the ratio is increased greatly by an increase in the ionic strength of the solution or by a more tightly bound counterion, especially one containing an alkyl chain of six or more carbon atoms.

5. The ratio is increased for POE nonionics, with increase in the length of the POE chain at constant hydrophobic chain length and decreased with increase in the length of the hydrophobic chain at constant POE chain length.
6. The ratio is decreased by an increase in temperature in the 10–40°C range.

Some of these factors affect Γ_m and the CMC/C_{20} ratio in parallel fashion (i.e., they increase both or decrease both); some in opposing fashion. When the effects are parallel, we can readily predict the resulting change in the effectiveness of surface tension reduction; when they are opposed, it is difficult to do so. Thus, increase in the length of the hydrophobic group in ionic surfactants has little effect on either Γ_m or the CMC/C_{20} ratio, and we can therefore expect that an increase in the length of the hydrophobic group will have little effect on their effectiveness of surface tension reduction.

On the other hand, the introduction of some branching into the hydrophobic group increases the CMC/C_{20} ratio but has little effect on Γ_m. We can therefore expect that the introduction of branching into the hydrophobic group will make the surfactant a more effective surface tension reducer. This is seen in the isomeric *p*-dodecylbenzene sulfonates (Figure 5.4), where the isomers with branched alkyl chains, although less efficient reducers of the surface tension than the isomer with the straight alkyl chain, reduce the surface tension to lower values than does the latter.

Table 5.2 lists some experimental values of Γ_m, CMC/C_{20}, and Π_{CMC}. The experimental Π_{CMC} values are very close to those calculated from the Γ_m and

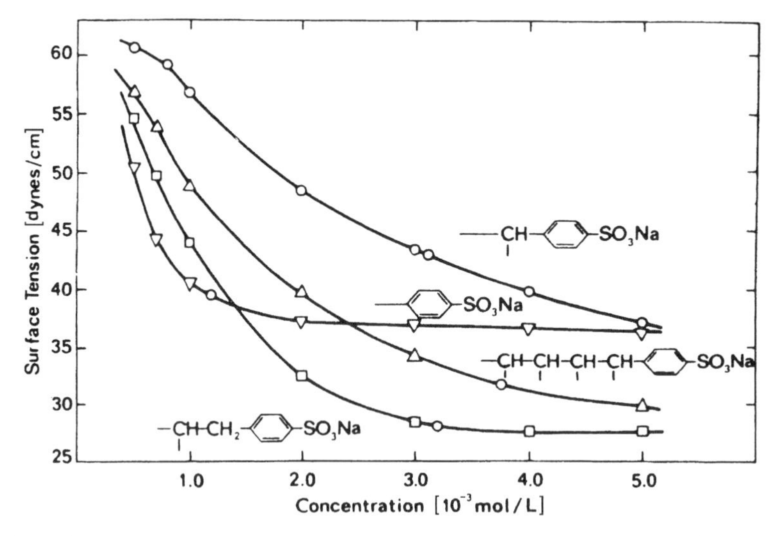

FIGURE 5.4 Surface tension of aqueous solutions of isomeric *p*-dodecylbenzenesulfonates at 75°C as a function of their concentration. Reprinted with permission from Greiss (1955).

TABLE 5.2 Values of Γ_m, CMC/C_{20}, and Π_{CMC} in Aqueous Medium

Surfactant	Temp. (°C)	$\Gamma_m \times 10^{10}$ mol/cm^2	CMC/C_{20}	Π_{CMC} (dyn/cm)	Reference
Anionics					
$C_{10}H_{21}OCH_2COO^-Na^+$ (0.1 *M* NaCl, pH 10.5)	30	5.4	4.9	$40._5$	Tsubone and Rosen (2001)
$C_{11}H_{23}CON(CH_3)CH_2COO^-Na^+$ (pH 10.5)	30	2.1	3.5	$32._9$	Tsubone and Rosen (2001)
$C_{11}H_{23}CON(CH_3)CH_2COO^-Na^+$ (0.1 *M* NaCl, pH 10.5)	30	2.9	6.5	$32._5$	Tsubone and Rosen (2001)
$C_{11}H_{23}CON(C_4H_9)CH_2COO^-$ Na^+	25	1.55	9.3	$36._8$	Zhu et al. (1998a)
$C_{11}H_{23}CON(C_4H_9)CH_2COO^-Na^+$ ("hard river" water, I.S. = 6.6×10^{-3} *M*)	25	2.90	28.8	$43._9$	Zhu et al. (1998a
$C_{11}H_{23}CON(CH_3)CH_2CH_2COO^-Na^+$ (pH 10.5)	30	1.6	3.7	$30._6$	Tsubone and Rosen (2001)
$C_{11}H_{23}CON(CH_3)CH_2CH_2COO^-Na^+$ (0.1 *M* NaCl, pH 10.5)	30	2.5	6.9	$31._5$	Tsubone and Rosen (2001)
$C_{13}H_{27}CON(C_3H_7)CH_2COO^-Na^+$	25	1.58	12.0	$39._2$	Zhu et al. (1998a)
$C_{13}H_{27}CON(C_3H_7)CH_2COO^-Na^+$ ("Hard river" water, I.S. = 6.6×10^{-3} *M*)	25	3.50	14.1	$42._9$	Zhu et al. (1998a)
$C_{10}H_{21}SO_3^-Na^+$	10	3.4	2.4	$33._0$	Dahanayake et al. (1986)
$C_{10}H_{21}SO_3^-Na^+$	25	3.3	2.1	$31._0$	Dahanayake et al. (1986)
$C_{10}H_{21}SO_3^-Na^+$	40	3.05	1.8	$29._2$	Dahanayake et al. (1986)
$C_{10}H_{21}SO_3^-Na^+$ (in 0.1 *M* NaCl)	25	3.85	4.1	$32._6$	Dahanayake et al. (1986)
$C_{10}H_{21}SO_3^-Na^+$ (in 0.5 *M* NaCl)	25	4.2	5.4	$37._1$	Dahanayake et al. (1986)
$C_{12}H_{25}SO_3^-Na^+$	25	2.9	2.8	$33._0$	Dahanayake et al. (1986)
$C_{12}H_{25}SO_3^-Na^+$	60	2.5	1.92	29	Rosen and Solash (1969)
$C_{12}H_{25}SO_3^-Na^+$ ("hard river" water, I.S. = 6.6×10^{-3} *M*)	25	2.34	9.97	$36._2$	Rosen et al. (1996)
$C_{12}H_{25}SO_3^-Na^+$ (in 0.1 *M* NaCl)	25	3.8	5.9	$36._4$	Dahanayake et al. (1986)
$C_{12}H_{25}SO_3^-Na^+$ (in 0.5 *M* NaCl)	40	3.6	6.8	$39._0$	Dahanayake et al. (1986)
$C_{12}H_{25}SO_3^-K^+$	25	3.3	2.38	34	Rosen (1974)
$C_{16}H_{33}SO_3^-K^+$	60	2.9	2.4	33	Rosen and Solash (1969)
$C_8H_{17}SO_4^-Na^+$ (heptane-H_2O)	50	2.3	4.0	39	Kling and Lange (1957)
$C_{10}H_{21}SO_4^-Na^+$	27	2.9	2.56	32	Dreger et al., 1944)
$C_{10}H_{21}SO_4^-Na^+$ (heptane-H_2O)	50	2.3	4.4	39	Kling and Lange (1957)

(*Continued*)

TABLE 5.2 (*Continued*)

Surfactant	Temp. (°C)	$\Gamma_m \times 10^{10}$ mol/cm^2	CMC/C_{20}	Π_{CMC} (dyn/cm)	Reference
branched $C_{12}H_{25}SO_4^-Na^+$	25	1.7	11.3	40.$_1$	Varadaraj et al. (1992)
branched $C_{12}H_{25}SO_4^-Na^{+a}$ (in 0.1 *M* NaCl)	25	3.3	15.2	42.$_7$	Varadaraj et al. (1992)
$C_{12}H_{25}SO_4^-Na^+$	25	3.2	2.6	32.$_5$	Dahanayake et al. (1986)
$C_{12}H_{25}SO_4^-Na^+$ (in 0.1 *M* NaCl)	25	4.0	6.0	38.$_0$	Dahanayake et al. (1986)
$C_{12}H_{25}SO_4^-Na^+$	25	3.2	2.6	32.$_5$	Dahanayake et al. (1986)
$C_{12}H_{25}SO_4^-Na^+$ (H_2O-octane)	25	3.3	4.7	42.$_8$	Rehfeld (1967)
$C_{12}H_{25}SO_4^-Na^+$ (H_2O-heptadecane)	25	3.3	4.8	42.$_5$	Rehfeld (1967)
$C_{12}H_{25}SO_4^-Na^+$ (H_2O-cyclohexane)	25	3.1	4.9	43.$_2$	Rehfeld (1967)
$C_{12}H_{25}SO_4^-Na^+$ (H_2O-benzene)	25	2.3	2.2	29.$_1$	Rehfeld (1967)
$C_{12}H_{25}SO_4^-Na^+$ (H_2O-1-hexene)	25	2.5	1.5	25.$_8$	Rehfeld (1967)
$C_{12}H_{25}SO_4^-Na^+$	60	2.6	1.74	28	Rosen and Solash (1969)
$C_{14}H_{29}SO_4^-Na^+$	25	—	2.6	37.$_2$	Lange and Schwuger (1968)
$C_{14}H_{29}SO_4^-Na^+$ (heptane-H_2O)	50	3.0	4.5	43	Kling and Lange (1957)
$C_{16}H_{33}SO_4^-Na^+$	60	3.3	2.5	35	Rosen and Solash (1969)
$C_{16}H_{33}SO_4^-Na^+$ (heptane-H_2O)	50	2.6	5.0	43.$_5$	Kling and Lange (1957)
$C_{18}H_{37}SO_4^-Na^+$ (heptane-H_2O)	50	2.5	5.0	44	Kling and Lange (1957)
$C_{10}H_{21}OCH_2CH_2SO_3^-Na^+$	25	3.2	2.0	30.$_8$	Dahanayake et al. (1986)
$C_{10}H_{21}OCH_2CH_2SO_3^-Na^+$ (in 0.1 *M* NaCl)	25	3.85	4.5	34.$_7$	Dahanayake et al. (1986)
$C_{10}H_{21}OCH_2CH_2SO_3^-Na^+$ (in 0.5 *M* NaCl)	25	4.3	7.1	39.$_0$	Dahanayake et al. (1986)
$C_{12}H_{25}OC_2H_4SO_4^-Na^+$	25	2.9	2.6	32.$_8$	Dahanayake et al. (1986)
$C_{12}H_{25}OC_2H_4SO_4^-Na^+$ (“hard river” water, I.S. = 6.6×10^{-3} *M*)	25	3.59	10.2	40.$_8$	Rosen et al. (1996)
$C_{12}H_{25}OC_2H_4SO_4^-Na^+$ (in 0.1 *M* NaCl)	25	3.8	7.3	38.$_6$	Dahanayake et al. (1986)
$C_{12}H_{25}OC_2H_4SO_4^-Na^+$ (in 0.5 *M* NaCl)	25	4.4	8.3	42.$_4$	Dahanayake et al. (1986)
$C_{12}H_{25}(OC_2H_4)_2SO_4^-Na^+$	10	2.8	2.8	32.$_6$	Dahanayake et al. (1986)
$C_{12}H_{25}(OC_2H_4)_2SO_4^-Na^+$	25	2.6	2.5	30.$_6$	Dahanayake et al. (1986)
$C_{12}H_{25}(OC_2H_4)_2SO_4^-Na^+$	40	2.5	2.0	28.$_6$	Dahanayake et al. (1986)
$C_{12}H_{25}(OC_2H_4)_2SO_4^-Na^+$ (“hard river” water, I.S. = 6.6×10^{-3} *M*)	25	3.24	11.5	39.$_0$	Rosen et al. (1996)
$C_{12}H_{25}(OC_2H_4)_2SO_4^-Na^+$ (in 0.5 *M* NaCl)	25	3.5	6.7	36.$_5$	Dahanayake et al. (1986)
$C_{12}H_{25}(OC_2H_4)_2SO_4^-Na^+$ (in 0.5 *M* NaCl)	25	3.8	10.0	40.$_2$	Dahanayake et al. (1986)

$C_{12}H_{25}(OC_2H_4)_2SO_4^-Na^+$ ("hard river" water, I.S. = 6.6×10^{-3} *M*)	25	2.41	10.5	$33._4$	Rosen et al. (1996)
$C_4H_9OC_{12}H_{24}SO_4^-Na^+$	25	1.1	4.2	28	Livingston and Drogin (1955)
$C_{14}H_{29}OC_2H_4SO_4^-Na^+$	25	2.1	8.8	40	Livingston and Drogin (1955)
$C_{14}H_{29}OC_2H_4SO_4^-Na^+$ ("hard river" water, I.S. = 6.6×10^{-3} *M*)	25	3.91	7.9	$40._0$	Rosen et al. (1996)
$C_4H_9CH(C_2H_5)CH_2OOCCH(SO_3^-Na^+CH_2COOCH_2CH(C_2H_5)C_4H_9$ ("hard river" water, I.S. = 6.6×10^{-3} *M*)	25	2.28	151.	$47._0$	Rosen et al. (1996)
$C_{11}H_{23}CON(CH_3)CH_2CH_2SO_3^-Na^+$ (pH 10.5)	30	2.2	2.0	$27._2$	Tsubone and Rosen (2001)
$C_{11}H_{23}CON(CH_3)CH_2CH_2SO_3^-Na^+$ (0.1 *M* NaCl, pH 10.5)	30	3.0	5.5	$31._7$	Tsubone and Rosen (2001)
$C_8H_{17}C_6H_4SO_3^-Na^+$	70	2.6	1.36	$24._7$	Lange (1964)
p-$C_9H_{19}C_6H_4SO_3^-Na^+$	75	1.8	1.3	23	Greiss (1955)
$C_{10}H_{21}C_6H_4SO_3^-Na^+$	70	3.2	1.33	$25._4$	Lange (1964)
p-$C_{10}H_{21}C_6H_4SO_3^-Na^+$	75	2.1	1.4	23.5	Greiss (1955)
$C_{11}H_{23}$-2-$C_6H_4SO_3^-Na^+$ ("hard river" water, I.S. = 6.6×10^{-3} *M*)	30	3.69	9.7	$40._0$	Zhu et al. (1998b)
p-Sodium 1,3,5,7-tetramethyl-(*n*-octyl)-1-benzenesulfonate	75	2.4	2.5	32	Greiss (1955)
$C_{12}H_{25}$-2-$C_6H_4SO_3^-Na^+$ ("hard river" water, I.S. = 6.6×10^{-3} *M*)	30	4.16	5.0	$35._6$	Zhu et al. (1998b)
$C_{12}H_{25}$-4-$C_6H_4SO_3^-Na^+$ ("hard river" water, I.S. = 6.6×10^{-3} *M*)	30	3.44	17.4	$43._8$	Zhu et al. (1998b)
p-$C_6H_{13}CH(C_4H_9)CH_2C_6H_4SO_3^-Na^+$	75	2.85	3.2	35	Greiss (1955)
p-$C_6H_{13}CH(C_5H_{11})C_6H_4SO_3^-Na^+$	75	2.1	>1.7	>26	Greiss (1955)
$C_{12}H_{25}$-6-$C_6H_4SO_3^-Na^+$ ("hard river" water, I.S. = 6.6×10^{-3} *M*)	30	3.15	21.5	$44._5$	Zhu et al. (1998b)
$C_{12}H_{25}C_6H_4SO_3^-Na^+$	70	3.7	1.33	$25._8$	Lange (1964)
$C_{12}H_{25}C_6H_4SO_3^-Na^+$ (0.1 *M* NaCl)	25	3.6	11.6	$41._9$	Murphy et al. (1990)
p-$C_{12}H_{25}C_6H_4SO_3^-Na^+$	75	2.8	1.6	24	Greiss (1955)
$C_{13}H_{27}$-2-$C_6H_4SO_3^-Na^+$ ("hard river" water, I.S. = 6.6×10^{-3} *M*)	30	4.05	3.1	$30._7$	Zhu et al. (1998b)
$C_{13}H_{27}$-5-$C_6H_4SO_3^-Na^+$ ("hard river" water, I.S. = 6.6×10^{-3} *M*)	30	3.58	15.8	$44._1$	Zhu et al. (1998b)
$C_{13}H_{27}$-5-$C_6H_4SO_3^-Na^+$	30	2.15	7.6	$39._0$	Zhu et al. (1998b)
$C_{14}H_{29}C_6H_4SO_3^-Na^+$	70	2.7	1.53	$26._5$	Lange (1964)
p-$C_{14}H_{29}C_6H_4SO_3^-Na^+$	70	2.2	1.6	$24._5$	Greiss (1955)
$C_{16}H_{33}C_6H_4SO_3^-Na^+$	70	1.9	1.93	$27._8$	Lange (1964)
$C_{16}H_{33}$-8-$C_6H_4SO_3^-Na^+$	45	1.61	14.4	$42._5$	Lascaux et al. (1983)
n-$C_7F_{15}COO^-Na^+$	25	4.0	9.4	$47._4$	Shinoda et al. (1972)

(*Continued*)

TABLE 5.2 (*Continued*)

Surfactant	Temp. (°C)	$\Gamma_m \times 10^{10}$ mol/cm^2	CMC/C_{20}	Π_{CMC} (dyn/cm)	Reference
n-$C_7F_{15}COO^-K^+$	25	3.9	9.3	$51._4$	Shinoda et al. (1972)
$(CF_3)_2CF(CF_2)_4COO^-Na^+$	25	2.8	11.2	$51._8$	Shinoda et al. (1972)
n-$C_8F_{17}SO_3^-Li^+$	25	3.0	10.0	$42._2$	Shinoda et al. (1972)
$C_4F_9CH_2OOCCH_2CH(SO_3^-Na^+)OOCCH_2C_4F_9$	30	3.0	-	$53._5$	Downer et al. (1999)
Cationics					
$C_{10}H_{21}N(CH_3)_3^+Br^-$ (in 0.1 *M* NaCl)	25	3.39	2.7	$30._4$	Li et al. (2001)
$C_{12}H_{25}N(CH_3)_3^+Br^-$ ("hard river" water, I.S. = 6.6×10^{-3} *M*)	25	2.72	3.99	$33._9$	Rosen et al. (1996)
$C_{12}H_{25}N(CH_3)_3^+Cl^-$	25	4.39	2.95	$31._5$	Li et al. (2001)
$C_{14}H_{29}N(CH_3)_3^+Br^-$	30	2.7	2.1	31	Venable and Nauman (1964)
$C_{14}H_{29}N(CH_3)_3^+Br^-$ ("hard river" water, I.S. = 6.6×10^{-3} *M*)	25	3.18	6.45	$34._6$	Rosen et al. (1996)
$C_{14}H_{29}N(C_3H_7)_3^+Br^-$	30	1.9	2.4	29	Venable and Nauman (1964)
$C_{16}H_{33}N(CH_3)_3^+Cl^-$ (in 0.1 *M* NaCl)	25	3.4	10.0	38	Caskey and Barlage (1971)
$C_{10}H_{21}Pyr^+Br^-$	25	2.01	3.97	$31._7$	Rosen et al. (1996)
$C_{12}H_{25}Pyr^+Br^-$	10	3.5	2.7	$34._6$	Rosen et al. (1982b)
$C_{12}H_{25}Pyr^+Br^-$	25	3.3	2.5	$32._9$	Rosen et al. (1982b)
$C_{12}H_{25}Pyr^+Br^-$	40	3.2	2.1	$30._8$	Rosen et al. (1982b)
$C_{12}H_{25}Pyr^+Br^-$ (in 0.1 *M* NaBr)	25	3.5	6.9	$35._2$	Rosen et al. (1982b)
$C_{12}H_{25}Pyr^+Br^-$ (in 0.1 *M* NaBr)	25	3.5	8.9	$37._2$	Rosen et al. (1982b)
$C_{12}H_{25}Pyr^+Cl^-$	10	2.7	2.3	$29._6$	Rosen et al. (1982b)
$C_{12}H_{25}Pyr^+Cl^-$	25	2.7	2.0	$28._3$	Rosen et al. (1982b)
$C_{12}H_{25}Pyr^+Cl^-$	40	2.6	1.8	$26._9$	Rosen et al. (1982b)
$C_{12}H_{25}Pyr^+Cl^-$ (in 0.1 *M* NaCl)	25	3.0	4.6	$30._4$	Rosen et al. (1982b)
$C_{12}H_{25}Pyr^+Cl^-$ (in 0.1 *M* NaCl)	25	3.1	5.5	$32._8$	Rosen et al. (1982b)
$C_{14}H_{29}Pyr^+Br^-$	30	2.8	2.2	31	Venable and Nauman (1964)
$C_{12}H_{25}N^+H_2CH_2CH_2OHCl^-$	25	1.93	7.0	31	Omar and Abdel-Khalek (1997)
$C_{12}H_{25}N^+H(CH_2CH_2OH)_2Cl^-$	25	2.49	7.3	32	Omar and Abdel-Khalek (1997)
$C_{12}H_{25}N^+(CH_2CH_2OH)_3Cl^-$	25	2.91	5.6	34	Omar and Abdel-Khalek (1997)

Anionic–Cationic Salts					
$CH_3SO_4^-.^+N(CH_3)_3C_{12}H_{25}$	25	2.70[b]	2.7	33.$_5$	Lange and Schwuger (1971)
$C_{12}H_{25}SO_4^-.^+N(CH_3)_3C_{12}H_{25}$	25	2.85[b]	3.4	37.$_5$	Lange and Schwuger (1971)
$C_{12}H_{25}SO_4^-.^+N(CH_3)_3C_{12}H_{25}$	25	2.63[b]	2.7	33.$_0$	Lange and Schwuger (1971)
$C_4H_9SO_4^-.^+N(CH_3)_3C_{10}H_{21}$	25	2.50[b]	7.0	44.$_2$	Lange and Schwuger (1971)
$C_{10}H_{21}SO_4^-.^+N(CH_3)_3C_4H_9$	25	2.85[b]	3.4	37.$_5$	Lange and Schwuger (1971)
$C_6H_{13}SO_4^-.^+N(CH_3)_3C_8H_{17}$	25	2.53[b]	10.4	49.$_8$	Lange and Schwuger (1971)
$C_8H_{17}SO_4^-.^+N(CH_3)_3C_6H_{13}$	25	2.50[b]	7.0	44.$_2$	Lange and Schwuger (1971)
$C_4H_9SO_4^-.^+N(CH_3)_3C_{12}H_{25}$	25	2.67[b]	5.3	42.$_0$	Lange and Schwuger (1971)
$C_6H_{13}SO_4^-.^+N(CH_3)_3C_{12}H_{25}$	25	2.58[b]	10.0	49.$_5$	Lange and Schwuger (1971)
$C_8H_{17}SO_4^-.^+N(CH_3)_3C_{12}H_{25}$	25	2.72[b]	9.6	50.$_6$	Lange and Schwuger (1971)
$C_{10}H_{21}SO_4^-.C_{10}H_{21}N(CH_3)_3^+$	25	2.9[b]	9.1	50	Corkill et al. (1963)
$C_{12}H_{25}SO_4^-.^+N(CH_3)_3C_{12}H_{25}$	25	2.74[b]	9.6	50.$_8$	Lange and Schwuger (1971)
$C_{12}H_{25}SO_3^-.^+HON(CH_3)_2C_{12}H_{25}$	25	2.14[b]	13.6	48.$_5$	Rosen et al. (1964)
Nonionics					
$C_8H_{17}CHOHCH_2OH$	25	5.1	9.6	48.$_6$	Kwan and Rosen (1980)
$C_8H_{17}CHOHCH_2CH_2OH$	25	5.3	8.9	48.$_4$	Kwan and Rosen (1980)
$C_{10}H_{21}CHOHCH_2OH$	25	6.3	6.5	49.$_3$[c]	Kwan and Rosen (1980)
$C_{10}H_{23}CHOHCH_2CH_2OH$	25	5.8	6.8	48.$_3$[c]	Kwan and Rosen (1980)
$C_{12}H_{25}CHOHCH_2CH_2OH$	25	5.1	7.7	45.$_5$	Kwan and Rosen (1980)
Decyl-β-D-glucoside (in 0.1 *M* NaCl, pH = 9)	25	4.18	11.1	44.$_2$	Li et al. (2001)
Decyl- β -D-maltoside (in 0.1 *M* NaCl, pH = 9)	25	3.37	6.5	35.$_7$	Li et al. (2001)
Dodecy1- β -D-maltoside (in 0.1 *M* NaCl, pH = 9)	25	3.67	7.1	37.$_3$	Li et al. (2001)
$C_6H_{13}(OC_2H_4)_6OH$	25	2.7	21.5	40	Mulley and Metcalf (1962); Elworthy and Florence (1964)
$C_8H_{17}OCH_2CH_2OH$	25	5.2	7.2	45.$_0$	Shinoda et al. (1959)
$C_8H_{17}(OC_2H_4)_5OH$ (in 0.1 *M* NaCl)	25	3.46	8.4	38.$_3$	Varadaraj et al. (1991)
$C_{10}H_{21}(OC_2H_4)_6OH$	25	3.0	17.0	42	Carless et al. (1964); Corkill et al. (1964)

(*Continued*)

TABLE 5.2 (*Continued*)

Surfactant	Temp. (°C)	$\Gamma_m \times 10^{10}$ mol/cm^2	CMC/C_{20}	Π_{CMC} (dyn/cm)	Reference
$C_{10}H_{21}(OC_2H_4)_6OH$ (in "hard river" water, I.S. = 6.6×10^{-3} *M*)	25	2.83	16.2	$39._4$	Rosen et al. (1996)
$C_{10}H_{21}(OC_2H_4)_8OH$	25	2.38	16.7	$36._4$	Meguro et al. (1981)
$C_{12}H_{25}(OC_2H_4)_3OH$	25	3.98	11.4	$44._1$	Rosen et al. (1996)
$C_{12}H_{25}$ $(OC_2H_4)_4OH$	25	3.63	13.7	$43._4$	Rosen et al. (1982a)
$C_{12}H_{25}(OC_2H_4)_4OH$ (H_2O-hexadecane)	25	3.16	16.8[d]	$52._1$	Rosen and Murphy (1991)
$C_{12}H_{25}(OC_2H_4)_5OH$	25	3.33	15.0	$41._5$	Rosen et al. (1982a)
$C_{12}H_{25}(OC_2H_4)_5OH$ (in 0.1 *M* NaCl)	25	3.31	18.5	$41._5$	Varadaraj et al. (1991)
$C_{12}H_{25}(OC_2H_4)_6OH$	25	3.7	9.6	41	Carless et al. (1964); Corkill et al. (1964)
$C_{12}H_{25}(OC_2H_4)_6OH$ (in "hard river" water, I.S. = 6.6×10^{-3} *M*)	25	3.19	12.8	$40._2$	Rosen et al. (1996)
$C_{12}H_{25}(OC_2H_4)_7OH$	25	2.90	14.9	$38._3$	Rosen et al. (1982a)
$C_{12}H_{25}(OC_2H_4)_8OH$	10	2.56	17.5	$37._4$	Rosen et al. (1982a)
$C_{12}H_{25}(OC_2H_4)_8OH$	25	2.52	17.3	$37._2$	Rosen et al. (1982a)
$C_{12}H_{25}(OC_2H_4)_8OH$	40	2.46	15.4	$37._3$	Rosen et al. (1982a)
$C_{12}H_{25}(OC_2H_4)_8OH$ (H_2O-hexadecane)	25	2.64	17.5[d]	$48._7$	Rosen and Murphy (1991)
$C_{12}H_{25}(OC_2H_4)_8OH$ (H_2O-heptane)	25	2.62	18.6[d]	$48._5$	Rosen and Murphy (1991)
$C_{12}H_{25}(OC_2H_4)_9OH$	23	2.3	17.0	36	Lange (1965)
$C_{12}H_{25}(OC_2H_4)_{12}OH$	23	1.9	11.8	32	Lange (1965)
6-branched $C_{13}H_{17}(OC_2H_4)_3OH$ (in 0.1 *M* NaCl)	25	2.87	35.7	$45._5$	Varadaraj et al. (1991)
$C_{13}H_{27}(OC_2H_4)_3OH$ (in 0.1 *M* NaCl)	25	3.89	8.8	$40._9$	Varadaraj et al. (1991)
$C_{13}H_{27}(OC_2H_4)_8OH$	25	2.78	11.3	$36._7$	Meguro et al. (1981)
$C_{14}H_{29}(OC_2H_4)_6OH$ (in "hard river" water, I.S. = 6.6×10^{-3} *M*)	25	3.34	10.5	$39._6$	Rosen et al. (1996)
$C_{14}H_{29}(OC_2H_4)_8OH$	25	3.43	8.4	$38._0$	Meguro et al. (1981)
$C_{14}H_{29}(OC_2H_4)_8OH$ (in "hard river" water, I.S. = 6.6×10^{-3} *M*)	25	2.67	13.8	$37._1$	Rosen et al. (1996)
$C_{15}H_{31}(OC_2H_4)_8OH$	25	3.59	7.1	$37._4$	Meguro et al. (1981)
$C_{16}H_{33}(OC_2H_4)_6OH$	25	4.4	6.3	40	Corkill et al. (1961); Elworthy and Florence (1964)
$C_{16}H_{23}(OC_2H_4)_6OH$ (in "hard river" water, I.S. = 6.6×10^{-3} M)	25	3.23	12.7	$40._1$	Rosen et al. (1996)
$C_{16}H_{33}(OC_2H_4)_7OH$	25	3.8	8.3	39	Elworthy and MacFarlane (1962)

$C_{16}H_{33}(OC_2H_4)_9OH$	25	3.1	7.8	36	Elworthy and MacFarlane (1962)
$C_{16}H_{33}(OC_2H_4)_{12}OH$	25	2.3	8.5	33	Elworthy and MacFarlane (1962)
$C_{16}H_{33}(OC_2H_4)_{15}OH$	25	2.1	8.9	32	Elworthy and MacFarlane (1962)
$C_{16}H_{33}\ (OC_2H_4)_{21}OH$	25	1.4	8.0	27	Elworthy and MacFarlane (1962)
p-t-$C_8H_{17}C_6H_4(OC_2H_4)_7OH$	25	2.9	22.9	42	Crook et al. (1963, 1964)
p-t-$C_8H_{17}C_6H_4(OC_2H_4)_8OH$	25	2.6	21.4	40	Crook et al. (1963, 1964)
p-t-$C_8H_{17}C_6H_4(OC_2H_4)_9OH$	25	2.5	18.6	$38._5$	Crook et al. (1963, 1964)
p-t-$C_8H_{17}C_6H_4(OC_2H_4)_{10}OH$	25	2.2	17.4	37	Crook et al. (1963, 1964)
$C_9H_{19}C_6H_4(OC_2H_4)_{10}OH$[e]	25	2.95	13.5	41	Schick et al. (1962)
$C_9H_{19}C_6H_4(OC_2H_4)_{15}OH$[e]	25	2.4	12.9	$35._5$	Schick et al. (1962)
$C_9H_{19}C_6H_4(OC_2H_4)_{30}OH$[e]	25	1.9	12.3	31	Schick et al. (1962)
$C_{11}H_{23}CON(CH_2CH_2OH)_2$	25	3.75	6.3	$37._1$	Rosen et al. (1964)
$C_{10}H_{21}CON(CH_3)CH_2(CHOH)_4CH_2OH$ (in 0.1 *M* NaCl)	25	3.80	10.5	$41._4$	Zhu et al. (1999)
$C_{11}H_{23}CONH(C_2H_4O)_4H$	23	3.4	–	$41._3$	Kjellin et al. (2002)
$C_{11}H_{23}CON(CH_3)CH_2CHOHCH_2OH$ (in 0.1 *M* NaCl)	25	4.34	10.9	$46._2$	Zhu et al. (1999)
$C_{11}H_{23}CON(CH_3)CH_2(CHOH)_3CH_2OH$ (in 0.1 *M* NaCl)	25	4.29	9.8	$44._7$	Zhu et al. (1999)
$C_{11}H_{23}CON(CH_3)CH_2(CHOH)_4CH_2OH$ (in 0.1 *M* NaCl)	25	4.10	8.7	$42._3$	Zhu et al. (1999)
$C_{12}H_{25}CON(CH_3)CH_2(CHOH)_4CH_2OH$ (in 0.1 *M* NaCl)	25	4.60	7.8	$43._9$	Zhu et al. (1999)
$C_{13}H_{27}CON(CH_3)CH_2(CHOH)_4CH_2OH$ (in 0.1 *M* NaCl)	25	4.68	4.0	$36._0$	Zhu et al. (1999)
$C_{10}H_{21}N(CH_3)CO(CHOH)_4CH_2OH$	20	3.96	5.2	$36._1$	Burczyk et al. (2001)
$C_{12}H_{25}N(CH_3)CO(CHOH)_4CH_2OH$	20	3.99	8.8	$37._6$	Burczyk et al. (2001)
$C_{14}H_{29}N(CH_3)CO(CHOH)_4CH_2OH$	20	3.97	8.5	$37._8$	Burczyk et al. (2001)
$C_{16}H_{33}N(CH_3)CO(CHOH)_4CH_2OH$	20	3.65	10.1	$38._5$	Burczyk et al. (2001)
$C_{18}H_{37}N(CH_3)CO(CHOH)_4CH_2OH$	20	3.97	8.1	$39._7$	Burczyk et al. (2001)
$C_6F_{13}C_2H_4SC_2H_4(OC_2H_4)_2OH$	25	4.74	—	54	Matos et al. (1989)
$C_6F_{13}C_2H_4SC_2H_4(OC_2H_4)_3OH$	25	4.46	—	$53._4$	Matos et al. (1989)
$C_6F_{13}C_2H_4SC_2H_4(OC_2H_4)_5OH$	25	3.56	—	54	Matos et al. (1989)
$C_6F_{13}C_2H_4SC_2H_4(OC_2H_4)_7OH$	25	3.19	—	51	Matos et al. (1989)

(*Continued*)

TABLE 5.2 (*Continued*)

Surfactant	Temp. (°C)	$\Gamma_m \times 10^{10}$ mol/cm^2	CMC/C_{20}	Π_{CMC} (dyn/cm)	Reference
$(CH_3)_3SiO[Si(CH_3)_2O]_3Si(CH_3)_2CH_2(C_2H_4O)_{8.2}CH_3$	25	3.4	37	50	Kanner et al. (1967)
$(CH_3)_3SiO[Si(CH_3)_2O]_3Si(CH_3)_2CH_2(C_2H_4O)_{12.8}CH_3$	25	4.2	19.5	51	Kanner et al. (1967)
$(CH_3)_3SiO[Si(CH_3)_2O]_3Si(CH_3)_2CH_2(C_2H_4O)_{17.3}CH_3$	25	4.2	17.4	$50._5$	Kanner et al. (1967)
$(CH_3)_3SiO[Si(CH_3)_2O]_9Si(CH_3)_2CH_2(C_2H_4O)_{17.3}CH_3$	25	3.6	11.8	42	Kanner et al. (1967)
Zwitterionics					
$C_{10}H_{21}N^+(CH_3)_2COO^-$	23	4.15	7.0	$39._7$	Beckett and Woodward (1963)
$C_{12}H_{25}N^+(CH_3)_2CH_2COO^-$	23	3.57	6.5	$36._5$	Beckett and Woodward (1963)
$C_{14}H_{29}N^+(CH_3)_2CH_2COO^-$	23	3.53	7.5	$37._5$	Beckett and Woodward (1963)
$C_{16}H_{33}N^+(CH_3)_2CH_2COO^-$	23	4.13	6.9	$39._7$	Beckett and Woodward (1963)
$C_{10}H_{21}CH(Pyr^+)COO^-$	25	3.59	3.90	$32._1$	Zhao and Rosen (1984)
$C_{12}H_{33}CH(Pyr^+)COO^-$	25	3.57	5.66	$35._0$	Zhao and Rosen (1984)
$C_{14}H_{29}CH(Pyr^+)COO^-$	40	3.40	6.16	$36._0$	Zhao and Rosen (1984)
$C_{10}H_{21}N^+(CH_2C_6H_5)(CH_3)CH_2COO^-$	25	2.91	12.0	$38._0$	Dahanayake and Rosen (1984)
$C_{12}H_{25}N^+(CH_2C_6H_5)(CH_3)CH_2COO^-$	25	2.86	14.4	$39._0$	Dahanayake and Rosen (1984)
$C_{12}H_{25}N^+(CH_2C_6H_5)(CH_3)CH_2COO^-$ (in 0.1 *M* NaCl, pH 5.7)	25	3.1	15.1	$39._9$	Rosen and Sulthana (2001)
$C_{12}H_{25}N^+(CH_2C_6H_5)(CH_3)CH_2COO^-$ (H_2O-heptane)	25	2.81	—	$48._4$	Murphy and Rosen (1988)
$C_{12}H_{25}N^+(CH_2C_6H_5)(CH_3)CH_2COO^-$ (H_2O-hexadecane)	25	2.90	—	$48._6$	Murphy and Rosen (1988)
$C_{12}H_{25}N^+(CH_2C_6H_5)(CH_3)CH_2COO^-$ (H_2O-toluene)	25	2.22	—	$35._8$	Murphy and Rosen (1988)
$C_{10}H_{24}N^+(CH_2C_6H_5)(CH_3)CH_2CH_2SO_3^-$	40	2.59	11.0	$33._8$	Dahanayake and Rosen (1984)

[a] From dodecyl alcohol with 4,4 methyl branches in the molecule.
[b] Since there are two chains in each molecule, the number of *hydrophobic chains* per centimeter is twice the value of Γ_m.
[c] Below the Krafft point; supersaturated solution.
[d] CMC/C_{30} value.
[e] Hydrophilic head is not homogeneous, but distribution of POE chain lengths is reduced by molecular distillation.
I.S., ionic strength of the solution; Pyr^+, pyridinium.

CMC/C_{20} values and Equation 5.3. For surfactants with hydrocarbon-chain hydrophobic groups, the most effective surface tension reducers (largest Π_{CMC} values) are (1) nonionic compounds having small hydrophilic head groups and (2) anionic–cationic salts where both hydrophobic chains contain six carbon atoms or more, especially when both chains are approximately of the same length. Because of the small hydrophilic groups and the absence of ionic repulsive forces at the aqueous solution–air interface, both of these types of surfactants have their hydrophobic groups closely packed at the interface (large Γ_m values) and relatively large CMC/C_{20} ratios, producing large Π_{CMC} values. Examples of the first type are the 1,2- and 1,3-alkanediols; of the latter, the salts $C_{12}H_{25}SO_4^-.^+N(CH_3)_3C_{12}H_{25}$ and $C_{12}H_{25}SO_3^-.^+HON(CHS_3)_2\text{-}C_{12}H_{25}$.

The replacement of the usual small, inorganic counterion in ionic surfactants by an organic straight-chain one that is itself surface-active (e.g., in $C_{12}H_{25}SO_4^-.C_{12}H_{25}N(CH_3)_3^+$ produces an ion pair that is strongly adsorbed at the aqueous solution–air interface. The mutual neutralization of charge in the ion pair results in (1) close packing at the interface (30.3×10^{-2} nm^2 per hydrophobic chain in this particular case) and (2) a high value for the CMC/C_{20} ratio, similar to that found in nonionic surfactants. Similar types of compounds, such as $C_{12}H_{25}N(CH_3)_2OH^+.C_{12}H_{25}SO_3^-$, are formed when long-chain amine oxides are added to anionic detergent compositions and are the basis for the foam-stabilization properties of amine oxides in these compositions.

The replacement of the usual hydrocarbon-chain hydrophobic group by a silicone- or fluorocarbon-chain hydrophobic group produces a large increase in the CMC/C_{20} ratio even, in the case of perfluoro compounds, when the surfactant is ionic. The large values of the CMC/C_{20} ratio in these cases may be due to steric barriers associated with the packing of these bulky chains into the micelle. The Γ_m values for these compounds (Table 5.2) are also high, even when ionic. The combination of large CMC/C_{20} ratios and large Γ_m values puts these compounds among the best surface tension reducers in aqueous media.

The addition of neutral electrolyte to an aqueous solution of an ionic surfactant, as mentioned above (Chapter 3, Section V), produces a much greater increase in adsorption at the aqueous solution–air interface than in micellization. Thus, the value of C_{20} is reduced more than the CMC, with the result that the CMC/C_{20} value is increased. In addition, Γ_m is increased by the increase in electrolyte. The increased values for Γ_m and CMC/C_{20} cause an increase in effectiveness.

The effect of an increase in the size of the hydrophilic head, without significant change in its nature, can be observed by comparing $C_{14}H_{29}N^+(CH_3)_3Br^-$ (or tetradecylpyridinium bromide) with $C_{14}H_{29}N^+(C_3H_7)_3Br^-$ (Table 5.2). The increase in the size of the three short alkyl groups surrounding the nitrogen results in a larger cross-sectional area of the molecule at the interface and therefore a smaller value of Γ_m. The log CMC/C_{20} ratio, however, shows only a small change in all three compounds, with the result that Π_{CMC} is reduced with increase in the size of the hydrophilic head.

For POE nonionics with the same (C_{12}) hydrophobic group, increase in the length of the POE chain from 1 to about 8 units causes a decrease in Γ_m but an increase in the CMC/C_{20} ratio. The change in Γ_m is greater than the change in log CMC/C_{20} and, as a result, the effectiveness of surface tension reduction decreases with increase in the length of the POE chain over this range. Above 8 OE units, there is little change in log CMC/C_{20} and a small decrease in Γ_m with increase in the OE content of the molecule, with the result that there is a continued small decrease in surface tension reduction effectiveness as the OE content is increased.

On the other hand, at constant POE content, an increase in the length of the hydrophobic group causes an increase in the value of Γ_m but an almost equal decrease in log CMC/C_{20}. As a result, as in the case of ionic surfactants, there is very little change in the surface tension reduction effectiveness of POE nonionic with increase in the length of the hydrophobic group.

For both ionic and POE nonionics, as the temperature is increased, there is a decrease in both Γ_m and the CMC/C_{20} ratios. As a result, although the surface *tension* of the solution may be reduced to a lower value by increase in the temperature, the surface tension reduction *effectiveness*, Π_{CMC} ($= \gamma_0 - \gamma_{CMC}$, where γ_0 is the surface tension of the pure solvent at that temperature), is always reduced by increase in temperature.

In contrast to their marked effect on the efficiency with which a POE nonionic reduces the surface tension of water, water structure promoters and breakers seem to show almost no effect on the effectiveness with which it reduces the tension (Schwuger, 1971).

Surfactants with hydrocarbon-chain hydrophobic groups generally do not lower the surface tension of alkanes, since any orientation of adsorbed surfactant of this type at the air–alkane surface would not reduce the surface free energy. However, fluorinated surfactants can adsorb and orient at the hydrocarbon–air surface to reduce the free energy there. Fluorinated surfactants of the type $C_6H_5CF(CF_3)O[CF_2CF(CF_3)O]_mC_3F_7$ have been observed to reduce the surface tension of *m*-xylene (28 mN/m) to 10 mN/m (Abe et al., 1992).

III. LIQUID–LIQUID INTERFACIAL TENSION REDUCTION

The reduction of the tension at an interface by a surfactant in aqueous solution when a second liquid phase is present may be considerably more complex than when that second phase is absent, that is, when the interface is a surface. If the second liquid phase is a nonpolar one in which the surfactant has almost no solubility, then adsorption of the surfactant at the aqueous solution–nonpolar liquid interface closely resembles that at the aqueous solution–air interface, and those factors that determine the efficiency and effectiveness of surface tension reduction affect interfacial tension reduction in a similar manner (Chapter 2, Section IIIC, E). When the nonpolar liquid phase is a saturated hydrocarbon, both the efficiency and effectiveness of interfacial tension reduction by the surfactant at the aqueous solution–hydrocarbon interface are greater than at the aqueous solution–air interface, as measured by pC_{20} and Π_{CMC}, respectively. The replacement of air as the second phase by a saturated hydrocarbon increases the tendency of the surfactant to adsorb at the interface, while the tendency to form micelles is not affected significantly. This results in an increase in the CMC/C_{20} ratio. Since the value of Γ_m, the effectiveness of adsorption (Chapter 2, Section IIIC), is not affected significantly by the presence of the saturated hydrocarbon, the increase in the value of Π_{CMC} is due mainly to this increase in the CMC/C_{20} ratio. When the hydrocarbon is a short-chain unsaturated or aromatic hydrocarbon, however, the Π_{CMC} value is smaller than when the second phase is air. Here, the effect is due mainly to the decrease in the value of Γ_m in the presence of these types of hydrocarbons. Both the tendency to adsorb and the tendency to form micelles are increased slightly in the presence of these types of hydrocarbons, but in almost equal amounts, resulting in little change in the CMC/C_{20} ratio.*

On the other hand, if the surfactant has appreciable solubility in both liquids, then very different factors may determine the value of the interfacial tension. Although low liquid–liquid interfacial tension is important in promoting emulsification (Chapter 8) and in the removal of oily soil by detergents (Chapter 10), advances in our knowledge of the factors governing the reduction at that interface stem from the intense interest in enhanced oil recovery by use of surfactant solutions.

A. Ultralow Interfacial Tension

For displacement of the oil in the pores and capillaries of petroleum reservoir rock, an aqueous solution–oil interfacial tension of $\approx 10^{-3}$ mN/m (dyn/cm) is generally required. To reach so low a value, the value of γ_{ab} (Equation 5.1), the interaction energy across the interface (Figure 5.1) must be large. This means that the nature of the material on both sides of the interface must be very similar. Since oil and water have very different natures, a situation where both sides of an oil–water interface (*O/W*) can have similar natures can occur only when both sides of the interface have similar concentrations of surfactant, oil, and water. There are a number of ways in which such a situation can be created.

In our discussion of the effect of temperature on solubilization capacity (Chapter 4, Section IB7), we mentioned that when the temperature of an aqueous micellar solution W_D of a POE nonionic surfactant is increased, its

* At the hydrocarbon-aqueous solution interface, the CMC/C_{30} ratio, where C_{30} is the molar surfactant concentration in the aqueous phase needed to produce a 30 mN/m (dyn/cm) reduction in the interfacial tension, is a better determinant of Π_{CMC} than CMC/C_{20}, since the interfacial tension-log C curve is often not linear from the CMC to surface pressures as low as 20 mN/m (dyn/cm).

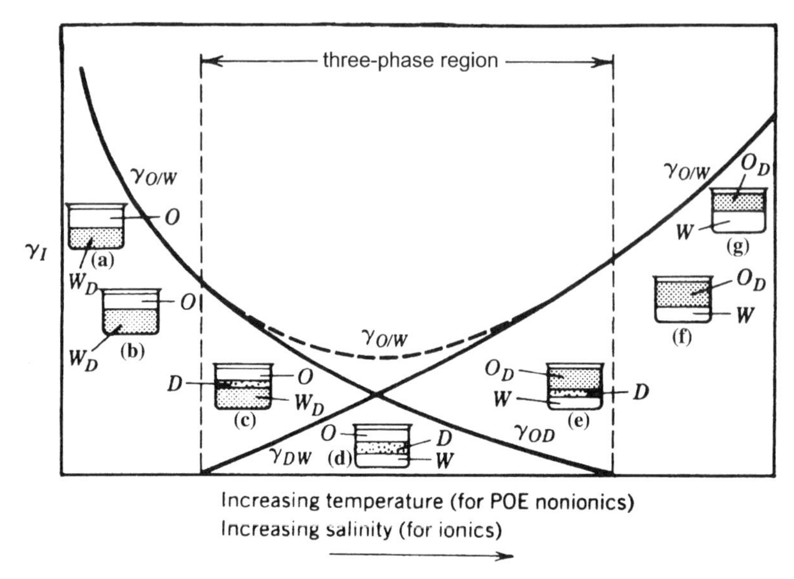

FIGURE 5.5 Effect of molecular environmental conditions on interfacial tension and phase volumes. Shaded phases indicate locations of the surfactant.

solubilization of nonpolar material O increases due to the increased dehydration of the POE chains, which increases the lipophilic character of the surfactant. If this occurs for POE nonionics of the proper structure in the presence of excess nonpolar material, the volume of the aqueous phase W_D increases and that of the nonpolar phase oil decreases as the temperature increases (Figure 5.5a,b). This is accompanied by a decrease in the tension γ_{OW} at the O/W interface. With further increase in temperature, the POE chains become more and more dehydrated, the surfactant becomes more lipophilic, and more and more nonpolar material oil is solubilized into the increasingly asymmetric micelles. When the vicinity of the cloud point of the nonionic is reached, the surfactant micelles, together with solubilized material, will start to separate from W_D as a separate phase D. If excess oil is still present, the system now contains three phases (Shinoda and Saito, 1968): excess oil; a phase D, the so-called middle phase, containing surfactant together with solubilized water and oil; and an aqueous phase W_D (Figure 5.5c).

The O/W_D interface is now replaced by a D/W_D interface, whose interfacial tension γ_{DW} is close to zero. At this point, there is also an O/D interface, whose interfacial tension γ_{OD} is low. As the temperature continues to increase, more and more surfactant micelles separate out of W_D, carrying with them solubilized O and W. The volume of W_D decreases and that of the middle phase D increases; γ_{OD} continues to decrease and γ_{DW} increases. Separation of the surfactant micelles from it has converted the aqueous phase into one (W) that contains only a small amount of unmicellized surfactant (Figure 5.5d). When

D is very small in the three-phase region, γ_{OW} is approximately the sum of γ_{DW} and γ_{OD}.

As the surfactant continues to become more and more lipophilic with increase in temperature, a point is reached at which the micelles start to invert (Figure 4.4) and these dissolve in the excess O, carrying with them solubilized water and forming a reversed micellar solution O_D. This is accompanied by an increase in O_D and a decrease in D to a very small volume (Figure 5.5e). γ_{OD} is close to zero; γ_{DW} continues to increase. Eventually, all of D dissolves in O_D, leaving only W (Figure 5.5f). At this point, the DW interface disappears and Γ_{DW}, still low, is replaced by γ_{OW}. With further increase in temperature, the surfactant becomes even more lipophilic, the solubilization capacity of the inverted micelles decreases, more W separates out, and γ_{OW} increases (Figure 5.5g).

The maximum volume of the surfactant (middle) phase D at the temperature where all three phases exist is dependent upon the percentage of surfactant in the system. If the percentage is very small, the surfactant phase may not be visible to the naked eye, and the system may appear to contain only two phases; if the percentage of surfactant is large, the aqueous and nonpolar phases may be completely solubilized in the surfactant phase, and the system may contain only one phase. In the latter case, the system is called a *microemulsion* (Chapter 8, Section II).

The structure of the surfactant phase D in equilibrium with both excess W and O phases has been a subject of considerable interest and speculation (Shinoda and Friberg, 1975; Huh, 1979; Shinoda, 1983). Data indicate that, at the point of minimum interfacial tension against both W and O phases, this phase may not be homogeneous, but may consist of a mixture of normal and reverse micelles, since gradients of concentration, density, and other properties appear upon standing for some time under normal gravity, or upon centrifugation (Hwan et al., 1979; Rosen and Li, 1984; Zhao et al., 1984; Good et al., 1986). It has also been suggested that the structure is bicontinuous (Scriven, 1977). If the surface-active material(s) present have long, straight hydrophobic groups, rodlike cylindrical or lamellar micelles may be present (Fowkes et al., 1985).

Temperature change in POE nonionic systems is not the only method of producing these phase changes and ultralow interfacial tension. For ionic surfactants of the proper structure, the addition of electrolyte, such as NaCl, with its consequent reduction of the electrical interactions of the ionic head groups, can cause the surfactant to change from hydrophilic to lyophilic. With increasing "salinity," such systems may show changes in phases, solubilization, and interfacial tension similar to those shown by POE nonionics with temperature change. The addition of hydrophilic or lipophilic polar compounds (*cosurfactants*) can also change the hydrophilic or lipophilic character of the system, its solubilization of water or oil, and the interfacial tension.

From the above discussion, it should be apparent that for POE nonionics, there is a particular temperature where the hydrophilic and lipophilic characters of the surfactant "balance" each other and γ_{OW} is at, or close to, its minimum value. It is usually defined operationally, for example, as the temperature where the surfactant phase solubilizes equal volumes of water and nonpolar material or the temperature at which an emulsion (Chapter 8) of the surfactant, water, and nonpolar material inverts. In the latter case, it is known as the *phase-inversion temperature* (PIT) (Chapter 8, Section IVB). Similarly, there is an electrolyte content at which the hydrophilic and lipophilic characters of ionic surfactants balance. The point at which equal volumes of water and nonpolar material are solubilized into the surfactant is known as the *optimal salinity* (Healy and Reed, 1974) and has been extensively investigated for enhanced oil recovery (Healy and Reed, 1977; Hedges and Glinsmann, 1979; Nelson, 1980). The optimal salinity or PIT is at or close to the point where the parameter $V_H/l_c a_0$ (Chapter 3, Section II) equals 1 and lamellar normal and reverse micelles are readily interconvertable.

The larger the volume of water (V_W) (or nonpolar material V_0) solubilized into the surfactant phase relative to its volume V_S, the lower the interfacial tensions γ_{DW}, γ_{OD}, and γ_{OW} (Robbins, 1974; Healy et al., 1976). This is understandable, since for both normal and reverse micelles, the interfacial tension against the second liquid phase decreases as the amount of second phase solubilized increases. The greater the amount solubilized in the presence of excess solubilizate, the more closely the natures of the two phases approach each other.

The Winsor ratio R (Winsor, 1948, 1968) is convenient for relating changes in the hydrophilic solvent W, the lipophilic solvent oil, and the surfactant C to interfacial tensions and phase volumes, and for explaining them in terms of the molecular interactions involved (Bourrel and Chambu, 1983; Bourrel et al., 1984). It is based upon the relative tendencies of the system to solubilize water and oil. The ratio,

$$R = \frac{A_{CO} - A_{OO} - A_{ll}}{A_{CW} - A_{WW} - A_{hh}} \tag{5.4}$$

measures the solubilization capacity of the surfactant micelles for W relative to that for O. A_{CO} and A_{CW} are the interaction strengths per unit area of interface of C with oil and water, respectively, promoting solubilization of the *other* liquid phase; A_{OO} and A_{WW} are the respective self-interaction strengths of the solvent molecules in oil and water, respectively, opposing solubilization into them; A_{ll} and A_{hh} are the strengths of the self-interactions between the lipophilic and hydrophilic portions, respectively, of the surfactant molecules, also opposing solubilization. When $R << 1$, the micelles solubilize oil much more readily than W, and a Type I system forms (Figure 5.5a,b); when $R >> 1$, they solubilize water much more readily than oil, and a Type II forms (Figure 5.5f,g). When $R \cong 1$, Type III or IV systems form, depending upon the magnitude of the numerator (or denominator). Type III is a three-phase system (Figure 5.5d); Type IV is a one-phase microemulsion (Chapter 8, Section II). When $R \cong 1$, the larger the value of the numerator (or denominator) of the expression for R, the greater the solubilization capacity for water (or oil), and consequently, the greater the tendency to form a Type IV system. R is therefore a semiquantitative method of measuring the balance between the hydrophilic and lipophilic characters of the surfactant *in the particular system in which it finds itself.**

The Winsor R parameter and the Mitchell–Ninham $V_H/l_C a_O$ parameter are related to each other in that both specify that when the value of the parameter exceeds 1, normal micelles in aqueous media will be converted into reverse micelles in the presence of excess nonpolar solvent. The former concept bases this on molecular interactions, the latter on molecular geometry.

The lowest interfacial tension values are produced when $R \cong 1$ and the value of the numerator (or denominator) in the expression for R is greatest. This produces the largest V_W/V_S, and V_H/V_S ratios. To reduce γ_{OW}, then, R should be made to approach 1, in the case where $R < 1$[†] by increasing the value of the numerator in the case where $R > 1$** increasing the denominator, rather than decreasing the numerator.

The value of the numerator in Equation 5.4 can be increased by increasing the value of A_{CO}, the interaction of the surfactant with oil, and/or by decreasing the values of A_{OO} and A_{ll}, the self-interaction of the lipophilic solvent molecules and of the lipophilic portions of the surfactant, respectively. A_{CO} can be increased by increasing the length of the lipophilic group of the surfactant, although this simultaneously increases, to a smaller extent, the value of A_{ll}. A_{CO} can also be increased by adding a moderately lipophilic nonionic *cosurfactant* (e.g., an alcohol, amide, or amine of intermediate chain length) or a *lipophilic linker* (Chapter 8, Section II) to an ionic surfactant or by any additive that increases the packing of the surfactant at the interface (since A_{CO} is an interaction strength per unit area of interface). When oil is an alkane, A_{CO} can be decreased by a decrease in the alkane chain length.

The value of the denominator can be increased by increasing the value of A_{CW} by the interaction of the surfactant with W, and/or by decreasing the values of A_{WW} and A_{hh}, the self-interaction of the hydrophilic solvent molecules and the hydrophilic portions of the surfactant, respectively. A_{CW} can be increased in the case of POE nonionics by increase in the length of the POE chain, although this simultaneously increases the value of A_{hh} slightly. It can also be increased by addition of a hydrophilic linker. All of these changes have been shown to decrease the value of A_{OW} when they bring the value of R closer to 1 (Healy et al., 1976; Salter, 1977; Bourrel et al., 1980; Shinoda and Shibata, 1986; Verzaro et al., 1984; Valint et al., 1987). In addition, from Equation 5.4, surfactants with large hydrophobic groups (large A_{CO} values) and

* When $R = 1$ and the A_{ll} and A_{hh} interactions are large, liquid crystals or gels may form (Bourrel et al., 1984).

[†] Determined by observing the type of system produced. When $R < 1$, a Type I system is formed; when $R > 1$, a Type II system is formed.

large hydrophilic groups (large A_{CW} values) should show lower interfacial tension values than similar-type surfactants of lower molecular weight with the same balance of hydrophilic and lipophilic groups. This has been confirmed experimentally (Kunieda and Shinoda, 1982; Barakat et al., 1983).

From the above, a surfactant capable of being both an efficient and an effective γ_{OW} reducer should have a balanced structure (R ≅ 1) in the system and under the conditions of use, with a considerable amount of both hydrophilic and lipophilic character (large values of A_{CW} and A_{CO}). The value of $R \cong 1$ and large values of A_{CW} and A_{CO} will cause reduction of γ_{OW} to a very low value, that is, make it an effective reducer of γ_{OW}. A surfactant of this type will also have limited solubility in a hydrophilic solvent because of the large lipophilic (hydrophobic) portion of the molecule and limited solubility in a lipophilic solvent because of the large hydrophilic (lipophobic) portion of the molecule. This limited solubility in both liquids will cause the surfactant to adsorb strongly at the interface and consequently to be a very efficient reducer of γ_{OW}.

The addition of an alcohol that adsorbs at the interface, such as *n*-pentanol, decreases A_{CW} by increasing the interfacial area per surfactant molecule. The addition of electrolyte, in the case of an ionic surfactant, decreases A_{CW} and increases A_{hh}. All these changes result in an increase in the value of R.

Another approach to obtaining ultralow interfacial tension is via the microemulsion *solubility parameter at optimum formulation* (Chapter 8, Section II).

IV. DYNAMIC SURFACE TENSION REDUCTION

A. Dynamic Regions

In many interfacial processes, such as in high-speed wetting of textile, paper, and other substrates (Chapter 6, Section IIC), or in foaming (Chapter 7), equilibrium conditions are not attained. In such cases, the dynamic surface tension (surface tension as a function of time) of the surfactant is a more important factor in determining the performance of the surfactant in the process than its equilibrium surface tension. With the introduction of simple instruments (such as the maximum bubble pressure apparatus, which measures the pressure and bubble rate of gas fed through a capillary), a considerable body of research data has accumulated on the dynamic surface tension of surfactant solution during the past decade or so.

The typical plot of the change in surface tension with time (Figure 5.6) contains four regions: an induction region (I), a rapid fall region (II), a meso-equilibrium region (III), and equilibrium (IV). Equation 5.5 (Hua and Rosen, 1988) fits the three dynamic regions (I–III) of this plot:

$$\gamma_t = \gamma_m + (\gamma_0 - \gamma_m)\Big/\left[1 + (t/t^*)^n\right], \tag{5.5}$$

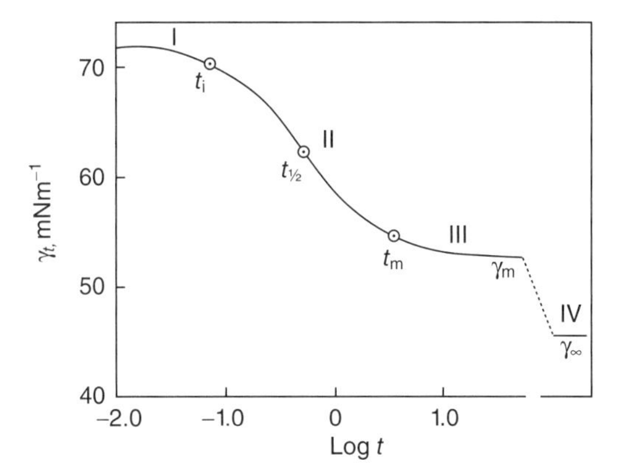

FIGURE 5.6 Generalized dynamic surface tension, γ_t versus log time, t, curve: region I, induction; region II, rapid fall; region III, meso-equilibrium; region IV, equilibrium. Reprinted with permission from Hua and Rosen (1988).

where γ_t is the surface tension of the surfactant solution at time t, γ_m is the meso-equilibrium surface tension (where γ_t shows only a small change with time), and γ_0 is the surface tension of the pure solvent. Equation 5.5 can be converted to its logarithmic form

$$\log(\gamma_0 - \gamma_t) - \log(\gamma_t - \gamma_m) = n \log t - n \log t^* \tag{5.6}$$

to calculate the values of n and t^*. The value of t^* is the time required for γ_t to reach half of the value between γ_0 and γ_m, and is related to the surfactant concentration. As the surfactant concentration increases, t^* decreases. From Equation 5.5, by differentiation, t^* is also the time at which, at constant surfactant concentration, the rate of surface tension change with $\log t$ (Equation 5.7) reaches its maximum value (Hua and Rosen, 1991):

$$(\delta\gamma_\tau / \delta \log t)_{\mathrm{max.c}} = 0.576n(\gamma_0 - \gamma_m). \tag{5.7}$$

Here, n is a constant related to the molecular structure of the surfactant. It has been suggested (Gao and Rosen, 1995) that n is related to the difference between the energies of adsorption and desorption of the surfactant. Some values of n are listed in Table 5.3. From the data, it is apparent that the value of n increases with increase in the hydrophobicity of the surfactant, thus increasing with (1) increase in the NaCl concentration of the solution for anionic surfactants (due to compression of the electrical double layer [Chapter 2, Section I]); (2) increase in the length of the hydrophobic group; (3) increase in the pH of the solution for the amine oxide, $C_{14}H_{29}N(CH_3)_2O$, which decreases its tendency to pick up a proton and become cationic; and (4) decrease in the number of OE units in POE nonionics. It decreases with branching of the

TABLE 5.3 Values of *n* (Equations 5.5 and 5.6) at 25°C

Compound	Medium	n	Reference
$C_{12}H_{25}SO_3^-Na^+$	0.1 M NaCl	0.9_0	Hua and Rosen (1991)
$C_{12}H_{25}OCH_2CH_2SO_4^-Na^+$	0.1 M NaCl	0.9_3	Gao and Rosen (1995)
$C_{12}H_{25}OCH_2CH_2SO_4^-Na^+$	0.5 M NaCl	1.0_5	Gao and Rosen (1995)
$C_{12}H_{25}(OCH_2CH_2)_2SO_4^-Na^+$	0.1 M NaCl	0.8_7	Gao and Rosen (1995)
$C_{12}H_{25}(OCH_2CH_2)_2SO_4^-Na^+$	0.5 M NaCl	0.9_8	Gao and Rosen (1995)
Branched $C_{16}H_{33}(OC_2H_4)_5SO_4^-Na^+$	0.1 M NaCl	0.9_9	Hua and Rosen (1991)
Linear $C_{16}H_{33}(OC_2H_4)_5SO_4^-Na^+$	0.1 M NaCl	1.4_5	Hua and Rosen (1991)
$C_4H_9CH(C_2H_5)CH_2COOCH(SO_3^-Na^+)$ -$CH_2COOCH_2CH(C_2H_5)C_4H_9$	0.1 M NaCl	1.6_6	Hua and Rosen (1991)
$[C_8H_{17}N^+(CH_3)_2CH_2]_2C_6H_4 \cdot 2Br^-$	0.1 M NaBr	1.1_5	Rosen and Song (1996)
$[C_{10}H_{21}N^+(CH_3)_2CH_2]_2C_6H_4 \cdot 2Br^-$	0.1 M NaBr	1.1	Rosen and Song (1996)
$[C_{12}H_{25}N^+(CH_3)_2CH_2]_2C_6H_4 \cdot 2Cl^-$	0.1 M NaCl	1.5	Rosen and Song (1996)
$[C_{12}H_{25}N^+(CH_3)_2CH_2]_2CHOH \cdot 2Cl^-$	0.1 M NaCl	1.8	Rosen and Song (1996)
N-octyl-2-pyrrolidinone	H_2O	0.7_3	Hua and Rosen (1991)
N-decyl-2-pyrrolidinone	H_2O	0.9_8	Hua and Rosen (1991)
N-dodecyl-2-pyrrolidinone	H_2O	1.5_4	Hua and Rosen (1991)
$C_{12}H_{25}(OC_2H_4)_4OH$	H_2O	1.0_6	Gao and Rosen (1995)
$C_{12}H_{25}(OC_2H_4)_7OH$	H_2O	0.9_6	Hua and Rosen (1991)
$C_{12}H_{25}(OC_2H_4)_7OH$	4 M urea	0.7_8	Hua and Rosen (1991)
$C_{12}H_{25}(OC_2H_4)_8OH$	H_2O	0.8_6	Gao and Rosen (1995)
$C_{12}H_{25}(OC_2H_4)_{10}OH$	H_2O	0.7_1	Gao and Rosen (1995)
$C_{12}H_{25}(OC_2H_4)_{11}OH$	H_2O	0.6_1	Tamura et al. (1995)
$C_{10}H_{21}N^+(CH_3)(CH_2C_6H_5)CH_2COO^-$	H_2O, pH 9	1.1_5	Hua and Rosen (1991)
$C_{12}H_{25}N^+(CH_3)(CH_2C_6H_5)CH_2COO^-$	H_2O, pH 9.0	1.4_0	Gao and Rosen (1995)
$C_{14}H_{29}N^+(CH_3)(CH_2C_6H_5)CH_2COO^-$	H_2O, pH 9.0	1.5_0	Gao and Rosen (1995)
$[C_{14}H_{29}N^+(CH_3)_2CH_2]_2CHOH \cdot 2Cl^-$	0.1 M NaCl	3.1	Rosen and Song (1996)
$C_{14}H_{29}N(CH_3)_2O$	H_2O, pH 3.0	0.9_5	Gao and Rosen (1995)
$C_{14}H_{29}N(CH_3)_2O$	H_2O, pH 9.5	1.1_6	Gao and Rosen (1995)

hydrophobic group in isomeric surfactants and with the addition of a structure breaker (urea) to the water. From Equation 5.6, the maximum rate of surface tension change with $\log t$, at constant surfactant concentration, increases with n, the hydrophobicity of the surfactant molecule.

The time, t_i, for the induction period (region I) to end is an important factor in determining the surface tension as a function of time, since only when that period ends does the surface tension start to fall rapidly. The value of t_i has been shown (Gao and Rosen, 1995; Rosen and Song, 1996) to be related to the surface coverage of the air–aqueous solution interface and to the apparent diffusion coefficient, D_{ap}, of the surfactant, calculated by use of the short-time approximation of the Ward–Tordai equation (Ward and Tordai, 1946) for diffusion-controlled adsorption (Equation 5.8):

$$\Gamma_t = 2(D_{\mathrm{ap}}/\pi)^{1/2} Ct^{1/2}. \tag{5.8}$$

From this,

$$\ln t_i = 2\ln(\Gamma_i / C) + \ln(\pi / 4D_{\mathrm{ap}}). \tag{5.9}$$

From Equation 5.9, the greater the amount of surfactant, Γ_i, at the surface at the end of the induction period, and the smaller the apparent diffusion coefficient of the surfactant, the longer the time, t_i, needed for the surface tension to start decreasing rapidly. Calculation of surface coverages at time t_i (Gao and Rosen, 1995; Rosen and Song, 1996) for the surfactants in Table 5.3 and others has shown that the rapid fall of surface tension starts when two-thirds of the maximum coverage, Γ_m (Chapter 2, Section IIIC) of the surface has been attained. Consequently, at the same use concentration, surfactants that have smaller values, that is, larger areas per molecule at the air–aqueous solution interface, meaning less molecules needed to attain this degree of coverage, and those that have larger apparent diffusion coefficients (D_{ap} Section IVB), meaning faster diffusion of these molecules to their interface, should show shorter induction times (faster reduction of surface tension). This explains why branched-chain surfactant molecules (which have larger a_m^s values [Table 2.2]) and small molecules (which have larger D_{ap} values) reduce surface tension faster than linear, larger surfactant molecules. This is consistent with their use as textile wetting agents (Chapter 6, Section IIC).

B. Apparent Diffusion Coefficients of Surfactants

As mentioned above, the value of t_i has been shown to be related to the coverage of the air–aqueous solution interface by the surfactant and to its apparent diffusion coefficient, D_{ap} (Equation 5.9). To calculate the values of D_{ap} at short times, Equation 5.10 (Bendure, 1971), based upon the short-time approximation equation of Ward and Tordai (Equation 5.8), and using dynamic short-time surface tension data, may be used:

$$(\gamma_0 - \gamma_t)/C = 2\,RT(D_{ap}/\pi)^{1/2} t^{1/2}. \tag{5.10}$$

At constant surfactant concentration, c, in the solution, a plot of $(\gamma_0 - \gamma_t)$ versus $t^{1/2}$ should be linear, if adsorption is diffusion-controlled (generally true, for simple-structured surfactants) and permits evaluation of D_{ap} from the slope of the plot.

Apparent diffusion coefficients may also be calculated from longer-time dynamic surface tension data by use of Equation 5.11 (Joos et al., 1992):

$$t \xrightarrow{\gamma_t} \infty = \gamma_e + n\left(RT\Gamma^2 / C\right)\left(7\pi / 12D_{\mathrm{ap}}t\right)^{1/2}, \tag{5.11}$$

where Γ is calculated from the Gibbs adsorption equation (Equation 2.19a), n is the constant in that equation, and γ_e is the surface tension at infinite time (close to the equilibrium surface tension value). For solutions of constant surfactant concentration, C, a plot of γ_t, versus $t^{-1/2}$ should be linear if adsorption is diffusion-controlled and permits evaluation of D_{ap} again from the slope of the plot. The value of γ_e can be evaluated from the y-axis intercept and should be close to the value of the equilibrium surface tension.

Values of D_{ap} for simple, conventional surfactants (with a single hydrophilic head group and a single hydrophobic group) are of the order of 10^{-6} $\mathrm{cm^2/s}$. The values decrease with increase in the alkyl chain length of the hydrophobic group and with the degree of hydration of the hydrophilic group. The value increases with branching of the alkyl chain compared to that of the isomeric straight-chain compound.

Interaction between two surfactants, producing an increase in the molecular weight of the complex, decreases the value of D_{ap} considerably (Gao and Rosen, 1994; Rosen and Gao, 1995). When interaction between the two surfactants is weak, the surface tension at short times ($t < 1$s) is close to that of the component with the lower surface tension value at that time; at longer times, it is closer to that of the component that has the lower equilibrium tension. When interaction is strong, the surface tension at short times is greater than that of either component (Gao and Rosen, 1994).

A fair correlation has been found (Smith, 2000) between diffusion coefficients and wetting times for cotton twill tape (using a modified Draves wetting test, Chapter 6, Section IIC).

REFERENCES

Abe, M., K. Morikawa, K. Ogino, H. Sawada, T. Matsumoto, and M. Nakayama (1992) *Langmuir* **8**, 763.

Barakat, Y., L. N. Fortney, R. S. Schechter, W. H. Wade, and S. H. Yir (1983) *J. Colloid Interface Sci.* **92**, 561.

Becher, P., *Emulsions: Theory and Practice*, 2nd ed., Reinhold, New York, 1965.

Beckett, A. H. and R. J. Woodward (1963) *J. Pharm. Pharmacol.* **15**, 422.

Bendure, R. L. (1971) *J. Colloid Interface Sci.* **35**, 238.

Bourrel, M. and C. Chambu (1983) *Soc. Pet. Eng. J.* **2**, 327.

Bourrel, M., J. L. Salager, R. S. Schechter, and W. H. Wade (1980) *J. Colloid Interface Sci.* **75**, 451.

Bourrel, M., F. Verzaro, and C. Chambu, SPE 12674, presented at 4th DOE/SPE Symp. on EOR, Tulsa, OK, April 1984.

Burczyk, R., K. A. Wilk, A. Sokolowski, and L. Syper (2001) *J. Colloid Interface Sci.* **240**, 552.

Carless, J. E., R. A. Challis, and B. A. Mulley (1964) *J. Colloid Sci.* **19**, 201.

Caskey, J. A. and W. B., Jr. Barlage (1971) *J. Colloid Interface Sci.* **35**, 46.

Corkill, J. M., J. F. Goodman, and R. H. Ottewill (1961) *Trans. Faraday Soc.* **57**, 1627.

Corkill, J. M., J. F. Goodman, C. R. Ogden, and J. R. Tate (1963) *Proc. R. Soc.* **273**, 84.

Corkill, J. M., J. F. Goodman, and S. P. Harrold (1964) *Trans. Faraday Soc.* **60**, 202.

Crook, E. H., D. B. Fordyce, and G. F. Trebbi (1963) *J. Phys. Chem.* **67**, 1987.

Crook, E. H., G. F. Trebbi, and D. B. Fordyce (1964) *J. Phys. Chem.* **68**, 3592.

Dahanayake, M. and M. J. Rosen, in *Structure/Performance Relationships in Surfactants*, M. J. Rosen, (ed.), ACS Symp. series 253, American Chemical Society, Washington, DC, 1984, p. 49.

Dahanayake, M., A. W. Cohen, and M. J. Rosen (1986) *J. Phys. Chem.* **90**, 2413.

Davey, T. M., W. A. Ducker, A. R. Hayman, and J. Simpson (1998) *Langmuir* **14**, 3210.

Downer, A., J. Eastoe, A. R. Pitt, E. A. Simister, and J. Penfold (1999) *Langmuir* **15**, 7591.

Dreger, E. E., G. I. Keim, G. D. Miles, L. Shedlovsky, and J. Ross (1944) *Ind. Eng. Chem.* **36**, 610.

Elworthy, P. H. and A. T. Florence (1964) *Kolloid-Z. Z. Polym.* **195**, 23.

Elworthy, P. H. and C. B. MacFarlane (1962) *J. Pharm. Pharmacol.* **14**, 100.

Fowkes, F. M., J. O. Carrali, and J. A. Sohara, in *Macro- and Microemulsions*, D. O. Shah (ed.), ACS Symp. Series 272, American Chemical Society, Washington, DC, 1985, pp. 173–183.

Gao, T. and M. J. Rosen (1994) *J. Am. Oil. Chem. Soc.* **71**, 771.

Gao, T. and M. J. Rosen (1995) *J. Colloid Interface Sci.* **172**, 242.

Good, R. J., C. J. van Oss, J. T. Ha, and M. Cheng (1986) *Colloids Surf.* **20**, 187.

Gotte, E. (1969) *Fette, Seifen, Anstrichmi* **71**, 219.

Greiss, W. (1955) *Fette, Seifen, Anstrichmi* **57**, 24, 168, 236.

Gu, T. and J. Sjeblom (1992) *Colloids Surfs.* **64**, 39.

Healy, R. N. and R. L. Reed (1974) *Soc. Pet. Eng. J.* **14**, 491.

Healy, R. N. and R. L. Reed (1977) *Soc. Pet. Eng. J.* **17**, 129.

Healy, R. N., R. L. Reed, and D. G. Stenmark (1976) *Soc. Pet. Eng. J.* **16**, 147.

Hedges, J. H. and G. R. Glinsmann, SPE 8324, presented at 54th Annu. Tech. Conf., SPE, Las Vegas, NM, September 1979.

Hikota, T., K. Morohara, and K. Meguro (1970) *Bull. Chem. Soc. Jpn.* **43**, 3913.

Hua, X. Y. and M. J. Rosen (1988) *J. Colloid Interface Sci.* **125**, 652.

Hua, X. Y. and M. J. Rosen (1991) *J. Colloid Interface Sci.* **141**, 180.

Huh, C. (1979) *J. Colloid Interface Sci.* **71**, 408.

Hwan, R. N., C. A. Miller, and T. Fort (1979) *J. Colloid Interface Sci.* **68**, 221.

Joos, P., J. P. Fang, and G. Semen (1992) *J. Colloid Interface Sci.* **151**, 144.

Kanner, B., W. G. Reid, and I. H. Peterson (1967) *Ind. Eng. Chem., Prod. Res. Dev.* **6**, 88.

Kjellin, U. R. M., P. M. Claesson, and P. Linse (2002) *Langmuir* **18**, 6745.

Kling, W. and H. Lange, 2nd Int. Congr. Surface Activity, London, September 1957, I, p. 295.

Kunieda, H. and K. Shinoda (1982) *Bull. Chem. Soc. Jpn.* **55**, 1777.

Kwan, C. C. and M. J. Rosen (1980) *J. Phys. Chem.* **84**, 547.

Lange, H., 4th Int. Congr. Surface-Active Substances, Brussels, Belgium, September 1964, II, P. 497.

Lange, H. (1965) *Kolloid-Z.* **201**, 131.

Lange, H. and M. J. Schwuger (1968) *Kolloid Z. Z. Polym.* **223**, 145.

Lange, H. and M. J. Schwuger (1971) *Kolloid Z Z. Polym.* **243**, 120.

Lascaux, M. P., O. Dusart, R. Granet, and S. Pickarski (1983) *J. Chim. Phys.* **80**, 615.

Li, F., M. J. Rosen, and S. B. Sulthawa (2001) *Langmuir* **17**, 1037.

Livingston, J. R. and R. Drogin (1955) *J. Am. Oil Chem. Soc.* **42**, 720.

Matos, S. L., J.-C. Ravey, and G. Serratrice (1989) *J. Colloid Interface Sci.* **128**, 341.

Meguro, K., Y. Takasawa, N. Kawahashi, Y. Tabata, and M. Ueno (1981) *J. Colloid Interface Sci.* **83**, 50.

Mulley, B. A. and A. D. Metcalf (1962) *J. Colloid Sci.* **17**, 523.

Murphy, D. S. and M. J. Rosen (1988) *J. Phys. Chem.* **92**, 2870.

Murphy, D. S., Z. H. Zhu, X. Y. Hua, and M. J. Rosen (1990) *J. Am. Oil Chem. Soc.* **67**, 197.

Nelson, R. S., SPE 8824, presented at 1st Joint SPE/DOE Symp. on EOR, Tulsa, OK, April 1980.

Ohbu, K., M. Fujiwara, and Y. Abe (1998) *Colloid Polym. Sci.* **109**, 85.

Omar, A. M. A. and N. A. Abdel-Khalek (1997) *Tenside Surf. Det.* **34**, 178.

Raison, M., 2nd Int. Congr. Surface Activity, Butterworths, London, 1957, p. 422.

Rehfeld, S. J. (1967) *J. Phys. Chem.* **71**, 738.

Robbins, M. L., presented at 48th Natl. Colloid Symp., Austin, Texas, June 1974.

Rosen, M. J. (1974) *J. Am. Oil Chem. Soc.* **51**, 461.

Rosen, M. J. (1976) *J. Colloid Interface Sci.* **56**, 320.

Rosen, M. J. and T. Gao (1995) *J Colloid Interface Sci.* **173**, 42.

Rosen, M. J. and X. Y. Hua (1990) *J. Colloid Interface Sci.* **139**, 397.

Rosen, M. J. and Z.-P. Li (1984) *J. Colloid Interface Sci.* **97**, 456.

Rosen, M. J. and D. S. Murphy (1991) *Langmuir* **7**, 2630.

Rosen, M. J. and J. Solash (1969) *J. Am. Oil Chem. Soc.* **46**, 399.

Rosen, M. J. and L. D. Song (1996) *J. Colloid Interface Sci.* **179**, 261.

Rosen, M. J. and S. B. Sulthana (2001) *J. Colloid Interface Sci.* **239**, 528.

Rosen, M. J., D. Friedman, and M. Gross (1964) *J. Phys. Chem.* **68**, 3219.

Rosen, M. J., A. W. Cohen, M. Dahanayake, and X. Y. Hua (1982a) *J. Phys. Chem.* **86**, 541.

Rosen, M. J., M. Dahanayake, and A. W. Cohen (1982b) *Colloids Surf.* **5**, 159.

Rosen, M. J., Y.-P. Zhu, and S. W. Morrall (1996) *J. Chem. Eng. Data* **41**, 1160.

Salter, S. J., SPE 6843, presented at 52nd Annual Fall Technical Conference, SPE of AIME, Denver, CO, October 9–12, 1977.

Schick, M. J., S. M. Atlas, and F. R. Eirich (1962) *J. Phys. Chem.* **66**, 1325.

Schwuger, M. J. (1971) *Ber. Bunsenes. Phys. Chem.* **75**, 167.

Scriven, L. E., in *Micellization, Solubilization, and Microemulsions*, Vol. 2, K. L. Mittal (ed.), Plenum, New York, 1977, p. 877.

Shinoda, K. (1983) *Progr. Colloid Polym. Sci.* **68**, 1.

Shinoda, K. and S. Friberg (1975) *Adv. Colloid Interface Sci.* **4**, 281.

Shinoda, K. and K. Masio (1960) *J. Phys. Chem.* **64**, 54.

Shinoda, K. and H. Saito (1968) *J. Colloid Interface Sci.* **26**, 70.

Shinoda, K. and Y. Shibata (1986) *Colloids Surf* **19**, 185.

Shinoda, K., T. Yamanaka, and K. Kiwashita (1959) *J. Phys. Chem.* **63**, 648.

Shinoda, K., M. Hato, and T. Hayashi (1972) *J. Phys. Chem.* **76**, 909.

Smith, D. L. (2000) *J. Surfactants Deterg.* **3**, 483.

Smith, F. D., A. J. Stirton, and M. V. Nunez Ponzoa (1966) *J. Am. Oil Chem. Soc.* **43**, 501.

Tamura, T., Y. Kaneko, and M. Ohyama (1995) *J. Colloid Interface Sci.* **173**, 493.

Tsubone, K. and M. J. Rosen (2001) *J. Colloid Interface Sci.* **244**, 394.

Ueno, M., S. Yamamoto, and K. Meguro (1974) *J. Am. Oil Chem. Soc.* **51**, 373.

Valint, P. L., J. Bock, M. W. Kim, M. L. Robbins, P. Steyn, and S. Zushma (1987) *Colloids Surf.* **26**, 191.

Varadaraj, R., J. Bock, P. Geissler, S. Zushma, N. Brons, and T. Colletti (1991) *J. Colloid Interface Sci.* **147**, 396.

Varadaraj, R., J. Bock, S. Zushma, and N. Brons (1992) *Langmuir* **8**, 14.

Venable, R. L. and R. V. Nauman (1964) *J. Phys. Chem.* **68**, 3498.

Verzaro, F., M. Bourrel, and C. Chambu, in *Surfactants in Solution*, Vol. 6, K. L. Mittal and P. Bothorel (eds.), Plenum, New York, 1984, pp. 1137–1157.

Ward, A. F. H. and L. Tordai (1946) *J. Chem. Phys.* **14**, 453.

Weers, J. G., J. E. Rathman, F. U. Axe, C. A. Crichlow, L. D. Foland, D. R. Schening, R. J. Wiersema, and A. G. Zielske (1991) *Langmuir* **7**, 854.

Weil, J. K., A. J. Stirton, R. G. Bistline, and E. W. Maurer (1959) *J. Am. Oil Chem. Soc.* **36**, 241.

Weil, J. K., F. S. Smith, A. J. Stirton, and R. G. Bistline, Jr. (1963) *J. Am. Oil Chem. Soc.* **40**, 538.

Weil, J. K., A. J. Stirton, and E. A. Barr (1966) *J. Am. Oil Chem. Soc.* **43**, 157.

Winsor, P. A. (1948) *Trans. Faraday Soc.* **44**, 376.

Winsor, P. A. (1968) *Chem. Rev.* **68**, 1.

Zhao, F. and M. J. Rosen (1984) *J. Phys. Chem.* **88**, 6041.

Zhao, F., M. J. Rosen, and N.-L. Yang (1984) *Colloids Surf.* **11**, 97.

Zhu, Y.-P., M. J. Rosen, and S. W. Morrall (1998a) *J. Surfactants Deterg.* **1**, 1.

Zhu, Y.-P., M. J. Rosen, S. W. Morrall, and J. Tolls (1998b) *J. Surfactants Deterg.* **1**, 187.

Zhu, Y.-P., M. J. Rosen, P. K. Vinson, and S. W. Morrall (1999) *J. Surfactants Deterg.* **2**, 357.

PROBLEMS

5.1 Indicate, in the table below, the effect of each of the following changes on the surface tension reduction effectiveness Π_{CMC} of the surfactant in aqueous solution. Use symbols: + = increase; – = decrease; 0 = little or no effect; ? = effect not clearly known.

Change	Effect

- **(a)** Increase in the length of the hydrophobic group
- **(b)** Replacement of straight chain hydrophobic group by isomeric branched chain.
- **(c)** For ionics, increase in the electrolyte content of the aqueous solution
- **(d)** Increase in the temperature of the solution

5.2 Predict the effect of each of the following changes on the value of the Winsor ratio, R:

- **(a)** Increase in the length of the hydrophobic group of the surfactant
- **(b)** Increase in the length of the POE chain of a nonionic surfactant
- **(c)** Replacement of n-hexane by n-octane as the hydrocarbon phase
- **(d)** Addition of n-pentanol to the system
- **(e)** Addition of NaCl to the system

5.3 Account for the following observations: The value of n in Equations 5.5 and 5.6: (a) increases with decrease in the number of oxyethylene units in nonionic surfactants with the same number of carbon atoms in their alkyl chains; (b) decreases when 4 M urea is added to the water.

5.4 A nonionic surfactant has a minimum area per molecule, a_m^s, value of 60 Å^2 in water. Its CMC value is 2×10^{-4} and its pC_{20} value is 4.8.

- **(a)** Estimate the surface tension of its aqueous solution at the CMC, γ_{CMC}, at 25°C, in mN/m (dyn/cm).
- **(b)** If the above values are the same at 40°C, what would the γ_{CMC} value be?

5.5 The adsorption density of an ionic surfactant on an ionic solid in water at a given salinity was determined to be 3.5×10^{-10} mol/cm^2. (a) Calculate the head group area of the surfactant molecule in the adsorbed state at the solid–liquid interface. (b) If the head group area measured for the same surfactant at water–air interface is 37.5 Å^2/molecule and corresponds to a monolayer, what is the average thickness of the adsorbed layer at the solid–liquid interface?

5.6 How is the Mitchell-Ninham parameter related to the Winsor *R* parameter in explaining the nature of aggregates in different media?

5.7 What are the characteristic features of dynamic surface tension and static surface tension? Explain and interpret the regions in the typical curves obtained for them.

6 Wetting and Its Modification by Surfactants

Wetting in its most general sense is the displacement from a surface of one fluid by another. Wetting, therefore, always involves three phases, at least two of which are fluids: a gas and two immiscible liquids, or a solid and two immiscible liquids, or a gas, a liquid, and a solid, or even three immiscible liquids. Commonly, however, the term *wetting* is applied to the displacement of air from a liquid or solid surface by water or an aqueous solution, and we restrict our discussion for the most part to those situations. The term *wetting agent* is applied to any substance that increases the ability of water or an aqueous solution to displace air from a liquid or solid surface. Wetting is a process involving surfaces and interfaces, and the modification of the wetting power of water is a surface property shown to some degree by all surface-active agents, although the extent to which they exhibit this phenomenon varies greatly. When the surface to be wet is small, as in the wetting of nongranular, nonporous solids (*hard surface wetting*), equilibrium conditions or conditions close to it can be attained during the wetting process and the free energy changes involved in the process determine the degree of wetting attained. On the other hand, when the surface to be wet is large, as in the wetting of porous or textile surfaces or finely powdered solids, equilibrium conditions are often not reached during the time allowed for wetting, and the degree of wetting is determined by the kinetics rather than the thermodynamics of the wetting process.

I. WETTING EQUILIBRIA

Three types of wetting have been distinguished (Osterhof and Bartell, 1930): (1) spreading wetting, (2) adhesional wetting, and (3) immersional wetting. The equilibria involved in these phenomena are well known.

Surfactants and Interfacial Phenomena, Fourth Edition. Milton J. Rosen and Joy T. Kunjappu.

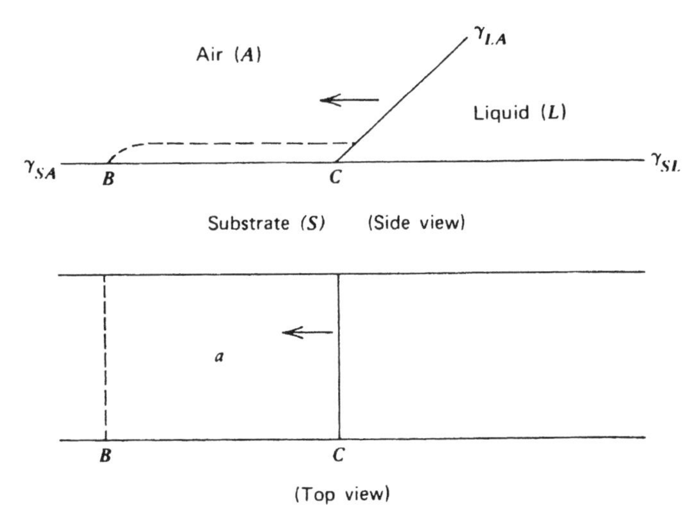

FIGURE 6.1 Spreading wetting.

A. Spreading Wetting

In spreading wetting (Figure 6.1), a liquid in contact with a substrate spreads over the substrate and displaces another fluid, such as air, from the surface. For the spreading to occur spontaneously, the surface free energy of the system must decrease during the spreading process. When the area of an interface increases, the surface free energy at that interface increases; when the area decreases, the surface free energy decreases. If the liquid L in Figure 6.1 spreads from C to B, covering an area a, then the decrease in surface free energy of the system due to decrease in area of the substrate–air interface is $a \times \gamma_{SA}$, where γ_{SA} is the interfacial free energy per unit area of the substrate in equilibrium with liquid-saturated air above it. At the same time, the free energy of the system has been increased because of the increase in liquid–substrate and liquid–air interfaces. The increase in surface free energy of the system due to the increase in the liquid–substrate interface is $a \times \gamma_{SL}$ (where γ_{SL} is the interfacial free energy per unit area at the liquid–substrate interface), and since the liquid–air interface has also been increased by area a, the increase in surface free energy due to increase in this interface is $a \times \gamma_{LA}$, where γ_{LA} is the surface tension of L. The total decrease in surface free energy per unit area of the system due to the spreading wetting, $-\Delta G_W/a$, is therefore $\gamma_{SA} - (\gamma_{SL} + \gamma_{LA})$. If the quantity $\gamma_{SA} - (\gamma_{SL} + \gamma_{LA})$ is positive, the system decreases in surface free energy during the spreading process, and the process can then occur spontaneously.

The quantity $\gamma_{SA} - (\gamma_{SL} + \gamma_{LA})$ is then a measure of the *driving force* behind the spreading process and is usually called the *spreading coefficient* $S_{L/S}$. If $S_{L/S}$, as defined by

$$S_{L/S} = \gamma_{SA} - (\gamma_{SL} + \gamma_{LA}) \tag{6.1}$$

is positive, spreading can occur spontaneously; if $S_{L/S}$ is negative, the liquid will not spread spontaneously over the substrate.

When a thin layer of liquid L_1 is being spread over a second liquid L_2 as substrate, $S_{L_1/L_2} = \gamma_{L_2A} - (\gamma_{L_1L_2} + \gamma_{L_1A})$, and the value of S can be obtained directly by measuring the surface tensions of the two liquids and the interfacial tension between them. However, this gives the initial spreading coefficient. The two phases will quickly become saturated with each other in the vicinity of the interface, and the equilibrium spreading coefficient will be that based on the tensions of the mutually saturated phases, which may be very different. As an example, the surface tensions of pure water and pure benzene, each at 20°C, are 72.8 and 28.9 dyn/cm, respectively, and their interfacial tension is 35.0 dyn/cm. Using these figures, the initial spreading coefficient of benzene on water is 72.8 – (28.9 + 35.0) = 8.9 dyn/cm. This means that benzene will initially spread spontaneously over water. However, the surface tensions of water saturated with benzene and benzene saturated with water at 20°C are 62.2 and 28.8 dyn/cm, respectively. Thus, the spreading coefficient a short time after both phases come in contact with each other becomes 62.2 – (28.8 + 35.0) = –1.4, and spontaneous spreading ceases to occur. The benzene retracts to a lens after the initial spreading.

Since the spreading coefficient involves only the surface tensions of the two liquids (in the case of one liquid spreading over another) and the interfacial tension between them, if we have a method for determining the interfacial tension between the two liquids from their respective surface tensions, we can calculate the spreading coefficient without additional experimental data and predict whether spreading will occur spontaneously. Good and Girifalco (1960) and Girifalco and Good (1957) have suggested a method of doing this. According to them, $\gamma_{L_1L_2} = \gamma_{L_1A} + \gamma_{L_2A} - 2\Phi\sqrt{\gamma_{L_1A}\gamma_{L_2A}}$, where Φ is an empirical factor measuring the degree of interaction between L_1 and L_2. Since $S_{L_1/L_2} = \gamma_{L_2A} - (\gamma_{L_1A} + \gamma_{L_2/L_2})$, by substituting for $\gamma_{L_1L_2}$ we obtain

$$\begin{aligned} S_{L_1/L_2} &= \gamma_{L_2A} - \left(\gamma_{L_1A} + \gamma_{L_1A} + \gamma_{L_2A} - 2\Phi\sqrt{\gamma_{L_1A}\gamma_{L_2A}}\right) \\ &= 2\left(\Phi\sqrt{\gamma_{L_1A}\gamma_{L_2A}} - \gamma_{L_1A}\right) \\ &= 2\gamma_{L_1A}\left(\Phi\sqrt{\gamma_{L_2A}/\gamma_{L_1A}} - 1\right) \end{aligned} \tag{6.2a}$$

In systems where there is no strong interaction between L_1 and L_2, Φ is less than 1. Thus, in those systems, γ_{L_1A} must be less than γ_{L_2A} for spreading to occur spontaneously (i.e., for the spreading coefficient to be positive, the spreading liquid must have a lower surface tension than the liquid over which it is spreading). The same is assumed to be true if the substrate over which the liquid is spreading is a solid:

$$S_{L/S} = 2\left(\Phi\sqrt{\gamma_{LA}\gamma_{SA}} - \gamma_{LA}\right) = 2\gamma_{LA}\left(\Phi\sqrt{\gamma_{SA}/\gamma_{LA}} - 1\right). \qquad (6.2b)$$

This concept of a *critical surface tension* for spreading on low-energy surfaces is one that was developed by Zisman and coworkers (Fox and Zisman, 1950; Shafrin and Zisman, 1960; Zisman, 1964). They demonstrated that, at least for low-energy substrates, in order to wet the substrate, the surface tension of the wetting liquid must not exceed a certain critical value that is characteristic of the particular substrate.

High-melting solids such as silica and most metals have high surface free energies ranging from several thousand to several hundred mJ/m^2 ($ergs/cm^2$). Low-melting solids, such as organic polymers, waxes, and covalent compounds, in general have surface free energies ranging from 100 to 25 mJ/m^2 ($ergs/cm^2$). Since nearly all liquids other than liquid metals have surface tensions of less than 75 mN/m (dyn/cm) (i.e., surface free energies of <75 (mJ/m^2)), they usually spread readily on metallic or siliceous surfaces but may not spread on low-melting solids.

1. The Contact Angle When the substrate is a solid, the spreading coefficient is usually evaluated by indirect means, since surface and interfacial tensions of solids cannot easily be measured directly. The method of doing this involves measuring the contact angle the substrate makes with the liquid in question.

The contact angle θ that the liquid makes when it is at equilibrium with the other phases in contact with it is related to the interfacial free energies per unit area of those phases. When the liquid is at equilibrium with the other two phases, gas and solid substrate, we can diagram the contact angle θ as shown in Figure 6.2. For a small reversible change in the position of the liquid on the surface so as to cause an increase, Δa, in the L/S interfacial area, there is a corresponding decrease Δa in the area of the S/A interface and an increase in the L/A interface of $\Delta a \cos\theta$. Thus, $\Delta G_W = -\gamma SA\Delta a + \gamma_{LS}\Delta a + \gamma_{LA}\Delta a \cos\theta$. As $\Delta a \to 0$, $\Delta G \to 0$ $\therefore$ $\gamma_{LA} da \cos\theta + \gamma_{LS} da - \gamma_{SA} da$. Therefore,

$$\gamma_{LA} \cos\theta = \gamma_{SA} - \gamma_{SL} \qquad (6.3)$$

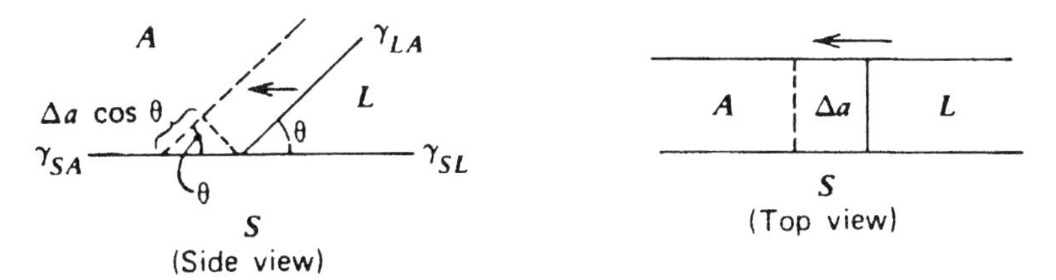

FIGURE 6.2 Contact angle.

or

$$\cos\theta = \frac{\gamma_{SA} - \gamma_{SL}}{\gamma_{LA}}. \tag{6.4}$$

Equation 6.3 is generally called *Young's equation* and the quantity $\gamma_{LA}\cos\theta$, the *adhesion tension* (Bartell and Bartell, 1934). Note that γ_{SA}, the interfacial tension in equilibrium with the gas and liquid phases in the system, is *not* γ_S, the free energy per unit area of the solid in a vacuum, but $\gamma_S - \pi$, where π is the reduction in interfacial free energy per unit area of S resulting from adsorption of vapor of L; that is,

$$\pi = \gamma_S - \gamma_{SA}.$$

If the contact angle is larger than 0°, then the spreading coefficient cannot be positive or zero. Since $S_{L/S} = \gamma_{SA} - (\gamma_{SL} + \gamma_{LA}) = \gamma_{SA} - \gamma_{SL} - \gamma_{LA}$ and $\gamma_{SA} - \gamma_{LA}\cos\theta$ when $\theta > 0°$, substituting $\gamma_{LA}\cos\theta$ for $\gamma_{SA} - \gamma_{SL}$ yields

$$\begin{aligned} S_{L/S} &= \gamma_{LA}\cos\theta - \gamma_{LA} \\ &= \gamma_{LA}(\cos\theta - 1) \end{aligned} \tag{6.5}$$

When θ is finite, $(\cos\theta - 1))$ is always negative, and $S_{L/S}$, too, is always negative. If the contact angle is 0°, then $S_{L/S}$ may be zero or positive. In either case here, complete spreading wetting occurs.

When the solid substrate is a nonpolar, low-energy surface, the contact angle can be used to determine the surface (excess) concentration of the surfactant at the solid–liquid interface Γ_{SL}.

From Young's equation,

$$\frac{d(\gamma_{LA}\cos\theta)}{d\ln C} = \frac{d(\gamma_{SA})}{d\ln C} - \frac{d(\gamma_{SL})}{d\ln C}. \tag{6.6}$$

If we assume that, for a low-energy surface, γ_{SA} does not change with change in the liquid phase surfactant concentration, that is, $d(\gamma_{SA})/d\ln C = 0$, then

$$\frac{d(\gamma_{LA}\cos\theta)}{d\ln C} = \frac{-d(\gamma_{SL})}{d\ln C}. \tag{6.7}$$

The Gibbs adsorption equation (Equations 2.19, 2.19a) at the solid–liquid interface can be written as

$$-d\gamma_{LS} = (1, n)RT\Gamma_{LS}d\ln C$$

from which

$$\frac{d(\gamma_{LA}\cos\theta)}{d\ln C} = (1, n)RT\Gamma_{LS}. \tag{6.8}$$

The value of Γ_{LS} can, therefore, under these conditions, be determined from the slope of a $\gamma_{LA}\cos\theta - \ln C(= 2.303\log C)$ plot at constant temperature.

2. Measurement of the Contact Angle Contact angles are measured on macroscopic, smooth, nonporous, planar substrates by merely placing a droplet of the liquid or solution on the substrate and determining the contact angle by any of a number of techniques (Adamson and Gast, 1997, p. 362). The contact angle can be measured directly by use of a microscope fitted with a goniometer eyepiece or by photographing the droplet. However, obtaining a valid, reproducible contact angle is more complicated and difficult than it appears for a number of reasons:

1. Contamination of the droplet by adsorption of impurities from the gas phase tends to reduce θ if γ_{LA} and/or γ_{SL} is reduced and γ_{SA} remains more or less constant.
2. A solid surface, even when apparently smooth, may have impurities and defects that vary from place to place on the surface and from sample to sample. Roughness reduces θ when the value on a smooth surface is $<90°$ and increases it when the value there is $<90°$.
3. The contact angle may show hysteresis. In this case, the advancing contact angle will always be greater than the receding contact angle, sometimes differing by as much as 60°. Contact angle hysteresis is always present when the surface is not clean or when it contains a considerable amount of impurity. However, even when the surface is clean and the substrate is pure, it may show hysteresis. For example, stearic acid becomes more wettable (shows a smaller contact angle) after being contacted with water. The explanation has been advanced that there is a change in orientation of the surface molecules in the presence of water, with more of the molecules becoming oriented with their carboxylic acid groups facing the water, thus decreasing the interfacial free energy. Other reasons for low receding angles are penetration of the wetting liquid into the substrate, removal of an adsorbed surface film from the substrate by the wetting liquid, and microscopic surface roughness.

Contact angles on finely divided solids are more difficult to measure, but are often more desired and more important than those on large solid surfaces. One method of obtaining such contact angles is to pack the powder into a glass tube and measure the rate of penetration of the liquid into it (Bruil and van Aartsen, 1974). The distance of penetration l in time t of a liquid of surface tension γ_{LA} and viscosity η is given by the modified Washburn equation (Washburn, 1921):

$$l^2 = \frac{(kr)t\gamma_{LA}\cos\theta}{2\eta}, \tag{6.9}$$

where r is the mean equivalent radius of the capillary passages through the powder and k is a constant to allow for the tortuous path through them. The (kr) product depends on the packing of the powder. When the powder is packed to the same bulk density, (kr) is assumed to be constant. The (kr) product is evaluated by passing through the powder a pure liquid of known γ_{LA} whose contact angle is known or assumed to be 0°. A limitation of the method is the assumption that (kr) will not change with change in the nature of the wetting liquid. This is justified only when the particle size of the powder is not changed by flocculation or dispersion produced by the passage through it of the surfactant solution.

This method is not reliable for dilute solutions of surfactants in many cases since it depends upon knowing the (constant) value of γ_{LA}. If adsorption of the surfactant onto the solid decreases its solution phase concentration to a value below the CMC, then γ_{LA} will change, and it will be impossible to determine θ accurately.

Adsorption of the surfactant onto the solid also makes this an unreliable method for determining the wetting effectiveness of dilute surfactant solution for powdered solids. Because of the small ratio of solution volume to solid–liquid interface, solutions that contain highly surface-active material that adsorbs well at the solid–liquid interface are rapidly depleted of surfactant and may penetrate more slowly than solutions of weakly surface-active material.

Another method of measuring the contact angles of powders involves measuring the height h of a drop of the wetting liquid on a cake of the powder, prepared by compressing it in a mold (Heertjes and Kossen, 1967). The contact angle is obtained from

$$\cos\theta = 1 - \sqrt{\frac{1}{3(1-\varepsilon)(1/Bh^2 - 1/2)}} \text{ for } \theta < 90° \tag{6.10}$$

$$\cos\theta = -1 + \sqrt{\frac{2}{3(1-\varepsilon)}\left(\frac{2}{Bh^2} - 1\right)} \text{ for } \theta > 90° \tag{6.11}$$

where $B = \rho_L g/2\gamma_{LA}$ for the wetting liquid of density ρ_L, γ_{LA} = the surface tension, g = the gravitational constant, and ε = the porosity of the cake. This method assumes that the powder consists of identical spheres.

B. Adhesional Wetting

In spreading wetting, a liquid in contact with a substrate and another fluid increases its area of contact with the substrate at the expense of the second

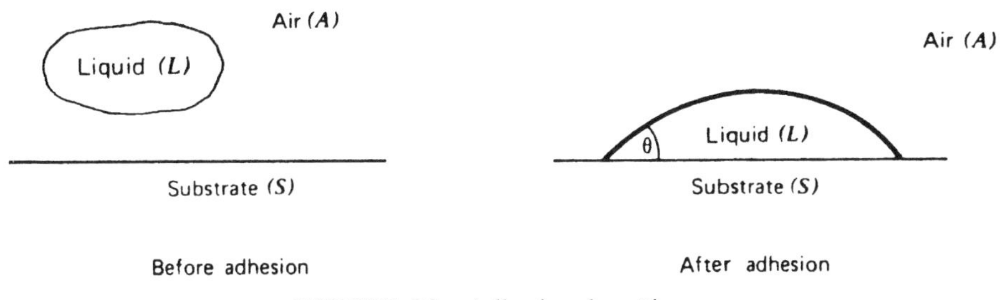

FIGURE 6.3 Adhesional wetting.

fluid. In adhesional wetting, a liquid *not* originally in contact with a substrate makes contact with that substrate and adheres to it. We can diagram this process as shown in Figure 6.3. In this case, the change in surface free energy is $-\Delta G_W = a(\gamma_{SA} + \gamma_{LA} - \gamma_{SL})$, where a is the surface area of the substrate in contact with (an equal) surface area of the liquid after adhesion and the "driving force" of this type of wetting phenomenon is $\gamma_{SA} + \gamma_{LA} - \gamma_{SL}$. This quantity is known as the *work of adhesion*, W_a, the reversible work required to separate the unit area of liquid from the substrate:

$$W_a = \gamma_{SA} + \gamma_{LA} - \gamma_{SL}. \tag{6.12}$$

The equation is from Dupré (1869). In this process, any reduction of the interfacial tension between substrate and the wetting liquid results in an increased tendency for adhesion to occur, but reduction of either the surface tension of the liquid or the surface tension of the substrate *decreases* the tendency of adhesion to occur. This accounts for the poor adhesion of substances to low-energy surfaces, especially when the natures of substance and substrate are very different (i.e., γ_{SL}, is large).

If the contact angle θ between liquid, substrate, and air, after adhesion, measured in the liquid, is finite (Figures 6.2 and 6.3), we can write, as before,

$$\gamma_{LA} \cos\theta - \gamma_{SA} - \gamma_{SL}. \tag{6.3}$$

Substituting for $\gamma_{SA} - \gamma_{SL}$ in Equation 6.12,

$$W_a = \gamma_{LA} \cos\theta + \gamma_{LA} = \gamma_{LA}(\cos\theta + 1), \tag{6.13}$$

from which it is apparent that an increase in the surface tension of the wetting liquid always causes increased adhesional wetting, whereas an increase in the contact angle obtained after wetting may or may not indicate a decreased tendency for adhesion to occur. If the increase in the contact angle (and consequent decrease of $\cos\theta$) reflects an increase in γ_{SL}, there is a diminished tendency to adhere; if it reflects merely an increase in γ_{LA}, there is increased tendency to adhere. The driving force in adhesional wetting can never be

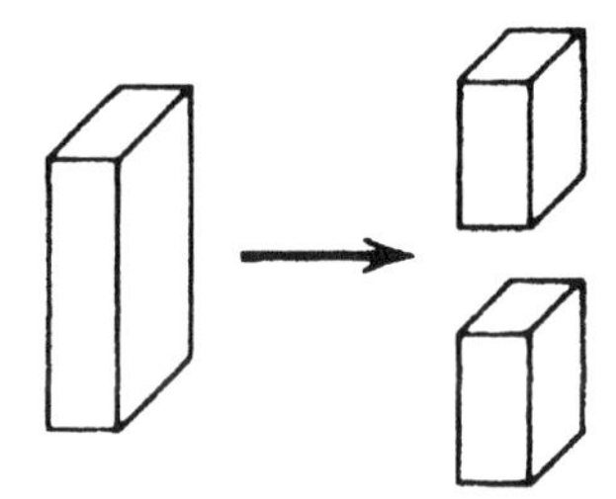

FIGURE 6.4 Work of adhesion.

negative and is equal to zero only when the contact angle is 180°, which is never achieved in practice. Since Equation 6.13 involves directly measurable quantities, γ_{LA} and θ, the driving force behind this type of wetting is readily evaluated.

Thc work of self-adhesion of a liquid is known as the *work of cohesion*, $W_c = 2\gamma_{LA}$. It is the work required to produce two unit areas of interface from an original unbroken column of the liquid or the $(-\Delta G_W)/a$ when the two columns are joined (Figure 6.4).

The difference between the work of adhesion of the liquid for the substrate and its work of cohesion equals the spreading coefficient $S_{L/S}$:

$$\begin{aligned} W_a - W_c &= \gamma_{SA} - \gamma_{SL} + \gamma_{LA} - 2\gamma_{LA} \\ &= \gamma_{SA} - \gamma_{SL} - \gamma_{LA} \\ &= S_{L/S} \end{aligned} \tag{6.14}$$

Therefore, if $W_a > W_c$, the spreading coefficient is positive, $\theta = 0°$, and the liquid spreads spontaneously over the substrate to form a thin film. If $W_a < W_c$, the spreading coefficient is negative, θ is greater than zero, and the liquid does not spread over the substrate but forms droplets or lenses with a finite contact angle.

When the work of adhesion equals the work of cohesion,

$$\gamma_{LA}(\cos\theta + 1) = 2\gamma_{LA}$$

or

$$\gamma_{SA} - \gamma_{SL} + \gamma_{LA} - 2\gamma_{LA}$$

and

$$\cos\theta + 1, \theta = 0^0 \text{ and } S_{L/S} = 0$$

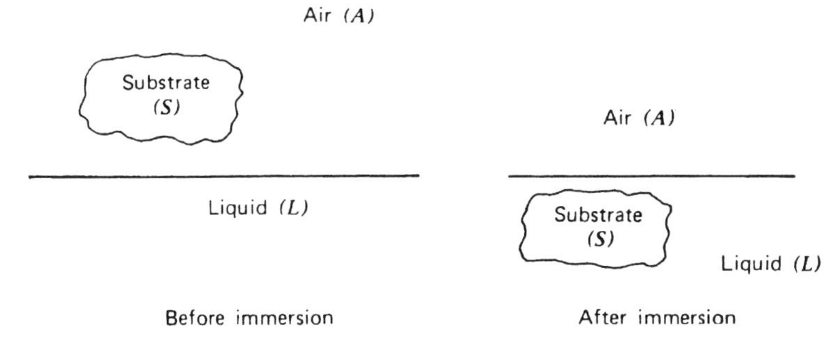

FIGURE 6.5 Immersional wetting.

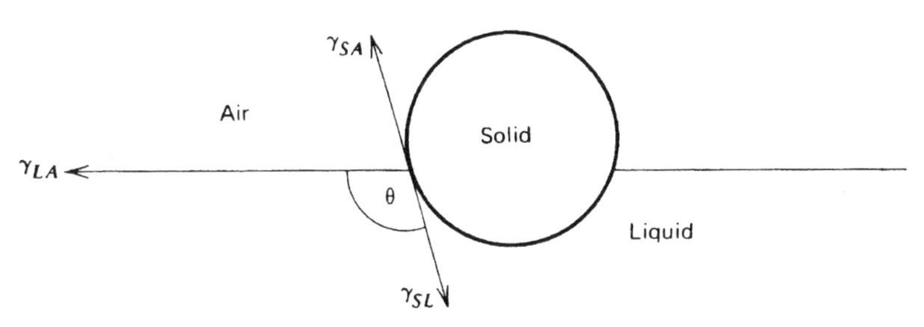

FIGURE 6.6 Contact angle of partially immersed solid.

C. Immersional Wetting

A third type of wetting is *immersional wetting*, where a substrate not previously in contact with a liquid is immersed completely by the liquid (Figure 6.5). In this case, the surface free energy change per unit area is

$$-\Delta G_w / a = \gamma_{SA} - \gamma_{SL} \tag{6.15}$$

and the driving force of the wetting phenomenon in this case is the quantity

$$(\gamma_{SA} - \gamma_{LA}).$$

If immersion of the solid in the wetting liquid gives a finite equilibrium contact angle θ, that is, $\theta > 0°$, then $(\gamma_{SA} - \gamma_{SL})$ equals $\gamma_{LA} \cos\theta$. We can determine $(\gamma_{SA} - \gamma_{SL})$, therefore, by observing the contact angle the solid makes with the liquid–air interface (Figure 6.6). If θ is $>90°$, then $\gamma_{SA} - \gamma_{SL}$ is negative; if $\theta < 90°$, then $\gamma_{SA} - \gamma_{SL}$, is positive.

However, since spreading occurs during the immersion process, the value of the spreading coefficient $S_{L/S} = \gamma_{SA} - \gamma_{SL} - \gamma_{LA}$ (Equation 6.1) determines the ease of complete immersion. When the spreading coefficient is ≥ 0 (i.e., $\gamma_{SA} - \gamma_{SL} \geq\geq \gamma_{LA}$), then, from Equation 6.4, $\theta = 0°$ and complete immersion is spontaneous; when the spreading coefficient is negative ($\gamma_{SA} - \gamma_{SL} < \gamma_{LA}$), then θ is finite, and work must be done to immerse the solid completely. In this latter case, to achieve spontaneous immersion of the solid, the values of any or all of the three quantities, γ_{SA}, γ_{SL}, and γ_{LA} must be changed in such a manner as to make the spreading coefficient greater than zero.

The depth of immersion of the solid in the wetting liquid is determined by the contact angle θ; the smaller the value of θ, the greater the depth of immersion. When $\theta = 0°$, immersion is complete. Here again, the relation of $\gamma_{SA} - \gamma_{SL}$ to γ_{LA}, as given in Equation 6.4, determines the value of θ.

When $\theta \leq 0°$, $\gamma_{SA} - \gamma_{SL}$ cannot be determined from the contact angle. Another experimentally determinable quantity, the heat of immersion ΔH^i, which is the heat change measured calorimetrically when the substrate is immersed in the wetting liquid, is then often used as a measure of immersional wetting. The heat of immersion per unit area of substrate is related to the surface free energy change per unit area due to immersional wetting by the relation $\Delta G_w/a = \Delta H^i/a - T\Delta S^i/a$, and is therefore equal to it only when the entropy change per unit area due to immersional wetting $\Delta S^i/a$ is insignificant.

The surface free energy change per unit area or driving force for the three different types of wetting can be expressed as

Spreading:

$$\frac{-\Delta G_w}{a} = \gamma_{SA} - \gamma_{LA} - \gamma_{SL} = S_{L/S}$$

Adhesion:

$$\frac{-\Delta G_w}{a} = \gamma_{SA} + \gamma_{LA} - \gamma_{SL} = W_a$$

Immersion:

$$\frac{-\Delta G_w}{a} = \gamma_{SA} - \gamma_{SL} (= \gamma_{SL} \cos\theta \text{ when } \theta > 0^0)$$

From these expressions we can see that in *all* wetting processes, reduction of the interfacial tension between substrate and the wetting liquid, γ_{SL}, is beneficial, but that reduction of the surface tension of the liquid γ_{LA} per se is not always of benefit.

D. Adsorption and Wetting

A convenient method of analyzing the relation of adsorption to equilibrium wetting has been developed by Lucassen-Reynders (1963). Combination of the Gibbs adsorption Equation 2.19 with Young's Equation 6.3 yields

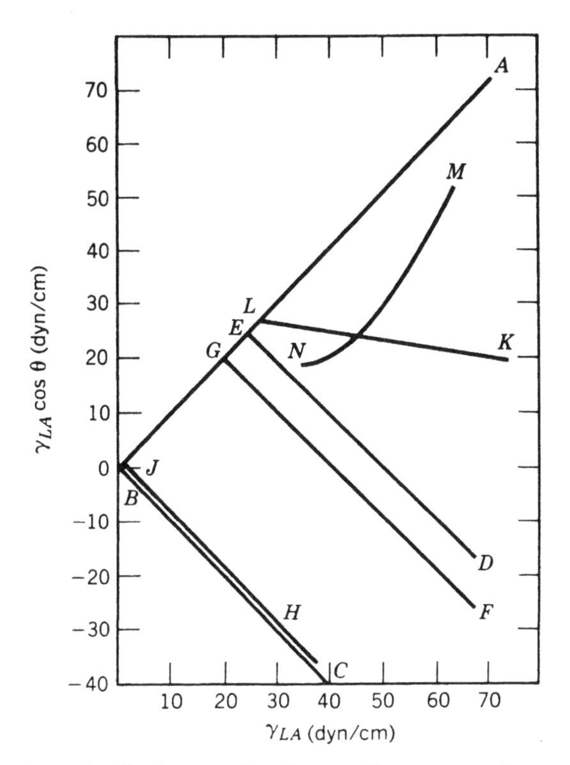

FIGURE 6.7 Plots of adhesion tension ($\gamma_{LA} \cos\theta$) versus surface tension (γ_{LA}) for surfactant solutions on different substrates: *DE*, paraffin in the presence of air; *FG*, Teflon in the presence of air; *HJ*, paraffin or Teflon in the presence of mineral oil (plot of $\gamma_{LO} \cos\theta$ vs. γ_{LO}); *KL*, anionic surfactant solution on nylon or polymethyl methacrylate in the presence of air; *MN*, mineral in the presence of air.

$$\frac{d(\gamma_{LA} \cos\theta)}{d\gamma_{LA}} = \frac{\Gamma_{SA} - \Gamma_{SL}}{\Gamma_{LA}}. \tag{6.16}$$

The slope of a plot of $\gamma_{LA} \cos\theta$, the *adhesion tension* (Equation 6.3), versus γ_{LA} consequently provides information on the surface (excess) concentrations of the surfactant at the three interfaces (Padday, 1967; Bargeman and Van Voorst Vader, 1973; Pyter et al., 1982).

Some typical $\gamma_{LA} \cos\theta$ versus γ_{LA} plots for surfactant solutions on solid substrates are shown diagrammatically in Figure 6.7. In this type of plot, the solution wets the substrate completely ($\theta = 0°$) when the curve reaches line *AB* (points *G*, *E*, *L)*; it dewets it completely ($\theta = 180°$) when it reaches line *BC*. Solutions that produce contact angles between 0° and 180° have points that fall in the region between those two lines (e.g., points *D*, *F*, *K*, *M*, *N*). When the slope of the plot is negative, wetting is improved by the presence of the surfactant; when it is positive, wetting is impaired by its presence.

On low-surface-energy (hydrophobic) solids such as paraffin and Teflon, data for several types of surfactants show constant slopes very close to –1 (lines *DE*, *EG*, and *HJ*), both for solid-aqueous-air systems and solid-aqueous-mineral oil systems (Bernett and Zisman, 1959; Bargeman and Van Voorst Vader, 1973; Pyter et al., 1982). This is usually taken to indicate that Γ_{SA} is close to 0 and $\Gamma_{SL/\Gamma_{LA}} \cong 1$. This is reasonable since, as shown in Chapter 2, Section IIIC, the effectiveness of adsorption at the nonpolar liquid (heptane)–aqueous solution interface is not very different from that at the air–aqueous solution interface, and we would therefore expect approximately equal adsorption at nonpolar solid–aqueous and air–aqueous interfaces. On these nonpolar solid surfaces, then, $\gamma_{LA} \cos\theta \cong -\gamma_{LA}$ + constant (Bargeman and Van Voorst Vader, 1973), and complete wetting is therefore achieved ($\theta = 0°$) when $\gamma_{LA} \cong$ constant/2. Thus, for paraffin wax in the presence of air (line *DE*), $\gamma_{LA} \cos\theta \cong -\gamma_{LA} + 49.4$, meaning that γ_{LA} must be decreased to ≅25 dyn/cm for complete wetting to occur. For Teflon in the presence of air (line *FG*), the constant is 40.6, and complete wetting occurs when $\gamma_{LA} \cong 20$ dyn/cm. When mineral oil replaces air on both of these solids (line *HJ*), the constant is almost zero, indicating the γ_{ow} must be very low for complete wetting to occur.

On negatively charged polar organic surfaces, such as polymethyl methacrylate or nylon, data by Pyter and coworkers (Pyter et al., 1982) show negative slopes close to zero for anionic surfactants (line *KL*), indicating either Γ_{SA} approaching Γ_{SL}, or, if Γ_{SA} remains close to zero, very low values of Γ_{SL}/Γ_{LA}, with the authors favoring the latter alternative. The solid–liquid interfacial tension γ_{SL} of a negatively charged solid is much lower than that of a nonpolar solid against water. Low adsorption of anionic surfactants at the aqueous interface with a negatively charged polar organic solid is consequently reasonable since adsorption would place either the hydrophobic group of the surfactant or its negatively charged hydrophilic group adjacent to the negatively charged polar surface, neither of which would be expected to reduce the solid–liquid interfacial tension. On the other hand, when the solid phase is nonpolar, the interfacial tension at the solid–aqueous solution interface is high and adsorption of the surfactant with its nonpolar hydrophobic group oriented toward the nonpolar solid and its hydrophilic group oriented toward the water would reduce γ_{SL}.

On high-energy surfaces with both positively and negatively charged sites, $\gamma_{LA} \cos\theta$ versus γ_{LA} plots show large positive slopes (curve *MN*), especially at low surfactant concentrations (high γ_{LA} values), indicating that $\Gamma_{SA} >> \Gamma_{SL}$ (Finch and Smith, 1973; Aronson et al., 1978; Bargava et al., 1978). This is presumably because direct adsorption onto oppositely charged sites on the solid substrate of ionic surfactants via their hydrophilic heads is occurring (Chapter 2, Section IIA and F). Wetting is impaired and the solid becomes more hydrophobic. This is the basis for the flotation process for ore beneficiation. At higher surfactant concentrations, the slope may become negative due to increased adsorption of surfactant at the now hydrophobic solid–aqueous solution interface, making $\Gamma_{SL} > \Gamma_{SA}$.

Perfluoroalkyl chain surfactants are much poorer wetting agents than alkyl chain surfactants for both paraffin and polymethyl methacrylate surfaces (Pyter et al., 1982). One explanation may be the mutual "phobicity" of alkyl and perfluoroalkyl chains, causing perfluoroalkyl chain surfactants to be adsorbed more poorly than alkyl chain surfactants at these solid–aqueous solution interfaces.

II. MODIFICATION OF WETTING BY SURFACTANTS

A. General Considerations

Since water has a rather high surface tension, 72 mN/m(dyn/cm) (reflecting the high intermolecular attractions of water molecules), it does not spontaneously spread over covalent solids that have surface free energies of less than 72 mJ/m^2(erg/cm^2). The addition of a surface-active agent to water, to modify the interfacial tensions of the system, is therefore often necessary to enable water to wet a solid or liquid surface. For water to wet a substrate spontaneously, the spreading coefficient $S_{W/S} = \gamma_{SA} - (\gamma_{SW} + \gamma_{WA})$ must be positive (Equation 6.1). The addition of a surface-active agent to the water, by reducing the surface tension of the water γ_{WA} and perhaps also the interfacial tension between the water and the substrate γ_{SW}, may cause the spreading coefficient to have a positive value and make spreading or immersion spontaneous.

However, the addition of a surface-active agent to water does not *always* increase its wetting power. Under certain conditions, the addition of a surface-active agent to the water may make spreading more difficult. In cases where the substrate is porous and can be considered to consist of a mass of capillaries, the pressure causing movement of liquid into the capillaries because of the curvature of the liquid surface is given by

$$\Delta P = \frac{2\gamma_{LA}\cos\theta}{r} = \frac{2(\gamma_{SA} - \gamma_{SL})}{r}, \tag{6.17}$$

where r is the equivalent radius of the capillaries and θ the contact angle at the air–liquid–substrate interface. When the contact angle is greater than 0°, ΔP depends only on the quantity $(\gamma_{SA} - \gamma_{SL})$, and any reduction only in γ_{LA} as a result of the addition of a surfactant to the system (i.e., without any change in γ_{SL}) merely causes a corresponding increase in $\cos\theta$, with ΔP remaining unchanged. However, if θ is already 0°, then

$$\Delta P = \frac{2\gamma_{LA}}{r}, \tag{6.18}$$

and any reduction of γ_{LA} by the surfactant in the system *decreases* the tendency of the liquid to move into the capillaries of the substrate.

Another situation in which the addition of a surface-active agent to water *decreases* its wetting power is when adsorption of the surfactant at the substrate–liquid interface occurs in such fashion that the amphipathic surfactant molecules are oriented with their polar ends toward the substrate and their hydrophobic tails toward the water. Adsorption in this manner can occur with ionic or polar substrates when there is strong interaction between the hydrophilic groups in the surfactant and the ionic or polar sites on the substrate. Such adsorption makes the surface of the substrate more nonpolar. The resulting increase in the interfacial tension between the substrate and the aqueous solution γ_{SL}, results in a *decrease* in the spreading coefficient. Furthermore, since the surfactant molecules are strongly adsorbed at the substrate surface in this case, if the wetting liquid retracts from the substrate because of the decrease in spreading coefficient, it will expose a substate with a surface free energy that has been decreased by adsorption of the surfactant. Thus, portions of the substrate that have already been wet by the solution will become even more difficult to wet (more hydrophobic) than they were originally. Cationic surfactants are adsorbed in this manner onto negatively charged solid surfaces, such as quartz, cellulose textile fibers, or glass, and render them more difficult to wet with aqueous solutions than they were originally, and more easily wet by nonpolar materials. This phenomenon is the basis for ore "flotation" processes (Somasundaran, 1972).

B. Hard Surface (Equilibrium) Wetting

In hard surface wetting, the substrate to be wet is a nonporous, nongranular solid or a nonporous film, and since the area to be wet is relatively small, conditions close to equilibrium are usually attained, and the thermodynamics of the wetting process is a major factor determining the extent of wetting.

The effectiveness of a surfactant in modifying the wetting properties of a liquid for this process can be evaluated by determining the spreading coefficient $S_{L/S}$ of the surfactant solution on the substrate at a given temperature. This can be done by measuring the surface tension γ_{LA} of the surfactant solution and the contact angle the solution makes with the substrate. The less negative the spreading coefficient, the more effective the wetting agent.

When the spreading coefficient is negative, the contact angle is finite and spreading is not complete. Since $S_{L/S} = \gamma_{SA} - \gamma_{SL} - \gamma_{LA}$ (Equation 6.1) and $\gamma_{SA} - \gamma_{SL} = \gamma_{LA} \cos\theta$ (Equation 6.3) in this case, $S_{L/S} = \gamma_{LA} (\cos\theta - 1)$ (Equation 6.5). When the spreading coefficient is zero, spreading wetting is complete, since $S_{L/S} = \gamma_{LA} (\cos\theta - 1) = 0$, and thus $\theta = 0°$. In this latter case, since $S_{L/S} = \gamma_{SA} - \gamma_{SL} - \gamma_{LA} = 0$, then $\gamma_{LA} = \gamma_{SA} - \gamma_{SL}$, meaning that the surface tension of the spreading liquid must be reduced to a value equal to $\gamma_{SA} - \gamma_{SL}$ for spreading to be complete.

On nonpolar (low-energy) surfaces, Zisman (1964) has shown that complete wetting can occur only when the surface tension of the wetting liquid has been reduced to a critical value γ_c characteristic of the substrate (e.g., about 31 mN/m for polyethylene and about 18 mN/m for Teflon), thus equating γ_c with $\gamma_{SA} - \gamma_{SL}$. γ_c is obtained by measuring the contact angles that liquids of different surface tension produce on a particular substrate. For many liquids, a plot of $\cos\theta$ versus γ_{LA} is linear. Extrapolation of this linear plot to $\cos\theta = 1$, that is, $\theta = 0°$, yields γ_c. A critical surface tension for complete wetting γ_c, which is constant for a particular substrate irrespective of the nature of the wetting agent, requires that $\gamma_{SA} - \gamma_{SL}$ be reduced to the same value in all cases. This is possible only on low-energy surfaces, where the following two necessary conditions may be obtained: (1) γ_{SA}, the free energy per unit area of the surface in equilibrium with the liquid-saturated air above it, may be equated with γ_s, the free energy per unit area of the surface free of adsorbed material (only on low-energy surfaces is it likely that γ_s will not be reduced by the adsorption of surfactant or solvent molecules), and (2) adsorption of the surfactant at the substrate–solution interface occurs with the same orientation and degree of packing, producing the same value of γ_{SL} (and thus a constant value for $\gamma_{SA} - \gamma_{SL}$), as the surface tension of the spreading liquid approaches γ_c.

Thus, for example, in the presence of some highly fluorinated carboxylic acids and their salts, the value γ_c for polyethylene is decreased from its usual value of almost 31 mN/m to about 20 mN/m (Bernett and Zisman, 1959) by adsorption of the fluorinated surfactants onto the polyethylene surface, with the result that solutions of these surfactants having surface tensions less than the normal γ_c for polyethylene do not spread on it. The requirement that the surface tension of the wetting liquid be reduced by the surfactant to some critical value characteristic of the substrate is thus a necessary, but not sufficient, condition for complete spreading wetting. A surfactant solution whose surface tension is above the critical tension for the substrate does not produce complete wetting, but a solution whose surface tension is below the critical tension for the substrate may or may not produce complete wetting (Schwarz and Reid, 1964).

On nonpolar surfaces, any structural or other factor that results in a decrease in γ_{LA} (Chapter 5) decreases the contact angle and improves wetting. Thus, the addition of a water structure-breaking additive (*N*-methylacetamide), which increases the surface tension of an aqueous solution of sodium dodecyl sulfate, causes an increase in its contact angle on polyethylene, whereas the addition of a water structure promoter (fructose, xylose), which decreases the surface tension of the surfactant solution, reduces its contact angle (Schwuger, 1971).

On ionic solid surfaces, if an aqueous wetting liquid contains surfactant ions of charge opposite to that of the surface, they will generally be adsorbed on the surface with their ionic hydrophilic heads oriented toward the solid and their hydrophobic groups oriented toward the aqueous phase (Chapter 2, Section IIC). Increase in the concentration of surfactant in the aqueous phase will then result in decreasing γ_{SA} and/or increasing γ_{SL}, thus decreasing wetting of the solid surface (decrease in $\cos\theta$), in spite of decreasing values for γ_{LA}, until the charge on the solid surface has been neutralized by adsorption of oppositely charged surfactant ions. Once the surface charge has been neutralized, further adsorption of surfactant ions will generally occur, with their hydrophobic groups oriented toward the surface and their hydrophilic heads toward the aqueous phase. This will result in γ_{SL} being reduced, and wetting will improve as the bulk concentration of the surfactant increases.

Ionic solid surfaces in contact with aqueous solutions containing surfactant ions of charge similar to that of the surface generally show only small adsorption of surfactant ions onto the similarly charged solid surface. As a result, γ_{SL} can be expected to show little change with change in the bulk concentration of surfactant, and any improvement in wetting with increase in the latter is due mainly to the decrease in the value of γ_{LA}.

A simple experimental method of evaluating the performance of a surfactant as a wetting agent for a liquid on a hard surface consists of measuring the area to which a drop of the surfactant-containing liquid spreads in a given amount of time and then comparing it with the area to which the pure liquid spreads in the same time. For example, a film of the surface to be wetted may be placed on a horizontal planar glass plate (10 cm × 10 cm) which has four small (1 cm^2) pieces of glass at each corner. Using a microsyringe, a 20-μL drop of the surfactant solution is placed on the film. The time is measured and another 10 cm × 10 cm glass square is immediately placed over the four pieces of glass at the corners, so that it is parallel to the film. After a given amount of time (e.g., 3 min), an outline of the spread solution is traced on the top glass. This area is then retraced on standard white paper from which it is cut and weighed. With the assumption that the paper has a constant mass per unit area, the exact spreading area is then calculated from the mass of a piece of the same paper of known area (Wu and Rosen, 2002). The ratio of the area spread by the surfactant solution to the area spread by the same volume of pure liquid under the same conditions is called the *spreading factor* (SF). Some SF values are listed in Table 6.3 (Section IV).

C. Textile (Nonequilibrium) Wetting

Textiles have large surface areas, and consequently equilibrium conditions are seldom attained in the times allowed for wetting in practical processes. As a result, the rate of wetting of the surface is generally a more important factor than wetting equilibria in determining the suitability of a surfactant as a wetting agent for a particular system, and evaluation of the surfactant is generally by some kinetic test. The performance of the surfactant can be evaluated by determining (1) the efficiency of the wetting agent, that is, the minimum concentration of the surfactant that will produce a given amount of wetting in a certain time at a specific temperature, (2) the effectiveness of the wetting agent, that is, the minimum wetting time that can be achieved by the surfactant in a given system, regardless of the surfactant concentration used, or (3) the wetting time at a fixed concentration of surfactant in a given system at a specific temperature. The performance rating of surfactants relative to each other may vary with the method of evaluation and with the temperature at which the evaluation is done, and it is therefore necessary to specify these conditions. Most frequently, the third method given above is used to evaluate performance, since it requires only one determination, generally by use of 0.1% solutions of surfactant at 25°C.

The most commonly used practical test for textile wetting ability is probably the Draves test (Draves, 1939), in which a 5-g skein of gray, naturally waxed cotton yarn (54-in. loops containing 120 threads) is attached to a 3-g hook and totally immersed in a tall cylinder of surfactant solution by means of a weight tied to the hook by a thread. The surfactant solution displaces the air in the skein by a spreading wetting process, and when sufficient air has been displaced, the skein suddenly sinks in the cylinder. The better the wetting agent, the shorter the time required for sinking to occur. This test is widely used, since it approximates important use conditions for wetting agents in the textile industry.

The physicochemical basis for this test was investigated by Fowkes (1953), who explained the well-known observation that the log of the wetting time (WOT) is a linear function of the log of the bulk phase concentration C_1 of the surfactant when the latter is used at concentrations below its CMC. Fowkes stated that the penetration of the surfactant into the cotton proceeds at a rate that is a function of the contact angle 0 of the advancing liquid front such that

$$\log \mathrm{WOT} = A - B\cos\theta, \tag{6.19}$$

where A and B are empirical constants. He also showed that for this system $\cos\theta$ is a linear function of γ, the surface tension at the wetting front, and that for surfactants used at concentrations below their CMCs where their surface excess concentration Γ_1 is a constant, γ is a linear function of log C_1; that is, the Gibbs adsorption equation in the form

$$d\gamma = -2.303RT\Gamma_1 d\log C_1 \tag{2.19}$$

can be integrated. Therefore,

$$\log \mathrm{WOT} = A^1 - B^1 \log C_1, \tag{6.20}$$

where A^1 and B^1 are empirical constants.

When adsorption of the surfactant onto the substrate is strong, however, Fowkes found that the rate of wetting was determined not by the bulk phase concentration of the surfactant, but by the rate of diffusion of the surfactant to the wetting front. In this case the concentration of surfactant present at the advancing liquid front was so depleted by adsorption that the surface tension (or contact angle) there, and consequently the wetting time, were determined solely by the rate at which new surfactant arrived at the front.

For a nonionic surfactant that is used at a concentration considerably above its CMC under diffusion-controlled conditions, the Fowkes (1953) equation for the factors determining WOT can be transformed (Cohen and Rosen, 1981) to

$$\log \mathrm{WOT} = K - 2\log(C - C^1) - \log D, \tag{6.21}$$

where K depends upon the physical characteristics of the skein being wetted and upon the surface area per gram of adsorbed surfactant at the air–aqueous solution interface S^1; C is the initial surfactant concentration, in g/L, in the aqueous phase; C^1 is the surfactant concentration, in g/L, at the skein–solution interface required to produce a given WOT; and D is the apparent diffusion coefficient of the surfactant.

For the wetting of similar skeins by a particular surfactant at $C >> C^1$, a log–log plot of WOT versus C should be linear with a slope of −2 if the diffusion coefficient D remains constant with change in C. Diffusion data available (Fowkes, 1953) for solutions of individual nonionics considerably below their cloud points but above their CMC indicate that D increases by a factor of about 2 in the concentration range between 0.25 and 1.0 g/L. This should yield a slope of −1.5 for a log–log plot of WOT versus C under these conditions. Wetting data for a series of well-purified individual nonionic surfactants of structure $C_{12}H_{25}(OC_2H_4)_xOH$, where x = 4, 5, 6, 7, or 8, yielded linear log–log plots with slopes close to that value in that initial surfactant concentration range, indicating the $C >> C^1$ in that range and that wetting is diffusion controlled for these compounds under these conditions.

For various nonionic surfactants at concentrations considerably above their CMC, the WOT at a given temperature depends also on their diffusion coefficients D and on their surface areas per gram of adsorbed surfactant S^1 at the air–aqueous solution interface. Thus, from Equation 6.21

$$\frac{\mathrm{WOT}_1}{\mathrm{WOT}_2} = \frac{D_2}{D_1}\frac{C_2 - C_2^1}{C - C_1^1}\left(\frac{MW_1}{a_1^S}\right)^2\left(\frac{a_2^S}{MW_2}\right)^2 \tag{6.22}$$

since $S^1 = Na^s/MW$, where a^s is the molecular area of the surfactant of molecular weight (MW) at the air–aqueous solution interface at that temperature and N = Avogadro's number.

When nonionic surfactants are compared at similar temperatures and similar initial surfactant concentrations in the aqueous phase such that $C >> C^1$, then

$$\frac{\mathrm{WOT}_1}{\mathrm{WOT}_2} = \frac{D_2}{D_1}\left(\frac{MW_1}{a_1^S}\right)^2\left(\frac{a_2^S}{MW_2}\right)^2. \tag{6.23}$$

Equation 6.23 indicates that, under these conditions, the shortest wetting times will be shown by those surfactants with the largest diffusion coefficients, largest surface areas per molecule at the air–aqueous solution interface, and smallest MWs. This is consistent with the finding (Chapter 5, Section IV) that surfactant molecules with larger areas per molecule at the air–aqueous solution interface and larger diffusion coefficients show shorter induction times in dynamic surface tension reduction, that is, show lower surface tensions at short times.

This may explain why short-chain surfactants with highly branched hydrophobic groups show such good wetting properties (see below). Since values of a^s are readily obtained (Chapter 2, Section IIIB), Equation 6.23 can also be used to obtain relative diffusion coeffficients from wetting times (Cohen and Rosen, 1981).

Since the rate of wetting is a function of the surface tension of the wetting front, the wetting power of a surfactant in wetting tests is a function of the concentration of molecularly dispersed material at the front, and those structural factors in the surfactant molecule that inhibit micelle formation and that increase the rate of diffusion of the monomeric surfactant to the interface should enhance its performance as a wetting agent. Some evidence that diffusion of surfactants is decreased by increase in the length of the (straight) alkyl chain and increased by branching in isomer alkyl chains has been supplied by Schwuger (1982). Data by Longsworth (1953) on the diffusion coefficients of amino acids indicate that the diffusion coefficient in aqueous solution is decreased by increase in the degree of hydration of the molecule and by increase in the chain length of the alkyl group. Branching of the alkyl group gave a more compact structure and a greater diffusion coefficient compared to the isomeric straight-chain compound. This is consistent with data showing that the fastest textile wetting is given by surfactants having relatively short, highly branched hydrophobic groups and that the highly hydrated polyoxyethylene (POE) nonionics are not as rapid textile wetting agents as the less hydrated anionics.

A fairly good correlation has been observed between the wetting time on cotton skeins, measured by the Draves test, and the dynamic surface tension (Chapter 5, Section IV) at 1 s. To achieve a 1-s dynamic surface tension value, γ_{1s}, that does not change much with decrease in surfactant concentration, a bulk phase surfactant concentration of at least 5×10^{-4} M is required (Rosen and Hua, 1990).

Wetting times for some surfactants, using the Draves test at 25°C, with a 3-g hook and a 5-g cotton skein, are listed in Table 6.1.

In water at 25°C containing no more than 300 ppm $CaCO_3$, and at a surfactant concentration of 0.1%, ionic surfactants having a terminal hydrophilic group seem to show optimum wetting when the hydrophobic group has an effective length of about 12–14 carbon atoms. When the hydrophilic group is centrally located, optimum wetting appears to be shown when the hydrophobic group has an effective length of almost 15 carbon atoms. The CMCs of ionic surfactants of these types generally fall in the range $1\text{–}8 \times 10^{-3}$ M. At very low concentrations, as in the case of surface tension, longer-chain compounds often perform better than shorter ones (Komor and Beiswanger, 1966), presumably because of their greater efficiency in reducing surface tension. However, at higher concentrations, the shorter-chain compounds

TABLE 6.1 Wetting Times of Surfactants in Aqueous Solution (Draves Test, 25°C)

Compound	Concentration (%)	Wetting Times (s)		Reference
		Distilled Water	300 ppm $CaCO_3$	
$C_{10}H_{21}SO_3^-Na^+$	0.1	>300		Dahanayake (1985)
$C_{10}H_{21}SO_3^-Na^+$	0.1	65		Dahanayake (1985)
$C_{12}H_{25}SO_3^-Na^+$	0.1 (in 0.1 *M* NaCl)	28		Dahanayake (1985)
n-$C_{12}H_{25}SO_4^-Na^+$	0.02	>300		Weil et al. (1960)
n-$C_{12}H_{25}SO_4^-Na^+$	0.05	39.9		Weil et al. (1960)
$C_{12}H_{25}SO_4^-Na^+$	0.1	7.5		Weil et al. (1960); Dahanayake (1985)
n-$C_{14}H_{29}SO_4^-Na^+$	0.10	12[a]		Weil et al. (1954)
n-$C_{16}H_{33}SO_4^-Na^+$	0.10	59[a]		Weil et al. (1954)
n-$C_{18}H_{37}SO_4^-Na^+$	0.10	280[a]		Weil et al. (1954)
Na oleyl sulfate	0.10	19[a]		Weil et al. (1954)
Na elaidyl sulfate	0.10	20[a]		Weil et al. (1954)
sec-n-$C_{13}H_{27}SO_4^-Na^+$	0.063	180+		Livingston et al. (1965)
(random-$SO_4^-Na^+$ positions)	0.125	11.6		Livingston et al. (1965)
sec-n-$C_{14}H_{29}SO_4^-Na^+$	0.063	180+		Livingston et al. (1965)
(random- $SO_4^-Na^+$ positions)	0.125	7.0		Livingston et al. (1965)
sec-n-$C_{15}H_{31}SO_4^-Na^+$	0.063	14.0		Livingston et al. (1965)
(random-$SO_4^-Na^+$ positions)	0.125	7.0		Livingston et al. (1965)
sec-n-$C_{16}H_{33}SO_4^-Na^+$	0.063	22		Livingston et al. (1965)
(random-$SO_4^-Na^+$ positions)	0.125	9		Livingston et al. (1965)
sec-n-$C_{17}H_{35}SO_4^-Na^+$	0.063	25		Livingston et al. (1965)
(random-$SO_4^-Na^+$ positions)	0.125	9		Livingston et al. (1965)
sec-n-$C_{18}H_{37}SO_4^-Na^+$	0.063	39		Livingston et al. (1965)
(random-$SO_4^-Na^+$ positions)	0.125	26		Livingston et al. (1965)
$C_{12}H_{25}OC_2H_4SO_4^-Na^+$	0.1	6		Dahanayake (1985)
$C_{12}H_{25}OC_2H_4SO_4^-Na^+$	0.088 (in 0.1 *M* NaCl)	6		Dahanayake (1985)
$C_{12}H_{25}(OC_2H_4)_2SO_4^-Na^+$	0.1	11		Dahanayake (1985)
$C_{12}H_{25}(OC_2H_4)_2SO_4^-Na^+$	0.1 (in 0.1 *M* NaCl)	13		Dahanayake (1985)

$C_{12}H_{25}C_6H_4(OC_2H_4)_2SO_3^-Na^+$ (from polypropylene tetramer)	0.125	6.9		Livingston et al. (1965)
n-$C_{10}H_{21}CH(CH_3)C_6H_4SO_3^-Na^+$	0.10	10.3	80	Smith et al. (1966)
n-$C_{12}H_{25}CH(CH_3)C_6H_4SO_3^-Na^+$	0.10	30	>300	Smith et al. (1966)
n-$C_{14}H_{29}CH(CH_3)C_6H_4SO_3^-Na^+$	0.10	155	>300	Smith et al. (1966)
$C_{10}DADS^b$	0.1	431		Rosen and Zhu (1993)
$C_{12}H_{25}(C_2H_4)_3OH$ + $C_{10}DADS$ mixture, 8:2 (w/w)	0.1	14.5		Rosen and Zhu (1993)
$C_{12}H_{25}OCH_2CH(SO_4^-Na^+)CH_3$	0.10	6		Weil et al. (1966)
$C_4H_9CH(C_2H_5)CH_2OOCCH_2CH(SO_3^-Na^+)COOCH_2CH(C_2H_5)C_4H_9$	0.025	20.1		Weil et al. (1960)
$C_4H_9CH(C_2H_5)CH_2OOCCH_2CH(SO_3^-Na^+)COOCH_2CH(C_2H_5)C_4H_9$	0.05	6.3		Weil et al. (1960)
$C_4H_9CH(C_2H_5)CH_2OOCCH_2CH(SO_3^-Na^+)COOCH_2CH(C_2H_5)C_4H_9$	0.10	1.9		Weil et al. (1960)
$C_7H_{15}CH(SO_3^-Na^+)COOC_5H_{11}$	0.10	12.1	5.3	Stirton et al. (1962b)
$C_8H_{17}CH(SO_3^-Na^+)COOC_8H_{17}$	10 mM	<2[c]		Ohbu et al. (1998)
$C_{10}H_{21}CH(SO_3^-Na^+)COOC_{10}H_{21}$	10 mM	<2[c]		Ohbu et al. (1998)
$C_{12}H_{25}CH(SO_3^-Na^+)COOCH_3$	10 mM	34		Ohbu et al. (1998)
$C_{12}H_{25}CH(SO_3^-Na^+)COOC_3H_7$	0.10	5.0	3.8	Stirton et al. (1962b)
$C_{12}H_{25}CH(SO_3^-Na^+)COOCH_3$	0.10	25	16	Stirton et al. (1962b)
$C_8H_{17}C(C_4H_9)(SO_3^-Na^+)COOCH_3$	0.10	13.3	5.2	Stirton et al. (1962b)
$C_8H_{17}C(C_6H_{13})(SO_3^-Na^+)COOCH_3$	0.10	1.3	3.7	Stirton et al. (1962b)
$C_8H_{17}C(C_8H_{17})(SO_3^-Na^+)COOCH_3$	0.10	2.8	3.8	Stirton et al. (1962b)
$C_7H_{15}CH(SO_3^-Na^+)COOC_6H_{13}$	0.10	2.2	1.4	Stirton et al. (1962b)
$C_7H_{15}CH(SO_3^-Na^+)COOC_7H_{15}$	0.10	0.0	3.0	Stirton et al. (1962b)
$C_7H_{15}CH(SO_3^-Na^+)COOC_8H_{17}$	0.025	15.4		Weil et al. (1960)
$C_7H_{15}CH(SO_3^-Na^+)COOC_8H_{17}$	0.05	5.0		Weil et al. (1960)
$C_7H_{15}CH(SO_3^-Na^+)COOC_8H_{17}$	0.10	1.5	10.8	Stirton et al. (1962b)
$C_7H_{15}CH(SO_3^-Na^+)COOCH(CH_3)COOC_6H_{13}$	0.10	1.3	4.5	Stirton et al. (1962a)

(*Continued*)

TABLE 6.1 (*Continued*)

Compound	Concentration (%)	Wetting Times (s)		Reference
		Distilled Water	300 ppm $CaCO_3$	
$C_7H_{15}CH(SO_3^-Na^+)COOCH_2CH(C_2H_5)C_4H_9$	0.10	0.0	4.5	Stirton et al. (1962a)
$C_7H_{15}CH(SO_3^-Na^+)COOC_9H_{19}$	0.10	3.8	33.1	Stirton et al. (1962a)
$C_2H_5CH(SO_3^-Na^+)COOC_{12}H_{25}$	0.10	5.5	4.4	Stirton et al. (1962a)
$C_{10}H_{21}CH(SO_3^-Na^+)COOC_4H_9$	0.10	5.5	4.4	Stirton et al. (1962b)
$C_{10}H_{21}CH(SO_3^-Na^+)COOC_5H_{11}$	0.10	1.6	4.9	Stirton et al. (1962b)
$C_{12}H_{25}Pyr^+Cl$	0.1	250		Dahanayake (1985)
$C_{12}H_{25}N^+(CH_3)(CH_2C_6H_5)CH_2COO^-$	0.1	250		Dahanayake (1985)
N-dodecyl-2-pyrrolid(in)one	0.1	131		Rosen and Zhu (1993)
$C_8H_{17}CHOHCH_2OH$	0.047	7		Rosen and Kwan (1979)
$C_8H_{17}CHOHCH_2CH_2OH$	0.041	8		Rosen and Kwan (1979)
$C_8H_{17}(OC_2H_4)_2OH$	0.1	5		Weil et al. (1979)
$C_8H_{17}(OC_2H_4)_3OH$	0.1	22		Weil et al. (1979)
$C_{10}H_{21}(OC_2H_4)_2OH$	0.1	10		Weil et al. (1979)
$C_{10}H_{21}(OC_2H_4)_3OH$	0.1	4		Weil et al. (1979)
$C_{10}H_{21}(OC_2H_4)_4OH$	0.1	5		Weil et al. (1979)
$C_{12}H_{25}(OC_2H_4)_3OH$	0.1	129		Rosen and Zhu (1993)
$C_{12}H_{25}(OC_2H_4)_4OH$	0.1	$4._8, 6._0$(10°C) $13._5$(40°C)		Cohen and Rosen (1981)
$C_{12}H_{25}(OC_2H_4)_{5.1}OH^c$	0.1	14		Cohen and Rosen (1981)
$C_{12}H_{25}(OC_2H_4)_6OH$	0.1	$3._9, 5._5$(10°C) $4._1$(40°C) $9._5$(60°C)		Cohen and Rosen (1981)
$C_{12}H_{25}(OC_2H_4)_7OH$	0.1	$5._9, 10._5$(10°C) $4._0$(40°C) $3._8$(60°C)		Cohen and Rosen (1981)
$C_{12}H_{25}(OC_2H_4)_8OH$	0.1	$8._3$, 18(10°C) $6._3$(40°C) $4._0$(60°C)		Cohen and Rosen (1981)
$C_{12}H_{25}(OC_2H_4)_{9.6}OH^c$	0.1	11		Weil et al. (1979)
p-t-$C_8H_{17}C_6H_4(OC_2H_4)_4OH$ (normal EO distribution)	0.05	50		Crook et al. (1964)

p-t-$C_8H_{17}C_6H_4(OC_2H_4)_5OH$	0.05	25	Crook et al. (1964)
p-t-$C_8H_{17}C_6H_4(OC_2H_4)_8OH$	0.05	~25	Crook et al. (1964)
p-t-$C_8H_{17}C_6H_4(OC_2H_4)_9OH$	0.05	25	Crook et al. (1964)
p-t-$C_8H_{17}C_6H_4(OC_2H_4)_{10}OH$	0.05	30	Crook et al. (1964)
p-t-$C_8H_{17}C_6H_4(OC_2H_4)_{12}OH$	0.05	50	Crook et al. (1964)
Igepal CO-630	0.05	27	Komor and Beiswanger (1966)
($C_9H_{19}C_6H_4(OC_2H_4)_9OH$)	0.10	12	Komor and Beiswanger (1966)
Igepal CO-710	0.05	33	Komor and Beiswanger (1966)
($C_9H_{19}C_6H_4(OC_2H_4)_{10\text{-}11}OH$)	0.10	15	Komor and Beiswanger (1966)
Igepal CO-730	0.05	>50	Komor and Beiswanger (1966)
($C_9H_{19}C_6H_4(OC_2H_4)_{15}OH$)	0.10	>50	Komor and Beiswanger (1966)
($C_9H_{19}C_6H_4(OC_2H_4)_{15}OH$)	0.05 (70^{0}C)	37	Komor and Beiswanger (1966)
($C_9H_{19}C_6H_4(OC_2H_4)_{15}OH$)	0.10 (70°C)	17	Komor and Beiswanger (1966)
$C_7H_{15}CO(OC_2H_4)_2OH$	0.1	>300	Weil et al. (1979
$C_7H_{15}CO(OC_2H_4)_4OH$	0.1	>300	Weil et al. (1979)
$C_8H_{17}CO(OC_2H_4)_2OH$	0.1	72	Weil et al. (1979)
$C_8H_{17}CO(OC_2H_4)_3OH$	0.1	6	Weil et al. (1979)
$C_8H_{17}CO(OC_2H_4)_4OH$	0.1	48	Weil et al. (1979)
$C_9H_{19}CO(OC_2H_4)_2OH$	0.1	41	Weil et al. (1979)
$C_9H_{19}CO(OC_2H_4)_3OH$	0.1	7	Weil et al. (1979)
$C_9H_{19}CO(OC_2H_4)_2OH$	0.1	11	Weil et al. (1979)
$C_9H_{19}CO(OC_2H_4)_5OH$	0.1	12	Weil et al. (1979)
$C_{11}H_{23}CO(OC_2H_4)_4OH$	0.1	23	Weil et al. (1979)
$C_{11}H_{23}CO(OC_2H_4)_5OH$	0.1	7	Weil et al. (1979)
$C_{11}H_{23}CO(OC_2H_4)_6OH$	0.1	34	Weil et al. (1979)
$C_{13}H_{27}CO(OC_2H_4)_5OH$	0.1	52	Weil et al. (1979)
$C_{13}H_{27}CO(OC_2H_4)_6OH$	0.1	21	Weil et al. (1979)
$C_{10}H_{19}(\Delta 10-11)CO(OC_2H_4)_3OH$	0.1	9	Weil et al. (1979)
$C_{10}H_{19}(\Delta 10-11)CO(OC_2H_4)_4OH$	0.1	6	Weil et al. (1979)
$C_{10}H_{19}(\Delta 10-11)CO(OC_2H_4)_5OH$	0.1	7	Weil et al. (1979)

(*Continued*)

TABLE 6.1 (*Continued*)

Compound	Concentration (%)	Wetting Times (s)		Reference
		Distilled Water	300 ppm $CaCO_3$	
$C_7H_{15}CO(OC_2H_4)_4OCH_3$	0.1	>300		Weil et al. (1979)
$C_7H_{15}CO(OC_2H_4)_5OCH_3$	0.1	>300		Weil et al. (1979)
$C_8H_{17}CO(OC_2H_4)_4OCH_3$	0.1	248		Weil et al. (1979)
$C_8H_{17}CO(OC_2H_4)_5OCH_3$	0.1	12		Weil et al. (1979)
$C_8H_{17}CO(OC_2H_4)_6OCH_3$	0.1	23		Weil et al., (1979)
$C_9H_{19}CO(OC_2H_4)_4OCH_3$	0.1	24		Weil et al. (1979)
$C_9H_{19}CO(OC_2H_4)_5OCH_3$	0.1	9		Weil et al. (1979)
$C_9H_{19}CO(OC_2H_4)_7OCH_3$	0.1	9		Weil et al. (1979)
$C_{11}H_{23}CO(OC_2H_4)_5OCH_3$	0.1	12		Weil et al. (1979)
$C_{11}H_{23}CO(OC_2H_4)_6OCH_3$	0.1	9		Weil et al. (1979)
$C_{11}H_{23}CO(OC_2H_4)_7OCH_3$	0.1	9		Weil et al. (1979)
$C_{13}H_{27}CO(OC_2H_4)_5OCH_3$	0.1	64		Weil et al. (1979)
$C_{13}H_{27}CO(OC_2H_4)_6OCH_3$	0.1	17		Weil et al. (1979)
$C_{13}H_{27}CO(OC_2H_4)_7OCH_3$	0.1	17		Weil et al. (1979)
Dimethylhexadecynediol +15 mol EO	0.05	11		Leeds et al. (1965)
	0.1	9		Leeds et al. (1965)
Tetramethyldecynediol +5 mol EO	0.1	24		Leeds et al. (1965)

[a] $1\frac{1}{4}$ in. binding tape, 1-g hook, and 40-g anchor (Shapiro, L. [1950] *Am. Dyestuff Reptr.* **39**, 38–45, 62).
[b] $(C_{10}H_{21})_2C_6H_2(SO_3^-Na^+)OC_6H_4SO_3^-Na^+$ ·
[c] 10 × 10 × 2 mm wool felt strip.
[d] Mixture with average EO number shown.
Pyr^+, pyridinium.

appear to become more effective and reach lower minimum wetting times than the long-chain ones. Thus, for aqueous solutions of sodium alkyl sulfates at 0.10% concentration, the order of increasing wetting time is tetradecyl < dodecyl < oleyl << hexadecyl << octadecyl, but at 0.15% concentration, the order is dodecyl < tetradecyl < oleyl, and this is the order of minimum wetting times attained at any concentration (Stirton et al., 1952; Weil et al., 1954).

As the temperature of the water is raised, the chain length for optimum wetting by ionic surfactants generally increases, probably because of the increased solubility of the surfactant at higher temperatures and its consequent lesser tendency to migrate to the interface. Thus, at 60°C, $C_{16}H_{33}SO_4^-Na^+$ shows better wetting than $C_{12}H_{25}SO_4^-Na^+$ (Weil et al., 1959) and in *p*-(*n*-alkyl) benzenesulfonates, the *n*-dodecyl compound (equivalent length = 15.5 carbons) gives the most rapid wetting (Greiss, 1955).

For determining the effective length of the hydrophobic groups, a carbon atom on a branch attached to the main chain appears to be equivalent to about two-thirds of a carbon atom on the main chain, one between an ionic hydrophilic group and a polar group in the molecule to about one-half of a carbon atom on the main chain, and a phenyl group to three and one-half carbon atoms in a straight chain. The ester linkage –COO– appears not to contribute to the length of the hydrophobic group.

Surfactants with a centrally located hydrophilic group are especially good textile wetting agents, particularly when they have branched hydrophobic groups, presumably because of their larger area/molecule at the air–aqueous solution interface and their more rapid diffusion to and orientation at the wetting front. For probably similar reasons, *o*-sulfonated alkylbenzenes and *N*-acylanilides are better wetting agents than the corresponding *p*-sulfonates (Shirolkar and Venkataraman, 1941; Gray et al., 1965). In compounds such as $RCH(R')SO_4^-M^+$ (Dreger et al., 1944; Püschel, 1966; Götte, 1969), $RCH(R')CH_2SO_4^-Na^+$ (Machemer, 1959), $RCH(R')C_6H_4SO_3^-Na^+$ (Baumgartner, 1954), and $RCH(SO_3^-M^+)COOR'$ (Weil et al., 1960; Stirton et al., 1962a, 1962b), wetting appears to improve as R and R′ approach each other in equivalent length. The excellent wetting properties of sulfated castor oil are attributed to the presence in the product of sulfated glyceryl ricinoleates with centrally located sulfate groups. Other structures that produce good wetting agents are $RC(R')(OH)C{\equiv}C(OH)(R')R$ and $ROOCH_2CH(SO_3^-Na^+)COOR$.

The introduction of a second ionic hydrophilic group into the molecule is generally unfavorable to wetting power (Götte and Schwuger, 1969). Thus, α-sulfocarboxylic acids and monoalkylsulfosuccinic acid esters are better wetting agents at acid pHs than at alkaline pHs; in sulfated ricinoleates at alkaline pHs, those in which the carboxylic acid group is esterified show better wetting properties than those in which it is not. Similarly, the introduction of oxyethylene (OE) groups between the hydrophilic and hydrophobic groups in the surfactant molecule in compounds of structure $R(OC_2H_4)_xSO_4^-Na^+$ where R is $C_{16}H_{33}$ or $C_{18}H_{37}$ and $x = 1$–4, is unfavorable to wetting power, with the wetting time increasing with the introduction of each additional OE group (Weil et al., 1959).

A study of the wetting power of well-purified individual POE nonionics (Cohen and Rosen, 1981) showed that individual surfactants of this type, $C_{12}H_{25}(OC_2H_4)_xOH$, are better wetting agents than a Poisson distribution mixture with the same average number of OE units. Temperature increase caused an increase in the wetting power of individual POE nonionics until the cloud point of the compound was approached. When the cloud point of the surfactant was exceeded, the wetting power markedly diminished. Best wetting was obtained when the wetting temperature was 10–30°C below the cloud point of the compound.

For commercial nonionic POE alcohols, alkylphenols, and mercaptans, wetting times go through a minimum with increase in the number of OE units in the POE chain, and optimum wetting power is generally shown by those surfactants whose cloud points are just above the temperature at which the wetting test is conducted (Komor and Beiswanger, 1966). In distilled water at 25°C, materials having an effective hydrophobic chain length of 10–11 carbon atoms and a POE chain of 6–8 OE units appear be the best wetting agents (Crook et al., 1964). POE alcohols and mercaptans appear to be better wetting agents than the corresponding POE fatty acids (Wrigley et al., 1957).

In POE polyoxypropylene block copolymers, wetting power appears to improve as the MW of the polyoxypropylene portion of the molecule is increased and to be at a maximum when the molecule contains the minimum OE content consistent with solubility in the aqueous phase at the use temperature.

In a study of POE straight-chain amines (Ikeda et al., 1984), the best wetting among isomeric materials was shown by compounds with two OE groups of approximately equal OE content attached to the nitrogen. Here, again, a more compact branched structure produced better wetting.

The importance of the absence of undesirable reaction products and impurities for good wetting power by aqueous surfactant solutions has been pointed out in a series of articles by Micich and Linfield (1984, 1985, 1986) on wetting agents for hydrophobic soils. Water-insoluble unreacted starting material and other reaction products markedly increased wetting times for cotton skeins (Draves test) and drop penetration times on hydrophobic soil. In a series of ethenoxylated secondary amides, $RCON(R^1)(EO)_XH$, best results were obtained when both hydrophobic groups (R, R^1) were branched and each contained 7–8 carbon atoms. This structure, with the hydrophilic group in the center of the molecule and an effective hydrophobic chain length of 12–14 carbon atoms, is typical of excellent wetting agents. When the compounds were synthesized in such manner as to minimize unreacted water-insoluble starting material and reaction products, almost instantaneous wetting and rewetting of both skeins and hydrophobic soil was obtained with compounds of this type whose cloud points were in the neighborhood of the wetting temperature.

TABLE 6.2 Wetting Efficiencies of Surfactants in Aqueous Solution (Concentrations for 25 s Sinking Time at 25°C in Draves Test)

Surfactant	Concentration (%)	Reference
$t\text{-}C_8H_{17}S(C_2H_4O)_2H$	0.098	Olin (1951)
$t\text{-}C_8H_{17}S(C_2H_4O)_{2.93}H$	0.084	Olin (1951)
$t\text{-}C_8H_{17}S(C_2H_4O)_{3.92}H$	0.102	Olin (1951)
$t\text{-}C_8H_{17}S(C_2H_4O)_{6.86}H$	0.175	Olin (1951)
$t\text{-}C_{12}H_{25}S(C_2H_4O)_{7.85}H$	0.074	Olin (1951)
$t\text{-}C_{12}H_{25}S(C_2H_4O)_{8.97}H$	0.051	Olin (1951)
$t\text{-}C_{12}H_{25}S(C_2H_4O)_{10.02}H$	0.046	Olin (1951)
$t\text{-}C_{12}H_{25}S(C_2H_4O)_{11.03}H$	0.047	Olin (1951)
$t\text{-}C_{12}H_{25}S(C_2H_4O)_{12.25}H$	0.052	Olin (1951)
$t\text{-}C_{14}H_{29}S(C_2H_4O)_{7.98}H$	0.132	Olin (1951)
$t\text{-}C_{14}H_{29}S(C_2H_4O)_{9.00}H$	0.135	Olin (1951)
$t\text{-}C_{14}H_{29}S(C_2H_4O)_{10.98}H$	0.113	Olin (1951)
$t\text{-}C_{14}H_{29}S(C_2H_4O)_{12.11}H$	0.108	Olin (1951)
$t\text{-}C_{14}H_{29}S(C_2H_4O)_{13.13}H$	0.135	Olin (1951)
$n\text{-}C_{11}H_{23}O(C_2H_4O)_8H$	0.035	Komor and Beiswanger (1966)
$n\text{-}C_{12}H_{25}O(C_2H_4O)_{10}H$	0.046	Komor and Beiswanger (1966)
$p\text{-}n\text{-}C_{10}H_{21}C_6H_4O(C_2H_4O)_{11}H$	0.054	Komor and Beiswanger (1966)
iso-$C_8H_{17}O(C_2H_4O)_4H$	0.13	Komor and Beiswanger (1966)
oxo-$C_{10}H_{21}O(C_2H_4O)_{10}H$	0.095	Komor and Beiswanger (1966)
Igepal CO-610 (nonylphenol + 8–9 mol EO)	0.05–0.06	GAF (1965)
Igepal CO-630 (nonylphenol + 9 mol EO)	0.03–0.05	GAF (1965)
Igepal CO-710 (nonylphenol + 10–11 mol EO)	0.04–0.07	GAF (1965)
Igepal CO-730 (nonylphenol + 15 mol EO)	0.14–0.16	GAF (1965)

Work on the wetting of hydrophobic sand by POE nonionics showed that both wetting rate and wetting effectiveness (measured by weight gain of the sand column) were greater for branched hydrophobic chain compounds than for linear ones (Varadaraj et al., 1994).

Wetting efficiencies of some nonionic surfactants are given in Table 6.2.

D. Effect of Additives

The electrolyte content of the aqueous phase has a considerable effect on the wetting time of ionic surfactants, reflecting its effect on the reduction of surface tension by the surfactant, its solubility in water, and its CMC. Electrolytes that decrease the surface tension of the surfactant solution (Chapter 5, Section III), such as Na_2SO_4, NaCl, and KCl, increase its wetting power (Gerault, 1964). When the aqueous phase contains added electrolyte or additional hardness, optimum wetting with anionic surfactants is generally obtained when the hydrophobic group is somewhat shorter than the optimum length for pure water. For solutions containing high concentrations of electrolyte, hydrophobic groups with lengths as short as seven to eight carbon atoms are effective.

The addition of long-chain alcohols to aqueous anionic and nonionic surfactant solutions is reported to increase their wetting power (Gerault, 1964; Bland and Winchester, 1968) and the addition of metal soaps, especially those of the alkaline earth metals, is stated to increase the wetting power of dodecylpyridinium chloride solutions (Suzuki, 1967).

III. SYNERGY IN WETTING BY MIXTURES OF SURFACTANTS

As will be described in Chapter 11, the interaction of two different types of surfactants with each other, either in mixed monolayers at an interface or in mixed micelles in aqueous solution, can result in synergistic enhancement of their interfacial properties. Such an enhancement can result in improved *performance properties*, such as wetting, foaming, solubilization, and so on.

In some cases, enhancement of (textile) wetting is the result of synergistic interaction producing a decrease in dynamic and equilibrium surface tension values (Section IIA) (Zhu et al., 1989). In other cases, enhancement of wetting is due to solubilization by a water-soluble surfactant of a water-insoluble but highly surface-active surfactant (Rosen and Zhu, 1993). Surfactants with limited solubility in water (less than 0.25 g/L) generally show poor textile wetting power, sometimes in spite of low equilibrium surface tension values. When some of these surfactants are mixed with a water-soluble surfactant that can interact with them to solubilize them into the aqueous phase, the wetting times of the mixture decrease, sometimes dramatically. Thus, short-chain alcohols or alkylphenol ethoxylates with just a few OE groups in the molecule, or *n*-dodecyl-2-pyrrolidinones, which show low solubility in water and poor textile wetting power, become excellent wetting agents when mixed with various anionic surfactants that can solubilize them into the water phase. The addition of a POE nonionic surfactant has been shown to increase the wetting power of some anionics and to diminish the wetting power of a cationic surfactant (Biswas and Mukherji, 1960). This is attributed to an increase by the nonionic in the mobility of the anionics and a decrease by it in the mobility of the cationic, resulting in more rapid diffusion of the former and slower diffusion by the latter to the wetting front.

IV. SUPERSPREADING (SUPERWETTING)

It is difficult to wet highly hydrophobic hard surfaces (e.g., with contact angles against water >100°) even with surfactant solutions. The areas to which the solutions spread (see method in Section IIB) is often just a small multiple of the area spread by pure water. Some data are shown in Table 6.3.

TABLE 6.3 Spreading Factors of Aqueous Solutions of Some Surfactants and Their Mixtures on Polyethylene Film at 25°C (0.1% conc., 3-min time)[a]

System	Medium	SF[b]	Reference
$C_{12}H_{25}SO_4^-Na^+$	H_2O	5	Rosen and Wu (2002)
$(CH_3)_2CHC_6H_4SO_3^-Na^+$	H_2O	5	Rosen and Wu (2002)
$C_4H_9OC_6H_4SO_3^-Na^+$	H_2O	5	Zhou et al. (2003)
Na diamylsulfosuccinate	H_2O	5	Rosen and Wu (2002)
Na di(2-ethylhexyl)sulfosuccinate	H_2O	8	Rosen and Wu (2002)
$C_{10}Pyr^+Br^-$	H_2O	5	Rosen and Wu (2002)
$C_{12}N^+Me_3Cl^-$	H_2O	3	Rosen and Wu (2002)
$C_{10}N^+(CH_3)_2O^-$	H_2O	3	Rosen and Wu (2002)
$C_2H_5C(CH_3)(OH)C{\equiv}CC(CH_3)(OH)\text{-}C_2H_5$[d]	H_2O	5	Rosen and Wu (2002)
$(CH_3)_2CHCH_2CH(CH_3)CH_2C\text{-}(OC_2H_4)_8CH_2CH(CH_3)_2$[d]	H_2O	65	Rosen and Wu (2002)
tert-$C_8H_{17}C_6H_4(OC_2H_4)_5OH$[d]	H_2O	15	Zhou et al. (2003)
tert-$C_8H_{17}C_6H_4(OC_2H_4)_5OH$[d]	0.1 *M* NaCl	10	Zhou et al. (2003)
N-hexylpyrrolid(in)one	Aqueous phosphate buffer, pH = 7.0	5	Wu and Rosen (2002)
N-(2-ethylhexyl)pyrrolid(in)one	Aqueous phosphate buffer, pH = 7.0	5	Wu and Rosen (2002)
N-octylpyrrolid(in)one	Aqueous phosphate buffer, pH = 7.0	5	Wu and Rosen (2002)
N-octylpyrrolid(in)one	H_2O	15	Zhou et al. (2003)
N-octylpyrrolid(in)one	0.1 *M* NaC1	10	Zhou et al. (2003)
N-dodecylpyrrolid(in)one	H_2O	45	Zhou et al. (2003)
N-dodecylpyrrolid(in)one	0.1 *M* NaC1	30	Zhou et al. (2003)
L 77[c]	Aqueous phosphate buffer, pH = 7.0	150	Wu and Rosen (2002)
L 77[c]-*N*-hexylpyrrolid(in)one	Aqueous phosphate buffer, pH = 7.0	210	Wu and Rosen (2002)
L 77[c]-*N*-(ethylhexyl)pyrrolid(in)one	Aqueous phosphate buffer, pH = 7.0	235	Wu and Rosen (2002)
L 77[c]-*N*-octylpyrrolid(in)one	Aqueous phosphate buffer, pH = 7.0	210	Wu and Rosen (2002)
n-$C_4H_9OC_6H_4SO_3^-Na^+$-*N*-dodecylpyrrolid(in)one	H_2O	80	Zhou et al. (2003)
n-$C_4H_9OC_6H_4SO_3^-Na^+$-*N*-dodecylpyrrolid(in)one	0.1 *M* NaCl	85	Zhou et al. (2003)
t-$C_8H_{17}C_6H_4(OC_2H_4)_5OH$[d]-*N*-octylpyrrolid(in)one	H_2O	100	Zhou et al. (2003)
t-$C_8H_{17}C_6H_4(OC_2H_4)_5OH$[d]-*N*-octylpyrrolid(in)one	0.1 *M*	130	Zhou et al. (2003)

[a] Values are approximate and depend upon the humidity.
[b] For mixtures, the largest value obtained for any ratio of the two surfactants.
[c] L 77 = commercial $(CH_3)_3SiOSi(CH_3)[CH_2CH_2(CH_2CH_2O)_{8.5}CH_3]OSi(CH_3)_3$.
[d] Commercial material.

Some POE trisiloxanes, for example, $[(CH_3)_3SiO]_2Si(CH_3)[CH_2(CH_2CH_2O)_{8.5}CH_3]$, on the other hand, spread readily to much greater areas on these highly hydrophobic areas. This phenomenon has been called *superspreading* or *superwetting* and has attracted considerable interest (Ananthapadmanabhan et al., 1990; He et al., 1993; Lin et al., 1993; Zhu et al., 1994; Gentle and Snow, 1995; Rosen and Song, 1996; Stoebe et al., 1996, 1997; Svitova et al., 1996). This superspreading is usually attributed to the ability of trisiloxane surfactants to decrease the surface tension of their aqueous solutions to 20–21 mN/m, significantly lower than the 25 mN/m minimum attainable with hydrocarbon chain surfactants. However, mixtures of certain short-hydrocarbon chain surfactants with the POE trisiloxane mentioned above show even greater superspreading ability than the latter (Rosen and Song, 1996). This synergistic effect (Table 6.3) has been shown (Rosen and Wu, 2001; Wu and Rosen, 2002) to be due not to a further reduction of the surface tension by the mixture, but to greater adsorption of the mixture at the aqueous solution–hydrophobic solid interface than at the aqueous solution–air interface.

For a series of *N*-alkyl-2-pyrrolidinones that produce enhanced superspreading of the POE trisiloxane mentioned above on polyethylene film, it has been shown (Rosen and Wu, 2001) that the addition of the alkylpyrrolidinone to the trisiloxane surfactant produces little or no increase in the total surfactant at the hydrophobic solid–air or aqueous solution–air interfaces, but a considerable increase in the total surfactant adsorption at the hydrophobic solid–aqueous solution interface. This enhanced adsorption of surfactant at the aqueous solution–solid interface relative to that at the aqueous solution–air interface produces a decrease in the surfactant concentration at the air–solution interface in the thin precursor film at the wetting front (Figure 6.8).

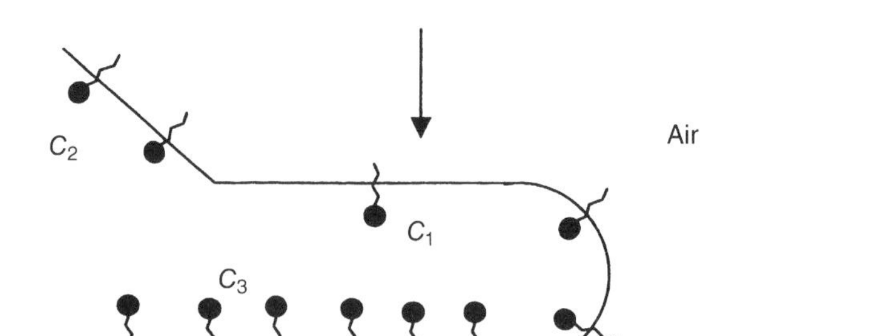

FIGURE 6.8 Precursor film of aqueous surfactant solution on a hydrophobic substrate, with surfactant concentration $C_1 < C_2 < C_3$ at the interfaces, and γ in the precursor film greater than in the bulk surfactant solution, due to the greater adsorption of the surfactant at the solid–liquid than at the air–liquid interface.

This results in a surface tension gradient in the precursor film promoting movement of the aqueous phase to the wetting front.

The order of increased surfactant adsorption on the solid produced by the different alkyl pyrrolidinones parallels the order of their enhancement of superspreading. In addition, it was shown (Wu and Rosen, 2002) (1) that the change in the spreading coefficient (Equation 6.1) parallels enhancement of superspreading and (2) that the order of increased attractive molecular interaction between the different alkylpyrrolidinones and the trisiloxane surfactant at the hydrophobic solid–aqueous solution interface, as measured by the interaction parameter β^{σ}_{SL} (Chapter 11): n-butyl < n-cyclohexyl < n-octyl < n-hexyl < 2-ethylhexyl, is exactly the same order as that of their enhancement of the superspreading.

Recently, it has been found that aqueous solutions of two different hydrocarbon-chain surfactants can also show superspreading on highly hydrophobic substrates (Rosen and Wu, 2002; Zhou et al., 2003). In these mixtures, the two different hydrocarbon-chain surfactants also interact to produce synergistic enhancement of the total surfactant adsorption at the hydrophobic solid–aqueous solution interface relative to that at the air–aqueous solution interface, and this is accompanied by an enhanced rate of reduction of the contact angle (Zhou et al., 2003). SF values for these mixtures are also listed in Table 6.3.

REFERENCES

Adamson, A. W. and A. P. Gast, *Physical Chemistry of Surfaces*, 6th ed., Wiley, New York, 1997.

Ananthapadmanabhan, K. P., E. D. Goddard, and P. Chandar (1990) *Colloids Surf.* **44**, 281.

Aronson, M. P., M. F. Petko, and H. M. Princen (1978) *J. Colloid Interface Sci.* **65**, 296.

Bargava, A., A. V. Francis, and A. K. Biswas (1978) *J. Colloid Interface Sci.* **64**, 214.

Bargeman, D. and F. Van Voorst Vader (1973) *J. Colloid Interface Sci.* **42**, 467.

Bartell, F. E. and L. S. Bartell (1934) *J. Am. Chem. Soc.* **56**, 2205.

Baumgartner, F. (1954) *Ind. Eng. Chem.* **46**, 1349.

Bernett, M. K. and W. A. Zisman (1959) *J. Phys. Chem.* **63**, 1911.

Biswas, A. K. and B. K. Mukherji (1960) *J. Appl. Chem.* **10**, 73.

Bland, P. and J. M. Winchester, *Proc. 5th Int. Congr. Surface Activity*, Barcelona, 1968, III, p. 325.

Bruil, H. G. and J. J. van Aartsen (1974) *Colloid Polym. Sci.* **252**, 32.

Cohen, A. W. and M. J. Rosen (1981) *J. Am. Oil Chem. Soc.* **58**, 1062.

Crook, E. H., D. B. Fordyce, and G. F. Trebbi (1964) *J. Am. Oil Chem. Soc.* **41**, 231.

Dahanayake, M., Surface chemistry fundamentals: thermodynamic and surface properties of highly purified model surfactants. Doctoral dissertation, City University of New York, 1985.

Draves, C. Z. (1939) *Am. Dyestuff Rep.* **28**, 425.

Dreger, E. E., G. I. Keim, G. D. Miles, L. Shedlovsky, and J. Ross (1944) *Ind. Eng. Chem.* **36**, 610.

Dupré, A., Theorie Mecanique de la Chaleur, Paris, 1869.

Finch, J. A. and G. W. Smith (1973) *J. Colloid Interface Sci.* **44**, 387.

Fowkes, F. M. (1953) *J. Phys. Chem.* **57**, 98.

Fox, H. W. and W. A. Zisman (1950) *J. Colloid Sci.* **5**, 514.

General Aniline and Film Corp. (GAF), Tech. Bull. 7543-002, 1965.

Gentle, T. E. and S. A. Snow (1995) *Langmuir* **11**, 2905.

Gerault, A., Fed. Assoc. Techniciens Ind. Peintures, Vernis, Emaux Encres Imprimerie Europe Continentale Congr. **7**, 119 (1964).

Girifalco, L. A. and R. J. Good (1957) *J. Phys. Chem.* **61**, 904.

Good, R. J. and L. A. Girifalco (1960) *J. Phys. Chem.* **64**, 561.

Götte, E. (1969) *Fette, Seifen, Anstrichemi* **71**, 219.

Götte, E. and M. J. Schwuger (1969) *Tenside* **6**, 131.

Gray, F. W., I. J. Krems, and J. F. Gerecht (1965) *J. Am. Oil. Chem. Soc.* **42**, 998.

Greiss, W. (1955) *Fette, Seifen, Anstrichmi* **57**, 24, 168, 236.

He, M., R. M. Hill, Z. Lin, L. E. Scriven, and H. T. Davis (1993) *J. Phys. Chem.* **97**, 8820.

Heertjes, P. M. and N. W. F. Kossen (1967) *Powder Tech.* **1**, 33.

Ikeda, I., A. Itoh, P.-L. Kuo, and M. Okahara (1984) *Tenside Dtrgts.* **21**, 252.

Komor, J. A. and J. P. G. Beiswanger (1966) *J. Am. Oil Chem. Soc.* **43**, 435.

Leeds, M. W., R. J. Redeschi, S. J. Dumovich, and A. W. Casey (1965) *Ind. Eng. Chem. Prod. Res. Dev.* **4**, 236.

Lin, Z., M. He, H. T. Davis, L. E. Scriven, and S. A. Snow (1993) *J. Phys. Chem.* **97**, 3571.

Livingston, J. R., R. Drogin, and R. J. Kelly (1965) *Ind. Eng. Chem. Prod. Res. Dev.* **4**, 28.

Longsworth, L. G. (1953) *J. Am. Chem. Soc.* **75**, 5705.

Lucassen-Reynders, E. H. (1963) *J. Phys. Chem.* **67**, 969.

Machemer, H. (1959) *Melliand Textilber* **40**, 56, 174.

Micich, T. J. and W. M. Linfield (1984) *J. Am. Oil Chem. Soc.* **61**, 591.

Micich, T. J. and W. M. Linfield (1985) *J. Am. Oil Chem. Soc.* **62**, 912.

Micich, T. J. and W. M. Linfield (1986) *J. Am. Oil Chem. Soc.* **63**, 1385.

Ohbu, K., M. Fujiwara, and Y. Abe (1998) *Progr. Colloid Polym. Sci.* **109**, 85.

Olin, J. F. (to Sharples Chemicals, Inc.), U.S. 2,565,986 (1951).

Osterhof, H. J. and F. E. Bartell (1930) *J. Phys. Chem.* **34**, 1399.

Padday, J. F., Wetting, S. C. I. Monograph No. 25, Soc. Chem. Ind., London, 1967, p.234.

Püschel, F. (1966) *Tenside* **3**, 71.

Pyter, R. A., G. Zografi, and P. Mukerjee (1982) *J. Colloid Interface Sci.* **89**, 144.

Rosen, M. J. and X. Y. Hua (1990) *J. Colloid Interface Sci.* **139**, 397.

Rosen, M. J. and C.-C. Kwan, Surface Active Agents Soc. Chem. Ind., London, 1979, pp. 99–105.

Rosen, M. J. and L. D. Song (1996) *Langmuir* **12**, 4945.

Rosen, M. J. and Y. Wu (2001) *Langmuir* **17**, 7296.

Rosen, M. J. and Y. Wu, U.S. Patent Appl. Serial No. 10/318,321, Dec. 12, 2002, Enhancement of the wetting of Hydrophobic Surfaces by Aqueous Surfactant Solutions.

Rosen, M. J. and Z. H. Zhu (1993) *J. Am. Oil Chem. Soc.* **70**, 65.

Schwarz, E. G. and W. G. Reid (1964) *Ind. Eng. Chem.* **56**, 9, 26.

Schwuger, M. J. (1971) *Ber. Bunsenes. Phys. Chem.* **75**, 167.

Schwuger, M. J. (1982) *J. Am. Oil Chem. Soc.* **59**, 258.

Shafrin, E. G. and W. A. Zisman (1960) *J. Phys. Chem.* **64**, 519.

Shirolkar, G. V. and K. Venkataraman (1941) *J. Soc. Dyers Colour.* **57**, 41.

Smith, F. D., A. J. Stirton, and M. V. Nunez-Ponzoa (1966) *J. Am. Oil Chem. Soc.* **43**, 501.

Somasundaran, P. (1972) *Separ. Purtf Methods* **1**, 117.

Stirton, A. J., J. K. Weil, A. A. Stawitzke, and S. James (1952) *J. Am. Oil. Chem. Soc.* **29**, 198.

Stirton, A. J., R. G. Bistline, J. K. Weil, and W. C. Ault (1962a) *J. Am. Oil Chem. Soc.* **39**, 55.

Stirton, A. J., R. G. Bistline, J. K. Weil, W. C. Ault, and E. W. Maurer (1962b) *J. Am. Oil Chem. Soc.* **39**, 128.

Stoebe, T., Z. Lin, R. M. Hill, M. D. Ward, and H. T. Davis (1996) *Langmuir* **12**, 337.

Stoebe, T., Z. Lin, R. M. Hill, M. D. Ward, and H. T. Davis (1997) *Langmuir* **13**, 7270, 7276, 7282.

Suzuki, H. (1967) *Yukagaku* **16**, 667. [C. A. 68, 41326h (1968)].

Svitova, T. F., H. Hoffmann, and R. M. Hill (1996) *Langmuir* **12**, 1712.

Varadaraj, R., J. Bock, N. Brons, and S. Zushma (1994) *J. Colloid Interface Sci.* **167**, 207.

Washburn, E. W. (1921) *Phys. Rev.* **17**, 273.

Weil, J. K., A. J. Stirton, and R. G. Bistline (1954) *J. Am. Oil Chem. Soc.* **31**, 444.

Weil, J. K., A. J. Stirton, R. G. Bistline, and E. W. Maurer (1959) *J. Am. Oil Chem. Soc.* **36**, 241.

Weil, J. K., A. J. Stirton, R. G. Bistline, and W. C. Ault (1960) *J. Am. Oil Chem. Soc.* **37**, 679.

Weil, J. K., A. J. Stirton, and E. A. Barr (1966) *J. Am. Oil Chem. Soc.* **43**, 157.

Weil, J. K., R. E. Koos, W. M. Linfield, and N. Parris (1979) *J. Am. Oil Chem. Soc.* **56**, 873.

Wrigley, A. N., F. D. Smith, and A. J. Stirton (1957) *J. Am. Oil Chem. Soc.* **34**, 39.
Wu, Y. and M. J. Rosen (2002) *Langmuir* **18**, 2205.
Zhou, Q., Y. Wu, and M. J. Rosen (2003) *Langmuir* **19**, 7955.
Zhu, Z. H., D. Yang, and M. J. Rosen (1989) *J. Am. Oil Chem. Soc.* **66**, 998.
Zhu, S., W. G. Miller, L. E. Scriven, and H. T. Davis (1994) *Coll. Surf. A* **90**, 63.
Zisman, W. A., *Advances in Chemistry*, No. 43, American Chemical Society, Washington, DC, 1964.

PROBLEMS

6.1 **(a)** The Good–Girifalco factor ϕ, which is a measure of the degree of interaction between the two phases in contact at an interface, varies from 0.5 when interaction is minimal to about 1.1 when interaction is strong. What can you conclude regarding the strength of the interaction of water ($\gamma_{LA} = 72$ mN/m (dyn/cm) at 25°C) and a liquid, X, whose surface tension is 20 mN/m (dyn/cm) at 25°C if the interfacial tension between them at that temperature is 45 dyn/cm?

(b) What is the value of the spreading coefficient in this system?

6.2 For low-energy surfaces, γ_c, the critical surface tension for wetting, is often equated with γ_{SA}. What is the implication of this?

6.3 Water at 25°C ($\gamma_c = 72$ mN/m (dyn/cm)) makes a contact angle of 102° on a solid substrate. The addition of a surfactant to the water decreases the surface tension to 40 mN/m (dyn/cm) and the contact angle to 30°. Calculate the change in the work of adhesion of the water to the substrate as a result of the surfactant addition.

6.4 Suggest a reason for the observation that POE nonionics often show shorter wetting times than anionics on hydrophobic substrates but show longer wetting times than anionics on cellulosic substrates.

6.5 Without consulting the tables, place the following surfactants in order of decreasing wetting time for a cellulosic substrate at alkaline pH:

(a) $C_{12}H_{25}CH(SO_3^-Na^+)COOC_4H_9$
(b) $C_{16}H_{33}CH(SO_3^-Na^+)COOCH_3$
(c) $C_{12}H_{25}N^+(CH_3)_3Cl^-$
(d) $C_{16}H_{21}CH(SO_3^-Na^+)COOC_{10}H_{21}$
(e) $C_{16}H_{33}CH(SO_3^-Na^+)COOC_4H_9$

6.6 **(a)** For what types of surfactants and low-energy surfaces can the assumption be made that the $d(\gamma_{SA})/d\ln C$ in Equation 6.6 equals zero?

(b) Give examples of systems for which this may not be true.

6.7 Design (synthetic scheme is optional) novel surfactant structures that have

- **(a)** maximum wetting
- **(b)** minimum wetting for individual hydrophobic and hydrophilic surfaces in polar and nonpolar media. (Awareness of current literature is highly recommended.)

7 Foaming and Antifoaming by Aqueous Solutions of Surfactants

Foam is produced when air or some other gas is introduced beneath the surface of a liquid that expands to enclose the gas with a film of liquid. Foam has a more or less stable honeycomb structure of gas cells whose walls consist of thin liquid films with approximately plane parallel sides. These two-sided films are called the *lamellae* of the foam. Where three or more gas bubbles meet, the lamellae are curved, concave to the gas cells, forming what is called the *Plateau border* or *Gibbs triangles* (Figure 7.1).

The pressure difference across a curved interface due to the surface or interfacial tension of the solution is given by the Laplace equation:

$$\Delta P = \gamma\left(\frac{1}{R_1} + \frac{1}{R_2}\right) \quad (7.1)$$

where R_1 and R_2 are the radii of curvature of the interface. Since the curvature in the lamellae is greatest in the Plateau borders, there is a greater pressure across the interface in these regions than elsewhere in the foam. Since the gas pressure inside an individual gas cell is everywhere the same, the liquid pressure inside the lamella at the highly curved Plateau border (B) must be lower than in the adjacent, less curved regions (A) of the Plateau area. This causes drainage of the liquid from the lamellae into the Plateau borders. In a column of foam, liquid also drains as a result of hydrostatic pressure, with the result that lamellae are thinnest in the upper region of the column and thickest in the lower region. Foams are destroyed when the liquid drains out from between the two parallel surfaces of the lamella, causing it to get progressively thinner. When it reaches a critical thickness (50–100 A), the film collapses.

Absolutely pure liquids do not foam. Foam is also not pronounced in mixtures of similar types of materials (e.g., aqueous solutions of hydrophilic substances). Bubbles of gas introduced beneath the surface of an absolutely pure liquid rupture immediately on contacting each other or escape from the liquid as fast as the liquid can drain away from them. For true foaming to occur, the presence of a solute capable of being adsorbed at the L/G interface is required.

Surfactants and Interfacial Phenomena, Fourth Edition. Milton J. Rosen and Joy T. Kunjappu.

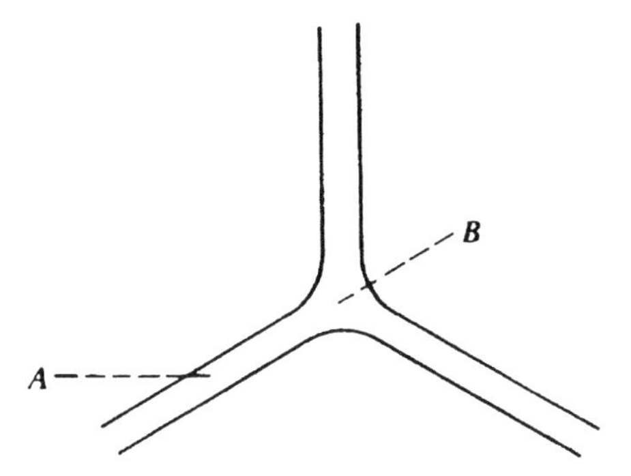

FIGURE 7.1 Plateau border at point of meeting of three bubbles.

The presence of this surface-active solute produces lamellae between the gas cells of the foam that have adsorbed monomolecular films of surfactant molecules on both sides at the *L/G* interface. These adsorbed films provide the system with the property that distinguishes foaming from nonfoaming systems—the ability of the former to resist excessive *localized* thinning of the lamella surrounding the bubbles—while general thinning of the lamella proceeds. This property, which is generally known as *film elasticity*, is a necessary condition for the production of foam; however, it is not sufficient for the formation of a *persistent* foam; the foam formed may subsequently prove persistent or transient. Persistent foams (often called *metastable foams*, to distinguish them from *transient* or unstable foams, since no foams are thermodynamically stable) are produced when some mechanism exists to prevent rupture of the lamella after most of the liquid has drained out of it. The lifetimes of persistent foams are measured in hours or days, whereas the lifetimes of transient foams are of the order of a few seconds to a few tens of seconds (less than a minute).

I. THEORIES OF FILM ELASTICITY

For a liquid to foam (persistently or transiently), the liquid membrane surrounding the bubbles must possess a special form of elasticity such that any applied stresses that tend toward local thinning or stretching of the membrane are rapidly opposed and counterbalanced by restoring forces generated during the initial displacement of the material of the foam film. That is, the very process of stretching or thinning must produce forces that tend to counteract stretching or thinning. Moreover, these restoring forces must increase with the amount of displacement of the film, like the stretching of a rubber band. This film elasticity is possible only if a surface-active solute is present.

Theories concerning the mechanisms of operation of this film elasticity depend on two observations concerning the surface tension of aqueous solutions of surface-active solutes: (1) its increase in value with decrease in concentration of the surface-active solute, at concentrations of the latter below

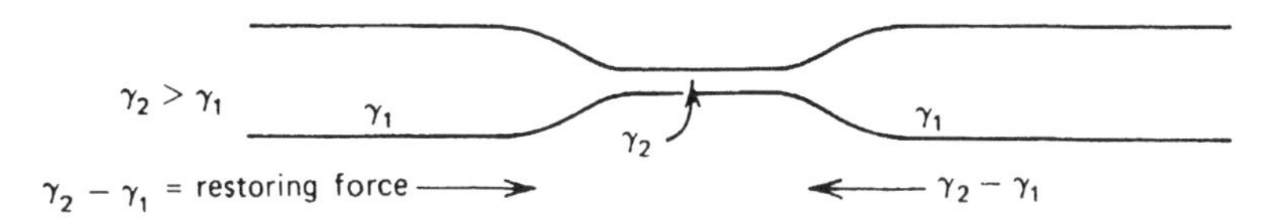

FIGURE 7.2 Stretched portion of foam lamella, illustrating the mechanism of film elasticity.

the critical micelle concentration (CMC), and (2) the time required for it to obtain its equilibrium value (the fact that the initial value of these surface tension at a new surface is always greater than the equilibrium value). The theory based on the first effect, the change in surface tension with change in concentration of the surface-active solute, is known as the *Gibbs effect* (Gibbs, 1878); that based on the second, the change in surface tension with time, is known as the *Marangoni effect* (Marangoni, 1872). The two theories are complementary and provide mechanisms for the operation of film elasticity under different conditions (Kitchener and Cooper, 1959).

Both film elasticity theories postulate that elasticity is due to the local increase in surface tension with extension of the film, that is, $d\gamma/dA = +$. As a local spot in the film thins and stretches and the area of the film in that region (Figure 7.2) increases, its surface tension increases and a gradient of tension is set up that causes liquid to flow toward the thinned spot from the thicker portions around it. The thinning spot thereby automatically draws liquid from its perimeter and prevents further thinning of the film. In addition, the movement of surface material carries with it underlying material that helps "heal" and thicken a thinned spot by a surface transport mechanism (Ewers and Sutherland, 1952). The theories differ in that the Marangoni theory explains this increase on the basis of the *instantaneous* value of γ, whereas the Gibbs theory explains it on the basis of the *equilibrium* value of γ.

The Marangoni effect is significant only in dilute solution and within a limited concentration range. The amount of solute adsorbed at a new surface in the absence of stirring or an energy barrier to adsorption is given by (Ward and Tordai, 1946)

$$n = 2\left(\frac{D}{\pi}\right)^{1/2} ct^{1/2}\frac{N}{1000}, \tag{7.2}$$

where n = number of molecules/cm^2,

D = bulk diffusion, coefficient cm^2/s,

c = bulk concentration, in mol/L,

t = time, in seconds,

N = Avogadro's number.

The times involved in foam production are in the range of 0.001–0.1 s; the value of D/π for the usual surface-active solute (with a hydrocarbon chain of

12–18 carbon atoms) is of the order of 1×10^{-6} cm^2/s; equilibrium surface concentrations for this same type of material are about 2×10^{14} molecules/cm^2. Thus, if the time needed to replace the solute at a new surface is to be no shorter than the time required to produce the foam, the concentration of the solution should not exceed, in the absence of an energy barrier to desorption,

$$c = \frac{n \times 1000}{2\,(D/\pi)^{1/2} t^{1/2} N} = \frac{2 \times 10^{17}}{2(1 \times 10^{-3})(1 \times 10^{-1}) \times 6 \times 10^{23}} = (1.7 \times 10^{-3} M)$$

(Here a mean time of 0.01 s for the production of the foam is used.)

On the other hand, if the solution is too dilute, then the surface tension of the solution will approach that of the pure solvent, and then the restoring force, which is the difference between the surface tension of the clean surface (than of the pure solvent) and the equilibrium surface tension of the solution, will be too small to withstand the usual thermal and mechanical shocks. Thus, according to this mechanism, there should be an optimum concentration for maximum foaming in any solution producing transient foams. (In these solutions, the foam stabilization effects are much less important than the foam-producing effects, and therefore the latter can be measured more or less independently of the former.) This maximum in the foam volume-concentration curve of solution producing transient foams has been well verified experimentally.

From the discussion of dynamic surface tension (Chapter 5, Section IV), the maximum rate of reduction of surface tension occurs when $t = t^*$, the time required for the surface tension to reach half of the value between that of the solvent, γ_0, and the meso-equilibrium surface tension value, γ_m. From Equation 5.4, it has been shown (Rosen et al., 1991) that

$$(-\delta\gamma/\delta t)_{t=t^*} = n(\gamma_0 - \gamma_m)/4t^* \qquad (7.3)$$

where n is the constant (Table 5.3), essentially independent of surfactant concentration, that increases with increasing tendency of the surfactant to adsorb on the surface. This implies that the parameter $n(\gamma_0 - \gamma_m)/t^*$ should be related to the foaming of the surfactant solution.

This has been shown to be valid for a series of pure and commercial polyoxyethylene (POE) dodecyl alcohols, POE nonylphenols, and POE alcohol sulfates (Varadaraj et al., 1990; Rosen et al., 1991; Tamura et al., 1995).

The Gibbs theory of film elasticity postulates that, on thinning and expansion of local areas of the lamellae, the rise in surface tension is due to depletion of the solute from the underlying layer of solution in the interior of the lamella. This theory is based on the assumption that in thin films, the length along the surface of the film is so much greater than the distance normal to the surface that equilibrium may be established normal to the surface much more rapidly than along the surface. The lamella may thus be regarded as consisting of individual sections of constant volume and constant solute content within which equilibrium is attained following some change in the surface of the lamella. If that section of the lamella is stretched, the surface area increases and the thickness decreases. However, if the film is thin, the concentration of solute in the solution beneath the surface may be insufficient to maintain the surface concentration of the surfactant as the film expands, and the surface tension of this section will increase. This effect is also significant only within a certain concentration range of the solute. If the concentration is very low, the change in surface tension with concentration will be too small and the consequent tension gradient insufficient to prevent further thinning and eventual rupture of the film. On the other hand, if the concentration is too far above the CMC of the solute, the change in surface tension with increase in the area of the film will also be too small to prevent rupture of the film because the surface tension does not vary much, if at all, with concentration change above the CMC and because the large reserve of surfactant in the solution will prevent significant change in the surface tension unless the film becomes very thin. The Gibbs effect, therefore, also accounts for the fact that in transient foams, foaming power goes through a maximum as the concentration of the solute is varied.

The Gibbs effect can be evaluated quantitatively. Gibbs defined a coefficient of surface elasticity E as the stress divided by the strain per unit area, E $[2d\gamma/(dA/A)]$. The greater the value of E, the greater the ability of the film to withstand shocks on thinning.

On the basis of the previously described model in which each section of the lamella is considered to act as an independent unit of constant volume containing a fixed amount of surfactant, E is given by the expression (Sheludko, 1966; Rosen, 1967)

$$E = \frac{4\Gamma^2 RT}{h_b C + 2\Gamma\left(1 - \frac{\Gamma}{\Gamma_m}\right)} \tag{7.4}$$

where h_b = thickness of the bulk solution in the lamella,

Γ = surface concentration of the surfactant, in mol/cm^2,

Γ_m = surface concentration of the surfactant, in mol/cm^2, at surface "saturation,"

C = bulk concentration of the surfactant, in mol/cm^3.

This expression implies, as expected, that E increases as the thickness of the bulk solution in the lamella or the bulk concentration of the surfactant in the lamella decreases. It also implies a very great dependence on the surface excess concentration of the surfactant, Γ, and indicates that if Γ is zero, there is no film elasticity.

From Equation 7.4 the thickness of the bulk solution in the lamella h_b at which the surface elasticity coefficient E becomes significant can be calculated. For surfactant concentrations greater than about one-third the CMC, Γ/Γ_m is very close to 1; thus, the second term in the denominator may be disregarded without significant error, and

$$E = \frac{4\Gamma^2 RT}{h_b C}. \tag{7.5}$$

The surface elasticity, therefore, decreases when either the thickness of the lamella or the bulk concentration of surfactant in it increases. Moreover, since Γ is proportional (Equation 2.25) to $(\delta\gamma/\delta \log C)_T$, the elasticity is very sensitive to change in the surface tension of the solution with change in the bulk phase concentration of the surfactant. Surface excess concentrations Γ for surfactants usually fall in the range $1\text{–}4 \times 10^{-10}$ mol/cm^2. Thus, at 27°C (300 K), since $R = 8.3 \times 10^7$ ergs/mol/deg,

$$E = \frac{4(1-16)\times 10^{-20} \times 8.3 \times 10^7 \times 300}{h_b C} = \frac{(1-16)\times 10^{-9}}{h_b C}$$

For E to be 10 dyn or more, h_b must be $10^{-3}\text{–}10^{-4}$ cm when $C = 1 \times 10^{-6}$ mol/cm^3 (1×10^{-3} M); $10^{-4}\text{–}10^{-5}$ cm when $C = 1 \times 10^{-5}$ mol/cm^3 (1×10^{-2} M), and so on.

II. FACTORS DETERMINING FOAM PERSISTENCE

For a foam to be persistent, mechanisms must be present to retard the loss of liquid and gas from the foam and to prevent rupture of the lamellae when they are subjected to mechanical shock or when a certain critical thickness is reached.

A. Drainage of Liquid in the Lamellae

The extent and rate of drainage of surplus solution from the interior of the lamellae is one of the important factors determining foam stability, since drainage causes thinning of the film, and when the film reaches a critical thickness (50–100 Å), the film may rupture spontaneously. Drainage of the film occurs under two influences: gravity and pressure difference.

Drainage by gravity is important mainly in very thick lamellae, such as are present when the foam is first formed; drainage under the influence of pressure difference is more important when the lamellae are thin. The bulk viscosity of the foaming solution is a major factor in determining the rate of drainage by gravity in thick lamellae. Electrolyte or organic additives (Chapter 3, Section III) that increase the bulk viscosity of the foaming solution decrease the rate of drainage of the liquid in the lamellae. Polymeric thickeners are often added to increase the bulk viscosity when very stable foams are desired. The formation of a viscous liquid crystalline phase (Chapter 3, Section IIC) in the bulk solution in the lamellae at certain concentrations above the CMC may increase the stability of the foam by retarding drainage. When the lamellae become thin, because of the drainage of liquid out of the interior of the lamellae, the viscosity of the remaining liquid is greatly affected by the oriented monolayers comprising the external surfaces of the lamellae. The orienting forces caused by these monolayers are transmitted to a significant depth in thin lamellae through successive polarization of the underlying layers of water molecules. In films 1000 Å thick, the viscosity of the water has been shown to be twice that of normal water, and in films 200 Å thick, five times that of normal water.

Drainage by pressure difference is due to the differences in curvature of the surface of the lamella. As mentioned earlier (Figure 7.1), at the intersection of three or more air bubbles, the lamella has a greater curvature than at the boundary between only two air sacs. The pressure difference causing drainage of liquid into the Plateau border at point B from point A outside it is given by the expression $\Delta P = \gamma(1/R_B + 1/R_A)$, where R_B and R_A are the radii of curvature of the lamella at points B and A, respectively. The greater the difference between R_A and R_B (i.e., the larger the bubble size in the foam) and the greater the surface tension of the solution in the lamella, the greater the pressure difference causing drainage.

B. Diffusion of Gas through the Lamellae

Another factor determining the stability of foams is the rate of diffusion of gas from one bubble to another through the lamella separating them. The rate of diffusion q of a gas between two bubbles of different radii, R_1 and R_2, is given by the equation

$$q = -\mathrm{J}A\Delta P \tag{7.6}$$

where J = the permeability of the diffusion path,

A = the effective perpendicular area through which diffusion occurs between the bubbles,

ΔP = the difference in gas pressures of the two bubbles,

$$= 2\gamma\left(\frac{1}{R_1} - \frac{1}{R_2}\right) \tag{7.7}$$

γ = the surface tension of the solution.

The right hand side of the equation is to be multiplied by 2 if both surfaces of the two bubbles are considered.

Since the negative sign in the equation indicates that diffusion is in the direction of pressure decrease and the gas pressure in small bubbles is higher than in larger bubbles ($\Delta P = 2\gamma/R$), large bubbles tend to grow at the expense of smaller ones. This growth may change the character of the foam completely from an initial one of small, spherical air cells to one containing large polyhedral cells. This change to large polyhedral cells increases the curvature in the Plateau borders and increases the forces causing drainage into these borders. This growth also may necessitate a rearrangement of the bubbles in the foam, with the consequent possibility of rupture of the lamellae at some points because of mechanical shocks occasioned by the rearrangement.

The value of J in the preceding diffusion equation depends on the resistance to gas transfer of the two interfaces and the liquid between them. Data indicate that this transfer of gas takes place through aqueous pores between the surfactant molecules in the surface films of the lamellae (Princen et al., 1967). Closer packing of the surfactant molecules in the film would therefore be expected to decrease the rate of diffusion of the gas between bubbles. Consistent with this, interfacial resistance to gas diffusion has been shown to increase with increase in the number of carbon atoms in the hydrophobic group of the surfactant and with decrease in the molecular mass of the hydrophilic group (Caskey and Barlage, 1972). The addition of certain concentrations of lauryl alcohol has been shown to decrease sharply the permeability of surface films of sodium lauryl sulfate, presumably because of condensation of the lauryl sulfate film by the lauryl alcohol.

C. Surface Viscosity

Qualitatively, in a number of cases, foam stability has been correlated with viscosity of the surface film, but the relation is not really clear. There are stable foams in which the viscosity of the surface film is not particularly high and viscous monolayers that do not produce particularly stable foams. However, it appears well accepted that if the viscosity of the surface film is either very low (a "gaseous" monomolecular film) or very high (a "solid" monomolecular film), the foam produced will be unstable. In both of these cases film elasticity is low. In addition, too high a surface viscosity can slow down self-healing of thinned spots in the film by the surface transport mechanism.

D. The Existence and Thickness of the Electrical Double Layer

Factors that may prevent the thinning of foam films (at least in the case of ionic surfactants) are electrostatic repulsion between the two sides of the film and the high osmotic pressure due to the large concentration of counterions present. These factors have been suggested by the existence of persistent foam in cases where the film is known *not* to have great surface viscosity—and this is true of foaming solutions of purified surface-active agents, where it is known that the surface films are not particularly coherent. In these cases, it has been suggested that when the film becomes very thin (<0.2 μm or 200 nm), stability is obtained chiefly because of the electrical repulsion between the ionic double layers associated with the adsorbed ionic surfactant on the two sides of the liquid film. Since the addition of electrolyte to the foaming solution causes compression of the electrical double layers associated with the surface films, such addition decreases their mutual repulsion. This is believed to account for the decreased thickness of liquid films with increase in their electrolyte content (Davies and Rideal, 1963) and for the decreased stability of many foams on the addition of electrolyte.

In summary then, the factors promoting foaming in aqueous surfactant solutions are (1) low equilibrium surface tension, (2) moderate rate of attaining equilibrium surface tension, (3) large surface concentration of surfactant, (4) high bulk viscosity, (5) moderate surface viscosity, and (6) electrical repulsion between the two sides of the foam lamella. The first three promote film elasticity; the last three promote foam persistence.

III. THE RELATION OF SURFACTANT CHEMICAL STRUCTURE TO FOAMING IN AQUEOUS SOLUTION

In foaming, as in other surface properties, correlations between surfactant structure and foaming in aqueous solution require a distinction between the efficiency of the surfactant, its bulk phase concentration required to produce a significant amount of foam, and its effectiveness, the maximum foam height obtained with the surfactant solution regardless of its concentration. A distinction must also be made between foam production, measured by the height of the foam initially produced, and foam stability, the height after a given amount of time. Therefore, in comparing the foaming properties of different surfactants, the term *foaming ability* must be clearly defined. In addition, such conditions as the method used to produce the foam, the temperature of the solution, the hardness of the water used, and its electrolyte content must all be specified. Since most of the foaming data on surfactants with clearly defined structures have been obtained by use of the Ross–Miles method (Ross and Miles, 1953), the structural correlations discussed here are based mainly on data obtained by that method.

In the Ross–Miles method, 200 mL of a solution of surfactant contained in a pipette of specified dimensions with a 2.9-mm-i.d. orifice is allowed to fall 90 cm onto 50 mL of the same solution contained in a cylindrical vessel maintained at a given temperature (often 60°C) by means of a water jacket. The height of the foam produced in the cylindrical vessel is read immediately after all the solution has run out of the pipette (initial foam height) and then again after a given amount of time (generally, 5 min).

A somewhat related but different method of measuring foam height and foam stability has been suggested by Lunkenheimer and Malysa (2003). The surfactant solution (50 mL) is poured into a 42-mm-i.d. glass cylinder with a fritted glass G-2 disc at the bottom, with a syringe attached to the bottom by means of a stopcock. Gas (50 or 100 mL) is introduced manually via the syringe and stopcock, in a fixed time (e.g., 20 s), into the bottom of the column and the stopcock closed. The initial heights of the foam generated and the solution column are measured. The changes in foam height and solution level are measured as a function of time.

A. Efficiency as a Foaming Agent

Foam height generally increases with increase in surfactant concentration below the CMC until the neighborhood of the CMC is reached, where foam height reaches a maximum value or increases slowly to a maximum value somewhat above the CMC. Thus, the CMC of a surfactant is a good measure of its efficiency as a foaming agent; the lower the CMC, the more efficient the surfactant as a foamer. Those structural factors that produce a lower CMC—for example, increased length of the hydrophobic group—would therefore be expected to increase the efficiency of the surfactant as a foaming agent. The addition of neutral electrolyte (which decreases the CMC of the surfactant) increases the efficiency of ionic surfactants. Table 7.1 lists the bulk phase concentrations at which foam height reaches a maximum for some aqueous solutions of surfactants, together with their CMCs at the same temperature at which the foaming data were obtained. It is apparent that surfactants with longer hydrophobic groups are more efficient but not necessarily more effective, foaming agents. Since the Ross–Miles foaming test is usually done at a 0.25% surfactant concentration, equivalent to about 8×10^{-3} *M* for most surfactants, only those materials having CMCs greater than that will not have reached their maximum foam volume at that concentration.

B. Effectiveness as a Foaming Agent

The effectiveness of a surfactant as a foaming agent appears to depend both on its effectiveness in reducing the surface tension of the foaming solution and on the magnitude of its intermolecular cohesive forces. The volume of foam produced when a given amount of work is done on an aqueous solution of surfactant to create foam depends on the surface tension of the solution, since the minimum amount of work required to produce the foam is $\gamma.\Delta A$, the product of the surface tension and the change in the area of the liquid–gas interface as a result of the foaming. The lower the surface tension of the aqueous solution, the greater appears to be the volume of foam of the same average bubble size produced by a given amount of work under the same foaming conditions (Rosen and Solash, 1969). It has also been suggested (Dreger et al., 1944) that the rate of attainment of surface tension reduction may also be a factor in determining the effectiveness of a surfactant as a foaming agent. Therefore, branched-chain surfactants and those containing centrally located hydrophobic groups, which are believed to diffuse rapidly to

TABLE 7.1 Foaming Efficiency of Aqueous Surfactant Solutions (Ross–Miles Method[a])

Surfactant	Temperature (°C)	Concentration (*M*) to Reach Maximum Foam Height	CMC	Height (mm)	Reference
p-$C_8H_{17}\ C_6H_4SO_3^-Na^+$	60	13×10^{-3}	16×10^{-3}	165	Gray et al. (1955)
p-$C_{10}H_{21}C_6H_4SO_3^-Na^+$	60	4.5×10^{-3}	3×10^{-3}	185	Gray et al. (1955)
o-$C_{12}H_{25}C_6H_4SO_3^-Na^+$	60	4×10^{-3}	3×10^{-3}	205	Gray et al. (1955)
p-$C_{12}H_{25}C_6H_4SO_3^-Na^+$	60	4×10^{-3}	1.2×10^{-3}	200	Gray et al. (1955)
o-$C_{11}H_{23}CH(CH_3)C_6H_4SO_3^-Na^+$	60	8×10^{-3}	—	195	Gray et al. (1965)
p-$C_{11}H_{23}CH(CH_3)C_6H_4SO_3^-Na^+$	60	8×10^{-3}	5×10^{-3}	215	Gray et al. (1955)
p-$C_7H_{15}CH(C_4H_9)C_6H_4SO_3^-Na^+$	60	7×10^{-3}	4×10^{-3}	230	Gray et al. (1955)
$C_5H_{11}CH(C_5H_{11})SO_4^-Na^+$	60	10×10^{-3}	83×10^{-3}	130	Dreger et al. (1944)
$C_{12}H_{25}SO_3^-Na^+$	60	11×10^{-3}	13×10^{-3}	210	Rosen and Solash (1969)
$C_{12}H_{25}SO_4^-Na^+$	46	5×10^{-3}	9×10^{-3}	205	Dreger et al. (1944)
$C_{11}H_{23}CH(CH_3)SO_4^-Na^+$	46	5×10^{-3}	6.5×10^{-3}	205	Dreger et al. (1944)
$C_6H_{13}CH(C_6H_{13})SO_4^-Na^+$	46	$>15 \times 10^{-3}$	19×10^{-3}	220	Dreger et al. (1944)
$C_{14}H_{29}SO_3^-K^+$	60	3×10^{-3}	3×10^{-3}	217	Rosen and Solash (1969)
$C_{14}H_{29}SO_4^-Na^+$	46	3×10^{-3}	2.3×10^{-3}	225	Dreger et al. (1944)
$C_{13}H_{27}CH(CH_3)SO_4^-Na^+$	46	3×10^{-3}	1.7×10^{-3}	220	Dreger et al. (1944)
$C_7H_{15}CH(C_7H_{15})SO_4^-Na^+$	46	5×10^{-3}	6.7×10^{-3}	240	Dreger et al. (1944)
$C_{16}H_{33}SO_3^-K^+$	60	0.8×10^{-3}	0.9×10^{-3}	233	Rosen and Solash (1969)
$C_{16}H_{33}SO_4^-Na^+$	60	0.8×10^{-3}	0.7×10^{-3}	220	Rosen and Solash (1969)
$C_{15}H_{31}CH(CH_3)SO_4^-Na^+$	46	$<1 \times 10^{-3}$	0.5×10^{-3}	212	Dreger et al. (1944)
$C_8H_{17}CH(C_8H_{17})SO_4^-Na^+$	46	4×10^{-3}	2.3×10^{-3}	245	Dreger et al. (1944)
p-$C_9H_{19}CH(CH_3)C_6H_4SO_3^-Na^+$	60	13×10^{-3}	—	190	Gray et al. (1965)
p-$C_{13}H_{27}CH(CH_3)C_6H_4SO_3^-Na^+$	60	4×10^{-3}	—	175	Gray et al. (1965)
p-$C_{15}H_{31}CH(CH_3)C_6H_4SO_3^-Na^+$	60	0.7×10^{-3}	—	126	Gray et al. (1965)

[a] Ross and Miles (1953).

the interface, would be expected to produce higher volumes of initial foam. However, not only must the surfactant produce the foam, it must also maintain it—the foam must have appreciable stability. This should require an interfacial film with sufficient cohesion to impart elasticity and mechanical strength to the liquid lamellae enclosing the gas in the foam. Since interchain cohesion increases with increase in the length of the hydrophobic group, this may account for the observation that foam height often goes through a maximum with increase in the length of the chain. Too short a chain probably produces insufficient cohesiveness, whereas too great a length produces too much rigidity for good film elasticity (or too low a solubility in water).

Shah and coworkers (Shah, 1998; Jha et al., 1999) have pointed out the relationship between micellar stability and foaming effectiveness. Micellar stability is inversely related to foaming ability, since very stable micelles are less capable of providing the flux of surfactants necessary to stabilize the new air–solution interface created during the foaming process. Thus, POE glycols, which decrease the stability of micelles of sodium dodecyl sulfate, increase the foaming of aqueous solutions of the latter (Dhara and Shah, 2001). For sodium dodecyl sulfate–alkyl trimethylammonium bromide mixtures, maximum micellar stability is observed for the sodium dodecyl sulfate–dodecyl trimethylammonium bromide mixtures when the alkyl chains of the two surfactants are of equal length, reflecting the maximum interaction obtained (Table 11.1). These mixtures, because of the close packing at both air–water interface and in the micelles, showed minimum surface tension, maximum surface viscosity, maximum foam stability, but minimum foam height (Patist et al., 1997). This concept is consistent with the explanation (below) offered by Dupré et al. (1960) for the marked decrease in foaming observed when POE nonionics reach their cloud points.

Since branched-chain surfactants and those with centrally located hydrophilic groups can depress the surface tension of water to lower values than isomeric straight-chain compounds or those with terminally located hydrophilic groups (Chapter 5, Section II), the former types of compounds would be expected to show higher initial foam heights than the latter. However, since hydrophobic groups with branches have weaker intermolecular cohesive forces than straight-chain ones, the former would be expected to show less foam stability. The result of these two opposing factors is that when the hydrophilic group of a straight-chain surfactant is moved from a terminal to a more central position in the molecule, foam heights generally increase, provided that the materials are all compared above their CMCs where foaming is at a maximum. This is necessary here because the shift of the hydrophilic group to a more central position in the molecule causes an increase in the CMC of the surfactant with a resulting decrease in its efficiency as a foaming agent. Surfactants with highly branched chains, on the other hand, generally show lower foam heights than isomeric straight-chain materials, except where the length of the hydrophobic group becomes too long for straight-chain compounds to have adequate water solubility for good foaming (e.g., >16 carbon atoms at

40°C). Presumably for a similar reason, 2, 5-di-*n*-alkylbenzenesulfonates show lower foam heights and stabilities than the corresponding *p*-*n*-alkylbenzenesulfonates (Kölbel et al., 1960b). Since branched-chain hydrophobic groups show greater water solubility than straight-chain ones and intermolecular cohesive forces increase with increase in chain length, good foaming at 40°C can be obtained with branched-chain surfactants containing up to 20 carbon atoms and foam heights in the C_{20} branched compounds appear to exceed those obtained with any shorter straight-chain compounds (Kölbel et al., 1960a).

In ionic surfactants, the effectiveness of foaming appears to depend also on the nature of the counterion, those with smaller counterions showing greater initial foam heights and foam stabilities. Thus, in the dodecyl sulfate series, the effectiveness decreases with increased size of the counterion in the order $NH_4^+ > (CH_3)_4N^+ > (C_2H_5)_4N^+ > (C_4H_9)_4N^+$ (Kondo et al., 1960).

The poor foaming of aqueous solutions of cationic surfactants in this and similar foaming tests may be due not to some inherent lack of foaming ability, but to the dewetting of the walls of the glass foaming apparatus as a result of adsorption of the cationic surfactant onto it with its hydrophobic group oriented toward the aqueous phase, causing foam rupture.

Table 7.2 lists the foaming effectiveness of some surfactants in aqueous solutions, as well as some data on their (short-term) stability.

In distilled water at room temperature, sodium alkyl sulfates and soaps with saturated straight-chain hydrophobic groups containing 12–14 carbon atoms seem to show the best foaming capacities (Broich, 1966); at higher temperatures, homologous materials with somewhat longer chains give optimum foaming. Thus, at 60°C, saturated straight-chain alkyl sulfates containing 16 carbon atoms, palmitate soaps, dodecyl- and tetradecylbenzenesulfonates (hydrophobic groups equivalent to 15.5–17.5 carbon chains), and α-sulfoesters containing 16–17 carbon atoms show maximum foaming power (Weil et al., 1954, 1966; Gray et al., 1955; Kölbel and Kuhn, 1959; Stirton et al., 1962; Micich et al., 1966). Near the boiling point, C_{18} compounds are best. Since interchain cohesion must overcome thermal agitation of the molecules, which increases with increase in temperature, it is to be expected that optimum chain lengths should increase with increase in temperature. The disodium salts of α-sulfocarboxylic acids produce much less foam than the monosodium salts of α-sulfoesters, presumably because increased electrostatic repulsion between hydrophilic groups counters interchain cohesive forces.

In hard water, somewhat shorter anionic compounds seem to give optimum foaming, probably because of the greater cohesiveness of anionic surface films in the presence of Ca^{2+}. Thus, in 300 ppm $CaCO_3$ solution at 60°C, C_{12}–C_{14} saturated straight-chain alkyl sulfates show the highest foaming capacities (Weil et al., 1954). A similar progressive shift to shorter-chain lengths for optimum foam stability with increase in water hardness was found in a dishwashing study at 46°C in the presence of triglyceride soil (Matheson and Matson, 1983).

TABLE 7.2 Foaming Effectiveness of Aqueous Surfactant Solutions (Ross–Miles Method[a])

Surfactant	Concentration (%)	Temp. (°C)	Foam Height (mm): Distilled Water, Initial	Foam Height (mm): Distilled Water, After Time (min)	Foam Height (mm): 300 ppm $CaCO_3$ Initial	Reference
$C_{15}H_{31}COO^-N^+$	0.25(pH 10.7)	60	236	232 (5)		Rosen and Zhu (1988)
$C_{10}H_{21}SO_3^-Na^+$	0.68	60	160	5 (5)		Dahanayake (1985)
$C_{12}H_{25}SO_3^-Na^+$	0.32	60	190	125 (5)		Dahanayake (1985)
$C_{14}H_{29}SO_3^-Na^+$	0.11	60	—	214 (1)	—	Rosen and Solash (1969)
$C_{16}H_{33}SO_3^-K^+$	0.033	60	—	233 (1)	—	Rosen and Solash (1969)
$C_{12}H_{25}SO_4^-Na^+$	0.25	60	220	200 (5)	240[b]	Weil et al. (1954, 1966); Dahanayake (1985)
$C_{14}H_{29}SO_4^-Na^+$	0.25	60	231	184 (5)	246[b]	Weil et al. (1954)
$C_{16}H_{33}SO_4^-Na^+$	0.25	60	245	240 (5)	178[b]	Weil et al. (1954, 1966)
$C_{18}H_{37}SO_4^-Na^+$	0.25	60	227	227 (5)	151[b]	Weil et al. (1954)
Sodium oleyl sulfate	0.25	60	246	240 (5)	226[b]	Weil et al. (1954)
Sodium elaidyl sulfate	0.25	60	243	241 (5)	202[b]	Weil et al. (1954)
$C_{12}H_{25}OC_2H_4SO_4^-Na^+$	0.14	60	246	241 (5)		Dahanayake (1985)
$C_{12}H_{25}(OC_2H_4)_2SO_4^-Na^+$	0.11	60	180	131 (5)	—	Dahanayake (1985)
$C_{12}H_{25}OCH_2CH(CH_3)SO_4^-Na^+$	0.25	60	200	—	—	Weil et al. (1966)
$C_{14}H_{29}OCH_2CH(CH_3)SO_4^-Na^+$	0.25	60	215	—	—	Weil et al. (1966)
$C_{14}H_{29}[OCH_2CH(CH_3)]_2SO_4^-Na^+$	0.25	60	210			Weil et al. (1966)
$C_{16}H_{33}OCH_2CH(CH_3)SO_4^-Na^+$	0.25	60	200	—	—	Weil et al. (1966)
$C_{18}H_{37}OCH_2CH(CH_3)SO_4^-Na^+$	0.25	60	160	—	—	Weil et al. (1966)
$C_{18}H_{37}OCH_2CH_2SO_4^-Na^+$	0.25	60	160	—	—	Weil et al. (1966)
o-$C_8H_{17}C_6H_4SO_3^-Na^+$	0.15	60	148	—	—	Gray et al. (1965)
p-$C_8H_{27}C_6H_4SO_3^-Na^+$	0.15	60	134	—	—	Gray et al. (1965)
p-$C_8H_{17}C_6H_4SO_3^-Na^+$	0.25	60	150	—	—	Gray et al. (1955)

(Continued)

TABLE 7.2 (*Continued*)

Surfactant	Concentration (%)	Temp. (°C)	Foam Height (mm)			Reference
			Distilled Water			
			Initial	After Time (min)	300 ppm $CaCO_3$ Initial	
o-$C_9H_{19}CH(CH_3)C_6H_4SO_3^-Na^+$	0.15	60	165	—	—	Gray et al. (1965)
p-$C_9H_{19}CH(CH_3)C_6H_4SO_3^-Na^+$	0.15	60	162	—	—	Gray et al. (1965)
o-$C_{12}H_{25}C_6H_4SO_3^-Na^+$	0.15	60	206	—	—	Gray et al. (1965)
o-$C_{12}H_{25}C_6H_4SO_3^-Na^+$	0.25	60	208	—	—	Gray et al. (1955)
p-$C_{12}H_{25}C_6H_4SO_3^-Na^+$	0.15	60	201	—	—	Gray et al. (1965)
$C_{10}H_{21}CH(CH_3)C_6H_4SO_3^-Na^+$	0.25	60	—	—	245	Smith et al. (1966)
o-$C_{11}H_{23}CH(CH_3)C_6H_4SO_3^-Na^+$	0.15	60	190	—	—	Gray et al. (1965)
p-$C_{11}H_{23}CH(CH_3)C_6H_4SO_3^-Na^+$	0.15	60	210	—	—	Gray et al. (1955)
p-$C_{11}H_{23}CH(CH_3)C_6H_4SO_3^-Na^+$	0.25	60	218	—	—	Gray et al. (1955)
p-$C_7H_{15}CH(C_4H_9)C_6H_4SO_3^-Na^+$	0.15	60	219	—	—	Gray et al. (1955)
p-$C_7H_{15}CH(C_4H_9)C_6H_4SO_3^-Na^+$	0.25	60	230	—	—	Gray et al. (1955)
$C_{12}H_{25}CH(CH_3)C_6H_4SO_3^-Na^+$	0.25	60	—	—	80	Smith et al. (1966)
$C_{14}H_{29}CH(CH_3)C_6H_4SO_3^-Na^+$	0.25	60	—	—	10	Smith et al. (1966)
o-$C_{15}H_{31}CH(CH_3)C_6H_4SO_3^-Na^+$	0.15	60	105	—	—	Gray et al. (1965)
p-$C_{15}H_{31}CH(CH_3)C_6H_4SO_3^-Na^+$	0.15	60	129	—	—	Gray et al. (1965)
$C_{16}H_{33}CH(CH_3)C_6H_4SO_3^-Na^+$	0.25	60	—	—	0	Smith et al. (1966)
$CH_3CH(SO_3^-Na^+)COOC_{14}H_{29}$	0.25	60	220	—	240	Stirton et al. (1962)
$C_2H_5CH(SO_3^-Na^+)COOC_{12}H_{25}$	0.25	60	220	—	225	Stirton et al. (1962)
$C_7H_{15}CH(SO_3^-Na^+)COOC_8H_{17}$	0.25	60	—	—	185	Weil et al. (1960)
$C_{10}H_{21}CH(SO_3^-Na^+)COOC_4H_9$	0.25	60	220	—	230	Stirton et al. (1962)
$C_{10}H_{21}CH(SO_3^-Na^+)COOC_5H_{11}$	0.25	60	220	—	235	Stirton et al. (1962)
$C_{14}H_{29}CH(SO_3^-Na^+)COOCH_3$	0.25	60	210	200 (5)	225	Stirton et al. (1962)
$C_{14}H_{29}CH(SO_3^-Na^+)COO^-Na^+$	0.25	60	175	165 (5)	125	Micich et al. (1966)
$C_{14}H_{29}CH(SO_3^-Na^+)COOC_2H_5$	0.25	60	210	—	215	Stirton et al. (1962)
$C_{13}H_{27}C(CH_3)(SO_3^-Na^+)COOCH_3$	0.25	60	180	160 (5)	200	Micich et al. (1966)
$C_{16}H_{33}C(CH_3)(SO_3^-Na^+)COOCH_3$	0.25	60	175	165 (5)	35	Micich et al. (1966)
$C_{18}H_{37}C(CH_3)(SO_3^-Na^+)COOCH_3$	0.25	60	140	130 (5)	30	Micich et al. (1966)

$C_8H_{17}C(C_8H_{17})(SO_3^-Na^+)COOCH_3$	0.25	60	210	200 (5)	215	Micich et al. (1966)
$C_8H_{17}C(C_8H_{17})(SO_3^-Na^+)COO^-Na^+$	0.25	60	0	0	95	Micich et al. (1966)
$C_8H_{17}C(C_6H_{13})(SO_3^-Na^+)COOCH_3$	0.25	60	204	190 (5)	213	Micich et al. (1966)
$C_8H_{17}C(C_4H_9)(SO_3^-Na^+)COOCH_3$	0.25	60	170	5 (5)	200	Micich et al. (1966)
$C_{12}H_{25}Pyr^+Br^-$	0.37	60	135	3 (5)	—	Dahanayake (1985)
$C_{10}H_{21}N^+(CH_3)(CH_2C_6H_5)CH_2COO^-$	0.14	60	35	2 (5)	—	Dahanayake (1985)
$C_{12}H_{25}N^+(CH_3)(CH_2C_6H_5)CH_2COO^-$	0.018	60	50	2 (5)	—	Dahanayake (1985)
$C_{12}H_{25}N^+(CH_3)_2CH_2COO^-$[c]	0.25 (pH 5.8)	60	199	29 (5)	—	Rosen and Zhu (1988)
$C_{12}H_{25}N^+(CH_3)_2CH_2COO^-$[c]	0.25 (pH 9.3)	60	197	34 (5)	—	Rosen and Zhu (1988)
$C_{12}H_{25}(OC_2H_4)_{10}OH$[c]	0.25	60	168	26 (5)	—	Rosen and Zhu (1988)
$C_{12}H_{25}O(C_2H_4O)_{15}H$[c]	0.25	60	—	—	197	Wrigley et al. (1957)
$C_{12}H_{25}O(C_2H_4O)_{20}H$[c]	0.25	60	—	—	195	Wrigley et al. (1957)
$C_{12}H_{25}O(C_2H_4O)_{33}H$[c]	0.25	60	—	—	180	Wrigley et al. (1957)
$C_{16}H_{33}O(C_2H_4O)_{15}H$[c]	0.25	60	—	—	153	Wrigley et al. (1957)
$C_{16}H_{33}O(C_2H_4O)_{20}H$[c]	0.25	60	—	—	167	Wrigley et al. (1957)
$C_{16}H_{33}O(C_2H_4O)_{30}H$[c]	0.25	60	—	—	149	Wrigley et al. (1957)
$C_{18}H_{37}O(C_2H_4O)_{15}H$[c]	0.25	60	—	—	165	Wrigley et al. (1957)
$C_{18}H_{37}O(C_2H_4O)_{21}H$[c]	0.25	60	—	—	152	Wrigley et al. (1957)
$C_{18}H_{37}O(C_2H_4O)_{30}H$[c]	0.25	60	—	—	115	Wrigley et al. (1957)
$C_{18}H_{35}O(C_2H_4O)_{15}H$[c,d]	0.25	60	—	—	140	Wrigley et al. (1957)
$C_{18}H_{35}O(C_2H_4O)_{20}H$[c,d]	0.25	60	—	—	160	Wrigley et al. (1957)
$C_{18}H_{35}O(C_2H_4O)_{31}H$[c,d]	0.25	60	—	—	140	Wrigley et al. (1957)
$t\text{-}C_9H_{19}C_6H_4O(C_2H_4O)_8H$[c]	0.10	25	55	45 (5)	—	GAF Corp. (1965)
$t\text{-}C_9H_{19}C_6H_4O(C_2H_4O)_9H$[c]	0.10	25	80	60 (5)	—	GAF Corp. (1965)
$t\text{-}C_9H_{19}C_6H_4O(C_2H_4O)_{10-11}H$[c]	0.10	25	110	80 (5)	—	GAF Corp. (1965)
$t\text{-}C_9H_{19}C_6H_4O(C_2H_4O)_{13}H$[c]	0.10	25	130	110 (5)	—	GAF Corp. (1965)
$t\text{-}C_9H_{19}C_6H_4O(C_2H_4O)_{20}H$[c]	0.10	25	120	110 (5)	—	GAF Corp. (1965)

[a] Ross and Miles (1953).
[b] 0.1. in 100 ppm $CaCO_3$.
[c] Commercial-type material.
[d] From oleyl alcohol.
Pyr^+, pyridinium.

POE nonionic surfactants generally produce less foam and much less stable foam than ionic surfactants in aqueous media. These effects are probably due to the larger surface area per molecule and the absence of highly charged surface films in these foams. Conversion of these materials to their corresponding sulfates generally increases their foaming ability. In POE nonionics, both foam stability and foam volume reach a maximum at a particular oxyethylene chain length and then decrease (Schick and Beyer, 1963). This is ascribed to a maximum in intermolecular cohesive forces in the adsorbed film as the oxyethylene content increases. Van der Waals forces between surfactant molecules decrease with increasing oxyethylene content, since the area per molecule at the surface increases with this change. However, the POE chain is believed to be coiled in the aqueous phase, and the cohesive forces due to intra- and intermolecular hydrogen bonding are stated to pass through a maximum with increasing oxyethylene content. The summation of the van der Waals and hydrogen bonding cohesive forces consequently passes through a maximum as the oxyethylene content of the molecule is increased. In 300 ppm $CaCO_3$ solution at 60°C, POE alcohols appear to be considerably better foaming agents than POE fatty acids. Immediate foam heights for POE *n*-dodecanol are higher than those for corresponding hexadecanol, octadecanol, or oleyl alcohol derivatives. Optimum oxyethylene content in these cases is at 15–20 mol of ethylene oxide per mole of hydrophobe (Wrigley et al., 1957). There appears to be no significant difference in foaming properties between POE linear primary alcohols and secondary alcohols. In distilled water at 25°C the optimum oxyethylene content for nonylphenol derivates is about 13 mol of ethylene oxide per mole of hydrophobe (GAF Corp., 1965). Homogeneous (single-species) POE materials show higher initial foam heights but lower foam stabilities than commercial materials of the same nominal structure (Crook et al., 1964).

The replacement of a straight-alkyl-chain hydrophobic group in POE nonionics by a cycloalkyl or 1-alkylcyclohexyl group with the same number of carbon atoms produces little or no decrease in initial foam volume but a marked decrease in foam stability. Somewhat similar effects are produced when the single-alkyl-chain hydrophobic group is replaced by two or three alkyl chains containing the same total number of carbon atoms (Kuwamura et al., 1979). The magnitude of the effect appears to decrease with increase in the number of carbon atoms in the hydrophobic portion and in the length of the POE chain.

The foam of POE nonionics decreases markedly at or above their cloud points. This has been attributed to a rate effect, the cloud point being marked by the aggregation of the dehydrated micelles into larger aggregates. Diffusion of surfactant molecules from these aggregated micelles to the newly created interface involved in bubble formation might be much slower than from the smaller, more highly hydrated micelles, thus decreasing the stabilization of the liquid lamellae in the forming foam (Dupré et al., 1960).

C. Low-Foaming Surfactants

In many industrial processes, it is often useful to add surfactants that can show certain types of surface activity without producing much foam. For example, in paper-making or textile dyeing processes that involve the high-speed movement of belts of material through an aqueous bath, surfactants are added that promote the wetting of the material passing through the bath. However, if the surfactant produces foam when the material is moved rapidly through the bath, then the foam bubbles will adhere to the surface of the material and blemish it. In processes such as these, consequently, low-foaming or nonfoaming surfactants are used.

Low-foaming surfactants can be produced by changing the structure of the surfactant molecule so that it retains its surface activity but produces an unstable foam. We mentioned above that if the surfactant is rapidly diffusing, it can destroy the elasticity of the surface film and thus prevent or minimize foaming.

Therefore, replacing a large, straight-chain hydrophobic group with an isomeric branched-chain one and positioning the hydrophilic group in a central, rather than a terminal, position in the molecule can reduce the foaming properties of the surfactant, while retaining, if not increasing, its surface activity.

Another method of decreasing the foaming of surfactants is to structure the surfactant molecule so that it has a large area/molecule at the liquid–air interface, thus forming a loosely packed noncoherent film that produces unstable foam. This can be accomplished by putting a second hydrophilic group into the molecule some distance from the first one, thus forcing the entire molecule between the two hydrophilic groups to lie flat in the interface. Another way of doing this is to use for the hydrophobic group a relatively short, highly branched or cis-unsaturated alkyl group rather than a long, straight, saturated one or by using a polyoxypropylene chain as part of the hydrophobic group. This type of modification, however, is sometimes not effective if the hydrophilic head already has a sizable cross-sectional area (as in POE nonionics). A third way of increasing the surface area/molecule of surfactant is to put a second hydrophobic group into the molecule, preferably of different size or shape from that of the first hydrophobic group, at some distance from the first one. Thus, high-foaming POE nonionics can be coverted to lower-foaming ones by "capping" the -OH of the POE chain with a short alkyl group or by replacing the terminal -OH group by -Cl. Capping the -OH group or replacing it by -Cl also decreases the cloud point of the POE nonionic and, above the cloud point, may result in the separation of a separate surfactant phase that can act as a foam breaker. Foaming decreases with increase in the length of the alkyl cap from CH_3 to C_4H_9 (Pryce et al., 1984). A fourth method is to put two bulky hydrophilic groups (e.g., POE chains) on the same carbon atom, thereby causing them to extend in different directions, increasing the area per molecule at the surface.

TABLE 7.3 Structures of Some Very Low-Foaming Surfactants

Structure		Reference
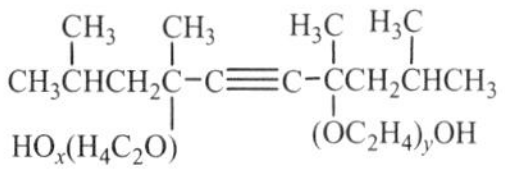 $CH_3CHCH_2C(CH_3)(O(C_2H_4O)_xH)-C\equiv C-C(CH_3)(CH_2CHCH_3CH_3)(OC_2H_4)_yOH$ (i.e., CH_3, CH_3 / $CH_3CHCH_2C-C\equiv C-CCH_2CHCH_3$ / $HO_x(H_4C_2O)$, $(OC_2H_4)_yOH$; H_3C, H_3C)	$x + y \leq 4$	Leeds et al. (1965)
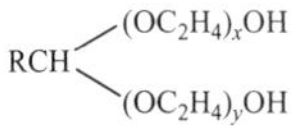 $RCH<^{(OC_2H_4)_xOH}_{(OC_2H_4)_yOH}$	$R < C_{11}, x = y \leq 5$	Kuwamura and Takahashi (1972)
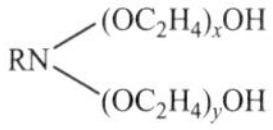 $RN<^{(OC_2H_4)_xOH}_{(OC_2H_4)_yOH}$	$R = C_{10}, x = y \leq 3$	Ikeda et al. (1984)
$HO(C_2H_4O)_x(CH_2)_{12}(OC_2H_4)_yOH$	$x + y \leq 12$	Takahashi et al. (1975)
$HO(C_2H_4O)_x(CH_2CH_2CH_2CH_2O)_y(C_2H_4O)_zH$	$y \leq 27, x + z \leq 82$	Kuwamura et al. (1971)
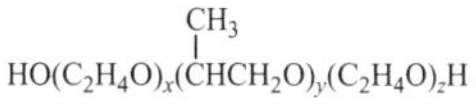 $HO(C_2H_4O)_x(CH(CH_3)CH_2O)_y(C_2H_4O)_zH$	$y = 35, x + z = 45$	Kuwamura et al. (1971)
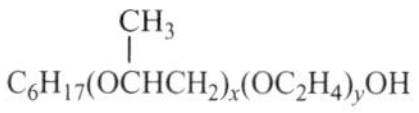 $C_6H_{17}(OCH(CH_3)CH_2)_x(OC_2H_4)_yOH$	$x = y \sim 10$	Kucharski (1974)

Structures of some very low-foaming surfactants are listed in Table 7.3. These materials all produce foam that disappears completely, or almost completely, within a few minutes.

IV. FOAM-STABILIZING ORGANIC ADDITIVES

The foaming properties of surfactant solutions can be modifed greatly by the presence or addition of other organic materials. Solutions that show excellent foaming properties can be converted to low- or nonfoaming material, and those that show poor foaming properties can be converted to high-foaming products by the addition of small amounts of the proper additive. Because of its practical importance, this method of modifying foaming properties has been extensively used and investigated.

Additives that increase the rate of attainment of surface tension equilibrium act as foam inhibitors by decreasing film elasticity, while those that decrease the rate of attainment of that equilibrium act as foam stabilizers. By decreasing the CMC of the surfactant solution and thereby lowering the activity of the monomeric surfactant in solution, an additive may decrease the rate of migration of the surfactant to the surface and the rate of attainment of surface tension equilibrium, with consequent increase in foam stability. On the other hand, additives that cause the breakdown of micelles, with the consequent increase in the activity of the monomeric surfactant, increase the rate of attainment of surface tension equilibrium and decrease foaming (Ross and Hauk, 1958). Another mechanism by means of which additives can act as foam stabilizers is by increasing the mechanical strength of foam films. The surface films produced by solutions of highly purified surfactants are often weakly coherent, containing molecules that are relatively widely spaced because of the mutual repulsion of the oriented polar heads. These films are mechanically weak and nonviscous. When they constitute the interfacial film in the lamellae of a foam, liquid drains rapidly from the lamellae. The addition of the proper additive to this type of film can convert it to a closer-packed, more coherent one of high surface viscosity, which is slow-draining and produces a much more stable foam. Such slow-draining films can be produced by additives (e.g., linear alcohols of intermediate or long chain length) that form liquid crystalline structures with the surfactant (Maner et al., 1982). On the other hand, additives that destroy liquid crystalline structures (e.g., short- or branched-chain alcohols) promote drainage and decrease foam.

Since micelles can solubilize organic additives and thereby remove them from the interface, much larger amounts of foam-stabilizing additives are required to stabilize the foam of aqueous solutions above their CMC than below their CMC.

The most effective additives for increasing the stability of the foam produced by surfactant solutions appear to be long-chain, often water-insoluble, polar compounds with straight-chain hydrocarbon groups of approximately the same length as the hydrophobic group of the surfactant. Examples are lauryl alcohol for use with sodium dodecyl sulfate, *N*,*N*-bis(hydroxyethyl) lauramide for use with dodecylbenzenesulfonate, lauric acid for use with potassium laurate, and *N*,*N*-dimethyldodecylamine oxide for use with dodecylbenzenesulfonate and other anionics.

Studies of the effectiveness of these additives in stabilizing the foam of various types of anionic surfactants indicate that the foam produced by straight-chain surfactants is more susceptible to stabilization than that produced by branched-chain materials. The order of susceptibility to foam stabilization is as follows: primary alkyl sulfates > 2-*n*-alkanesulfonates > secondary alkyl sulfates > *n*-alkylbenzenesulfonates > branched-chain alkylbenzenesulfonates (Sawyer and Fowkes, 1958). This is exactly the order of decreasing van der Waals interaction with an adjacent compound containing a straight-chain hydrocarbon group. Moreover, the most effective foam-stabilizing compounds are those that lower the CMC of the surfactant solution considerably (Schick and Fowkes, 1957). Since the CMC of a surfactant in aqueous solution is not lowered significantly by solubilization of the material into the interior of the micelle, but only by solubilization between the surfactant molecules in the outer portion of the micellar core, the so-called palisade layer (Chapter 3, Sections IVC and D), it appears that the additive operates by penetrating into the surface film and organizing the surfactant molecules into a condensed structure by orienting itself between the molecules of surface-active agent in the film in a manner similar to that in the palisade layer of a micelle.

The increased cohesion of the resulting film may be due to the presence of a nonionic, polar "buffer" between the mutually repelling ionic heads of the surfactant molecules to which both ionic heads are attracted by ion–dipole interactions, whereas the hydrocarbon portions of all the molecules are held together by van der Waals forces. This would account for the greater susceptibility of surfactants having straight-chain, compared to branched-chain, hydrophobic groups to foam boosting and for the greater effectiveness of additives having straight-chain, compared to branched-chain, hydrophobic groups.

Polar additives may also increase foam stabilization by solubilizing foamicidal oils (Schick and Fowkes, 1957), since micelles containing solubilized polar additives (Chapter 4, Section IB4) have increased solubilization power for nonpolar materials.

The nature of the polar group in these additives is important. It has been found (Sawyer and Fowkes, 1958) that the order of effectiveness in these additives is *N*-polar substituted amides > unsubstituted amides > sulfolanyl ethers > glycerol ethers > primary alcohols. This order may be that of decreasing ability to form hydrogen bonds with the adjacent surfactant and water molecules since film viscosity increases greatly where hydrogen bonding between adjacent molecules is possible. The OH group in an alcohol is *not* capable sterically of forming direct bonds with adjacent molecules containing only OH groups, whereas the -CONH- group is capable of direct bonding with adjacent molecules. Also, foam stabilization is greater for those additives containing more than one polar group capable of forming hydrogen bonds. The explanation given for this is that the multiple hydrogen bonds with water prevent the polar additives from being forced out from between the surfactant molecules and into the interior of the micelles in the bulk phase.

Another foam stabilizer for anionic surfactants, *N*,*N*-dimethyldodecylamine oxide, appears to operate in a somewhat different manner. Here it has been shown (Kolp et al., 1963; Rosen et al., 1964) that interaction occurs between the protonated amine oxide cation, $RN(CH_3)_2OH^+$, and the surfactant anion, yielding a product that has been isolated, $RN(CH_3)_2OH^+ . {}^-O_3SR$, in which cation and anion are very strongly hydrogen-bonded via the H^+ of the cation. This compound is much more surface active than either the amine oxide or the anionic surfactant and adsorbs strongly at the air–water interface to form a closely packed film (Rosen et al., 1964). Similarly, salts of long-chain amines and alkyl sulfonates of equal chain length, for example, $C_{10}H_{21}SO_3^- . {}^+N(CH_3)_3C_{10}H_{21}$, have been shown to produce unusually stable thin aqueous films because strong electrical attraction between the cationic and anionic surface-active ions promotes formation of a close-packed surface film (Corkill et al., 1963).

As described in Chapter 4, Section IB5, surfactants interact with polymers to form complexes, the strength of the surfactant–polymer interaction being dependent upon the chemical structures of the polymer and surfactant. These complexes adsorb at the air–aqueous solution interface, causing reduction in surface tension and increase in surface viscosity, with resulting changes in foaming effectiveness and foam stability.

The effect of the adsorbed surfactant–polymer complex on the rheology of the air–aqueous solution interface is easily detected by the "talc particle" test (Regismond et al., 1997). A small quantity of calcined talc powder is sprinkled on the surface of the aqueous solution in a 10-cm petri dish. A gentle current of air is directed tangentially to the talc particles for 1–2 s and then removed. The observed movement is noted in the following categories: fluid (F), viscous (V), gel (G) (= almost no flow), solid (S) (= no flow), and viscoelastic (VE) (= net movement, with some recovery upon removal of air current).

The effect of interaction of sodium dodecyl sulfate with the polymer, polyvinylpyrrolidone, on the foaming of aqueous solutions of the former has been investigated by Folmer and Kronberg (2000). Depending upon the surfactant and polymer concentrations, the foaming can either be decreased or increased. Foaming increases when surface and/or bulk viscosities are increased by the surfactant–polymer concentration; it decreases when surfactant–polymer interaction in the bulk phase causes desorption of them from the air–aqueous solution interface.

V. ANTIFOAMING

When undesirable foaming of a solution cannot be reduced sufficiently by replacing the surfactant with a lower foaming one, or when the foam is caused partially or entirely by nonsurfactant components of the solution, then antifoaming agents are used to reduce the foam. Antifoaming agents act in various ways:

1. *By Removing Surface-Active Material from the Bubble Surface.* The decreased foaming shown by surfactant solutions in the presence of certain types of soil is often due to this mechanism: surfactant removal from the surface by adsorption onto or dissolution in the soil (Princen and Goddard, 1972). Finely divided hydrophobic silica particles dispersed in silicone oil are effective anifoaming agents and are believed to act in this fashion by adsorbing surfactant molecules from the bubble surface and carrying them into the aqueous phase (Kulkarni et al., 1979). Hydrophobic particles also destabilize foam by forming lenses at the Plateau borders of the foam, promoting dewetting of the film lamellae and causing bubble coalescence (Wang et al., 1999).
2. *By Replacing the Foam-Producing Surface Film with an Entirely Different Type of Film that Is Less Capable of Producing Foam.* One method of doing this is by swamping the surface with rapidly diffusing noncohesive molecules of limited solubility in the solution. These must produce a surface tension low enough so that they can spread spontaneously over the existing film (i.e., their spreading coefficient over the surface, $S_{L/S} = \gamma_{SA} - \gamma_{SL} - \gamma_{LA}$ [Equation 6.1], must be positive). The rapidly diffusing molecules at the surface produce a surface film with little or no

elasticity, since transient surface tension gradients producing film elasticity (Section I) are rapidly destroyed by the fast-diffusing molecules. Some wetting agents with limited solubility in water, for example, tertiary acetylenic glycols, act in this manner. Ethyl ether ($\gamma = 17$ mN/m; dyn/cm) and isoamyl alcohol ($\gamma = 23$ mN/m; dyn/cm) are believed to act as foam breakers by reducing the surface tension in local areas to exceptionally low values, thereby causing these areas to be thinned rapidly to the breaking point by the pull of the surrounding higher-tension regions (Okazaki and Sasaki, 1960). Another method is by replacing the elastic surface film with a brittle, close-packed surface film. Calcium salts of long-chain fatty acids (stearic and palmitic) do this with the foam of sodium dodecylbenzenesulfonate or sodium lauryl sulfate by displacing it from the surface film and replacing it wholly or in part by calcium soap molecules that form a "solid," brittle film having no elasticity. This calcium soap film consequently produces an unstable foam. If the calcium soap can form a true mixed film with the surfactant, the foam is *not* destroyed by the calcium soap (Peper, 1958).

3. *By Promoting Drainage in the Foam Lamellae.* Tributyl phosphate is believed to act as an antifoaming agent in this manner by reducing surface viscosity sharply. It has a large cross-sectional area at the aqueous solution–air interface. By intercalating between the surfactant molecules in the interfacial film, it reduces the cohesive forces between them and consequently the surface viscosity. Symmetrical tetraalkylammonium ions may also act in this manner to destabilize the foam of sodium lauryl sulfate solutions. The surface viscosity decreases, accompanying increase in the area per surfactant molecule at the air–aqueous solution as the alkyl chain length of the quaternary ammonium ion increases, with resultant decrease in foam stability. Tetrapentylammonium bromide is particularly effective as a foam destabilizer (Blute et al., 1994).

VI. FOAMING OF AQUEOUS DISPERSIONS OF FINELY DIVIDED SOLIDS

When the aqueous system contains finely divided solids, then foaming of the system may be influenced greatly by the nature of the dispersed solid particles. If the particles have a surface that is hydrophobic, and if the particles are divided finely enough, then the particles may adsorb onto the surface of any air bubbles introduced into the system and stabilize them against coalescence. They adsorb at the air–solid interface from the aqueous system because their solid–aqueous solution interfacial tension, γ_{SL}, is high, and their solid–(nonpolar) air interfacial tension, γ_{SA}, is low because of their nonpolar surface. Consequently, their contact angle, θ, with the aqueous phase, from Equation 6.3

$$\gamma_{LA} \cos\theta = \gamma_{SA} - \gamma_{SL}$$

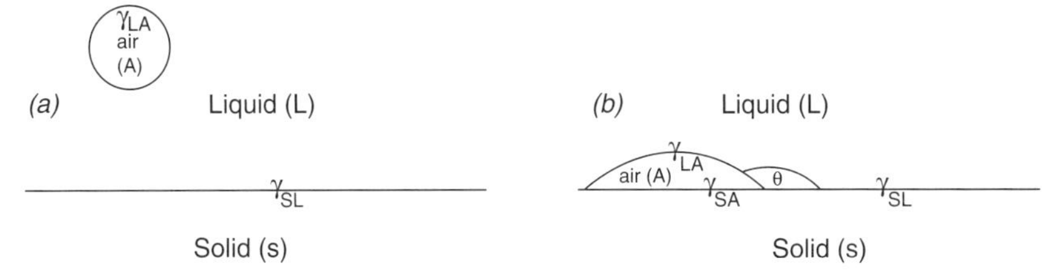

FIGURE 7.3 (a) Before adhesion of the solid to the air. (b) After adhesion of the solid to the air. Note the large value of θ measured in the liquid.

is high, and their contact angle with the air is consequently low. The work of adhesion of the solid to the air

$$W_a = \gamma_{LA} + \gamma_{SL} - \gamma_{SA} \tag{7.8}$$

as shown in Figure 7.3, is consequently large, and adhesion of the solid to the air is strong. The air bubbles in this case are stabilized not by a film of liquid, but by a film of adsorbed solid.

This type of adsorption is the basis for a number of important industrial processes, notably the separation of mineral ores by "froth flotations" (Somasundaran and Ananthapadmanabhan, 1979), the de-inking of waste paper (Turai, 1982), the ultrapurification of fine powders for the chemical and ceramic industries (Moudgil and Behl, 1991), and the production of "foamed" concrete. In the last case, for the concrete to entrain air, it is not even necessary for the liquid phase to show any foaming. In most of the processes used above, surfactants, such as salts of long-chain carboxylic acids or long-chain amines, that adsorb with their polar or ionic heads oriented toward the solid and their hydrophobic groups oriented away from it, are used to make the surface of the solid hydrophobic.

VII. FOAMING AND ANTIFOAMING IN ORGANIC MEDIA

Although many of the examples that had been described pertain to the formation of foams in aqueous media, foaming in nonaqueous media is of practical concern in diverse applications. The petroleum industry faces foaming problem because the natural surface-active agent asphaltene tends to stabilize foam. Despite the concern over the use of volatile organic compounds, many industrial formulations such as inks, paints, and cosmetics still use organic media to disperse the constituents. Chemical manufacturers also have to tackle the nuisance of foams during distillations from organic solvents.

The low surface tension of organic solvents is a prime foam-stabilizing source as it favors large interfacial areas. Increasing the bulk viscosity tends to stabilize the foam by reducing the rate of film thinning. Formulators control the order of addition of high viscosity constituents (thickeners) in a formulation, and these are added only after the foaming problem is tackled.

Effective antifoaming agents have been developed to control the foam in organic media. The earlier use of hydrocarbon-based mineral oils, particulates such as silica, fatty esters, and waxes to combat foams in organic media was replaced to a great extent by silicone-based antifoaming agents such as polydimethylsiloxane (PDMS, structure below) with hundreds of repeating units (see Chapter 1, Section 3C8). The efficacy of such antifoaming agents can be enhanced by admixture with hydrophobic particles.

$$H_3C{-}\overset{H_3C}{\underset{H_3C}{SiO}}{-}\left[\overset{CH_3}{\underset{CH_3}{SiO}}\right]_x{-}\overset{CH_3}{\underset{CH_3}{Si}}{-}CH_3$$

REFERENCES

Blute, I., M. Jansson, S. G. Oh, and D. O. Shah (1994) *J. Am. Oil Chem. Soc.* **71**, 41.

Broich, F. (1966) *Seifen-Ole-Fette-Wachse* **92**, 853.

Caskey, J. A. and W. B., Jr. Barlage (1972) *J. Colloid Interface Sci.* **41**, 52.

Corkill, J. M., J. F. Goodman, C. P. Ogden, and J. R. Tate (1963) *Proc. R. Soc.* **273**, 84.

Crook, E. H., D. B. Fordyce, and G. F. Trebbi (1964) *J. Am. Oil Chem. Soc.* **41**, 231.

Dahanayake, M., Surface chemistry fundamentals: thermodynamic and surface properties of highly purified model surfactants. Doctoral dissertation, City University of New York, 1985.

Davies, J. T. and E. K. Rideal, *Interfacial Phenomena*, 2nd ed., Academic, New York, 1963.

Dhara, D. and D. O. Shah (2001) *Langmuir* **17**, 7233.

Dreger, E. E., G. I. Keim, G. D. Miles, L. Shedlovsky, and J. Ross (1944) *Ind. Eng. Chem.* **36**, 610.

Dupré, J., R. E. Wolfram, and D. R. Fordyce (1960) *Soap Chem. Specs.* **36** (2), 55; (3), 55.

Ewers, W. E. and K. L. Sutherland (1952) *Aust. J. Sci. Res.* **A5**, 697.

Folmer, B. M. and B. Kronberg (2000) *Langmuir* **16**, 5987.

GAF Corp., Tech. Bull. 7543-002, 1965.

Gibbs, J. W. (1878) *Trans. Conn. Acad.* **3**, 343; Collected works 1, 269, 300, Longmans, Green, New York, 1931.

Gray, F. W., J. F. Gerecht, and I. J. Krems (1955) *J. Org. Chem.* **20**, 511.

Gray, F. W., I. J. Krems, and J. F. Gerecht (1965) *J. Am. Oil Chem. Soc.* **42**, 998.

Ikeda, I., A. Itoh, P.-L. Kuo, and M. Okahara (1984) *Tenside Detgts.* **21**, 252.

Jha, B. K., A. Patist, and D. O. Shah (1999) *Langmuir* **15**, 3042.

Kitchener, J. A. and C. F. Cooper (1959) *Quart. Rev.* **13**, 71.

Kölbel, H. and P. Kuhn (1959) *Angew. Chem.* **71**, 211.

Kölbel, H., D. Klamann, and P. Kurzendorfer, *Proc. 3rd Int. Congr. Surface-Active Substances,* Cologne, 1960a, I, p. 1.

Kölbel, H., D. Klamann, and E. Wagner, *Proc. 3rd Int. Congr. Surface-Active Substances,* Cologne, 1960b, I, p. 27.

Kolp, D. G., R. G. Laughlin, F. R. Krause, and R. E. Zimmerer (1963) *J. Phys. Chem.* **67**, 51.

Kondo, T., K. Meguro, and S. Sukigara (1960) *Yukagaku* **9**, 63. [C. A. **54**, 21797 (1960)].

Kucharski, S. (1974) *Tenside Detgts.* **11**, 101.

Kulkarni, R. D., E. D. Goddard, and M. R. Rosen (1979) *J. Soc. Cosmet. Chem.* **30**, 105.

Kuwamura, T. and H. Takahashi (1972) *Bull. Chem. Soc. Jpn.* **45**, 617.

Kuwamura, T., H. Takahashi, and T. Hatori (1971) *J. Am. Oil Chem. Soc.* **48**, 29.

Kuwamura, T., M. Akimaru, H. Takahashi, and M. Arai (1979) *Rep. Asahi Glass Found. Ind. Tech.* **35**, 45.

Leeds, M. W., R. J. Tedeschi, S. J. Dumovich, and A. W. Casey (1965) *Ind. Eng. Chem. Prod. Res. Dev.* **4**, 236.

Lunkenheimer, K. and K. Malysa (2003) *J. Surfactants Deterg.* **6**, 69.

Maner, E. D., S. V. Sazdanova, A. A. Rao, and D. T. Wasan (1982) *J. Disp. Sci. Tech.* **3**, 435.

Marangoni, G. (1872) *Il Nuovo Cimento* **2**, 239.

Matheson, K. L. and T. P. Matson (1983) *J. Am. Oil Chem. Soc.* **60**, 1693.

Micich, T. J., E. A. Diamond, R. G. Bistline, A. J. Stirton, and W. C. Ault (1966) *J. Am. Oil Chem. Soc.* **43**, 539.

Moudgil, B. M. and S. Behl, in *Surfactants in Solution*, Vol. 11, K. L. Mittal and D. O. Shah (eds.), Plenum, New York, 1991, p. 457.

Okazaki, S. and T. Sasaki (1960) *Bull. Chem. Soc. Jpn.* **33**, 564.

Patist, A., V. Chhabra, R. Pagidipati, and D. O. Shah (1997) *Langmuir* **13**, 432.

Peper, H. (1958) *J. Colloid Sci.* **13**, 199.

Princen, H. M. and E. D. Goddard (1972) *J. Colloid Interface Sci.* **38**, 523.

Princen, H. M., J. Th. G. Overbeek, and S. G. Mason (1967) *J. Colloid Interface Sci.* **24**, 125.

Pryce, A., R. Hatton, M. Bell, and P. Lees, Proc. World Surfactants Congr., Munich, May 1984, Kurle Druck and Verlag, Geinhausen, FRG, III, p. 51.

Regismond, S. T. A., K. D. Gracie, F. M. Winnik, and E. D. Goddard (1997) *Langmuir* **13**, 5558.

Rosen, M. J. (1967) *J. Colloid Interface Sci.* **24**, 279.

Rosen, M. J. and J. Solash (1969) *J. Am. Oil Chem. Soc.* **46**, 399.

Rosen, M. J. and Z. H. Zhu (1988) *J. Am. Oil Chem. Soc.* **65**, 663.

Rosen, M. J., D. Friedman, and M. Gross (1964) *J. Phys. Chem.* **68**, 3219.

Rosen, M. J., X. Y. Hua, and Z. H. Zhu, in *Surfactants in Solution*, Vol. II, K. L. Mittal and D. O. Shah (eds.), Plenum, New York, 1991, p. 315.

Ross, J. and G. D. Miles, Am. Soc. for Testing Materials, Method D1173-53, Philadelphia, PA, 1953; *Oil Soap* **18**, 99 (1941).

Ross, S. and R. M. Hauk (1958) *J. Phys. Chem.* **62**, 1260.

Sawyer, W. M. and F. M. Fowkes (1958) *J. Phys. Chem.* **62**, 159.

Schick, M. J. and E. A. Beyer (1963) *J. Am. Oil Chem. Soc.* **40**, 66.

Schick, M. J. and F. M. Fowkes (1957) *J. Phys. Chem.* **61**, 1062.

Shah, D. O., in *Micelles, Microemulsions and Monolayers*, D. O. Shah (ed.), Marcel Dekker, New York, 1998, pp. 1–52.

Sheludko, A., *Colloid Chemistry*, Elsevier, Amsterdam, 1966.

Smith, F. D., A. J. Stirton, and M. V. Nunez-Ponzoa (1966) *J. Am. Oil Chem. Soc.* **43**, 501.

Somasundaran, P. and K. P. Ananthapadmanabhan, in *Solution Chemistry of Surfactants*, Vol. 2, K. L. Mittal, (ed.), Plennum, New York, 1979, p. 777.

Stirton, A. J., R. G. Bistline, J. K. Weil, W. C. Ault, and E. W. Maurer (1962) *J. Am. Oil Chem. Soc.* **39**, 128.

Takahashi, H., T. Fujiwara, and T. Kuwamura (1975) *Yukagaku* **24**, 36.

Tamura, T., Y. Kaneko, and M. Ohyama (1995) *J. Colloid Interface Sci.* **173**, 493.

Turai, L. L., in *Solution Behavior of Surfactants*, Vol. 2, K. L. Mittal and E. J. Fendler (eds.), Plenum, New York, 1982, p. 1381.

Varadaraj, R., J. Bock, P. Valint, S. Zushma, and N. Brons (1990) *J. Colloid Interface Sci.* **140**, 31.

Wang, G., R. Pelton, A. Hrymak, N. Shawatafy, and Y. M. Heng (1999) *Langmuir* **15**, 2202.

Ward, A. F. H. and L. Tordai (1946) *J. Chem. Phys.* **14**, 453.

Weil, J. K., A. J. Stirton, and R. G. Bistline (1954) *J. Am. Oil Chem. Soc.* **31**, 444.

Weil, J. K., A. J. Stirton, R. G. Bistline, and W. C. Ault (1960) *J. Am. Oil Chem. Soc.* **37**, 679.

Weil, J. K., A. J. Stirton, and E. A. Barr (1966) *J. Am. Oil Chem. Soc.* **43**, 157.

Wrigley, A. N., F. D. Smith, and A. J. Stirton (1957) *J. Am. Oil Chem. Soc.* **34**, 39.

PROBLEMS

7.1 Explain why film elasticity is greatest in the region of the CMC.

7.2 Discuss two properties of surfactants that account for the existence of film elasticity.

7.3 Describe two different mechanisms by means of which antifoams operate.

7.4 Give structural formulas for three different types of low-foaming surfactants, indicating the structural characteristics that cause them to foam poorly.

7.5 Calculate the time it would take for the surface concentration to reach a value of 2×10^{-10} mol/cm^2 from a 1×10^{-2} M solution of surfactant in the absence of stirring or an energy barrier to adsorption. Assume the bulk diffusion constant of the surfactant to be 2×10^{6} cm^2/s.

7.6 Explain the observation that aqueous solutions of branched-chain surfactants show higher initial foam heights but poorer foam stability than their linear-chain isomeric surfactant based upon

(a) their equilibrium interfacial properties.

(b) their dynamic interfacial properties.

7.7 Aqueous solutions of some surfactants show good foaming at room temperature but very poor foaming at higher temperatures.

(a) Suggest a type of surfactant that shows this foaming behavior and explain the behavior.

(b) At what temperature would you expect to observe this behavior?

7.8 Design (synthetic scheme is optional) novel surfactant structures that have

(a) maximum foaming and/or antifoaming

(b) minimum foaming and/or antifoaming properties in aqueous/organic solvents. (Awareness of current literature is highly recommended.)

7.9 After reading Chapters 6 and 7, can you establish the connection between surfactant structure of a particular class and its general effectiveness as a wetting agent, foaming agent, or an antifoaming agent?

8 Emulsification by Surfactants

Emulsification—the formation of emulsions from two immiscible liquid phases—is probably the most versatile property of surface-active agents for practical applications and, as a result, has been extensively studied. Paints, polishes, pesticides, metal cutting oils, margarine, ice cream, cosmetics, metal cleaners, nanoparticles, and textile processing oils all have emulsion technology in common, that is, they are either emulsions *per se* or use emulsions in their preparations, or are used in emulsified form. Since there are a number of books, chapters of books, and review articles devoted to emulsions and emulsification (Sjoblom, 1996; Solans and Kunieda, 1996; Becher, 2001; Boutonnet et al., 2008; McClements, 2010), the discussion here covers only those aspects of emulsification that bear on the role of surfactants in this phenomenon.

An *emulsion* is a significantly stable suspension of particles of liquid of a certain size within a second, immiscible liquid. The term *significantly stable* means relative to the intended use and may range from a few minutes to a few years. Investigators in this field distinguish between three different types of emulsions, based upon the size of the dispersed particles: (1) *macroemulsions*, the most well-known type, opaque emulsions with particles >400 nm (0.4 μm), easily visible under a microscope; (2) *microemulsions*, transparent dispersions with particles <100 nm (0.1 μm) in size; and (3) *nanoemulsions (miniemulsions)*, a type that is blue-white, with particle sizes between those of the first two types (100–400 nm [0.1–0.4 μm]). *Multiple emulsions* (Matsumoto et al., 1976), in which the dispersed particles are themselves emulsions, have been the subject of considerable investigation.

Two immiscible, pure liquids cannot form an emulsion. For a suspension of one liquid in another to be stable enough to be classified as an emulsion, a third component must be present to stabilize the system. The third component is called the *emulsifying agent*, and it is usually a surface-active agent, although not necessarily of the type that is usually considered a surface-active agent (finely divided solids, for example, may act as emulsifying agents). The emulsifying agent, if of the conventional type, need not be an individual substance; in fact, the most effective emulsifying agents are usually mixtures of two or more substances, as we will see.

Surfactants and Interfacial Phenomena, Fourth Edition. Milton J. Rosen and Joy T. Kunjappu.

I. MACROEMULSIONS

Macroemulsions are of two types, based on the nature of the dispersed phase: oil-in-water (*O/W*) and water-in-oil (*W/O*). The *O/W* type is a dispersion of a water-immiscible liquid or solution, always called the *oil* (*0*), regardless of its nature, in an aqueous phase (*W*). The oil is, in this case, the "discontinuous" (inner) phase; the aqueous phase is the "continuous" (outer) phase. The *W/O* type is a dispersion of water or an aqueous solution (*W*) in a water-immiscible liquid *(0)*. The type of emulsion formed by the water and the oil depends primarily on the nature of the emulsifying agent and, to some extent, on the process used in preparing the emulsion and the relative proportions of oil and water present. In general, *O/W* emulsions are produced by emulsifying agents that are more soluble in the water than in the oil phase, whereas *W/O* emulsions are produced by emulsifying agents that are more soluble in the oil than in the water phase. This is known as the *Bancroft rule* (Bancroft, 1913). *O/W* and *W/O* emulsions are not in thermodynamic equilibrium with each other; one type is usually inherently more stable than the other for a particular emulsifying agent at a given concentration under a given set of conditions. However, one type can be converted to the other by changing conditions. This is called *inversion* of the emulsion.

These two types of emulsions are easily distinguished: (1) An emulsion can readily be diluted with more of the outer phase, but not as easily with the inner phase. Consequently O/*W* emulsions disperse readily in water; *W/O* ones do not, but they do disperse readily in oil. This method works best on dilute emulsions. (2) *O/W* emulsions have electrical conductivities similar to that of the water phase; *W/O* emulsions do not conduct current significantly. (3) *W/O* emulsions will be colored by oil-soluble dyes, whereas *O/W* emulsions show the color faintly, if at all, but will be colored by water-soluble dyes. (4) If the two phases have different refractive indices, microscopic examination of the droplets will determine their nature. A droplet, on focusing upward, will appear brighter if its refractive index is greater than the continuous phase and darker if its refractive index is less than that of the continuous phase. This clearly identifies the substance in the droplet if one knows the relative refractive indices of the two phases. (5) In filter paper tests, a drop of an *O/W* emulsion produces an immediate wide, moist area; a drop of a *W/O* emulsion does not. If the filter paper is first impregnated with 20% cobaltous chloride solution and dried before the test, the area around the drop immediately turns pink if the emulsion is *O/W* and remains blue (shows no color change) if it is *W/O* (Tronnier and Bussins, 1960).

There are three similarities between macroemulsions and foams: (1) They both consist of a dispersion of an immiscible state of matter in a liquid phase. Foams are dispersions of a gas in a liquid; emulsions are dispersions of a liquid in a second immiscible liquid. (2) The tension γ_I at the relevant interface is always greater than zero, and since there is a marked increase in interfacial area ΔA during the process (of emulsification or foaming), the minimum work involved is the product of the interfacial tension and the increase in interfacial area ($W_{min} = \Delta A \times \gamma_I$). (3) The system will spontaneously revert to two bulk phases unless there is an interfacial film present that produces steric and/or electrical barriers to coalescence of the dispersed phase.

On the other hand, there are two significant differences between macroemulsions and foams: (1) The surfactants in the interfacial film of a foam cannot dissolve in the dispersed (gas) phase, while in a macroemulsion the solubility of the surfactants in the liquid being dispersed is a major factor determining the stability of the emulsion. (2) In macroemulsions, both oil and water can serve as the continuous phase, that is, both *O/W* and *W/O* emulsions are commonly encountered, while in foams, only the liquid acts as the continuous phase.

A. Formation

In the formation of macroemulsions, one of the two immiscible liquids is broken up into particles that are dispersed in the second liquid. Since the interfacial tension between two immiscible pure liquids is always greater than zero, this dispersion of the inner liquid, which produces a tremendous increase in the area of the interface between them, results in a correspondingly large increase in the interfacial free energy of the system. The emulsion produced is consequently highly unstable thermodynamically relative to the two bulk phases separated by a minimum area interface. It is for this reason that two immiscible liquids, when pure, cannot form an emulsion. The function of the emulsifying agent is to stabilize this basically unstable system for a sufficient time so that it can perform some function. This the emulsifying agent does by adsorption at the liquid–liquid interface as an oriented interfacial film. This oriented film performs two functions: (1) It reduces the interfacial tension between the two liquids and consequently the thermodynamic instability of the system resulting from the increase in the interfacial area between the two phases. (2) It decreases the rate of coalescence of the dispersed liquid particles by forming mechanical, steric, and/or electrical barriers around them. The steric and electrical barriers inhibit the close approach of one particle to another. The mechanical barrier increases the resistance of the dispersed particles to mechanical shock and prevents them from coalescing when they do collide. In the formation of macroemulsions, the reduction of interfacial tension reduces the amount of mechanical work required to break the inner phase into dispersed particles. In the case of microemulsions, the interfacial tension is reduced, at least temporarily, to such a low value that emulsification can occur spontaneously.

B. Factors Determining Stability

The term *stability*, when applied to macroemulsions used for practical applications, usually refers to the resistence of emulsions to the coalescence of their dispersed droplets. The mere rising or settling of the droplets (*creaming*) because of a difference in density between them and the continuous phase is usually not considered instability. Flocculation or coagulation of the dispersed particles, *without* coalescence of the liquid interior of the particles, although a form of instability, is not considered as serious a sign of instability as coalescence or *breaking* of the emulsion. The factors determining flocculation of the dispersed droplets in an emulsion are the same as those that bear on the flocculation of solid particles in a dispersion (Chapter 9). For a detailed discussion of flocculation in macroemulsions, see Kitchener and Mussellwhite (1968).

The rate of coalescence of the droplets in a macroemulsion is stated to be the only quantitative measure of its stability (Boyd et al., 1972). It can be measured by counting the number of droplets per unit volume of the emulsion as a function of time in a haemocytometer cell under a microscope (Sherman, 1968) or by means of a Coulter centrifugal photosedimentometer (Groves et al., 1964; Freshwater et al., 1966).

The rate at which the droplets of a macroemulsion coalesce to form larger droplets and eventually break the emulsion has been found to depend on a number of factors: (1) the physical nature of the interfacial film, (2) the existence of an electrical or steric barrier on the droplets, (3) the viscosity of the continuous phase, (4) the size distribution of the droplets, (5) the phase volume ratio, and (6) the temperature.

1. Physical Nature of the Interfacial Film The droplets of dispersed liquid in an emulsion are in constant motion, and therefore, there are frequent collisions between them. If, on collision, the interfacial film surrounding the two colliding droplets in a macroemulsion ruptures, the two droplets will coalesce to form a larger one, since this results in a decrease in the free energy of the system. If this process continues, the dispersed phase will separate from the emulsion, and it will break. The mechanical strength of the interfacial film is therefore one of the prime factors determining macroemulsion stability.

For maximum mechanical stability, the interfacial film resulting from the adsorbed surfactants should be condensed, with strong lateral intermolecular forces, and should exhibit high film elasticity. The liquid film between two colliding droplets in an emulsion is similar to the liquid lamella between two adjacent air sacs in a foam (Chapter 7) and shows film elasticity for the same reasons (Gibbs and Marangoni effects).

Since highly purified surfactants generally produce interfacial films that are not close-packed (Table 2.2) and hence not mechanically strong, good emulsifying agents are usually a mixture of two or more surfactants rather than an individual surfactant. A commonly used combination consists of a water-soluble surfactant and an oil-soluble one. The oil-soluble surfactant, which generally has a long, straight hydrophobic group and a hydrophilic head that is only slightly polar, increases the lateral interaction between the surface-active molecules in the interfacial film and condenses it to one that is mechanically stronger than in its absence. Thus, the addition of an amount of lauryl alcohol sufficient to produce a close-packed monomolecular film increases the stability of sodium lauryl sulfate emulsions, as does the addition of NaCl, which, by compressing the electrical double layer, decreases the electrostatic repulsions between the ionic heads and allows the hydrophobic chains of the surfactant to approach each other more closely (Table 2.2). Consistent with this, polyoxyethylene (POE) alcohol emulsifying agents that have a broader distribution of POE chains produce more stable *O/W* emulsions than those with a narrower distribution. They also are stable over a larger temperature range (Saito et al., 1990).

An example of an oil-soluble surfactant and a water-soluble surfactant that are commonly used together as the emulsifying agent for many applications is a sorbitol ester and a POE sorbitol ester. Because of the greater interaction of the POE sorbitol derivative with the aqueous phase, its hydrophilic group extends further into the water than that of the nonoxyethylenated ester, and this is believed to permit the hydrophobic groups of the two materials to approach each other more closely in the interfacial film and to interact more strongly than when each surfactant is present by itself (Boyd et al., 1972). Figure 8.1 illustrates this complex formation at the interface.

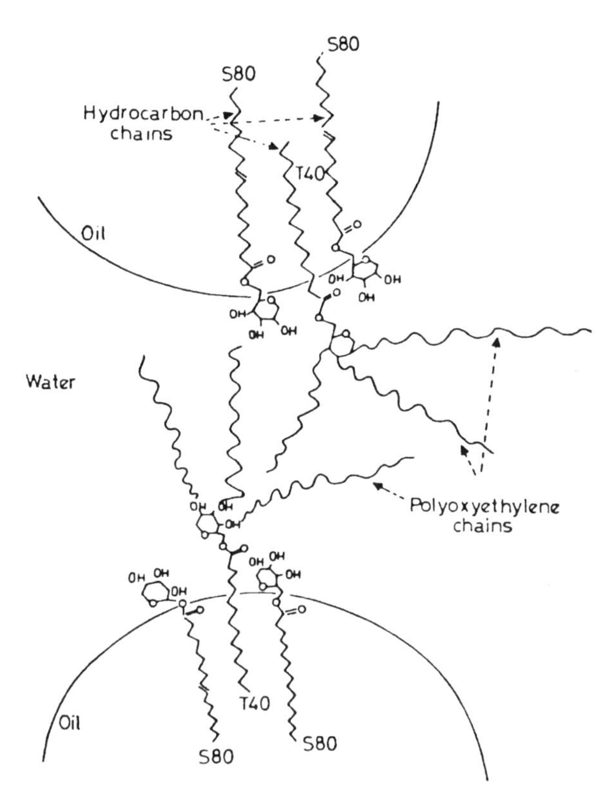

FIGURE 8.1 Complex formation between a Span (S80) and a Tween (T40) at the oil–water interface. Reprinted with permission from Boyd et al. (1972).

Liquid crystal formation can also stabilize the emulsion. By accumulating at the interface surrounding the dispersed particles, liquid crystals surround the particles with a high-viscosity region that resists the coalescence of individual droplets and also acts as a steric barrier (see Section B2) preventing the dispersed particles from approaching each other closely enough for van der Waals forces of attraction (Chapter 9, Section I) to operate (Friberg, 1976).

The films surround the droplets in *W/O* macroemulsions, in particular, must be very strong, and these films are believed to be of the solid-condensed type (Schulman and Cockbain, 1940; Ford and Furmidge, 1966) characterized by very strong lateral intermolecular forces and well-developed orientation of the film with respect to the interface, which confers a good deal of rigidity to the film. This type of film is necessary since the water droplets in a *W/O* emulsion carry little or no charge and therefore have no electrical barrier to coalescence (discussed in the following section). It is therefore mainly the mechanical strength of the interfacial film that prevents coalescence of the droplets in *W/O* macroemulsions, and to survive under the constant bombardment by neighboring droplets, the film must have unusual strength. The great rigidity of the film in these *W/O* emulsions is evidenced by the irregular shape of the water droplets in them, in contrast to the spherical shape of oil droplets in *O/W* emulsions.

2. Existence of an Electrical or Steric Barrier to Coalescence on the Dispersed Droplets The presence of a charge on the dispersed droplets constitutes an electrical barrier to the close approach of two particles to each other. This is believed to be a significant factor only in *O/W* emulsions. In *O/W* emulsions, the source of the charge on the dispersed droplets is the adsorbed layer of surfactant with its hydrophilic group oriented toward the water phase. In emulsions stabilized by ionic surfactants, the sign of the charge on the dispersed droplets is always that of the amphipathic ion. In emulsions stabilized by nonionic surfactants, the charge on the dispersed phase may arise either from adsorption of ions from the aqueous phase or from frictional contact between droplets and the aqueous phase. In the latter case, the phase with the higher dielectric constant is charged positively. In *W/O* emulsions, there is very little charge, if any, on the dispersed particles, and experimental data indicate no correlation between stability and any charge present. In fact, for water-in-benzene emulsions stabilized by oleate soaps of polyvalent metals, an *anticorrelation* was found between zeta potential and stability against coalescence. The true stabilizers in these systems are probably insoluble basic metal oleates produced by hydrolysis of the original metal oleates. Those metal oleates that do not stabilize water-in-benzene emulsions show no hydrolysis and have the highest zeta potentials. The hydrolysis products, if insoluble in both phases, accumulate at the interface and prevent the formation of an electrical double layer in the oil phase. Their accumulation at the interface stabilizes the *W/O* emulsion, since these basic metal oleates are preferentially wetted by the benzene and, in addition, form an interfacial film or layer that mechanically prevents coalescence of the water droplets (Albers et al., 1959). Hydrophobic solid particles stabilize *W/O* emulsions, while hydrophilic solid particles stabilize *O/W* emulsions (Aveyard and Clint, 2003).

The presence of groupings in the interfacial film that may be forced into higher energy arrangements by the close approach to each other of two dispersed droplets constitutes a steric barrier to such approach. Highly hydrated hydrophilic groups on the surfactants constituting the interfacial film in *O/W* emulsions, which may be forced to dehydrate on the close approach to each other of the two dispersed droplets, or long POE chains, which may be forced out of their usual coiled arrangement in water by such approach, may constitute such a barrier. In *W/O* emulsions, long alkyl groups extending into the oil phase from surfactants constituting the interfacial film may produce such a steric barrier.

3. Viscosity of the Continuous Phase An increase in the viscosity η of the continuous phase reduces the diffusion coefficient D of the droplets, since, for spherical droplets,

$$D = \frac{kT}{6\pi\eta a}, \tag{8.1}$$

where k = the Boltzmann constant, T = the absolute temperature, and a = the radius of the droplets.

As the diffusion constant is reduced, the frequency of collision of the droplets and their rate of coalescence are reduced. The viscosity of the external phase is increased as the number of suspended particles increases, and this is one of the reasons that many emulsions are more stable in concentrated form than when diluted. The viscosity of the external phase in emulsions is often increased by the addition of special ingredients for this purpose, such as natural and synthetic "thickening" agents. Friberg (1969) has pointed out the importance of the presence of liquid crystalline phases (Chapter 3, Section IIC) in stabilizing emulsions. At certain concentrations of oil, water, and emulsifying agent, liquid crystalline mesophases that increase the viscosity of the continuous phase may be formed. These can increase the stability of the macroemulsion greatly.

4. Size Distribution of Droplets A factor influencing the rate of coalescence of the droplets is the size distribution. The smaller the *range* of sizes, the more stable the emulsion. Since larger particles have less interfacial surface per unit volume than smaller droplets, in macroemulsions they are thermodynamically more stable than the smaller droplets and tend to grow at the expense of the smaller ones. If this process continues, the emulsion eventually breaks. An emulsion with a fairly uniform size distribution is therefore more stable than one with the same average particle size having a wider distribution of sizes.

5. Phase Volume Ratio As the volume of the dispersed phase in a macroemulsion increases, the interfacial film expands further and further to surround the droplets of dispersed material, and the basic instability of the system increases. As the volume of the dispersed phase increased beyond that of the continuous phase, the type of emulsion (*O/W*) or (*W/O*) becomes basically more and more unstable relative to the other type of emulsion, since the area of the interface that is now enclosing the dispersed phase is larger than that which would be needed to enclose the continuous phase. It often happens, therefore, that the emulsion inverts as more and more of the dispersed phase is added. If the emulsifying agent is so unbalanced as to strongly favor only the original type of emulsion, it may not invert, and may instead form a multiple emulsion, either *W/O/W* or *O/W/O*—the former type when it normally favors *W/O*, the latter type when it normally favors *O/W* (see Section ID).

6. Temperature A change in temperature causes changes in the interfacial tension between the two phases, in the nature and viscosity of the interfacial film, in the relative solubility of the emulsifying agent in the two phases, in the vapor pressures and viscosities of the liquid phases, and in the thermal agitation of the dispersed particles. Therefore, temperature changes usually cause considerable changes in the stability of emulsions; they may invert the emulsion or cause it to break. Emulsifying agents are usually most effective when near the point of minimum solubility in the solvent in which they are dissolved, since at that point they are most surface-active. Since the solubility of the emulsifying agent usually changes with temperature change, stability of the emulsion usually also changes because of this. Finally, anything that disturbs the interface decreases its stability, and the increased vapor pressure resulting from an increase in temperature causes an increased flow of molecules through the interface, with a resulting decrease in stability.

A quantitative expression for the rate of coalescence of droplets in a macroemulsion, which includes most of the factors discussed previously, was developed by Davies and Rideal (1963), based on the von Smoluchowski (1916) theory of the coagulation of colloids.

The rate of diffusion-controlled coalescence of spherical particles in a disperse system as a result of collisions has been shown by von Smoluchowski to be proportional to the collision radius of the particles, the diffusion coefficient, and the square of the concentration of the particles:

$$\frac{-dn}{dt} = 4\pi Drn^2, \tag{8.2}$$

where D = diffusion coefficient, r = collision radius (distance between centers when coalescence begins), and n = number of particles per cm^3.

This assumes that every collision is effective in decreasing the number of particles. In the presence of an energy barrier to coalescence E, which is present in all dispersed systems,

$$\frac{-dn}{dt} = 4\pi Drn^2 e^{-E/kT} \tag{8.3}$$

On integration at constant temperature,

$$\frac{1}{n} = 4\pi Drte^{-E/kT} + \text{constant} \tag{8.4}$$

From the Einstein equation,

$$D = \frac{kT}{6\pi\eta \text{a}}, \tag{8.1}$$

where a is the average radius of the particles, and if we assume that coalescence occurs on contact (i.e., when $r = 2\text{a}$), then

$$\frac{1}{n} = 4\pi \frac{kT}{6\pi\eta \text{a}} 2ate^{-E/kT} + constant \tag{8.5}$$

$$= \frac{4kT}{3\eta} te^{-E/kT} + constant. \tag{8.6}$$

A plot of $1/n$ versus t (n is determined by counting the particles per unit volume of the emulsion under a microscope) then permits the evaluation of E, since the slope of the curve equals

$$\frac{4kT}{3\eta} e^{-E/kT}$$

and k, T, and η are all known constants. It should be noted, however, that E may vary as the size or the number of particles in the emulsion changes.

If we define the mean volume of a particle V = V/n, where V = the volume fraction of the dispersed phase (i.e., the volume per cm^3 of the emulsion), then

$$\bar{V} = \frac{4}{3} \frac{VkT}{\eta} te^{-E/kT} + constant. \tag{8.7}$$

Differentiating this expression yields an expression for the rate of coalescence of the particles and thus for the stability of the emulsion:

$$\frac{d\bar{V}}{dt} = \frac{4}{3} \frac{VkT}{\eta} e^{-E/kt} \tag{8.8}$$

$$= Ae^{-E/kt}. \tag{8.9}$$

A is a constant for a particular system, called the *collision factor*. The effect of the surfactants used as the emulsifying agent is seen in the value of E, the energy barrier to coalescence, which includes both mechanical and electrical barriers.

C. Inversion

Macroemulsions may be changed from *W/O* to *O/W* and vice versa by varying some of the emulsification conditions: (1) the order of addition of the phases (by adding the water to the oil plus emulsifier, a *W/O* emulsion may be obtained, whereas the addition of oil to the same emulsifier plus water may produce an *O/W* emulsion); (2) the nature of the emulsifier (making the emulsifier more oil-soluble tends to produce a *W/O* emulsion, whereas making it more water-soluble tends to produce an *O/W* emulsion); (3) the phase volume ratio (increasing the ratio of oil to water tends to produce a *W/O* emulsion and vice versa); (4) the phase in which the emulsifying agent is dissolved (placing the more hydrophilic of the surfactants used as the emulsifying agent in the aqueous phase appears to favor *O/W* emulsion formation); (5) the temperature of the system (as the temperature of an *O/W* emulsion stabilized with a POE nonionic surfactant is increased, the surfactant becomes more hydrophobic and the emulsion may invert to *W/O)*; on the other hand, some emulsions stabilized by ionic surfactants may invert to *W/O* on cooling; (6) the electrolyte or other additive content (the addition of strong electrolyte to *O/W* emulsions stabilized by ionic surfactants may invert them to *W/O* by decreasing the electrical potential on the dispersed particles and by increasing interaction between the surfactant ions and counterions (thereby making them less hydrophilic); the addition of long-chain alcohols or fatty acids may invert an *O/W* emulsion to *W/O* by making the combination of surfactants acting as emulsifying agents more hydrophobic).

In the process of inverting an *O/W* emulsion to a *W/O* emulsion, any charge on the dispersed oil particles must be removed and an interlinked, solid condensed film formed from the original interfacial film The process has been represented diagrammatically as shown in Figure 8.2 (Schulman and Cockbain, 1940). According to this mechanism, the charged film in the *O/W* emulsion is neutralized and the oil droplets tend to coagulate to form the continuous phase. The trapped water is surrounded by an interfacial film that realigns to form irregularly shaped droplets of water stabilized by a rigid, uncharged film. The result is a *W/O* emulsion.

D. Multiple Emulsions

There has been considerable interest in multiple emulsions, in part because of their potential as a means of (1) delivering drugs to specified targets in the body without the possible deleterious effects of these drugs on other organs and (2) prolonging the release of drugs that have a short biological half-life. Both *W/O/W* and *O/W/O* emulsions exist. In the first type (Figure 8.3a), the water-immiscible liquid (*O*) globules that are suspended in the aqueous (*W*)

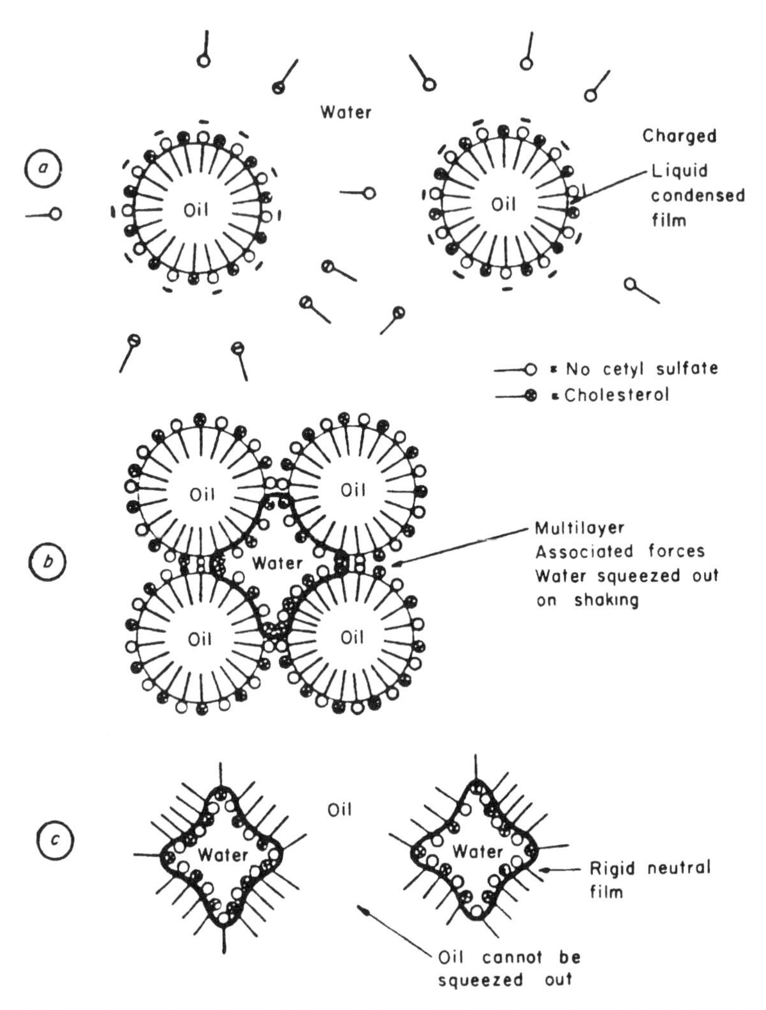

FIGURE 8.2 Inversion of an *O/W* emulsion stabilized by an interfacial film of sodium cetyl sulfate and cholesterol to a *W/O* emulsion upon addition of polyvalent cations. Adsorption of the cations neutralizes the negative charge on the oil droplets, thus allowing them to coalesce. Reprinted with permission from Schulman and Cockbain (1940).

phase themselves contain dispersed globules of an aqueous solution; in O/*W*/*O* emulsions (Figure 8.3b), the globules of aqueous solution suspended in the oil phase contain dispersed oil particles. It is believed that *W/O/W* formation is a mesophase preceding complete inversion of *W/O* to *O/W* emulsions (Matsumoto et al., 1985).

Multiple emulsions of the *W/O/W* type are generally prepared by a two-step procedure (Matsumoto et al., 1976; Garti et al., 1983; Magdassi et al., 1984): a preformed *W/O* emulsion is added slowly, with stiring, to an aqueous solution

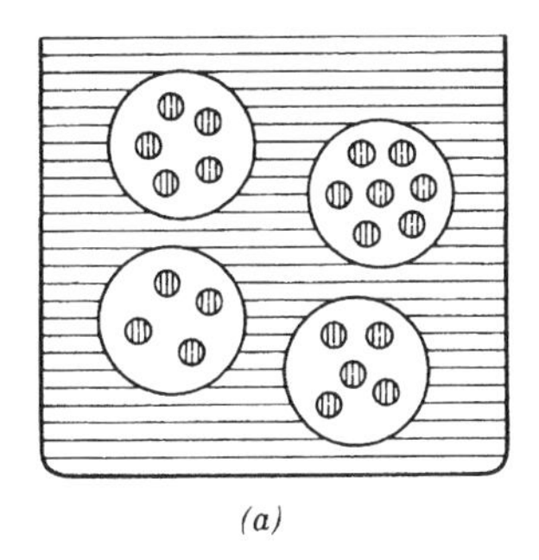

(a)

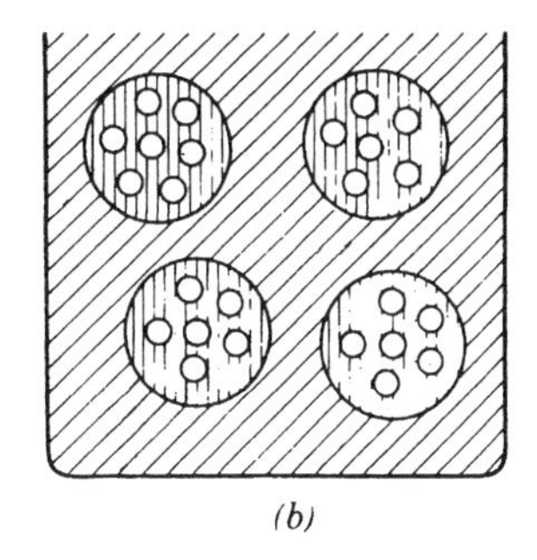

(b)

FIGURE 8.3 Multiple emulsions: (a) *W/O/W* emulsion: ▥, inner *W* phase; □, inner *O* phase; ▤, outer *W* phase. (b) *O/W/O* emulsion: □, inner *O* phase; ▥, inner *W* phase; ▨, outer *O* phase.

containing a hydrophilic emulsifying agent. However, it is possible to form them by a one-step procedure (Matsumoto, 1983) if a dilute aqueous solution of the hydrophilic emulsifying agent (e.g., 1/20–1/50 the concentration of the lipophilic emulsifying agent) is used as the aqueous phase in forming the *W/O* emulsion. Inversion occurred at a water-to-oil volume ratio greater than 0.7, yielding a mixture of *O/W* and *W/O/W* emulsions. For the formation of *W/O/W* emulsions by this technique, a close-packed interfacial film in the *W/O* emulsion is needed.

In general, to obtain good yields of *W/O/W* emulsions, a high concentration of the lipophilic emulsifying agent in the oil phase during the preparation of the *W/O* emulsion and a low concentration of the hydrophilic emulsifier in the aqueous phase during the formation of the *W/O/W* emulsion are required. In some cases, to get >90% yields of *W/O/W* emulsions, the concentration of the lipophilic emulsifying agent in the oil phase had to exceed 30% and be 10–60 times that of the hydrophilic emulsifier (Matsumoto et al., 1976). The presence of anionic surfactant in the hydrophilic emulsifier produced greater stability in the *W/O/W* emulsion (Garti et al., 1983; Matsumoto, 1983), as did the addition of a protein (bovine serum albumin) to the aqueous inner phase (Omotosho et al., 1986).

E. Theories of Emulsion Type

1. Qualitative Theories All qualitative theories explaining the formation of *O/W* and *W/O* emulsions are based on the empirical Bancroft rule. Some investigators believe that the interfacial region produced by the adsorption and orientation of the surface-active molecules at the liquid–liquid interface can have different interfacial tensions (or interfacial pressures) on either of its two sides; that is, the interfacial tension between the hydrophilic ends of the surfactant molecules and the water phase molecules (or the interfacial pressure between the hydrophilic heads) can be different from the interfacial tension between the hydrophobic ends of the surfactant and the oil phase

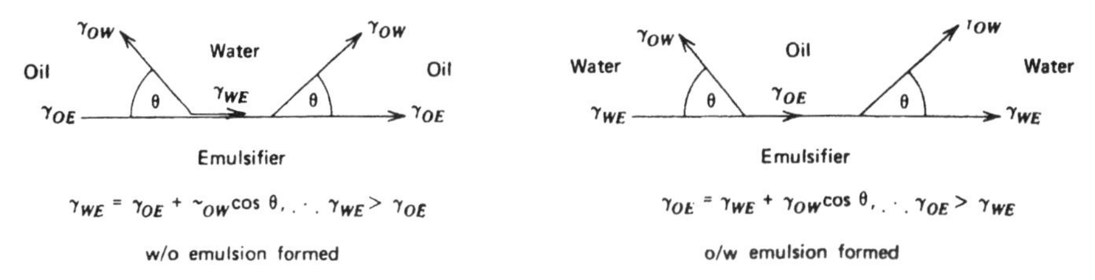

FIGURE 8.4 Effect of contact angle on emulsion type.

molecules (or the interfacial pressure between the hydrophobic ends). In the formation of the emulsion, the interfacial region would tend to curve so as to shorten the area of the side with the greater interfacial tension (or lower interfacial pressure), thus minimizing the interfacial free energy. If the oil-hydrophobic end tension were greater (or interfacial pressure lower) than the water-hydrophilic end tension, then the former side would be shortened, causing the film to be concave toward the oil, resulting in the enclosure of the oil by the water and therefore forming an *O/W* emulsion. On the other hand, if the water-hydrophilic end tension were greater (or interfacial pressure lower) than the oil-hydrophobic end tension, then the former side would be shortened, causing the film to be concave toward the water, forming a *W/O* emulsion. A preferentially oil-soluble emulsifying agent would, of course, produce a lower interfacial tension (or greater interfacial pressure) at the oil interface, yielding a *W/O* emulsion; preferentially water-soluble emulsifying agent would produce a lower interfacial tension (or greater interfacial pressure) at the water interface, yielding an *O/W* emulsion.

Other investigators (Schulman and Leja, 1954) explain the formation of these two types of emulsions on the basis of the difference in contact angles at the oil–water–emulsifier boundary (Figure 8.4). If, at the contact between oil, water, and emulsifier, the oil contact angle (the contact angle, measured in the oil phase) is less than 90°, then the oil surface is concave toward the water, producing a *W/O* emulsion. On the other hand, if at the same oil-water-emulsifier contact, the water contact angle is less than 90°, then the water surface is concave toward the oil, producing an *O/W* emulsion. Note, however, that if the oil contact angle is <90°, then $\gamma_{OE} < \gamma_{WE}$ (i.e., the emulsifier is more hydrophobic than hydrophilic). If the water contact angle is <90°, then $\gamma_{WE} < \gamma_{OE}$, and the emulsifying agent is more hydrophilic than hydrophobic. Thus, emulsifying agents with mainly hydrophilic character produce *O/W* emulsions, whereas those with mainly hydrophobic character produce *W/O* emulsions. This relation is *quantitatively* useful only for emulsifying agents, such as certain solids, that dissolve in neither the oil nor the water phase, or for emulsifying agents adsorbed on solids insoluble in both phases. If the emulsifying agent dissolves in either of the two phases, or in both, its contact angle with each phase in which it dissolves is nonexistent and therefore cannot be measured. Qualitatively, however, it follows that if an emulsifying agent is soluble in only one of the phases, its contact angle with that phase is zero, and therefore less than that which it makes with the phase in which it is not soluble; therefore, the phase in which it is soluble is the continuous phase in the emulsion. Thus, emulsifying agents that are predominantly oil-soluble form *W/O* emulsions, whereas those that are predominantly water-soluble form *O/W* emulsions.

This concept was tested (Schulman and Leja, 1954) for $BaSO_4$-stabilized emulsions containing surface-active agents. When the contact angle measured in the aqueous phase was slightly greater than 90°, a *W/O* emulsion was formed; when it was slightly less than 90°, an *O/W* emulsion was formed. If the angle was much greater than 90°, $BaSO_4$ particles dispersed in the oil phase; if much lower than 90°, they dispersed in the water phase. In the last two cases, the emulsions broke.

Studies on the coalescence of oil and water droplets at the oil–water interface confirm the conclusion that the dispersed phase will be the one that makes the higher contact angle with the emulsifier and have suggested still another factor in the formation of these two types of emulsions. It is believed that stability of the droplets in an external liquid medium is determined mainly by the ability or lack of ability of the material constituting the droplets to wet the adsorbed film between the droplets and the external medium (Cockbain and McRoberts, 1953). If the material in the droplets can wet the interfacial film (which means that a low contact angle exists between the emulsifier comprising the film and the material in the droplet), then the droplets will coalesce and the emulsion will be unstable. If the material in the droplets cannot wet the interfacial film (i.e., a high contact angle exists between emulsifier and material in the droplet), then it will be difficult for the droplets to coalesce, since it is difficult for the external phase to wet the area between the two droplets, and the emulsion will be stable. Therefore, the more stable type of emulsion will be that in which the droplets contain the phase that wets the emulsifier less (i.e., the one that makes the higher contact angle with the emulsifier).

2. Kinetic Theory of Macroemulsion Type Davies (1957) developed a quantitative theory of macroemulsion type relating the type of emulsion formed to the kinetics of coalescence (Equations 8.8 and 8.9) of the two types of droplets present: oil droplets and water droplets. According to this theory, the type of macroemulsion formed when oil and water are agitated together in the presence of an emulsifying agent is due to the relative rates of the two competing processes: (1) coalescence of oil droplets and (2) coalescence of water droplets. Agitation is presumed to break simultaneously both the oil and the water phases into droplets, with the emulsifying agent being adsorbed at the interface around these droplets. The phase that becomes the continuous one is that which has the faster rate of coalescence. If the rate of coalescence of the water droplets is much greater than that of the oil droplets, then an *O/W* emulsion forms; if the rate of coalescence of the oil droplets is much greater than that of the water droplets, then a *W/O* emulsion forms. When the rates of coalescence of the two phases are similar, the phase of larger volume becomes the outer phase.

In general, hydrophilic groups in the interfacial film constitute a barrier to the coalescence of oil droplets, whereas hydrophobic groups in the interfacial film constitute a barrier to the coalescence of water droplets. Hence, an interfacial film that is predominantly hydrophilic tends to form *O/W* emulsions, whereas one that is predominantly hydrophobic tends to produce *W/O* emulsions.

According to Davies, a rate of 10^{-2} times the collision factor (i.e., $Ae^{-E/kT} = 10^{-2}$ A in Equation 8.9) is a fast rate of coalescence, corresponding to complete coalescence of that phase within an hour, whereas a rate of 10^{-5} A is a very slow rate, corresponding to a stability of the order of several months for that phase in a dispersed form. Therefore, if the rate of coalescence of one phase is of the order of 10^{-5} A and the rate of coalescence of the other phase is considerably faster, then a stable emulsion will be formed with the phase having the slower rate as the dispersed phase. On the other hand, if the rate of coalescence of both phases is of the order of 10^{-2} A, then both phases will coalesce rapidly and the emulsion will break, regardless of which phase has the slower rate.

If some substance is added to an emulsion or if some condition is varied, which changes the rates of coalescence of the two phases in such a manner that the rate of coalescence of the continuous phase is decreased very considerably (e.g., to the order of 10^{-5} A) and the rate of coalescence of the dispersed phase is increased very considerably (e.g., to the order of 10^{-2} A), then the emulsion, on agitation, inverts and the two phases switch roles in the emulsion.

II. MICROEMULSIONS

Although microemulsions (called *soluble oils* at that time) have been produced commercially since the 1930s, significant understanding of their nature has been acquired only during the past few decades, mainly as a result of the intense interest generated in them by laboratory and field tests that showed that they could increase the recovery of petroleum from reservoir rock. This is due to the ultralow interfacial tensions (Chapter 5, Section IIIA) attained at the microemulsions–petroleum interface, a prerequisite for the displacement of the residual petroleum in the capillaries of the rock. There has also been considerable recent interest in microemulsions of fluorocarbons as a result of the exceptionally high solubility of O_2 in these systems and their consequent potential as O_2 carriers in cases of circulatory dysfunction (Mathis et al., 1984). In addition, microemulsions are used in the preparation of solid nanoparticles (Barette, 1992), in foods and beverages (Dungan et al., 1997), and as reaction media for organic syntheses (Schomacker, 1992).

Microemulsions are transparent, thermodynamically stable dispersions containing two immiscible liquids with particles of 10–100 nm (0.01–0.1 pm) diameter that are generally obtained upon mixing the ingredients gently. They differ markedly from both macro- and miniemulsions in this respect, since these two types depend upon intense agitation for their formation. Microemulsions may be water-external (*O/W*), oil-external (*W/O*), or both, that is, bicontinuous (Mishra et al., 1989).

Whether one considers a microemulsion to be a solution in one liquid of micelles swollen by a solubilized second liquid or a dispersion of tiny droplets of one liquid in a second liquid, the interfacial tension of the microemulsion against both of these liquids must be close to zero. In the first case, the system is one-phase and therefore has no interface against either liquid as long as the micelles are capable of solubilizing more of the second liquid. In the second case, the interfacial area is so large that an exceedingly low interfacial tension must be present to permit formation of the microemulsion with so little work. In addition, the interfacial region must be highly flexible, either to permit the large curvature required to surround exceedingly small particles or to allow the easy transition from oil-continuous to water-continuous structures that is characteristic of microemulsions.

It is generally accepted that the clear, fluid, middle (surfactant) phase between a nonpolar phase (*O*) and an aqueous phase (*W*) in a three-phase system (Chapter 5, Section III) is a microemulsion; if the concentration of surfactant is increased, the middle phase incorporates both the (oil and water) phases into a single (microemulsion) phase. As discussed in Chapter 5, Section IIIA, the guidelines for formation of this middle phase indicate the conditions for forming microemulsions. The Winsor R ratio (Equation 5.3) measures the solubilization capacity for water relative to oil. Modification of that ratio by changing the structure of the surfactant, changing the temperature of the system, adding a cosurfactant, or adding electrolyte can change the solubilization capacity of the system for either water or oil, or both (Verzaro et al., 1984).

Microemulsions are generally prepared with more than one surfactant or with a mixture of a surfactant and a cosurfactant (e.g., a polar compound of intermediate chain length*), although cosurfactant-free microemulsions have been prepared (Holmberg and Osterberg, 1986). The combination is usually required to provide the proper balance between hydrophilic and lipophilic properties for the required oil and water phases under the conditions of use. This balance can be determined experimentally by mixing the oil and water phases in the desired proportions with the surfactant–cosurfactant combination and noting whether a Winsor Type I, II, III, or IV system (Figure 5.5) is obtained. It is advisable to use graduated vessels for this purpose so that the volumes of the phases can be measured. The surfactant–cosurfactant combination is then adjusted to bring the Winsor R value close to 1. If a three-phase system is finally obtained instead of a one-phase microemulsion, the

*Long chain polar compounds are generally not desirable as cosurfactants since they tend to form liquid crystalline structures that may increase the viscosity of the system and the rigidity of the interface.

concentration of surfactant–cosurfactant mixture can be increased until both water and oil phases disappear by solubilization into the surfactant phase.

As a result of the intensive research done on microemulsions in connection with enhancing the recovery of petroleum from oil reservoir rock, a number of methods have been developed for determining the conditions under which a microemulsion is formed. Thus, for microemulsions with anionic surfactants, Equation 8.10 has been suggested (Salager et al., 1979) to develop the optimum formulation (when the equation equals zero):

$$\ln S - kACN - f(C_A) + \sigma - \alpha_T\ \Delta T = 0, \tag{8.10}$$

where S is the salinity of the aqueous phase in weight % NaCl,

ACN is the carbon number (or equivalent) of the alkane used,

$f(C_A)$ is a function of the alcohol concentration used as a cosurfactant,

σ is a parameter characteristic of the chemical structure of the surfactant used (which increases linearly with the hydrophilic chain length),

ΔT is the temperature deviation from a reference temperature (25°C),

k and α_T are empirical constants.

Equation 8.10 takes into consideration the formulation ingredients that may be necessary to produce a microemulsion with the ultralow interfacial tension (Chapter 5, Section IIIA) required for enhancing the recovery of petroleum from the reservoir rock.

From the equation, it is apparent that as the number of carbon atoms in the alkane (ACN) to be emulsified increases, or as the number of carbon atoms in the hydrophobic group of the surfactant (σ) decreases, the salinity (S) of the solution must be increased to yield a microemulsion.

Equation 8.10, useful in defining the conditions for forming microemulsions from anionics, involves some empirical parameters that can be beneficially interpreted to understand the physical chemistry of amphiphilic layers. Acosta (Acosta et al., 2003, 2008) has redefined σ as the normalized net curvature of the surfactant at reference conditions: $S = 1$ g/100 mL of NaCl which makes $\ln S = 0$; $ACN = 0$ for a hydrocarbon like benzene; since no cosurfactant is used and if the reference surfactant is sodium dihexyl sulfosuccinate that forms a microemulsion unassisted by a cosurfactant, $C_A = 0$; $T = 25°$, which makes $\Delta T = 0$; thus $\ln S = \sigma$), and renamed it as the *characteristic curvature* (C_c). A negative C_c value indicates a tendency to form normal micelles and a positive value, reverse micelles, under these conditions. A simple linear equation for determining C_c of an anionic surfactant is obtained (Equation 8.11; Acosta et al., 2008) by applying Equation 8.10 for a mixture of two anionic surfactants and by approximating ACN as zero for the microemulsion formation condition of $HLD_{mix} = 0$:

$$ln\frac{S^*}{S_1^*} = (Cc_1 - Cc_2)X_2, \tag{8.11}$$

where S^* and S_1^* are the salinities for the reference surfactant (sodium dihexyl sulfosuccinate, for which $Cc_1 = -0.92$) and the investigated surfactant, respectively, and Cc_1 and Cc_2 are their characteristic curvatures, and X_2 is the mole fraction of the investigated surfactant. Thus, the slope of the plot of optimum salinities, S^*/S_1^*, against the mole fraction X_2 of the tested surfactant readily yields its characteristic curvature.

An equation analogous to Equation 8.10 has been developed (Anton et al., 1997) for cationic surfactant-based microemulsions.

For POE nonionic surfactant-based microemulsions, Equation 8.11 has been suggested (Bourrel et al., 1980):

$$\alpha - EON + bS - kACN - \Phi(\mathrm{C_A}) + \alpha_\mathrm{T}\ \Delta\mathrm{T} = 0, \tag{8.12}$$

where α is characteristic of the surfactant hydrophobe (and increases with the number of carbons in it),

EON is the average number of oxyethylene groups in the surfactant hydrophilic group,

S, *ACN*, C_A, and ΔT are the same as in Equation 8.10, and *b*, *k*, *Φ*, and α_T are again empirical constants.

Here, it is apparent that as the number of oxyethylene groups *(EON)* in the surfactant molecule increases, the salinity *(S)* must be increased to yield a microemulsion. In contrast to Equation 8.10, where the sign of the term $\alpha_T\Delta T$ is negative, in Equation 8.12, the sign of the $\alpha_T\Delta_T$ term is positive. This is because ionic surfactants become more water-soluble (more hydrophilic) as the temperature is raised, and the left-hand side of Equation 8.10 should reflect this by becoming less positive (or more negative) with temperature increase. On the other hand, POE nonionics, because of dehydration of their oxyethylene groups as the temperature increases (Chapter 4, Section IIIC), become less water-soluble (more hydrophobic) with this change, and the left-hand side of Equation 8.12 should reflect this by becoming more positive.

The volume of the oil phase (in milliliter) solubilized per gram of surfactant used at the conditions where Equations 8.10 or 8.12 are equal to zero ("optimum salinity") is called the *solubilization parameter* at optimum formulation and symbolized by SP^*. The interfacial tension under these conditions, γ^*, is inversely proportional to the SP^*, and $\gamma^* = K/(SP^*)^2$ (Chun, 1979). Consequently, to obtain the lowest interfacial tension (Chapter 5, Section IIIA), the value of SP^* should be maximized.

Lipophilic linkers (Salager et al., 1998) and hydrophilic linkers (Uchiyama et al., 2000; Acosta et al., 2002) are used to increase the value of SP^* and decrease γ^*. Lipophilic linkers are long-chained alcohols (above C_8) and their low oxyethylenation products that increase the surfactant–oil interaction. The most effective ones have hydrophobic chain lengths that are an average of the hydrophobic chain length of the surfactant and the chain length of the alkane oil. Hydrophilic linkers increase the surfactant–water interaction. Examples are mono- and dimethylnaphthalene sulfonates and sodium octanoate.

III. NANOEMULSIONS

These are also known as *miniemulsions* (Ugelstad et al., 1973; El-Asser et al., 1977, 1984; Grimm et al., 1983; Brouwer et al., 1986), *finely dispersed emulsions* (Sagitani, 1981), or *ultrafine emulsions* (Nakajima et al., 1993). They are blue-white semiopaque emulsions of 100–400 nm (0.1–0.4 μm) droplet size. The emulsifier is generally 1–3% of the oil phase, in contrast to the 15–30% in microemulsions, and is a mixture of an ionic surfactant and a cosurfactant, where the latter is generally a long-chain alcohol. The chain length of the cosurfactant is at least 12 carbons, in contrast to the considerably shorter lengths used in microemulsions. Nanoemulsions are used in the preparation of polymer latices, in cosmetics (where their translucent or transparent appearance makes them especially attractive), and in pharmaceutical drug delivery systems. Nanoemulsions of the *O/W* type are prepared by stirring a mixture of the surfactant and cosurfactant in water for at least an hour to produce a mixed micellar solution. The phase inversion temperature (PIT) method (Section IVB below) is commonly used for their preparation (Forster, 1997). The initial location of the (cosurfactant) fatty alcohol in the aqueous phase is essential for successful emulsification. The mechanism suggested for nano-emulsion formation is swelling of mixed micellar structures by solubilized solvent, followed by breakdown of these swollen structures to tiny droplets of <400 nm diameter. The gain in entropy of mixing during net transfer of fatty alcohol from aqueous phase to oil phase is suggested as the driving force for their formation (Brouwer et al., 1986). Nanoemulsions do not cream or settle because Brownian movement is larger than gravity effects on particles <1 μm.

For styrene nanoemulsions prepared with 10^{-2} *M* sodium lauryl sulfate and a 1:1 molar ratio of ionic surfactant : fatty alcohol, the order of decreasing stability with fatty alcohols of different chain length is $C_{16} > C_{18} > C_{14} > C_{12} > C_{10}$. For sodium lauryl sulfate-C_{16} alcohol mixtures, the order of decreasing stability with different sodium lauryl sulfate : fatty alcohol ratios is 1:3 > 1:2 > 1:1 > 1:6 > 1:0.5. The 1:3 and 1:2 ratios produce emulsions with stabilities >1 month. The presence of rodlike liquid crystalline structures at 1:1 to 1:3 ionic surfactant : fatty alcohol ratios is believed to be essential for the preparation of a stable nanoemulsion. (El-Asser et al., 1984).

This method has been used to prepare nanoemulsions of such polymers as cellulose esters and epoxy resins, similar to latexes produced by emulsion polymerization. The nanoemulsions are prepared by direct emulsification of solutions of the polymers in organic solvents, followed by removal of the organic solvent by steam distillation under reduced pressure. The nanoemulsions produced in this fashion had stabilities >1 year.

IV. SELECTION OF SURFACTANTS AS EMULSIFYING AGENTS

Correlations between the chemical structure of surface-active agents and their emulsifying power are complicated by the fact that both phases, oil and water, are of variable composition. This is in contrast to such phenomena as foaming or wetting, where one phase (the air) is more or less constant and some specific correlations can be made between structure and activity. Moreover, the concentration at which the emulsifying agent is used determines not only its emulsifying power, but even the type of emulsion (*O/W* or *W/O*) formed. As a result of this necessity of taking into consideration the composition of the two phases and the concentration of the emulsifying agent, it is not possible to rate specific surfactants as general emulsifying agents in any particular order. However, there are some general guidelines that can be helpful in the selection of surfactants as emulsifying agents. In general, for a surfactant to act as an emulsifier: (1) It must show good surface activity and produce a low interfacial tension *in the particular system in which it is to be used*. This means that it must have a tendency to migrate to the interface, rather than to remain dissolved in either one of the bulk phases. It must therefore have a balance of lyophilic and lyophobic groups such that it will distort the structure of both bulk phases to some extent, although not necessarily equally. Too great a solubility in either bulk phase will make its usefulness dubious. (2) It must form, at the interface, either by itself or with other adsorbed molecules that are present there, an interfacial film that is condensed because of lateral interactions between the molecules comprising the interfacial film. This means that for *O/W* macroemulsions, the hydrophobic groups in the interfacial film should have strong lateral interactions; for *W/O* macroemulsions, the hydrophilic groups should interact strongly. (3) It must migrate to the interface at a rate such that the interfacial tension is reduced to a low value in the time during which the emulsion is being produced. Since the rate of migration to the interface of a particular surfactant usually varies, depending on whether it is placed in the oil or in the water phase before the emulsification process, its emulsifying behavior often depends on the phase in which it is placed prior to emulsification.

Two other very general guidelines have already been established on the basis of the previous discussion: (1) Emulsifying agents that are preferentially oil-soluble form *W/O* emulsions and (2) a mixture of a preferentially oil-soluble surface-active agent and a preferentially water-soluble one often produces better and more stable emulsions than an individual surfactant. To these can be added a third guideline, which takes into consideration the nature of the oil phase: (3) The more polar the oil phase, the more hydrophilic the emulsifier should be; the more nonpolar the oil to be emulsified, the more lipophilic the emulsifier should be. This generalization is the basis for a number of methods of minimizing the work of selecting the most suitable emulsifying agent or combination of emulsifying agents for a particular system.

A. The Hydrophile–Lipophile Balance (HLB) Method

A frequently used method is known as the *HLB method*. In this method (Griffin, 1949), a number (0–40) indicative of emulsification behavior and related to the balance between the hydrophilic and lipophilic (hydrophobic) portions of the molecule* has been assigned to many commercial emulsifying agents. (In some cases, the HLB number is calculated from the structure of the molecule; in other cases, it is based on experimental emulsification data). In addition, a similar range of numbers has been assigned to various substances that are frequently emulsified, such as oils, lanolin, paraffin wax, xylene, carbon tetrachloride, and so on. These numbers are generally based on the emulsification experience† rather than on structural considerations. Then an emulsifying agent—or better still, a combination of emulsifying agents—is selected whose HLB number is approximately the same as that of the ingredients to be emulsified. If there are a number of ingredients to be emulsified simultaneously, the weighted average of the assigned numbers corresponding to the percentage composition of the mixture of ingredients is used. As in the case of the ingredients to be emulsified, when a combination of emulsifying agents of different HLB values is used, the HLB number of the mixture is the weighted average of the individual HLB numbers.

For example, if a mixture of 20% paraffin wax (HLB = 10) and 80% aromatic mineral oil (HLB = 13) is to be emulsified, then the HLB number of the emulsifying agent combination should be $(10 \times 0.20) + (13 \times 0.80) = 12.4$. For this purpose, a mixture of 60% of POE lauryl alcohol made from 23 mol of ethylene oxide (HLB = 16.9) and 40% of POE cetyl alcohol made from 2 mol of ethylene oxide (HLB = 5.3) could be tried:

$$\mathrm{HLB} = (16.9 \times 0.60) + (5.3 \times 0.40) = 12.2.$$

To determine the *optimum* emulsifier combination, however, various mixtures of other *types* of emulsifying agents with the same weighted average HLB number must then be tried to determine which structural types of emulsifying agents give the best results with this particular combination of emulsion ingredients, since the HLB number is indicative only of the type of emulsion to be expected, not the efficiency or effectiveness with which it will be accomplished (Griffin, 1954; Becher, 1973). For *O/W* emulsions stabilized with POE nonionics, emulsion stability increases with increase in the length of the POE chain; for *W/O* emulsions, with length of the hydrophobic group (Shinoda et al., 1971).

* Becher (1984) has pointed out the relation between HLB number and the $V_H/l\ a$ *parameter* (chapter 3, Section IIA);

† For determining the HLB of an oil of unknown HLB value, see Becher (1973, p. 84).

As expected from the definition of the HLB value, materials with high HLB values are *O/W* emulsifiers, and materials with low HLB value are *W/O* emulsifiers. An HLB value of 3–6 is the recommended range for *W/O* emulsification; 8–18 is recommended for *O/W* emulsification. Since the requirements for emulsification of a particular ingredient differ markedly, depending on whether the ingredient is the dispersed phase (*O/W* emulsion) or the continuous phase (*W/O* emulsion), each ingredient has a different HLB value, depending on which phase of the final emulsion it will become. Thus, paraffinic mineral oil has an HLB value of 11 for emulsification as the dispersed phase in an *O/W* emulsion and a value of 4 as the continuous phase in a *W/O* emulsion.

The HLB value for some types of nonionic surface-active agents can be calculated from their structural groupings (Griffin, 1954). Thus, for fatty acid esters of many polyhydric alcohols,

$$\mathrm{HLB} = 20\left(1 - \frac{S}{A}\right), \tag{8.13}$$

where S is the saponification number of the ester and A is the acid number of the fatty acid used in the ester. For example, glyceryl monostearate has $S = 161$, A = 198, and hence HLB = 3.8. For esters for which good saponification data are not readily obtainable, the following formula can be used:

$$\mathrm{HLB} = \frac{E + P}{5}, \tag{8.14}$$

where E is the weight percentage of oxyethylene content and P is the weight percentage of polyol content. For materials where a POE chain is the only hydrophilic group, this reduces to

$$\mathrm{HLB} = \frac{E}{5}. \tag{8.15}$$

Thus, a POE cetyl alcohol made from 20 mol of ethylene oxide (77% oxyethylene) would have a calculated HLB of 15.4.

A commonly used general formula for nonionics is

$$20\mathrm{x}\frac{M_H}{M_H + M_L}, \tag{8.16}$$

where MH is the formula weight of the hydrophilic portion of the molecule and M_L is the formula weight of the lipophilic (hydrophobic) portion of the molecule.

The water solubility of the surfactant can be used to obtain a rough approximation of its HLB value (Becher, 2001).

Behavior in Water	HLB Range
No dispersibility	1–4
Poor dispersion	3–6
Milky dispersion after vigorous agitation	6–8
Stable milky dispersion (upper end almost translucent)	8–10
From translucent to clear	10–13
Clear solution	13+

There have been numerous attempts to determine HLB numbers from other fundamental properties of surfactants, for example, from cloud points of nonionics (Schott, 1969), from CMCs (Lin et al., 1973), from gas chromatography retention times (Becher and Birkmeier, 1964; Petrowski and Vanatta, 1973), from NMR spectra of nonionics (Ben-et and Tatarsky, 1972), from partial molal volumes (Marszall, 1973), from solubility parameters (Hayashi, 1967; McDonald, 1970; Beerbower and Hill, 1971), from quality control data (Pasquali et al., 2010), and from QSPR (quantitative structure–performance relationship) studies (Chen et al., 2009). Although relations have been developed between many of these quantities and HLB values calculated from structural groups in the molecule, particularly in the case of nonionic surfactants, there are few or no data showing that the HLB values calculated in these fashions are indicative of actual emulsion behavior.

It has become apparent that although the HLB method is useful as a rough guide to emulsifier selection, it has serious limitations. Although, as mentioned previously, the HLB number of a surfactant is indicative of neither its efficiency (the required concentration of the emulsifying agent) nor its effectiveness (the stability of the emulsion), but only of the type of emulsion that can be expected from it, data have accumulated that show that even this is not reliably related to the HLB number. It has been pointed out (Shinoda, 1968; Boyd et al., 1972; Kloet and Schramm, 2002) that a single surfactant can produce either an *O/W* or a *W/O* emulsion, depending on the temperature at which the emulsion is prepared, the shear rate, or, at high oil concentrations, and depending on the ratio of surfactant to oil. *O/W* emulsions can be prepared with certain surfactants over the entire range of HLB numbers from 2 to 17.

B. The PIT Method

A major disadvantage of the HLB method of selecting surfactants as emulsifying agents for a particular system is that it makes no allowance for the change in HLB value with change in the conditions for emulsification (temperature, nature of the oil and water phases, presence of cosurfactants or other additives). For example, we saw in Chapter 5, Section IIIA, that when the temperature is raised, the degree of hydration of a POE nonionic surfactant decreases and the surfactant becomes less hydrophilic. Consequently, its HLB must decrease. An *O/W* emulsion made with a POE nonionic surfactant may invert to a *W/O* emulsion when the temperature is raised; a *W/O* emulsion may invert to an *O/W* emulsion when the temperature is lowered. The temperature in the middle of the three-phase region at which inversion occurs is known as the PIT and is the temperature, as we have seen in Chapter 5, Section IIIA, at which the hydrophilic and lipophilic tendencies of the surfactant (or surfactant–cosurfactant mixture) "balance" in that particular system of oil and water phases. There is also a very good linear relationship between the PIT and the cloud points (Chapter 4, Section IIIB) of various types of POE nonionic surfactants when the system is saturated with the oil phase (Shinoda and Arai, 1964).

Since the oil–water interfacial tension is at a minimum at the PIT, emulsions made at this temperature should have the finest particle size. The minimum work needed to create the emulsions is the product of the interfacial tension and the increase in interfacial area ($W_{min} = \gamma_I \times \Delta A$) and, for a given amount of mechanical work expended, AA should be a *maximum* at the temperature. Since the particle size diminishes as AA increases for a given amount of mechanical work expended, the particle size should be at a minimum at the PIT. This is the basis for a method of selecting surfactants as emulsifying agents for a particular system, the PIT method (Shinoda and Arai, 1964, 1965; Shinoda, 1968). This method is applicable only to emulsions that show inversion at a particular temperature.

According to this method, an emulsion made with equal weights of oil and aqueous phases and 3–5% of surfactant is heated and shaken at different temperatures and the temperature at which the emulsion inverts from *O/W* to *W/O*, or vice versa, is determined. A suitable emulsifier for an *O/W* emulsion should yield a PIT 20–60°C higher than the storage temperature of the emulsion; for a *W/O* emulsion, a PIT 10–40°C lower than the storage temperature is recommended (although PITs cannot be determined below 0°C).

For optimum stability, Shinoda and Saito (1969) suggest "emulsification by the PIT method," in which the emulsion is prepared at a temperature 2–4°C below the PIT and then cooled down to the storage temperature (for *O/W* emulsions). This is because an emulsion prepared near the PIT has a very fine average particle size but is not very stable to coalescence. Cooling it down to a temperature considerably below the PIT increases its stability without significantly increasing its average particle size.

The PIT is affected by the HLB and the concentration of the surfactant, the polarity of the oil phase, the phase ratio of the bulk phases and the presence of additives in them, and the distribution of POE chain lengths in POE nonionics (Shinoda, 1968; Mitsui et al., 1970). The PIT appears to be an almost linear function of the HLB value of the surfactant for a given set of emulsification conditions; the higher the HLB value, the greater the PIT. This is to be expected, since the larger the ratio of the hydrophilic to the lipophilic moiety in the surfactant molecule, the higher the temperature required to dehydrate it to the point where its structure is balanced. When the distribution of POE chain lengths in an emulsion stabilized by a POE nonionic surfactant is broad, its PIT is higher and its stability greater than when the distribution is narrow (Shinoda et al., 1971).

For a POE surfactant with a given HLB value, as the polarity of the oil phase decreases, the PIT increases. (The surfactant must be made more lipophilic to match the decreased polarity of the oil.) Thus, to keep the PIT constant (and hence a constant emulsifying power balance), the surfactant used must have a lower HLB value as the polarity of the oil phase decreases. The PIT of an emulsion made from binary mixture of oils is the weighted average, by volume, of the PITs of the emulsion made from the individual oils, using the same emulsifying agent (Arai and Shinoda, 1967):

$$\mathrm{PIT}_{(\mathrm{mix})} = \mathrm{PIT}_{\mathrm{A}} \cdot \phi_{\mathrm{A}} + \mathrm{PIT}_{\mathrm{B}} \cdot \phi_{\mathrm{B}} \tag{8.17}$$

where ϕ_A and ϕ_B are the volume fractions of oils *A* and *B* used in the emulsion.

The PIT appears to reach a constant value at 3–5% surfactant concentration when a POE nonionic containing a single POE chain length is used. When there is a distribution of POE chain lengths in the surfactant, the PIT decreases very sharply with increase in the concentration of the surfactant when the degree of oxyethylenation is low and less sharply when the degree of oxyethylenation is high.

As the oil–water ratio increases in an emulsion with a fixed surfactant concentration, the PIT increases. However, fixed ratios of surfactant to oil give the same PIT, even when the oil–water ratio varies. The higher the surfactant–oil ratio, the lower the PIT.

Additives, such a paraffin, that decrease the polarity of the oil phase increase the PIT, whereas those, such as oleic acid or lauryl alcohol, that increase its polarity lower the PIT. The addition of salts to the aqueous phase decreases the PIT of emulsions made with POE nonionics (Shinoda and Takeda, 1970).

Since the PIT of a hydrocarbon–water emulsion stabilized with a POE nonionic surfactant is, as might be expected, related to the cloud point of an aqueous solution of the nonionic saturated with that hydrocarbon (Chapter 4), these effects on the PIT of emulsions stabilized by POE nonionics are readily understood. As mentioned in the discussion (Chapter 4, Section IIIB) of the effect of solubilizate on the cloud points of POE nonionics, long-chain aliphatic hydrocarbons that are solubilized in the inner core of the micelle increase the cloud point, whereas short-chain aromatic hydrocarbons and polar materials that are solubilized between the POE chains decrease it. They have the same effect on the PIT: long-chain aliphatic hydrocarbons increase the PIT and therefore tend to form stable *O/W* emulsions, whereas short-chain aromatics and polar additives decrease it and tend to form stable *W/O* emulsions (Shinoda and Arai, 1964). An increase in the length of the POE chain increases the cloud point and the PIT, and consequently increases the tendency to form *O/W* emulsions, consistent with the generalization that the more water-soluble the emulsifier, the greater its tendency to form *O/W* emulsions.

C. The Hydrophilic Lipophilic Deviation (HLD) Method

The method developed originally for microemulsion formulation (Section II) has been adapted (Salager et al., 1983, 2000) to macroemulsion formation. In this method, the value of the left-hand side of Equation 8.10 or 8.12 is called the HLD. When the value equals zero, as in Section II, a microemulsion is formed; when the value is positive, a *W/O* macroemulsion is preferentially formed; when it is negative, an *O/W* macroemulsion is preferentially formed. The HLD is similar in nature to the Winsor *R* ratio (Equation 5.3) in that when the HLD is larger than, smaller than, or equal to O, *R* is larger than, smaller than, or equal to 1. The value of the HLD method is that, on a qualitative basis, it takes into consideration the other components of the system (salinity, cosurfactant, alkane chain length, temperature, and hydrophilic and hydrophobic groups of the surfactant). On the other hand, on a quantitative basis, it requires the experimental evaluation of a number of empirical constants.

V. DEMULSIFICATION

In some processes, the emulsification of two liquid phases is an undersirable phenonmenon. This often occurs when two immiscible phases are mixed together with considerable agitation, as in industrial extraction processes. However, probably the most important case of undersirable emulsification is in the recovery of petroleum from oil reservoirs. Crude oil always is associated with water or brine in the reservoir and also contains natural emulsifying agents, such as asphaltenes and resins. These, particularly the asphaltenes, together with other components in the petroleum, such as the resins and waxes, form a thick, viscous interfacial film around water droplets, with their polar groups oriented toward the water and their nonpolar groups toward the oil. This interfacial film is highly viscous, producing very stable, viscous *W/O* emulsions. To break these emulsions and separate the petroleum from the water in them, various techniques are used, notably the addition of surfactants called *demulsifiers* or *demulsifying agents*. Demulsification and demulsifiers in petroleum recovery have been discussed by Angle (2001) and Sjoblom et al. (2001), respectively, but there have been few systematic studies (Shetty et al., 1992; Bhardwaj and Hartland, 1993).

The demulsifiers used at various times may be classified as (Mikula and Munoz, 2000):

1. soaps, salts of naphthenic acids, aromatic and alkylaromatic sulphonates (1920s)
2. petroleum sulfonates, mahogany soaps, oxidized castor oil, and sulfosuccinic acid esters (1930s)
3. ethoxylates of fatty acids, fatty alcohols, and alkylphenols (1935)
4. ethylene oxide/propylene oxide copolymers, *p*-alkylphenol formaldehyde resins with ethylene/propylene oxides modifications (1950)
5. amine oxalkylates (1965)
6. oxalkylated, cyclic *p*-alkylphenol formaldehyde resins, and complex modifications (1976)
7. polyesteramines and blends (1896)
8. the later ones (see below)

With time, the effective concentration of demulsifier was reduced from 1000 ppm in the 1920s to ~5 ppm in 1986.

Mechanisms involved in the demulsification by surfactants of petroleum *W/O* emulsions include, (1) adsorption of the surfactant at the oil–water interface, (2) change in the nature of the interfacial film from a highly hydrophobic one to a less hydrophobic one (and, consequently, one more wettable by water), (3) reduction of the viscosity of the interfacial film by penetration of the surfactant into it, and (4) displacement of the original *W/O* emulsion stabilizers, particularly the asphaltenes, from the interface into the oil phase.

Since the chemical composition of the crude oil and the natural emulsifying agents contained in it vary greatly, depending upon the material from which it was formed and the conditions of its formation, no one surfactant demulsifier can be used. Instead, a "chemical cocktail" is used, containing different surfactants to perform the required functions. These include wetting agents, such as di(2-ethylhexyl) sulfoscuccinate, and various polymeric surfactants, such as POE polyoxypropylenes and POE alkylphenol-formaldehyde polymers. The structure of the POE (and polyoxypropylenated) material can be "tailored" to meet the different composition of the petroleum.

Investigations on the demulsification mechanism performed on asphaltene-stabilized petroleum by functional variations in ethoxylated polyalkylphenol formaldehyde surfactants (Al-Sabagh et al., 2009) highlight the importance of water : oil ratios, surfactant concentration, surfactant molecular weight, ethylene oxide content, alkyl chain length, and asphaltene content. Different novel structures were also tested as demulsifiers (triblock copolymers of poly(ethylene)oxide and poly(dimethyl)siloxane [Le Follotec et al., 2010]; hydrolyzed fatty oils adducted with maleic anhydride and modified by esterification with polyethylene glycols or ethyleneoxide-propyleneoxide block copolymers [El-Ghazawy et al., 2010]; magnetic amphiphilic composites based on carbon nanotubes and nanofibers grown on an inorganic matrix [Oliveira et al., 2010]; *silicone* based [Dalmazzone and Noïk, 2005]; ionic liquid based

[Guzman-Lucero et al., 2010]). In some cases, the chemically initiated demulsification has been accelerated by external stimuli like microwave radiation (Xia et al., 2010).

REFERENCES

Acosta, E., H. Uchiyama, D. A. Sabatini, and J. H. Harwell (2002) *J. Surfactants Deterg.* **5**, 151.

Acosta, E., E. Szekeres, D. A. Sabatini, and J. H. Harwell (2003) *Langmuir* **19**, 186.

Acosta, E. J., J. S. Yuan, and A. S. Bhakta (2008) *J. Surfact. Deterg.* **11**, 145.

Albers, W., J. Th. G. Overbeek (1959) *J. Colloid Sci.* **14**, 501, 510.

Al-Sabagh, A. M., M. R. N. El-Din, S. A.-E. Fotouh, and N. M. Nasser (2009) *J. Disp. Sci. Technol.* **30**, 267.

Angle, C. W., in *Encyclopedia of Emulsion Technology*, J. Sjoblom (ed.), Marcel Dekker, New York, 2001, chapter 24.

Anton, R. E., N. Garces, and A. Yajure (1997) *J. Disp. Sci. Technol.* **18**, 539.

Arai, H. and K. Shinoda (1967) *J. Colloid Interface Sci.* **25**, 396.

Aveyard, R. and J. H. Clint, in *Adsorption and Aggregation of Surfactants in Solution*, K. L. Mittal and D. O. Shah (eds.), Marcel Dekker, New York, 2003, p. 76.

Bancroft, W. D. (1913) *J. Phys. Chem.* **17**, 514.

Barette, D., Memoir de Licence, FUNDP, Namur, Belgium, 1992.

Becher, P., in *Pesticide formulations*, W. Van Valkenburg (ed.), Marcel Dekker, New York, 1973, p. 84 and 85.

Becher, P. (1984) *J. Disp. Sci. Tech.* **5**, 81.

Becher, P., *Emulsions. Theory and Practice*, 3rd ed., American Chemical Society, Washington, DC, 2001.

Becher, P. and R. L. Birkmeier (1964) *J. Am. Chem. Soc.* **41**, 169.

Beerbower, A. and M. Hill, *McCutcheon's Detergents and Emulsifiers Annual*, Allured Publ. Co., Ridgewood, NJ, 1971.

Ben-et, G. and D. Tatarsky (1972) *J. Am. Oil Chem. Soc.* **49**, 499.

Bhardwaj, A. and S. Hartland (1993) *J. Disp. Sci. Technol.* **14**, 541.

Bourrel, M., J. L. Salager, R. S. Schechter, and W. H. Wade (1980) *J. Colloid Interface Sci.* **75**, 451.

Boutonnet, M., S. Logdberg, and E. E. Svensson (2008) *Curr. Opin. Colloid Interface Sci.* **13**, 270.

Boyd, J., C. Parkinson, and P. Sherman (1972) *J. Colloid Interface Sci.* **41**, 359.

Brouwer, W. M., M. S. El-Asser, and J. W. Vanderhoff (1986) *Colloids Surf.* **21**, 69.

Chen, M. L., Z. W. Wang, and H. J. Duan (2009) *J. Disp. Sci. Tech.* **30**, 1481.

Chun, H. (1979) *J. Colloid Interface Sci.* **71**, 408.

Cockbain, E. G. and T. S. McRoberts (1953) *J. Colloid Sci.* **8**, 440.

Dalmazzone, C. and C. Noïk (2005) *SPE J.* **10**, 44.

Davies, J. T., 2nd Int. Congr. Surface Activity, London, 1957, 1, p. 426.

Davies, J. T. and E. K. Rideal, *Interfacial Phenomena*, 2nd ed., Academic, New York, 1963, Chap. 8.

Dungan, S. R., C. Solans, and H. Kunieda (eds.), in *Industrial Applications of Microemulsions*, Marcel Dekker, New York, 1997, p. 147.

El-Asser, M. S., S. C. Misra, J. W. Vanderhoff, and J. A. Manson (1977) *J. Coatings Tech.* **49**, 71.

El-Asser, M. S., C. D. Lack, Y. T. Choi, T. I. Min, J. W. Vanderhoff, and F. M. Fowkes (1984) *Colloids Surf.* **12**, 79.

El-Ghazawy, R. A., A. M. Al-Sabagh, N. G. Kandile, and M. N. El-Din (2010) *J. Disp. Sci. Technol.* **31**, 1423.

Ford, R. E. and C. G. L. Furmidge (1966) *J. Colloid Interface Sci.* **22**, 331.

Forster, T., in *Surfactants in Cosmetics*, M. Rieger and L. D. Rhein (eds.), Marcel Dekker, New York, 1997, p. 105.

Freshwater, D. C., B. Scarlett, and M. J. Groves (1966) *Am. Cosmet. Perfum.* **81**, 43.

Friberg, S. (1969) *J. Colloid Interface Sci.* **29**, 155.

Friberg, S. (1976) *J. Colloid Interface Sci.* **55**, 614.

Garti, N., M. Frenkel, and R. Shwartz (1983) *J. Disp. Sci. Tech.* **4**, 237.

Griffin, W. C. (1949) *J. Soc. Cosmet. Chem.* **1**, 311.

Griffin, W. C. (1954) *J. Soc. Cosmet. Chem.* **5**, 249.

Grimm, W. L., T. I. Min, M. S. El-Asser, and J. W. Vanderhoff (1983) *J. Colloid Interface Sci.* **94**, 531.

Groves, M. J., B. H. Kaye, and B. Scarlett (1964) *Be Chem. Eng.* **9**, 742.

Guzman-Lucero, D., P. Flores, T. Rojo, and R. Martı′nez-Palou (2010) *Energy Fuels* **24**, 3610.

Hayashi, S. (1967) *Yukagaku* **16**, 554.

Holmberg, K. and E. Osterberg (1986) *J. Disp. Sci. Tech.* **7**, 299.

Kitchener, J. A. and P. R. Mussellwhite, *Emulsion Science*, P. Sherman (ed.), Academic, New York, 1968, p. 96ff.

Kloet, J. V. and L. L. Schramm (2002) *J. Surfactants Deterg.* **5**, 19.

Le Follotec, A., I. Pezron, C. Noik, C. Dalmazzone, and L. Metlas-Komunjer (2010) *Coll. Surf. A- Physicochem. Eng. Aspects* **365**, 162.

Lin, I. J., J. P. Friend, and Y. Zimmels (1973) *J. Colloid Interface Sci.* **45**, 378.

Magdassi, S., M. Frenke, and N. Garti (1984) *J Disp. Sci. Tech.* **5**, 49.

Marszall, L. (1973) *J. Pharm. Pharmacol.* **25**, 254.

Mathis, G., P. Leempoel, J. C. Ravey, C. Selve, and J. J. Delpuech (1984) *J. Am. Chem. Soc.* **106**, 6162.

Matsumoto, S. (1983) *J. Colloid Interface Sci.* **94**, 362.

Matsumoto, S., Y. Kita, and D. Yonezawa (1976) *J. Colloid Interface Sci.* **57**, 353.

Matsumoto, S., Y. Koh, and A. Michiura (1985) *J Disp. Sci. Tech.* **6**, 507.

McClements, D. J. (2010) *Ann. Rev. Food Sci. Technol.* **1**, 241.

McDonald, C. (1970) *Can. J. Pharm. Sci.* **5**, 81.

Mikula, R. J. and V. A. Munoz, in *Surfactants: Fundamentals and Applications in the Petroleum Industry*, L. L. Schram (ed.), Cambridge University Press, Cambridge, UK, 2000, p. 54.

Mishra, B. K., B. S. Valaulikar, J. T. Kunjappu, and C. Manohar (1989) *J. Colloid Interface Sci.* **127**, 373.

Mitsui, T., Y. Machida, and F. Harusawa (1970) *Bull. Chem. Soc. Jpn.* **43**, 3044.

Nakajima, H., H. Tomomasa, and M. Okabe, Proc. First World Emulsion Conf, Paris, Vol. 1, p. 1, 1993.

Oliveira, A. A. S., I. F. Teixeira, L. P. Ribeiro, J. C. Tristao, C. Juliana, A. Dias, R. M. Lago, and M. Rochel (2010) *J. Braz. Chem. Soc.* **21**, 2184.

Omotosho, J. A., T. K. Law, T. L. Whateley, and A. T. Florence (1986) *Colloid Surf.* **20**, 133.

Pasquali, R. C., N. Sacco, and C. Bregni (2010) *J. Disp. Sci. Tech.* **31**, 479.

Petrowski, G. E. and J. R. Vanatta (1973) *J. Am. Oil Chem. Soc.* **50**, 284.

Prince, L. M. Presented before 7th N. E. regional meeting, American Chemical Society, Albany, New York, August 9 1976.

Sagitani, H. (1981) *J. Am. Oil Chem. Soc.* **58**, 738.

Saito, Y., T. Sato, and I. Anazawa (1990) *J. Am. Oil Chem. Soc.* **67**, 145.

Salager, J. L., L. Morgan, R. S. Schechter, W. H. Wade, and E. Vasquez (1979) *Soc. Petrol. Eng. J.* **19**, 107.

Salager, J. L., M. Minana-Perez, M. Perez-Sanchez, M. Ranfrey-Gouveia, and C. I. Rojas (1983) *J. Disp. Sci. Technol.* **4**, 313.

Salager, J. L., A. Graciaa, and J. Lachaise (1998) *J. Surfactants Deterg.* **1**, 403.

Salager, J. L., N. Marquez, A. Graciaa, and J. Lachaise (2000) *Langmuir* **16**, 5534.

Schomacker, R. (1992) *Nachr. Chem. Tech. lab.* **40**, 1344.

Schott, H. (1969) *J. Pharm. Sci.* **58**, 1443.

Schulman, J. and J. Leja (1954) *Trans. Faraday Soc.* **50**, 598.

Schulman, J. H. and E. G. Cockbain (1940) *Trans. Faraday Sco.* **36**, 661.

Sherman, P. (ed.), *Emulsion Science*, Academic, New York, 1968.

Shetty, C. A., A. D. Nikolov, and D. T. Wasan (1992) *J. Disp. Sci. Technol.* **13**, 121.

Shinoda, K., Proc. 5th Int. Congr. Detergency, Balcelona, Vol. II, p. 275, September 1968.

Shinoda, K. and H. Arai (1964) *J. Phys. Chem.* **68**, 3485.

Shinoda, K. and H. Arai (1965) *J. Colloid Sci.* **20**, 93.

Shinoda, K. and H. Saito (1969) *J. Colloid Interface Sci.* **30**, 258.

Shinoda, K. and H. Takeda (1970) *J. Colloid Interface Sci.* **32**, 642.

Shinoda, K., H. Saito, and H. Arai (1971) *J. Colloid Interface Sci.* **35**, 624.

Sjoblom, J., *Emulsions and Emulsion Stability*, Marcel Dekker, New York, 1996.

Sjoblom, J., E. E. Johnsen, A. Westvik, M.-H. Ese, J. Djuve, I. H. Auflem, and H. Kallevik, in *Encyclopedia of Emulsion Technology*, J. Sjoblom (ed.), Marcel Dekker, New York, 2001, chapter 25.

von Smoluchowski, M. (1916) *Phys. Z.* **17**, 557, 585; Z. Phys. Chem. 92, 129 (1917).

Solans, C. and H. Kunieda, *Industrial Applications of Microemulsions*, Marcel Dekker, New York, 1996.

Tronnier, H. and H. Bussins (1960) *Seifen-Ole-Fette-Wachse* **86**, 747.

Uchiyama, H., E. Acosta, D. A. Sabatini, and J. H. Harwell (2000) *Ind. Eng. Chem.* **39**, 2704.

Ugelstad, J., M. S. El-Asser, and J. W. Vanderhoff (1973) *J. Polym. Sci. Polym. Lett.* **11**, 503.

Verzaro, F., M. Bourrel, and C. Chambu, in *Surfactants in Solution*, Vol 6, K. L. Mittal and P. Bothorel (eds.), Plenum, New York, 1984, pp. 1137–1157.

Xia, L., K. Gong, S. Wang, J. Li, and D. Yang (2010) *J. Disp. Sci. Technol.* **31**, 1574.

PROBLEMS

8.1 List four different ways of distinguishing *O/W* from *W/O* macroemulsions.

8.2 Describe or give the characteristic properties of each of the following:

- **(a)** macroemulsion
- **(b)** nanoemulsion
- **(c)** microemulsion
- **(d)** multiple emulsion

8.3 Discuss the changes in interfacial tension that occur in the conversion of an *O/W* macroemulsion stabilized by a POE nonionic surfactant to a *W/O* macroemulsion upon raising the temperature above the cloud point.

8.4 Explain the relationship between γ_{OE}, γ_{WE}, spreading coefficient, and emulsion type.

8.5 An oil has an HLB of 10 for *O/W* emulsification. Calculate the percentages of $C_{12}H_{25}(OC_2H_4)_2OH$ and $C_{12}H_{25}(OC_2H_4)_8OH$ that should be used in attempting to emulsify this oil with a mixture of these two surfactants.

8.6 **(a)** Describe the effect of the following changes on the tendency of a system of a POE nonionic surfactant, an alkane, and water, to form an *O/W* emulsion:

1. Increase in the temperature from 25°C to 40°C
2. Change in the alkane from *n*-octane to *n*-dodecane
3. Increase in the number of carbon atoms of the hydrophobic group of the surfactant.

(b) Describe the effect in each case if the surfactant is an anionic surfactant.

8.7 Suggest and explain conditions under which the HLB value for a particular surfactant will vary

(a) for a POE nonionic surfactant

(b) for an ionic surfactant

8.8 List the important mechanisms involved in the demulsification of a W/O emulsion by surfactants. Can you draw visualizing cartoon pictures to describe the changes associated with these mechanisms?

9 Dispersion and Aggregation of Solids in Liquid Media by Surfactants

In many products and processes it is important to obtain significantly stable, uniform dispersions of finely divided solids. Paints, pharmaceutical preparations, drilling muds for oil wells, pigments, and dyestuffs are commonly used as suspensions of finely divided solids in some liquid medium.

However, when a preformed, finely divided solid is immersed in a liquid, it often does not form a stable dispersion. Many of the particles remain attached (aggregated) in the form of clumps, and those particles that do disperse in the liquid very often clump together again to form larger aggregates that settle out of the suspension. In addition, even when the particles do disperse in the liquid, the dispersion may be viscous or thin, the particles may remain dispersed for different lengths of time, and the sensitivity of the dispersions to molecular environmental conditions (pH, temperature, additives) may vary greatly. Before discussing the role of surfactants in these systems and the relation of the structure of the surfactant to its performance as a dispersing agent, it is necessary to review the forces between particles in these suspensions, since these forces, together with the particle size and shape and the volume of the dispersed phase, determine the properties of the suspension.

I. INTERPARTICLE FORCES

Tadros (1986) describes four types of interparticle forces: hard sphere, soft (electrostatic), van der Waals, and steric. Hard-sphere interactions, which are repulsive, become significant only when particles approach each other at distances slightly less than twice the hard-sphere radius. They are not commonly encountered.

Surfactants and Interfacial Phenomena, Fourth Edition. Milton J. Rosen and Joy T. Kunjappu.

A. Soft (Electrostatic) and van der Waals Forces: Derjaguin and Landau and Verwey and Overbeek (DLVO) Theory

The soft (electrostatic) and van der Waals interparticle forces are described in the well-established theory of the stability of lyophobic dispersions (colloidal dispersions of particles that are not surrounded by solvent layers). This theory was developed independently by Derjaguin and Landau (1941) and Verwey and Overbeek (1948) and therefore is called the *DLVO theory*. It assumes a balance between repulsive and attractive potential energies of interaction of the dispersed particles. Repulsive interactions are believed to be due either to the similarly charged electrical double layers surrounding the particles or to particle–solvent interactions. Attractive interactions are believed to be due mainly to the van der Waals forces between the particles. To disperse the particles, the repulsive interactions must be increased to the point where they overcome the attractive interactions; to aggregate the particles, the reverse must be done.

The total potential energy of interaction V is the sum of the potential energy of attraction V_A and that of repulsion V_R:

$$V = V_A + V_R \tag{9.1}$$

The potential energy of attraction in a vacuum for similar spherical particles of radius a whose centers are separated by a distance R is given by the expression (Hamaker, 1937)

$$V_A = \frac{-\mathrm{A}a}{12H} \tag{9.2}$$

where A is the Hamaker (van der Waals) constant and H is the nearest distance between the surfaces of the particles (= $R - 2\mathrm{a}$) when H is small ($R/a \leq 5$). The attractive potential energy is always negative because its value at infinity is zero and decreases as the particles approach each other.

In a liquid dispersion medium, A must be replaced by an effective Hamaker constant,

$$A_{eff} = \left(\sqrt{\mathrm{A}_2} - \sqrt{\mathrm{A}_1}\right)^2 \tag{9.3}$$

where A_2 and A_1 are the Hamaker constants for the particles and the dispersion medium, respectively (Vold, 1961). As the particles and the dispersion medium become more similar in nature, A_2 and A_1 become closer in magnitude and $\mathrm{A}_{\mathrm{eff}}$ becomes smaller. This results in a smaller attractive potential energy between the particles.

The potential energy of repulsion V_R depends on the size and shape of the dispersed particles, the distance between them, their surface potential Ψ_0, the dielectric constant ε_r of the dispersing liquid, and the effectiveness thickness of the electrical double layer 1/K (Chapter 2, Section I), where

$$\frac{1}{\kappa} = \left(\frac{\varepsilon_r \varepsilon_0}{4\pi F^2 \sum_i C_i Z_i^2} \right)^{1/2} \tag{2.1}$$

For two spherical particles (Lyklema, 1968) of radius a, when $a/(1/\kappa)$ (= κa) << 1, that is, small particles and a relatively thick electrical double layer,

$$V_R = \frac{\varepsilon_r a^2 \Psi_0^2}{R} e^{-\kappa H} \tag{9.4}$$

When $a/(1/\kappa)$ (= κa) >> 1, that is, large particles and a relatively thin electrical double layer,

$$V_R = \frac{\varepsilon_r a \Psi_0^2}{2} \ln(1 + e^{-\kappa H}) \tag{9.5}$$

The potential energy of repulsion is always positive, since its value at infinity is zero and increases as the particles approach each other.

Typical plots of V_A and V_R as a function of the distance H between the particles are shown in Figure 9.1, together with the plot of the total energy of interaction V, the sum of V_A and V_R. The particles tend to aggregate at those distances where the attractive potential energy is greater than the repulsive energy and V becomes negative.

The form of the curve for the total potential energy of interaction V depends on the ratio of the particle size to the thickness of the electrical double layer $a/(1/\kappa) = \kappa a$ (Figure 9.2), the electrolyte concentration (Figure 9.3), and the surface potential Ψ_0 (Figure 9.4).

When κa >> 1 (i.e., when the ratio of particle size to thickness of the electrical double layer is very large), the curve for V (Figure 9.2b) shows a secondary minimum (S) at a relatively large distance of separation between the particles in addition to the primary minimum (P). Particles may therefore aggregate at a relatively large distance between the particles. This type of aggregation is sometimes called *flocculation* to distinguish it from aggregation in the primary minimum, which is termed *coagulation.* Since the depth of the secondary minimum is rather shallow, flocculation of this type is easily reversible and the particles can be freed by agitation. Particles larger than a few micrometers, especially flat ones, may show this phenomenon.

The effect on V of the addition of electrolyte to the (aqueous) dispersion medium and the consequent compression of the double layer is shown in Figure 9.3. With increase in the concentration of indifferent electrolyte, κ increases and the energy barrier to coagulation (V_{max}) decreases and may even disappear, consistent with the known coagulation of lyophobic colloidal dispersions by electrolyte. Figure 9.4, illustrating the effect of the surface potential of the particles on V, indicates that the energy barrier to coagulation increases with increase in the surface potential. The effect of adsorption of

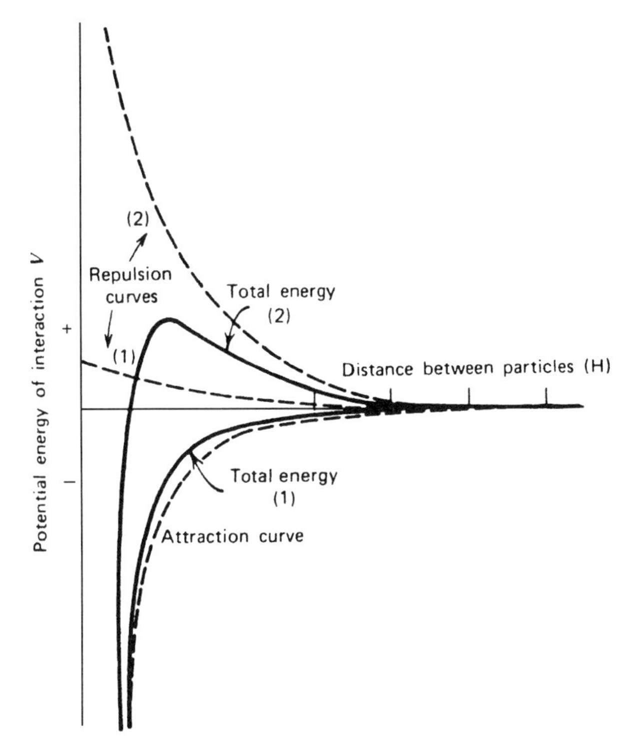

FIGURE 9.1 Total interaction energy curves (obtained by summation of attraction and repulsion curves) for two repulsion curves of different heights.

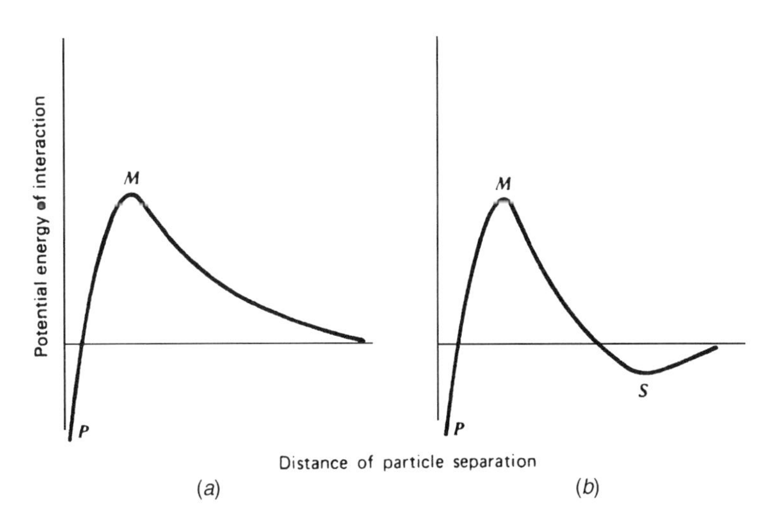

FIGURE 9.2 Potential energy of interaction as a function of distance of particle separation and ratio of particle size to thickness of the electrical double layer, $a/(1/\kappa) = \kappa a$. (a) $\kappa a << 1$; (b) $\kappa a >> 1$.

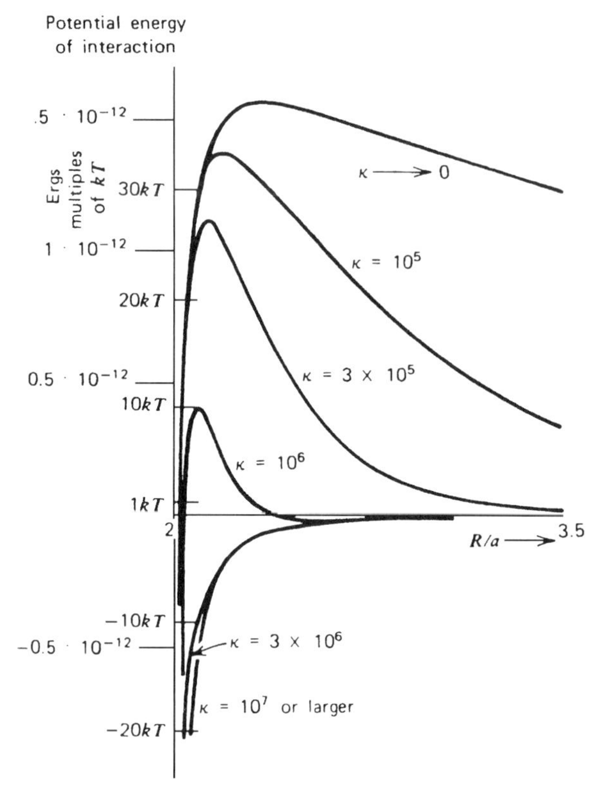

FIGURE 9.3 Influence of electrolyte concentration (as measured by κ) on the total potential energy of interaction of two spherical particles. Reprinted with permission from J. Th. G. Overbeek in *Colloid Science,* Vol. 1, H. Kruyt (ed.), Elsevier, Amsterdam, 1952, Chap. 6, p. 276.

surfactant ions onto the particle surface is apparent. When adsorption results in an increase in the potential of the particle at the Stern layer, the stability of the dispersion is increased; when it results in a decrease in that potential, the stability of the dispersion is lowered. Since the range of thermal energies for dispersed particles may go as high as 10 kT, an energy barrier of greater than *15* kT is usually considered necessary for a stable dispersion.

The stability of a colloidal dispersion is usually measured by determining the rate of change in the number of particles n during the early stages of aggregation. The rate of diffusion-controlled coalescence of spherical particles in a disperse system as a result of collisions in the absence of any energy barrier to coalesence is given by the von Smoluchowski equation:

$$\frac{-dn}{dt} = 4\pi Drn^2. \tag{8.2}$$

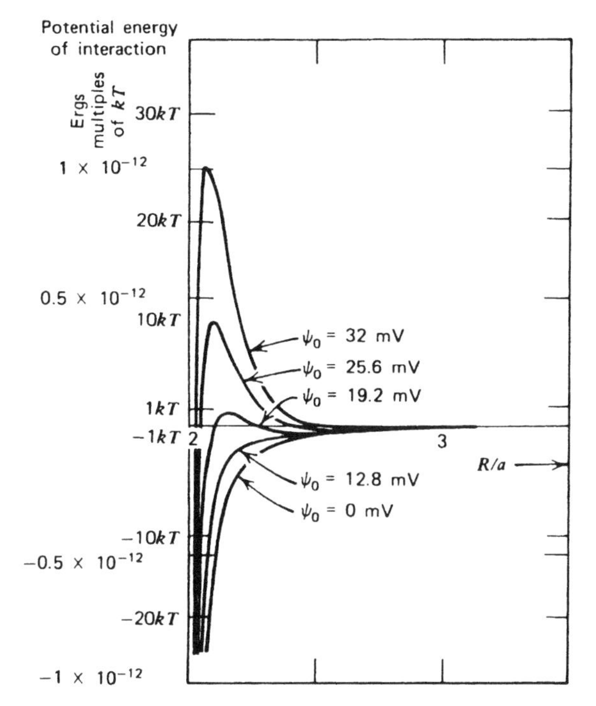

FIGURE 9.4 Influence of the surface potential Ψ_0 on the total potential energy of interaction of two spherical particles. Reprinted with permission from J. Th. G. Overbeek in *Colloid Science,* Vol. 1, H. Kruyt (ed.), Elsevier, Amsterdam, 1952, Chap. 6, p. 277.

Since, from the Einstein equation, $D = kT/6\pi\eta a$ (Equation 8.1), and $r = 2a$,

$$\frac{-dn}{dt} = \frac{4KT}{3\eta}n^2 = K_0 n^2 \frac{-dn}{dt} = \frac{4KT}{3\eta}n^2 = K_0 n^2, \tag{9.6}$$

where K_0 is the rate (constant) for diffusion-controlled coalescence. Experimentally determined rate constants K_0 for coalescence in the absence of an electrical barrier to aggregation can be determined by adding electrolyte to the dispersion until no further rate increase is obtained (Parfitt and Wharton, 1972).

In the presence of an energy barrier V_{max}, to coalescence,

$$\frac{-dn}{dt} \propto K_0 n^2 e^{-V_{max}/kT} = Kn^2, \tag{9.7}$$

where K is the rate of (slow) coalescence in the presence of an energy barrier. The stability W of the dispersion is defined as the ratio of the rate constants in the absence and presence of an energy barrier, respectively:

$$W_{st} = \frac{K_0}{K} \propto e^{V_{max}/kT}. \tag{9.8}$$

Log W_{st} is therefore a linear function of V_{max}/kT and is usually plotted against some function of the concentration of an additive to show its effect on the energy barrier to aggregation and hence on the stability of the dispersion. From theoretical considerations, log W_{st} has been shown (Fuchs, 1934; Reerink and Overbeek, 1954) to be an approximately linear function of the log of the concentration of the electrolyte in the liquid phase during the initial phase of slow coagulation at constant surface potential. Values of W_{st}, are usually calculated by determining the change in particle concentration with time during the initial period of aggregation, either directly by counting particles per unit volume under the microscope (Garvey and Tadros, 1972) or ultramicroscope (Ottewill and Shaw, 1966; Parfitt and Wharton, 1972), in a manner similar to that used for determining the stability of emulsions (Chapter 8, Section IB), or indirectly from measurements of optical density using a spectrophotometer (Ottewill and Rastogi, 1960; Ottewill and Shaw, 1966).

When absolute values of W_{st} are not required and the optical density D is proportional to n, measurements of optical density as a function of time may be used to show the effect of some additive on the stability of the dispersion. Thus $log(dt/dD)_{t \to 0}$ is plotted versus the log of the concentration of added surfactants (Watanabe, 1960) to show their effect on some AgI sols. Values of $(dt/dD)_{t \to 0}$ are obtained from the reciprocal of the initial slope of a plot of D versus time. Alternatively, the rate constant K may be determined (McGown and Parfitt, 1966) from the slope of a linear plot of $1/n$ versus t (Equation 8.4).

From the preceding discussion of the DLVO theory and Equations 9.1–9.5 and 9.8, it is apparent that the stability of a lyophobic dispersion is a function of the particle radius and surface potential, the ionic strength and dielectric constant of the dispersing medium, the value of the Hamaker constant, and the temperature. Stability is increased by increase in the particle radius or surface potential or in the dielectric constant of the medium and by decrease in the effective Hamaker constant, the ionic strength of the dispersing liquid, or the temperature.

1. Limitations of the DLVO Theory The effect of a surfactant on the stability of a lyophobic dispersion, according to DLVO theory, is therefore limited to its effect on the surface potential of the dispersed particles, the effective Hamaker constant, and the ionic strength of the dispersing liquid (in the case of ionic surfactants). Since surfactants are generally used at very low concentrations, the main effect of ionic surfactants would be expected to be on the surface potential of the dispersed particles, and this is observed experimentally. The addition of an ionic surfactant to a dispersion and its adsorption onto the dispersed particles generally increases the stability of dispersions whose particles are of the same sign as the surfactant and decreases the stability of those whose particles are of opposite sign. However, the situation is sometimes considerably more complex (see below). In the case of nonionics, the DLVO theory limits their effect to a change in the effective Hamaker constant, and although this may account for part of their effect, it is probably insufficient to account for the very large increase in stability produced by many POE nonionics.

Therefore, although the DLVO theory is very useful in predicting the effect of ionic surfactants on electrical barriers to aggregation, to fully understand the effects of surfactants on dispersion stability, other factors must also be considered. They include the following: (1) Adsorption of the surfactant onto dispersed particles that are larger than colloidal in size may change the contact angle (Chapter 2, Section IIF) they make with the dispersing liquid. This change may affect the stability of the dispersion. An increase in the contact angle may cause the particles to flocculate from the dispersion or to float to the surface. A decrease in the contact angle may increase dispersibility (Parfitt and Wharton, 1972). (2) Surfactants that are polymeric or that have long POE chains may form nonelectrical steric barriers to aggregation in aqueous media. The presence of these barriers, which are not covered by the DLVO theory, may increase the stability of dispersions, even when electrical barriers are reduced or absent. (3) In liquids of low dielectric constant, electrical barriers to aggregation are largely absent. In spite of this, stable dispersions of solids in these liquids can be prepared by use of surfactants that produce steric barriers to aggregation. (4) There is currently no accepted experimental method for measuring the potential at the Stern layer of the dispersed particles. The zeta potential, which is often used to estimate that potential, merely indicates the electrical potential at the plane of shear (the distance from the charged surface where the solvated particle and the solvent move with respect to each other). For highly solvated particles in particular, this may be quite different from the Stem layer potential.

The limitations of the DLVO theory have been investigated further and elaborated in many research publications. For example, the dispersion stability of oil-in-water emulsions stabilized in tetradecyltrimethylammonium bromide as a function of salinity underwent transitions from aggregation to dispersion, and such a variation in stability was explained by the predominance of non-DLVO surface forces, parsed into structural, hydration, and thermal fluctuation forces (Petkov et al., 1998). Ninham (1999) critically analyzed the "limitations" of the discussions on the limiting non-DLVO forces in explaining dispersion stability, the inconsistencies and insufficiencies in their description, and the need to consider the effect of dissolved gases in those systems.

The dispersion stability of nonspherical, crystalline (β-form) copper phthalocyanine particles, a well-known pigment in inks and paints with a blue color and cyan shade, was explained by a non-DLVO approach, by incorporating two types of DLVO treatments (one for spherical particles and one for parallel cubical faces) in which a nondimensionless quantity, defined as the ratio of electrostatic double-layer energy to Hamaker constant, signified the repulsive and attractive forces of the original DLVO theory (Dong et al., 2010). The two approaches used above integrated the strength and weakness of the original DLVO theory in water, as the DLVO model for spheres overpredicted the stability, while the model for cubes underpredicted the stability. The Fuchs–Smoluchowski stability ratio (McGown and Parfitt, 1967), needed for this study, was determined from dynamic light scattering data and the Rayleigh–Debye–Gans scattering theory.

B. Steric Forces

As mentioned above, dispersions of solids in liquids can be stabilized by steric barriers and in the absence or presence of electrical barriers. Such barriers can be produced when portions (lyophilic chains) of molecules adsorbed onto the surfaces of the solid particles extend into the liquid phase and interact with each other. These interactions (Tadros, 1986) produce two effects: (1) a mixing effect and (2) an entropic effect. The mixing effect is due to solvent–chain interactions and the high concentration of chains in the region of overlap. This effect becomes significant when adjacent particles approach each other to slightly less than twice the thickness of the adsorbed layer on the particles. It depends greatly upon the relative strengths of solvent–chain and chain–chain interactions. When the solvent–chain interaction is stronger than the chain–chain interaction, the free energy of the system is increased when the regions containing the extended portions of the adsorbed molecules overlap, and an energy barrier is produced to a closer approach. When the chain–chain interaction is greater than the solvent–chain interaction, the free energy is decreased when the regions overlap, and attraction rather than repulsion occurs. The entropic effect is due to restriction of the motion of the chains extending into the liquid phase when adjacent particles approach each other closely. This effect becomes particularly important when the separation between particle surfaces becomes less than the thickness of the adsorbed layer.

Both effects increase with increase in the number of adsorbed chains per unit of surface area on the dispersed particles and with the length of the chains extending into the liquid phase. However, there is an optimum chain length for maximum stabilization, since the possibility of flocculation also increases with chain length. In cases where the nature of the liquid phase can be varied, steric stabilization is best when one group of the adsorbed molecule has only limited solubility in the liquid phase, thereby promoting its adsorption onto the solid to be dispersed, while the other (long) group has good compatibility or interaction with the liquid phase, assisting its extension into it (Lee et al., 1986). Some examples may serve to illustrate the application of both DLVO theory and steric factors to the explanation of stability changes in dispersions.

1. The addition of a cationic surfactant (Figure 9.5) to a negatively charged colloidal dispersion (Ottewill and Rastogi, 1960) at first decreased the zeta potential of the dispersed particles and the stability of the dispersion

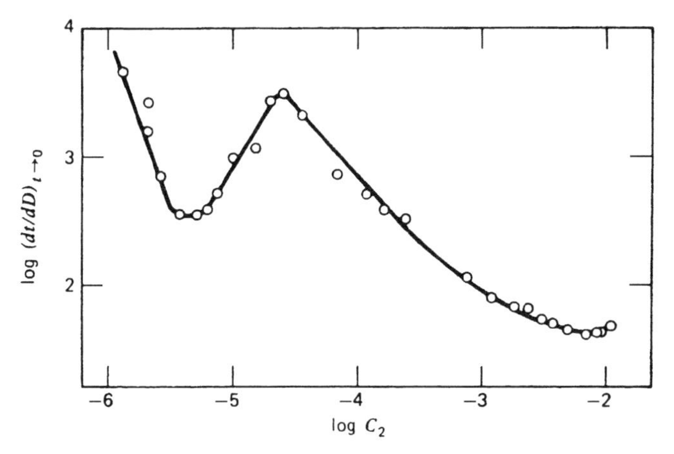

FIGURE 9.5 Stability [log$(dt/dD)_{t\to 0}$] of a negatively charged colloid dispersion of AgI as a function of log C_1, the concentration of added cationic surfactant, dodecylpyridinium bromide. Reprinted with permission from R. H. Ottewill and M. C. Rastogi (1960) *Trans. Faraday. Soc.* 56, 866.

until the zeta potential and the potential at the Stern layer had been reduced to zero, at which point stability reached a minimum. With further addition of surfactant, however, the stability increased again. This was because further adsorption of surfactant by the dispersed particles beyond the point of zero charge caused them to acquire an electrical potential of positive sign. At still higher concentrations of the cationic surfactant, the stability of the dispersion decreased again, although the zeta potential continued to become more positive, the decrease in stability this time being due to compression of the electrical double layer by the increased concentration of ionic surfactant.

2. In another investigation (Garvey and Tadros, 1972), adsorption of a polymeric ionic surfactant onto dispersed particles of the same sign initially caused an increase in both the stability of the dispersion and the zeta potential of the particles. Still higher concentrations of surfactant, however, resulted in a considerable decrease in the zeta potential of the particles. At these lower potentials, moreover, the stability of the dispersion showed a sharp increase. Here the explanation was that the decrease in the zeta potential was the result of a shift in the plane of shear (at which the zeta potential is measured) away from the surface of the particles resulting from the thicker layer of adsorbed polymeric surfactant at the higher liquid phase concentrations. This thicker adsorbed layer constituted a steric barrier to the aggregation of the particles and sharply increased the stability of the dispersion.
3. The addition of a POE nonionic surfactant to an aqueous dispersion whose particles carried a small negative charge increased the stability of

the dispersion to flocculation by a polyvalent cation (Ottewill and Walker, 1968). The stability increased sharply as adsorption of the nonionic surfactant onto the particles approached a close-packed vertical monolayer. Stability at this point was very high even when the electrical double layer was compressed by large amounts of electrolyte or the charge removed by lowering the pH of the dispersion. The high stability at this point was attributed to the closely packed, strongly solvated POE chains. Close approach of the dispersed particles to each other would require desolvation of these chains, which consequently constituted an energy barrier to aggregation of the particles. This high stability may also be due, at least in part, to a decrease in the effective Hamaker constant (Equation 9.3). The adsorption onto a dispersed particle of a layer that is chemically more similar to the solvent than to the particle itself causes a decrease in the effective Hamaker constant. The attraction between particles decreases rapidly with increase in the thickness of this adsorbed layer (Vold, 1961). Since the POE chain of the surfactant is highly hydrated, it is to be expected that adsorption of a POE nonionic surfactant onto the dispersed particles will decrease the effective Hamaker constant. As the adsorbed surfactant becomes more closely packed, the thickness of the adsorbed layer increases greatly and the attraction between the dispersed particles should show a sharp decrease. The stability of the dispersions at this point should consequently be very high.

II. ROLE OF THE SURFACTANT IN THE DISPERSION PROCESS

The dispersal of a solid in a liquid has been described as a three-stage process (Parfitt and Picton, 1968; Parfitt and Wharton, 1972): (1) wetting the powder and displacing trapped air, (2) deaggregation or fragmentation of the particle clusters, (3) prevention of reaggregation of the dispersed particles.

A. Wetting of the Powder

For a liquid to disperse a finely divided solid, it must first wet each particle cluster completely. This will involve, at least in the final stages of wetting, a spreading type of wetting, in which the air is completely displaced from the surface by the wetting medium. The driving force for this process, as we have seen, is the spreading coefficient:

$$S_{L/S} = \gamma_{SA} - \gamma_{SL} - \gamma_{LA} \tag{6.1}$$

and for spontaneously spreading wetting, this quantity must be positive, producing a contact angle of 0°. Wetting agents are therefore added to the liquid to lower γ_{SL}, and/or γ_{LA} by adsorption at those interfaces, especially in those cases where a liquid of high γ_{LA}, such as water, is the dispersion medium.

Reduction of the contact angle has been correlated with increased dispersibility in aqueous medium (Parfitt and Wharton, 1972).

B. Deaggregation of Fragmentation of Particle Clusters

Once the particle clusters have been wet by the suspending liquid, they must be dispersed in it. This may be accomplished by the surface-active agent in two ways:

1. By being adsorbed in "microcracks" in the solid, it may reduce the mechanical work needed to fragment the particles of solid (Rebinder, 1947). These microcracks are believed to be formed in crystals under stress, but are self-healing and disappear when the stress is removed. The adsorption of surface-active agents onto the surfaces of these microcracks may increase their depth and reduce their self-healing ability and thus reduce the energy required to rupture solid particles mechanically.
2. The adsorption of an ionic surfactant onto the particles in a cluster may cause the individual particles in the cluster to acquire an electrical charge of similar sign, resulting in their mutual repulsion and dispersion in the liquid phase.

C. Prevention of Reaggregation

Once the solid has been dispersed in the liquid, it is necessary to prevent the individual dispersed particles from coming together once again to form aggregates. Reduction of the thermodynamic instability of the dispersion ($\gamma_{SL} \times \Delta A$, where ΔA is the increase in interfacial area as a result of the dispersion) relative to the aggregated state can be reduced, although probably not eliminated, by adsorption of surfactants there in such a fashion as to reduce γ_{SL}. In the case of aqueous dispersions, this means adsorption of the surfactant with the hydrophilic group oriented toward the aqueous phase. The tendency for the dispersed particles to aggregate can be further reduced by adsorption of surfactant onto the dispersed particles in such a fashion as to increase or produce energy barriers to aggregation. These energy barriers, examples of which have been described earlier, may be electrical or nonelectrical in nature. In both cases, solvation of the lyophilic heads probably plays an important, not fully understood role in stabilizing the dispersion.

III. COAGULATION OR FLOCCULATION OF DISPERSED SOLIDS BY SURFACTANTS

Surfactants may be used not only to disperse solids in liquid media, but also to coagulate or flocculate solids already dispersed in liquid media. This may be accomplished with surface-active agents by a number of different mechanisms.

A. Neutralization or Reduction of the Potential at the Stern Layer of the Dispersed Particles

Electrostatic attraction of surface-active ions to oppositely charged sites on the surface of the dispersed particles results in a lowering of the electrical energy barrier to the close approach of two particles to each other, thereby promoting coagulation. If the surfactant ion contains only one hydrophilic (ionic) group, electrostatic attraction of that group to the oppositely charged site on the surface of the particle may, in addition, result in adsorption of the surfactant ion with its hydrophobic group oriented toward the liquid phase. In aqueous media this will cause an increase in the solid–liquid interfacial tension and an increase in the contact angle that the liquid makes with the particle (increased water repellence), with the result that the solid tends either to flocculate from the medium or to be expelled to the air–solution interface. The concentrations of surfactants needed to coagulate hydrophobic colloidal dispersions are orders of magnitude less than those of inorganic ions of the same charge, longer-chain surfactants being more efficient than shorter-chain ones (Ottewill and Rastogi, 1960). On the other hand, orientation of the surfactant in this fashion may make the particles more dispersible in nonaqueous media (e.g., in the "flushing" of pigments (Moilliet, 1955)).

Surfactants that at low concentrations cause flocculation in aqueous media by this mechanism may act as deflocculators at higher concentrations. This phenomenon is due to adsorption of additional surfactant molecules by interaction of their hydrophobic groups with those of the previously adsorbed surfactant molecules and with their hydrophilic groups oriented toward the aqueous phase (Figure 2.12, Chapter 2, Section IIC). Adsorption of this additional surfactant probably occurs only after the potential at the Stern layer has been completely neutralized by adsorption of the oppositely charged surfactant (i.e., only after the point of zero charge has been reached). Adsorption of this additional surfactant consequently produces a potential of the same sign as the surfactant ion (opposite in sign to that of the original potential at the Stern layer), which helps to redisperse the particle. This additional adsorbed surfactant appears to be much more easily removed than the originally adsorbed layer by reducing the concentration of the surfactant in the dispersion. Dilution of the dispersion therefore may reaggregate the particles.

If the surfactant is one that has two or more hydrophilic (ionic) groups at different points in the molecule, then adsorption of the surfactant onto an oppositely charged surface may be with one hydrophilic group oriented toward the surface and with the other(s) oriented toward the aqueous phase. In this case, adsorption of the surfactant will neutralize or reduce the electrical potential of the surface but may not cause flocculation of the particles from the aqueous phase.

B. Bridging

Flocculation by a bridging mechanism may occur in two ways: (1) A long (usually polymeric) surfactant molecule containing functional groups at various points in the molecule that can adsorb onto sites on the surface of adjacent particles may attach itself to two or more dispersed particles, thereby binding them together in a loose arrangement. This type of bridging appears to occur when adsorption of the surfactant onto the surface of the dispersed particles is low, thus providing ample sites for attachment of surfactant molecules extending from other particles (Kitchener, 1972). Thus, adsorbed molecules that act as steric stabilizers when their particle surface coverage is high can act as flocculants when their surface coverage is low and surface sites for adsorption are available on adjacent particles. Bridging by this mechanism generally reaches a maximum at about one-half total surface coverage. (2) When surfactant molecules are adsorbed onto dispersed particles in such a fashion that the adsorbed molecules extend into the liquid phase and these extended portions are capable of interacting with each other, bridging may occur by interaction of the extended portions attached to different particles. This type of bridging may occur with long polymeric surfactant molecules when surface coverage by the adsorbed molecules is so high that sites for attachment by the previously described bridging mechanism are scarce. As mentioned above, this type of bridging is prone to occur when the liquid phase is not a good solvent for the portion of the adsorbed molecules extending into it and is minimized when there is strong interaction between those portions and the liquid phase. Therefore, it may also occur with the usual type of surfactant (containing one terminal hydrophilic group and a hydrophobic group) when adsorption onto the dispersed particles is with the hydrophobic groups oriented toward the aqueous phase and surface coverage is so high that the hydrophobic group is forced to extend into the aqueous phase. Under these conditions, hydrophobic groups from two dispersed particles may come together to reduce their surface energies, thereby bridging the two particles (Somasundaran et al., 1966). In either of these cases, the energy released upon bridging must be greater than the energy required to desolvate the bridging groups. Consequently, strong interaction of the extended groups with the solvent may inhibit bridging.

C. Reversible Flocculation

This technique is useful when it is desired to flocculate an aqueous dispersion temporarily (e.g., for ease in filtration, handling, or storage), but in a condition where it can readily be deflocculated. The particles are first treated with an ionic surfactant that imparts a potential to the particles sufficiently high to disperse them. Then the dispersion is treated with sufficient easily soluble electrolyte to compress the electrical double layer surrounding the particles to the point where flocculation occurs. Subsequent dilution of the flocculated material (and consequent reduction of the electrolyte concentration) redisperses it when desired (Stewart and Bunbury, 1935).

Reversible flocculation of aqueous dispersions stabilized sterically with POE nonionics can be accomplished by increasing the temperature. With increase in temperature, the hydrogen bonding of the POE chains to water is reduced and the chains tend to aggregate, flocculating the dispersion. Upon reducing the temperature, the chains hydrate again and the particles redisperse.

IV. THE RELATION OF SURFACTANT CHEMICAL STRUCTURE TO DISPERSING PROPERTIES

For the purpose of this discussion, the term *dispersing properties* is used for those properties of a surfactant that enable it to adsorb onto a solid particle and to produce by this adsorption energy barriers of sufficient height to disperse the particle in a (generally aqueous) liquid medium. Surfactants sharing these properties are termed *dispersing agents.* Thus, although wetting of the solid particle by the dispersing liquid is a necessary first step in the dispersion process, a surfactant that produces only wetting of the particle surface without raising energy barriers of sufficient height to disperse the particle is not demonstrating dispersing properties in this system. It is acting merely as a wetting agent. On the other hand, a surfactant that does not promote wetting of the particle surface yet produces energy barriers of sufficient height for dispersion of the particle is considered as demonstrating dispersion properties. Of course, there are surfactants that demonstrate both wetting and dispersing power in a particular system, but wetting agents are often added to dispersing agents to compensate for their lack of wetting power. This discussion is therefore restricted to those structural features that bear on the ability of the surfactant to form energy barriers to aggregation.

A. Aqueous Dispersions

For the formation of electrical barriers to aggregation, ionic surfactants are generally used. When the solid to be dispersed is essentially nonpolar (e.g., hydrophobic carbon) and the dispersing medium is aqueous, conventional surfactants (containing one terminal hydrophilic group and a long hydrophobic group) may be used, since adsorption of the surface-active ion onto the essentially uncharged solid particles causes them all to acquire a charge of the same sign and to repel each other. An electrical barrier to aggregation will then have been formed. In addition, the adsorbed surfactant ions will be oriented with their hydrophobic groups toward the nonpolar particle and their hydrophilic heads toward the aqueous phase, producing a lowering of the solid–liquid interfacial tension. Since the efficiency of adsorption (Chapter 2, Section IID) in this case increases with increase in the length of the hydrophobic group, longer-chain compounds can be expected to be more efficient dispersing agents for this type of particle than shorter-chain ones.

When the solid to be dispersed is charged, however, conventional surfactants may not be useful. If a conventional surfactant of opposite charge is used, then flocculation rather than dispersion will probably occur until the charge on the particles has been neutralized. Only then may dispersion, caused by adsorption of additional surfactant ions onto the now uncharged particles, occur. Therefore, this is generally not an efficient method of forming the dispersion. On the other hand, if a conventional surfactant of the same sign as the particle is used, the situation is not improved much. Although adsorption of the surfactant ion may increase the electrical energy barrier to aggregation, and generally occurs with the ionic hydrophilic head oriented away from the similarly charged particle surface (and hence oriented toward the aqueous phase), repulsion between the adsorbing surfactant ion and the similarly charged particle inhibits adsorption. Only at relatively high concentrations of the surfactant in the aqueous phase is adsorption sufficiently high to stabilize the dispersion.

As a result, ionic dispersing agents for use with charged or polar solids in aqueous medium usually have ionic groups at various positions in the surfactant molecule and hydrophobic groups containing polarizable structures such as aromatic rings or ether linkages rather than saturated hydrocarbon chains. The multiple ionic groups probably serve a number of purposes: (1) They inhibit adsorption of the surfactant molecule with the hydrophobic group facing the aqueous phase. On oppositely charged particles, one of the multiple ionic groups in the surfactant molecule can be adsorbed onto an oppositely charged site while another may be oriented toward the aqueous phase, thus preventing adsorption of the surfactant with its hydrophobic group facing the aqueous phase and the consequent tendency of the dispersion to flocculate. (2) They increase the efficiency of the surfactant molecule in producing an electrical barrier to aggregation. The larger the number of ionic charges of similar sign per molecule, the greater the increase in the electrical barrier per adsorbed molecule on similarly charged particles and the greater the neutralization of charge leading to formation of an electrical barrier of sign similar to that of the surfactant on oppositely charged particles. (3) They permit extension of the surfactant molecule into the aqueous phase (thus creating a steric barrier to coalescence) without an increase in the free energy of the system. The decrease in free energy resulting from hydration of the ionic hydrophilic groups may compensate for the free energy increase due to the increased contact of the hydrophobic group with the aqueous phase.

The difference in the properties of aqueous dispersions treated with oppositely charged surfactants containing a single hydrophilic group and a hydrophobic group or containing two hydrophilic groups at opposite ends of the hydrophobic group has been discussed by Esumi et al. (1986). The dispersions were, in all cases, flocculated by the addition of the oppositely charged surfactant. However, when the surfactant used had a single hydrophilic group, the flocculated particles were easily dispersed ("flushed") into toluene. When it had two hydrophilic groups at opposite ends of the hydrophobic group, the flocculated material could not be dispersed in toluene, but formed a film at the toluene–water interface. In the first case, the flocculated particles were lipophilic because of the orientation of the hydrophobic groups of thead-sorbed surfactant molecules toward the aqueous phase; in the second case, each hydrophobic group extending into the aqueous phase had a terminal hydrophilic group which prevented the particles from becoming lipophilic.

The polarizable structures in the hydrophobic group of the dispersing agent offer sites by means of which the surfactant can interact with charged sites on the particle size and consequently adsorb via its hydrophobic group. It has been shown, for example, that alumina adsorbs surfactants onto its surface by polarization of the π electrons in the adsorbate (Snyder, 1968). This gives the adsorbed surfactant molecules the proper orientation for acting as dispersing agents in aqueous media.

The adsorption from aqueous solution of surfactants with two hydrophilic and two hydrophobic groups (gemini surfactants, Chapter 12) onto oppositely charged sites on solid particle surfaces—cationic geminis onto clay particles (Li and Rosen, 2000), anionic geminis onto limestone particles (Rosen and Li, 2001)—results in one hydrophilic group oriented toward the solid surface and the second oriented toward the aqueous phase. The solid particles are dispersed in both cases.

Examples of commonly used dispersing agents containing multiple ionic groups and aromatic hydrophobic groups are β-naphthalene–sulfonic acid–formaldehyde condensates and ligninsulfonates (Chapter 1, Section IA2).

Polyelectrolytes prepared from ionic monomers are often excellent dispersing agents for solids in aqueous media. Their multiple ionic groups can impart high surface charges to the solid particles onto which they adsorb. When the tendency to adsorb onto the surface of a solid particle of an individual functional group attached to the backbone of the polymer is low, the number of such groups in the macromolecule must be large enough that the total adsorption energy of the molecule is sufficient to anchor it firmly to the particle surface. Homopolymers, consequently, are not as versatile as copolymers, since the former have a more limited range of substrates to which they anchor well, especially when the molecular weight of the polymer is low. Copolymers of monomers with different structural characteristics yield products that can adsorb strongly to a wider variety of substrates. Thus, acrylic acid or maleic anhydride copolymerized with styrene yields dispersing agents, with aromatic nuclei attached to the backbone of the polymer, which can adsorb onto a wide range of substrates. For nonpolar substrates, short-chain monomers, such as acrylic acid, are copolymerized with long-chain monomers, for example, lauryl methacrylate, to increase the binding energy of the dispersing agent to the particle surface (Buscall and Corner, 1986).

With increase in the number of hydrophilic groups per molecule of dispersing agent, there is often an increase in its solubility in water, and this may cause a decrease in its adsorption onto a particular particle surface (Garvey and Tadros, 1972), especially when the interaction between the surfactant and the particle surface is weak. In some cases, therefore, the adsorption of the dispersing agent onto a particle surface and its dispersing power for it may go through a maximum, with increase in the number of ionic groups in the surfactant molecule. Thus, in the preparation of aqueous dispersions of dyestuffs (Prazak, 1970), hydrophobic dyestuffs, which would be expected to interact strongly with the hydrophobic groups of ligninsulfonate dispersing agents, produce dispersions that are stable to heat when the ligninsulfonate is highly sulfonated. Hydrophilic dyestuffs, on the other hand, which would not be expected to interact strongly with this type of dispersing agent, form dispersions with it that are not stable to heating. A less highly sulfonated ligninsulfonate, however, produces dispersions that are heat stable with these hydrophilic dyestuffs. Presumably, the high solubility of the highly sulfonated dispersing agent at the elevated temperature removes it from the surface of the hydrophilic, but not the hydrophobic, dyestuff. In order to get equal heat stability with the hydrophilic dyestuffs, a less soluble (less sulfonated) dispersing agent must be used.

Particles can also be dispersed in the aqueous medium by the use of steric barriers. For this purpose, both ionic and nonionic surfactants can serve as steric stabilizers. As discussed previously, steric barriers to aggregation are produced when the adsorbed surfactant molecules extend chains into the aqueous phase and inhibit the close approach of two particles to each other. As mentioned in Section IB, steric stabilization increases with increase in the length of the chains extending into the liquid phase. Consequently, polymeric surfactants, both ionic and nonionic, are commonly used as steric stabilizers since the length of the chain extending into the liquid phase can often be increased conveniently by increasing the degree of polymerization. Surfactants with ionic groups distributed along the length of molecule can, as mentioned previously, produce such steric barriers, and their effectiveness in doing so increases with the distance into the aqueous phase to which the molecules can extend. Hence, longer compounds are more effective than shorter ones (Garvey and Tadros, 1972), provided that increased solubility in the aqueous phase does not decrease their adsorption onto the particle surface significantly. Nonionic surfactants of the POE type are excellent dispersing agents for many purposes because their highly hydrated POE chains extend into the aqueous phase in the form of coils that present excellent steric barriers to aggregation. In addition, the thick layer of hydrated oxyethylene groups similar in nature to the aqueous phase would be expected to produce a considerable decrease in the effective Hamaker constant (Equation 9.3) and a consequent sharp decrease in the van der Waals attraction between particles. For the adsorbed layer to be an effective steric barrier, its thickness must generally exceed 25 Å. For a POE chain in aqueous medium, this is usually attained when there are more than 20 EO units in the chain, although stable aqueous dispersion of solid particles by POE nonionics with much smaller POE chains has been observed. Thus, ferric oxyhydroxide (β-FeOOH), precipitated by the hydrolysis of ferric chloride in the presence of a POE C_{12}-C_{15} alcohol with an average of 4 oxyethylene units, was obtained as a dispersion of nano-sized particles stable for several months (O'Sullivan et al., 1994).

Block and graft polymers are widely used as steric stabilizers. Since the two blocks are separated in the molecule, they can be designed chemically, by use of the proper functional groups and degree of polymerization, for optimum efficiency and effectiveness. One block should be designed to adsorb strongly onto the particle surface (and also have limited solubility in the liquid phase), the other block(s) to extend into the liquid phase (good compatibility with and/or interaction with the liquid phase). One commonly used type is the polyoxyethylene-polyoxypropylene (POE-POP) block copolymer, made from ethylene oxide and propylene oxide. For dispersions in aqueous media, products of structure

$$H(OCH_2CH_2)_x[OCH(CH_3)CH_2]_y(OCH_2CH_2)_zOH$$

are used. The central POP block, $[OCH(CH_3)CH_2]_y$, which is not soluble in water, is attached to the surface of the solid particle, while the water-soluble POE chains extend into the aqueous phase as random coils and produce a steric barrier to the close approach of adjacent particles. Another type that is used has the structure

$$H[OCH(CH_3)CH_2]_x(OCH_2CH_2)_y[OCH(CH_3)CH_2]_zOH.$$

Here, the POE block is central, surrounded by POP chains. This type is most effective in nonaqueous liquids in which the POE block has limited solubility and the POP blocks have good solubility, causing the former to adsorb efficiently onto the solid particles, while the latter extend into the liquid phase to produce the steric barrier. In both of these types, steric stabilization increases with increase in the length of the chain extending into the liquid phase. The increase in solubility with increase in the length of the POE chain can, in this type of compound, readily be compensated for by increasing the length of the POP hydrophobic groups. Thus, the most effective dispersing agents of this type would be expected to be those in which both the POE and POP chains are long.

When the particles to be dispersed are hydrophilic, then adsorption of conventional POE nonionic surfactants occurs with the POE chains oriented toward the hydrophilic particle surface and the hydrophobic chains oriented toward the aqueous phase (Glazman et al., 1986). Stabilization of the dispersion is then achieved by bilayer formation, with the hydrophobic groups of the two surfactant layers oriented toward and associated with each other, while the POE chains of the second layer are oriented toward the aqueous phase. Consistent with this explanation is the lack of any stabilization effect until the surfactant concentration in the aqueous phase is considerably above the concentration required for hydrophobic group association, that is, the critical micelle concentration (CMC).

B. Nonaqueous Dispersions

In nonaqueous media of low dielectric constant, electrical barriers to aggregation are usually ineffective and steric barriers are generally required to disperse solid particles. As the dielectric constant of the dispersing medium increases, electrical barriers become more significant. The steric barriers may arise, as the particles approach each other, either (1) from the energy required to desolvate, the portions of the adsorbed surfactant molecules extending into the dispersing medium, or (2) from the decrease in the entropy of the system as these portions of the adsorbed molecules are restricted in their movement or arrangement by the close approach of two particles. The effective Hamaker constant and the consequent attraction between particles can also be reduced by the use of molecules that, on adsorption, extend out from the particle surface lyophilic groups of a nature similar to that of the dispersing liquid.

Thus, the dispersion of carbon in aliphatic hydrocarbons is improved by the addition of alkylbenzenes (van der Waarden, 1950, 1951). The benzene rings are presumably adsorbed onto the surface of the carbon and the aliphatic chains extend into the dispersing liquid. Increase in the length and the number of alkyl groups attached to the benzene nucleus increases the stability of the dispersion. In somewhat similar fashion, the dispersion of two ionic solids (halite and sylvite) in nonpolar solvents was improved by the addition of long-chain amines (Bischoff, 1960). An increase in the chain length of the amine increased its efficiency as a dispersing agent.

A mechanism for electrical charging of solid particles in nonaqueous media has been proposed, involving acid–base interaction between neutral particle and neutral adsorbed dispersing agent. Charge separation between them occurs when the charged dispersing agent is desorbed and incorporated into bulky reverse micelles in the nonaqueous phase with the charged sites in the interior of the micelle. Acidic or basic polymers are consequently effective dispersing agents for solid particles in nonaqueous media.

C. Design of New Dispersants

The need for new dispersants with optimal molecular structural features frequently arises because many diverse novel materials developed require special dispersants to maintain its dispersion stability. Aside from regular monomeric surfactant dispersants, there is a growing trend toward developing polymeric dispersants that control the surface properties of particles, especially their dispersion characteristics. A typical polymeric dispersant contains either a hydrophobic backbone and hydrophilic side chains, or a hydrophilic backbone and hydrophobic side chains. Many examples exist in the natural and synthetic worlds, such as proteins and carbohydrates (see Chapter 13 and Table 13.1), and modified EO or PE surfactants (see Chapter 1). Polymeric surfactants also serve as viscosity modifiers in industrial dispersions. Paints, inks (including ink-jet and electronic inks), nanocarbon dispersions, and biological and pharmaceutical dispersions, all require specific dispersants.

Moreover, stringent restrictions imposed by government agencies on limiting the use of volatile organic compounds (VOCs), have resulted in a resurgence in polymeric surfactants, because many of the traditional organic solvents that come under VOCs must be replaced by water. Polymeric dispersants such as acrylates (Kunjappu, 1999, 2001) were the earliest ones to be developed to support the high solid/liquid ratio aqueous media, such as the dispersion of organic pigments (Kunjappu, 2000) in water. Nitroxide-mediated radical polymerization of acrylic block copolymers (Auschra et al., 2002) produced dispersants that were tested on pigments.

The importance of functional group modifiers in controlling the polarity of these molecules depending on their location in the molecule, and their influence on the packing density and orientation at the interface, have been pivotal in defining the molecular architecture and regulatory behavior of polymeric dispersants (Chevalier, 2002).

There still exist several special low-volume applications, where organic solvents are deemed to be essential. Specific polymeric dispersants required in preparing stable dispersions of single-walled carbon nanotubes (SWCNT), have been investigated (Kim et al., 2007), using monomers of 3-hexylthiophene to produce oligomers that differ in the number and regioregularity of head groups, and the head-to-tail ratios of the hexyl group. These dispersants were found to be effective even at very low concentrations, imparting long-term stability of single-walled nanotube (SWNT) dispersions in liquids such as 1,2-dichloroethane, *N*,*N*-dimethylformamide, and *N*-methyl-2-pyrrolidone. These liquid dispersing media helped enhance the properties of SWNT (nano) dispersions because of the superior mechanical strength, electrical conductivity, and chemical stability of SWNT.

REFERENCES

Auschra, C., E. Eckstein, A. Mühlebach, M. O. Zink, and F. Rime (2002) *Prog. Org. Coat.* **45**, 83.

Bischoff, E. (1960) *Kolloid-Z.* **168**, 8.

Buscall, R. and T. Corner (1986) *Colloids Surf.* **17**, 39.

Chevalier, Y. (2002) *Curr. Opin. Colloid Interface Sci.* **7**, 3.

Derjaguin, B. and L. Landau (1941) *Acta Physicochim.* **14**, 633.

Dong, J., D. S. Corti, E. I. Franses, Y. Zhao, H. T. Ng, and E. Hanson (2010) *Langmuir* **26**, 6995.

Esumi, K., K. Yamada, T. Sugawara, and K. Meguro (1986) *Bull. Chem. Soc. Japan* **59**, 697.

Fuchs, N. *Z. Phys.* **89**, 736 (1934).

Garvey, M. J. and T. F. Tadros, *Proc. 6th Int. Congr.* Surface-Active Substances, Zurich, p. 715, 1972.

Glazman, Y. M., G. D. Botsaris, and P. Dansky (1986) *Colloids Surf.* **21**, 431.

Hamaker, C. H. (1937) *Physica* **4**, 1058.

Kim, K. K., S. M. Yoon, J. Y. Choi, J. Lee, B. K. Kim, J. M. Kim, J. H. Lee, U. Paik, M. H. Park, C. W. Yang, K. H. An, Y. Chung, and Y. H. Lee (2007) *Adv. Funct. Mater.* **17**, 1775.

Kitchener, J. A. (1972) *Be Polym. J.* **4**, 217.

Kunjappu, J. T. (1999) Ink World February, pp. 40–45.

Kunjappu, J. T. (2000) Ink World Paints and Coatings Industry, Septmeber, pp. 124–133.

Kunjappu, J. T. *Essays in Ink Chemistry (For Paints and Coatings Too)*, Nova Science Publishers, New York, 2001.

Lee, H., R. Pober, and P. Calvert (1986) *J. Colloid Interface Sci.* **110**, 144.

Li, F. and M. J. Rosen (2000) *J. Colloid Interface Sci.* **224**, 265.

Lyklema, J. (1968) *Adv. Colloid Interface Sci.* **2**, 67.

McGown, D. N. L. and G. D. Parfitt (1966) *Disc. Faraday Soc.* **42**, 225.

McGown, D. N. L. and G. D. Parfitt (1967) *J. Phys. Chem.* **71**, 449.

Moilliet, J. L. (1955) *J. Oil Colour Chem. Asoc.* **38**, 463.

Ninham, B. W. (1999) *Adv. Colloid Interface Sci.* **83**, 1.

O'Sullivan, E. C., A. J. I. Ward, and T. Budd (1994) *Langmuir* **10**, 2985.

Ottewill, R. H. and M. C. Rastogi (1960) *Trans. Faraday Soc.* **56**, 866, 880.

Ottewill, R. H. and J. N. Shaw (1966) *Disc. Faraday Soc.* **42**, 154.

Ottewill, R. H. and T. Walker (1968) *Kolloid-Z. Z. Polym* **227**, 108.

Parfitt, G. D. and N. H. Picton (1968) *Trans. Faraday Soc.* **64**, 1955.

Parfitt, G. D. and D. G. Wharton (1972) *J. Colloid Interface Sci.* **38**, 431.

Petkov, J., J. Se′ne′chal, F. Guimberteau, and F. Leal-Calderon (1998) *Langmuir* **14**, 4011.

Prazak, G. (1970) *Am. Dyestuff Rep.* **59**, 44.

Rebinder, P. (1947) *Nature* **159**, 866.

Reerink, H. and J. T. G. Overbeek (1954) *Disc. Faraday Soc.* **18**, 74.

Rosen, M. J. and F. Li (2001) *J. Colloid Interface Sci.* **234**, 418.

Snyder, L. R. (1968) *J. Phys. Chem.* **72**, 489.

Somasundaran, P., T. W. Healy, and D. W. Fuerstenau (1966) *J. Colloid Interface Sci.* **22**, 599.

Stewart, A. and H. M. Bunbury (1935) *Trans. Faraday Soc.* **31**, 214.

Tadros, T. F. (1986) *Colloids Surf.* **18**, 137.

van der Waarden, M. (1950) *J. Colloid Sci.* **5**, 317.

van der Waarden, M. (1951) *J. Colloid Sci.* **6**, 443.

Verwey, E. and J. T. G. Overbeek *Theory of the Stability of Lyophobic Colloids*, Elsevier, Amsterdam, 1948.

Vold, M. J. (1961) *J. Colloid Sci.* **16**, 1.

Watanabe, A. (1960) *Bull. Inst. Chem. Res. Kyoto Univ.* **38**, 179.

PROBLEMS

9.1 List three different ways of increasing the stability of a dispersion of an ionic solid in a liquid.

9.2 Discuss two different mechanisms by which the POE chain in a nonionic surfactant adsorbed on a finely divided hydrophobic substance can help stabilize a dispersion of that substance in water.

9.3 Explain why Ca^{2+} is much more effective than Na^{+}, at the same molar concentration in the solution phase, as a flocculant for an aqueous dispersion stabilized by an anionic surfactant.

9.4 Which one of the following, at the same molar concentration in the solution phase, would be expected to be the most effective stabilizer for a dispersion of a positively charged hydrophilic solid in heptane?

(a) $C_{12}H_{25}\ (OC_2H_4)_2OH$
(b) $C_{12}H_{25}\ (OC_2H_4)_8OH$
(c) $C_{12}H_{25}N^{+}(CH_3)_3Cl^{-}$
(d) Sodium ligninsulfonate

9.5 A dispersion of an ionic solid in aqueous medium is precipitated by small amounts of $C_{12}H_{25}SO_4^{-}Na^{+}$ or $C_{12}H_{25}(OC_2H_4)_{10}OH$, but is unaffected by the addition of $C_{12}H_{25}N^{+}(CH_3)_3Cl^{-}$. What conclusions regarding the solid can be drawn from these data?

9.6 Discuss the effect of the following surfactants on an aqueous dispersion of a water-insoluble salt of a polyvalent metal whose particles are positively charged:

(a) A small amount of $C_{12}H_{25}SO_4^{-}Na^{+}$
(b) A small amount of $C_{12}H_{25}(OC_2H_4)_{10}OH$
(c) A small amount of $C_{12}H_{25}N^{+}(CH_3)_3Cl^{-}$
(d) A large amount of $C_{12}H_{25}C_6H_4SO_3^{-}Na^{+}$

9.7 Explain why a gemini surfactant (with two hydrophobic and two hydrophilic groups in the molecule, Chapter 12) is a much better dispersing agent than a similar surfactant with only one hydrophobic and one hydrophilic group in the molecule for oppositely charged solids but not much better for similarly charged solids.

9.8 Find the typical values of the Hamaker constant in the literature for a particle that can form a lyophobic dispersion. What is the significance of the magnitude of this constant in controlling the dispersion stability?

9.9 What are the limitations of the DLVO theory in the context of dispersion stability of colloids in surfactant solutions? Read any recent article from the literature on this topic and critically evaluate the developments in this area.

9.10 Assume that you are asked to design a new surfactant candidate to increase the stability of a lyophobic dispersion. What will be your important considerations and general strategy for this?

10 Detergency and Its Modification by Surfactants

Since detergency is by far the largest single use for surfactants, there is a voluminous literature on the subject. In spite of that, it is only in recent years that a real understanding of the factors involved in the cleaning process has started to emerge and a great deal about the subject still remains obscure. This is undoubtedly due to the complexity of the cleaning process and the large variations in soils and substrates encountered. Since several books have been devoted to various aspects of detergency (Cutler and Kissa, 1986; Lai, 1997; Showell, 1998; Broze, 1999; Friedli, 2001), this discussion will cover only those areas that are pertinent to the role of surfactants in the cleaning process.

The term *detergency*, as used to describe a property of surface-active agents, has a special meaning. As a general term, it means cleaning power, but no surfactant by itself can clean a surface. The term *detergency*, when applied to a surface-active agent, means the special property it has of enhancing the cleaning power of a liquid. This it accomplishes by a combination of effects involving adsorption at interfaces, alteration of interfacial tensions, solubilization, emulsification, and the formation and dissipation of surface charges.

I. MECHANISMS OF THE CLEANING PROCESS

Three elements are present in every cleaning process: (1) the substrate (the surface that is to be cleaned), (2) the soil (the material that is to be removed from the substrate in the cleaning process), and (3) the cleaning solution or "bath" (the liquid that is applied to the substrate to remove the soil). The difficulty in developing a unified mechanism for the cleaning process lies in the almost infinite variety of the first two elements—the substrate and the soil. The substrate may vary from an impervious, smooth, hard surface like that of a glass plate to a soft, porous, complex surface like that of a piece of cotton or wool yarn. The soil may be liquid or solid (usually a combination of both), ionic or nonpolar, finely or coarsely ground, inert or reactive toward the clean-

Surfactants and Interfacial Phenomena, Fourth Edition. Milton J. Rosen and Joy T. Kunjappu.

ing bath. As a result of this great variability of substrate and soils, there is no one single mechanism of detergency, but rather a number of different mechanisms, depending on the nature of substrate and soil. The bath used is generally a solution of various materials, collectively called the *detergent*, in the cleaning liquid. Except in the case of dry cleaning, which is covered later in this section, the liquid in the bath is water.

In general, cleaning consists essentially of two processes: (1) removal of the soil from the substrate and (2) suspension of the soil in the bath and prevention of its redeposition. This second process is just as important as the first, since it prevents redeposition of the soil onto another part of the substrate.

A. Removal of Soil from Substrate

Soils are attached to substrates by various types of forces and as a result are removed from them by different mechanisms. This discussion is restricted to the removal of soils by mechanisms in which surfactants play a major role; it will not cover the removal of soil by mechanical work, or chemical reagents such as bleaches, reducing agents, or enzymes. Substances that are chemisorbed via covalent bond formation can generally be removed only by chemical means that destroy those bonds (e.g., by use of oxidizing agents or enzymes); soils that can be removed by the use of surfactants are generally attached by physical adsorption (van der Waals forces, dipole interactions) or by electrostatic forces. The removal of soil by surfactants generally involves their adsorption onto the soil and substrate surfaces from the cleaning bath (Schwuger, 1982). This adsorption changes interfacial tensions and/or electrical potentials at the soil–bath and substrate–bath interfaces in such a manner as to enhance the removal of the soil by the bath.

Soils whose removal from substrates can be enhanced by the presence of surfactants in the cleaning bath are generally classified according to the mechanisms by which they are removed. Liquid soils, which may contain skin fats (sebum), fatty acids, mineral and vegetable oils, fatty alcohols, and the liquid components found in cosmetic materials, are generally removed by the *roll-back* mechanism. Solid soils may consist either of organic solids, such as mineral or vegetable waxes, that can be liquefied by the application of heat or the action of additives, or of particulate matter, such as carbon, iron oxide, or clay particles, that cannot be liquefied. The former are generally removed, after liquefaction, by the roll-back mechanism; the latter, by the production of repulsive electrical potentials on soil and substrate surfaces.

Evidence is accumulating that maximum detergency may be associated with the presence of an insoluble, surfactant-rich phase—the middle phase discussed in Chapter 5, Section IIIA. Thus, soap forms dispersed particles of an insoluble surfactant-rich product in hard water in the presence of certain surfactants known as *lime soap dispersing agents* (LSDAs), described in Section II. This product, although insoluble in the cleaning bath, shows high surface activity and high detergency (Weil et al., 1976). Commercial linear alkylbenzenesulfonate (LAS) in hard water in the presence of certain polyoxyethylene (POE) nonionics forms a suspension of insoluble particles that solubilizes mineral oil (Smith et al., 1985). The suspension shows better detergency than LAS by itself in water of the same hardness (Smith et al., 1985). In oily soil detergency by POE nonionics (Section IB2 below), maximum detergency is obtained 15–30°C above the cloud point of the nonionic, where particles of a surfactant-rich phase are present in the cleaning bath.

1. Removal of Liquid Soil Removal of liquid (oily) soil by aqueous baths is accomplished mainly by a *roll-back* or *roll-up* mechanism in which the contact angle that the liquid soil makes with the substrate is increased by adsorption of surfactant from the cleaning bath.

Figure 6.3 illustrates the situation of a liquid soil adhering to a substrate in the presence of air. The reversible work to remove the liquid oily soil O from the substrate, the work of adhesion W_a, (Equations 6.12 and 6.13) is given by the expressions

$$W_{O/S(A)} = \gamma_{SA} + \gamma_{OA} - \gamma_{SO} \quad (10.1)$$
$$= \gamma_{OA}(\cos\theta + 1), \quad (10.2)$$

where θ is the contact angle, measured in the liquid soil phase, at the soil substrate–air junction. Figure 10.1 illustrates the situation where the air is replaced by a cleaning bath. The work of adhesion of the liquid soil for the substrate is now given by the expression

$$W_{O/S(B)} = \gamma_{SB} + \gamma_{OB} - \gamma_{SO} \quad (10.3)$$
$$= \gamma_{OB}(\cos\theta + 1) \quad (10.4)$$

and the contact angle by the expression

$$\cos\theta = \frac{\gamma_{SB} - \gamma_{SO}}{\gamma_{OB}}. \quad (10.5)$$

When surfactants of the proper structure are present in the bath, they will adsorb at the substrate–bath (SB) and liquid soil–bath (OB) interfaces in such

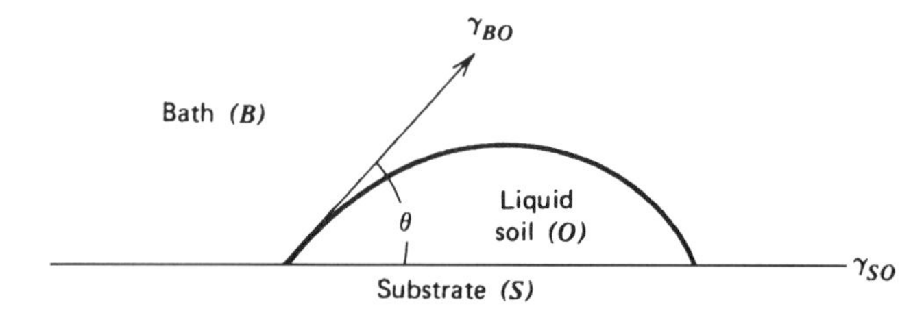

FIGURE 10.1 Contact angle at the bath–liquid soil–substrate junction.

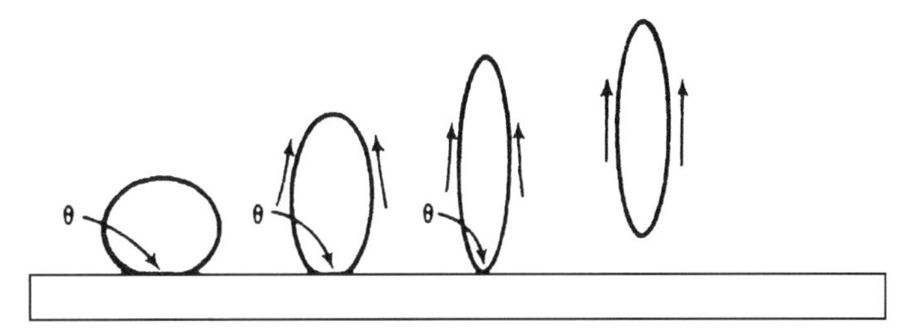

FIGURE 10.2 Complete removal of oil droplets from substrate by hydraulic currents (arrows) when θ remains constant at >90°. Reprinted with permission from A. M. Schwartz, in *Surface and Colloid Science*, E. Matijevic (ed.), Vol. 5, Wiley, New York, 1972, p. 212.

a fashion (i.e., with the hydrophilic group oriented toward the aqueous bath) as to reduce γ_{SB} and y_{OB}, with consequent reduction in the work to remove the soil from the substrate. Reduction in γ_{SB} will also cause a decrease in cos θ and an increase in θ, resulting in the observed roll-back of the liquid soil. Many investigators of oily soil removal, both on textile and hard surfaces, have found that reduction of γ_{0B} (Dillan et al., 1979; Matson and Smith, 1980; Pierce and Trowbridge, 1980; Aronson et al., 1983; Dillan, 1984) and/or increase in θ, measured in the oily soil phase (Rubingh and Jones, 1982), correlates well with increase in detergency. In some cases, this low γ_{0B} value may be associated with the separation of an insoluble surfactant-rich phase (Chapter 5, Section III). In many cases, γ_{SB} is reduced to the point where $\gamma_{SB} - \gamma_{S0}$ is negative, with resulting increase in θ to a value greater than 90°. Such a situation is illustrated in Figure 10.2.

If the contact angle is 180°, the bath will spontaneously completely displace the liquid soil from the substrate; if the contact angle is less than 180° but more than 90°, the soil will not be displaced spontaneously but can be removed by hydraulic currents in the bath (Figure 10.2) (Schwartz, 1972). When the contact angle is less than 90°, at least part of the oily soil will remain attached to the substrate, even when it is subjected to the hydraulic currents of the bath (Figure 10.3) (Schwartz, 1971, 1972), and mechanical work or some other mechanism (e.g., solubilization, see below) is required to remove the residual soil from the substrate.

In high-speed spray cleaning, a critical factor is the dynamic surface tension reduction of the surfactant solution (Chapter 5, Section IV), rather than its equilibrium surface tension value, since under these cleaning conditions equilibrium values are not attained. Surfactants that reduce the surface tension to the lowest values in short times exhibit the best soil removal (Prieto et al., 1996).

2. Removal of Solid Soil *Liquefiable Soil* The first stage in the removal of this type of soil is believed to be liquefaction of the soil (Cox, 1986).

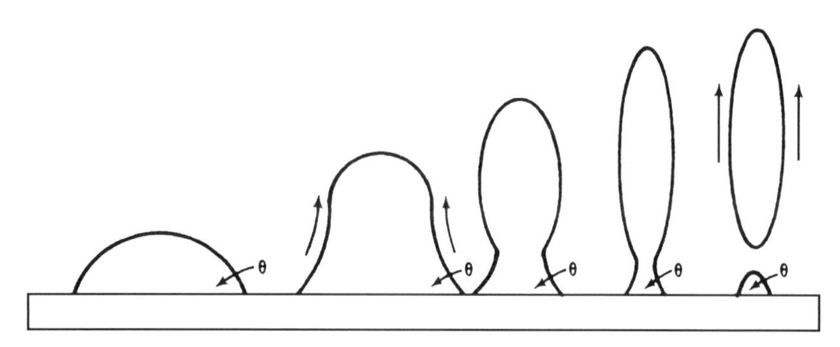

FIGURE 10.3 Rupture and incomplete removal of large oil droplets by hydraulic currents (arrows) when θ remains constant at <90°. A small droplet remains attached to the substrate. Reprinted with permission from A. M. Schwartz, in *Surface and Colloid Science*, E. Matijevic (ed.), Vol. 5, Wiley, New York, 1972, p. 211.

Penetration of the soil by surfactant (and associated water molecules) from the cleaning bath with resulting liquefaction may be a key process in the removal of this type of soil (Cox and Matson, 1984). In cases where penetration of the solid soil by surfactants or other additives does not occur, an increase in the temperature of the cleaning process may result in its liquefaction. The liquefied soil is then removed by the roll-back mechanism described above (Cox et al., 1987).

Particulate Soil Removal of particulate solid soil by aqueous baths is accomplished by the following mechanisms:

1. *Wetting of the Substrate and the Soil Particles by the Bath.* Adhesion of small solid particles to solid substrate is greatly diminished by immersion in water because of interaction of the water with substrate and particles. The presence of water results in the formation of electrical double layers at the substrate–liquid and particle–liquid interfaces. These electrical double layers almost always result in charges of similar sign on substrate and particle with a resulting mutual repulsion that, when superimposed on the preexisting van der Waals attraction, causes a diminution of the net adhesion (Chapter 9). In addition, water may cause the substrate surface, especially if it is of a fibrous nature, to hydrate and swell, resulting in an increase in the distance between particle and substrate.

   The tendency of the bath B to spread over the soil particle P or the substrate S is given by the spreading coefficients (Chapter 6), $S_{B/P}$ and $S_{B/S}$, respectively, where

$$S_{B/P} = \gamma_{PA} - \gamma_{PB} - \gamma_{AB} \tag{10.6}$$

and

$$S_{B/S} = \gamma_{SA} - \gamma_{SB} - \gamma_{AB}. \tag{10.7}$$

(Subscripts PA, SA, and AB refer to the particle–air, substrate–air, and air–bath interfaces, respectively.) If the spreading coefficient is positive, then spreading occurs spontaneously; if not, mechanical work must be done to wet the surface completely. Adsorption of surfactant from the bath at the air–bath interface or onto soil or substrate (with the hydrophilic group oriented toward the bath) can reduce γ_{AB}, γ_{PB}, or γ_{SB}, respectively, and thereby increase the tendency to wet the soil and/or substrate. Since the soil particle or the substrate is often hydrophobic, with the consequence that γ_{PA} or γ_{SA}, respectively, is low, mechanical work is often necessary to wet both the soil and the substrate completely with the bath. This is one of the reasons that washing is always accompanied by some mechanical work.

2. *Adsorption of Surfactant and Other Bath Components (e.g., Inorganic Ions) at the Substrate–Liquid and Particle–Liquid Interfaces.* This causes a decrease in the work required to remove the particle from the substrate, since the free energy change per unit area involved in this process is the work of adhesion W_a (Chapter 6, Section IB), given in this case by the expression

$$W_a = \gamma_{SB} + \gamma_{PB} - \gamma_{SP}. \tag{10.8}$$

Adsorption of surfactants at these interfaces can result in a decrease in γ_{SB} and γ_{PB}, with a consequent decrease in the work required to cause removal of the particle from the substrate.

3. However, the major mechanism by means of which particulate soil is removed from the substrate by nonmechanical means is probably through the increase *in the negative electrical potentials at the Stern layers* (Chapter 2, Section I, and Chapter 9) *of both soil and substrate on adsorption onto them of anions from the bath.* As might be expected, anionic surfactants in the bath are particularly effective for increasing negative potentials on both substrate and particulate soil, although inorganic anions in the bath, especially polyvalent ones, are also useful for this purpose. This increase in the negative potentials of both substrate and soil increases their mutual repulsion (i.e., the energy barrier for removal of the soil from the substrate is decreased and, at the same time, the energy barrier for soil redeposition is increased [Kling and Lange, 1960]).

Since the adsorption of nonionics onto soil or substrate does not significantly increase its electrical potential at the Stern layer, this mechanism of soil removal is probably not a major one for nonionics, and nonionics are generally not as effective as anionics for the removal of particulate soil (Albin et al., 1973). On the other hand, they appear to be very effective for producing steric barriers (see below) for the prevention of soil redeposition.

The reduced work of adhesion between soil and substrate caused by adsorption onto them of surfactant molecules with their hydrophilic groups oriented toward the bath, the reduced van der Waals attraction resulting from hydration of these hydrophilic groups (Chapter 9), and the increased electrostatic repulsion caused by the increase in magnitude of the electrical potentials of similar sign on soil and substrate all facilitate the separation of soil and substrate. However, mechanical work is almost always required to remove solid soil from the substrate. Larger particles are removed more readily than smaller particles. In small particles, the ratio of the area of true contact A_0 of the particle with the substrate to total surface area A is high. Any noninertial force tending to remove the particle is proportional to $A - A_0$, whereas the force holding the particle to the substrate is proportional to A_0 (Schwartz, 1972). Thus, a greater force per unit area is required to remove a small particle than a large one. In addition, the streaming velocity resulting from mechanical agitation approaches zero near the surface of the substrate, and therefore small particles encounter only the smaller velocities. Soil particles smaller than 0.1 μm cannot be removed from fibrous textile material at all (Lange, 1967).

B. Suspension of the Soil in the Bath and Prevention of Redeposition

Suspension of the soil in the bath and the prevention of redeposition are also accomplished by different mechanisms, depending on the nature of the soil.

1. Solid Particulate Soil: Formation of Electrical and Steric Barriers; Soil Release Agents The formation of electrical and steric barriers is probably the most important mechanism by which solid soil is suspended in the bath and prevented from redepositing on the substrate. Adsorption of similarly charged (almost always negative) surfactant or inorganic ions from the bath onto the detached soil particles increases their electrical potentials at the Stern layer, causing mutual repulsions and preventing agglomeration (Chapter 9, Section IA). Adsorption of POE nonionic surfactants with their hydrated POE chains oriented toward the bath also prevents agglomeration of soil particles by reducing van der Waals attraction between them and by producing a steric barrier to their close approach to each other (Chapter 9, Section IB).

Adsorption in similar fashion of other bath components onto the substrate or soil can also produce electrical and steric barriers to the close approach of soil particles to the substrate, thus inhibiting or preventing the redeposition of soil particles. Special components, called *soil release agents* or *antiredeposition agents*, are often added to the bath for this purpose. These are generally polymeric materials that by adsorption onto the fabric or soil produce a steric and sometimes also an electrical barrier to the close approach of soil particles (Trost, 1963).

Thus, sodium carboxymethylcellulose, used in laundry detergents, adsorbs onto cotton and increases its negative charge, thereby enhancing its repulsion of (negatively charged) soil. Polyacrylates are also used for this purpose

(Bertleff et al., 1998). POE terephthalate polyesters are used for polyester fabrics. They adsorb on the (hydrophobic) polyester with their POE groups oriented toward the aqueous bath phase, making the fabric surface hydrophilic and causing it to repel oily soil particles (O'Lenick, 1999). For polyester or polyester/cotton fabrics, methylhydroxyethylcellulose is also very effective (Carrion Fite, 1992).

2. Liquid Oily Soil *Solubilization* Solubilization has long been known to be a major factor in the removal of oily soil and its retention by the bath. This is based upon the observation (Ginn and Harris, 1961; Mankowich, 1961) that oily soil removal from both hard and textile surfaces becomes significant only above the critical micelle concentration (CMC) for nonionics and even for some avionics having low CMCs, and reaches its maximum only at several times the CMC. A considerable amount of research has been devoted to the removal of oily soil by POE nonionic surfactants, particularly from polyester or polyester/cotton (Dillan et al., 1979; Pierce and Trowbridge, 1980; Benson, 1982, 1985, 1986; Dillan, 1984; Raney and Miller, 1987; Miller and Raney, 1993). Optimal oily soil detergency has been correlated (Benson, 1986; Raney and Benson, 1990; Raney, 1991) with the phase-inversion temperature (PIT) (Chapter 8, Section IVB), both for POE nonionic surfactants and for POE nonionic–anionic mixtures. As discussed in Chapter 5, Section III, γ_{OW} reaches its minimum value at the PIT. Moreover, the solubilization of nonpolar material (Chapter 4, Section IB7) increases markedly with separation of the surfactant-rich middle phase. Thus, conditions are optimum at the PIT for the removal of oily soil by the roll-back mechanism and its retention in the bath via solubilization. On the other hand, a relationship has been found between the interfacial tension in the *supersolubilization region*, that is, in the oil-swollen aqueous micellar region close to the three-phase region (Figure 5.5), and oily soil detergency (Tongcumpou et al., 2003). Although the interfacial tension is not at its minimum and the solubilization parameter (Chapter 8, Section II) is not at its maximum at that point, the removal of the oily soil investigated was close to the maximum and avoided the complications of a three phase system.

Liquid crystalline phase or microemulsion formation between surfactant, water, and oily soil accompanies oily soil removal from hydrophobic fabrics such as polyester (Raney and Miller, 1987; Yatagai et al., 1990). It has been suggested (Miller and Raney, 1993) that maximum soil removal occurs not by solubilization into ordinary micelles, but into the liquid crystal phases or microemulsions that develop above the cloud point of the POE nonionic.

The extent of solubilization of the oily soil depends on the chemical structure of the surfactant, its concentration in the bath, and the temperature (Chapter 4, Section IB). At low bath concentrations only a relatively small amount of oily soil can be solubilized, whereas at high surfactant concentrations (10–100 times the CMC), solubilization is more similar to microemulsion formation (Chapter 8, Section II) and the high concentration of surfactant can accommodate a much larger amount of oily matter (Schwartz, 1972). With ionic surfactants, the use concentration is generally not much above the CMC; consequently, solubilization is almost always insufficient to suspend all the oily soil. When insufficient surfactant is present to solubilize all the oily soil, the remainder is probably suspended in the bath by macroemulsification. Antiredeposition agents, such as the POE terephthate polyesters mentioned in Section IB1 above, help prevent redeposition of suspended oily soil particles.

Macroemulsification For macroemulsification to be important, it is imperative that the interfacial tension between oily soil droplets and bath be low, so that emulsification can be accomplished with very little mechanical work. Here adsorption of surfactants at the oily soil–bath interface, with consequent lowering of the interfacial tensions, may play an important role. Emulsification was found to become a major factor when alkaline builders were added to a cleaning bath containing POE nonionic surfactant and the soil was mineral oil containing 5% oleic acid (Dillan et al., 1979). It is also involved in the suspension of liquefiable solid soil (Cox et al., 1987).

The ability of the bath to emulsify the oily soil is, however, in itself insufficient to keep all the soil from redepositing on the substrate (Schwartz, 1972). When the emulsified oil droplets impinge on the substrate, some of them may adhere to it in part, with the adhering portion tending to assume the equilibrium contact angle, unless the latter is 180° (i.e., unless complete oily soil removal by roll-back has been attained). This is in contrast to solubilization, which can result in complete removal of the oily soil from the substrate.

Mere dispersion of soil particles in the bath appears to be insufficient to accomplish effective cleaning. There appears to be no correlation between detergency and dispersing power of the bath. Surfactants with excellent dispersing powers are often very poor detergents and vice versa. On the other hand, increased adsorption onto substrate and soil, in the case of anionics and nonionics, and solubilizing power, in the case nonionics and fatty soil, appear to correlate well with detergency.

C. Skin Irritation (see Chapter 1, Section IIIB)

Skin irritation is an important factor in the selection of surfactants for use in cleaning materials that may contact the skin. The adsorption of monomeric surfactant from the cleaning product onto charged sites on the skin results in protein denaturation. All types of surfactants have been shown to produce skin protein denaturation in the order anionics > cationics > zwitterionics > amine oxides > POE nonionics (Miyazawa et al., 1984; Ohbu et al., 1984; Rhein et al., 1986). The order for anionics is sodium lauryl sulfate > C_{12}LAS > sodium laurate > AOS ≈ NaC_{12} AES (Kastner, 1980). No denaturation occurs with $C_{12}H_{25}(OC_2H_4)_6SO_4^-Na^+$ or $C_{12}H_{25}(OC_2H_4)_8OH$ (Ohbu et al., 1984).

The skin irritancy of anionics can be diminished by the addition of positively charged materials such as protein hydrolysates (Taves et al., 1986) or long-chain amine oxides that interact with the anionic and decrease its tendency to adsorb onto the skin, or by polymers that interact with them (Chapter 1, Section IVB5) to reduce the CMC and, consequently, the concentration of monomeric anionic surfactant (Goddard, 1994), since it is the latter that produces the skin irritation.

D. Dry Cleaning

Here the bath liquid used is not water but a hydrocarbon or chlorinated hydrocarbon. However, water in small amounts is always present in these systems and is a most essential ingredient.

Since oily soil is completely removed by the solvent, the main task for surfactants and other components of the bath is probably that of inhibiting redeposition of solid soil particles that are freed from the substrate when oily soil binding them to it is dissolved by the bath liquid. The rather high interfacial tension between textile fibers and the solvents used in dry cleaning promotes such soil redeposition. No clear generalizations have been deduced regarding the influence of type or concentration of surfactant on this process (Lange, 1967). Charge on the particles appears to play no role in the stabilization of the dispersion, probably because the potentials in the electrical double layers adjacent to the surfaces decay very rapidly in the nonpolar solvent. (Because of the very low dielectric constant, counterions are held very close to the surfaces.) However, there seems to be a correlation between adsorption of the surfactants present onto the substrate and soil particles and the prevention of redeposition of solid soil (von Hornuff and Mauer, 1972). Surfactants are probably adsorbed with polar heads oriented toward the substrate and soil and hydrophobic chains oriented toward the nonpolar solvent. This manner of adsorption produces a steric barrier to agglomeration or redeposition of the particles, since any close approach to another surface will constrain the free mobility of the hydrophobic chains. This adsorption of surfactants appears to be increased by small amounts of water in the solvent that hydrate soil and substrate surfaces.

Water-soluble soil (sodium chloride, sugar) appears to be removed by solubilization (Chapter 4, Section II) into free water in the interior of surfactant micelles in the solvent. Surfactant micelles in nonpolar solvents are formed with the polar heads oriented into the interior of the micelle. Water is added to the dry-cleaning solvent and is solubilized into the interior of these micelles. Some of this water in the interior is bound strongly to the polar heads of the surfactants in the interior of the micelle and some is essentially free water. Studies (Aebi and Wiebush, 1959) have shown that it is the free water that dissolves water-soluble soil rather than the bound water. In the absence of any free water in the solvent, water-soluble soil is not removed to any significant extent. The water-soluble soil appears to be removed from fibrous surfaces by a process involving hydration of the soil followed by solubilization (Mönch, 1960; Rieker and Kurz, 1973).

II. EFFECT OF WATER HARDNESS

The presence of polyvalent cations, notably Ca^{2+} and Mg^{2+}, in the bath water is invariably detrimental to the cleaning process for a number of reasons:

1. Adsorption of polyvalent cations onto the negatively charged substrate and soil reduces their electrical potentials, thus impeding soil removal and facilitating its redeposition. The detrimental effect attributed to this has been noted also in detergency studies involving only nonionic surfactants (Porter, 1967; Schwuger, 1971).
2. Polyvalent cations can act as linkages between negatively charged substrate and negatively charged soil, thus promoting soil redeposition (deJong, 1966). They can also act as linkages between the negatively charged hydrophilic groups of anionic surfactants and the negatively charged soil or substrate, causing adsorption of the former with their hydrophilic groups oriented toward the latter and their hydrophobic groups toward the bath. Adsorption with this orientation results in increases in γ_{SB} and γ_{PB}, the interfacial tensions at the substrate–bath and soil–bath interfaces, increasing the work of adhesion and impeding wetting and oily soil roll-back.
3. Adsorption of polyvalent cations onto solid soil particles dispersed in the bath can reduce their (negative) electrical potentials and cause them to flocculate and redeposit onto the substrate.
4. At high polyvalent cation concentrations, the corresponding metal salts of anionic surfactants and other anions (e.g., phosphates, silicates) in the bath may precipitate onto the substrate. In some cases, this may mask the presence of soil on the substrate (Rutkowski, 1971) or produce other deleterious effects (Vance, 1969; Brysson et al., 1971).

A. Builders

In addition to surfactants, a number of other materials are present in formulated laundry detergents. Among these are materials called builders. Their main purpose is to counter the detrimental effects of polyvalent cations on detergency. Polyvalent cations are introduced into the wash bath mainly by water hardness but may also come from soil or substrate. In addition, builders serve to increase the detersive efficiency and effectiveness of surfactants and to supplement their beneficial effects on soil removal.

Builders perform the following primary functions, in the order of decreasing importance:

1. *Sequestration, Precipitation, or Ion Exchange.* These are the three mechanisms by which builders reduce the concentration of polyvalent cations in the wash bath. Excellent sequestration is provided by sodium and

potassium polyphosphates, especially the tripolyphosphates which for decades were the builders of choice in laundry detergents. However, they are responsible for *eutrophication* (overfertilization of stagnant bodies of water) with adverse effects on aquatic organisms. As a result, use of polyphosphates in U.S. household detergents is limited to automatic dishwashing detergents. Precipitation by sodium carbonate effectively removes polyvalent cations from the wash liquor, but the insoluble calcium compounds which result can present an unsightly precipitate on washed goods. Sodium aluminosilicates (e.g., Zeolite A) physically trap polyvalent cations and exchange them for sodium ions. This builder is insoluble in water and is not suitable for liquid detergents. At present, the builder system in powdered laundry detergents consists of zeolite, carbonate, and low levels of polycarboxylate cobuilders.

Interest in effective biodegradable, nonphosphate sequestering builders that can also be used in liquid detergents continues to be active. Sodium citrate is the principal small-molecule commercial polycarboxylate builder at present even though it is only moderately effective. Other small molecules—polycarboxylates, ethylenediamino disuccinate, and tartrate mono/disuccinate—have been tested but have not attained large-scale usage.

Polymeric polycarboxylates, such as polyacrylates and acrylate–maleate copolymers, are finding usage as cobuilders in zeolite–carbonate builder systems. Polymers are finding increasing application in detergent formulations as dispersing agents, soil release agents, antiwrinkling additives, dye transfer inhibitors, fabric care additives, and other functionalities (Bertleff et al., 1998).

2. *Deflocculation and Dispersion of Particulate Soils.* This is accomplished by adsorption of the builder onto soil particles with a consequent increase in their negative electrical potentials, thus increasing their mutual repulsion. For this purpose polyphosphate and polycarboxylate ions, with their multiple negative charges, are particularly suited. Inorganic salts, in general, by decreasing the solubility of surfactants in the bath, promote their adsorption onto substrate and soil particles, and thereby increase their efficiency and effectiveness as soil dispersants.
3. *Alkalinity and Buffering.* High pH increases the negative potentials at soil and substrate and promotes cleaning. Buffering is necessary to prevent soil and substrate components from lowering the pH, with consequent lowering of surface potentials. Sodium carbonate is particularly effective for this purpose.

Current laundry detergents in powder form contain 8–25% surfactant and 30–80% builders. The builders are mainly inorganic salts, used at fairly high percentages, but a few organic polymeric materials are also used, at low percentages. Sodium polyacrylates have been recommended for use with sodium carbonate as builder. The polyacrylates prevent precipitation of insoluble carbonates (Nagarajan, 1985).

In addition to these primary functions, some builders are used for special purposes. Sodium silicates are used to prevent corrosion of aluminium parts in washing machines (they form a protective aluminium silicate coating), to prevent overglaze corrosion on china and, in powder detergents, as a structural agent to yield a crisp, nonsticky product. Organic polymers called *antiredeposition agents* are used, at low concentrations, to prevent redeposition of soil onto the substrate. Sodium carboxymethylcellulose is used at concentrations below 2% and in alkaline medium to prevent redeposition of soil onto cellulosic fibers. It adsorbs via H bonding to the cellulosic material and produces a (negatively charged) electrical barrier to the deposition of soil. It performs poorly on more hydrophobic synthetic fabrics, such as polyester, presumably because of poor adsorption. On such substrates, nonionic cellulose derivatives, such as hydroxyethyl-, 2-hydroxypropyl-, and 3-hydroxybutylcellulose, have been suggested as antiredeposition agents. The latter compound was reported to be the best of the three for use on polyester (Greminger et al., 1978). POE–polyoxypropylene copolymers (Chapter 9, Section VA) can also be used as antiredeposition agents on polyester. To be effective for this purpose, the molecules must adsorb onto the polyester via the polyoxypropylene group, leaving the POE chains free to extend into the aqueous phase and form a steric barrier against soil redeposition. For effective protection, the thickness of the adsorbed layer should exceed 25 Å (Gresser, 1985).

B. LSDAs

LSDAs are surfactants that enable soap to act as an effective laundry detergent in hard water without the deposition of insoluble calcium soap. For a surfactant to act as an LSDA, it must possess a bulky hydrophilic group (e.g., an ester, ether, amido, or amino linkage between the terminal hydrophilic group and the hydrophobic group) and a straight-chain hydrophobic group. It is believed that, in the presence of hardness ions (Ca^{2+}, Mg^{2+}), the soap and LSDA form a mixed micelle that shows high surface activity, including detergency. The bulky hydrophilic group of the LSDA forces the mixed micelle, with its hydrophilic groups oriented toward the aqueous phase, to retain its convex curvature (Linfield, 1978) toward the water. Soap micelles by themselves are believed to invert in hard water, with their hydrophobic groups oriented toward the aqueous phase, producing insoluble lime soaps (Stirton et al., 1965).

An extensive investigation of tallow-derived surfactants as LSDA in soap formulations for laundry detergents revealed (Linfield, 1978) that anionic and particularly zwitterionic surfactants are the best surfactant types for use as LSDA. POE nonionics are very effective LSDA but have a deleterious effect on the detergency of soap, while cationics form water-insoluble salts with soap. Among the anionics, an *N*-methyl tauride, $RCON(CH_3)CH_2CH_2SO_3^-Na^+$ (Noble et al., 1972), a sulfated alkanolamide, $RCONHCH_2CH_2OCH_2CH_2OSO_3^-Na^+$ (Weil et al., 1970), and a sulfated POE alcohol, $R(OCH_2CH_2)_3OSO_3^-Na^+$ (Weil et al., 1966; Bistline et al., 1972), all based on a tallow-derived hydrophobic group, were found to be the most effective. Zwitterionic surfactants of the sulfobetaine type were found to be an even more effective LSDA. Although a simple betaine, $RN(CH_3)_2CH_2COO^-$, showed only fair lime soap dispersing properties and poor detergency in a soap formulation, an amidosulfobetaine, $RCONH(CH_2)_3N^+(CH_3)_2CH_2CH_2CH_2SO_3^-$ (Parris et al., 1973, 1976), was the best LSDA among the materials studied. The corresponding sulfated material, $RCONH(CH_2)_3N^+(CH_3)_2CH_2CH_2CH_2OSO_3^-$ (Parris et al., 1976), and *N*-alkylsulfobetaine, $RN^+(CH_3)_2CH_2CH_2CH_2SO_3^-$ (Parris et al., 1973), were also very effective LSDA and showed even better detergency in a soap formulation. A coconut oil-derived amido hydroxysulfobetaine $RCONH(CH_2)_3N^+CH_3)_2CH_2CHOHCH_2SO_3^-$, showed excellent detergency in soap formulations in water of 1000 ppm hardness (Noble and Linfield, 1980).

III. FABRIC SOFTENERS

Fabric softeners have two major functions: (1) to impart a soft *feel* to dried fabrics and (2) to reduce static cling. They also reduce drying time and thus extend the life of tumble-dried garments by reducing mechanically induced fiber damage. Their development, including environmental considerations and chemical structures, has been reviewed by Levinson (1999). Originally designed for use in the final rinse cycle of washing machines, fabric softeners have been modified for use in the drying cycle of automatic dryers. They impart softness by adsorbing onto the (negatively charged) fabrics via their positively charged hydrophilic head groups, with their hydrophobic groups oriented away from the surface, reducing γ_{SA} and the work of adhesion of water to the substrate. This reduces the shrinkage (reduction of substrate surface area) and resulting hard feel that accompanies the removal of water from the substrate. They reduce static cling by reducing the negative static charge present on most surfaces.

Currently used fabric softeners are all cationic surfactants of structural types (Puchta et al., 1993; Levinson, 1999; Friedli et al., 2001, 2002):

I. *Dialkyl Dimethylammonium Salts*

$$R_2N^+(CH_3)_2X^- \text{ } (X^- \text{ is } Cl^- \text{ or } CH_3SO_4^-)$$

II. *Polyoxyethylenated Diamido Quaternary Ammonium Salts*

$$(RCONHCH_2CH_2)_2N^+(CH_3)(CH_2CH_2O)_xHX^- \text{ } (X^- \text{ is usually } CH_3SO_4^-)$$

III. *Amido Imidazolinium Salts*

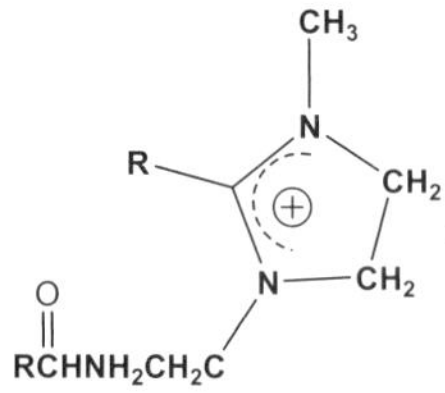

CH_3
N
R
CH_2
X^- (X^- is usually $CH_3SO_4^-$)
N—CH_2
O
‖
$RCHNH_2CH_2C$

IV. *Ester Quaternary Ammonium Salts (Ester Quats)*

$$(RCOOCH_2CH_2)2N^+(CH_3)(CH_2CH_2OH)\cdot CH_3SO_4^-$$

$$RCOOCH_2CH(OOCR)CH_2N^+(CH_3)_3Cl^-$$

V. *Amidoester Quaternary Ammonium Salts (Amidoester Quat)*

$$RCOOCH_2CH_2N^+(CH_3)_2CH_2CH_2CH_2NHCORCl^-$$

R is based upon tallow or hydrogenated tallow.

All of the above types are used as rinse cycle additives. Shorter-chain hydrophobic groups show little softening and unsaturated chains produce a dry, rather than a slick, feel. The Type I materials with saturated tallow chains give excellent softening performance but are now not used in Europe because of biodegradability considerations. For use in automatic clothes dryers, Type I in the form of the CH_3SO_4 salt is generally used, since the Cl^- may release corrosive hydrogen chloride. Type III materials are often used as drying cycle additives because of the difficulty of formulating them as rinse cycle additives. Type IV materials (ester quats) are now commonly used in Europe because of their excellent biodegradability (see Chapter 1, Section IIIA). It appears that two ester groups are needed for ready biodegradability (Friedli et al., 2001). Palm stearine fatty acids (60% C_{16} and 40% C_{18}) were found to give softening performance similar to that of tallow acids in IV A-type (triethanolamine-based) ester quat formulations. The addition of monoalkyl trimethylammonium halides increased the dye transfer inhibition properties of the formulation (Friedli et al., 2002).

Since these materials are all cationics that operate by adsorbing strongly onto fabric surfaces via their positively charged hydrophilic groups, when formulated with anionic surfactants for use in the wash cycle, both the softening properties of the cationic and the detergency properties of the anionic are reduced because of anionic–cationic interaction. When formulated with POE nonionics, the detergency of the POE nonionic appears not to be impaired, but the softening properties of the cationic are significantly reduced (Williams, 1981). An investigation of the use of alkyldimethylamine oxides as additives to ditallowdimethylammonium chloride fabric softeners found that octadecyldimethylamine oxides exhibited synergistic behavior with the quaternary ammonium compound in its softening effect on cotton towels and its prevention of static buildup on polyester fabrics (Crutcher et al., 1992). For fabric softening, the order of decreasing effectiveness is, consequently, rinse cycle > dryer cycle > wash cycle addition. Rinse cycle additives of Type I based upon hydrogenated tallow are the most effective. For static control, the order of decreasing effectiveness is dryer cycle > wash cycle > rinse cycle.

IV. THE RELATION OF THE CHEMICAL STRUCTURE OF THE SURFACTANT TO ITS DETERGENCY*

Correlations between the chemical structure of the surfactant and its detergency are complicated by the differing soils and substrates to be cleaned, by the amount and nature of builders present, by the temperature and hardness of the water used in the bath, and by the different mechanisms by means of which soils are removed. Correlations are therefore valid only when many of these variables are specified and controlled.

A. Effect of Soil and Substrate

1. Oily Soil Nonpolar soil has been found to be removed from hydrophobic substrates (e.g., polyester) more effectively by POE nonionics than by anionics (Fort et al., 1968; McGuire and Matson, 1975), and investigations of this type of soil removal have concentrated on the use of POE nonionics. POE nonionics have also been found (Rutkowski, 1971) to remove oily soils and prevent their redeposition at lower bath concentrations than anionics (i.e., nonionic surfactants are more efficient for these purposes than anionics). The greater efficiency of nonionics in soil removal is presumably due to their lower CMCs; in the prevention of soil redeposition it is probably due to their greater surface coverage per molecule when adsorbed on substrate and soil.

As mentioned above, maximum oily soil removal from polyester substrates by POE nonionics is obtained when the PIT of the surfactant in the presence of that soil is close to the wash temperature. Since the PIT decreases with decrease in the EO content of the POE nonionic surfactant, it is to be expected that as the wash temperature is decreased, the EO content of the surfactant showing optimum oily soil removal will decrease. Thus, for single homogeneous surfactants, $C_{12}H_{25}\,(OC_2H_4)_xOH$, maximum cetane detergency at 30°C was shown by the 4EO compound (PIT = 30°C), at 50°C by the 5EO compound (PIT = 52°C), and at 65°C by the 6EO compound (Benson, 1986). The

* As mentioned previously, the term *detergency* as used here refers to the power of the surfactant to enhance the cleaning power of the bath liquid. Except for dry cleaning (Section IVD), the bath liquid referred to in this discussion is water.

detergency of the 5EO compound at 30°C could be increased by additives that decreased its cloud point (and PIT).

In addition, as the wash temperature decreases, the chain length of the hydrophobic group for optimum oily soil removal appears to decrease. Thus, for oily soil removal from polyester/cotton by blends of homogeneous 3EO and 8EO nonionics having similar cloud points, the order of maximum oil removal at 70° was $C_{14} = C_{12} > C_{10}$, at 38°C it was $C_{10} = C_{12} > C_{14}$, and at 24°C it was $C_{10} > C_{12} > C_{14}$. The difference has been ascribed to the rate of solubilization of the soil, since the rate for these surfactant blends decreases with increase in the length of the hydrophobe (Benson, 1982).

For commercial POE nonionics with different types of hydrophobic groups of approximately equivalent chain length and the same degree of oxyethylenation (9 mol EO), the order of decreasing nonpolar soil removal from polyester/cotton was nonylphenol adduct > secondary C_{11}–C_{15} alcohol adduct > linear primary C_{12}–C_{15} alcohol adduct. This was the order of decreased effectiveness of equilibrium γ_{OW} reduction and of reduced rate of γ_{OW} reduction (Dillan, 1984).

Nonionics have been shown also to be more effective than ionics in the removal of oily soil from relatively nonpolar substrates (polyester, nylon). On cotton, however, a relatively hydrophilic fiber, anionics can outperform nonionics in detergency, and both of these are superior to cationics (Fort et al., 1968). The effects here may be due to differences in the orientation of adsorption of the different types of surfactants on the different substrates. On nonpolar substrates and soils, POE nonionics are adsorbed (Chapter 2) from aqueous solution via dispersion forces or hydrophobic bonding with their hydrophobic POE groups oriented toward the adsorbent and their hydrophilic POE groups toward the bath. Adsorption of the surfactant in this fashion on the substrate lowers the substrate–bath interfacial tension γ_{SB} and facilitates soil removal (Equation 10.3); adsorption in this fashion on both substrate and soil produces a steric barrier that inhibits soil redeposition.

On a cellulose substrate, on the other hand, adsorption–desorption data (Waag, 1968) indicate that POE nonionics can be adsorbed, at least partly, by hydrogen bonding between the hydroxyl groups of the cellulose and the ether linkages of the hydrophilic POE chain. This results in orientation of the surfactant with its hydrophilic group toward the substrate and its hydrophobic group toward the bath. Adsorption of the surfactant onto the cellulose substrate in this fashion makes the latter more hydrophobic and increases γ_{SB}, impeding removal of oily soil and facilitating its redeposition. This may account for the poorer washing performance of nonionics on cotton than on nonpolar substrates.

This unfavorable orientation of adsorption may also account for the even poorer performance of cationics on cotton. Since cotton acquires a negative charge at neutral or alkaline pHs, cationic surfactants may be adsorbed onto it by electrostatic attraction between the negatively charged sites on the fiber and the positively charged hydrophilic groups of the surfactant, with the hydrophobic groups of the adsorbed surfactant molecules oriented toward the bath. This orientation of the cationic surfactant onto the cotton will make it more hydrophobic and increase γ_{SB}, impeding the removal of oily soil and facilitating its redeposition.

Anionic surfactants, by contrast, although not adsorbed well onto negatively charged cotton except at relatively high bath concentrations of surfactant, can adsorb onto it only with their negatively charged hydrophilic groups oriented away from the similar charged substrate, and toward the bath, thereby increasing its hydrophilic character and decreasing γ_{SB}, facilitating soil removal and inhibiting soil redeposition. These considerations may underlie the observation by several investigators (Fort et al., 1966; Gordon et al., 1967; Spangler et al., 1967; Rutkowski, 1971) that for the removal of oily soil, nonionics are best for nonpolar substrates and anionics are best for cellulose substrates.

Geol (1998) has suggested guidelines for the optimal removal of oily soil from 65:35 polyester/cotton fabric by POE nonionic–anionic mixtures, based on the observation that maximum oily soil removal from this type of substrate is obtained at the PIT of the system, where the oil–water interfacial tension is at a minimum. Since the addition of anionic surfactant or increase in the POE content of the nonionic produces an increase in the PIT of the system, while the addition of salting-out electrolyte (NaCl, Na_2CO_3, $Na_5P_3O_{10}$) or decrease in the POE content of the nonionic reduces the PIT, these two opposing tendencies can be used to "tune" the system so that the PIT approximates the wash temperature. Thus, at a fixed level of electrolyte in the system, increase in the anionic/nonionic ratio, which increases the PIT, can be compensated for by use of a nonionic with a lower POE content. Increase in the electrolyte content, which decreases the PIT, can be compensated for by increase in the POE content of the nonionic or by increase in the anionic/nonionic ratio. Detergency results with C_{12} LAS–POE lauryl alcohol mixtures in the presence of $Na_2CO_3/Na_5P_3O_{10}$ mixtures, using artificial sebum as oily soil, were in accordance with these concepts.

Studies of solid hydrophobic soil removal from hard surfaces (Cox and Matson, 1984; Cox, 1986; Cox et al., 1987) indicate that liquefaction of the soil, involving penetration by surfactant and associated water molecules, is a key step in the removal of this type of soil. POE nonionics with short-chain (C_6, C_8) hydrophobic groups gave better performance than longer-chain (C_{12}) materials with the same percentage of EO. Performance also increased with decrease in EO content, in the 50–80% range. These effects are believed to be due to faster penetration by the shorter-chain materials. A C_{13} LAS performed better than shorter-chain homologs and its performance was improved in the presence of Mg^{2+}. The effects here are ascribed to better soil emulsification by the longer-chain LAS and better penetration in the presence of Mg^{2+}.

2. Particulate Soil Particulate soil is removed better from both cotton and Dacron–cotton blends by anionics than by nonionics (Albin et al., 1973). Here, increase in the electrical potentials on soil and substrate is probably the major mechanism by which this type of soil is removed and dispersed in the bath. Nonionics are consequently frequently less effective for removing this soil from the substrate, although they may be as effective and generally are more efficient than anionics in preventing its redeposition onto substrates (Rutkowski, 1968; Schott, 1968). Optimum particulate soil removal by POE nonionics from polyester/cotton requires a longer chain hydrophobic group and a higher EO content than oily soil removal (Vreugdenhil and Kok, 1984). Schwuger (1982) has pointed out that surfactants that adsorb equally well onto a hydrophobic soil and show equal detergency for this type of soil may show differences in adsorption for a hydrophilic soil. This may account for differences in detergency shown for this latter type of soil or for mixed soil (below).

Cationics, again, show poor detergency, since most substrates and particulate soils acquire negative potentials when contacted with aqueous baths at neutral or alkaline pHs. Adsorption of the positively charged surfactant ions onto substrate and soil decreases their (negative) electrical potentials, making more difficult the removal of soil and facilitating its redeposition.

3. Mixed Soil Mixed soil, which contains both oily and particulate matter, is commonly used in laundering investigations, since it approximates the composition of the soil found in clothing. A comparison of the washing performance of sodium alkyl sulfates and sodium methyl α-sulfocarboxylates has indicated that only when two different surfactants show equally good adsorption onto the fabric and all the components of the soil do they show similar washing properties. Surfactants that adsorb less strongly than others onto the textile fiber or onto some major component of the soil show poorer detergency (Schwuger, 1971). However, good adsorption by the surfactant onto a substrate and soil is not sufficient to ensure good washing properties. The surfactant must also be adsorbed with the proper orientation for soil removal, with its hydrophilic group oriented toward the aqueous bath phase. Thus, in a study comparing the laundering properties at 49°C of unbuilt sodium dodecylbenzenesulfonate with those of a nonionic POE isooctylphenol using a mixed soil, it was found (Rutkowski, 1971) that the nonionic was more efficient than the anionic both in removing soil from polyester fabric and in preventing soil redeposition on it, but it was not more effective than the anionic at higher concentrations in either respect. On cotton, however, although the nonionic was still somewhat more efficient but not more effective than the anionic in removing soil, it was not more efficient and was considerably less effective than the anionic in preventing soil redeposition. The poorer performance of the nonionic in preventing soil redeposition on the cotton may again be due to its adsorption, at least in part, onto the fiber via hydrogen bonding of the POE group with the hydrophobic groups oriented toward the bath, thus providing sites for soil redeposition.

In some cases, this orientation may be difficult to predict. Thus, a study of the adsorption of a series of *n*-alkyl sulfates and POE nonylphenols onto isotactic polypropylene fiber and of the washing properties of the same surfactants for the fiber using a mixed soil indicated that, although the adsorption of sodium hexadecyl sulfate onto the fiber was greater than that of a POE nonylphenol with a chain of 10 oxyethylene groups, the detergency of the former for the fiber was much lower. This was in spite of the fact that both compounds have equally good, similar emulsifying and solubilizing properties for this soil. Closer investigation of the adsorption of the two surfactants revealed that the anionic was adsorbed by strong, essentially irreversible interaction of its hydrophilic group with sites on the fiber (possibly polyvalent cations from the catalyst used in the polymerization), whereas the nonionic was adsorbed in normal, reversible fashion via its hydrophobic group (Schwuger, 1971).

B. Effect of the Hydrophobic Group of the Surfactant

Since both the extent of adsorption of the surfactant onto substrate and soil and its orientation with the hydrophobic group toward the adsorbents are of major importance in both soil removal and the prevention of soil redeposition, it is to be expected that changes in the length of the hydrophobic group will result in changes in detergency. Since an increase in the length of the hydrophobic group results in an increase in its efficiency of adsorption from aqueous solution (Chapter 2, Section II) and an increase in its tendency to adsorb via its hydrophobic group, whereas branching of the hydrophobic group or a centrally located hydrophilic group decreases the efficiency of adsorption, these probably account for the general observation that good detergents generally have a long, straight hydrophobic group and a hydrophilic group that is located either terminally or close to one end of the surfactant molecule. Numerous studies have indicated that detergency increases with increase in the length of the hydrophobic group, subject to solubility limits, and with movement of the hydrophilic group from a central to a more terminal position in the molecule (Kölbel and Kuhn, 1959; Burgess et al., 1964; Hellsten, 1965; Finger et al., 1967; Schwuger, 1982). Thus, in distilled water, compounds with straight hydrophobic groups are better detergents than their branched-chain isomers, fatty acid soaps generally are better than rosin soaps, and C_{16} and C_{18} fatty acid soaps are better than C_{12} and C_{14} soaps. The same effects are noticed in the alkyl sulfate and alkylbenzenesulfonate series in distilled water. In the latter series, parasulfonates appear to be better than ortho and monoalkyl better than isomeric dialkyl (Kölbel et al., 1960; Ginn and Harris, 1961).

There is one very important limitation on this increase of detergency with hydrophobic chain length, and that is the solubility of the surfactant in the cleaning bath. Particularly for ionic-type surfactants, the solubility of the surfactant in aqueous media decreases rapidly with increased length of the hydrophobic portion of the molecule, and precipitation of the surfactant, especially by any polyvalent cations present in the system, causes a marked decrease in detergency. Thus, optimum detergency is generally shown by the longest straight-chain surfactants whose solubility in the aqueous bath *under use conditions* is sufficient to prevent their precipitation onto the substrate in the presence of polyvalent cations. As the hardness of the bath water increases, optimum detergency appears to shift to shorter-chain homologs. Since solubility of ionic straight-chain surfactants in water generally increases with movement of the hydrophilic group from a terminal to a more central position in the molecule, when the surfactant with terminal hydrophilic group is either too insoluble or too sensitive to polyvalent cations for effective detergency, isomers with centrally located hydrophilic groups may show superior detergency (Rubinfeld et al., 1965). Thus, commercial linear dodecylbenzenesulfonate, which is produced with the phenylsulfonate group at the 2-, 3-, and other internal positions of the dodecyl group, rather than at the 1-position, is superior to the latter in detergency. 1-Phenylsulfonate, although superior to the other isomers in hot distilled water, has so slight a solubility at ordinary temperatures and is so sensitive to the presence of hard water cations (Ca^{2+}, Mg^{2+}) that it is not usable under normal washing conditions. As the bath temperature increases and the solubility of ionic surfactants in the bath increases, the length of the hydrophobic chain for optimum detergency increases (Matson, 1963).

In POE nonionics containing the same number of oxyethylene units, increase in the length of the hydrophobic group increases the efficiency of oily soil removal by decreasing the CMC and hence the concentration at which solubilization commences. Optimum detergency increases with increase in the chain length of the hydrophobe to a maximum that again is dependent on the temperature of the bath.

C. Effect of the Hydrophilic Group of the Surfactant

From our previous discussion, it is apparent that the charge on an ionic surfactant plays an important role in detergency. Because of the unfavorable (for detergency) orientation of the surfactant resulting from electrostatic attraction of its hydrophilic group to oppositely charged sites on the substrate or soil, ionic surfactants cannot be used efficiently for the removal of soil from oppositely charged substrates. Thus, cationics perform poorly on negatively charged substrates, especially at alkaline pHs, whereas anionics would not be expected to perform as well as cationics for the removal of soil from positively charged substrates at acidic pHs.

In POE nonionics, an increase in the number of oxyethylene groups in the hydrophilic POE chain appears to decrease the efficiency of adsorption of the surfactant onto most materials (Chapter 2), and this is sometimes accompanied by a decrease in detergency. For example, the detergency of wool at 30°C by a fixed molar concentration of POE nonylphenol in distilled water decreases with increase in the number of oxyethylene groups from 9 to 20 (Schwuger, 1971). This is consistent with adsorption studies that show that the greater the adsorption of these nonionics on wool, the greater the detergency (Kame et al., 1963).

On the other hand, the detergency of isotactic polypropylene at 90°C by these same surfactants in distilled water increases with increase in the number of oxyethylene units in the POE chain to a maximum at 12 and then decreases (Schwuger, 1971). The major factor involved here is probably the PIT of the surfactant, which increases with increase in the number of oxyethylene groups in the molecule. Detergency is optimum in the vicinity of the PIT, presumably because solubilization of oily soil by the surfactant increases markedly there.

A similar detergency maximum at almost the same oxyethylene content has been observed in the removal of oily soil from metal surfaces using similar surfactants in an alkaline, built formulation (Komor, 1969). The maximum here is at 68% oxyethylene (about 11 oxyethylene units per nonylphenol) at bath temperatures from 40 to 80°C. For a series of polyoxyethylenated nonrandom linear alkylphenols with C_8–C_{18} alkyl chains, optimum removal of sebum soil from cotton at 49°C and 50 and 300 ppm water hardness was obtained at 63–68% oxyethylene content (Smithson, 1966). A study of the removal of oily soil from cotton and permanent press cloths, and of clay from permanent press cloths by commercial POE alcohols, showed that POE C_{12}–C_{14} alcohols with 60% or greater ethylene oxide content achieved the best soil removal (Cox, 1989).

Studies of the soil removal properties of polyoxyethylenated straight-chain primary alcohols on cotton and Dacron–cotton permapress fabric indicate that this detergency maximum with change in the number of oxyethylene units in the POE chain is also shown on these fabrics. In liquid no-phosphate formulations built only with diethanolamine to provide an alkaline pH, optimum removal of both sebum and clay soils from Dacron–cotton permapress at 49°C in 150 ppm hard water occurs with about 5, 9, and 10 oxyethylene units for POE C_{9-11}, C_{12-15}, and C_{16-18} alcohol mixtures, respectively. For removal of the same soils from cotton at the same temperature, the optimum POE chain lengths are about two oxyethylene units larger (Albin et al., 1973).

The effect of changing the hydrophilic group from nonionic to anionic can be seen by comparing the soil removal properties of these same POE alcohols with two series of anionics made from the same hydrophobes, either by sulfating the alcohol mixture directly or after polyoxyethylenation with 3 or 6 mol of ethylene oxide. Using the same liquid no-phosphate formulation and the same laundering conditions at 49°C in 150 ppm hard water, the following results were obtained (Albin et al., 1973):

1. Both series of anionics made from the C_{12-15} alcohol mixture showed better detergency than the corresponding surfactants made from the C_{16-18} alcohol mixture; the corresponding materials made from the C_{9-11} alcohol mixture were poorest.
2. The directly sulfated materials (alkyl sulfates) made from the C_{9-11} and C_{16-18} alcohol mixtures were always poorer, in most cases considerably poorer, than the best nonionics made from the same alcohol mixtures.

3. Generally, there was no significant difference in the performance of sulfated surfactants polyoxyethylenated with 3 mol of ethylene oxide and the corresponding materials with six oxyethylene units.
4. Polyoxyethylenation of the C_{9-11} and C_{16-18} alcohol mixtures prior to sulfation generally improved their detergency considerably; in the case of the C_{12-15} alcohol mixture, polyoxyethylenation prior to sulfation reduced its performance on Dacron–cotton permapress slightly and improved its performance on cotton.
5. The best nonionics (made from the C_{12-15} alcohol mixture) were better than any of the anionics for both sebum and clay removal from Dacron–cotton permapress, but were not as good as the best anionics (sulfated POE C_{12-15} alcohols) for sebum or clay removal from cotton.

Comparable results were obtained in formulations containing sodium silicate as a builder together with 0–45% sodium tripolyphosphate, using 250 ppm hard water and a bath temperature of 49°C (Illman et al., 1971). A nonionic surfactant prepared by polyoxyethylenation of a C_{12-15} alcohol mixture with 9–11 mol of ethylene oxide generally showed similar detergency to an anionic prepared by sulfation of a C_{12-15} alcohol mixture previously polyoxyethylenated with 3 mol of ethylene oxide at all percentages of sodium tripolyphosphate, and both were considerably superior to a linear tridecylbenzenesulfonate and a sulfated C_{16-18} alcohol mixture. The nonionic was somewhat better than the sulfated POE alcohol for removing nonpolar fatty soil from Dacron–cotton permapress, and the reverse was true for the removal of polar soil from Dacron-cotton permapress and carbon soil from cotton, but similar results for the two surfactants were obtained for clay removal from both Dacron-cotton permapress and cotton, and polar and nonpolar fatty soil from cotton.

D. Dry Cleaning

Surfactants used as detergents in dry cleaning must, of course, be soluble in the solvent used as the bath liquid. They are often added as solutions in some suitable solvent. Surfactants used for this purpose include solvent-soluble petroleum sulfonates, sodium and amine salts of alkylarylsulfonates, sodium sulfosuccinates, POE phosphate esters, sorbitan esters, POE amides, and POE alkylphenols (Martin, 1965).

There seem to have been few systematic investigations of the effect of the chemical structure of the surfactant on the cleaning properties of the bath. The hydrophilic group of the surfactant appears to play a more important role than the hydrophobic (Kajl, 1960; Lange, 1967). In view of the two main mechanisms by means of which surfactants aid the cleaning process—(1) adsorption via the hydrophilic group onto soil and substrate to prevent redeposition of solid soil and (2) solubilization of water-soluble soil by water held between the hydrophilic groups in the interior of the micelles (Chapter 4, Section

II)—the importance of the hydrophilic group is not unexpected. The function of the hydrophobic group appears to be that of producing the steric barrier to the aggregation of solid soil particles dispersed in the bath. Consistent with this, C_{16-18} straight chains appear to be the most effective for this purpose Wedell 1960).

V. BIOSURFACTANTS AND ENZYMES IN DETERGENT FORMULATIONS

A detergent contains several major and minor components—surfactants, builders, bleaching-agents, enzymes, fillers, and minor additives like stabilizers for enzymes, optical whitening agents, antifoams, and perfumes, to name a few. However, the major component of a detergent is the surfactant. Surfactants end up in ponds and aquifers through the laundry waste and cause detrimental environmental effects and toxicity to living systems. Hence, there is a resurgent effort to incorporate nontoxic and biodegradable surfactants in detergent formulations. Biosurfactants (Chapter 13, Section I) and enzymes are being included in modern detergents and investigated so as to identify useful candidates from renewable and biological sources.

Biosurfactants from two strains of *Bacillus subtilis* (Mukherjee, 2007) have been isolated that showed good thermal and high pH stability, two of the essential qualities of a laundry detergent. These cyclic lipopeptide surfactants (Table 13.1) are compatible with ionic and nonionic surfactants traditionally used in the detergent formulations, and show acceptable surface properties at 80°C for up to 60 min at a pH range of 7.0–12.0. The detergency effect of these biosurfactants was evaluated from their ability to remove vegetable oil and blood stains from cotton fabrics. The high biodegradability, lower toxicity, and minimal ecological impact of these microbial surfactants make them environmentally and ecologically safe materials. Biosurfactants compatible with laundry detergents have also been isolated and studied from food wastes (Savarino et al., 2010) and soil samples (Moradian et al., 2009).

Enzymes are mostly proteinaceous compounds with a variety of biological functions, which are available from natural and synthetic sources. Currently used industrial enzymes function mainly by their *hydrolytic* action, *proteases* being the dominant enzyme type (used in detergents and dairy industry) and *carbohydrases*, in the form of *amylases* and *cellulases*, being the second largest enzyme class (used in detergents, and starch, textile, and baking industries). Some of the important enzyme classes (and their applications) (Kirk et al., 2002) currently used in detergents are *protease* (removal of protein stains); *amylase* (removal of starch stains); *lipase* (removal of lipid stains); *cellulase* (for cleaning, color clarification, and antiredeposition on cotton); and the new enzyme class *mannanase* (removal of mannanan stains, i.e., reappearing stains). Recent developments in this area are centered on producing improved genetic versions of these enzymes so that they operate at lower temperatures (to save energy in the use of washing machines and dishwashers) and still retain the surfactant character, and remain stable at higher alkaline pHs. A *trypsin*-based enzyme, derived from the intestine of striped sea bream (*lithognathus mormyrus*), for use in laundry detergents has been investigated (Ali et al., 2009) with reference to its alkaline stability, enzyme activity, and useful temperature range.

VI. NANODETERGENTS (SEE CHAPTER 14, SECTION IIIF)

REFERENCES

Aebi, C. M. and J. R. Wiebush (1959) *J. Colloid Sci.* **14**, 161.

Albin, T. B., D. W. Bisacchi, J. C. Illman, W. T. Shebs, and H. Stupel Paper presented before Am. Oil Chem. Soc., Chicago, Sept. 18 1973.

Ali, N. E., N. Hmidet, A. Bougatef, R. Nasri, and M. Nasri (2009) *J. Agricult. Food Chem.* **57**, 10943.

Aronson, M. P., M. L. Gum, and E. D. Goddard (1983) *J. Am. Oil Chem. Soc.* **60**, 1333.

Benson, H. L. Presented 73rd Annu. Am. Oil Chem. Soc. Meet., Toronto, Canada, May 4 1982.

Benson, H. L. Presented 77th Annu. Am. Oil Chem. Soc. Meet., Honolulu, Hawaii, May 14–18 1986.

Benson, H. L., K. R. Cox, and J. E. Zweig (1985) *Soap/Cosmetics/Chem. Specs.* **3**, 35.

Bertleff, W., P. Neumann, R. Baur, and D. Kiessling (1998) *J. Surfactants Deterg.* **1**, 419.

Bistline, R. G., Jr., W. R. Noble, J. K. Weil, and W. M. Linfield (1972) *J. Am. Oil Chem. Soc.* **49**, 63.

Broze, G. (ed.), *Handbook of Detergents, Part A, Properties*, Marcel Dekker, New York, 1999.

Brysson, R. J., B. Piccolo, and A. M. Walker, 1971 *Text. Res. J.* **41**, 86.

Burgess, J., G. R. Edwards, and M. W. Lindsay, Chem. Phys. Appl. Surface Active Subs., Proc. Int. Congr. Surface Active Subs., 4th, Sept. 1964, III, 153.

Carrion Fite, F. (1992) *J., Tenside Surf Det.* **29**, 213.

Cox, M. F. (1986) *J. Am. Oil Chem. Soc.* **63**, 559.

Cox, M. F. (1989) *J. Am. Oil Chem. Soc.* **66**, 367.

Cox, M. F. and T. P. Matson (1984) *J. Am. Oil Chem. Soc.* **61**, 1273.

Cox, M. F., D. L. Smith, and G. L. Russell (1987) *J. Am. Oil Chem. Soc.* **64**, 273.

Crutcher, T., K. R. Smith, J. E. Borland, J. Sauer, and J. W. Previne (1992) *J. Am. Oil Chem. Soc.* **69**, 682.

Cutler, W. G. and E. Kissa (eds.), *Detergency: Theory and Technology*, Marcel Dekker, New York, 1986.

deJong, A. L. (1966) *Textiles* **25**, 242.

Dillan, K. W. (1984) *J. Am. Oil Chem. Soc.* **61**, 1278.

Dillan, K. W., E. D. Goddard, and D. A. M. Kenzie (1979) *J. Am. Oil Chem. Soc.* **56**, 59.
Durham, K. (1956) *J. Appl. Chem.* **6**, 153.

Finger, B. M., G. A. Gillies, G. M. Hartwig, W. W. Ryder, and W. M. Sawyer (1967) *J. Am. Oil Chem. Soc.* **44**, 525.

Fort, T., H. R. Billica, and T. H. Grindstaff (1966) *Textile Res. J.* **36**, 99.

Fort, T., H. R. Billica, and T. H. Grindstaff (1968) *J. Am. Oil Chem. Soc.* **45**, 354.

Friedli, F. E. (ed.), *Detergency of Specialty Surfactants*, Marcel Dekker, New York, 2001.

Friedli, F. E., R. Keys, C. J. Toney, O. Portwood, D. Whittlinger, and M. Doerr (2001) *J. Surfactants Deterg.* **4**, 401.

Friedli, F. E., H. J. Koehle, M. Fender, M. Watts, R. Keys, P. Frank, C. J. Toney, and M. Doerr (2002) *J. Surfactants Deterg.* **5**, 211.

Geol, S. K. (1998) *J. Surfactants Deterg.* **1**, 213.

Ginn, M. E. and J. C. Harris (1961) *J. Am. Oil Chem. Soc.* **38**, 605.

Goddard, E. D. (1994) *J. Am. Oil Chem. Soc.* **71**, 1.

Gordon, B. E., J. Roddewig, and W. T. Shebs (1967) *J. Am. Oil Chem. Soc.* **44**, 289.

Greminger, G. K., A. S. Teot, and N. Sarkar (1978) *J. Am. Oil Chem. Soc.* **55**, 122.

Gresser, R. (1985) *Tenside Detergents* **22**, 178.

Hellsten, M., in *Surface Chemistry*, P. Ekwall et al. (eds.), Academic, New York, 1965, p. 123.

Illman, J. C., T. B. Albin, and H. Stupel. Paper presented at Am. Oil. Chem. Soc. Short Course, Lake Placid, NY, June 14, 1971.

Kajl, M. Vortraege Originalfassung Int. Kongr. Grenzflaechenaktive Stolle, 3rd, Cologne, 1960, **4**, 187 (Publ. 1961).

Kame, M., Y. Danjo, S. Kishima, and H. Koda (1963) *Yukagaku* **12**, 223. [C.A. 59, 11715g (1963)].

Kastner, W. in *Anionic Surfactants, Biochemistry, Toxicology, Dermatology*, C. Gloxhuber (ed.), Marcel Dekker, New York, 1980, pp. 139–307.

Kirk, O., T. V. Borchert, and C. C. Fuglsang (2002) *Curr. Opin. Biotech.* **13**, 345.

Kling, W. and H. Lange (1960) *J. Am. Oil Chem. Soc.* **37**, 30.

Kölbel, H. and P. Kuhn (1959) *Angew. Chem.* **71**, 211.

Kölbel, H., D. Klamann, and P. KurzenclÖrfer *Proc 3rd mt. Congr. Surface Active Substances*, Cologne, I. p. 1, 1960.

Komor, J. A. Am. Spec. Mfr. Ass., *Proc. Annu. Meet.* 1969 (Publ. 1970) **56**, 81.

Lai, K-Y. (ed.) *Liquid Detergents*, Marcel Dekker, New York, 1997.

Lange, H. in *Solvent Properties of Surfactant Solution*, K. Shinoda (ed.), Marcel Dekker, New York, 1967, Chap. 4.

Levinson, M. L. (1999) *J. Surfactants Deterg.* **2**, 223.

Linfield, W. M. (1978) *J. Am. Oil Chem. Soc.* **55**, 87.

Mankowich, A. M. (1961) *J. Am. Oil Chem. Soc.* **38**, 589.

Martin, A. R. *Kirk-Othmer Encyclopedia of Chemical Technology*, 2nd ed., Vol. 7, John Wiley, New York, 1965, pp. 307–326.

Matson, T. P. (1963) *Soap Chem. Specialties* **39**, 52, 91, 95, 97, 100.

Matson, T. P. and G. D. Smith Presented 71st Annu. Am. Oil Chem. Soc. Meet., New York, April 27–May 1 1980.

McGuire, S. E. and T. P. Matson (1975) *J. Am. Oil Chem. Soc.* **52**, 411.

Miller, C. A. and K. H. Raney (1993) *Colloids Surf.* **74**, 169.

Miyazawa, K., M. Ogawa, and T. Mitsui (1984) *Int. Cosmet. J. Sci.* **6**, 33.

Mönch, R. (1960) *Faserforschu. Textiltech.* **11**, 228.

Moradian, F., K. Khajeh, H. Naderi-Manesh, and M. Sadeghizadeh (2009) *Appl. Biochem. Biotech.* **159**, 33.

Mukherjee, A. K. (2007) *Lett. Appl. Microbiol.* **45**, 330.

Nagarajan, M. K. Presented 76th Annu. Am. Oil Chem. Soc. Meet., Philadelphia, PA, May 5–9 1985.

Noble, W. R. and W. M. Linfield (1980) *J. Am. Oil Chem. Soc.* **57**, 368.

Noble, W. R., R. G. Bistline, Jr., and W. M. Linfield (1972) *Soap Cosmet. Chem. Spec.* **48**, 38.

Ohbu, K., N. Jona, N. Miyajima, and M. Fukuda *Proc. 1st World Surfactants Congr*, Munich, Vol. 3, p. 317, 1984.

O'Lenick, A. J. (1999) *J. Surfactants Deterg.* **2**, 553.

Parris, N., J. K. Weil, and W. M. Linfield (1973) *J. Am. Oil Chem. Soc.* **50**, 509.

Parris, N., J. K. Weil, and W. M. Linfield (1976) *J. Am. Oil Chem. Soc.* **53**, 97.

Pierce, R. C. and J. R. Trowbridge Presented at the 71st Annu. Am. Oil Chem. Soc. Meet., New York, April 27–May 1 1980.

Porter, A. S. *Proc. Mt. Congr. Surface Active Substances*, 4th, Brussels, III, 1964, p. 187, (1967).

Prieto, N. E., W. Lilienthal, and P. L. Tortorici (1996) *J. Am. Oil Chem. Soc.* **73**, 9.

Puchta, R., P. Krings, and P. Sandkuhler (1993) *Tenside, Surf. Det.* **30**, 186.

Raney, K. H. (1991) *J. Am. Oil Chem. Soc.* **68**, 525.

Raney, K. H. and H. L. Benson (1990) *J. Am. Oil Chem. Soc.* **67**, 722.

Raney, K. H. and C. A. Miller (1987) *J. Colloid Interface Sci.* **119**, 537.

Rhein, L. D., C. R. Robbins, K. Fernec, and R. Cantore (1986) *J. Soc. Cosmet. Chem.* **37**, 125.

Rieker, J. and J. Kurz (1973) *Melliand Textilber. Mt.* **54**, 971.

Rubinfeld, J., E. M. Emery, and H. D. Cross (1965) *Ind. Eng. Chem., Prod. Res. Dev.* **4**, 33.

Rubingh, D. N. and T. Jones (1982) *Ind. Eng. Chem. Prod. Res. Dev.* **21**, 176.

Rutkowski, B. J. (1968) *J. Am. Oil Chem. Soc.* **45**, 266.

Rutkowski, B. J. Paper presented at Am. Oil Chem. Soc. Short Course, Lake Placid, NY, June 14 1971.

Savarino, P., E. Montoneri, G. Musso, and V. Boffa (2010) *J. Surfact. Deterg.* **13**, 59.

Schott, H. (1968) *J. Am. Oil Chem. Soc.* **45**, 414.

Schwartz, A. M. (1971) *J. Am. Oil Chem. Soc.* **48**, 566.

Schwartz, A. M. in *Surface and Colloid Science*, E. Matijevic (ed.), Wiley, New York, 1972, Vol. 5, pp. 195–244.

Schwuger, M. J. (1971) *Chem-Ing.-Tech.* **43**, 705.

Schwuger, M. J. (1982) *J. Am. Oil Chem. Soc.* **59**, 258, 265.

Showell, M. S. (ed.), *Powdered Detergents*, Marcel Dekker, New York, 1998.

Smith, D. L., K. L. Matheson, and M. F. Cox (1985) *J. Am. Oil Chem. Soc.* **62**, 1399.

Smithson, L. H. (1966) *J. Am. Oil Chem. Soc.* **43**, 568.

Spangler, W. G., R. C. Roga, and H. D. Cross (1967) *J. Am. Oil Chem. Soc.* **44**, 728.

Stirton, A. J., F. D. Smith, and J. K. Weil (1965) *J. Am. Oil Chem. Soc.* **42**, 114.

Taves, E. A., E. Eigen, V. Temnikov, and A. M. Kligman (1986) *J. Am. Oil Chem. Soc.* **63**, 574.

Tongcumpou, C., E. J. Acosta, L. B. Quencer, A. F. Joseph, J. F. Scamehorn, D. A. Sabatini, S. Chavadej, and N. Yanumet (2003) *J. Surfactants Deterg.* **6**, 205.

Trost, H. B. (1963) *J. Am. Oil Chem. Soc.* **40**, 669.

Vance, R. F. (1969) *J. Am. Oil Chem. Soc.* **46**, 639.

von Hornuff, G. and W. Mauer (1972) *Deut. Textitech.* **22**, 290.

Vreugdenhil, A. D. and R. Kok *Proc. World Surfactant Congr.* May 6–10, 1984, Munich, Kurle Verlag, Gelnhausen, Vol. 4, p. 24, 1984.

Waag, A. Chim., Phys., Appi. Prat. Ag. Surface, C. R. Congr. Int. Deterg. 5th, Barcelona, 1968 (Publ. 1959), III, p. 143.

Wedell, H. (1960) *Melliand Textilber* **41**, 845.

Weil, J. K., A. J. Stirton, and M. V. Nufiez-Ponzoa (1966) *J. Am. Oil Chem. Soc.* **43**, 603.

Weil, J. K., N. Parris, and A. J. Stirton (1970) *J. Am. Oil Chem. Soc.* **47**, 91.

Weil, J. K., C. J. Pierce, and W. M. Linfield (1976) *J. Am. Oil Chem. Soc.* **53**, 757.

Williams, J. A. presented at the, 72nd Annu. Am. Oil Chem. Soc. Meet. New Orleans, May 17–21 1981.

Yatagai, M., M. Komaki, T. Nakajima, and T. Hashimoto (1990) *J. Am. Oil Chem. Soc.* **67**, 154.

PROBLEMS

10.1 Explain why cationic surfactants, which ordinarily show poor detergency in aqueous media, can be used successfully as detergents at low pH.

10.2 Explain how the addition of a small amount of a cationic surfactant can increase the efficiency of an alkaline solution of an anionic surfactant in soil removal from a textile surface.

10.3 **(a)** What effect would adsorption of a surfactant onto a textile surface via its hydrophilic head have on the spreading coefficient of the bath on the textile surface?

(b) List two cases where this may occur.

10.4 Sodium sulfate is often found in laundry detergent powders based on sodium linear alkylbenzenesulfonate. Aside from acting as an inexpensive "filler," suggest other reasons for its presence and possible useful functions.

10.5 Considering the fact that skin is a proteinaceous material, explain what type of molecular interaction you would expect between the skin and a cationic surfactant. In light of this interaction, provide a reasonable description of protein denaturing.

10.6 Sections ID and IVD pertain to the dry cleaning process and the redeposition phenomenon. How would you link the surfactant structure with the redeposition process in this case?

10.7 What are lime soap dispersing agents (LSDAs)? Can you give an explanation for the observation that anionic and zwitterionic surfactants function as effective LSDAs?

10.8 What is the role of a builder in detergent formulations?

11 Molecular Interactions and Synergism in Mixtures of Two Surfactants

In most practical applications, mixtures of surfactants, rather than individual surfactants, are used. In some cases, this is involuntary since the commercial surfactants used, even when designated by the name of an individual surfactant, for example, sodium lauryl sulfate, are mixtures of surface-active materials as a result of the nonhomogeneous raw materials used in their manufacture and/or the presence of unreacted raw materials and manufactured by-products. In other cases, different types of surfactants are purposely mixed to improve the properties of the final product.

In most cases, when different types of surfactants are purposely mixed, what is sought is *synergism*, the condition in which the properties of the mixture are better than those attainable with the individual components by themselves. For example, a long-chain amine oxide is often added to a formulation based upon an anionic surfactant because the foaming properties of the mixture are better than those of either surfactant by itself.

Although the existence of synergistic relations between certain types of surfactants has been known and utilized for many years, the investigation of synergism in *quantitative* terms is a recent development based upon a simple, convenient method for measuring molecular interactions between surfactants. The molecular interactions between two different surfactants adsorbed at various interfaces are measured by a parameter, β, which indicates the nature and strength of those interactions. The value of the β-parameter is related to the free energy change upon mixing of the two surfactants [$\Delta G_{\text{mix}} = \beta X(1 - X)$ RT], where X is the mole fraction of the first surfactant in the mixture (on a surfactant-only basis) adsorbed at the interface and $(1 - X)$ is the mole fraction of the second surfactant.

The *regular solution* equation (Rubingh, 1979) for β, is $\beta = [W_{AB} - (W_{AA} + W_{BB}/2]$RT, where W_{AB} is the molecular interaction energy between the mixed surfactants, W_{AA} is the molecular interaction energy between the first surfactant before mixing with the second, W_{BB} is the molecular interaction energy

Surfactants and Interfacial Phenomena, Fourth Edition. Milton J. Rosen and Joy T. Kunjappu.

between the second surfactant before mixing with the first, R is the gas constant, and T, the absolute temperature, is a convenient way of understanding its meaning. For attractive interaction, the sign of W is negative; for repulsive interaction, it is positive. Thus, a negative β-value indicates that, upon mixing, the two surfactants experience either greater attraction or less repulsion than before mixing; a positive β-value indicates less attraction or greater repulsion upon mixing than before mixing. A value close to zero indicates little or no change in interactions upon mixing. Since in ionic surfactant-containing mixtures there is always repulsive interaction between the ionic surfactant molecules before mixing, the β-parameter is almost always negative if only because of the dilution effect upon mixing with a second surfactant, except for anionic–anionic mixtures. Steric effects contribute to the value of the β-parameter when there are variations in the size of the hydrophilic head group or in the branching of the hydrophobic groups of the two surfactants (Zhou and Rosen, 2003). From the relevant properties of the individual surfactants and the values of the molecular interaction parameters, it is possible to predict whether synergism will exist in a mixture of surfactants and, if so, the ratio of the materials at which synergism will be a maximum and the optimum value of the relevant surface property at that point. At the present time, mixtures containing only two surfactants have been investigated, although the method is theoretically (Holland and Rubingh, 1983) capable of handling any number of components. However, in any multicomponent system, the strongest interaction between two surfactants usually determines the properties of the entire system, and the evaluation of that one interaction will probably be sufficient to allow prediction of the properties of the mixture.

I. EVALUATION OF MOLECULAR INTERACTION PARAMETERS

The two fundamental properties of surfactants are monolayer formation at interfaces and micelle formation in solution; for surfactant mixtures, the characteristic phenomena are mixed monolayer formation at interfaces (Chapter 2, Section IIIG) and mixed micelle formation in solution (Chapter 3, Section VIII). The molecular interaction parameters for mixed monolayer formation by two different surfactants at an interface can be evaluated using Equations 11.1 and 11.2, which are based upon the application of the nonideal solution theory to the thermodynamics of the system (Rosen and Hua, 1982):

$$\frac{X_1^2 \ln\left(\alpha C_{12} / X_1 C_1^0\right)}{(1-X_1)^2 \ln\left[(1-\alpha) C_{12} / (1-X_1) C_2^0\right]} = 1 \tag{11.1}$$

and

$$\beta^\sigma = \frac{\ln\left(\alpha C_{12} / X_1 C_1^0\right)}{(1-X_1)^2} = 1, \tag{11.2}$$

where α is the mole fraction of surfactant 1 in the total surfactant in the solution phase; that is, the mole fraction of surfactant 2 equals $1-\alpha$; X_1 is the mole fraction of surfactant 1 in the total surfactant in the mixed monolayer; C_1^0, C_2^0, and C_{12} are the solution phase molar concentrations of surfactants 1, 2, and their mixture, respectively, required to produce a given surface tension value; and β^σ is the molecular interaction parameter for mixed monolayer formations at the aqueous solution–air interface.

For evaluating the molecular interaction parameters for mixed micelle formation by two different surfactants, Equations 11.3 and 11.4 (Rubingh, 1979) are used:

$$\frac{(X_1^M)^2 \ln(\alpha C_{12}^M / X_1^M C_1^M)}{(1-X_1^M)^2 \ln[(1-\alpha)C_{12}^M/(1-X_1^M)C_2^M]} = 1 \tag{11.3}$$

and

$$\beta^M = \frac{\ln(\alpha C_{12}^M / X_1^M C_1^M)}{(1-X_1^M)^2} = 1, \tag{11.4}$$

where C_1^M, C_2^M, and C_{12}^M are the critical micelle concentrations (CMCs) of individual surfactants 1 and 2 and their mixture at a given value of α, respectively; X_1^M is the mole fraction of surfactant 1 in the total surfactant in the mixed micelle; and β^M is a parameter that measures the nature and extent of the interaction between the two different surfactant molecules in the mixed micelle in aqueous solution. Equation 11.1 (or 11.3) is solved numerically for X_1 (or X_1^M), and substitution of this in Equation 11.2 (or 11.4) yields the value of β^σ (or β^M).

The determination of β^σ and β^M experimentally is shown in Figure 11.1. The surface tension–log surfactant concentration curves for each of the two individual surfactants in the system and at least one mixture of them at a fixed value of α must be determined. For calculating β^σ (the molecular interaction parameter for mixed monolayer formation at the aqueous solution–air interface), C_1^0, C_2^0, and C_{12}^0 are required; for β^M, the CMCs, C_1^M, C_2^M, and C_{12}^M, are needed.

A. Notes on the Use of Equations 11.1–11.4

In order to obtain valid β-parameter values, that is, values that do not change significantly with change in the ratio of the surfactant in the mixture, the following conditions must be met:

1. The two surfactants must be molecularly homogeneous and free from surface-active impurities.
2. Since Equations 11.1–11.4 neglect counterion effects, all solutions containing ionic surfactants should have the same total ionic strength, with a swamping excess of any counterions.

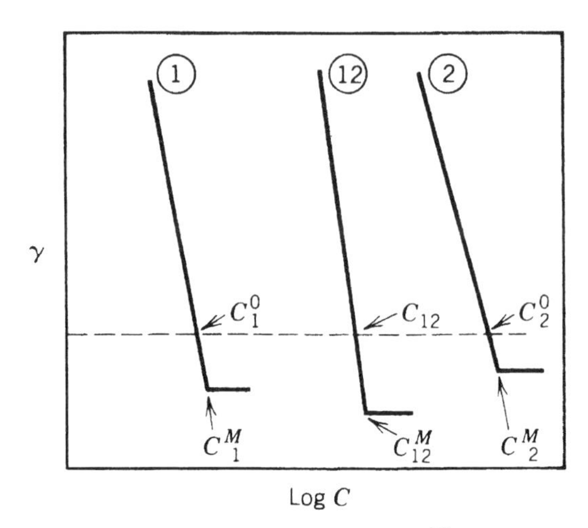

FIGURE 11.1 Experimental evaluation of β^σ or β^M. ①, individual surfactant 1; ②, individual surfactant 2; ⑫, mixture of surfactants 1 and 2 at a fixed mole fraction α in solution. Reprinted with permission from Rosen, M. J., in *Phenomena in Mixed Surfactant Systems*, J. F. Scamehorn (ed.), ACS Symposium Series 311, American Chemical Society, Washington, DC, 1986, p. 148.

3. Since the derivation of these equations is based upon the assumption that the mixed micelle or monolayer can be considered to contain only surfactants, these structures are considered to contain no free water. This is reasonable when the surfactant molecules are so closely packed (e.g., at their maximum surface excess concentration in the monolayer) that all the water present can be considered to be bound to the hydrophilic head groups. Because of this and because surfactant mixtures are generally used above their CMC, it is advisable to determine β^σ using C_1^0, C_2^0, and C_{12} values taken from the log C plots at such a value of γ that the slopes are all linear or almost so, and preferably at the lowest possible γ-value. For this purpose, it is permissible to extend the γ-log C plot to values above the CMC of one of the surfactants (see Figure 11.8a) by extrapolating linearly the portion of maximum slope just below the CMC. If the plot shows a decrease in slope close to the CMC, that portion should be ignored in extending the plot.

Since Equations 11.1 and 11.3 contain the terms $((X_1)^2/(1-X_1)^2)$ and $\left((X_1^M)^2/(1-X_1^M)^2\right)$, respectively, which change rapidly in value when X approaches 1 or 0, it is advisable to use for the surfactant mixtures α-values that yield X_1 or X_1^M values between 0.2 and 0.8. When the value is beyond these limits, small experimental errors may cause large errors in calculating X_1 or X_1^M, with consequent large deviations in the value of β^σ or β^M. Values of X_1 approximating 0.5 can be obtained by using an α-value equal to the ratio

$C_1^0/(C_1^0 + C_2^0)$, and values of X_1^M approximating 0.5 can be obtained by using an α-value equal to $C_1^M/(C_1^M + C_2^M)$. Regrettably, in much of the recent literature on β-parameters, these conditions have not been met.

The interaction parameters in the presence of a second liquid (hydrocarbon) phase, β_{LL}^{σ} and β_{LL}^{M}, for mixed monolayer formation at the aqueous solution–hydrocarbon interface and for mixed micelle formation in the aqueous phase, respectively, can be evaluated (Rosen and Murphy, 1986) by equations analogous to Equations 11.1, 11.2, 11.3, and 11.4, respectively. The necessary data are obtained from interfacial tension–concentration curves.

Interaction parameters for mixed monolayer formation at the aqueous solution–solid interface (β_{SL}^{σ}) can also be evaluated (Rosen and Gu, 1987), *in the case where the solid has a low-energy (hydrophobic) surface*, by equations analogous to Equations 11.1 and 11.2. In this case, quantities C_1^0, C_2^0, and C_{12} are the concentrations in aqueous solutions of surfactant 1, surfactant 2, and their mixture, respectively, *at the same value of the adhesion tension*, $\gamma_{LA}\cos\theta$, where γ_{LA} is the surface tension of the surfactant solution and θ is the contact angle measured in the aqueous phase on a smooth, nonporous planar surface of the hydrophobic solid (Figure 6.2). Alternatively, β_{SL}^{σ} values can be obtained from the adsorption isotherms of solutions of surfactant 1, surfactant 2, and at least one mixture of them at a fixed equilibrium value of α on the finely divided solid (Chapter 2, Section IIB).

From Equation 2.19a, $d_{\gamma SL} = -n\mathrm{RT}\Gamma_{SL} d\ln C$, and, integrating both sides of the equation,

$$\int_{\gamma_{SL}}^{\gamma_{SL}^0} d\gamma_{SL} = \gamma_{SL}^0 - \gamma_{SL} = \Pi_{SL} = RT\int_C^0 \Gamma_{SL} \cdot d\ln C \tag{11.5}$$

Integrating the area under the plot of adsorption, Γ_{SL} versus $\ln C$ (or $\log C$) (Figure 11.2) below the CMCs of the respective solutions, yields the value of Π_{SL}, the amount of solid–aqueous solution interfacial tension reduction. This is plotted (Figure 11.3) against ln (or log) C and values of C_1^0, C_2^0, and C_{12} selected, at the largest common value of Π_{SL}, for substitution into Equations 11.1 and 11.2 to evaluate β_{SL}^{σ}. Π_{SL} can also be evaluated (Zhu and Gu, 1991) directly from adsorption isotherm data by the use of equations

$$\Pi_{SL} = \Gamma_{\infty}\mathrm{RT}\ln(1 + C_1/a) \tag{11.6}$$

for monolayer adsorption and

$$\Pi_{SL} = \Gamma_{\infty}^{S}/n\mathrm{RT}\ln(1 + C_1^n/a) \tag{11.7}$$

for adsorption with surface aggregates, by use of Equations 2.7 or 2.8 to evaluate a and Γ_M^S, and Equation 2.12 to evaluate K (= $1/a$) and n.

The molecular interaction parameters evaluated using Equations 11.1–11.4, together with the properties of the individual surfactants (see Section III), are

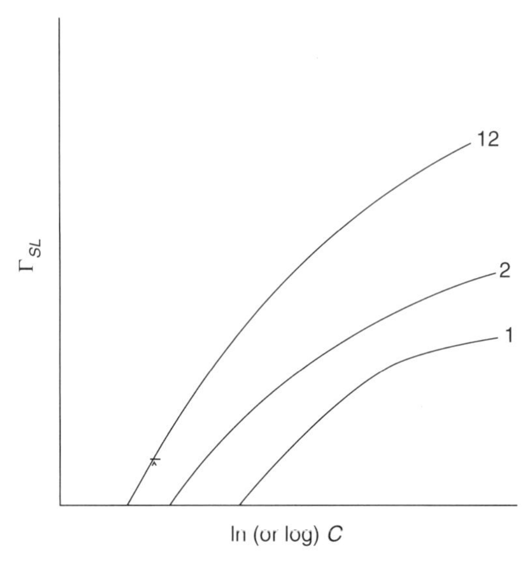

FIGURE 11.2 Plots of adsorption (Γ_{SL}) on a powdered, finely divided solid versus ln (or log) total surfactant concentration, C, of aqueous solutions of surfactants 1, 2, and a mixture of them, 12, at a fixed value of α, for the evaluation of Π_{SL}.

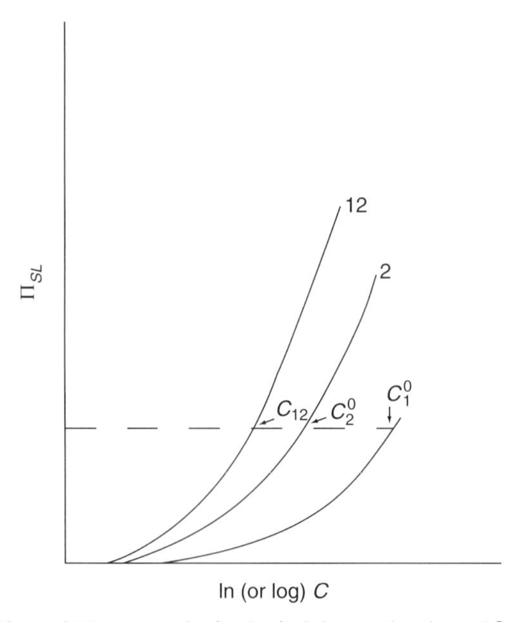

FIGURE 11.3 Plots of Π_{SL} versus ln (or log) C for evaluation of β_{SL} from adsorption isotherm data below the CMCs of the aqueous solutions of surfactants 1, 2, and a mixture of them (12) at a fixed value of α.

used to predict whether synergism of a particular type will occur when the two surfactants are mixed and, if so, the molar ratio of the two surfactants at which maximum synergism will exist and the relevant property of the mixture at that point. The particular interaction parameters used depend upon the nature of the interfacial phenomenon involved as described below.

II. EFFECT OF CHEMICAL STRUCTURE AND MOLECULAR ENVIRONMENT ON MOLECULAR INTERACTION PARAMETERS

A considerable number of molecular interaction parameters on well-characterized surfactant pairs have been measured during the past two decades. In addition, information on how the parameters change with variation in the chemical structures of the two surfactants and in their molecular environment (pH, temperature, and ionic strength of the solution) has accumulated. This permits the estimation of these parameters when it may not be convenient to determine them experimentally.

Table 11.1 lists values of the various types of molecular interaction parameters. Almost all of the mixtures show negative β-values, indicating more attraction (or less repulsion) after mixing than before. The only examples observed to date of surfactant pairs with positive β-values (greater repulsion or smaller attraction of the two components after mixing than before) are (1) anionic–anionic mixtures of sodium soaps (≥C14) with either commercial dodecylbenzenesulfonate (linear alkylbenzenesulfonate [LAS]) or sodium hexadecanesulfonate (Rosen, 1989) and (2) mixtures of hydrocarbon-chain and perfluorocarbon-chain surfactants of the same charge type (Zhao and Zhu, 1986). These latter types have been shown to form aggregated separate domains rather than mixed films or mixed micelles upon mixing (Kadi et al., 2002).

The interaction between the two surfactants is mainly due to electrostatic forces. The strength of attractive electrostatic interaction decreases in the order anionic–cationic > anionic–zwitterionic capable of accepting a proton > cation–zwitterionic capable of losing a proton > anionic polyoxyethylene (POE) nonionic > cationic–POE nonionic. Mixtures of surfactants of the same charge type (anionic–anionic, cationic–cationic, nonionic–nonionic, and zwitterionic–zwittenonic) show only a very weak interaction (negative β-values of 1 or less) at the aqueous solution–air interface, although they can show significant interaction at other interfaces.

The large negative β-values observed in the cases of two oppositely charged surfactants are consequently due to the attractive electrostatic interaction they experience after mixing. In the case of ionic–nonionic mixtures, however, where attractive electrostatic interaction after mixing may be expected to be weak, a major contribution to the negative β-values observed may be the reduction after mixing with the nonionic of the self-repulsion of the ionic surfactant before mixing, that is, a dilution effect (Zhou and Rosen, 2003).

TABLE 11.1 Values of Molecular Interaction Parameters

Mixture	Temperature (°C)	β^{σ}	β^{M}	Reference
Anionic–Anionic Mixtures				
$C_{13}COO^-Na^+$-LAS^-Na^{+a} (0.1 *M* NaCl pH 10.6)	60	+0.2	–0.6	Rosen and Zhu (1989)
$C_{15}COO^-Na^+$-$C_{12}SO_3^-Na^+$ (0.1 *M* NaCl pH 10.6)	60	–0.01	+02	Rosen and Zhu (1989)
$C_{15}COO^-Na^+$-LAS^-Na^{+a} (0.1 *M* NaCl pH 10.6)	60	+1.4	+0.7	Rosen and Zhu (1989)
$C_{15}COO^-Na^+$-$C_{16}SAS^-Na^{+b}$ (0.1 *M* NaCl pH 10.6)	60	–0.1	–0.7	Rosen and Zhu (1989)
$C_{15}COO^-Na^+$-$C_{16}SO_3^-Na^+$ (0.1 *M* NaCl pH 10.6)	60	+0.7	+0.7	Rosen and Zhu (1989)
$C_7F_{15}COO^-Na^+$-$C_{12}SO_4^-Na^+$ (0.1 *M* NaCl heptane–H_2O)	30	+0.8 (β^{σ}_{LL})	+0.3	Zhao and Zhu (1986)
$C_7F_{15}COO^-Na^+$-$C_{12}SO_4^-Na^+$ (0.1 *M* NaCl)	30	+2.0	—	Zhao and Zhu (1986)
$C_{12}SO_3^-Na^+$-LAS^-Na^{+a} (0.1 *M* NaCl)	25	–0.3	–0.3	Rosen and Zhu (1989)
$C_{12}SO_3^-Na^+$-AOT^-Na^{+c} (0.1 *M* NaCl)	25	–0.3	–0.5	Rosen and Zhu (1989)
Anionic–Cationic Mixtures				
$C_7F_{15}COO^-Na^+$-$C_7N^+Me_3Br^-$ (0.1 *M* NaCl)	30	–15.0	—	Zhao and Zhu (1986)
$C_5SO_3^-Na^+$-$C_{10}Pyr^+Cl^{-d}$ (0.01 *M* NaCl)	25	$-11_{.8}$	—	Goralczyk et al. (2003)
$C_5SO_3^-Na^+$-$C_{10}Pyr^+Cl^{-d}$ (0.03 *M* NaCl)	25	$-10_{.8}$	—	Goralczyk et al. (2003)
$C_5SO_3^-Na^+$-$C_{10}Pyr^-Cl^{-d}$ (0.03 *M* NaBr)	25	–8.2	—	Goralczyk et al. (2003)
$C_5SO_3^-Na^+$-$C_{10}Pyr^+Cl^{-d}$ (0.03 *M* NaI)	25	–5.5	—	Goralczyk et al. (2003)
$C_7SO_3^-Na^+$-$C_{10}Pyr^+Cl^{-d}$ (0.01 *M* NaCl)	25	$-15_{.4}$	—	Goralczyk et al. (2003)
$C_8SO_3^-Na^+$-$C_{14}N^+Me_3Br^-$ (0.1 *M* NaBr (aq.)–air)	25	$-13_{.5}$	—	Gu and Rosen (1989)
$C_8SO_3^-Na^+$-$C_{14}N^+Me_3Br^-$ (0.1 *M* NaBr (aq.)–PTFE[e])	25	$-10_{.8}$ (β^{σ}_{SL})	—	Gu and Rosen (1989)
$C_8SO_3^-Na^+$-$C_{14}N^+Me_3Br^-$ (0.1 *M* NaBr (aq.)–Parafilm)	25	$-11_{.2}$ (β^{σ}_{SL})	—	Gu and Rosen (1989)
$C_{10}SO_3^-Na^+$-$C_{12}N^+Me_3Br^-$	25	$-35_{.6}$	—	Rodakiewicz-Nowak (1982)
$C_{10}SO_3^-Na^+$-$C_{12}N^+Me_3Br^-$ (H_2O–PTFE[e])	25	$-28_{.8}$ (β^{σ}_{SL})	—	Gu and Rosen (1989)
$C_{10}SO_3^-Na^+$-$C_{12}N^+Me_3Br^-$ (H_2O–polyethylene)	25	$-26_{.6}$ (β^{σ}_{SL})	—	Gu and Rosen (1989)
$C_{10}SO_3^-Na^+$-$C_{12}N^+Me_3Br^-$ (0.1 *M* NaBr (aq.)–air)	25	$-19_{.6}$	—	Gu and Rosen (1989)
$C_{10}SO_3^-Na^+$-$C_{12}N^+Me_3Br^-$ (0.1 *M* NaBr (aq.)-PTFE[e])	25	$-14_{.1}$	—	Gu and Rosen (1989)
$C_{10}SO_3^-Na^+$-$C_{12}N^+Me_3Br^-$ (0.1 *M* NaBr (aq.)–Parafilm)	25	$-15_{.3}$	—	Gu and Rosen (1989)
$C_{10}SO_3^-Na^+$-$C_{12}Pyr^+Br^{-d}$ (0.1 *M* NaBr (aq.)–Parafilm)	25	$-19_{.7}$	—	Gu and Rosen (1989)
$C_{10}SO_3^-Na^+$-$C_{12}Pyr^+Br^{-d}$ (0.1 *M* NaBr (aq.)–PTFE[e])	25	$-14_{.2}$ (β^{σ}_{SL})	—	Gu and Rosen (1989)

$C_{10}SO_3^-Na^+$-$C_{12}Pyr^+Br^{-d}$ (0.1 *M* NaBr (aq.)–Parafilm)	25	$-15_{.5}$ (β^{σ}_{SL})	—	Gu and Rosen (1989)
$C_{12}SO_3^-Na^+$-$C_8Pyr^+Br^{-d}$ (0.1 *M* NaBr (aq.)–air)	25	$-19_{.5}$	—	Gu and Rosen (1989)
$C_{12}SO_3^-Na^+$-$C_8Pyr^+Br^{-d}$ (0.1 *M* NaBr (aq.)–PTFE[e])	25	$-14_{.1}$ (β^{σ}_{SL})	—	Gu and Rosen (1989)
$C_{12}SO_3^-Na^+$-$C_8Pyr^+Br^{-d}$ (0.1 *M* NaBr (aq.)–Parafilm)	25	$-15_{.3}$ (β^{σ}_{SL})	—	Gu and Rosen (1989)
$C_{12}SO_3^-Na^+$-$C_{10}Pyr^+Cl^{-d}$ (0.1 *M* NaCl)	25	$-33_{.2}$	—	Liu and Rosen (1996)
$C_8SO_4^-Na^+$-$C_8N^+Me_3Br^-$	25	$-14_{.2}$	$-10_{.2}$	Zhao et al. (1980)
$C_8SO_4^-Na^+$-$C_8N^+Me_3Br^-$ (0.1 *M* NaBr)	25	−14	−10	Liu and Rosen (1996)
$C_8SO_4^-Na^+$-$C_8Pyr^+Br^{-d}$	25	—	$-10_{.7}$	Li and Zhao (1992)
$C_8SO_4^-Na^+$-$C_8(OE)_3Pyr^+Cl^{-d}$	25	—	$-6_{.3}$	Li and Zhao (1992)
$C_8OESO_4^-Na^+$-$C_8Pyr^+Br^{-d}$	25	—	$-8_{.1}$	Li and Zhao (1992)
$C_8(OE)_3SO_4^-Na^+$-$C_8Pyr^+Br^{-d}$	25	—	$-4_{.4}$	Li and Zhao (1992)
$C_8(OE)_3SO_4^-Na^+$-$C_8(OE)_3Pyr^+Cl^-$	25	—	$-3_{.9}$	Li and Zhao (1992)
$C_8(OE)_3SO_4^-Na^+$-$C_{10}Pyr^+Cl^{-d}$	25	—	$-8_{.1}$	Li and Zhao (1992)
$C_8(OE)_3SO_4^-Na^+$-$C_{12}Pyr^+Br^{-d}$	25	—	$-10_{.4}$	Li and Zhao (1992)
$C_8(OE)_3SO_4^-Na^+$-$C_{14}Pyr^+Br^{-d}$	25	—	$-11_{.4}$	Li and Zhao (1992)
$C_{10}SO_4^-Na^+$-$C_{10}N^+Me_3Br^-$	25	—	$-18_{.5}$	Corkill and Goodman (1963)
$C_{10}SO_4^-Na^+$-$C_{10}N^+Me_3Br^{-d}$ (0.05 *M* NaBr)	23	—	$-13_{.2}$	Holland and Rubingh (1983)
$C_{12}SO_4^-Na^+$-$C_{12}N^+Me_3Br^-$	25	−27.8	$-25_{.5}$	Lucassen-Reynders et al. (1981)
$C_{12}SO_4^-Na^+$-$C_{12}N^+Me_3Br^-$ (H_2O–PTFE[e])	25	−30.6 (β^{σ}_{SL})	—	Gu and Rosen (1989)
$C_{12}SO_4^-Na^+$-$C_{12}N^+Me_3Br^-$ (H_2O–polyethylene)	25	−26.7 (β^{σ}_{SL})	—	Gu and Rosen (1989)
$C_{12}(OE)_3SO_4^-Na^+$-$C_{16}N^+Me_3Cl^-$	25	—	$-23_{.1}$	Esumi et al. (1994)
$C_{12}(OE)_5SO_4^-Na^+$-$C_{16}N^+Me_3Cl^-$	25	—	$-16_{.8}$	Esumi et al. (1994)
$C_{12}(OE)_3SO_4^-Na^+$-$C_8F_{17}CH_2CH(OH)CH_2N^+(CH_3)(C_2H_4OH)_2.Cl^-$	25	—	$-17_{.1}$	Esumi et al. (1994)
$C_{12}(OE)_5SO_4^-Na^+$-$C_8F_{17}CH_2CH(OH)CH_2N^+(CH_3)(C_2H_4OH)_2.Cl^-$	25	—	$-10_{.7}$	Esumi et al. (1994)

(*Continued*)

TABLE 11.1 (*Continued*)

Mixture	Temperature (°C)	β^{σ}	β^{M}	Reference
$C_4H_9\Phi SO_3^-Na^+$-$C_{16}N^+Me_3Br^-$	27	–9.9$_5$	—	Bhat (1999)
$(CH_3)_2CHCH_2\Phi SO_3^-Na^+$-$C_{16}N^+Me_3Br^-$	27	–9.4	—	Bhat (1999)
$(CH_3)_3C\Phi SO_3^-Na^+$-$C_{16}N^+Me_3Br^-$	27	–8.2	—	Bhat (1999)
C_{16}-2-Φ-$SO_3^-Na^+$-$C_{14}N^+Me_3Br^-$	50	—	–19.4	Bourrel et al. (1984)
C_{16}-4-Φ-$SO_3^-Na^+$-$C_{14}N^+Me_3Br^-$	50	—	–17.2	Bourrel et al. (1984)
C_{16}-6-Φ-$SO_3^-Na^+$-$C_{14}N^+Me_3Br^-$	50	—	–16.1	Bourrel et al. (1984)
C_{16}-8-Φ-$SO_3^-Na^+$-$C_{14}N^+Me_3Br^-$	50	—	–15.3	Bourrel et al. (1984)
C_{14}-7-Φ-$SO_3^-Na^+$-$C_{14}N^+Me_3Br^-$	50	—	–17.3	Bourrel et al. (1984)
C_{12}-6-Φ-$SO_3^-Na^+$-$C_{14}N^+Me_3Br^-$	50	—	–18.7	Bourrel et al. (1984)
C_{10}-5-Φ-$SO_3^-Na^+$-$C_{14}N^+Me_3Br^-$	50	—	–19.9	Bourrel et al. (1984)
$C_{12}SO_3^-Na^+$-$C_{14}N^+Me_3Br^-$	50	—	–20.0	Bourrel et al. (1984)
Anionic–Nonionic Mixture				
$C_7F_{15}COO^-Na^+$-C_8SOCH_3	25	–4.7	–3.2	Zhao and Zhu (1986)
$C_7F_{15}COO^-Li^+$-C_8-β-D-glucoside	25	—	–1.9	Esumi et al. (1996)
$C_{10}SO_3^-Na^+$-1,2-C_{12} diol (0.1 *M* NaCl)	25	–2.4	—	Rosen and Zhao (1983)
$C_{12}SO_3^-Na^+$-1,2-C_{10} diol (0.1 *M* NaCl)	25	–2.7$_5$	–1.3	Zhou and Rosen (2003)
$C_{12}SO_3^-Na^+$-1,2-C_{12} diol (0.1 *M* NaCl)	25	–3	–1.45	Rosen and Zhao (1983)
$C_{12}SO_3^-Na^+$-4,5-C_{10} diol (0.1 *M* NaCl)	25	–3.2	—	Zhou and Rosen (2003)
$C_{14}SO_3^-Na^+$-1,2-C_{12} diol (0.1 *M* NaCl)	25	–2.6	—	Rosen and Zhao (1983)
$C_{12}SO_3^-Na^+$-*N*-octyl-2-pyrrolid(in)one (H_2O–air)	25	–2.6	—	Rosen et al. (1989)
$C_{12}SO_3^-Na^+$-*N*-octyl-2-pyrrolid(in)one (H_2O–Parafilm)	25	–2.1 (β^{σ}_{SL})	—	Rosen and Zhao (1983)
$C_{12}SO_3^-Na^+$-*N*-octyl-2-pyrrolid(in)one (H_2O–PTFE[e])	25	–2.0 (β^{σ}_{SL})	—	Rosen (1989b)
$C_{12}SO_3^-Na^+$-*N*-octyl-2-pyrrolid(in)one (0.1 *M* NaCl (aq.)–air)	25	–3.1	—	Rosen (1989b)
$C_{12}SO_3^-Na^+$-*N*-octyl-2-pyrrolid(in)one (0.1 *M* NaCl (aq.)–Parafilm)	25	–2.9 (β^{σ}_{SL})	—	Rosen (1989b)

$C_{12}SO_3^-Na^+$-*N*-octyl-2-pyrrolid(in)one (0.1 *M* NaCl (aq.)–PTTE[e])	25	–2.5 (β^{σ}_{SL})	—	Rosen (1989b)
$C_{12}SO_3^-Na^+$-*N*-decyl-2-pyrrolid(in)one (0.1 *M* NaCl (aq.)–hexadecane)	25	–1.7 (β^{σ}_{SL})	—	Rosen (1989b)
$C_{12}SO_3^-Na^+$-*N*-octyl-2-pyrrolid(in)one (0.1 *M* NaCl (aq.)–hexadecane)	25	–2.3 (β^{σ}_{SL})	—	Rosen (1989b)
$C_{12}SO_3^-Na^+$-$C_{11}H_{23}CON(CH_3)CH_2(CHOH)_4CH_2OH$ (0.1 *M* NaCl)	25	–2.8	–1.8	Zhou and Rosen (2003)
$C_{12}SAS^-Na^{+b}$-$C_{12}(OE)_7OH$	25	–0.2	–1.0	Rosen and Zhu (1989)
$C_{12}SO_3^-Na^+$-$C_{12}(OE)_8OH$ (0.1 *M* NaCl)	25	–2.2	—	Rosen and Zhao (1983)
$C_{12}SO_3^-Na^+$-TMN6[f] (0.1 *M* NaCl)[f]	25	–1.7	–2.1	Zhou and Rosen (2003)
$C_{12}SO_3^-Na^+$-$C_{12}(OE)_4OH$ (0.1 *M* NaCl)	25	–1.6	–0.8	Zhou and Rosen (2003)
$C_{12}SO_3^-Na^+$-$C_{12}(OE)_7OH$ (0.1 *M* NaCl)	25	–1.7	–2.4	Zhou and Rosen (2003)
$C_{12}SO_3^-Na^{+b}$-$C_{12}(OE)_8OH$	25	–1.5	–3.4	Rosen and Zhao (1983)
$C_{12}SO_3^-Na^+$-$C_{12}(OE)_8OH$ (0.1 *M* NaCl)	25	–2.6	–3.1	Rosen and Zhao (1983)
$C_{12}SO_3^-Na^+$-$C_{12}(OE)_8OH$ (0.5 *M* NaCl)	25	–2.0	—	Rosen and Zhao (1983)
$C_{12}SO_3^-Na^+$-$C_{12}(OE)_8OH$ (0.1 *M* NaCl (aq.)–PTFE[e])	25	–2.1 (β^{σ}_{SL})	—	Gu and Rosen (1989)
$C_{12}SO_3^-Na^+$-$C_{12}(OE)_8OH$ (0.5 *M* NaCl (aq.)–PTFE[e])	25	–1.7 (β^{σ}_{SL})	—	Gu and Rosen (1989)
$C_{12}SO_3^-Na^+$-$C_{14}(OE)_4OH$ (0.1 *M* NaCl)	25	–1.1	–0.5	Zhou and Rosen (2003)
$C_{12}SO_3^-Na^+$-$C_{14}(OE)_8OH$ (0.1 *M* NaCl)	25	–1.4	–2.0	Zhou and Rosen (2003)
$C_{14}SO_3^-Na^+$-$C_{12}(OE)_8OH$ (0.1 *M* NaCl)	25	–2.3	—	Rosen and Zhao (1983)
$C_{10}SO_4^-Na^+$-$C_{12}(OE)_8OH$ (0.1 *M* NaCl)	25	–3.2	—	Rosen and Zhao (1983)
$C_{12}SO_4^-Na^+$-$C_8(OE)_4OH$	25	—	–3.1	Lange and Beck (1973)
$C_{12}SO_4^-Na^+$-$C_8(OE)_6OH$	25	—	–3.4	Lange and Beck (1973)
$C_{12}SO_4^-Na^+$-$C_8(OE)_{12}H$	25	—	–4.1	Lange and Beck (1973)
$C_{12}SO_4^-Na^+$-$C_{10}(OE)_4OH$ (5×10^{-4} *M* Na_2CO_3)	23	—	–3.6	Holland and Rubingh (1983)
$C_{12}SO_4^-Na^+$-$C_{12}(OE)_4OH$ (0.1 *M* NaCl)	25	–3.0	—	Huber (1991)
$C_{12}SO_4^-Na^+$-$C_{12}(OE)_4OH$ (0.1 *M* NaCl–PTFE[e])	25	–2.1 (β^{σ}_{SL})	—	Huber (1991)
$C_{12}SO_4^-Na^+$-$C_{12}(OE)_6OH$ (0.1 *M* NaCl)	25	–2.5	–3.4	Penfold et al. (1995) and Goloub et al. (2000)
$C_{12}SO_4^-Na^+$-$C_{12}(OE)_8OH$	25	–2.7	–4.1	Rosen and Zhao (1983)
$C_{12}SO_4^-Na^+$-$C_{12}(OE)_8OH$ (0.1 *M* NaCl)	25	–3.5	—	Rosen and Zhao (1983)

(*Continued*)

TABLE 11.1 (*Continued*)

Mixture	Temperature (°C)	β^{σ}	β^{M}	Reference
$C_{12}SO_4^-Na^+$-$C_{12}(OE)_8OH$ (0.1 *M* NaCl–PTFE[e])	25	−2.9 (β^{σ}_{SL})	—	Gu and Rosen (1989)
$C_{12}SO_4^-Na^+$-$C_{14}(OE)_8OH$ (0.5 *M* NaCl)	25	−3.3, 3.1	−3.0	Ingram (1980) and Rosen and Zhao (1983)
$C_{12}SO_4^-Na^+$-$C_{12}(OE)_8OH$ (0.5 *M* NaCl–PTFE[e])	25	−2.7 (β^{σ}_{SL})	—	Gu and Rosen (1989)
$C_{12}SO_4^-Na^+$-$C_{16}(OE)_{10}OH$[g]	30	−4.3	−6.6	Ogino et al. (1986)
$C_{12}SO_4^-Na^+$-$C_{16}(OE)_{20}OH$[g]	30	−4.3	−6.2	Ogino et al. (1986)
$C_{12}SO_4^-Na^+$-$C_{16}(OE)_{30}OH$[g]	30	—	−4.3	Ogino et al. (1986)
$C_{14}SO_4^-Na^+$-$C_{12}(OE)_8OH$ (0.1 *M* NaCl)	25	−3.2	—	Rosen and Zhao (1983)
$C_{12}(OE)SO_4^-Na^+$-C_{10}-β-glucoside (0.1 *M* NaCl, pH 5.7)	25	−1.8	−1.4	Rosen and Sulthana (2001)
$C_{12}(OE)SO_4^-Na^+$-C_{10}-β-maltoside (0.1 *M* NaCl, pH 5.7)	25	−1.5	−1.2	Rosen and Sulthana (2001)
$C_{12}(OE)SO_4^-Na^+$-C_{12}-β-maltoside (0.1 *M* NaCl, pH 5.7)	25	−1.4	−1.3	Rosen and Sulthana (2001)
$C_{12}(OE)SO_4^-Na^+$-C_{12}-2 : 1 (molar) C_{12} maltoside, C_{12} glucoside (0.1 *M* NaCl, pH 5.7)	25	−3.2	−3.2	Rosen and Sulthana (2001)
$C_{12}(OE)_2SO_4^-Na^+$-1,2-C_{10} diol (0.1 *M* NaCl)	25	−1.4	~0	Zhou and Rosen (2003)
$C_{12}(OE)_2SO_4^-Na^+$-$C_{11}H_{23}CON(CH_3)CH_2(CHOH)_4OH$ (0.1 *M* NaCl)	25	−1.8	−1.2	Zhou and Rosen (2003)
$C_{12}(OE)_2SO_4^-Na^+$-$C_8(OE)_4OH$	25	—	−1.6	Holland (1984)
$C_{12}(OE)_2SO_4^-Na^+$-TMN6[f] (0.1 *M* NaCl)	25	−1.6	−0.9	Zhou and Rosen (2003)
$C_{12}(OE)_2SO_4^-Na^+$-$C_{12}(OE)_4OH$ (0.1 *M* NaCl)	25	−1.4	−0.9	Zhou and Rosen (2003)
$C_{12}(OE)_2SO_4^-Na^+$-$C_{12}(OE)_6OH$ (0.1 *M* NaCl)	25	−1.5	-1.9_5	Zhou and Rosen (2003)
$C_{12}(OE)_2SO_4^-Na^{+b}$-$C_{12}(OE)_{10}OH^b$ (0.1 *M* NaCl)	25	−2.1	−2.3	Rosen and Zhu (1988)
C_{10}-3Φ$SO_3^-Na^+$-C_9Φ$(OE)_{10}OH$ (0.17 *M* NaCl)[g]	27	—	−1.5	Osborne-Lee et al. (1985)
LAS^-Na^{+a}-C_{10}-β-maltoside (0.1 *M* NaCl)	22	−1.9	−2.1	Liljekvist and Kronberg (2000)

LAS^-Na^{+a}-$C_{11}CON(C_2H_4OH)_2$ (0.1 *M* NaCl)	25	−2.4	−1.5	Rosen and Zhu (1988)
LAS^-Na^{+a}-*N*-Octyl-2-pyrrolid(in)one (0.005 *M* NaCl (aq.)–air)	25	−3.8	−2.3	Zhu et al. (1989)
LAS^-Na^{+a}-*N*-dodecyl-2-pyrrolid(in)one (0.005 *M* NaCl (aq.)–air)	25	−3.1	−1.7	Zhu et al. (1989)
LAS^-Na^{+a}-$C_{10}(OE)_8OH$ (0.1 *M* NaCl)	22	−4.8	−3.3	Liljekvist and Kronberg (2000)
LAS^-Na^{+a}-$C_{12}(OE)_{10}OH$ (0.1 *M* NaCl)	25	−2.4	−2.7	Rosen and Zhu (1988)
C_{12}-2-$\Phi SO_3^-Na^+$-$C_{12}(OE)_8OH$	25	−3.1	−5.2	C. Utarapichart and M. J. Rosen, unpublished data
C_{12}-2-$\Phi SO_3^-Na^+$-$C_{12}(OE)_8OH$ (0.005 *M* NaCl)	25	−4.0	−5.8	C. Utarapichart and M. J. Rosen, unpublished data
C_{12}-2-$\Phi SO_3^-Na^+$-$C_{12}(OE)_8OH$ (0.01 *M* NaCl)	25	−4.3	−5.4	C. Utarapichart and M. J. Rosen, unpublished data
C_{12}-2-$\Phi SO_3^-Na^+$-$C_{12}(OE)_8OH$ (0.01 *M* NaCl)	40	−3.4	−3.8	C. Utarapichart and M. J. Rosen, unpublished data
C_{12}-4-$\Phi SO_3^-Na^+$-$C_{12}(OE)_8OH$	25	−2.3	−5.1	C. Utarapichart and M. J. Rosen, unpublished data
C_{12}-4-$\Phi SO_3^-Na^+$-$C_{12}(OE)_8OH$ (0.005 *M* NaCl)	25	−3.9	−5.5	C. Utarapichart and M. J. Rosen, unpublished data
C_{12}-4-$\Phi SO_3^-Na^+$-$C_{12}(OE)_8OH$ (0.01 *M* NaCl)	25	−3.9	−5.0	C. Utarapichart and M. J. Rosen, unpublished data
C_{12}-4-$\Phi SO_3^-Na^+$-$C_{12}(OE)_8OH$ (0.1 *M* NaCl)	25	−3.5	−3.9	C. Utarapichart and M. J. Rosen, unpublished data
C_{12}-4-$\Phi SO_3^-Na^+$-$C_9\Phi(OE)_{50}OH^b$ (0.17 *M* NaCl)	27	—	−2.6	Osborne-Lee et al. (1985)
AOT^c-1,2-C_{10} diol (0.1 *M* NaCl)	25	−1.3	−1.2	Zhou and Rosen (2003)
AOT^c-TMN6[f] (0.1 *M* NaCl)	25	−0.5	−0.5	Zhou and Rosen (2003)
AOT^cNa^+-$C_{12}(OE)_5OH$	25	−0.9	−1.2	Chang et al. (1985)
AOT^c-$C_{12}(OE)_6OH$ (0.1 *M* NaCl)	25	−1.6	−1.5	Zhou and Rosen (2003)
AOT^cNa^+-$C_{12}(OE)_7OH$	25	−1.6	−1.9	Chang et al. (1985)
AOT^cNa^+-$C_{12}(OE)_8OH$	25	−2.6	−2.0	C. Utarapichart and M. J. Rosen, unpublished data

(*Continued*)

TABLE 11.1 (*Continued*)

Mixture	Temperature (°C)	β^σ	β^M	Reference
AOT[c]Na^+-$C_{12}(OE)_8OH$ (0.05 *N* NaCl)	25	−1.7	−3.6	C. Utarapichart and M. J. Rosen, unpublished data
AOT[c]Na^+-$C_{14}(OE)_8OH$ (0.1 *M* NaCl)	25	-2.0_5	−0.2	Zhou and Rosen (2003)
$C_{12}H_{25}CH(SO_3^-Na^+)COOCH_3$-$C_9H_{19}CON(CH_3)CH_2(CHOH)_4CH_2OH$	30	—	−2.1	Okano et al. (2000)
Anionic–Zwitterionic Mixtures				
$C_8F_{17}SO_3^-Li^+$-$C_6F_{13}C_2H_4SO_2NH(CH_2)_3N^+(CH_3)_2CH_2COO$	25	—	−8.3	Esumi and Ogawa (1993)
$C_{12}SO_3^-Na^+$-$C_{12}N^+H_2(CH_2)_2COO^-$ (0.1 *M* NaBr (aq.)–air, pH 5.8)	25	−4.2	−1.2	Rosen (1991)
$C_{12}SO_3^-Na^+$-$C_{12}N^+(B_z)(Me)CH_2COO^-$ (pH 5.0)	25	−6.9	−5.4	Rosen and Zhu (1984)
$C_{12}SO_3^-Na^+$-$C_{12}N^+(B_z)(Me)CH_2COO^-$ (pH 6.7)	25	−4.9	−4.4	Rosen and Zhu (1984)
$C_{12}SO_3^-Na^+$-$C_{12}N^+(B_z)(Me)CH_2COO^-$ (pH 9.3)	25	−2.9	−1.7	Rosen and Zhu (1984)
$C_{12}SO_3^-Na^+$-$C_{12}N^+(B_z)(Me)CH_2COO^-$ (0.1 *M* NaBr–PTFE[e], pH 5.8)	25	−6.2	—	Rosen and Gu (1987)
$C_{12}SO_3^-Na^+$-$C_{12}N^+(B_z)(Me)CH_2COO^-$ (0.1 *M* NaBr–Parafilm, pH 5.8)	25	−6.9	—	Gu and Rosen (1989)
$C_{12}SO_3^-Na^+$-$C_{12}N^+(B_z)(Me)CH_2COO^-$ (pH 5.8), hexadecane–water	25	−5.2 (β^σ_{LL})	−4.0 (β^M_{LL})	Rosen and Murphy (1989)
$C_{12}SO_3^-Na^+$-$C_{12}N^+(B_z)(Me)CH_2COO^-$ (pH 5.8), dodecane–water	25	−4.8 (β^σ_{LL})	−3.6 (β^M_{LL})	Rosen and Murphy (1989)
$C_{12}SO_3^-Na^+$-$C_{12}N^+(B_z)(Me)CH_2COO^-$ (pH 5.8), heptane–water	25	−4.7 (β^σ_{LL})	−4.0 (β^M_{LL})	Rosen and Murphy (1989)
$C_{12}SO_3^-Na^+$-$C_{12}N^+(B_z)(Me)CH_2COO^-$ (pH 5.8), isooctane–water	25	−4.4 (β^σ_{LL})	−4.0 (β^M_{LL})	Rosen and Murphy (1989)
$C_{12}SO_3^-Na^+$-$C_{12}N^+(B_z)(Me)CH_2COO^-$ (pH 5.8), cyclohexane–water	25	−5.0 (β^σ_{LL})	−4.2 (β^M_{LL})	Rosen and Murphy (1989)
$C_{12}SO_3^-Na^+$-$C_{12}N^+(B_z)(Me)CH_2COO^-$ (pH 5.8), toluene–water	25	−3.2 (β^σ_{LL})	2.1 (β^M_{LL})	Rosen and Murphy (1989)
$C_{12}SO_3^-Na^+$-$C_{10}N^+(B_z)(Me)C_2H_4SO_3^-$ (pH 6.6)	25	−2.5	—	Rosen and Zhu (1984)

$C_{12}SO_3^-Na^+$-$C_{14}N^+(CH_3)_2O^-$ (0.1 *M* NaCl (aq.)–air, pH 5.8)	25	–10.3	–7.8	Rosen et al. (1994)
$C_{12}SO_3^-Na^+$-$C_{14}N^+(CH_3)_2O^-$ (0.1 *M* NaCl (aq.)–air, pH 2.9)	25	–13.5	—	Rosen et al. (1994)
$C_{10}SO_4^-Na^+$-$C_{12}N^+H_2(CH_2)_2COO^-$	30	–13.4	–10.6	Tajima et al. (1979)
$C_{10}SO_4^-Na^+$-$C_{10}S^+(Me)O^-$	25	–4.3	–4.3	Zhu and Zhao (1988)
$C_{12}SO_4^-Na^+$-$C_{12}N^+H_2(CH_2)_2COO^-$	30	–15.7	–14.1	Tajima et al. (1979)
$C_{12}SO_4^-Na^+$-$C_{12}N^+(CH_3)_2O^-$	23	—	–7.0	Goloub et al. (2000)
$C_{12}SO_4^-Na^+$-$C_{10}S^+(Me)O^-$ (1×10^{-3} *M* Na_2CO_3)	24	—	–2.4	Holland and Rubingh (1983)
$C_{12}SO_4^-Na^+$-$C_{10}P^+(Me)_2O^-$ (1×10^{-3} *M* Na_2CO_3)	24	—	–3.7	Holland and Rubingh (1983)
$C_{12}SO_4^-Li^+$-$C_6F_{13}C_2H_4SO_2NH(CH_2)_3N^+(CH_3)_2CH_2COO^-$	25	—	0	Esumi and Ogawa (1993)
$C_{14}SO_4^-Na^+$-$C_{12}N^+H_2(CH_2)_2COO^-$	30	–15.5	–15.5	Tajima et al. (1979)
LAS^-Na^{+a}-$C_{12}N^+(Me)_2CH_2COO^-$ (0.1 *M* NaCl, pH 5.8)	25	–3.8	–2.9	Rosen and Zhu (1988)
LAS^-Na^{+a}-$C_{12}N^+(Me)_2CH_2COO^-$ (0.1 *M* NaCl, pH 9.3)	25	–2.8	–1.7	Rosen and Zhu (1988)
Cationic–Cationic Mixtures				
$C_{12}N^+$-Me_3Cl^--$C_{14}N^+Me_3Cl$	30	—	–0.8	Filipovic-Vincekovic et al. (1997)
Cationic–Nonionic Mixtures				
$C_{10}N^+Me_3Br^-$-C_{10}-β-glucoside (0.1 *M* NaCl, pH 9.0)	25	–1.2	–1.2	Li et al. 2001; Rosen and Sulthana (2001)
$C_{10}N^+Me_3Br^-$-C_{10}-β-maltoside (0.1 *M* NaCl, pH 9.0)	25	–0.3	–0.3	Li et al. 2001; Rosen and Sulthana (2001)
$C_{12}N^+Me_3Cl^-$-C_{12}-β-maltoside (0.1 *M* NaCl, pH 5.7)	25	–1.0	–0.8	Rosen and Sulthana (2001)
$C_{12}N^+Me_3Cl^-$-C_{12}-β-maltoside (0.1 *M* NaCl, pH 9.0)	25	–1.9	–1.5	Li et al. 2001; Rosen and Sulthana (2001)
$C_{12}N^+Me_3Cl^-$2:1 (molar) C_{12} maltoside, C_{12} glucoside (0.1 *M* NaCl, pH 9.0)	25	–2.8	–2.8	Rosen and Sulthana (2001)

(*Continued*)

TABLE 11.1 (***Continued***)

Mixture	Temperature (°C)	β^{σ}	β^{M}	Reference
$C_{14}N^{+}Me_3Br^{-}$-C_{12}-β-maltoside (0.1 *M* NaCl, pH 9.0)	25	−1.8	−1.3	Rosen and Sulthana (2001)
$C_{10}N^{+}Me_3Br^{-}$-$C_8(OE)_4OH$ (0.05 *M* NaBr)	23	—	−1.8	Holland and Rubingh, 1983
$C_{12}N^{+}Me_3Cl^{-}C_{12}(OE)_4OH$ (0.1 *M* NaCl)	25	−1.8	-0.3_5	Zhou and Rosen (2003)
$C_{12}N^{+}Me_3Cl^{-}C_{12}(OE)_5OH$	25	—	−1.0	Rubingh and Jones (1982)
$C_{12}N^{+}Me_3Cl^{-}C_{12}(OE)_7OH$ (0.1 *M* NaCl)	25	−1.8	−1.2	Zhou and Rosen (2003)
$C_{16}N^{+}Me_3Br^{-}C_{12}(OE)_5OH$	25	—	−3.0	Rubingh and Jones (1982)
$C_{16}N^{+}Me_3Cl^{-}C_{12}(OE)_8OH$ (0.1 *M* NaCl)	25	—	−3.1	Lange and Beck (1973)
$C_{20}N^{+}Me_3Cl^{-}$-$C_{12}(OE)_8OH$	25	—	−4.6	Lange and Beck (1973)
$C_{12}Pyr^{+}Br^{-d}$-*N*-octylpyrrolid(in)one (0.1 *M* NaBr, pH 5.9)	25	−1.6	—	Rosen (1991)
$C_{12}Pyr^{+}Br^{-}$- $C_{12}(OE)_8OH$	25	−1.0	—	Hua and Rosen (1982a)
$C_{12}Pyr^{+}Br^{-}$-$C_{12}(OE)_8OH$ (0.1 *M* NaBr)	25	−0.8	—	Rosen and Hua (1982)
$C_{12}Pyr^{+}Cl^{-}$-$C_{12}(OE)_8OH$	25	−2.8	—	Rosen and Zhao (1983)
$C_{12}Pyr^{+}Cl^{-}$-$C_{12}(OE)_8OH$ (0.1 *M* NaCl)	10	−2.5	—	Rosen and Zhao (1983)
$C_{12}Pyr^{+}Cl^{-}$-$C_{12}(OE)_8OH$ (0.1 *M* NaCl)	25	−2.2	—	Rosen and Zhao (1983)
$C_{12}Pyr^{+}Cl^{-}$-$C_{12}(OE)_8OH$ (0.1 *M* NaCl)	40	−2.0	—	Rosen and Zhao (1983)
$C_{12}Pyr^{+}Cl^{-}$-$C_{12}(OE)_8OH$ (0.5 *M* NaCl)	25	−1.5	—	Rosen and Zhao (1983)
$(C_{12})_2N^{+}Me_2Br^{-}$-$C_{12}(OE)_5OH$	25	—	−1.6	Rubingh and Jones (1982)
Cationic–Zwitterionic Mixtures				
$C_{10}N^{+}$-Me_3Br^{-}-$C_{10}S^{+}MeO^{-}$, pH 5.9	25	−0.6	−0.6	Zhu and Zhao (1988)
$C_{12}N^{+}$-Me_3Br^{-}-$C_{12}N^{+}(B_z)(Me)CH_2COO^{-}$	25	−1.3	−1.3	Rosen and Zhu (1984)
$C_{12}Pyr^{+}Br^{-d}$-$C_{12}N^{+}H_2CH_2CH_2COO^{-}$ (0.1 NaBr (aq.), pH 5.8)	25	−4.8	−3.4	Rosen (1991)

Nonionic–Nonionic Mixtures				
C_{10}-β-glucoside-C_{10}-β-maltoside (0.1 *M* NaCl, pH 9.0)	25	–0.3	–0.2	Rosen and Sulthana (2001)
C_{10}-β-glucoside-$C_{12}(OE)_7OH$	25	—	–0.04	Sierra and Svensson (1999)
C_{10}-β-maltoside-$C_{10}(OE)_8OH$ (0.1 *M* NaCl)	22	–0.5	–0.3	Liljekvist and Kronberg (2000)
C_{12}-β-maltoside-$C_{12}(OE)_7OH$ (0.1 *M* NaCl, pH 5.7)	25	–0.7	–0.05	Rosen and Sulthana (2001)
$C_{12}(OE)_3OH$-$C_{12}(OE)_8OH$	25	–0.2	—	Rosen and Hua (1982)
$C_{12}(OE)_3OH$-$C_{12}(OE)_8OH$ (H_2O–hexadecane)	25	–0.7 (β^{σ}_{LL})	–0.2 (β^{M}_{LL})	Rosen and Murphy (1991)
$C_{12}(OE)_4OH$-$C_{12}(OE)_8OH$ (0.1 *M* NaCl)	25	–0.3	—	Huber (1991)
$C_{12}(OE)_8OH$-$C_{12}(OE)_4OH$ (0.1 *M* NaCl–PTFE[e])	25	0.0 (β^{σ}_{LL})	—	Huber (1991)
$C_{10}F_{19}(OE)_9OH$-t-$C_8H_{17}C_6H_4(OE)_{10}OH$	25	+0.8	—	Zhao and Zhu (1986)
N-butyl-2-pyrrolid(in)one-$(CH_3)_3SiOSi(CH_3)[CH_2(CH_2CH_2O)_{8.5}CH_3]OSi(CH_3)_3$, pH 7.0	25	–0.4	—	Wu and Rosen (2002)
N-butyl-2-pyrrolid(in)one-$(CH_3)_3SiOSi(CH_3)[CH_2(CH_2CH_2O)_{8.5}CH_3]OSi(CH_3)_3$, on polyethylene, pH 7.0	25	–3.5 (β^{σ}_{SL})	—	Wu and Rosen (2002)
N-hexyl-2-pyrrolid(in)one-$(CH_3)_3SiOSi(CH_3)[CH_2(CH_2CH_2O)_{8.5}CH_3]OSi(CH_3)_3$, pH 7.0	25	–0.8	—	Wu and Rosen (2002)
N-hexyl-2-pyrrolid(in)one-$(CH_3)_3SiOSi(CH_3)[CH_2(CH_2CH_2O)_{8.5}CH_3]OSi(CH_3)_3$, on polyethylene, pH 7.0	25	–5.9 (β^{σ}_{SL})	—	Wu and Rosen (2002)
N-(2-ethylhexyl-2-pyrrolid(in)one-$(CH_3)_3SiOSi(CH_3)$-$[CH_2(CH_2CH_2O)_{8.5}CH_3]OSi(CH_3)_3$, pH 7.0	25	–0.7	—	Wu and Rosen (2002)
N-(2-ethylhexyl-2-pyrrolid(in)one-$(CH_3)_3SiOSi(CH_3)$-$[CH_2(CH_2CH_2O)_{8.5}CH_3]OSi(CH_3)_3$, on polyethylene, pH 7.0	25	–6.7 (β^{σ}_{SL})	—	Wu and Rosen (2002)
N-octyl-2-pyrrolid(in)one-$(CH_3)_3SiOSi(CH_3)[CH_2(CH_2CH_2O)_{8.5}CH_3]OSi(CH_3)_3$, pH 7.0	25	–0.4	—	Wu and Rosen (2002)

(*Continued*)

TABLE 11.1 (*Continued*)

Mixture	Temperature (°C)	β^σ	β^M	Reference
N-octyl-2-pyrrolid(in)one-$C_{12}(OE)_8OH$ (H_2O–hexadecane)	25	−0.5 (β^σ_{LL})	−0.1 (β^M_{SL})	Rosen and Murphy (1991)
N-octyl-2-pyrrolid(in)one-$(CH_3)_3SiOSi(CH_3)[CH_2(CH_2CH_2O)_{8.5}CH_3]OSi(CH_3)_3$, on polyethylene, pH 7.0	25	−5.4 (β^σ_{SL})	—	Wu and Rosen (2002)
N-decyl-2-pyrrolid(in)one-$(CH_3)_3SiOSi(CH_3)[CH_2(CH_2CH_2O)_{8.5}CH_3]OSi(CH_3)_3$, pH 7.0	25	+0.1	—	Wu and Rosen (2002)
N-decyl-2-pyrrolid(in)one-$(CH_3)_3SiOSi(CH_3)[CH_2(CH_2CH_2O)_{8.5}CH_3]OSi(CH_3)_3$, on polyethylene, pH 7.0	25	+1.2 (β^σ_{SL})	—	Wu and Rosen (2002)
N-dodecyl-2-pyrrolid(in)one-$C_{12}(OE)_8OH$ (H_2O-hexadecane)	25	−2.0 (β^σ_{LL})	−1.4 (β^M_{SL})	Rosen and Murphy (1991)
Nonionic–Zwitterionic Mixtures				
C_{12}-β-maltoside-$C_{12}N^+(B_z)(Me)CH_2COO^-$ (0.1 *M* NaCl, pH 5.7)	25	−1.7	−1.1	Rosen and Sulthana (2001)
2:1 (molar) C_{12}-maltoside-C_{12} glucoside $C_{12}N^+(B_z)(Me)CH_2COO^-$ (0.1 *M* NaCl, pH 5.7)	25	−2.7	−2.7	Rosen and Sulthana (2001)
$C_{10}(EO)_4OH$-$C_{12}N^+(Me)_2O^-$ (5×10^{-4} *M* Na_2CO_3)	23	—	−0.8	Holland and Rubingh (1983)
$C_{12}(EO)_6OH$-$C_{12}N^+(Me)_2O^-$ (pH 2)	23	—	−1.0	Goloub et al. (2000)
$C_{12}(EO)_6OH$-$C_{12}N^+(Me)_2O^-$ (pH 8)	23	—	−0.3	Goloub et al. (2000)
$C_{12}(EO)_8OH$-$C_{12}N^+(B_z)(Me)CH_2COO^-$	25	−0.6	−0.9	Rosen and Zhu (1984)

[a] LAS^-Na^+, commercial sodium C_{12} benzenesulfonate.
[b] SAS, commercial secondary alkanesulfonate.
[c] AOT^-Na^+, sodium di(2-ethylhexyl)sulfosuccinate.
[d] Pyr^+, pyridinium.
[e] PTFE, polytetrafluoroethylene.
[f] TMN6, commercial 2,4,8-trimethyl nonanol $(OC_2H_4)_8OH$.
[g] Commercial materials..

Except for some mixtures of anionics with POE nonionics that have about six or more oxyethylene groups, β^M values are less negative, at best equally negative, compared to their β^σ values for the same surfactant under the same conditions. This may be due to the greater difficulty of accommodating hydrophobic groups in the interior of a convex micelle compared to a planar interface.

Steric effects appear when either surfactant molecule of the mixture varies in the size of the head group or in the branching of the hydrophobic group. Thus, sodium tertiarybutylbenzenesulfonate interacts less strongly with cetyl trimethylammonium bromide than isobutylbenzenesulfonate, which in turn interacts less strongly than normal butylbenzenesulfonate (Bhat and Gaikar, 1999).

Branching near the hydrophilic head group, or an increase in its size decreases the negative values of both β^σ and β^M, with greater effect on β^M than on β^σ. Branching in the hydrophobic group appears to reduce the negative value mainly of β^M. On the other hand, an increase in the number of oxyethylene groups in a POE nonionic increases sharply the negative value of β^M in sodium anionic–POE nonionic mixtures. This effect is not seen in cationic–POE nonionic mixtures and may be due to the acquisition of a positive charge by the POE chain when it is large enough to complex the Na^+ of the anionic (Matsubara et al., 1999, 2001; Liljekvist and Kronberg, 2000; Rosen and Zhou, 2001; Zhou and Rosen, 2003). Also, it has been observed that bulky head groups in a surfactant are more readily accommodated at the surface of a convex micelle than at the planar air–solution interface (Matsuki et al., 1997).

The values of both β^σ and β^M become more negative as the chain lengths of the alkyl groups on the surfactants are increased. The negative value of β^σ appears to become larger as the alkyl chains approach each other in length. This appears not to be true for β^M, which becomes more negative with an increase in the total number of carbon atoms in the alkyl chains of the two surfactants. Zwitterionics that are capable of accepting a proton (amino carboxylates and amine oxides) interact with anionics by acquiring a net positive charge through the acceptance of a proton from the water. The resulting cationic conjugate acid interacts electrostatically with the anionic surfactant. An increase in the pH of the aqueous phase consequently causes a reduction in the strength of the attractive interaction between the two surfactants, as illustrated by the $C_{12}SO_3^-Na^+$-$C_{12}N^+(B_z)(Me)CH_2COO^-$ system. A decrease in the basicity of the zwitterionic, for example, [$C_{10}N^+(B_z)(Me)CH_2CH_2SO_3^-$ vs. $C_{12}N^+(B_z)(Me)CH_2COO^-$] at constant pH, also decreases the attractive interaction with an anionic. Zwitterionics that are capable of losing a proton and acquiring a negative charge interact significantly with cationic surfactants. *N*-Alkyl-*N*, *N*-dimethylamine oxides and *N*-alkyl-*N*-methylsulfoxides interact with anionic surfactants in a manner similar to that of other zwitterionics by accepting a proton from the water to form the cationic conjugate acid. Their interaction with cationic surfactants is far weaker (Zhu and Zhao, 1988) since these compounds cannot become solely anionic in nature.

In general, an increase in the electrolyte content of the aqueous phase produces a decrease in the negative value of β^σ. This is true even for ionic–POE nonionic mixtures, indicating that the interaction between them is, at least partly, electrostatic. For anionic–cationic mixtures, the decrease in the negative value of β^σ upon the addition of sodium halides was found to be NaI > NaBr > NaCl (Goralczyk et al., 2003), reflecting the order of decreasing tendency to neutralize the charge of the cationic surfactant (and, consequently, its attraction for the anionic). However, in the case of anionic–POE nonionic mixtures, an initial *increase* in the negative value of β^σ is observed when NaCl is added to the salt-free mixture. This has been attributed (Rosen and Zhao, 1983) to the complex formation between the Na^+ and the ether oxygens of the POE chain, resulting in its acquiring a positive charge that increases the strength of its interaction with an anionic surfactant. This effect is not observed in cationic–POE nonionic mixtures (acquisition of a positive charge by the POE chain would not increase its interaction with a cationic surfactant).

A temperature increase in the 10–40°C range generally causes a decrease in attractive interaction.

III. CONDITIONS FOR THE EXISTENCE OF SYNERGISM

Based upon the same nonideal solution theory used in the evaluation of molecular interaction parameters above, the conditions for the existence of synergism in various fundamental interfacial phenomena, that is, reduction of surface or interfacial tension, mixed micelle formation, have been derived mathematically. When synergism exists, the conditions at the point of maximum synergism, such as α^* (the mole fraction of surfactant 1 in the total surfactant in the solution phase), X^* (the mole fraction of surfactant 1 in the total surfactant at the interface), $C_{12,\min}^M$ (the minimum CMC of the mixture), and γ_{CMC}^* (the minimum surface tension of the mixture at its CMC), can all be determined from the values of the relevant molecular interaction parameters and properties of the individual surfactants.

However, it should be understood that, because of the assumptions and approximations used in the nonideal solution theory upon which these relations are based, the calculated values for conditions at the point of maximum synergism may only approximate the values found under experimental conditions and should be used mainly for estimation purposes. This is especially true when commercial surfactants are used that may contain surface-active materials (impurities) of a type different from that of the nominal surfactant. These may cause the molecular interaction parameters to have values somewhat different from those listed in Table 11.1 for the nominal surfactant. When such impurities are suspected, it is advisable to determine experimentally the values of the interaction parameters.

A. Synergism or Antagonism (Negative Synergism) in Surface or Interfacial Tension Reduction Efficiency

The efficiency of surface (or interfacial) tension reduction by a surfactant has been defined (Chapter 5, Section I) as the solution phase surfactant concentration required to produce a given surface (or interfacial) tension (reduction). Synergism, in this respect, is present in an aqueous system containing two surfactants when a given surface (or interfacial) tension can be attained at a total mixed surfactant concentration lower than that required of either surfactant by itself. Antagonism (negative synergism) is present when it is attained at a higher mixed surfactant concentration than that required of either surfactant by itself. Synergism and antagonism are illustrated in Figure 11.4.

From the relations upon which Equations 11.1 and 11.2 are based and the definition for synergism or antagonism (negative synergism) of this type, it has been shown mathematically (Hua and Rosen, 1982b, 1988) that the conditions for synergism or antagonism in surface tension reduction efficiency to exist are

Synergism	Antagonism
1. β^σ must be negative.	1. β^σ must be positive.
2. $\lvert\beta^\sigma\rvert > \lvert\ln(C_1^0/C_2^0)\rvert$	2. $\lvert\beta^\sigma\rvert > \lvert\ln(C_1^0/C_2^0)\rvert$

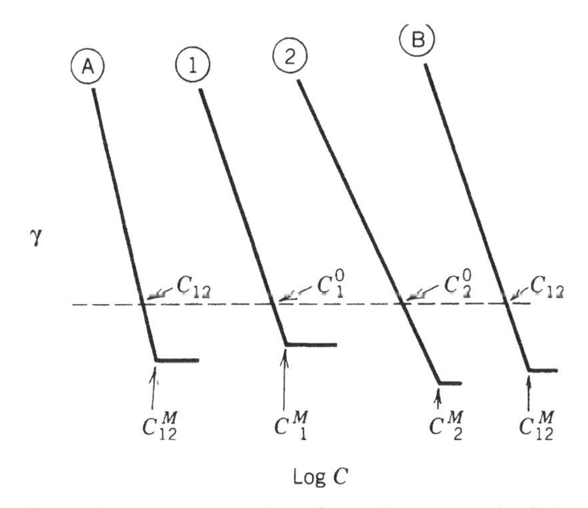

FIGURE 11.4 Synergism or antagonism (negative synergism) in surface tension reduction efficiency or in mixed micelle formation. ①, pure surfactant 1; ②, pure surfactant 2; Ⓐ, mixture of surfactants 1 and 2 at a given mole fraction α in the aqueous phase showing synergism ($C_{12} < C_1^0, C_2^0$ or $C_{12}^M < C_1^M, C_2^M$); Ⓑ, mixture of surfactants 1 and 2 at a given mole fraction α in the solution phase showing antagonism (negative synergism) ($C_{12} > C_1^0, C_2^0$ or $C_{12}^M > C_1^M$).

It is apparent from condition 2 that to increase the probability of synergism existing, the two surfactants selected for the mixture should have C_1^0 and C_2^0 values as close to each other as possible. When the values are equal, *any* value of β^σ (other than 0) will produce synergism or antagonism.

At the point of maximum synergism or maximum antagonism, that is, at the minimum or maximum aqueous phase total molar concentration, respectively, of mixed surfactant to produce a given surface tension, the mole fraction α^* of surfactant 1 in the solution phase (on a surfactant-only basis) equals its mole fraction X_1^* at the interface and is given by the relation

$$\alpha^* = \frac{\ln(C_1^0/C_2^0) + \beta^\sigma}{2\beta^\sigma}. \tag{11.8}$$

The minimum (or maximum) aqueous phase total molar concentration of mixed surfactants in the system to produce a given surface tension is

$$C_{12,\min} = C_1^0 \exp\left\{\beta^\sigma \left[\frac{\beta^\sigma - \ln(C_1^0/C_2^0)}{2\beta^\sigma}\right]^2\right\}. \tag{11.9}$$

From the above relation, the larger the negative value of β^σ, the smaller will be the value of $C_{12,\min}$; the larger its positive value, the greater the value of $C_{12,\max}$. Figure 11.5 illustrates the relations between $\log C_{12}$ and α in systems showing synergism or antagonism in surface tension reduction efficiency.

Analogous expressions have been derived for the existence of synergism in interfacial tension reduction efficiency at the liquid–liquid interface (Rosen and Murphy, 1986) and at the liquid–hydrophobic solid interface (Rosen and Gu, 1987), and for the conditions at the point of maximum synergism.

B. Synergism or Antagonism (Negative Synergism) in Mixed Micelle Formation in an Aqueous Medium

Synergism, in this respect, is present when the CMC in an aqueous medium of any mixture of two surfactants is smaller than that of either individual surfactant. Antagonism, in this respect, is present when the CMC of the mixture is larger than the CMC of either surfactant of the mixture. These are illustrated in Figure 11.4. From Equations 11.3 and 11.4 and the definition for this type of synergism or antagonism, the conditions for synergism or antagonism in this respect in a mixture containing two surfactants (in the absence of a second liquid phase) have been shown mathematically (Hua and Rosen, 1982b, 1988) to be

Synergism	Antagonism
1. β^M must be negative.	1. β^M must be positive.
2. $\lvert\beta^M\rvert > \lvert\ln(C_1^M/C_2^M)\rvert$	2. $\lvert\beta^M\rvert > \lvert\ln(C_1^M/C_2^M)\rvert$

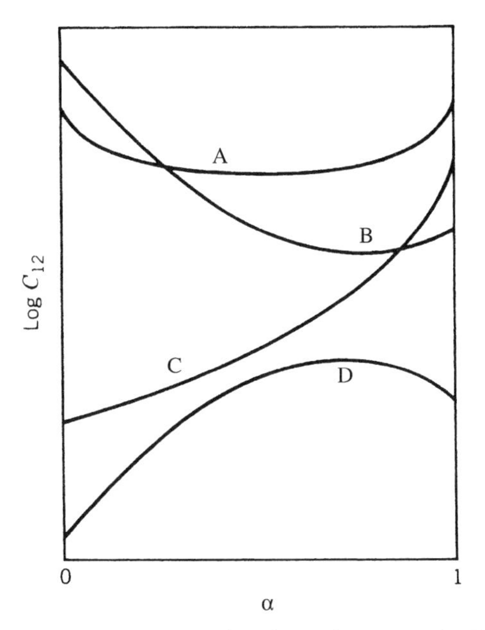

FIGURE 11.5 Synergism and antagonism (negative synergism) in surface tension reduction efficiency. Log C_{12} versus α curves illustrating (A) synergism when $\beta^\sigma < 0$, $|\ln C_1^0/C_2^0| \approx 0$; (B) synergism when $\beta^\sigma < 0$, $|\beta^\sigma| > |\ln C_1^0/C_2^0| > 0$; (C) no synergism when $\beta^\sigma < 0$, $|\beta^\sigma| < |\ln C_1^0/C_2^0|$; and (D) antagonism when $\beta^\sigma > 0$, $|\beta^\sigma| > |\ln C_1^0/C_2^0|$.

At the point of maximum synergism or antagonism, that is, where the CMC of the system is at a minimum or maximum, respectively, the mole fraction α^* of surfactant 1 in the solution phase (on a surfactant-only basis) equals its mole fraction $X_1^{M,*}$ in the mixed micelle and is given by the relation

$$\alpha^* = \frac{\ln(C_1^M/C_2^M)+\beta^M}{2\beta^M}. \tag{11.10}$$

The minimum (or maximum) CMC of the mixture is

$$C_{12,\min}^M = C_1^M \exp\left\{\beta^M\left[\frac{\beta^M - \ln(C_1^M/C_2^M)}{2\beta^M}\right]^2\right\}. \tag{11.11}$$

Figure 11.6 illustrates the relation of $\log C_{12}^M$ to α in some systems showing synergism in mixed micelle formation.

Analogous expressions have been derived (Rosen and Murphy, 1986) for the existence of synergism in mixed micelle formation in the presence of a second liquid phase and for the conditions at the point of maximum synergism.

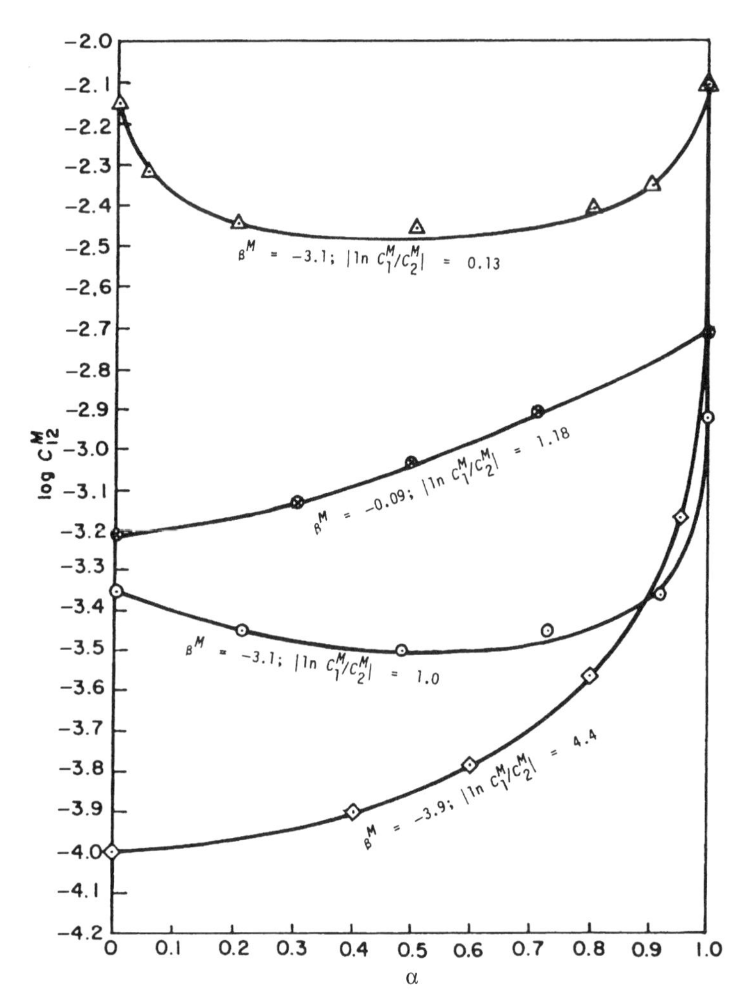

FIGURE 11.6 Synergism in mixed micelle formation for some binary surfactant mixtures. ◇, $C_{12}H_{25}SO_4^-Na^+/C_{12}H_{25}(OC_2H_4)_8OH$ mixtures in water at 25°C, showing no synergism; △, $C_{12}H_{25}SO_4^-Na^+/C_8H_{17}(OC_2H_4)_7OH$ mixtures in water at 25°C, showing synergism. Data from Lange, H. and Beck, K. H. (1973) *Kolloid Z. Z Polym.* **251**, 424. ⊙, $(C_{12}H_{25}SO_4^-)_2$ $M/C_{12}H_{25}(OC_2H_4)_{49}OH$ ($M = Zn^{2+}, Mn^{2+}, Cu^{2+}, Mg^{2+}$) mixtures in water at 30°C, showing synergism. Data from Nishioka, N. (1977) *J. Colloid Interface Sci.* **60**, 242. ⊗, $C_{10}H_{21}S(O)CH_3/C_{10}H_{21}(OC_2H_4)_3$ mixtures at 25°C, showing no synergism. Data from Ingram, B. T. and Luckhurst, A. H. W., in *Surface Active Agents*, Society of Chemical Industry, London, 1979, p. 89. Adapted with permission from Figure 3. Copyright 1982, American Oil Chemists' Society (from Rosen, M. J., in *Phenomena in Mixed Surfactant Systems*, J. F. Scamerhorn [ed.], ACS Symposium Series 311, American Chemical Society, Washington, DC, 1986, p. 151).

C. Synergism or Antagonism (Negative Synergism) in Surface or Interfacial Tension Reduction Effectiveness

Synergism of this type exists when the mixture of two surfactants at its CMC reaches a lower surface (or interfacial) tension γ_{12} value than that attained at the CMC of either individual surfactant (γ_1^{CMC}, γ_2^{CMC}); antagonism exists when it reaches a higher surface (or interfacial) tension γ_{12}^{CMC} value. These are illustrated in Figure 11.7. The conditions for synergism or antagonism in surface tension reduction effectiveness to occur (Zhu and Rosen, 1984; Hua and Rosen, 1988) are

Synergism	Antagonism
1. $\beta^\sigma - \beta^M$ must be negative.	1. $\beta^\sigma - \beta^M$ must be positive.
2. $\lvert\beta^\sigma - \beta^M\rvert > \left\lvert \ln\left(\frac{C_1^{0,CMC} C_2^M}{C_2^{0,CMC} C_1^M}\right)\right\rvert$	2. $\lvert\beta^\sigma - \beta^M\rvert > \left\lvert \ln\left(\frac{C_1^{0,CMC} C_2^M}{C_2^{0,CMC} C_1^M}\right)\right\rvert$

where $C_1^{0,CMC}$ and $C_2^{0,CMC}$ are the molar concentrations of surfactants 1 and 2, respectively, required to yield a surface tension equal to that of any mixture at its CMC.

It is apparent from condition 1 that synergism in surface tension reduction effectiveness can occur only when the attractive interaction between the two surfactants in the mixed monolayer at the aqueous solution–air interface

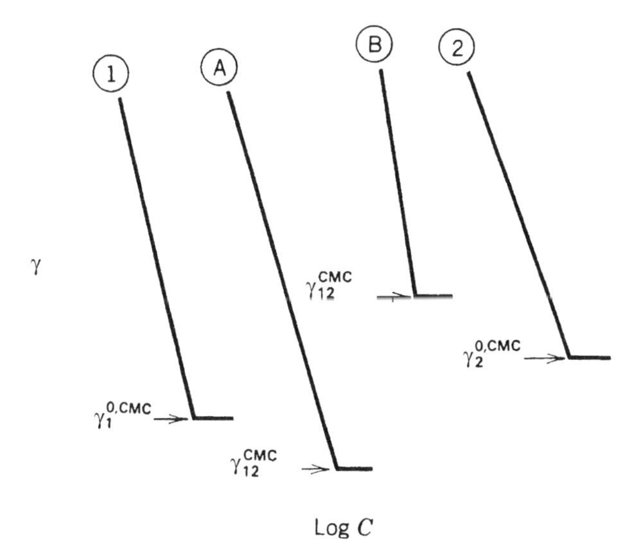

FIGURE 11.7 Synergism or, antagonism (negative synergism) in surface tension reduction effectiveness. ①, pure surfactant 1; ②, pure surfactant 2; Ⓐ, mixture of surfactants 1 and 2 at a given mole fraction α in the solution phase showing: synergism ($\gamma_{12}^{CMC} < \gamma_1^{CMC}, \gamma_2^{CMC}$); Ⓑ, mixture of surfactants 1 and 2 at a given mole fraction α in the solution phase showing antagonism ($\gamma_{12}^{CMC} > \gamma_1^{CMC}, \gamma_2^{CMC}$).

is stronger than that in the mixed micelle in the solution phase. When the attraction between the two surfactants in the mixed micelle is stronger than in the mixed monolayer, it is possible for antagonism of this type to occur.

When, from the values of β^σ and β^M, it is possible that the system may show synergism of this type, it is advisable (for the purpose of determining the values of $C_1^{0,\mathrm{CMC}}$ and $C_2^{0,\mathrm{CMC}}$ for testing condition 2 above) to extend the γ-log C plot of the surfactant having the larger γ_{CMC} value down to a γ-value equal to that of the other surfactant (with the smaller γ_{CMC}). To do this, the linear (or almost linear) portion of maximum slope below the CMC is extended downward (see Figure 11.8a); any portion of the plot close to the CMC showing a decrease in slope is ignored. The quantity $\left|\ln\left(C_1^{0,\mathrm{CMC}}/C_2^{0,\mathrm{CMC}}\right)\left(C_2^M/C_1^M\right)\right|$ then equals $\ln(C^{0,\mathrm{CMC}}/C^M)$ for the surfactant whose plot has been extended.

When it is possible that the system may show antagonism (negative synergism) (from the values of β^σ and β^M), it is advisable to use values of $C_1^{0,\mathrm{CMC}}$ and $C_2^{0,\mathrm{CMC}}$ at the γ_{CMC} of the surfactant having the larger surface tension value at its CMC (see Figure 11.8b). In this case, the quantity $\left|\ln\left(C_1^{0,\mathrm{CMC}}/C_2^{0,\mathrm{CMC}}\right)\left(C_2^M/C_1^M\right)\right|$

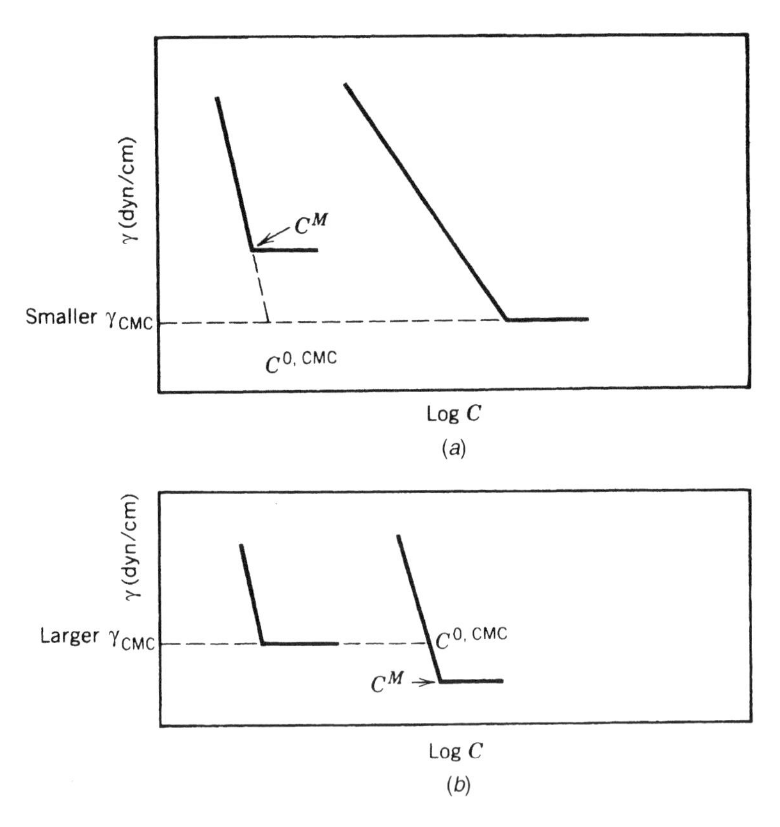

FIGURE 11.8 Evaluation of $[(C_1^{0,\mathrm{CMC}}/C_2^{0,\mathrm{CMC}})(C_2^M/C_1^M)]$ (a) for systems that may show synergism and (b) for systems that may show antagonism.

equals $\ln(C^M/C^{0,\text{CMC}})$ for the surfactant having the smaller surface tension value at its CMC.

At the point of maximum synergism or maximum antagonism in surface or interfacial tension reduction effectiveness, the composition of the mixed interfacial layer equals the composition of the mixed micelle; that is, $X_1^{*,E} = X_1^{M,*,E} \cdot \alpha^{*,E}$; the mole fraction of surfactant 1 in the solution phase (on a surfactant-only basis) at this point is obtained (Hua and Rosen, 1988) by solving Equation 11.12 numerically for $X_1^{*,E}$ and by substituting that value in Equation 11.13:

$$\frac{\gamma_1^{0,\text{CMC}} - K_1(\beta^\sigma - \beta^M)(1 - X_1^*)^2}{\gamma_2^{0,\text{CMC}} - K_2(\beta^\sigma - \beta^M)(1 - X_1^*)^2} = 1 \tag{11.12}$$

and

$$\alpha^{*,E} = \frac{\dfrac{C_1^M}{C_2^M} \cdot \dfrac{X_1^*}{1 - X_1^*} \exp[\beta^M(1 - 2X_1^*)]}{1 + \dfrac{C_1^M}{C_2^M} \cdot \dfrac{X_1^*}{1 - X_1^*} \exp[\beta^M(1 - 2X_1^*)]}, \tag{11.13}$$

where K_1 and K_2 are the slopes of the $\gamma - \ln C$ plots of the aqueous solutions of surfactants 1 and 2, respectively; $\gamma_1^{*,\text{CMC}}$ and $\gamma_2^{*,\text{CMC}}$ are the surface or interfacial tensions of surfactants 1 and 2, respectively, at their respective CMCs.

D. Selection of Surfactant Pairs for Optimal Interfacial Properties

1. *For Maximum Reduction of the CMC.* Select surfactant pairs with the largest negative β^M values (strongest attractive interaction in mixed micelle formation) (Hua and Rosen, 1982b). When interaction between the two surfactants is weak, that is, β^M is a small negative number, select a pair with approximately equal CMC values. The surfactant with the smaller CMC should always be used in a larger quantity than the one with the larger CMC.
2. *For Maximum Efficiency in Surface (or Interfacial) Tension Reduction.* Select surfactant pairs with the largest negative β^σ or β_{SL}^σ or β_{LL}^σ values. If one of the surfactants in the formulation is specified, then the second surfactant should, if possible, have a larger pC_{20} value (be more efficient) than the first. If β^σ (or β_{SL}^σ or β_{LL}^σ) for the surfactant pair is a small negative number (attractive interaction between them is weak), select surfactants with approximately equal pC_{20} values. If β^σ (or β_{SL}^σ or β_{LL}^σ) is a large negative number, use equimolar amounts of each surfactant to achieve maximum efficiency; otherwise, use a larger quantity of the more efficient surfactant (i.e., the one with the larger pC_{20} value).
3. *For Maximum Reduction of Surface (or Interfacial) Tension.* Select surfactant pairs with the largest negative β^σ (or β_{SL}^σ or β_{LL}^σ) $- \beta^M$ values. If

this quantity is only a small negative number, use, if possible, two surfactants with approximately equal γ-values at their CMC. When this is not possible, the surfactant with the higher γ-value at its CMC should, preferably, have the smaller area/molecule at the interface.

IV. THE RELATION BETWEEN SYNERGISM IN FUNDAMENTAL SURFACE PROPERTIES AND SYNERGISM IN SURFACTANT APPLICATIONS

The relations between synergism (or antagonism) in the fundamental properties of mixed monolayer formation at an interface or mixed micelle formation in solution and synergism in various practical applications of surfactants is still a relatively unexplored area. Some studies have probed this area, but much remains to be known.

A study that investigated a number of applications was of aqueous mixtures of commercial sodium dodecylbenzenesulfonate (LAS) and sulfated POE dodecyl alcohol. These mixtures show synergism in interfacial tension reduction effectiveness (both static and dynamic) against olive oil, with the degree of synergism increasing with an increase in the number of oxyethylene groups from one to four (Figure 11.9). When the sulfated alcohol is not oxyethylenated, that is, in LAS–sodium dodecyl sulfate mixtures, no synergistic interaction is observed. Synergism for these mixtures was observed in wetting of polyester, in emulsification of olive oil, in dishwashing, and in soil removal from wool (Figures 11.10 and 11.11), with the point of maximum synergism in all these phenomena being at approximately the same surfactant ratio as that observed for synergism in interfacial tension reduction effectiveness (Schwuger, 1984).

An investigation of synergism in foaming in aqueous media and its relation to synergism in the fundamental properties of surface tension reduction and mixed micelle formation showed that synergism (or negative synergism) in foaming effectiveness, measured by initial foam heights by the Ross–Miles technique (Chapter 7, Section III), is related to synergism (or antagonism) in surface tension reduction effectiveness (Rosen and Zhu, 1988). Binary mixtures of surfactants that lowered the surface tension to values below that attainable with the individual surfactants showed higher initial foam heights than those produced by the individual surfactants (Figure 11.12a). Maximum foam height was obtained at approximately the same mole ratio of the two surfactants that produced maximum synergism in surface tension reduction, and this ratio was in agreement with that calculated by the equations in Section III. A surfactant mixture that showed antagonism in surface tension reduction effectiveness (higher surface tension at the CMC of the mixture than that observed at the CMC of the individual surfactants by themselves) showed lower initial foam height than that produced by the individual surfactants (Figure 11.12b) at the same total surfactant concentration. There appeared to

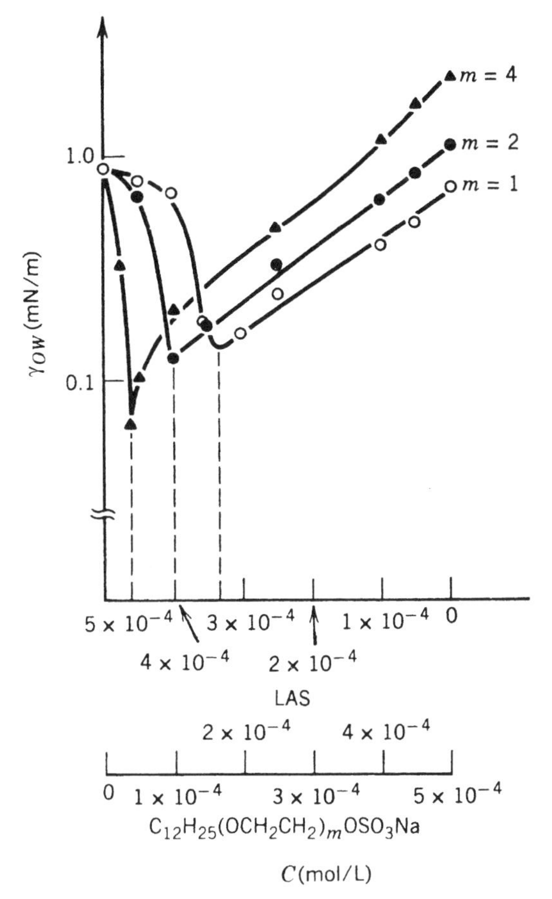

FIGURE 11.9 Olive oil–water interfacial tension for LAS–alkylether sulfate mixtures (purity: LAS, technical product; ether sulfates, 98.0–99.5%). Reprinted with permission from Schwuger, M. J., in *Structure/Performance Relationships in Surfactants*, M. J. Rosen (ed.), ACS Symposium Series 253, American Chemical Society, Washington, DC, 1984, p. 22.

be no relation between synergism in surface tension reduction efficiency or mixed micelle formation and foaming efficiency (surfactant concentration to produce a given amount of initial foam height).

Various mixtures of anionic and cationic surfactants, which interact strongly and show marked synergism in surface tension reduction efficiency, mixed micelle formation, and surface tension reduction effectiveness, show synergism in wetting at various interfaces. Thus, mixtures of sodium n-octyl sulfate and octyltrimethylammonium bromide, which interact very strongly in aqueous media (Table 11.1), show much better wetting properties for paraffin wax than the individual surfactants by themselves (Zhao et al., 1980). Aqueous solutions

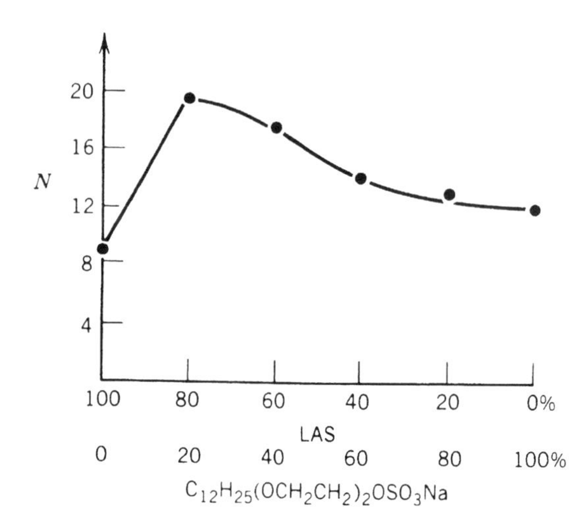

FIGURE 11.10 Dishwashing by LAS–dodecyl 2 EO sulfate mixtures. *N*, number of plates washed at 45°C (technical surfactants). Reprinted with permission from Schwuger, M. J., in *Structure/Performance Relationships in Surfactants*, M. J. Rosen (ed.), ACS Symposium Series 253, American Chemical Society, Washington, DC, 1984, p. 23.

of mixtures of sodium perfluorooctanoate and octyltrimethylammonium bromide, which show synergism in the three fundamental properties mentioned above, spread readily over kerosene and n-heptane surfaces, although aqueous solutions of the individual surfactants by themselves do not (Zhao and Zhu, 1983).

Interaction between the surface-active components in surfactant mixtures and with the solubilizate can both increase and decrease solubilization into the mixed micelles. Thus, the addition of small quantities of sodium dodecyl sulfate sharply decreases the solubilization of butobarbitone by micellar solutions of a commercial POE nonionic, $C_{12}H_{25}(OC_2H_4)_{23}OH$. The competitive interaction of the sodium dodecyl sulfate with the oxyethylene groups on the surface of the micelles of the nonionic surfactant is believed to be the cause of this phenomenon (Treiner et al., 1985). On the other hand, a mixture of sodium dodecyl sulfate and sorbitan monopalmitate in aqueous solution (Span 40) solubilized dimethylaminoazobenzene more than either surfactant by itself, with maximum solubilization observed at a 9:1 molar ratio of the anionic to the nonionic (Fukuda and Taniyama, 1958).

The solubilizing power of a tetradecylammonium bromide–sodium octane sulfonate mixture for equal amounts of water and hydrocarbon at 50°C is less than that of the individual components and decreases sharply as the surfactant proportions approach a 1:1 molar ratio. Here, the interaction between the two surfactants produces antagonism.

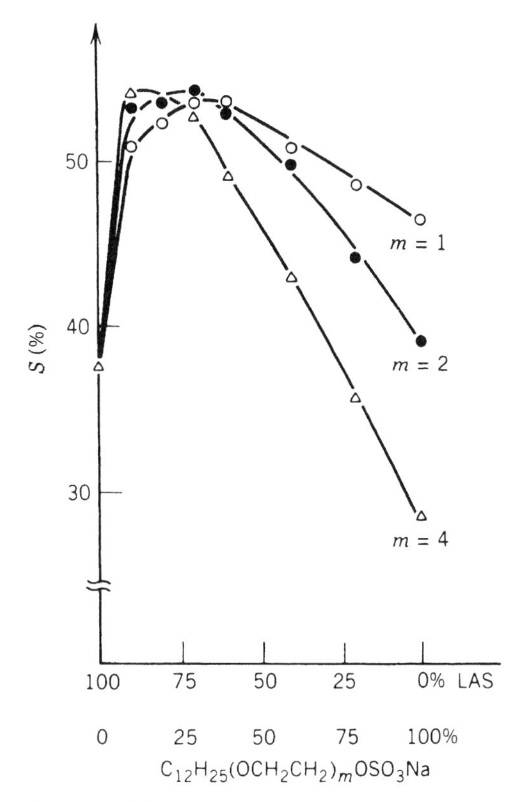

FIGURE 11.11 Soil removal from wool by LAS–alkylether sulfate mixtures (technical surfactants). Test conditions: 30°C, total surfactant concentration 5×10^{-3} mol/L, sebum–pigment mixture soil. Reprinted with permission from Schwuger, M. J. in *Structure/Performance Relationships in Surfactants*, M. J. Rosen (ed.), ACS Symposium Series 253, American Chemical Society, Washington, DC, 1984, p. 24.

The proposed explanation is based upon the Winsor *R* concept (Chapter 5, Section III), with the interaction between the surfactants producing a pseudo-nonionic complex that decreases the *Acw* value in the denominator of the *R* ratio, with a consequent decrease in the solubilization power (Bourrel et al., 1984).

Interaction between the two surfactants has also been shown to both increase and decrease their adsorption at various interfaces. The addition of a small amount (<20 mol %) of a POE nonionic to an anionic surfactant, sodium dodecyl sulfate, increased the adsorption of the anionic onto carbon at low surfactant concentrations. As the ratio of nonionic to anionic increased, this effect was diminished, and at a 1:1 molar ratio, the anionic was scarcely adsorbed. It was suggested that the inclusion of the POE nonionic in the adsorbed film on the carbon reduces electrical repulsion between adsorbed

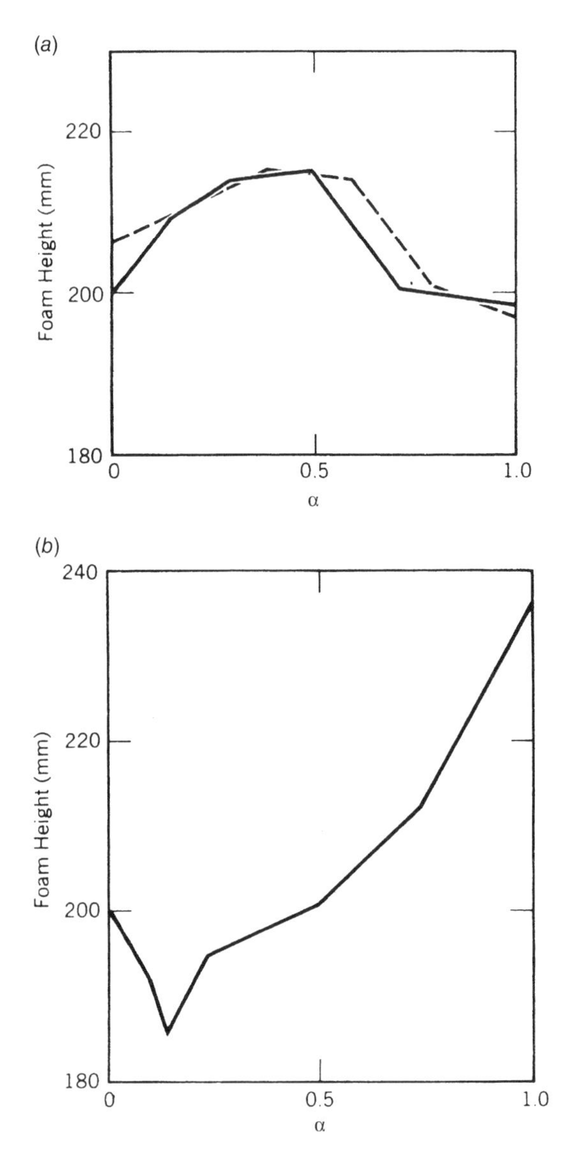

FIGURE 11.12 (a) Initial foam height versus mole fraction α of commercial sodium linear alkylbenzensulfonate (LAS) in the total surfactant in the aqueous phase in 0.25% LAS-dodecylbetaine mixtures (60°C, 0.1 *M* NaC1). —, pH 5.8; - - -, pH 9.3. (b) Initial foam height versus mole fraction α of C_{16} sodium soap in the total surfactant in the aqueous phase of 0.25% C_{16} soap–LAS mixtures (60°C, 0.1 *N* NaCl, pH 10.6). From Rosen, M. J. and Zhu, Z. H. (1988) *J. Am. Oil Chem. Soc.* **65**, 663.

surfactant molecules and also between them and the negatively charged carbon surface. The increased solution concentration of POE nonionic caused displacement of the anionic from the adsorbed film by the more surface-active nonionic (Schwuger and Smolka, 1977).

Interaction between two surfactants in aqueous solution producing synergism in foaming and decreased adsorption onto solid surfaces has been used to advantage in the separation of minerals. An alkyl sulfosuccinate–POE nonionic mixture that shows synergism in foaming and whose interaction results in decreased adsorption onto scheelite and calcite surfaces produced enhanced selectivity and recovery of scheelite by the flotation process (von Rybinski and Schwuger, 1986).

The addition of a second surfactant that interacts with the first surfactant to change various fundamental interfacial properties has been shown to affect detergency in several recent investigations (Schwuger, 1982; Aronson et al., 1983; Matson and Cox, 1984; Cox et al., 1985; Smith et al., 1985). Both synergistic and antagonistic effects were observed. Thus, significant improvement in the detergency by LAS of sebum soiled permanent press and cotton cloth at water hardness >100 ppm Ca^{2+} was obtained by replacing a portion of the LAS by a POE nonionic (Cox et al., 1985). Optimum detergency at 100°F was obtained with a C_{12}-C_{14} nonionic containing 70% ethylene oxide (EO) at a 1:4 nonionic/LAS ratio. Calculation of the mole fraction of the nonionic in the mixed micelle (by Equation 11.3) for this 1:4 nonionic–LAS mixture showed that the mixed micelle formed consisted predominantly of the nonionic. The improved detergency is consequently believed to be due to the nonionic acting as a micelle promotion agent, incorporating LAS into the mixed micelle and Ca^{2+} by counterion binding to the micellar surface, thus reducing the formation of $Ca(LAS)_2$ in the solution phase. The free LAS, on the other hand, is believed to be mainly responsible for the interfacial and detergency properties of the mixture.

On the other hand, the addition of minor amounts of LAS to a POE nonionic solution that showed rapid removal of mineral oil soil from a polyester surface decreased the rate of removal or inhibited its removal completely. The effect appeared to be related to its effect on the mineral oil-in-water (*O*/*W*) interfacial tension: As the *O*/*W* interfacial tension increased, the removal time of the oily soil increased; when the *O*/*W* interfacial tension exceeded a critical value, removal was completely inhibited (Aronson, 1983).

REFERENCES

Aronson, M. P., M. L. Gum, and E. D. Goddard (1983) *J. Am. Oil Chem. Soc.* **60**, 1333.

Bhat, M. and V. G. Gaikar (1999) *Langmuir* **15**, 4740.

Bourrel, M., D. Bernard, and A. Graciaa (1984) *Tenside Detergents* **21**, 311.

Chang, J. H., Y. Muto, K. Esumi, and K. Meguro (1985) *J. Am. Oil Chem. Soc.* **62**, 1709.

Corkill, M. and J. Goodman (1963) *Proc. R. Soc.* **273**, 84.

Cox, M. F., N. F. Borys, and T. P. Matson (1985) *J. Am. Oil Chem. Soc.* **62**, 1139.

Esumi, K. and M. Ogawa (1993) *Langmuir* **9**, 358.

Esumi, K., N. Nakamura, and K. Nagai (1994) *Langmuir* **10**, 4388.

Esumi, K., T. Arai, and K. Takasuji (1996) *Coll. Surf. A.* **111**, 231.

Filipovic-Vincekovic, N., I. Juranovic, and Z. Grahek (1997) *Coll. Surf. A.* **125**, 115.

Fukuda, K. and Y. Taniyama (1958) *Sci. Rep. Saitama Univ.* **3A**, 27. [C. A. 53, 10902i (1959)].

Goloub, T. P., R. J. Pugh, and B. V. Zhmud (2000) *J. Colloid Interface Sci.* **229**, 72.

Goralczyk, D., K. Hac, and P. Wydro (2003) *Coll. Surf. A: Physicochem. Eng. Aspects* **220**, 55.

Gu, B. and M. J. Rosen (1989) *J. Colloid Interface Sci.* **129**, 537.

Holland, P., in *Structure/Performance Relationships in Surfactants*, M. J. Rosen (ed.), ACS Symposium Series 253, American Chemical Society, Washington, DC, 1984, p. 141.

Holland, P. and D. N. Rubingh (1983) *J. Phys. Chem.* **87**, 1984.

Hua, X. Y. and M. J. Rosen (1982a) *J. Colloid Interface Sci.* **87**, 469.

Hua, X. Y. and M. J. Rosen (1982b) *J. Colloid Interface Sci.* **90**, 212.

Hua, X. Y. and M. J. Rosen (1988) *J. Colloid Interface Sci.* **125**, 730.

Huber, K. (1991) *J. Colloid Interface Sci.* **147**, 321.

Ingram, B. T. (1980) *Colloid Polym. Sci.* **258**, 191.

Kadi, M., P. Hansson, and M. Almgren (2002) *Langmuir* **18**, 9243.

Lange, H. and K. H. Beck (1973) *Kolloid Z. Z. Polym.* **251**, 424.

Li, F., M. J. Rosen, and S. B. Sulthana (2001) *Langmuir* **17**, 1037.

Li, X.-G. and G.-X. Zhao (1992) *Colloids Surfs.* **64**, 185.

Liljekvist, P. and B. Kronberg (2000) *J. Colloid Interface Sci.* **222**, 159.

Liu, L. and M. J. Rosen (1996) *J. Colloid Interface Sci.* **179**, 454.

Lucassen-Reynders, E. H., J. Lucassen, and D. Giles (1981) *J. Colloid Interface Sci.* **81**, 150.

Matson, T. P. and M. F. Cox (1984) *J. Am. Oil Chem. Soc.* **61**, 1270.

Matsubara, H., A. Ohta, M. Kameda, M. Villeneuve, N. Ikeda, and M. Aranoto (1999) *Langmuir* **15**, 5496.

Matsubara, H., S. Muroi, M. Kameda, N. Ikeda, A. Ohta, and M. Aranoto (2001) *Langmuir* **17**, 7752.

Matsuki, H., S. Hashimoto, S. Kaneshina, and Y. Yamanaka (1997) *Langmuir* **13**, 2687.

Ogino, K., T. Kakihara, H. Uchiyama, and M. Abe Presented 77th Annual Meeting, Am. Oil Chem. Soc., Honolulu, Hawaii, May 1986.

Okano, T., T. Tamura, Y. Abe, T. Tsuchida, S. Lee, and G. Sugihara (2000) *Langmuir* **16**, 1508.

Osborne-Lee, I., W. Schechter, R. S. Wade, and Y. Barakat (1985) *J. Colloid Interface Sci.* **108**, 60.

Penfold, J., E. Staples, L. Thompson, I. Tucker, J. Hines, R. K. Thomas, and J. R. Lu (1995) *Langmuir* **11**, 2496.

Rodakiewicz-Nowak, J. (1982) *J. Colloid Interface Sci.* **84**, 532.

Rosen, M. J. (1989) *J. Am. Oil Chem. Soc.* **66**, 1840.

Rosen, M. J. (1991) *Langmuir* **7**, 885.

Rosen, M. J. and B. Gu (1987) *Colloids Surf* **23**, 119.

Rosen, M. J. and B. Y. Zhu (1984) *J. Colloid Interface Sci.* **99**, 427.

Rosen, M. J. and D. S. Murphy (1986) *J. Colloid Interface Sci.* **110**, 224.

Rosen, M. J. and D. S. Murphy (1989) *J. Colloid Interface Sci.* **129**, 208.

Rosen, M. J. and D. S. Murphy (1991) *Langmuir* **7**, 2630.

Rosen, M. J. and F. Zhao (1983) *J. Colloid Interface Sci.* **95**, 443.

Rosen, M. J. and Q. Zhou (2001) *Langmuir* **17**, 3532.

Rosen, M. J. and S. B. Sulthana (2001) *J. Colloid Interface Sci.* **239**, 528.

Rosen, M. J. and X. Y. Hua (1982) *J. Colloid Interface Sci.* **86**, 164.

Rosen, M. J. and Z. H. Zhu (1988) *J. Am. Oil Chem. Soc.* **65**, 663.

Rosen, M. J. and Z. H. Zhu (1989) *J. Colloid Interface Sci.* **133**, 473.

Rosen, M. J., B. Gu, D. S. Murphy, and Z. H. Zhu (1989) *J. Colloid Interface Sci.* **129**, 468.

Rosen, M. J., T. Gao, Y. Nakatsuji, and A. Masuyama (1994) *Coll. Surf. A.: Physicochem. Eng. Aspects* **88**, 1.

Rubingh, D. N., in *Solution Chemistry of Surfactants*, K. L. Mittal (ed.), Vol. 1, Plenum, New York, 1979, pp. 337–354.

Rubingh, D. N. and T. Jones (1982) *Ind. Eng. Chem. Prod. Res. Dev.* **21**, 176.

Schwuger, M. J. (1982) *J. Am. Chem. Soc.* **59**, 265.

Schwuger, M. J., in *Structure/Performance Relationships in Surfactants*, M. J. Rosen (ed.), ACS Symposium Series 253, Amer. Chem. Soc., Washington, DC, 1984, p. 3.

Schwuger, M. J. and H. G. Smolka (1977) *Colloid Polym. Sci.* **255**, 589.

Sierra, M. L. and M. Svensson (1999) *Langmuir* **15**, 2301.

Smith, D. L., K. L. Matheson, and M. F. Cox (1985) *J. Am. Oil Chem. Soc.* **62**, 1399.

Tajima, K., A. Nakamura, and T. Tsutsui (1979) *Bull. Chem. Soc. Jpn.* **52**, 2060.

Treiner, C., C. Vaution, E. Miralles, and F. Puisieux (1985) *Colloids Surf.* **14**, 285.

von Rybinski, W. and M. J. Schwuger (1986) *Langmuir* **2**, 639.

Wu, Y. and M. J. Rosen (2002) *Langmuir* **18**, 2205.

Zhao, G.-X. and B. Y. Zhu (1983) *Colloid Polym. Sci.* **261**, 89.

Zhao, G.-X. and B. Y. Zhu, in *Phenomena in Mixed Surfactant Systems*, J. F. Scamehorn (ed.), ACS Symposium Series 311, American Chemical Society, Washington, DC, 1986, p. 184.

Zhao, G.-X., Y. Z. Chen, J. G. Ou, B. X. Tien, and Z. M. Huang (1980) *Hua Hsueh Hsueh Pao (Acta Chimica Sinica)* **38**, 409.

Zhou, Q. and M. J. Rosen (2003) *Langmuir* **19**, 4555.

Zhu, D. and G.-X. Zhao (1988) *Wuli Huaxue Xuebao Acta. Phys.-Chim. Sin.* **4**, 129.

Zhu, B. Y. and M. J. Rosen (1984) *J. Colloid Interface Sci.* **99**, 435.

Zhu, B. Y. and T. Gu (1991) *J. Chem. Soc., Faraday Trans.* **87**, 2745.

Zhu, Z. H., D. Yang, and M. J. Rosen (1989) *J. Am. Oil Chem. Soc.* **66**, 998.

PROBLEMS

11.1 **(a)** Surfactant A has a pC_{20} value of 3.00 in 0.1 *M* NaCl (aq.); surfactant B has a pC_{20} value of 3.60 in the same medium. The β^{σ} value for the mixture in 0.1 *M* NaCl is –2.80. Will a mixture of surfactants A and B in 0.1 *M* NaCl show synergism in surface tension reduction efficiency?

(b) If this system does show synergism of this type, calculate the values of α^* (the mole fraction of surfactant A in the mixture, on a surfactant-only basis, at the point of maximum synergism) and $C_{12,\min}$ (the minimum total molar surfactant concentration to yield a 20 dyn/cm reduction in the surface tension of the solvent).

11.2 Surfactants C and D have CMC values of 1.38×10^{-4} and 4.27×10^{-4} mol/L, respectively, in aqueous 0.1 *M* NaCl. This mixture, in the same medium, has a CMC value of 3.63×10^{-4} mol/L when the mole fraction α of surfactant C in the mixture is 0.181 (on a surfactant-only basis).

(a) Calculate β^M for a mixture of surfactants A and B.

(b) Will this mixed system exhibit synergism or antagonism in mixed micelle formation? If so, calculate the values of α^* and $C^M_{12,\min(\max)}$.

11.3 Surfactants C and D of Problem 2 individually reduce the surface tension of an aqueous 0.1 *M* NaCl solution to 30 dyn/cm when their respective molar concentrations are 9.1×10^{-4} and 3.98×10^{-4}. The mixture of them at $\alpha = 0.181$ in Problem 11.2 has a surface tension value of 30 dyn/cm when the total molar surfactant concentration is 3.47×10^{-4}. Will a mixture of surfactants C and D exhibit synergism or antagonism in surface tension reduction effectiveness?

11.4 Without consulting tables, place the following mixtures in order of increasing attractive interaction (increasing negative β^{σ} value) at the aqueous solution–air interface:

(a) $C_{12}H_{25}(OC_2H_4)_6OH$-$C_{12}H_{25}SO_3^-Na^+$ (H_2O)

(b) $C_{12}H_{25}(OC_2H_4)_6OH$-$C_{12}H_{25}(OC_2H_4)_{15}OH$ (0.1 *M* NaCl, H_2O)

(c) $C_{12}H_{25}SO_3^-Na^+$-$C_{12}H_{25}N^+(CH_3)_2CH_2COO^-$ (H_2O)

(d) $C_{12}H_{25}N^+(CH_3)_3Cl^-$-$C_{12}H_{25}N^+(CH_3)_2CH_2COO^-$ (H_2O)

11.5 Explain the β-values obtained for the following mixtures:

(a) $C_7F_{15}COO^-Na^+$-$C_{12}H_{25}SO_4^-Na^+$ (0.1 *M* NaCl, 30^0), $\beta^{\sigma} = +2.0$

(b) $C_5H_{11}SO_3^-Na^+$-$C_{10}H_{21}Pyr^+Cl^-$ (0.03 *M* NaCl, 25^0), $\beta^{\sigma} = -10.8$;
$C_5H_{11}SO_3^-Na^+$-$C_{10}H_{21}Pyr^+Cl^-$ (0.03 *M* NaCl, 25^0), $\beta^{\sigma} = -5.5$

(c) $C_{12}H_{25}SO_3^-Na^+$-$C_{12}H_{25}N^+(B_z)(Me)CH_2COO^-$ (pH = 5.0, 20^0), $\beta^M = -5.4$;

$C_{12}H_{25}SO_3^-Na^+$-$C_{12}H_{25}N^+(B_z)(Me)CH_2COO^-$ (pH = 9.3, 25^0) $\beta^M = -1.7$

(d) $C_{10}H_{21}SO_3^-Na^+$-$C_{12}H_{25}N^+(Me)_3Br^-$ (H_2O, 25^0), $\beta^\sigma = -35.6$;
$C_{10}H_{21}SO_3^-Na^+$-$C_{12}H_{25}N^+(Me)_3Br^-$ (0.1 *M* NaBr, 25^0), $\beta^\sigma = -19.6$

(e) $C_{10}H_{21}S^+(CH_3)O^- + C_{10}H_{21}N^+SO_4^-Na^+$ (pH = 5.9), $\beta^\sigma = -4.3$;
$C_{10}H_{21}S^+(CH_3)O^- + C_{10}H_{21}\ N^+(Me)_3Cl^-$ (pH = 5.9), $\beta^\sigma = -0.6$;

(f) $C_{12}H_{25}N^+H_2CH_2CH_2COO^- + C_{12}H_{25}SO_3^-Na^+$ (pH = 5.8), $\beta^\sigma = -4.2$;
$C_{12}H_{25}N^+H_2CH_2CH_2COO^- + C_{12}H_{25}Pyr^+Br^-$ (pH = 5.8), $\beta^\sigma = -4.8$

12 Gemini Surfactants

Gemini surfactants, sometimes also called *dimeric surfactants*, contain two hydrophobic groups (sometimes three) and two hydrophilic groups in the molecule, connected by a linkage (spacer) close to the hydrophilic groups (Figure 12.1).* They therefore have three structural elements—a hydrophilic group, a hydrophobic group, and their linkage—that may be varied to change the properties of the surfactant. There has been a considerable interest in these compounds, both academic and industrial, since it was pointed out (Rosen, 1993a) that the interfacial properties of these surfactants in aqueous media can be orders of magnitude greater than those of comparable conventional surfactants (i.e., surfactants with single but similar hydrophilic and hydrophobic groups). Hundreds of scientific papers and patents have appeared in the literature in the past decade and have been reviewed (Rosen and Tracy, 1998; Menger and Keiper, 2000a; Zana, 2002). All charge types of geminis have been synthesized and investigated: anionics, including dicarboxylates, disulfates, disulfonates, and diphosphates (e.g., Zhu et al., 1990; Menger and Littau, 1991; Rosen et al., 1992; Duivenvoorde et al., 1997); cationics (e.g., Devinsky et al., 1985; Zana et al. 1991); nonionics (e.g., Eastoe et al., 1994; Paddon-Jones et al., 2001; Xie and Feng, 2010); zwitterionics (e.g., Seredyuk et al., 2001); and a variety of structural types: alkylglucoside based (Castro et al., 2002), arginine based (Pinazo et al., 1999), gemini-like peptides (Damen et al., 2010), glucamide based (Eastoe et al., 1996), taurine based (Li et al., 2010), hydrocarboxylic acid based (Altenbach et al., 2010), sugar based (Johnsson et al., 2003), with unsaturated linkages (Menger et al., 2000b; Tatsumi et al., 2001), hydrolyzable (Tatsumi et al., 2000; Tehrani-Bagha et al., 2007), polymerizable (Abe et al., 2006), discotic liquid crystalline (Kumar et al., 2010), and with nonidentical head groups (Alami et al., 2002).

* If the linkage is not close to the hydrophilic groups, the unique properties mentioned below are not observed.

Surfactants and Interfacial Phenomena, Fourth Edition. Milton J. Rosen and Joy T. Kunjappu.

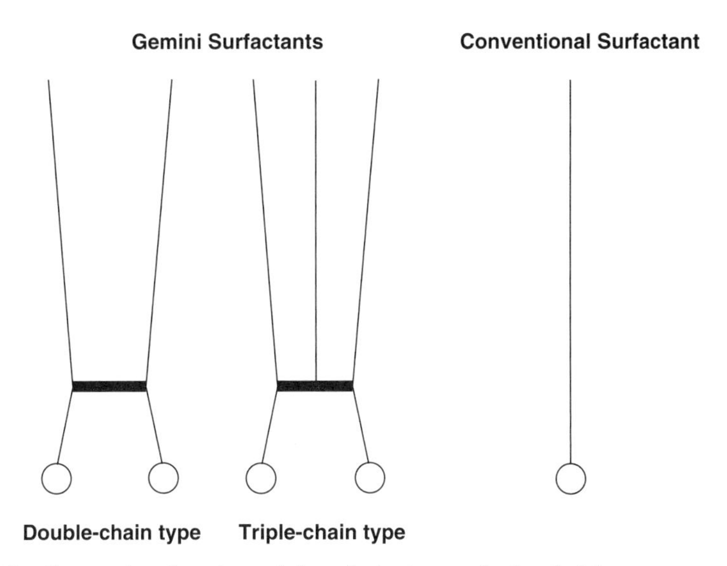

FIGURE 12.1 Conventional and gemini surfactants. —, hydrophobic group; ▬ , connecting group; ○, hydrophilic group.

I. FUNDAMENTAL PROPERTIES

Some examples of geminis, their C_{20} and critical micelle concentration (CMC) values, and those of comparable conventional surfactants, are shown in Table 12.1.

As can be seen from the data in Table 12.1, the C_{20} values, a measure of the efficiency of adsorption of the surfactant at the interface (Chapter 2, Section IIIE), can be two to three orders of magnitude smaller than the C_{20} values of comparable conventional surfactants, and their CMCs (Chapter 3, Section I) can be one to two orders of magnitude smaller than those of comparable conventional surfactants. The reason for this greater surface activity of geminis, compared to comparable conventional surfactants, is the larger total number of carbon atoms in the hydrophobic chains of the geminis. The larger the total number of carbon atoms in the surfactant molecule, the greater the distortion of the water structure of the aqueous phase and the greater the tendency to adsorb at the interfaces surrounding the aqueous phase or to form micelles in the aqueous phase, that is, greater surface activity (Chapter 1, Section II). This results in smaller C_{20} values (Chapter 2, Section IIIE) and smaller CMC values (Chapter 3, Section IV).

On the other hand, an increase in the total number of carbon atoms in the hydrophobic chain(s) of the surfactant molecule decreases the solubility of the surfactant in water and limits its surface activity. When the surfactant contains two hydrophilic groups, however, its solubility in water increases and the molecule can accommodate more carbon atoms in the hydrophobic groups without becoming water insoluble. Consequently, gemini surfactants are much more

TABLE 12.1 C_{20} and CMC Values of Gemini and Comparable Conventional Surfactants at 25°C

Compound	Medium	C_{20} (10^{-6}) M	CMC (10^{-6} M)	Reference
$[C_{10}H_{21}OCH_2CH(OCH_2COO^-Na^+)CH_2]_2O$	H_2O	4	84	Zhu et al. (1993)
$C_{11}H_{23}COO^-Na^+$	H_2O	5000	20,000	Zhu et al. (1993)
$[C_{10}H_{21}OCH_2CH(OCH_2CH_2CH_2SO_3^-Na^+]_2O$	H_2O	8	33	Zhu et al. (1991)
$C_{12}H_{25}SO_3^-Na^+$	H_2O	4400	9,800	Zhu et al. (1991)
$[C_{10}H_{21}OCH_2CH(SO_4^-Na^+)CH_2OCH_2]_2$	H_2O	1	13	Zhu et al. (1990)
$C_{12}H_{25}SO_4^-Na^+$	H_2O	3100	8,200	Zhu et al. (1990)
$[C_{12}H_{25}N^+(CH_3)_2CH_2]_2 \bullet 2Br^-$	H_2O	—	840	Zana et al. (1991)
$[C_{12}H_{25}N^+(CH_3)_2\ CH_2CH_2]_2 \bullet 2Br^-$	H_2O	—	1170	Pei et al. (2010)
$[C_{12}H_{25}\ N^+(CH_3)_2CH_2CHOH]_2 \bullet 2Br^-$	H_2O	129	700	Rosen and Liu (1996)
$C_{12}\ H_{25}\ N^+\ (CH_3)_3 \bullet Br^-$	H_2O	8000	16,000	Rosen and Liu (1996)
$[C_{12}H_{25}\ N^+(CH_3)_2CH_2]_2CHOH \bullet 2Cl^-$	0.1 M NaCl	0.9	9.6	Song and Rosen, 1996
$[C_{12}H_{25}N^+(CH_3)_2CH_2CHOH]_2 \bullet 2Br^-$	0.1 M NaCl	0.9	21	Rosen and Liu (1996)
$C_{12}H_{25}N^+\ (CH_3)_3 \bullet Cl^-$	0.1 M NaCl	1950	5,760	Li et al. (2001)
$(C_{11}H_{23}CONHCH[(CH_2)_3NHC(NH_2)_2^+]CONHCH_2)_2 \bullet 2C1^-$	H_2O	1.9	9.5	Perez et al. (1998)
$C_{11}H_{23}CONHCH[(CH_2)_3NHC(NH_2)_2^+]COOCH_3 \bullet Cl^-$	H_2O	630	6,000	Perez et al. (1998)
$\{C_{11}H_{23}CONHCH[(CH_2)_3NHC(NH_2)_2^+]CONHCH_2\}_2 \bullet 2C1^-$	0.01 M NaCl	1	9.2	Perez et al. (1998)
$C_{11}H_{23}CONHCH[(CH_2)_3NHC(NH_2)_2^+]COOCH_3 \bullet Cl^-$	0.01 M NaCl	50	270	Perez et al. (1998)

surface active than their comparable conventional surfactants, which have only half the number of carbon atoms in the molecule. In addition, geminis are more water soluble than comparable conventional surfactants and have much lower Krafft points (Chapter 5, Section IIA).

Trimeric and oligomeric surfactants have also been prepared (Zana et al., 1995; Sumida et al., 1998; In et al., 2000; Onitsuka et al., 2001). Their CMC values are even smaller than those of the analogous geminis. As the number of hydrophobic groups per molecule increases for gemini quaternary C_{12} ammonium compounds with polymethylene $(CH_2)_n$ spacers, their surface layers become more dense, their micellar microviscosity increases, and their micellar shape changes from spherical to wormlike, to branched wormlike, to ringlike. Zero-shear viscosity increases very rapidly to a maximum with an increase in the number of oligomers (In et al., 2000).

As in conventional surfactants, the C_{20} and CMC values decrease with an increase in the total number of carbon atoms in the molecule. However, unlike this behavior in conventional surfactants, when the number of carbon atoms in the alkyl chains of the geminis exceeds a certain value (about 14 carbon atoms per chain, the exact number dependent upon the gemini structure, temperature, and the electrolyte content of the aqueous phase), the linear relationship with the log of the C_{20} value (Chapter 2, Section IIIE) or with the log of the CMC (Chapter 3, Section IVA4) breaks down. The log C_{20} or log CMC values start to deviate more and more from this linear relationship, with the C_{20} and CMC values becoming larger than expected, that is, in the direction of decreased surface activity and micellization tendency. This deviation becomes larger and larger with an increase in the number of carbon atoms in the alkyl chains until the C_{20} and CMC values actually increase with this change. This behavior has been attributed (Menger and Littau, 1991) to the formation of small, non-surface-active aggregates (dimers, trimers, etc.) in the aqueous phase below the CMC, decreasing the concentration of the surface-active monomeric species and, consequently, the surface activity and micellization tendency. This has been confirmed by the calculation of the equilibrium constants for this self-association (Rosen and Liu, 1996; Song and Rosen, 1996) and by fluorescence spectroscopy (Mathias et al., 2001). This self-aggregation below the CMC to form small, non-surface-active aggregates when the alkyl chains of the geminis are long may be due to the exceptionally large free energy decrease resulting from hydrophobic bonding between adjacent gemini molecules, each containing two (or three) long hydrophobic chains.

The flexibility, length, and hydrophobicity (or hydrophilicity) of the linkage between the two hydrophilic groups of the gemini affect such properties as the C_{20}, CMC, and area/molecule at the air–aqueous solution interface values, sometimes considerably. For cationic geminis with the flexible, hydrophobic polymethylene, $(CH_2)_n$ linkage, the value of the CMC increases with an increase in the number of methylene groups to a maximum in the neighborhood of 6 (Devinsky et al., 1991; Zana et al., 1991) and the area/molecule at the air–aqueous solution interface to a maximum in the neighborhood of 10 methylene groups (Alami et al., 1993; Espert et al., 1998), and then they both decrease. This is believed to be due to the unfavorable orientation of the hydrophobic polymethylene linkage when it is in contact with the aqueous phase. Consequently, it is believed to penetrate into the micellar interior (in the case of the CMC) or to loop into the air (in the case of the area/molecule at the air–aqueous solution interface) when the linkage is sufficiently long. This decreases the CMC and the area/molecule at the interface, respectively. A rigid linkage, for example, $—CH_2C_6H_4CH_2—$, $—CH_2C{\equiv}C—CH_2—$, prevents this and consequently increases the CMC and the area/molecule at the interface. The smallest CMC, C_{20}, and area/molecule values are obtained when the linkage is short, flexible, and slightly hydrophilic. However, Zhu and coworkers (1993), Dreja and coworkers (1999), and Wettig and coworkers (2003), investigating the properties of geminis with hydrophilic spacers, have observed a monotonic increase in the area/molecule of the gemini at the air–aqueous solution interface with an increase in the length of the hydrophilic spacer, presumably due to the compatibility of the linkage with the aqueous phase.

A very useful property of geminis is their packing at various interfaces. At the air–aqueous solution interface, when the linkage between the two hydrophilic groups is small or hydrophilic and close to the hydrophilic groups, the hydrophobic groups of the gemini can be more closely packed (In et al., 2000) than those of the comparable conventional surfactant under the same conditions (temperature and ionic strength) in the aqueous phase. The packing of the chains is so close in some cases (Rosen et al., 1999; Onitsuka et al., 2001; Seredyuk et al., 2001; Tsubone et al., 2003a) that it appears that multilayers are formed. This closer packing makes the interfacial film of the geminis more coherent, and this is reflected in their superior foaming (Zhu et al., 1991, 1993; Ono et al., 1993) and their superior emulsifying properties (Briggs and Pitts, 1990; Dreja and Tieke, 1998). This close packing of the hydrophobic groups in geminis with short linkages close to their hydrophilic groups (indicated by small area/molecule values at the air–aqueous solution interface) results in a packing parameter (Chapter 3, Section II) indicative of cylindrical micelle formation in aqueous media and may account for the unusually high viscosities shown by some geminis (Zana and Talmon, 1993; Schmitt et al., 1995). The geminis $[C_{12}H_{25}N^+(CH_3)_2CH_2]_2{\cdot}2Br^-$ and $[C_{12}H_{25}N(CH_3)_2CH_2]_2CH_2{\cdot}2Br^-$ have been shown (Zana and Talmon, 1993; Danino et al., 1995) to have long, wormlike micelles in aqueous solution. The entanglement of these wormlike micelles produces the high viscosity of the aqueous phase.

Gemini surfactants also show unique adsorption at oppositely charged solid surfaces from aqueous media. Whereas conventional surfactants adsorb onto this type of solid from aqueous media with their hydrophilic groups oriented toward the oppositely charged solid surface (Figure 2.12) and their hydrophobic groups oriented toward the aqueous phase, making the solid, at least initially, more hydrophobic, geminis with short linkages adsorb onto these surfaces with one hydrophilic group oriented toward the solid and the other oriented toward the aqueous phase (Li and Rosen, 2000; Rosen and Sulthana, 2001), retaining the hydrophilic character of the solid. One of the effects of this is to make the solid more dispersible in the aqueous phase.

A standard free energy change upon micellization for ionic gemini surfactants with two hydrophilic groups and monovalent counterions can be calculated (Zana, 1996) by taking into account the degree of binding $(1 - \alpha)$ of the counterions to the micelle:

$$\begin{aligned} \Delta G^{\circ}_{\text{mic}} &= RT[1+2(1-\alpha)]\ln X_{\text{CMC}} \\ &= 2.3RT(3-2\alpha)\log X_{\text{CMC}}, \end{aligned} \tag{12.1}$$

where α is the degree of ionization of the gemini surfactant, measured by the ratio of the slopes of the plot of specific conductivity versus C above and below the CMC (Chapter 3, Section IVA3), and X_{CMC} is the mole fraction of the surfactant in the liquid phase at the CMC. The degree of binding of cationic geminis appears to be similar to those of comparable conventional surfactants (Zana et al., 1995) (Table 3.3).

II. INTERACTION WITH OTHER SURFACTANTS

Because of the double charge on ionic geminis, they interact more strongly with oppositely charged surfactants at interfaces or in mixed micelles than do singly charged (conventional) surfactants. The strength of the interaction between two different types of surfactants, both conventional and gemini, is measured by the so-called β-parameter (Chapter 11, Section I). The numerical value of the β-parameter (which is negative when greater attractive [or less repulsive] interaction occurs after mixing and is positive when there is greater repulsion [or less attraction] between the two different surfactants after mixing) depends upon the nature of the interface (liquid–air, liquid–liquid, or liquid–solid) at which they interact and also upon whether the interaction is between the different surfactants adsorbed in a mixed monolayer at an interface (β^{σ}) or between them in a mixed micelle in aqueous medium (β^{M}). The nature and strength of the interactions between the two different surfactants determine whether the mixtures of the two surfactants will exhibit synergism, antagonism (negative synergism), or ideal behavior in their interfacial properties (Chapter 11, Section III). Interaction parameters for mixtures of gemini surfactants or conventional surfactants with the same second surfactant indicate that the interactions are much stronger for the gemini surfactants than for their comparable conventional surfactants (Rosen, 1993a; Rosen et al., 1993b; Liu and Rosen, 1996; Li et al., 2001). Some data are shown in Table 12.2 together with interaction parameters for some comparable conventional surfactant mixtures. The data also indicate that interaction between the two different types of surfactants is much stronger in the mixed monolayer at the

TABLE 12.2 Interaction Parameters of Gemini Surfactants with Conventional Surfactants at 25°C

System	Medium	β^{σ}	β^{M}	Reference
Anionic–Cationic Mixtures				
$[C_8H_{17}N^+(CH_3)_2CH_2CHOH]_2 \bullet 2Br^-$—$C_{10}H_{21}SO_3^-Na^+$	0.1 *M* NaBr	−26	−12	Liu and Rosen (1996)
$[C_8H_{17}N^+(CH_3)_2CH_2CHOH]_2 \bullet 2Br^-$—$C_{12}H_{25}SO_3^-Na^+$	0.1 *M* NaBr	−30	−13	Liu and Rosen (1996)
$C_8Pyr^+Br^-$—$C_{12}H_{25}SO_3^-Na^+$	0.1 *M* NaBr	−19.5	—	Gu and Rosen (1989)
$[C_{10}H_{21}N^+(CH_3)_2CH_2CHOH]_2 \bullet 2Br^-$—$C_{10}H_{21}SO_3^-Na^+$	0.1 *M* NaBr	−34	−14	Liu and Rosen (1996)
$[C_{10}H_{21}N^+(CH_3)_2CH_2CHOH]_2 \bullet 2Br^-$—$C_{12}H_{25}SO_3^-Na^+$	0.1 *M* NaBr	−34	−18	Liu and Rosen (1996)
$[C_{10}H_{21}N^+(CH_3)_2CH_2CHOH]_2 \bullet 2Br^-$—$C_{12}H_{25}SO_3^-Na^+$	0.1 *M* NaCl	−40	−19	Liu and Rosen (1996)
$[C_8H_{17}N^+(CH_3)_2CH_2CHOH]_2 \bullet 2Br^-$—$C_{12}H_{25}(OC_2H_4)_4SO_4^-Na^+$	0.1 *M* NaBr	−28	—	Liu and Rosen (1996)
$[C_{10}H_{21}N^+(CH_3)_2CH_2CHOH]_2 \bullet 2Br^-$—$C_{12}H_{25}(OC_2H_4)_4SO_4^-Na^+$	0.1 *M* NaBr	−31	−11	Liu and Rosen (1996)
Anionic–Nonionic Mixtures				
$(C_{10}H_{21})_2C_6H_2(SO_3^-Na^+)OC_6H_4SO_3^-Na^+$—$C_{12}H_{25}(OC_2H_4)_7OH$	0.1 *M* NaCl	−6.9	−0.8	Rosen et al. (1993b)
$C_{10}H_{21}C_6H_3(SO_3^-Na^+)OC_6H_4SO_3^-Na^+$—$C_{12}H_{25}(OC_2H_4)_7OH$	0.1 *M* NaCl	−1.8	−0.9	Rosen et al. (1993b)
Anionic–Zwitterionic Mixtures				
$(C_{10}H_{21})_2C_6H_2(SO_3^-Na^+)OC_6H_4SO_3^-Na^+$—$C_{14}H_{29}N(CH_3)_2O$, pH = 6.0	0.1 *M* NaCl	−7.3	−2.4	Rosen et al. (1994)
$C_{10}H_{21}C_6H_3(SO_3^-Na^+)OC_6H_5$—$C_{14}H_{29}N(CH_3)_2O$, pH = 5.8	0.1 *M* NaCl	−4.7	−3.2	Rosen et al. (1993b)
Cationic–Nonionic Mixtures				
$[C_{10}H_{21}N^+(CH_3)_2CH_2CH_2]_2 \bullet 2Br^-$—decyl-β-glucoside (pH = 9)	0.1 *M* NaCl	−4.0	−1.9	Li et al. (2001)
$2C_{10}H_{21}N^+(CH_3)_3 \bullet 2Br^-$—decyl-β-glucoside (pH = 9)	0.1 *M* NaCl	−1.2	−1.2	Rosen and Li (2001)
$[C_{10}H_{21}N^+(CH_3)_2CH_2]_2CHOH \bullet 2Br^-$-decyl-β-glucoside (pH = 9)	0.1 *M* NaCl	−4.2	−1.2	Li et al. (2001)
$[C_{10}H_{21}N^+(CH_3)_2CH_2CH_2]_2 \bullet 2Br^-$-decyl-β-maltoside	0.1 *M* NaCl	−2.7	−1.9	Li et al. (2001)
$C_{10}H_{21}N^+(CH_3)_3 \bullet Br^-$-decyl-β-maltoside	0.1 *M* NaCl	−0.3	−0.3	Rosen and Li (2001)
$[C_{10}H_{21}N^+(CH_3)_2CH_2]_2CHOH \bullet 2Br^-$-decyl-β-glucoside	0.1 *M* NaCl	−4.2	−1.2	Li et al. (2001)
$[C_{10}H_{21}N^+(CH_3)_2CH_2]_2CHOH \bullet 2Br^-$-decyl-β-maltoside	0.1 *M* NaCl	−2.9	−1.4	Li et al. (2001)
$[C_{10}H_{21}N^+(CH_3)_2CH_2CHOH]_2 \bullet 2Br^-$-decyl-β-glucoside	0.1 *M* NaCl	−3.1	−1.4	Li et al. (2001)
$[C_{10}H_{21}N^+(CH_3)_2CH_2CHOH]_2 \bullet 2Br^-$-decyl-β-maltoside	0.1 *M* NaCl	−2.0	−1.7	Li et al. (2001)
$[C_{10}H_{21}N^+(CH_3)_2CH_2CH_2]_2O \bullet 2Br^-$-decyl-β-glucoside	0.1 *M* NaCl	−3.3	−1.5	Li et al. (2001)
$[C_{10}H_{21}N^+(CH_3)_2CH_2CH_2]_2O \bullet 2Br^-$-decyl-β-maltoside	0.1 *M* NaCl	−2.3	−1.7	Li et al. (2001)
$[C_{12}H_{25}N^+(CH_3)_2CH_2CH_2]_2 \bullet 2Br^-$-dodecyl-β-maltoside	0.1 *M* NaCl	−3.0	−2.2	Li et al. (2001)
$[C_{12}H_{25}N^+(CH_3)_2CH_2]_2 \bullet 2Br^-$-$C_{12}(OC_2H_4)_6OH$	H_2O	—	−2.2	Esumi et al. (1998)
$[C_{12}H_{25}N^+(CH_3)_3 \bullet Cl^-$ -$C_{12}(OC_2H_4)_5OH$	H_2O	—	−1.0	Rubingh and Jones (1982)

planar air–aqueous solution interface than in the convex mixed micelle. This is believed to be due to the greater difficulty of accommodating the two hydrophobic groups of the gemini in the interior of the convex micelle. This may also account for the observation mentioned above that, whereas the C_{20} values of gemini are often two to three orders of magnitude smaller than those of comparable conventional surfactants, their CMC values are only one to two orders of magnitude smaller.

An unexpected aspect of these interactions for ionic geminis, both in mixed monolayers and in mixed micelles, is that 1 mol of the doubly charged molecule interacts with only 1 mol of a second oppositely singly charged conventional surfactant, contrary to the expected 1:2 molar ratio. This is in contrast to what is observed with conventional surfactants, where they interact with oppositely charged surfactants in the expected 1:1 molar ratio and generally produce a water-insoluble product with no charge and little surface activity, unless there are additional hydrophilic groups in the molecules. The 1:1 interaction product of the doubly charged gemini and an oppositely singly charged conventional surfactant, on the other hand, has a net charge (of the same type as the gemini), is water soluble, and retains high surface activity. Gemini surfactants also interact more strongly than do their comparable conventional surfactants with water-soluble polymers. This is true for both neutral and oppositely charged polymers (Kastner and Zana, 1999; Pisarcik et al., 2000).

This much stronger interaction of geminis with other surfactants in mixed monolayers than in mixed micelles means that there is a strong possibility of synergism existing between them in effectiveness of surface or interfacial tension reduction since that is one of the requirements for synergism of this type (Chapter 11, Section IIIC). Synergism in surface or interfacial tension reduction effectiveness is very important in enhancing such performance properties as detergency, foaming, and wetting.

In a manner similar to the self-aggregation of geminis below their CMC to form small, non-surface-active aggregates when the alkyl chain exceeds a critical number of carbon atoms, described above, the interaction of ionic geminis with oppositely charged surfactants produces a marked decrease in surface activity when the combined alkyl carbon number of the two interacting surfactants (32 in the cases studied) exceeds a critical value (Liu and Rosen, 1996). Again, this is attributed to the formation of small, non-surface-active aggregates, in this case involving both oppositely charged surfactants. As with the shorter-chain geminis, the interaction product of the doubly charged gemini and the singly charged conventional surfactant is again 1:1 molar and water soluble.

Because of this interaction of only one of the ionic groups of the gemini with an oppositely charged conventional surfactant, tiny amounts of the gemini can be used to promote substantial growth of oppositely charged micelles by cross-linking them (Menger and Eliseev, 1995), with a consequent marked increase in the viscosity of the solution (Liu and Zhao, 2010).

III. PERFORMANCE PROPERTIES

The unique physicochemical properties described above result in gemini surfactants exhibiting some very desirable performance properties. Their very low C_{20} values make them remarkably efficient in reducing equilibrium surface tension, while their very low CMC values make them very efficient solubilizers (Chapter 4) of water-insoluble material since solubilization occurs only above the CMC. Thus, cationic geminis of the structure $[C_{12}H_{25}N(CH_3)_2]_2(CH_2)_n \bullet 2Br^-$ have been observed to solubilize more toluene, n-hexane, or styrene into water than their comparable conventional surfactants, particularly when n is small (Dam et al., 1996; Dreja and Tieke, 1998), and disodium didecyl diphenyl ether disulfonate has been found to solubilize water-insoluble nonionic surfactants more efficiently and more effectively than monosodium monodecyl diphenyl ether monosulfonate (Rosen et al., 1992). Their low CMC values also cause them to show very low skin irritation (which is associated with the concentration of monomeric surfactant in the aqueous phase) (Diz et al., 1994; Li et al., 1997; Okano et al., 1997; Kitsubi et al., 1998; Tracy et al., 1998). The double charge on the molecules of ionic geminis also make them better dispersing agents for finely divided solids in aqueous media (Chapter 9) than their comparable conventional surfactants.

When the alkyl chains of geminis are short and branched and the group linking the hydrophilic groups is short, geminis show excellent wetting properties. Tertiary acetylenic glycols of the structure $R_1R_2C(OH)C{\equiv}CC(OH)R_1R_2$, where R_1 is CH_3 and R_2 is an alkyl chain with two to four carbon atoms, and which are gemini-type surfactants that have been commercialized for decades, are excellent wetting agents. Gemini diamides of the structure $\{RN[(C_2H_4O)_xH]CO\}_2R^1$, where R is 2-ethylhexyl, $x = 4$, and R^1 is either —$(CH_2)_2$— or —CH=CH—, have been reported (Micich and Linfield, 1988) to be excellent hydrophobic soil wetting and rewetting agents.

As a result of the tighter packing of the hydrophobic groups of geminis when the linkage between the hydrophilic groups is small, as mentioned above, and the resulting more coherent surface film, the foaming of aqueous solutions of geminis of this type has been found in many cases to be superior to that of their comparable conventional surfactants (*monomers*). Both initial foam height and foam stability have been found to be significantly greater in several series of anionic geminis (Zhu et al., 1991, 1993; Ono et al., 1993; Kitsubi et al., 1998). Although conventional cationic surfactants of the alkyl trimethylammonium chloride structure show very little foam in aqueous solution, the analogous geminis, $[RN^+(CH_3)_2]_2(CH_2)_n \bullet 2Cl^-$, where $n = 2$ or 3 and R is C_2H_{25} or $C_{14}H_{29}$, show very high foam (Kim et al., 1996). When the length of the linkage between the hydrophilic groups was increased, both initial foam and foam stability decreased.

Studies of the removal of some pollutants from water by anionic and cationic geminis adsorbed on soil solids showed that they are both more efficient and more effective at removing the pollutants than their comparable conventional surfactants (Li and Rosen, 2000; Rosen and Sulthana, 2001). Quaternary ammonium surfactants often show strong antimicrobial activity, and an extensive study was made (Pavlikova et al., 1995) of the geminis of the structure $[RN^+(CH_3)_2CH_2OC(O)]_2(CH_2)_n$ to determine their activity. Maximum activity was shown when R is $C_{12}H_{25}$ and $n = 2$, and the antimicrobial activity of this gemini was far superior to the activity of single-chain commercial antimicrobials such as benzyldodecyldimethylammonium bromide. Another study (Diz et al., 1994) showed that diquaternary geminis of the structure $[C_{12}H_{25}N^+(CH_3)_2CH_2C(O)NHCH_2CH_2]_2{\bullet}2Cl^-$ or $[C_{12}H_{25}N^+(CH_3)_2CH_2C(O)NH(CH_2)S)_2{\bullet}2Cl^-$ showed greater antimicrobial activity against both gram-positive and gram-negative organisms and against *Candida albicans* than did hexadecyltrimethylammonium bromide. A gemini-type surfactant, *N*,*N*-di-*n*-hexadecyl-*N*,*N*-dihydroxyethylammonium bromide, was found (Banerjee et al., 1999) to be a more effective transfection (gene-delivering) agent than either the monohexadecyl or monohydroxyethyl analogue.

REFERENCES

Abe, M., K. Tsubone, T. Koike, K. Tsuchiya, T. Ohkubo, and H. Sakai (2006) *Langmuir* **22**, 8293.

Alami, E., G. Beinert, P. Marie, and R. Zana (1993) *Langmuir* **9**, 1465.

Alami, E., K. Holmberg, and J. Eastoe (2002) *J. Colloid Interface Sci.* **247**, 447.

Altenbach, H., R. Ihizane, B. Jakob, K. Lange, M. Schneider, Z. Yilmaz, and S. Nandi (2010) *J. Surfact. Deterg.* **13**, 399.

Banerjee, R., P. K. Das, G. V. Srilakshmi, A. Chaudhuri, and N. M. Rao (1999) *J. Med. Chem.* **42**, 92.

Briggs, C. B. and A. R. Pitts (1990) U.S. Patent 4,892,806 .

Castro, M. J. L., J. Kovensky, and A. F. Cirelli (2002) *Langmuir* **18**, 2477.

Dam, T., J. B. F. N. Engberts, J. Karthauser, S. Karabomi, and N. M. Van Os (1996) *Colloids Surf. A.* **118**, 41.

Damen, M., J. Aarbiou, S. F. M. van Dongen, R. M. Buijs-Offerman, P. P. Spijkers, M. van den Heuvel, K. Kvashnina, R. J. M. Nolte, B. J. Scholte, and M. C. Feiters (2010) *J. Control. Release* **145**, 33.

Danino, D., Y. Talmon, and R. Zana (1995) *Langmuir* **11**, 1448.

Devinsky, F., I. Marasova, and I. Lacko (1985) *J. Colloid Interface Sci.* **105**, 235.

Devinsky, F., J. Lacko, and T. Iman (1991) *J. Colloid Interface Sci.* **143**, 336.

Diz, M., A. Manresa, A. Pinazo, P. Erra, and M. R. Infante (1994) *J. Chem. Soc., Perkin Trans.* **2**, 1871.

Dreja, M. and B. Tieke (1998) *Langmuir* **14**, 800.

Dreja, M., W. Pyckhouf-Hintzen, H. Mays, and B. Tiecke (1999) *Langmuir* **15**, 391.

Duivenvoorde, F. L., M. C. Feiters, S. J. van der Gaast, and J. F. N. Engberts (1997) *Langmuir* **13**, 3737.

Eastoe, J., P. Rogueda, B. J. Harrison, A. M. Howe, and A. R. Pitt (1994) *Langmuir* **10**, 4429.

Eastoe, J., P. Rogueda, A. M. Howe, A. R. Pitt, and R. K. Heenan (1996) *Langmuir* **12**, 2701.

Espert, A., R. V. Klitzing, P. Poulin, A. Cohn, R. Zana, and D. Langevin (1998) *Langmuir* **14**, 1140.

Esumi, K., M. Miyazaki, T. Arai, and Y. Koide (1998) *Colloids Surfs. A.* **135**, 117.

Gu, B. and M. J. Rosen (1989) *J. Colloid Interface Sci.* **129**, 537.

In, M., V. Bec, O. Aguerre-Chariol, and R. Zana (2000) *Langmuir* **16**, 141.

Johnsson, M., A. Waganaer, M. C. A. Stuart, and J. F. N. Engberts (2003) *Langmuir* 19. **4609**.

Kastner, U. and R. Zana (1999) *J. Colloid Interface Sci.* **218**, 468.

Kim, T.-S., T. Kida, Y. Nakatsuji, T. Hirao, and I. Ikeda (1996) *J. Am. Oil Chem. Soc.* **73**, 907.

Kitsubi, T., M. Ono, K. Kita, Y. Fujikura, A. Nakano, M. Tosaka, K. Yahagi, S. Tamura, and K. Maruta (1998) U. S. Patent 5,714,457.

Kumar, S., S. K. Gupta, and S. Kumar (2010) *Tetrahedron Lett.* **51**, 5459.

Li, F. and M. J. Rosen (2000) *J. Colloid Interface Sci.* **224**, 265.

Li, F., M. J. Rosen, and S. B. Sulthana (2001) *Langmuir* **17**, 1037.

Li, J., M. Dahanayake, R. L. Reierson, and D. J. Tracy (1997) U.S. Patent 5,656,586.

Li, X., S. Zhao, Z. Hu, H. Zhu, and D. Cao (2010) *Tenside Surf. Detg.* **47**, 243.

Liu, J. and J. Zhao (2010) *J. Surfact. Deterg.* **13**, 83.

Liu, L. and M. J. Rosen (1996) *J. Colloid Interface Sci.* **179**, 454.

Mathias, J. H., M. J. Rosen, and L. Davenport (2001) *Langmuir* **17**, 6148.

Menger, F. M. and A. V. Eliseev (1995) *Langmuir* **11**, 1855.

Menger, F. M. and J. S. Keiper (2000a) *Angew. Chem. Int. Ed.* **39**, 1906.

Menger, F. M. and C. A. Littau (1991) *J. Am. Chem. Soc.* **113**, 1451.

Menger, F. M., J. S. Keiper, and V. Azov (2000b) *Langmuir* **16**, 2062.

Micich, T. J. and W. M. Linfield (1988) *J. Am. Oil Chem. Soc.* **65**, 820.

Okano, T., M. Fukuda, J. Tanabe, M. Ono, Y. Akabane, H. Takahashi, N. Egawa, T. Sakotani, H. Kanao, and Y. Yoneyanna (1997) U.S. Patent 5,681,803

Onitsuka, E., T. Yoshimura, Y. Koide, H. Shosenji, and K. Esumi (2001) *J. Oleo Sci.* **50**, 159.

Ono, D., T. Tanaka, A. Masuyama, Y. Nakatsuji, and M. Okahara (1993) *J. Jpn. Oil Chem. Soc. (Yukagaku)* **42**, 10.

Paddon-Jones, G., S. Regismond, K. Kwetkat, and R. Zana (2001) *J. Colloid Interface Sci.* **243**, 496.

Pavlikova, M., I. Lacko, F. Devinsky, and D. Mlynareik (1995) *Collect. Czech. Chem. Commun.* **60**, 1213.

Pei, X., J. Zhao, and R. Jiang (2010) *Colloid Polym. Sci.* **288**, 711.

Perez, L., A. Pinazo, M. J. Rosen, and M. R. Infante (1998) *Langmuir* **14**, 2307.

Pinazo, A., X. Win, L. Perez, M. R. Infante, and E. I. Frances (1999) *Langmuir* **15**, 3134.

Pisarcik, M., T. Imae, F. Devinsky, I. Lacko, and D. Bakos (2000) *J. Colloid Interface Sci.* **228**, 207.

Rosen, M. J. CHEMTECH, 1993a, March, pp. 30–33.

Rosen, M. J. and F. Li (2001) *J. Colloid Interface Sci.* **234**, 418.

Rosen, M. J. and L. Liu (1996) *J. Am. Oil Chem. Soc.* **73**, 885.

Rosen, M. J. and S. B. Sulthana (2001) *J. Colloid Interface Sci.* **239**, 528.

Rosen, M. J. and D. J. Tracy (1998) *J. Surfacts. Detgts.* **1**, 547.

Rosen, M. J., Z. H. Zhu, and X. Y. Hua (1992) *J. Amer. Oil Chem. Soc.* **69**, 30.

Rosen, M. J., Z. H. Zhu, and T. Gao (1993b) *J. Colloid Interface Sci.* **157**, 254.

Rosen, M. J., T. Gao, Y. Nakatsuji, and A. Masuyama (1994) *Colloids Surfaces A. Physicochem. Eng. Aspects* **88**, 1.

Rosen, M. J., J. H. Mathias, and L. Davenport (1999) *Langmuir* **15**, 7340.

Rubingh, D. N. and T. Jones (1982) *Ind. Eng. Chem. Prod. Res. Dev.* **21**, 176.

Schmitt, V., F. Schosseler, and F. Lequeux (1995) *Europhys. Lett.* **30**, 31.

Seredyuk, V., E. Alami, M. Nyden, K. Holmberg, A. V. Peresypkin, and F. M. Menger (2001) *Langmuir* **17**, 5160.

Song, L. D. and M. J. Rosen (1996) *Langmuir* **12**, 1149.

Sumida, Y., T. Oki, A. Masuyama, H. Maekawa, M. Nishiura, T. Kida, Y. Nakatsuji, I. Ikeda, and M. Nojima (1998) *Langmuir* **14**, 7450.

Tatsumi, T., W. Zhang, T. Kida, Y. Nakatsuji, D. Ono, T. Takeda, and I. Ikeda (2000) *J. Surfactants Deterg.* **3**, 167.

Tatsumi, T., W. Zhang, T. Kida, Y. Nakatsuji, D. Ono, T. Takeda, and I. Ikeda (2001) *J. Surfactants Deterg.* **4**, 279.

Tehrani-Bagha, A. R., H. Oskarsson, C. G. van Ginkel, and K. Holmberg (2007) *Colloid Interface Sci.* **312**, 444.

Tracy, D. J., R. Li, and J. M. Ricca (1998) U.S. Patent 5,710,121.

Tsubone, K., Y. Arakawa, and M. J. Rosen (2003a) *J Colloid Interface Sci.* **262**, 516.

Tsubone, K., T. Ogawa, and K. Mimura (2003b) *J. Surfactants Deterg.* **6**, 39.

Wettig, S. D., X. Li, and R. E. Verrall (2003) *Langmuir* **19**, 3666.

Xic, Z. and Y. Fcng (2010) *J. Surfact. Dcterg.* **13**, 51.

Zana, R. (1996) *Langmuir* **12**, 1208.

Zana, R. (2002) *Adv. Colloid Interface Sci.* **97**, 205.

Zana, R. and Y. Talmon (1993) *Nature* **362**, 228.

Zana, R., M. Benrraoa, and R. Rueff (1991) *Langmuir* **7**, 1072.

Zana, R., H. Levy, D. Papoutsi, and G. Beinert (1995) *Langmuir* **11**, 3694.

Zhu, Y.-P., A. Masuyama, and M. Okahara (1990) *J. Am. Oil Chem. Soc.* **67**, 459.

Zhu, Y.-P., A. Masuyama, T. Nagata, and M. Okahara (1991) *J. Jpn. Oil Chem. Soc. (Yukagaku)* **40**, 473.

Zhu, Y.-P., A. Masuyama, Y. Kobata, Y. Nakatsuji, M. Okahara, and M. J. Rosen (1993) *J. Colloid Interface Sci.* **158**, 40.

PROBLEMS

12.1 Explain the following observations regarding the properties of geminis, compared to those of conventional surfactants having similar, but single, hydrophilic and hydrophobic groups.

- **(a)** Their C_{20} values are generally two or three orders of magnitude smaller than those of the latter.
- **(b)** Their CMC values are generally one or two orders of magnitude smaller than those of the latter, not the two or three orders of magnitude observed in (a) (above) for the C_{20} values.
- **(c)** They wet substrates less rapidly than the latter.
- **(d)** They are better foaming and emulsifying agents than the latter.
- **(e)** They are better dispersing agents for oppositely charged, finely divided solids than the latter.

12.2 Suggest reasons for the fact that the unique properties of geminis decrease as the linkage (spacer) between the hydrophilic groups increases in length.

12.3 Explain why a mixture of two gemini surfactants shows more negative interaction β parameter for micellization than a mixture of regular surfactants of the same type.

13 Surfactants in Biology

Many biological systems are related either directly or indirectly to surfactant molecules. For example,

- Cell membranes are made up of lipid surfactant bilayered vesicular structures.
- The pulmonary mechanical motion in the lungs is dependent on certain biosurfactants, a deficiency of which causes the respiratory distress syndrome found in prematurely born babies.
- Biochemists have used surfactants for cell lysis.
- Surfactants can cause denaturation of proteins.
- Electrophoresis techniques that biologists commonly use are based upon surfactants.
- Some surfactants have biocidal action.
- Biotechnological processes make use of surfactants at several stages.
- Pharmaceutical formulation can involve surfactants.

I. BIOSURFACTANTS AND THEIR APPLICATION AREAS

Surfactants produced by microorganisms and other living systems are referred to as biosurfactants. They can produce low surface tensions and are characterized by low critical micelle concentration (CMC) values. They are effective in forming both water-in-oil (*W*/*O*) and oil-in-water (*O*/*W*) microemulsions. The hydrophilic moiety in biosurfactants is composed of anionic or cationic derivatives of amino acids or peptides, and nonionic derivatives of di- or polysaccharides; the hydrophobic portion is made up of saturated, unsaturated, or hydroxylated fatty acid chains, or hydrophobic peptides. Microorganisms are known to produce surfactants both from hydrocarbons and from water-soluble compounds (glucose, sucrose, ethanol, and glycerol). In spite of the somewhat higher production costs of biosurfactants, their biodegradability and renewable resource base make them attractive in a number of industrial areas such

Surfactants and Interfacial Phenomena, Fourth Edition. Milton J. Rosen and Joy T. Kunjappu.

as agriculture, building and construction, the food and beverage industries, industrial cleaning, the leather industry, emulsion polymerization, paper and metal industries, textile manufacture, cosmetic formulation, the pharmaceuticals, petroleum and petrochemical industries, and enhanced oil recovery (EOR).

Biosurfactants are also more acceptable than synthetic surfactants with respect to biodegradability and toxicity. The application of biosurfactants in bioremediation (the conservation of the natural environment by eliminating contaminants) and similar environmental applications stems from their low toxicity profile toward freshwater, marine, and terrestrial ecosystems (Finnerty, 1994). Biosurfactants have been employed to clean up oil tanks and for oil production (Van Dyke et al., 1991; Bordoloi and Konwar, 2008). The benign nature of biosurfactants makes them the choice in many cosmetic and pharmaceutical and agricultural formulations, and as antimicrobial and immunomodulatory agents (Singh et al., 2007). In various sectors of the food industry, they are used because of their efficacy as emulsifiers and emulsion stabilizers (Bognolo, 1999). Biosurfactants can also be effective general antimicrobials (Nitschkea and Costa, 2007). They have been investigated for their ability in gene transfection (process of introducing nucleic acids into cells by nonviral methods—a fundamental process in molecular and cell biology) and as convenient releasers of DNA in gene therapy applications (Ueno et al., 2007).

True biosurfactants, as opposed to their synthetic analogues, are produced *in vivo*. The main biochemical synthetic mechanism involves the interpolation of an oxygen or polar atom into the C–H bond of hydrocarbon-like molecules. For example, with the alkane hydroxylase AlkB of *Pseudomonas putida* GPo1, a membrane-spanning diiron oxygenase, the oxygenation step proceeds with the solubilization of hydrocarbons in the cell membrane's amphiphilic environment assisted by the coordination center of ionic iron in the hemoproteinous enzyme *cytochrome P450* (Austin et al., 2008). The relevant interfacial phenomenon is the reduction of surface tension at the phase boundary that permits the microorganisms to grow on water-immiscible substrates and makes the substrate more readily available for uptake (Fiechter, 1992). The microbes feed on various chemicals producing the biosurfactant molecules as by-products. The biosynthetic sequences may be reenacted also *in vitro*, where the nature of the broth medium containing nutrients influences the structure of the surfactant. For example, by adjusting the amino acid concentration in the media, we can obtain val-7 or leu-7 surfactin (see p. 475 for its variations), thus controlling the biosurfactant structure. (The most well-known member of surfactin has a ring in its structure consisting of iso-C_{15}-hydroxy [or amino] carboxylic acid and seven amino acids: L-Glu$^{(1)}$-L-Leu$^{(2)}$-D-Leu$^{(3)}$-L-Val$^{(4)}$-L-Asp$^{(5)}$-D-Leu$^{(6)}$-L-Leu$^{(7)}$. In val-7, the seventh amino acid in the peptide ring is valine.) The chemical structure of the biosurfactant produced by a species of microorganism is regulated by the nature of the carbon source, the nutritional limitations, aerobic or anaerobic environment, temperature, ionic strength, and pH.

TABLE 13.1 Various Classes of Microbial Biosurfactants and the Corresponding Producer Microorganisms

Surfactant Class	Producer Microorganism	Reference[a]
Glycolipids		Banat (1995), Desai and Banat (1997), Rosenberg and Ron (1999), Deleu and Paquot (2004), Kulakovskaya et al. (2005), Nitschkea and Costa (2007)
Rhamnolipids	*Pseudomonas aeruginosa*	
Trehalose lipids	*Rhodococcus erythropolis*	
	Arthobacter sp.	
Sophorolipids	*Candida bombicola*, *Candida apicola*	
Mannosylerythritol lipids	*Candida antartica*	
Ustilagic acid	*Pseudozyma fusiformata*	
Lipopeptides		
Surfactin/iturin/fengycin	*Bacillus subtilis*	
Viscosin	*Pseudomonas fluorescens*	
Lichenysin	*Bacillus licheniformis*	
Serrawettin	*Serratia marcescens*	
Phospholipids	*Acinetobacter* sp.	
	Corynebacterium lepus	
Fatty acids/neutral lipids		
Corynomicolic acids	*Corynebacterium insidibasseosum*	
Polymeric surfactants		
Emulsan	*Acinetobacter calcoaceticus*	
Alasan	*Acinetobacter radioresistens*	
Liposan	*Candida lipolytica*	
Lipomanan	*Candida tropicalis*	
Particulate biosurfactants		
Vesicles	*A. calcoaceticus*	

[a] Common references for all.

Table 13.1 shows the major biosurfactant classes produced by microorganisms. The chemical structures of some of these classes are shown in Table 13.2. These structures contain a number of structural moieties derived from fatty acids, carbohydrates, glycerides, peptides, and phospholipids.

Table 13.3 lists the names, abbreviations, and structures of the 20 essential amino acids used in representing the biosurfactant structures.

Table 13.4 lists some common industrial application areas of biosurfactants. Each application may use many different microorganisms or a mixture of biosurfactants derived from them that belong to multiple chemical classes.

Table 13.5 shows the representative surface tension values of some biosurfactants at their CMCs at ambient temperatures in water (Table 13.2 lists chemical structures).

TABLE 13.2 Structures of Some Biosurfactants

Rhamnolipids (Zhang and Miller, 1992)

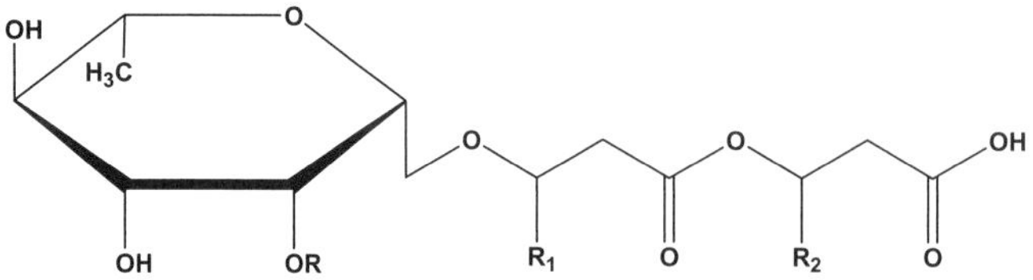

In a monorhamnolipid, R=H or R=CO-CH=CH-$(CH_2)_6$-CH_3; in a dirhamnolipid, R = rhamnosyl or R = rhamnosyl-O-CO-CH=CH$(CH_2)_6$-CH_3; R_1 and R_2 = C_7H_{15}

Trehalose lipid (Ortiz et al., 2009)

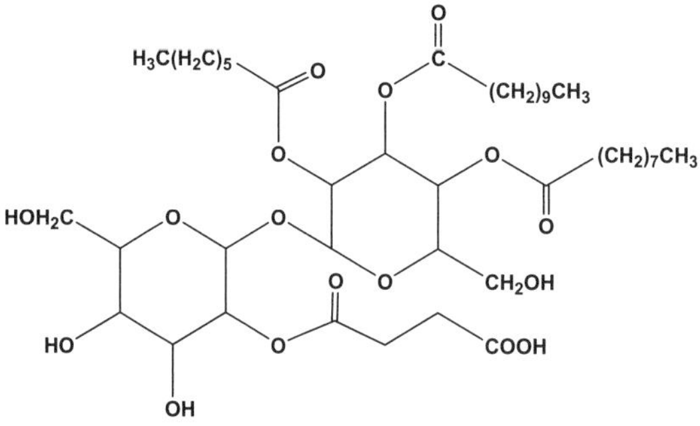

Sophorolipids (Fu et al., 2008)

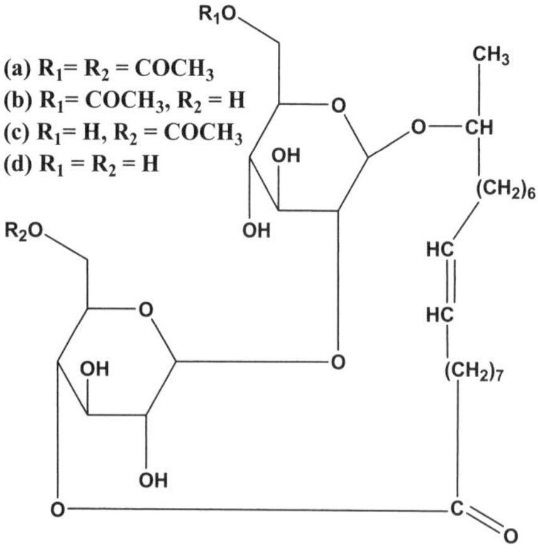

(a) $R_1 = R_2 = COCH_3$
(b) $R_1 = COCH_3$, $R_2 = H$
(c) $R_1 = H$, $R_2 = COCH_3$
(d) $R_1 = R_2 = H$

The main component is the lactone, in which R is as in (a). This also contains a component corresponding to the hydrolyzed structure of the lactone.

TABLE 13.2 (*Continued*)

Mannosylerythritol lipids (Jae et al., 2001)

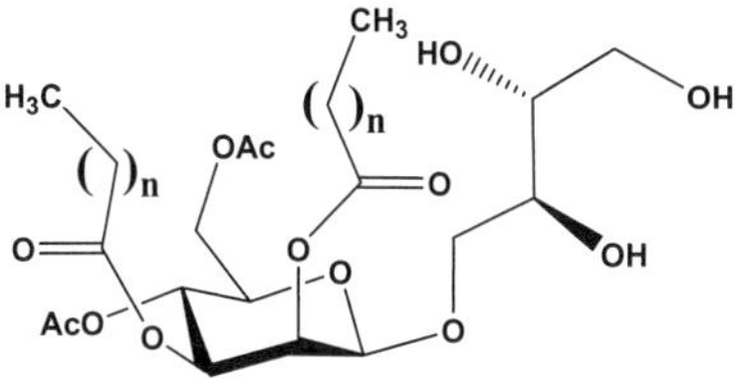

***n* = 6 - 10**

Surfactins (Deleu et al., 1999)

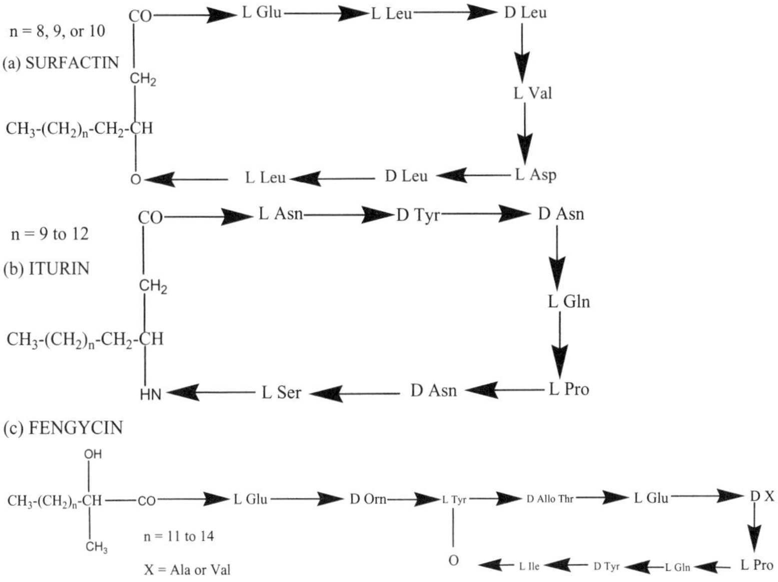

n = 8, 9, or 10

(a) SURFACTIN

CH_3-$(CH_2)_n$-CH_2-CH — CH_2 — CO → L Glu → L Leu → D Leu → L Val → L Asp → D Leu → L Leu → O

n = 9 to 12

(b) ITURIN

CH_3-$(CH_2)_n$-CH_2-CH — CH_2 — CO → L Asn → D Tyr → D Asn → L Gln → L Pro → D Asn → L Ser → HN

(c) FENGYCIN

CH_3-$(CH_2)_n$-CH(OH)(CH_3) — CO → L Glu → D Orn → L Tyr → D Allo Thr → L Glu → D X → L Pro → L Gln → D Tyr → L Ile → O

n = 11 to 14

X = Ala or Val

Viscosin (Saini et al., 2008)

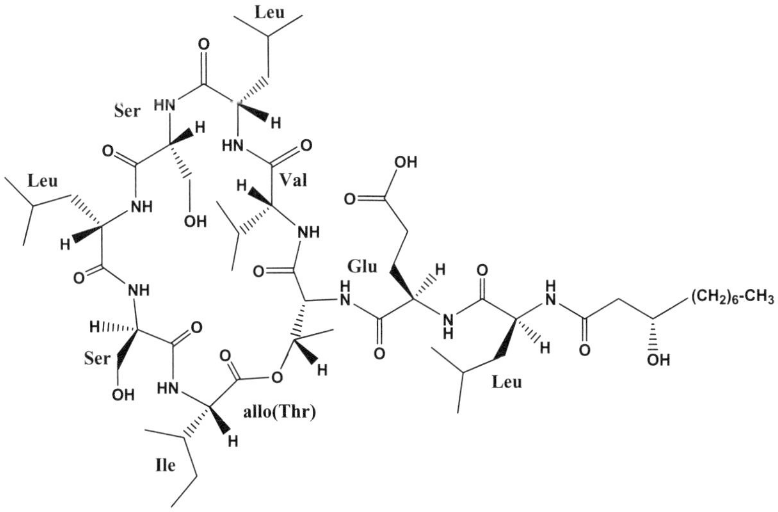

(*Continued*)

TABLE 13.2 (*Continued*)

Lichenysin (Michail et al., 1999)

Serrawettin (Matthew and Weibel, 2009)

Phospholipids: cardiolipin (Torregrossa et al., 1977)

R^1, R^2, R^3, and R^4 are miscellaneous alkyl groups in the fatty acid residues; for example, in mammalian cardiolipin, the alkyl groups are linoleoyl.

TABLE 13.2 (*Continued*)

Corynomicolic acids (Cooper et al., 1979; Silva et al., 1980)

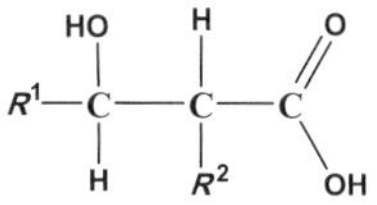

$R^1 = C_{40}H_{73}, C_{42}H_{77}, C_{44}H_{81}, C_{46}H_{85}, C_{48}H_{29}, C_{50}H_{93}$

$R^2 = C_{14}H_{29}, C_{16}H_{33}$

Emulsan (Kim et al., 1997)

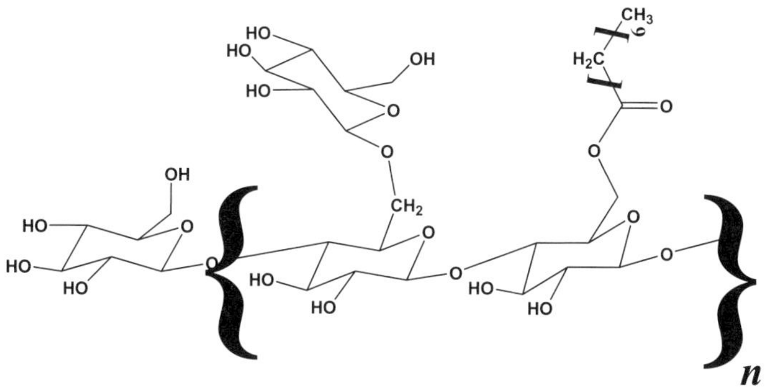

Molecular weight = 12,000–13,000 (n = 1000–1500)

Ustilagic acid (Kulakovskaya et al., 2005)

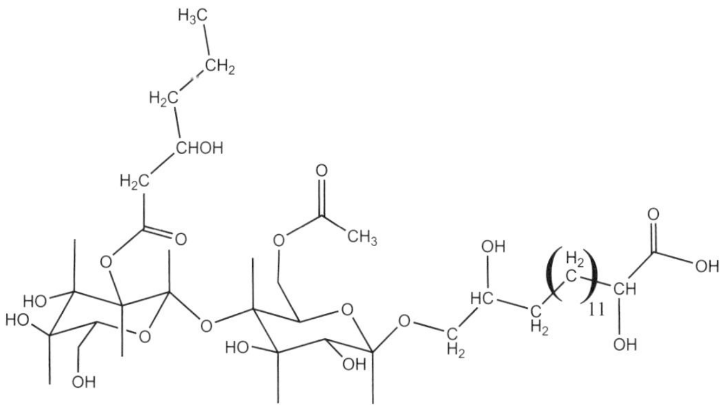

(*Continued*)

TABLE 13.2 (*Continued*)

Alasan (Walzer et al., 2006)

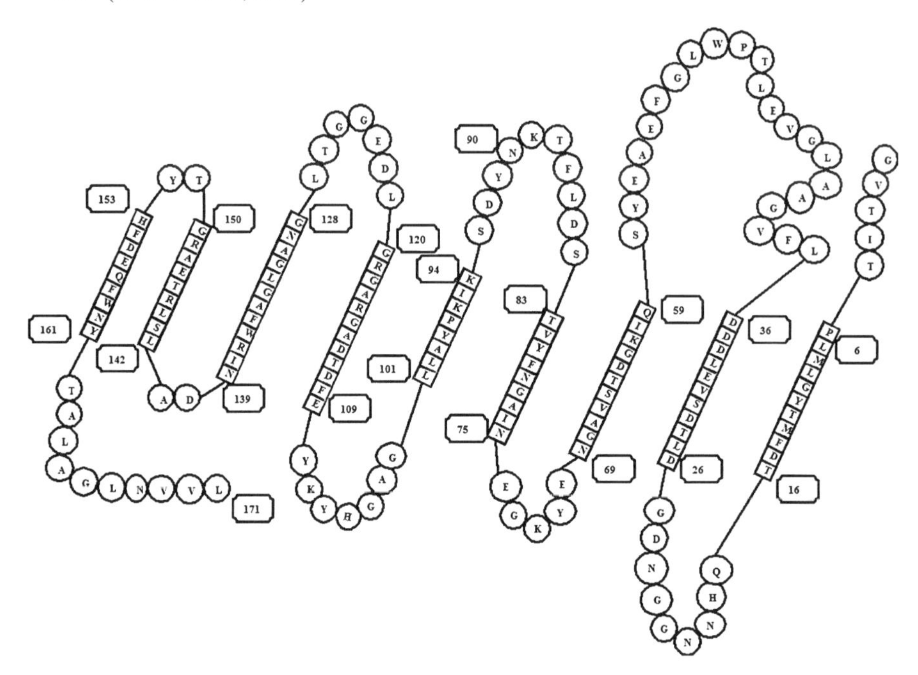

Alasan is a high-mass complex of proteins and polysaccharides. The major emulsification activity of this complex is associated with a 45-kDa protein (AlnA) (see Table 13.3 for abbreviations).

Liposan (Cirigliano and Carman, 1985)
Liposan is composed of approximately 83% carbohydrate and 17% protein. The carbohydrate portion is a complex heteropolysaccharide consisting of glucose, galactose, galactosamine, and galacturonic acid.

Lipomanan (Kordula'kova' et al., 2003)

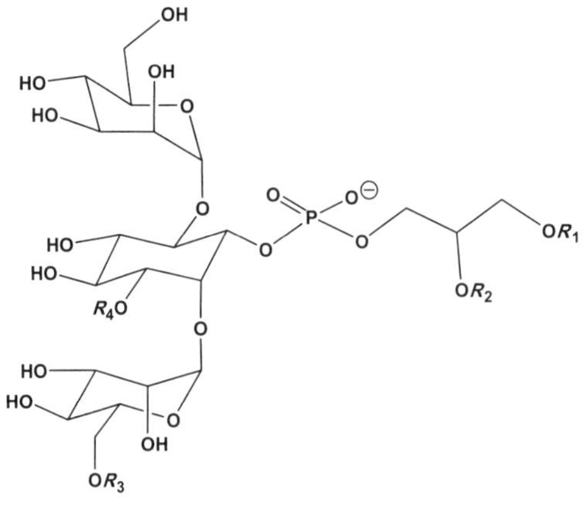

A precursor in the lipomannan biosynthetic pathway
$R_1 = C_{19}$ and $R_2 = C_{16}$, $R_3 = C_{16}/C_{19}$, and $R_4 = C_{16}/C_{18}/C_{19}$ fatty acyl groups.

TABLE 13.3 Amino Acids, Their Three- and One-Letter Abbreviations, and Structures

Name of Amino Acid	Abbreviation	Structure
Alanine	Ala, A	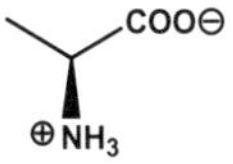
Asparagine	Asn, N	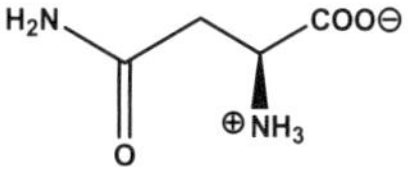
Aspartic acid	Asp, D	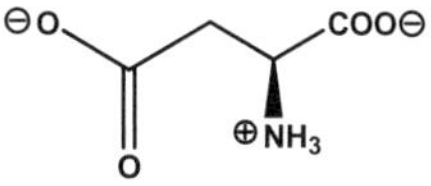
Arginine	Arg, R	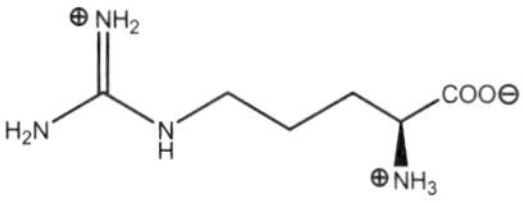
Cysteine	Cys, C	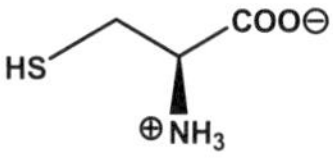
Glutamine	Gln, Q	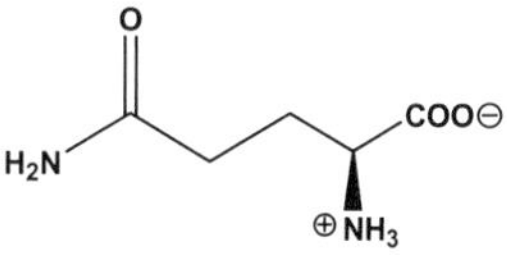
Glycine	Gly, G	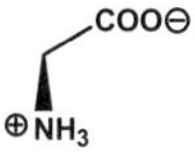
Glutamic acid	Glu, E	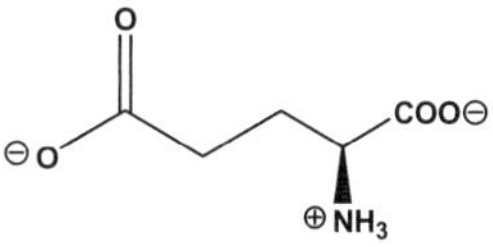
Histidine	His, H	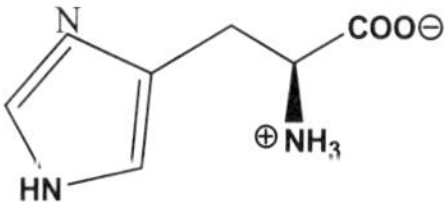
Isoleucine	Ile, I	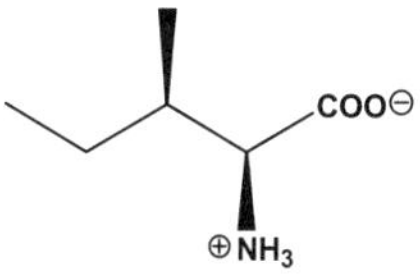
Lysine	Lys, K	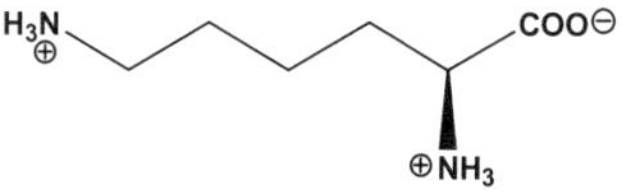
Leucine	Leu, L	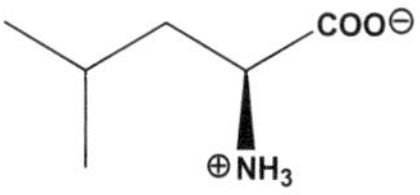

(*Continued*)

TABLE 13.3 (*Continued*)

Name of Amino Acid	Abbreviation	Structure
Methionine	Met, M	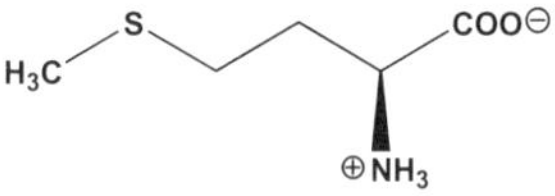
Phenylalanine	Phe, F	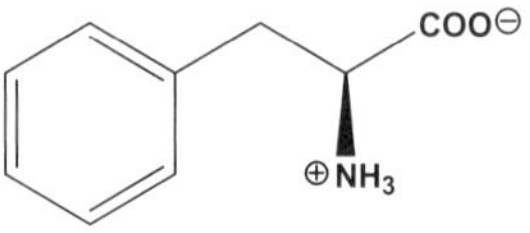
Proline	Pro, P	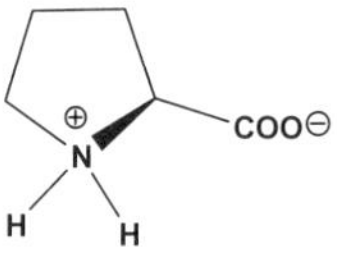
Serine	Ser, S	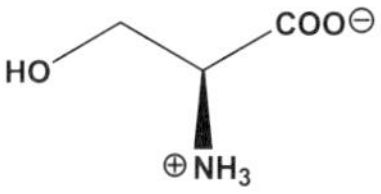
Tryptophan	Trp, W	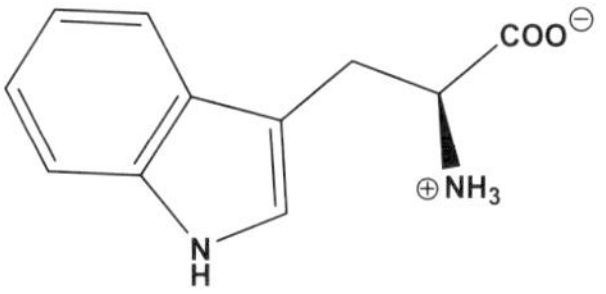
Threonine	Thr, T	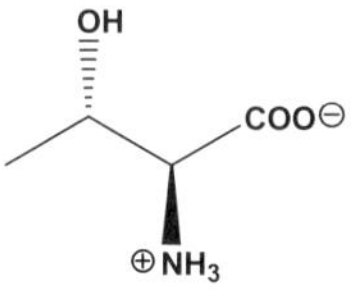
Tyrosine	Tyr, Y	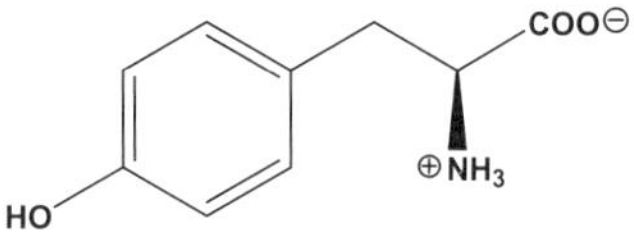
Valine	Val, V	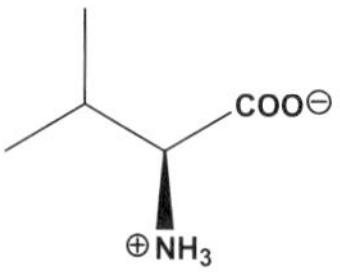

II. CELL MEMBRANES

The boundary of a cell is defined by an envelope called the cell membrane, which helps compartmentalize the contents of the cell. Within the cell, subcellular organelles such as the nucleus, mitochondria, and chloroplasts are endowed with their own membranes.

TABLE 13.4 Some Common Industrial Application Areas of Biosurfactants (Biotechnological Applications Are Elaborated in Table 13.6) (Singh et al., 2007)

Industry	Application/Surfactant	Role of Surfactants
Petroleum	Enhanced oil recovery (*trehalolipids*)	Improve oil drainage into well bore; stimulate release of oil entrapped by capillaries; control wetting of solid surfaces; reduce viscosity of oil pour point; lower interfacial tension; enhance oil dissolution
	De-emulsification (*polysaccharides*, *glycolipids*, *glycoproteins*, *phospholipids*, and *rhamnolipids*)	De-emulsification of oil emulsions, oil solubilization, viscosity reduction, wetting agent
Environmental	Bioremediation (*rhamnolipids*, *sophorolipids*, and *surfactin*)	Emulsification of hydrocarbons, lowering of interfacial tension, metal sequestration
	Soil remediation and flushing (*rhamnolipids*, *surfactin*)	Emulsification through adherence to hydrocarbons, dispersion, foaming agent, detergent, soil flushing
Food	Emulsification and de-emulsification and as functional ingredients (*lecithin and its derivatives*)	Emulsifier; solubilizer; demulsifier; suspension, wetting, foaming, defoaming, thickener, lubricating agent Interaction with lipids, proteins, and carbohydrates
Biological	Microbiological (e.g., *rhamnolipids*)	Cell mobility, cell–cell communication, nutrient accession, cell–cell competition, plant and animal pathogenesis, regulation of gene expression
	Pharmaceuticals and therapeutics (*rhamnolipids*, *lipopeptides*, *mannosylerythritol lipids*; *surfactin*)	Antibacterial, antifungal, antiviral agents; adhesive agents; immunomodulatory molecules; vaccines; gene therapy

(*Continued*)

TABLE 13.4 (*Continued*)

Industry	Application/Surfactant	Role of Surfactants
Agricultural	Biocontrol (*combination of microbes like fungus* Myrothecium verrucaria *with regular surfactants like Silwet L-77*)	Facilitation of biocontrol mechanisms of microbes such as "symbiotic" parasitism, "detrimental" antibiosis, competition, induced systemic acquired resistance (resistance response to an earlier exposure to a pathogen) and hypovirulence (reduced virulence of a pathogen)
Bioprocessing	Downstream processing (*performance of cells to secrete surface-active agents enhanced by conventional nonionic and ionic surfactants*)	Biocatalysis in aqueous two-phase systems and microemulsions, biotransformations, recovery of intracellular products, enhanced production of extracellular enzymes and fermentation products
Cosmetic health and beauty products	Health and beauty products (*sophorolipids*)	Emulsifiers, foaming agents, solubilizers, wetting agents, cleansers, antimicrobial agent, mediators of enzyme action
Paint industry	*Biodispersan*	Dispersing properties
Ceramic industry	*Heteropolysaccharides*	Dispersants
Biosensors (a device that detects an analyte by coupling biocomponent and physicochemical detection)	*Rhamnolipids, biodispersan*	Degradation of anionic surfactants, surfactants detection, rapid evaluation of surfactants in water Media
Mining	*Biodispersan*	Prevention of flocculation and better dispersion Stabilization of coal slurries

Membranes are made up of amphiphilic lipid molecules, which form bilayered structures. Glycerophospholipids and sphingolipids are two important bilayer-forming classes of lipids (Table 13.6). They are double-tailed surfactants with the potential to form bilayers, as can be seen from the $V_H/l_c a_0$ rule, explained in Chapter 3, Section IIA, where the principle called the packing parameter is elaborated.

Lipids are hydrophobic, sparingly water-soluble materials of biological origin that are soluble in organic solvents such as chloroform and methanol. Bilayer thickness is typically ~60 Å, as measured by electron microscopy and

TABLE 13.5 Surface Tension Values of Some Biosurfactants at Their CMCs

Name of Biosurfactant	Concentration (CMC)	Surface Tension (mN/m)	Reference
Trehalose dicorynornycolates	10^2 mg/L	43	Bognolo (1999)
Trehalose monocorynomycolates	1–10^2 mg/L	32	Bognolo (1999)
Rhamnolipid 1	1–10 mg/L	31	Bognolo (1999)
Rhamnolipid 3	10 mg/L	31	Bognolo (1999)
Sophorolipid	82.0 mg/L	37	Finnerty (1994)
Surfactin	11.0 mg/L	27	Finnerty (1994)
Trehalose-2,3,4,2′-tetraester	10.0 mg/L	26	Finnerty (1994)
Trehalose-di-, tetra-, hexa-, octaester	1500.0 mg/L	30	Finnerty (1994)
Glucose-6-mycolate	20.0 mg/L	40	Finnerty (1994)
Cellobiose-6-mycolate	4.0 mg/L	35	Finnerty (1994)
Maltotriose-6,6′,6″-trimycolate	10.0 mg/L	44	Finnerty (1994)
Ustilagic acid	20.0 mg/L	30	Finnerty (1994)

X-ray diffraction. The head group and the hydrocarbon chain of the membrane lipids each have an area of ~15 $Å^2$ (Kunjappu and Somasundaran, 1996).

A generalized schematic representation of a cell membrane along with the integral and transmembrane proteins is shown in Figure 13.1.

Based on the chemical class, the components of cell membranes may be categorized as follows (see Chart 13.1):

(a) Long-chain carboxylic acids containing 14–20 carbon atoms, which can be saturated, for example, palmitic acid–C16, stearic acid–C18, or unsaturated, for example, oleic acid–C18 (cis-9, double bond), linoleic acid–C18 (cis-9, and cis-12 double bonds)

(b) Esters of glycerol, e.g., 1-palmitoyl-2-linoleoyl-3-stearoylglycerol

(c) Glycerophospholipids, two OH groups in glycerol esterified with long-chain carboxylic acids and the third esterified with phosphoric acid or its derivatives; for example, phosphatidylcholine (lecithin) is a glycerophospholipid. Cardiolipins found in heart muscle are diphosphatidylglycerol.

(d) Sphingolipids, C18 aminoalcohol derivatives—e.g., sphingosine. Their N-acyl fatty acid derivatives are known as ceramides. Cerebroside, the simplest sphingoglycolipid, is a sugar-containing compound.

(e) Cholesterol, a component of animal plasma membranes and a derivative of cyclopentanoperhydrophenanthrene

TABLE 13.6 Biotechnological Processes with Biosurfactants and/or Added Synthetic Surfactants

Application	Surfactant	Physicochemical Processes	References
Soil bioremediation	Sophorose lipids (*Candida bombicola*), emulsan (*Acinetobacter*), rhamnolipids (*Pseudomonas aeruginosa*), surface-active extracts (*Mycobacterium avescens Ex 91*)	Substrate–surfactant interactions, emulsification by reducing surface and interfacial tension, dispersion, partial solubilization and partitioning of hydrocarbons between phases, biodegradation.	Banat et al. (2000)
Conversion of lignocelluloses to ethanol	Biosurfactants from *Trichoderma reesei Cel7A* (CBHI), and enhanced by added ionic and nonionic surfactants	Reduction in unfavorable adsorption of cellulose onto lignin and enhancement of cellulose bioconversion into sugars	Eriksson et al. (2002)
Microbial enhanced oil recovery (MEOR)	Various bacteria	Reduction in interfacial tension is achieved by microbial surfactants; biosurfactant degrades faster than synthetic analogues.	Rapp et al. (1979)
	Lipopeptides (*Bacillus licheniformis JF-2*, *Bacillus subtilis*) (high-temperature processes)		Rapp et al. (1979), Lin et al. (1994), Yakimov et al. (1997), Makkar and Cameotra (1998)
	Trehalolipids (*Nocardia rhodochrus*), uncharacterized surfactant mixture (*Desulfovibrio*), neutral lipids (*Clostridium*), heteropolysaccarides and lipopeptides (*Arthrobacter*)		Tanner et al. 1991
	Rhamnolipid (*Pseudomonas aeruginosa*), lichenysin (*B. licheniformis*), and *Desulfovibrio Desulfuricans* (see above)		Kosaric et al. (1987); Kosaric (1992)
	Emulsan (*Acinetobacter calcoaceticus*), alasan (*Acinetobacter radioresistens*), fatty acids and phospholipids (*Alteromonas* sp.), surfactin (*Bacillus subtilis*), corynomycolic acids (*Corynebacterium*), glycolipids (*Petrophilum*, *Torulopsis bombicola*), ceramides (*Sphingobacterium*)		Hayes et al. (1986)
	Emulsan (*A. calcoaceticus RAG-1*)	De-emulsification of petroleum emulsions formed during oil recovery; Emulsifying ability to remove oil from the bottom of the tank; easing transportation of crude through pipelines (as large as 2000 times reduction in viscosity of the oil)	Banat et al. (1991)

Enhanced metal extraction	Rusticyanin, a copper protein (*T. ferrooxidans* and *Sulfolobus acidocaldarius*)	Biosolubilization and biohydrometallurgy (metal recovery in aqueous solution assisted by microorganisms)	Somasundaran et al. (2001), Petrisor et al. (2007)
Paint industry	Biodispersan (*A. calcoaceticus A2*)	Stabilization of dispersions, probable viscosity modifier	Rosenberg and Ron (1998)
Coal processing	Lipopeptides (*Bacillus licheniformis*)	Solubilization of coal component of lignite	Polman et al. (1994)
Ceramic industry	Anionic polysaccharide with carboxylates (*Macrocystis pyrifera* and *Azotobacter vinelandii*)	Effective bonding between particles, thus controlling the size, shape, and texture of the material	Pellerin et al. (1992)
Biosensor	Lipopeptides (enzyme) (*Pseudomonas* and *Achromobacter*)	Degradation of anionic surfactants like sodium dodecyl sulfate, their detection in aqueous media	Taranova et al. (2002)
Agriculture	Polyoxyethylene (20) sorbitan monooleate	Destruction of microbes—for example, antagonistic effect on organisms like sweet potato whitefly (*Bemisia tabaci*) and *Camptotylus reuteri*	Jazzar and Hammad (2003)
Influence on extracellular products	Trichoderma Cellulase production enhanced by polyethylene glycol-(1,1 3,3-tetramethylbutyl)-phenyl ether	Promotion of protein secretion by modified interactions with the lipid components of cell membranes	Reese and Maguire (1969), Ron and Rosenberg (1969)
Recovery of intracellular products, for example, extraction of penicillin	Acylase (*Escherichia coli* in reverse micelles of sodium bis-(2-ethylhexyl) sulfosuccinate in water/hexane)	Cell permeabilization—the surfactant penetrates the inner phospholipid layer and increases the membrane permeability by disorganizing the phospholipids	Bansal-Mutalik and Gaikar (2003)
Applied biocatalysis	Lipase (*Rhizopus delemar* modified with the detergent didodecyl glucosyl glutamate)	Retention of enzyme activity in washing powders	Okazaki et al. (1997)
		Surfactants—enzyme complexes alter enzyme tertiary structure and regulate enzyme activity and specificity; useful in biocatalysts resistant to denaturation in organic solvents	
Microbial production of biosurfactants	Glycolipids (*Pseudomonas and Candida* sp.)	Incorporation of oxygen and other electronegative elements by biocatalysts in the cell membrane of the microbes is an important process in creating biosurfactants	de Lima et al. (2009)
	Sophorolipids (*Candida bombicola*, a nonpathogenic yeast)	Specific sites in the gene of this organism are involved in the surfactant synthesis	Van Bogaert et al. (2008)

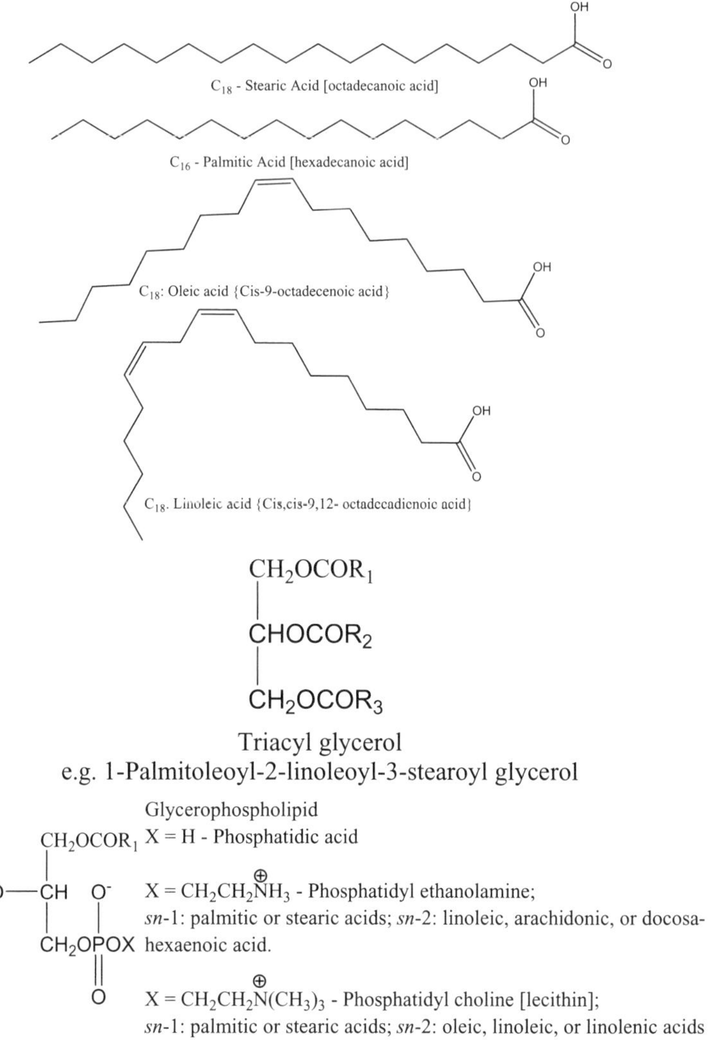

CHART 13.1 Structures of Some Representative Fatty Acids and Triglycerides Found in Cell Membranes.

III. SURFACTANTS IN CELL LYSIS

Cell lysis (breaking up [*lysis*] of cells) is an important process routinely used in isolating membrane proteins.

Surfactants can be effective in breaking cell membranes and opening cells. The three possible types of interactions of the membrane components—lipid:lipid, lipid:protein, and protein:protein—are disrupted by surfactants. Disruption is then followed by solubilization of the surfactant complex into the surfactant micelles (Schuck et al., 2003).

(*sn* stands for stereospecific numbering system—if the second OH in glycerol is written on the left side, then the top carbon is designated as *sn*-1, middle one as *sn*-2, and the bottom one as *sn*-3 according to the IUPC/IUB convention)

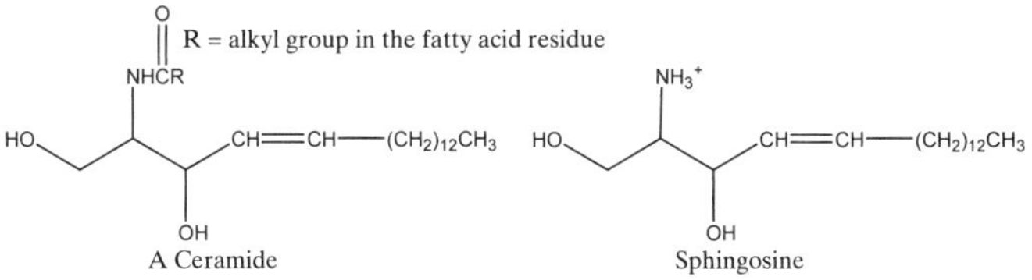

R = alkyl group in the fatty acid residue

NHCR HO CH=CH—$(CH_2)_{12}CH_3$ OH
A Ceramide

NH_3^+ HO CH=CH—$(CH_2)_{12}CH_3$ OH
Sphingosine

(in ceramide *proper* N-acyl group is palmitoyl)

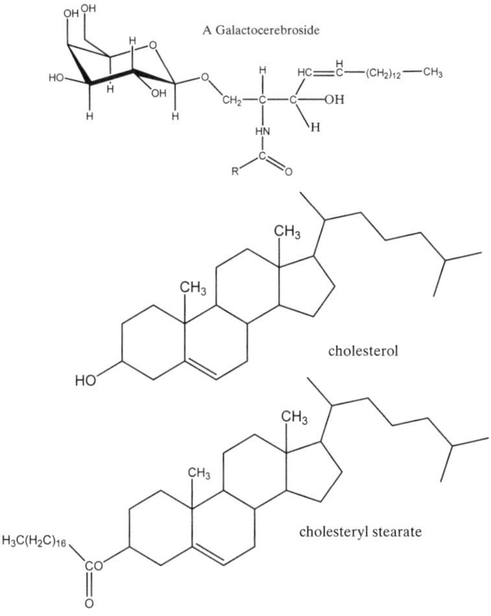

A Galactocerebroside

cholesterol

cholesteryl stearate

CHART 13.1 (*Continued*)

Externally added molecules adsorb cooperatively onto the transmembrane surface of the protein forming a monolayer-like assembly, which is in rapid equilibrium with aqueous monomers and protein-free micelles. Below the CMC level, the proteins aggregate and precipitate. The choice of surfactants, concentration, and conditions such as temperature, pH, duration of the process, and salt level can influence whether the protein will remain in its native state or will be denatured. Above the CMC, the micelles may assist in solubilizing the product of the surfactant's interaction with lipids and proteins or may further adsorb more of monomeric surfactant on the already formed

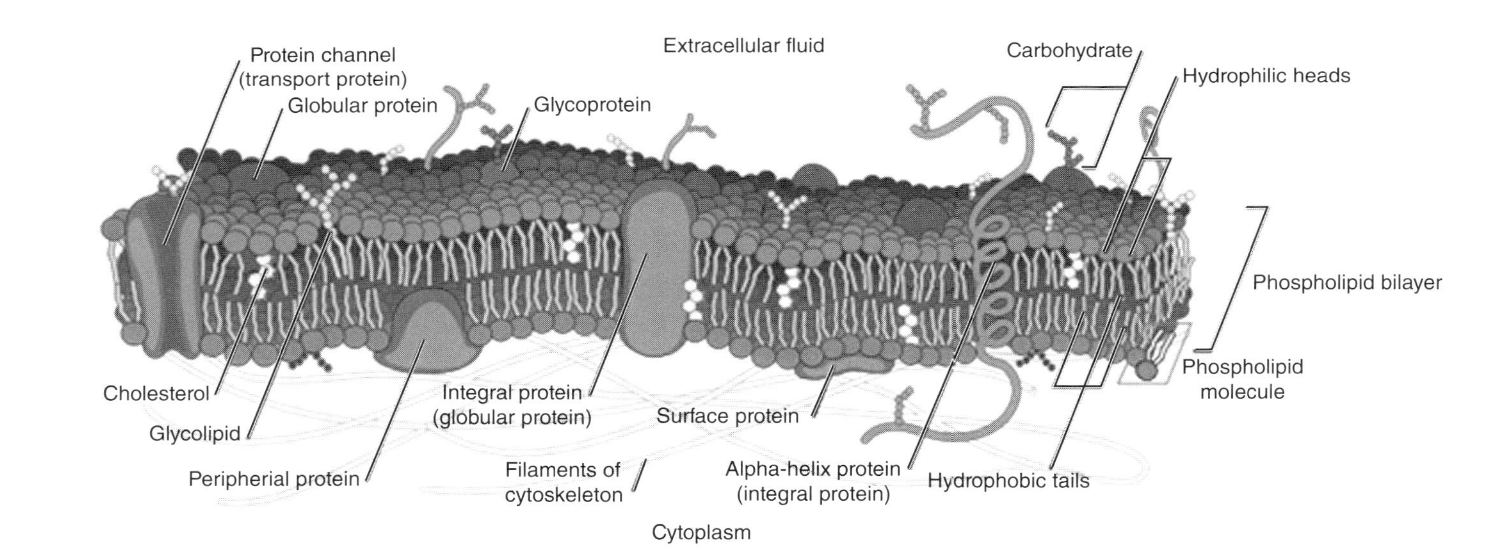

FIGURE 13.1 Schematic representation of a cell membrane.

monolayer. When the cell is broken, the cell content can ooze out, exposing nucleic acids.

Nonionic detergents such as polyethylene glycol p-(1,1,3,3-tetramethylbutyl)-phenyl ether (Chapter 1, Section IC1) cause less damage to the native conformational structure of proteins than ionic detergents. Obviously, the electrostatic interactions between the charged groups in the protein and the ionic surfactant modify the repulsions in certain segments of the protein and, consequently, its shape, much more strongly than in the case of a nonionic surfactant. The extent of denaturization is concentration dependent. At lower surfactant concentration, the binding is highly specific as a result of interaction with specific binding sites and noncooperative. At a higher surfactant concentration, the binding is highly nonspecific and cooperative, resulting in the destruction of the three-dimensional structure of the protein, which is called denaturation (Curry et al., 1998).

Lysis media based on polyethylene glycol p-(1,1,3,3-tetramethylbutyl)-phenyl ether are used in several protocols for cell lysis. A typical lysis solution contains 100 mM potassium phosphate (pH 7.8), 0.2% (v/v) polyethylene glycol p-(1,1,3,3-tetramethylbutyl)-phenyl ether, and 1 mM dithiothreitol (DTT). Polyethylene glycol p-(1,1,3,3-tetramethylbutyl)-phenyl ether/glycylglycine buffer is also popular for cell lysis.

The downside of using most surfactants in cell lysis is that the separated proteins tend to be denatured. Novel surfactant structures have been developed to avoid the denaturing tendency, which arises from the strong chemical bonding between the surfactant and the protein (Popota et al., 2003). The surfactants developed, for example, amphipols, avoid the strong interaction with the biopolymer, while still capable of retaining the solubilizing property. Amphipols are a class of amphiphilic polymeric surfactants with high solubilizing power, but the major binding force is noncovalent in character. In an amphipol (Figure 13.2), the hydrophobic backbone is connected to hydrophilic and highly charged groups, which in turn are tethered to hydrophobic groups (e.g., octyl). The sizes of the interacting polymer particles (the length of the chain and the area of the interacting groups) qualitatively match with the

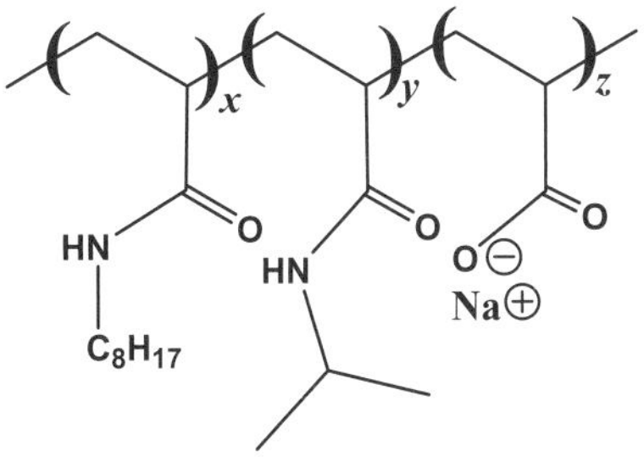

FIGURE 13.2 Structure of a typical polyacrylate-based amphipol polymer.

dimensions of the interacting groups on the membrane surface and the total area so that they can be linked at multiple points on the cell. Amphipols can stabilize many well-characterized integral membrane proteins in aqueous solution, maintaining their native stereospecific state (Tribet et al., 1996).

Special surfactants have also been used for making well-defined crystal structures for diffraction studies in order to elucidate the three-dimensional features of some proteins. Here again, the protein–surfactant complex is subject to the dichotomy that the surfactant–protein complex should be stable enough to remain solubilized in the micelles, but flexible enough to make reasonable contact with the crystal. A case in point is the light-harvesting membrane protein bacteriorhodopsin, isolated from *Halobacterium salinarum*. This has been studied (Essen et al., 1998) both by a sugar-based surfactant, n-octyl β-glucopyranoside (Chapter 1, Section IC9), and a tripod amphiphile (Figure 13.3) that is characterized by a tripod-like hydrophobic appendage of three short chains on a head group (Chae et al., 2008). Many novel surfactant systems have been designed to study membrane proteins that help throw light into membrane structural biology as, for example, zwitterionic phosphocholine detergents that have the ability to reconstitute membrane proteins (refolding of protein to mimic the native state).

A lipid bilayer structure with a spherical geometry constituting the external membrane of cells has been referred to as a *vesicle* (Chapter 3. Section IIA). Vesicles can be prepared *in vitro* by assembling lipid molecules such as dipalmitoyl phosphatidylcholine. Vesicles have the potential to mimic biological membranes (acting as microreactors for artificial photosynthesis and other

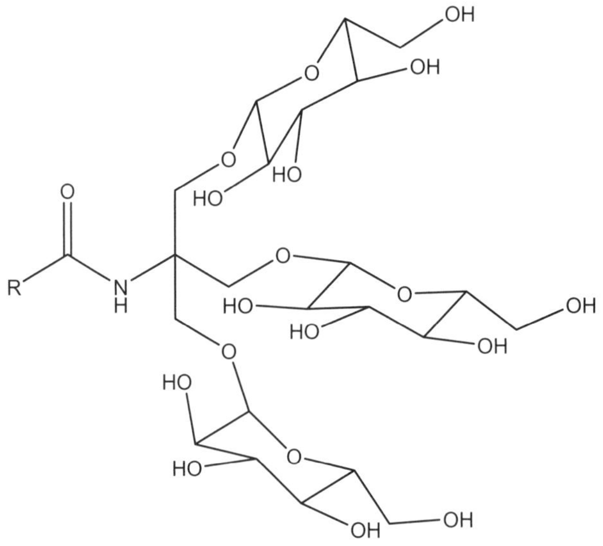

FIGURE 13.3 Structure of a tripod amphiphile.

chemical reactions) for drug delivery and in the design of electronic, photonic, and magnetic functional units. *Liposomes* are spherical bilayers created out of the same or similar material as the amphiphilic lipids in the cell membrane. Liposomes with a variety of structural features in the interior and exterior layers can be made, and they find application in targeted drug delivery and gene therapy. Liposomes being made from natural amphiphiles will not have any detrimental interaction with other cell components and act as inert carriers of medicine.

IV. PROTEIN DENATURING AND ELECTROPHORESIS WITH SURFACTANTS

Proteins in natural conformation are said to be in the *native* state. The weak nonbonding interactions that maintain the integrity of the native form is highly susceptible to changes in temperature, pH, and solvent. Even in micromolar concentrations, surfactants can perturb the protein structure, causing denaturation by hydrophobically associating with the nonpolar residues of the protein, which are responsible for the native structure.

The ability of sodium dodecyl sulfate (SDS, Chapter 1, Section IA3) to complex with proteins finds application in gel electrophoresis. Most proteins bind to SDS in the same ratio, 1.4 g of SDS per gram of protein, which is equivalent to one SDS molecule per two amino acid residues. The SDS charge overrides the protein charge, with the result that the protein–SDS complex tends to have the same charge/mass ratios and a similar shape. As a consequence of this, separation of proteins by gel permeation electrophoresis in an SDS-containing gel of polyacrylamide is possible on a mass basis (cf. charge/mass ratio as the basis of mass spectrometric separation). The principle of gel permeation chromatography (GPC) or size exclusion chromatography is that the pores of the gel retain smaller size polymers, and large size polymers are excluded.

V. PULMONARY SURFACTANTS

Pulmonary or lung surfactants are a mixture of biosurfactants [chiefly a lipoprotein complex (phospholipoprotein)] formed by alveolar cells, in the lung or outside of it. They consist of approximately 90% phospholipids and 10% proteins. The phospholipid fraction is primarily dipalmitoylphosphatidylcholine (DPPC) with smaller amounts of unsaturated cholines, phosphatidylglycerol (PG), phosphatidylinositol, phosphatidylethanolamine, and neutral lipids.

During the inhalation process, air drawn through the nasal cavity and pharynx passes down through the trachea and the bronchi to the tiniest bronchioles, which end in microscopic air sacs called the *alveoli*. The exchange of gases within the lungs is accelerated by pulmonary biosurfactants. They adsorb at the air–water interface of the alveoli with the polar groups of the lung surfactant facing toward water and the hydrophobic tails facing toward the air, reducing the surface tension at the air–liquid interface in the alveoli. The alveoli may be thought of as air bubbles in water. The equilibrium surface tension at the beginning of expiration has been experimentally determined to be ~23 mN/m (Floros, 2005). The lung surfactants help prevent the collapse of the lung at the end of expiration by reducing the surface tension at the air–aqueous solution interface, allowing the lung to inflate much more easily and considerably reducing the work of breathing.

A deficiency in lung biosurfactants is responsible for respiratory malfunctioning in some babies born prematurely. Pulmonary surfactants derived from other sources, as from the bronchoalveolar cow lung *lavage* (cleaning or rinsing process), can be used in treating some of the syndromes associated with these malfunctions.

During a lung injury, blood plasma leaks into air spaces and inactivates the lung surfactant by chemical interaction. Blood plasma contains dissolved polymeric proteins among other components, and the protein–surfactant interaction will, in effect, deprive the alveoli of the much needed lung surfactant. The lost lung surfactant can be replenished artificially. Commonly used pulmonary surfactants approved by the Food and Drug Administration (FDA) are either synthetic or derived from various animal sources. Additives like polyethylene glycol and dextran substantially lower the viscosity of the artificial lung surfactant solution (Lu et al., 2009). The composition of some of these is

(a) A mixture of DPPC with hexadecanol and tyloxapol (Figure 13.4) (a polymeric nonionic surfactant) added as spreading agents
(b) A natural extract of bovine lung surfactant, containing 88% phospholipid (about 50% DPPC) and proteins (about 1%), supplemented with synthetic palmitic acid, and tripalmitin (about 50% DPPC) and pro-

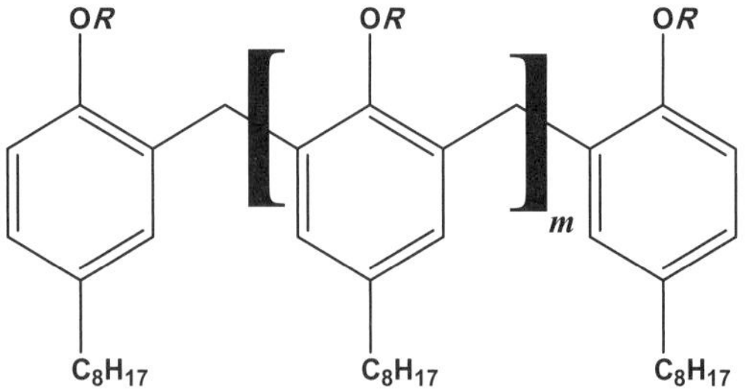

$R = CH_2CH_2O{-}[CH_2CH_2O]_n{-}OCH_2CH_2OH$

$m < 6$
$n = 6–8$

FIGURE 13.4 Molecular structure of tyloxapol.

teins (about 1%), supplemented with synthetic palmitic acid and tripalmitin

(c) A natural surfactant extract from porcine lungs, containing almost exclusively polar lipids, in particular phosphatidylcholine (about 70% of total phospholipid content), PG (about 30% of total phospholipid content), and about 1% of hydrophobic proteins.

(d) An extract from calf lung lavage fluid; the main component phospholipids are DPPC, PG, and some unsaturated phospholipids; the phospholipid concentration is 35 mg/mL (Lalchev et al., 2004; Fernando and Maturana, 2007).

A typical lung surfactant is usually formulated by mixing the animal organ extract with a synthetic analogue of the surfactant, for example, DPPC and PG. Tests on animal models have indicated that synthetic analogues of lung surfactants can be even more effective than their natural counterparts. These studies also reveal that an insufficiency in lung surfactants can be treated more effectively with a formulation than a surfactant inactivation from lung injury because the latter case produces several foreign components within the lung that retard the treatment.

Further investigations on synthetic analogues have been directed to special surfactant systems as, for example, fluorinated molecules. Fluorocarbons have better oxygen and carbon dioxide transportation capacity and superior surface activity than their protium analogues, and their use has been an area of active study in this context (Hlastala and Souders, 2001).

VI. SURFACTANTS IN BIOTECHNOLOGY

In biotechnology, a microorganism or a cell or its extract plays the lead role. Biotechnology is based on the function of biocatalysts that are small or large polypeptides. Each process or product requires a specific biocatalyst. Thus, the optimization of properties and the environment are essential steps in realizing any biotechnological process.

All biotechnological processes involving surfactants, at some stage or other, make use of one or more of the physicochemical properties of surfactants that have been described in previous chapters (Chapters 1–5). These are reduction in surface tension, aggregation behavior in solution, solubilizing ability, and adsorption at interfaces (solid–liquid, liquid–gas, and liquid–liquid). The stability of these systems depends on the thickness and compactness of the adsorbed layers of surfactants, which, in turn, depends upon the ionic, covalent, and noncovalent hydrophobic and hydrogen bond interactions. The interaction of polymers with surfactants and the stability of gas bubbles often recur as the underlying mechanistic theme in all these processes, where interfacial adsorption phenomena are often critical. Other surfactant-related concepts that are relevant to biotechnological applications are their properties as emulsifiers, demulsifiers, microemulsion formers, wetting and penetrating agents, detergents, foaming agents, thickening agents, metal sequestering agents, vesicle formers, and viscosity-reducing agents.

Some applications of surfactants in biotechnology are described below.

A. Mineral Engineering

One of the earliest industrial applications of surfactants was in the flotation process, where a mineral ore was separated from gangue by adsorbing surfactants on the surface, followed by carrying the hydrophobically modified particles to the top by attaching air bubbles onto them. When surfactants adsorb on the solid surface, the polar groups will be oriented toward the surface and the exterior will be replete with nonpolar hydrophobic tails. Such a coating of surfactant endows the solid with a high degree of hydrophobicity that expedites the attachment of the particles onto the (hydrophobic) gas bubbles, allowing the solid to float.

Bacterial mining has been a useful alternative to a purely chemical approach, where the processes may be operated under mild conditions (Petrisor et al., 2007). A specific system of interest is the mineral iron pyrite (FeS_2), the geochemical breakdown of which is through biological intervention. The overall chemical process can be represented as

$$4FeS_2 + 15O_2 + 2H_2O \rightarrow 2Fe_2(SO_4)_3 + 2H_2SO_4.$$

This reaction is accelerated a million times by the intervention of bacteria like *Thiobacillus ferrooxidans*. The bacteria physically adhere to the mineral surface and degrade the sulfide by the following sequence of reactions:

$$FeS_2\ (\text{solid}) + 3.5O_2 + H_2O \xrightarrow{\text{Bacteria}} Fe^{2+} + 2SO_4^{2-} + 2H^+$$

$$Fe^{2+} + 0.25O_2 + H^+ \xrightarrow{\text{Bacteria}} Fe^{3+} + 0.5H_2O.$$

Fe^{3+} further degrades the sulfide as

$$FeS_2 + 2Fe^{3+} + 3O_2 + 2H_2O \rightarrow 3Fe^{2+} + 2SO_4^{2-} + 4H^+.$$

The bacterial attachment to the mineral surface is related to the concentration of bacteria in the solution by the Langmuir isotherm (Chapter 2, Section IIB1). The rate of the process depends on a number of factors that include the physical and chemical characteristics of the mineral, the nature of the intervening microbial agent, and the ambient conditions such as pH and temperature.

Surface-active agents excreted by the microorganisms have a specific role in the mass transfer of the reactants. Here, the wetting of the mineral surface by the surfactants has been identified as a prominent step in the total process.

Adsorption on the mineral surface by surfactants leading to changes in the hydrophobicity of the mineral particles can also be important in these processes. The *Thiobacillus* species has been investigated in detail and some of the participating surfactants have been identified (see Table 13.6) (Torma and Boseker, 1982; Gupta and Mishra, 1983), which includes the profile of ornithine lipids (Dees and Shively, 1982).

B. Fermentation

In fermentation processes, the size and the stability of the gas bubbles are controlled by the presence of surfactants. The surfactants can be those generated from the microorganisms or deliberately added to the fermenting broth. Surfactants adsorb at the interfaces, bringing down the interfacial free energy. As a result, the gas–liquid interfacial area can be increased with a consequent decrease in the size of the gas bubbles and an increase in the fermentation rate. A concomitant decrease in the *mass transfer coefficient* (a transport parameter that connects the diffusion rate constant, the mass transfer area, and the concentration gradient) is compensated by the increased interfacial area.

Aeration is a critical factor in aerobic fermentation processes, where oxidation of organic compounds is the key step. Studies on turbine aerators show that low concentrations of SDS increase the interfacial area and the mass transfer coefficients (Benedek and Heideger, 1971).

C. Enzymatic Deinking

The consumption of paper has been escalating and environmental concerns recommend the recovery of used paper. Deinking is of prime concern. Nonylphenol ethoxylate surfactants have been used in the traditional deinking processes along with chemicals such as sodium hydroxide, hydrogen peroxide, sodium silicate, and chelating agents. The deinking process, as practiced, generates effluents that are not environmentally friendly. Enzymes derived from biological sources can replace many of the chemicals without compromising the quality of the fiber (Soni et al., 2008). Aside from reducing the use of classical chemicals, enzyme technology helps reduce energy consumption, produces stronger fibers, and eliminates unfriendly waste products. However, surfactants such as alkylphenol ethoxylates (Chapter 1, Section IC) can still be beneficial in the process. Synergism resulting from the combination of surfactant and enzyme mixtures is exploited in advanced deinking technology (Jobbins and Franks, 1997).

D. EOR and Oil Bioremediation

Biosurfactants produced from plants and microorganisms can have good interfacial properties, especially at hydrocarbon–water interfaces. Furthermore, they do not produce hazardous wastes, eliminating disposal problems. They function effectively under a variety of conditions, showing high stability at high temperatures, pH, and salt concentrations.

Biosurfactants find use in EOR and pollution control of oil spills. The central property on which these processes depend is the ability of the biosurfactant to emulsify a hydrocarbon–water mixture. An increase in the interfacial area due to the biosurfactants assists in the degradation of hydrocarbon in the environment (Banat, 1995).

E. Enzyme Activity in Surfactant Media

The performance of an enzyme is dependent on its environment, such as solvent and added chemicals. Surfactants can alter the enzyme activity. To illustrate, the activity of *cholesterol oxidase* from *Streptomyces hygroscopicus* and *Brevibacterium sterolicum* has been studied in the presence of nonionic surfactants polyoxyethylene glycol 400 dodecyl ether and polyethylene glycol-(1,1,3,3-tetramethylbutyl)-phenyl ether. Polyoxyethylene glycol 400 dodecyl ether showed a concentration dependence on the activity of this enzyme, an increase in activity at low levels, and a decrease at high levels. In contrast, polyethylene glycol-(1,1,3,3-tetramethylbutyl)-phenyl ether showed low activity irrespective of the concentration. The selective activation of enzymes in reverse micellar media has also been investigated for the two hydrolyzing enzymes, *protease* and *α-amylase*. In one study in the presence of surfactants of different classes, it was observed that the nonionic surfactant polyoxyethylene (20) sorbitan monooleate enhanced the activity of both enzymes; anionic surfactants, SDS, and sodium dioctyl sulfosuccinate enhanced the activity of protease but inhibited the activity of α-amylase, whereas the cationic surfactant, cetylpyridinium chloride, increased the activity of α-amylase and decreased the activity of protease (Gajjar et al., 1994).

F. Carbon Dioxide "Fixing" in Bioreactors

Fixing is the conversion of carbon dioxide into organic compounds such as carbohydrates, similar to photosynthesis. Carbon dioxide is considered to be a waste gas produced as the final product of combustion that contributes to global warming. It can be "fixed" to produce biomass as animal feed in bioreactors of microalgae. This process produces a variety of organic chemicals, including surfactants. The single cells of microalgae are particularly efficient at photosynthesis with the potential to fix high carbon dioxide concentrations from industrial exhaust gases.

G. Soil Remediation

Bioremediation of soil to get rid of accumulated surfactant waste has environmental importance. Microbial preparations have been developed for enhancing the rate of soil purification from surfactants (Gradova et al., 1996). Accelerated biodegradation of the surfactant occurs, giving rise to purified soils devoid of surfactants. Moreover, surfactants themselves are found to accelerate the degradation of carcinogenic polyaromatic hydrocarbons (PAH) in artificial soils made from resinous materials using axenic cultures (a culture of an organism that is entirely free of all other "contaminating" organisms of bacteria) in the presence of nonionic surfactants (Vacca et al., 2005).

H. Effluent Purification

Surfactant accumulation in aquifers contaminates drinking water sources. Biotechnological methods are being developed to remove surfactants in these cases. *Pseudomonas putida* strain TP-19 immobilized in polyvinyl alcohol cryogels showed a destruction activity of almost 100% (Turkovskaya et al., 1995). One group utilized a strain of *Pseudomonas desmolytica* to destroy zwitterionic surfactants such as amidobetaines (Chapter 1, Section ID1) (Taranova et al., 1990).

I. Surfactants in Horticulture

Plant shoot generation is an area of botanical importance. In this connection, the effect of polyoxyethylene-polyoxypropylene block copolymer nonionic surfactant on the growth has been studied in the culture of jute cotyledons (*Corchorus capsularis* L.). Control cotyledons did not produce shoots in the absence of the surfactant, highlighting the potential value of nonionic surfactants as growth-stimulating additives to plant culture media (Khatun et al., 1993). Other horticultural applications of surfactants include the use of ethylene oxide-based nonionic surfactants in enhancing the cuticular absorption of pesticides (Stevens and Bukovact, 1987).

J. Vesicle Manipulation

Genetic engineering through DNA encapsulation and liposome-mediated transfection can be achieved using vesicular aggregates. Surfactants, both synthetic, like polyethylene glycol–(1,1,3,3-tetramethylbutyl)-phenyl ether, octyl glucosides, and natural, like bile acids, influence the fusion of unilamellar vesicles of phosphatidylcholine (Goni and Alonso, 1988).

K. Genetic Engineering and Gene Therapy

DNA–surfactant complexes, for example, the DNA–cetyltrimethylammoniumm bromide complex, have been investigated in view of their potential biotechnological applications in gene therapy and other genetic engineering methods. Cationic surfactants form insoluble charge-neutralized complexes with DNA, and such neutral complexes can be encapsulated in surfactant micelles through solubilization. This adduct has the potential to function as a genetic information carrier (Kunjappu and Nair, 1992; Bonincontro et al., 2007).

Table 13.6 (p. 484) lists some of the applications, names of the biosurfactants, the relevant physicochemical principles, and references. The literature is replete with hundreds of microbial species, the biosurfactants in many of which are not well characterized; nevertheless, microbial extracts from them have beneficial effects in specific processes. In many instances, mixtures of bacterial culture and synthetic surfactants are used in these processes. The structures of most of the biosurfactants indicated in Table 13.6 may be found either in this chapter or in previous chapters along with the names of the microorganism producing the biosurfactant (see Tables 13.2 and 13.3).

REFERENCES

Austin, R. N., K. Luddy, K. Erickson, M. Pender-Cudlip, E. Bertrand, D. Deng, R. S. Buzdygon, J. B. van Beilen, and J. T. Groves (2008) *Angew. Chem. Int. Ed.* **47**, 5232.

Banat, I. M. (1995) *Bioresour. Technol.* **51**, 1.

Banat, I. M., N. Samarah, M. Murad, R. Horne, and S. Banerjee (1991) *World J. Microbiol. Biotechnol.* **7**, 80.

Banat, I. M., R. S. Makkar, and S. S. Cameotra (2000) *Appl. Microbiol. Biotechnol.* **53**, 495.

Bansal-Mutalik, R. and V. G. Gaikar (2003) *Enzyme Microb. Technol.* **32**, 14.

Benedek, A. and W. J. Heideger (1971) *Biotechnol. Bioeng.* **13**, 663.

Bognolo, G. (1999) *Colloids Surf A: Physicochem. Eng. Aspects* **152**, 41.

Bonincontro, A., C. L. Mesa, C. Proietti, and G. Risuleo (2007) *Biomacromolecules* **8**, 1824.

Bordoloi, N. K. and B. K. Konwar (2008) *Colloids Surf. B: Biointerfaces* **63**, 73.

Chae, P. S., M. J. Wander, A. P. Bowling, P. D. Laible, and S. H. Gellman (2008) *ChemBioChem* **9**, 1706.

Cirigliano, M. C. and G. M. Carman (1985) *Appl.Environ. Microbiol.* **50**, 846.

Cooper, D. G., J. E. Zajic, and D. F. Gerson (1979) *Appl.Environ. Microbiol.* **37**, 4.

Curry, S., H. Mandelcow, P. Brick, and N. Franks (1998) *Nat. Struct. Biol.* **5**, 827.

de Lima, C. J. B., E. J. Ribeiro, E. F. C. Servulo, M. M. Resende, and V. L. Cardoso (2009) *Appl. Biochem. Biotech.* **152**, 156.

Dees, C. and J. M. Shively (1982) *J. Bacteriol.* **149**, 798.

Deleu, M. and M. Paquot (2004) *C. R. Chim.* **7**, 641.

Deleu, M., H. Razafindralambo, Y. Popineau, P. Jacques, P. Thonart, and M. Paquot (1999) *Colloids Surf. A: Physicochem. Eng. Aspects* **152**, 3.

Desai, J. D. and I. M. Banat (1997) *Microbiol. Mol. Rev.* **61**, 47.

Eriksson, T., J. Boerjesson, and F. Tjerneld (2002) *Enzyme Microb. Technol.* **31**, 353.

Essen, L., R. Siegert, W. D. Lehmann, and D. Oesterhelt (1998) *Proc. Natl. Acad. Sci. U.S.A.* **95**, 11673.

Fernando, M. and A. Maturana (2007) *Clin. Perinatol.* **34**, 145.

Fiechter, A. (1992) *Trends Biotechnol.* **10**, 208.

Finnerty, W. R. (1994) *Curr. Opin. Biotech.* **5**, 291.

Floros, J. (2005) *Curr. Resp. Med. Rev.* **1**, 77.

Fu, S. L., S. R. Wallner, W. B. Bowne, M. D. Hagler, M. E. Zenilman, R. Gross, and M. H. Bluth (2008) *J. Surg. Res.* **148**, 77.

Gajjar, L., R. S. Dubey, and R. C. Srivastava (1994) *Appl. Biochem. Biotech.* **49**, 101.

Goni, F. M. and A. Alonso (1988) *Adv. Exp. Med. Biol.* **238**, 81.

Gradova, I. B., P. A. Kozhevin, N. L. Rabinovich, S. S. Korchmary, and D. Ul'yanov (1996) *Biotekhnologiya* **11**, 46.

Gupta, M. D. and A. K. Mishra, in *Recent Progress in Biohydrometallurgy*, G. Rossi and A. E. Torma, (eds.), Associazione Mineraria Sarda—09016, Iglesias, Italy, 1983, p. 1.

Hayes, M. E., E. Nestaas, and K. R. Hrebenar (1986) *Chemtech* **4**, 239.

Hlastala, M. and J. Souders (2001) *Am. J. Respir. Crit. Care Med.* **164**, 1.

Jae, H. I., T. Nakane, H. Yanagishita, T. Ikegami, and D. Kitamoto (2001) *BMC Biotechnol.* **1**, 5.

Jazzar, C. and E. A. Hammad (2003) *Bull Insectol.* **56**, 269.

Jobbins, J. M. and N. E. Franks (1997) *TAPPI J.* **80**, 73.

Khatun, A., L. Laouar, M. R. Davey, J. B. Power, B. J. Mulligan, and K. C. Lowe (1993) *Plant Cell Tissue Organ Cult.* **34**, 133.

Kim, P., D. Oh, S. Kim, and J. Kim (1997) *Biotechnol. Lett.* **19**, 457.

Kordula'kova', J., M. Gilleron, G. Puzo, P. J. Brennan, B. Gicquel, K. Mikušova', and M. Jackson (2003) *J. Biol. Chem.* **278**, 36285.

Kosaric, N. (1992) *Pure Appl. Chem.* **64**, 1731.

Kosaric, N., W. L. Cairns, and N. C. C. Gray, in *Biosurfactants and Biotechnology*, N. Kosaric, W. L. Cairns, and N. C. C. Gray (eds.), Marcel Dekker, New York, 1987, pp. 247–320.

Kulakovskaya, T. V., A. S. Shashkov, E. V. Kulakovskaya, and W. I. Golubev (2005) *FEMS Yeast Res.* **5**, 919.

Kunjappu, J. T. and C. K. K. Nair (1992) *Ind. J. Chem.* **31A**, 432.

Kunjappu, J. T. and P. Somasundaran (1996) *Colloids Surf. A.* **117**, 1.

Lalchev, Z., G. Georgiev, A. Jordanova, R. Todorov, E. Christova, and C. S. Vassilieff (2004) *Colloids Surf. B: Biointerfaces* **33**, 227.

Lin, S.-C., K. S. Carswell, M. M. Sharma, and G. Georgiou (1994) *Appl. Microbiol. Biotechnol.* **41**, 281.

Lu, K. W., J. Pérez-Gil, and H. William Taeusch (2009) *Biochim. et Biophys. Acta* **1788**, 632.

Makkar, R. S. and S. S. Cameotra (1998) *Ind. Microbiol. Biotechnol.* **20**, 48.

Matthew, F. C. and D. B. Weibel (2009) *Soft Matter* **5**, 1174.

Michail, M. Y., A. Wolf-Rainer, H. Meyer, L. Giuliano, and P. N. Golyshin (1999) *Biochim. et Biophys. Acta* **1438**, 273.

Nitschkea, M. and S. G. V. A. O. Costa (2007) *Trends Food Sci. Tech.* **18**, 252.

Okazaki, S., N. Kamiya, and M. Goto (1997) *Biotechnol. Prog.* **13**, 551.

Ortiz, A., J. A. Teruel, M. J. Espuny, A. Marqués, Á. Manresa, and F. J. Aranda (2009) *Chem. Phys. Lipids* **158**, 46.

Pellerin, N. B., J. T. Staley, T. Ren, G. L. Graf, D. R. Treadwell, and I. A. Aksay (1992) *Mater. Res. Soc. Symp.* **218**, 123.

Petrisor, I. G., I. Lazar, and T. F. Yen (2007) *Petroleum Sci. Tech.* **25**, 1347.

Polman, J. K., K. S. Miller, D. L. Stoner, and C. R. Brakenridge (1994) *J. Chem. Technol. Biotechnol.* **61**, 11.

Popota, J.-L., E. A. Berryb, D. Charvolina, C. Creuzenetc, C. Ebeld, D. M. Engelmane, M. Flötenmeyerf, F. Giustia, Y. Gohona, P. Hervéa, Q. Hongg, J. H. Lakeyg, K. Leonardf, H. A. Shumani, P. Timminsi, D. E. Warschawskia, F. Zitoa, M. Zoonensa, B. Puccij, and C. Tribet (2003) *CMLS, Cell. Mol. Life Sci.* **60**, 1559.

Rapp, P., H. Bock, V. Wray, and F. Wagner (1979) *J. Gen. Microbiol.* **115**, 491.

Reese, E. T. and A. Maguire (1969) *Appl. Microbiol.* 17. **242**.

Ron, E. Z. and E. Rosenberg (1969) *Curr. Opin. Biotechnol.* **13**, 249.

Rosenberg, E. and E. Z. Ron, in *Biopolymers from Renewable Resources*, D. L. Kaplan (ed.), Springer, Berlin, 1998, pp. 281–289.

Rosenberg, E. and E. Z. Ron (1999) *Appl. Microbiol. and Biotechnol.* **52**, 154.

Saini, H. S., B. E. Barraga′n-Huerta, A. L. ′n-Paler, J. E. Pemberton, R. R. Va′zquez, A. M. Burns, M. T. Marron, C. J. Seliga, A. A. L. Gunatilaka, and R. M. Maier (2008) *J. Nat. Prod.* **71**, 1011.

Schuck, S., M. Honsho, K. Ekroos, A. Shevchenko, and K. Simons (2003) *Proc. Nat. Acad. Sci.* **100**, 5795.

Silva, C. L., J. L. Gesztesi, M. C. Zupo, M. Breda, and T. Ioneda (1980) *Chem. Phys. Lipids* **26**, 197.

Singh, A., J. D. Van Hamme, and O. P. Ward (2007) *Biotechnol. Adv.* **25**, 99.

Somasundaran, P., N. Deo, and K. A. Natarajan, in *Mineral Biotechnology*, S. K. Kawatra and K. A. Natarajan, (eds.), Society for Mining, Metallurgy, and Exploration, Littleton, CO, 2001, p. 221.

Soni, R., A. Nazir, B. S. Chadha, and M. S. Saini (2008) *Bioresources* **3**, 234.

Stevens, P. J. G. and M. J. Bukovact (1987) *Pestic. Sci.* **20**, 19.

Tanner, R. S., E. O. Udegbunam, M. J. McInerney, and R. M. Knapp (1991) *Geomicrobiol. J.* **9**, 169.

Taranova, L., I. Semenchuk, T. Manolov, P. Iliasov, and A. Reshetilov (2002) *Biosens. Bioelectron.* **17**, 635.

Taranova, L. A., L. F. Ovcharov, and M. N. Rotmistrov (1990) *Biotekhnologiya* **4**, 31.

Torma, A. E. and K. Boseker (1982) *Prog. Ind. Microbiol.* **16**, 77.

Torregrossa, E., R. A. Makula, and W. R. Finnerty (1977) *J. Bacteriol.* **131**, 493.

Tribet, C., R. Audebert, and J. Popot (1996) *Proc. Natl. Acad. Sci. U S A* **93**, 15047.

Turkovskaya, O. V., L. V. Panchenko, O. V. Ignatov, and A. V. Tambovtsev (1995) *Khimiya I Tekhnologiya* **17**, 105.

Ueno, Y., Y. Inoh, T. Furuno, and N. Hirashima (2007) *J. Control. Release* **123**, 247.

Vacca, D. J., W. F. Bleam, and W. J. Hickey (2005) *Appl. Environ. Microbiol.* **71**, 3797.

Van Bogaert, N. A., S. L. De Maeseneire, D. Develter, W. Soetaert, and E. J. Vandamme (2008) *Yeast* **25**, 273.

Van Dyke, M. I., S. L. Gulley, H. Lee, and J. T. Trevors (1991) *Biotech. Adv.* **9**, 241.

Walzer, G., E. Rosenberg, and E. Z. Ron (2006) *Appl. Environ. Microbiol.* **61**, 3240.

Yakimov, M. M., M. M. Amor, M. Bock, K. Bodekaer, H. L. Fredrickson, and K. N. Timmis (1997) *J Petrol. Sci. Eng.* **18**, 147.

Zhang, Y. and R. M. Miller (1992) *Appl. Environ. Microbiol.* **58**, 3276.

PROBLEMS

13.1 Explain the connection between surfactants and the cell membrane.

13.2 How do surfactants influence the pulmonary function?

13.3 Explain any three biotechnological applications of surfactants, indicating their role in the selected processes.

13.4 How do the interactions of charged groups of proteins and the ionic and nonionic surfactants affect cell lysis?

13.5 Explain the use of surfactants in exploring the three-dimensional structure of proteins.

13.6 Identify a biological application of surfactants that is not discussed in this chapter, clearly bringing out the role of surfactants. You may have to search primary or secondary literature for this.

14 Surfactants in Nanotechnology

Nanoscience is the study of phenomena and manipulation of materials at atomic, molecular, and macromolecular scales, where properties differ significantly from those at a larger scale. *Nanotechnologies* are the design, characterization, production, and application of structures, devices, and systems by controlling shape and size at the nanometer level (Drexler, 1986; Anonymous, 2004).

In scientific terms, they encompass the dimension range 1–100 nm (nm = nanometer = 10^{-9} m). The above considerations clearly indicate that surfactant science, with its description of surfactant molecules in layers at interfaces (Chapter 2) and the formation of micelles, vesicles, and liquid crystals in solution (Chapter 3) fundamentally belongs to nanoscience.

Nanotechnology is the confluence of several disciplines that include physics, chemistry, biology, applied mathematics, and many engineering areas. It covers topics such as quantum dots, nanowires, nanotubes, nanorods, nanofilms, nano-self-assemblies, thin films, nanosize metals, semiconductors, biomaterials, oligomers, polymers, and nanodevices.

Nanotechnology means different things to different areas: In health science, it means many biomedical implant devices and the treatment of cancer with targeted medicines that act in a cellular level and the easing of clogging of arteries using nanodevices; in electronics, it means the development of various nanoparticle-based technology to enhance the performance and speed of microelectronics devices; in chemistry, among many applications, it refers to the synthesis of novel nanocatalysts; in biology, it includes the development of various nanobiotechnological processes and the mimicking of the biochemical processes that occur in this size range; in defense, it covers the production of lightweight but strong combat implements, among others; and for the surfactant scientist, it means the self-assembly of amphiphilic molecules on surfaces and their use in various devices and applications, as well as the use of surfactants in the preparation of nanoparticles and nanodispersions.

Surfactants and Interfacial Phenomena, Fourth Edition. Milton J. Rosen and Joy T. Kunjappu.

I. SPECIAL EFFECTS OF THE NANOSTATE

The ultralow particle size of nanomaterials provides large interfacial areas and the subsequent increase in the adsorption density of molecules that modify the surface behavior. Increase in surface area will produce an increase in chemical reactivity. This enables such fine particulate matter to function as efficient catalysts for batteries and fuel cells. Nanoparticles exhibit special electronic and optical effects. At this size scale (shape and volume), the wave properties of the electrons inside matter are affected and modified, manifested as quantum size effects. Such quantum effects are significant at a particle size below 50 nm and can be observed even at room temperature for sizes below 10 nm. Nanosized quantum dot particles of semiconductors like cadmium selenide and gold nanoparticles show a gradation in color. Chemists often demonstrate different shades of color from blue to red with metallic nanoparticles in test tubes as a manifestation of the optical quantum size effect. Absorption energy levels are modified as a result of size change, which affects the molecular orbital interactions in them. Theory supports a red shift in the absorption spectrum of gold nanoparticles with size reduction. This is because the wavelengths of the interacting light and the timescales of their interaction match the size range of the nanoparticles.

The small size range means consumption of less material. This aspect helps to compensate for the quantity of material and fuel (energy) consumption in the fabrication and application stages. A proportionate reduction in waste and pollution is also achieved.

Characterization of nanostructures requires microscopic techniques with atomic resolution capability. Novel techniques such as scanning tunneling microscopy (STM) and atomic force microscopy (AFM) can "see" the nanostructures in atomic resolution.

II. ROLE OF SURFACTANTS IN THE PREPARATION OF NANOSTRUCTURES

The basis for any nanoapplication is the creation of nanostructured materials. Surfactants can assist directly or indirectly in the synthesis of nanoparticles. Directly, some surface-active materials can be assembled as part of nanostructures. Since many nanoapplications are in the form of dispersions, the property of surfactants to function as effective dispersing agents find a direct use in nanotechnology. Included in this is the ability of surfactants to alter the surface properties of nanoparticles by adsorbing onto them, causing changes in electrical, hydrophobic, and related properties.

Two methods for building nanostructures can be visualized: "bottom-up" and "top-down" methods. In the bottom-up method, higher dimensions are reached starting with molecular dimension, and in the top-down method, nanoscopic dimension is reached starting with bulk solid state. These methods may be considered to be the counterparts of the condensation and dispersion methods used in the preparation of colloids.

A. Bottom-Up Methods

1. Surfactant Self-Assembly Self-assembly of amphiphilic molecules in general is an effective way to create nanostructures. A number of structures such as micelles and vesicles (Chapter 3) and films (Chapter 2) in the nanoscale to microscale range can be formed by this method.

Surfactant self-assemblies are examples of the bottom-up process because they always start with the solution phase below the critical micelle concentration (CMC), where the surfactant is present in the nonaggregated molecular state. At or above the CMC, the surfactant aggregates in aqueous solution (Chapter 3, Sections IIB and IIC) to form micelles, and by varying the surfactant concentration, they transform into cylindrical and bilayer structures. These are fundamental nano-self-assembly processes. The structural diversity in surfactant aggregates is clearly reflected in their phase diagrams (Chapter 3, Section IIC). The formation of fatty acid monolayers on a liquid surface, or their deposition onto glass or other surfaces (Chapter 2, Section IIB1), and the hydrophobization of polar solid surfaces by the adsorption of surfactants (Chapter 2, Section II) are typical examples of creating surfactant nanoassemblies at interfaces.

Surfactants can induce dramatic conformational changes in proteins. The self-assembly of proteins into hydrophobic and hydrophilic domains, assisted by surfactant-induced electrostatic and hydrophobic interactions, leads to nanostructures. Similar self-assembled natural structures are formed in nucleic acids (DNA and RNA) with the aid of amino acids and proteins (biological catalysts) that function with their hydrophobic and hydrophilic domains as surfactants. The nanostructures are stabilized by supramolecular forces (physical binding forces, such as hydrogen bonding, and interactions with dipoles and quadrupoles) that engender various conformations. The loss in entropic contribution accompanying the self-organization process is compensated by the enthalpic contribution arising from such stabilizing factors.

Nanoassemblies of chiral surfactants are useful as reactors for stereo-controlled organic synthesis and as media for separating chiral molecules that have special relevance to pharmaceutical sciences. Nano-self-assemblies of the two chiral isomers (Figure 14.1) of the surfactant (±)-2-dodecyl-β-D-glucoside have large differences in their thermotropic and lyotropic phase behavior, which is attributed to the differences in their molecular shapes as a result of their chirality (Boyd et al., 2002).

The causes for the nearly hundred times higher biological activity of L-ascorbic acid compared to its epimer, D-ascorbic acid (Figure 14.2), have been investigated by thermal studies on the nanostructures of the vitamin

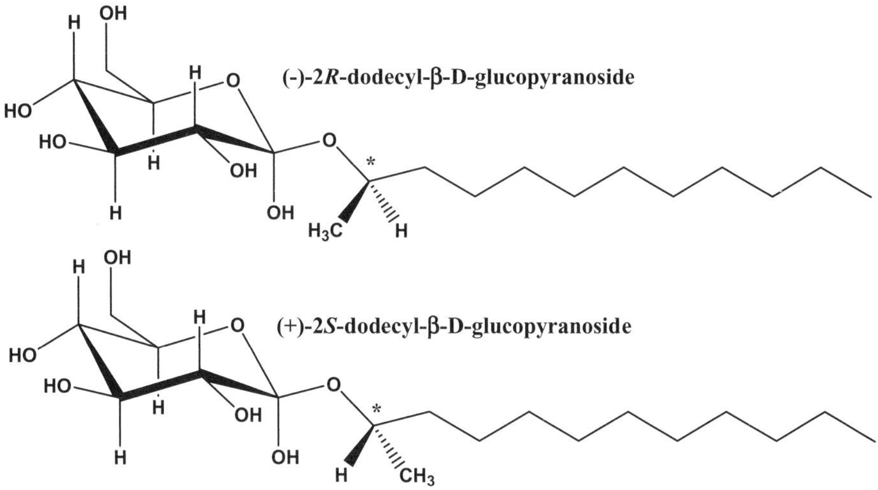

FIGURE 14.1 Structures of isomeric chiral surfactants; * shows the chiral centers (Boyd et al., 2002).

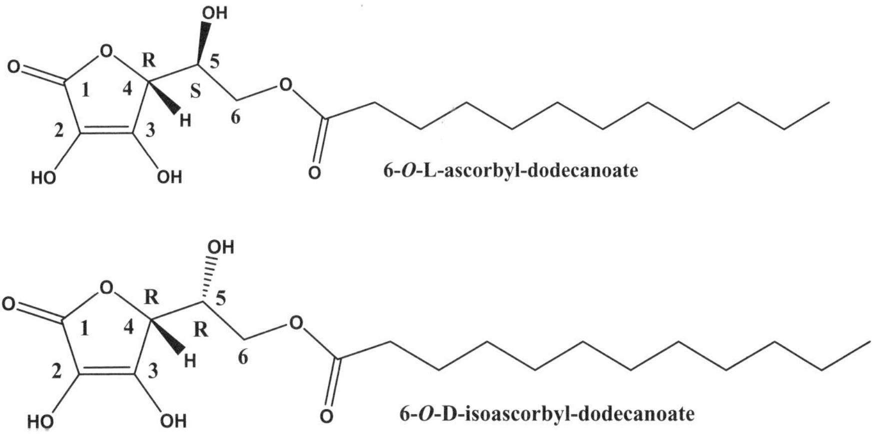

FIGURE 14.2 Structures of ascorbyl-dodecanoates (Nostro et al., 2009).

C-based surfactants, 6-*O*-L-ascorbyl-dodecanoate and 6-*O*-D-isoascorbyl-dodecanoate, and the role of stereochemistry on their intermolecular interactions has been invoked to explain the differences in biological activity (Nostro et al., 2009).

It is possible to build up supramolecular self-assembled structures layer by layer to produce macroscopic tubular structures. The amphiphilic macromolecule with hydrophobic hyperbranched poly(3-ethyl-3-oxetanemethanol) core (HBPO) and many hydrophilic polyethylene glycol arms (polyethylene oxide [PEO]) [(HBPO-star-PEO)] (Figure 14.3) at a concentration range of 10 mg/mL to 1 g/mL in acetone generates multiwalled tubes with diameters in

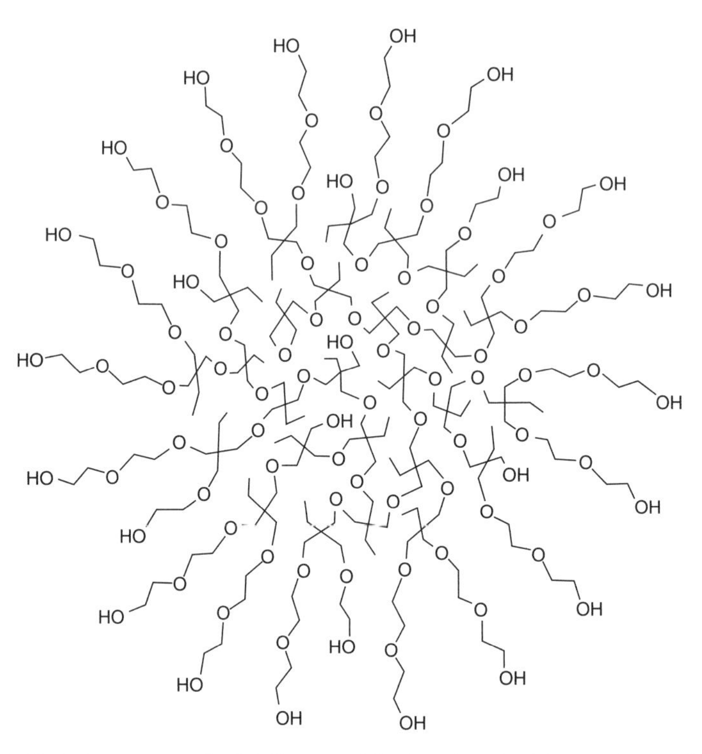

FIGURE 14.3 Structure of HBPO-star-PEO. Reprinted with permission from Yan, D., et al. (2004) *Science* **303**, 65. Copyright 2004, American Association for the Advancement of Science, USA.

millimeters and lengths in centimeters by the self-assembly process (Yan et al., 2004). The walls have a lamellar structure with a morphology in which ordered hydrophilic domains and amorphous, hydrophilic domains alternate. Only acetone was found to produce the tubules; water, alcohol, and esters did not support the self-assembly (Figure 14.4). Ionic strength and pH influenced the shape and structure of the tubules.

Complex vesicle-based polymeric surfactants can act as nanostructures. They can function as templates for colloidal particles. Nanoscopic vesicles of the block copolymer, polyethylene oxide-polyethylethylene (PEO-PEE), are formed in water as shown in Figure 14.5. Under cryo-transmission electron microscopy (TEM) (low-temperature TEM to directly image self-aggregation in liquids) preparation conditions, a coexistence of wormlike and spherical micelles of submicrometer dimension was observed and, by conventional methods of observation, giant vesicles are formed. The hydrophobic (ethylethylene) and hydrophilic (ethylene oxide) regions provide surfactant-like properties to this block copolymer (Figure 14.6) (Zasadzinski et al., 2001).

Nanoparticle self-assembly processes involving surfactants have been very popular in the electronics and photonics industries. For example, N-alkylthiolates and dialkylsulfides are reversibly attached to gold surfaces to yield compact,

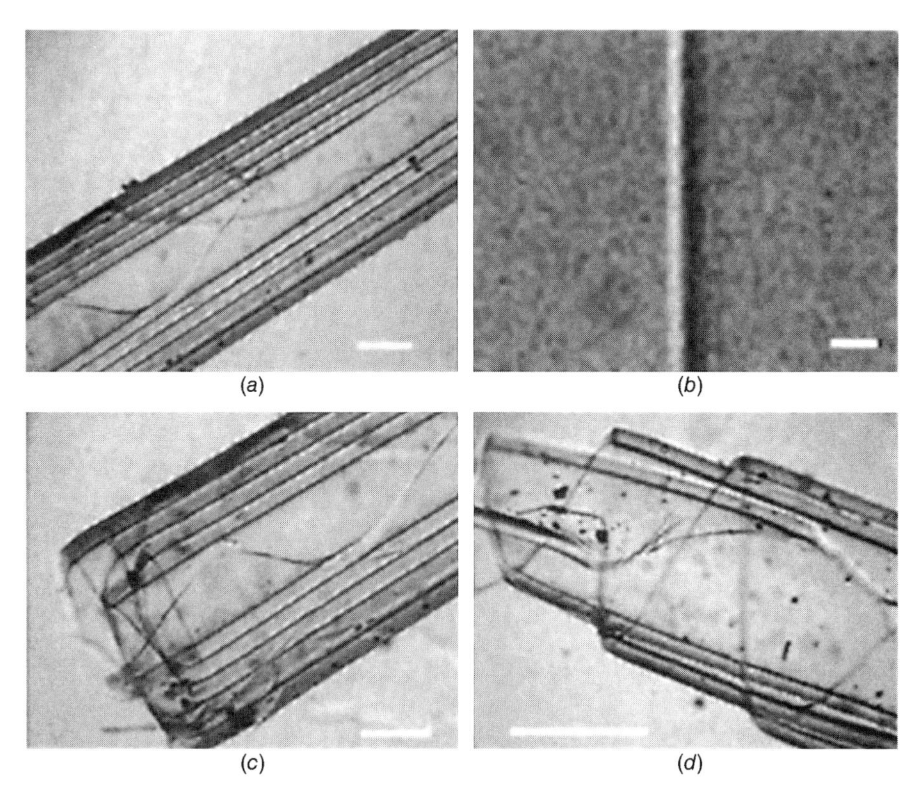

FIGURE 14.4 Optical microscopy images of the HBPO-star-PEO tubes in acetone: (a) A pentawalled self-assembly tube. One of the dark lines shows a single wall, and the space between two lines is vacant. (b) Image of a single wall. (c) An internal screw end of the self-assembly tube. It is a left-hand screw. (d) An external screw end of the self-assembly tube. It is a right-hand screw. Scale bars: 300 μm (a,c,d), 1 μm (b) Reprinted with permission from Yan, D., et al. (2004) *Science* **303**, 65. Copyright 2004, American Association for the Advancement of Science, USA.

orderly monolayers (usually referred to as self assembled monolayer [SAM]), which may be transformed or incorporated as molecular components in electronic devices. Close-packed monolayered clusters of alkylthiol molecules can encapsulate crystals of nanogold particles, with the particles interconnected covalently through alkyldithiols (Andres et al., 1996). Methods for creating and regulating crystalline superstructures of metallic and semiconducting nanoparticles have been a fertile area of research (Bognolo, 2003).

In a similar way, self-assembly of metallic nanoparticle arrays is achieved by DNA, where DNA acts as a programmable molecular template scaffolding (Xiao et al., 2002). Such precisely assembled structures provide a new basis for the manufacturing of future electronic circuitry, replacing current nanolithography. Oligonucleotide structures can be modified with surfactant-like alkylthiols connected to metallic particles like gold and assembled into nanostructures by hydrophobic, electrostatic, and dominant hydrogen bonding. This

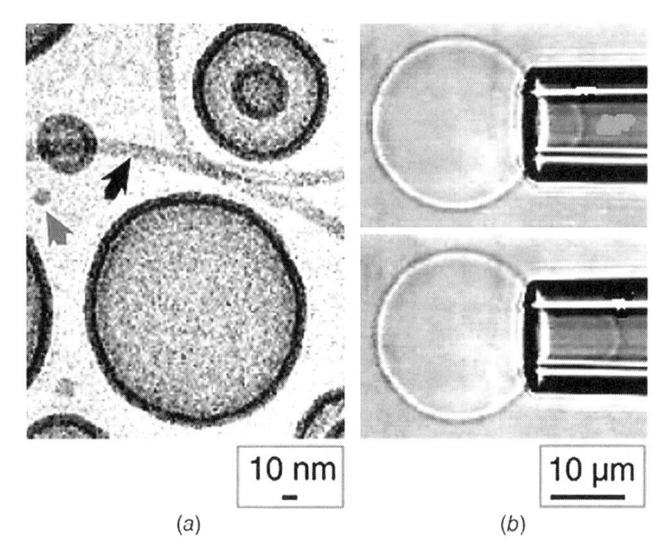

FIGURE 14.5 Vesicles formed in water from the block copolymer polyethylene oxide-polyethylethylene: (a) vesicles in coexistence with wormlike and spherical micelles under cryo-TEM preparation conditions; (b) giant vesicles formed by conventional methods. Note the difference in the magnification scales in (a) and (b). Reprinted with permission from Zasadzinski, J. A., et al. (2001) *Curr. Opin. Colloid Interf. Sci.* **6**, 85. Copyright 2001, Elsevier Limited, UK.

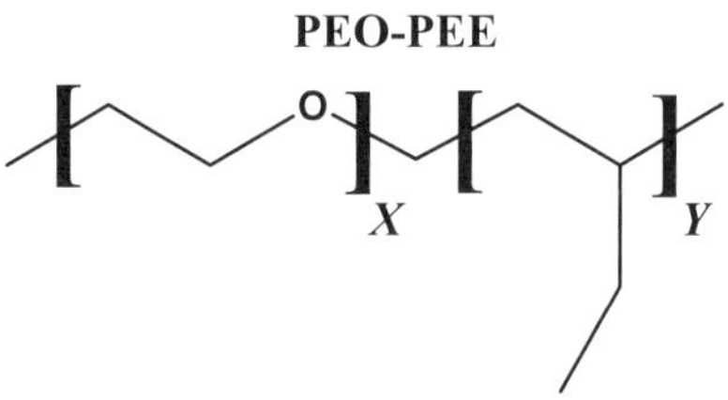

PEO-PEE

O X Y

FIGURE 14.6 Structure of the block copolymer, polyethylene oxide-polyethylethylene (PEO-PEE) (X = 40, Y = 37).

approach exploits the self-assembling ability through base pairing on the nanoscale that yields precise and regular arrays. Ordered DNA tile structures organize metallic particles of gold (1.4 nm in diameter) in highly ordered two-dimensional (2-D) arrays with regular interparticle spacings as shown in Figure 14.7, where the black circles represent gold nanoparticles hosted by DNA tiles.

2. Synthetic Processes Surface-active materials play an important role in the synthesis of nanoparticles with diverse applications as catalysts and magnetic and photonic materials that can be further manipulated into parts of nanodevices. A central theme in all these procedures is the ability of different surfactant self-assembly systems to host the reactive components that produce nanomaterials by functioning as confined media that direct the reactivity and

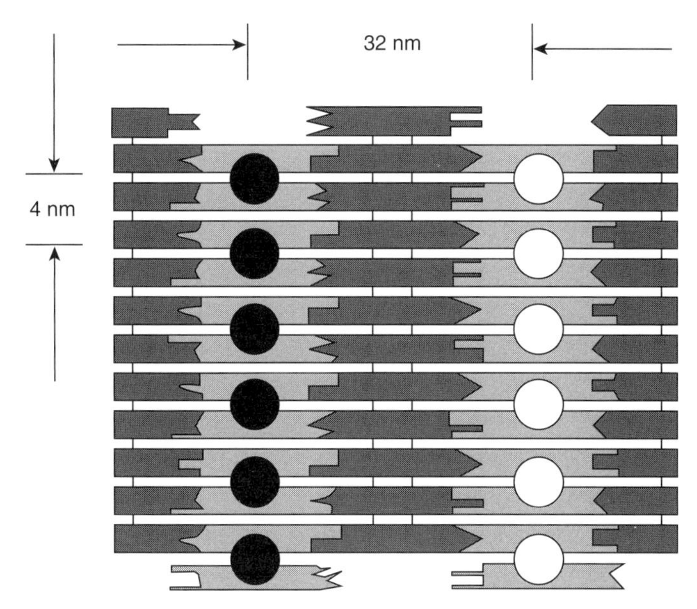

FIGURE 14.7 DNA tiles with metallic particles assembled from DNA–Au conjugates; the spacing between these features is 32 nm in the horizontal direction and 4 nm in the vertical direction. Reprinted with permission from Xiao, S., et al. (2002) *Nanoparticle Res.* **4**, 313. Copyright 2002, Springer.

selectivity of a process. For example, in the case of polymeric nanoparticles, kinetic control of the reaction leads to a material with the desired molecular weight and dispersity index.

Nanosized cadmium sulfide particles are obtained by mixing a solution of sodium sulfide incorporated in reverse micelles of sodium bis(2-ethylhexyl) sulfosuccinate in heptane, with a solution of cadmium perchlorate in the same reverse micelles. The nanocadmium sulfide particles formed are protected by the inverse micellar environment, which prevents the Ostwald ripening (building up of larger crystals from small ones) that is common in aqueous media (Lianos and Thomas, 1986). Here, although the surfactant is not an active component in the chemical sense, it provides an environment conducive to the sustenance of the desired size and morphology. Multilayered cadmium sulfide particle assemblies have also been prepared on gold nanoparticles by initially creating ~3-nm particles in the same surfactant medium and depositing it on gold nanoparticles coated with alkanethiols. A schematic of such a process is given in Figure 14.8, explaining the sequential nature of the process (Nakanishi et al., 1998).

Shifts in the absorption spectra of these particles are correlated with their particle size (optical quantum size effect). In Figure 14.9, the blue shift (i.e., spectral shift to lower wavelength region of light) correlates well with the decreasing size of the CdS particles in this case, and the particle size is proportional to the size of the water pool of the reverse micelle.

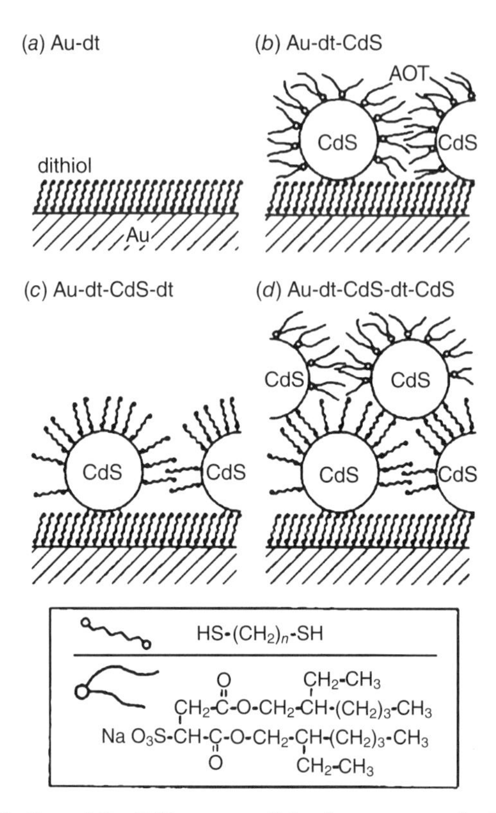

FIGURE 14.8 Binding of the CdS nanoparticles from reversed micelles onto gold via dithiol and the formation of an alternating layer-by-layer structure: (a) alkanedithiol (dt) adsorbed (self-assembled monolayer [SAM]) on a gold substrate (Au-dt); (b) CdS nanoparticles attached onto the SAM (Au-dt-CdS); (c) adsorption of dithiol layers onto CdS nanoparticles (Au-dt-CdSdt); and (d) formation of a second CdS-nanoparticle layer (Au-dt-CdS-dt-CdS). AOT, aerosol OT. Reprinted with permission from Nakanishi, T., et al. (1998) *J. Phys. Chem. B* **102**, 1571. Copyright 1998, American Chemical Society.

A similar synthetic strategy in reverse micelles with a water content of the reverse micelle, defined as [H_2O]/[surfactant] (six in this case), for the same surfactant, sodium bis(2-ethylhexyl)sulfosuccinate, in isooctane at a concentration of 0.1 mol/L, has been used to obtain ruthenium sulfide (RuS_2) nanoparticles immobilized on thiol-modified polystyrene particles. Hydrogen sulfide gas is injected into the surfactant solution of ruthenium chloride ($RuCl_3$) to get RuS_2 nanoparticles. Then, thiol-modified polystyrene is injected under mild stirring conditions to get the desired composite that exhibits photocatalytic activity in generating hydrogen gas from water. The thiol modification was performed by substituting the chloro group in chloromethylstyrene with the sulfhydryl group (Hirai et al., 2003).

Superhigh magnetic recording densities in the range of 100 GB (gigabyte)/in.^{2} to 1 TB (terabyte)/in.^{2} or more are sought in the magnetic recording

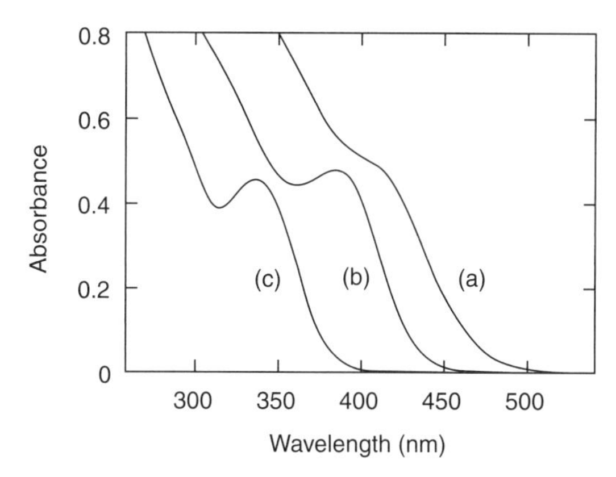

FIGURE 14.9 Absorption spectra of CdS nanoparticles in heptane prepared with different $[H_2O]/[surfactant]$ ratios: (a) 5.5, (b) 4.0, and (c) 2.7. Concentration of CdS: $\sim 5 \times 10^{-4}$ mol/L. Reprinted with permission from Nakanishi, T., et al. (1998) *J. Phys. Chem. B* **102**, 1571. Copyright 1998, American Chemical Society.

devices for which magnetic colloids (*ferrocolloids*) are developed. Monodisperse cobalt nanocrystals that can be assembled into magnetic superlattices can be prepared by reducing cobalt(II) chloride with dioctyl superhydride solution in the presence of the surfactant, oleic acid, and trialkyl phosphine. At higher temperatures, clusters of cobalt grow into nanosized single crystals. Micelles of oleic acid act as a confined space for this reaction. In a typical experiment, 1 mmol anhydrous cobalt chloride and 1 mmol oleic acid were mixed under nitrogen. At a temperature of 100°C, 3 mmol tributylphosphine was added, followed by an injection of 1 mL of 2 mmol dioctylether superhydride solution at 200°C. After 20 min, black metal particles were precipitated with ethanol at room temperature. The single crystalline cobalt particles formed (with a complex cubic structure that is related to the beta phase of elemental manganese) were converted to a hexagonal close-packed form by annealing the nanocrystals at 300°C. They were then deposited on substrates such as amorphous carbon-coated copper grids, silicon nitride membranes, and silicon (100) wafers by the evaporation of the carrier solvent, after redispersion in hexane stabilized by oleic acid into two- and three-dimensional self-assemblies of magnetic superlattices (Sun and Murray, 1999).

Magnetic iron particles stabilized by amphiphilic molecules similar to a double-chain surfactant have been prepared using thiodiglycolic acid and 4-vinylaniline. These nanoparticles provide another layer of regulation to the effectiveness of recording media in terms of data storage capacity and quality of recording. During the synthesis, thiodiglycolic acid and 4-vinylaniline were reacted with the nanoparticles formed by chemical coprecipitation of a 1:2 mixture of Fe^{2+} : Fe^{3+} with alkali under an inert environment. In effect, the coating of the two compounds produces the same effect as an adsorbed layer of an amphiphile (Figure 14.10) (Shamim et al., 2005).

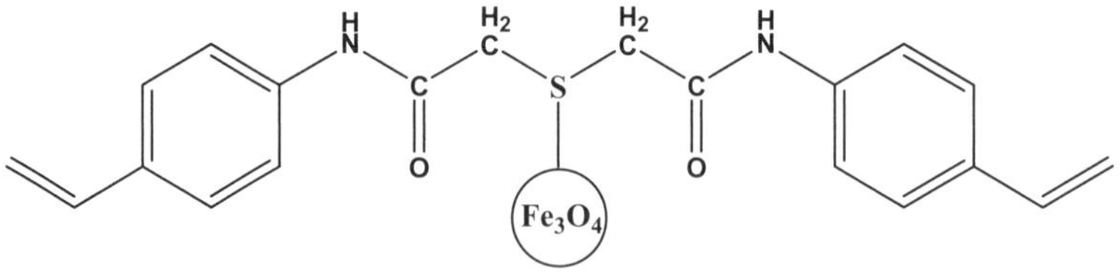

FIGURE 14.10 Structure of the surfactant-coated magnetic nanoparticles (Shamim et al., 2005).

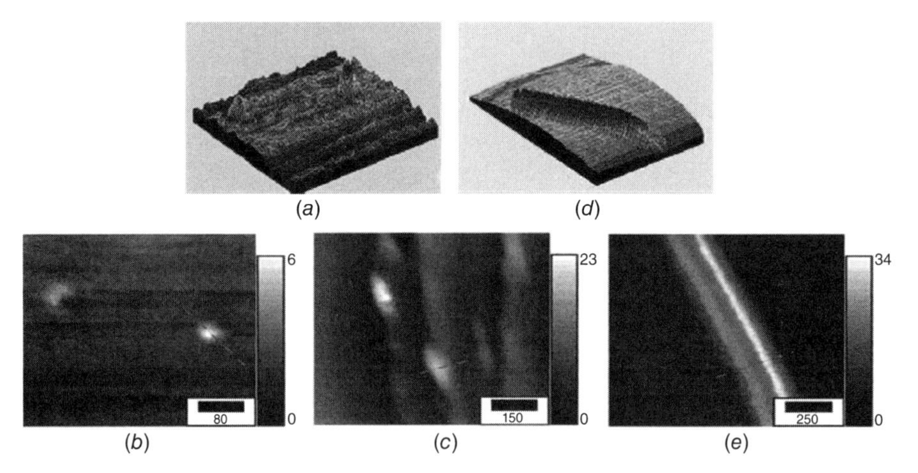

FIGURE 14.11 STM topographic images of individual iron nanoparticles and extended nanostructures synthesized in the mixed monolayer deposited onto the surface of a graphite substrate without external fields (a and b); under applied magnetic (c) and combined magnetic and electric fields (d and e). Note the shapes of the particles marked by an indicator bar, and the differences in the magnifications in (b), (c), and (d). Reprinted with permission from Khomutov, G. B., et al. (2002) *Colloids Surf.* **198–200**, 593. Copyright 2002, Elsevier.

Nanoparticles can also be prepared in a 2-D matrix of surfactant assembly on a Langmuir–Blodgett film at the liquid–air interface and deposited on a graphite substrate under individually or synergistically applied external electric and magnetic fields (Figure 14.11). Iron pentacarbonyl, ($Fe(CO)_5$), contained in a mixed Langmuir monolayer at the water–air interface with stearic acid as a surfactant matrix, when photochemically decomposed with ultraviolet (UV) light at 300 nm, generated iron-containing magnetic nanoparticles. The surfactant provided a 2-D assembly and prevented the aggregation of the particles. The monolayer was spread on the film balance from a chloroform solution at a pH of 5.6. Anisotropic properties could be induced in the particles by applying electrical and magnetic fields during photolysis (Khomutov et al., 2002).

Disk-shaped hcp (hexagonal close-packed) nanoparticles of cobalt with a controlled crystal structure have been obtained by rapid decomposition of

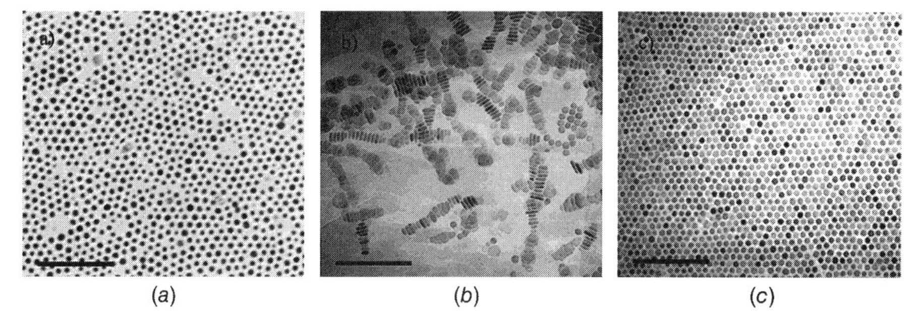

FIGURE 14.12 TEM pictures of Co nanocrystals (left to right) synthesized (a) after refluxing for 5 min in the presence of oleic acid, (b) in the presence of both oleic acid and trioctylphosphine oxide (TOPO) at 10 s, and (c) in the presence of both oleic acid and TOPO at 5 min. Bars are 100 nm. Reprinted with permission from Puntes, V. F., et al. (2002) *J. Am. Chem. Soc.* **124**, 12874. Copyright 2002, American Chemical Society.

cobalt carbonyl [$Co_2(CO)_8$] in the presence of alkylamines, phosphine oxide, and/or oleic acid surfactants, which improved the size and shape and nanocrystal stability (Puntes et al., 2002). Surfactant molecules adsorbed on a particular crystallographic face regulate the surface energy of the nanoparticles that, in turn, influences the direction of growth of crystals because the rate constants for the crystal growth depends exponentially upon its surface energy within the kinetic limit. The magnetic properties of the nanocrystals were measured by a SQUID (superconducting quantum interference device that works as a magnetometer and measures extremely weak magnetic fields), and morphological characteristics were measured by diffraction techniques and microscopy. The surfactant environment and the experimental conditions influenced the size distribution, shape control, and stability of the nanocrystals. As an example, by injecting $Co_2(CO)_8$ into hot ortho-dichlorobenzene in the presence of oleic acid, spherically shaped ε-Co nanocrystals were obtained with a relatively broad size distribution (10–20% variation in diameter) as in Figure 14.12a. But the presence of trioctylphosphine oxide (TOPO) leads to anisotropically shaped Co nanocrystals (disks) in the first 10 s as in Figure 14.12b. The disks then rapidly redissolve with increasing temperature. Minutes later, a very narrow size distribution of spherically shaped nanocrystals results with a 3–5% variation in diameter as in Figure 14.12c. TOPO acts both as a selective absorber that alters the relative growth rates of different faces of the crystals and as a molecule that enables size distribution focusing. A narrow size distribution may also be achieved by treating the dispersed fractions by sequential precipitation and centrifugation (Sun and Murray, 1999).

Surface-active polymers like poly(3-hexylthiophene) (P3HT), (Figure 14.13) in conjunction with nanorods of CdSe, have been used to fabricate hybrid nanorod–polymer solar cells, where the band gap has been tuned by controlling the nanorod length and radius such that the absorption spectrum of the cell overlaps with the solar energy spectrum. CdSe nanorods,

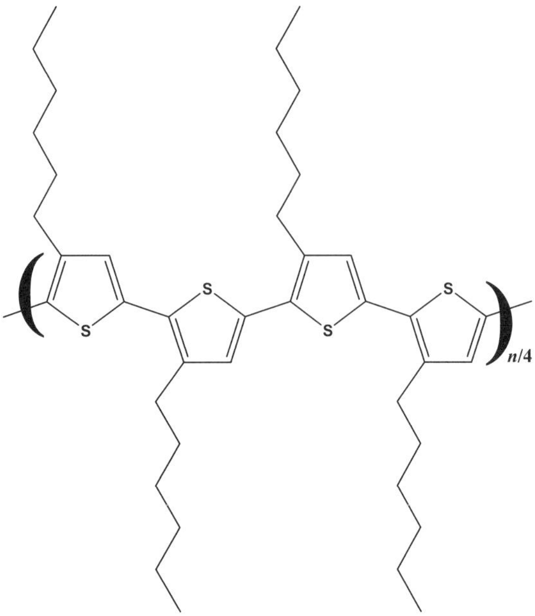

FIGURE 14.13 Structure of regioregular polymer P3HT.

in combination with the conjugated polymer, P3HT, create charge transfer junctions with high interfacial area. This is an important application of nanotechnology involving surfactant-like polymers, where their adsorption behavior is decisive in fine-tuning the particles' dimension. Further, the nanorod length fine-tunes the efficiency of electron transport through the thin film (Huynh et al., 2002). This procedure can still be rated as a bottom-up process because the polymer molecules are assembled onto nanorods from solution.

Composite structures of semiconductor nanocrystals of cadmium sulfide (Figure 14.14) were prepared by linking the nanostructures to metal surfaces (gold or aluminum) through SAMs of alkanedithiol for gold, and alkanethiol carboxylic acid for aluminum substrates. The metal substrates were either coatings of thin films by vapor deposition that were further glued to the glass slides using (3-mercaptopropyl)-trimethoxysilane or naked thin metallic blocks. The SAMs were formed on the gold substrate with 1,6-hexanedithiol, one thiol group linking to the gold film and the other to the semiconductor particle. For the aluminum substrate, 1-sulfhydryl-propane-3-carboxylic acid was the bifunctional amphiphile, which linked to aluminum through the carboxyl group and to cadmium sulfide particles through the thiol group. The cadmium sulfide crystals were either made in inverse micelles of dioctyl sulfosuccinate in heptane or in aqueous solutions by mixing cadmium chlorate and sodium sulfide solutions (Colvin et al., 1992).

Surfactant assemblies created by adsorption from solution can serve as templates to shape nanostructures of inorganic particles such as oxides, sul-

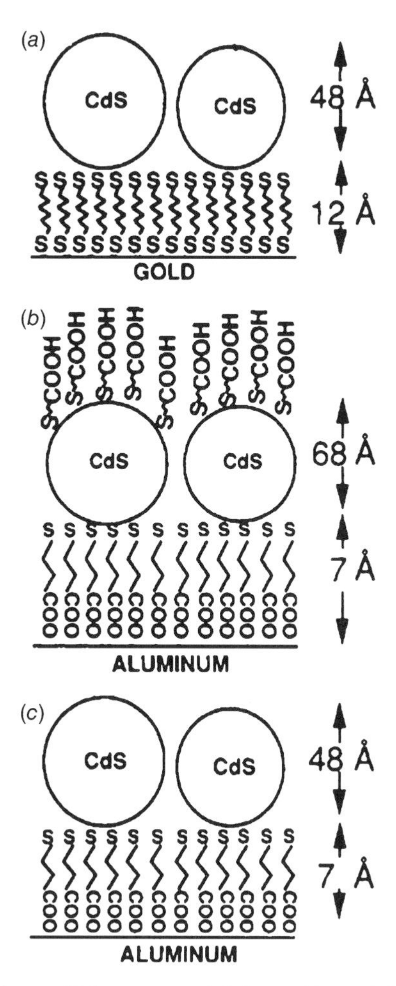

FIGURE 14.14 Cadmium sulfide nanocrystals bound to metal surfaces: (A) Cadmium sulfide from inverse micelles of sodium dioctyl sulfosuccinate in heptane bound to gold via 1,6-hexanedithiol. (B) Cadmium sulfide nanocrystals synthesized in water and coated with carboxylates bound to aluminum. (C) Cadmium sulfide from the same inverse micelles bound to aluminum via a thiocarboxylic acid. The lengths of the layers of the linker thiol molecules and the particles are marked in the diagram. Reprinted with permission from Colvin, V. L., et al. (1992) *J. Am. Chem. Soc.* **114**, 5221. Copyright 1992, American Chemical Society.

fides, and nitrides of metals. The procedure is typical of the one used in creating casts and molds for making multiple replicas using wax models of the figure. An initially carved wax model is enveloped with plaster and, upon removing the wax by melting, the cast remains. The resulting hollow shape can be filled with metals to get the replica. Extending this principle to the preparation of nanoparticles, the surfactant self-assembly created following appropriate procedures is either encased with the particulate materials or the gaps and voids

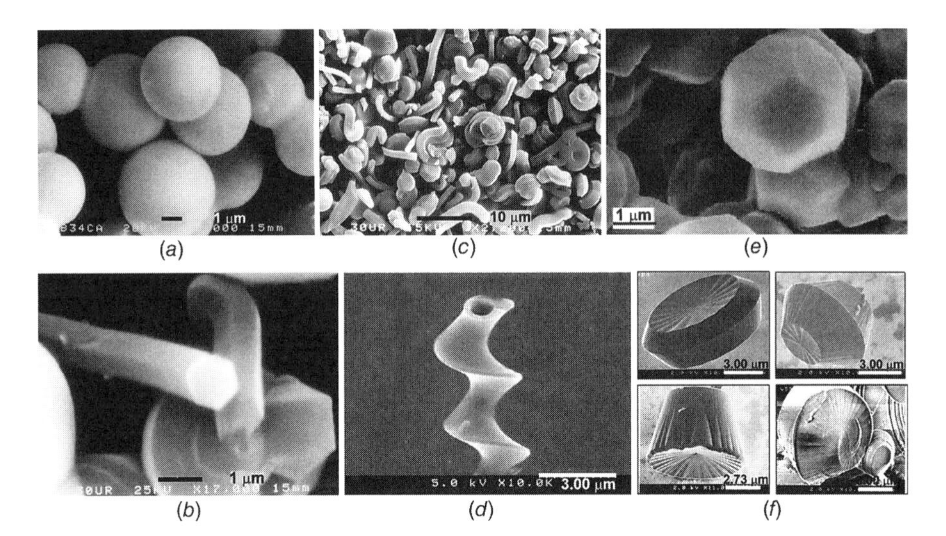

FIGURE 14.15 Scanning electron micrographs of surfactant-templated materials with different particle shapes: (a) spheres; (b) hexagonal rods; (c) shapes formed from twisted rods; (d) hollow helicoids; (e) single crystals; and (f) discoid, gyroids, and spiral shapes with radial patterns. Note the magnification scales on each segment. Reprinted with permission from Edler, K. J. (2004) *Philos. Trans. R. Soc. Lond. A* **362**, 2635. Copyright 2004, The Royal Society.

within these assemblies are filled with them. By burning out the surfactants at elevated temperatures, various nanostructured porous materials result that find application as catalysts or adsorbents, and in several areas of materials engineering. Figure 14.15 shows the scanning electron micrographs of surfactant-templated materials with different particle shapes: (1) spheres; (2) hexagonal rods; (3) shapes formed from twisted rods; (4) hollow helicoid; (5) single crystals; and (6) discoid, gyroids, and spiral shapes with radial patterns. Soap and sand are used as the construction tools for the nanotechnology, indicative of the link between surfactants, metal oxide particles, and nanotechnology (Edler, 2004).

The possible mechanisms for the organization of silica nanoparticles directed by the surfactant self-assembly may be visualized in the following scheme (Figure 14.16): gray indicates surfactant-filled regions, while white represents silica. (1) Micelle assembly mechanism: Micelles become coated with silica monomers, causing them to change shape. The coated micelles then assemble into particles. (2) Phase separation mechanism: Silica monomers polymerize to become oligomers; surfactant micelles adsorb onto the oligomers and phase separation occurs, forming disordered concentrated droplets. Nanostructure formation within the droplets drives the particles to transform from spheres into other shapes. Particles with shapes such as those shown could form by either mechanism.

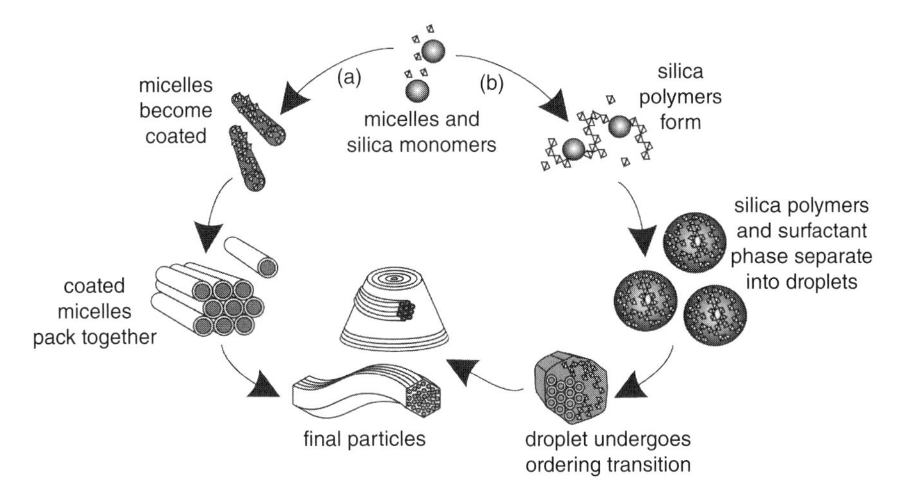

FIGURE 14.16 Mechanisms for the organization of silica into nanostructures in surfactant assembly. Reprinted with permission from Edler, K. J. (2004) *Philos. Trans. R. Soc. Lond. A* **362**, 2635. Copyright 2004, The Royal Society.

B. Top-Down Methods

Similar to the dispersion methods used in the preparation of colloids, solid-state processes are available for the preparation of nanoparticles. Top-down methods like laser ablation and mechanochemical synthesis can yield nanoparticles. In the former, highly focused, high-energy laser beams interact with the solid surface, chipping out nanoscaled particles; in the latter, high-energy collisions of the grinding media and reactant particles lead to nanoparticles during a ball milling process. In another process called physical vapor synthesis, a metal is heated to release vapors, which are then cooled with a gas to a liquid and then further cooled to nanoparticles. If the cooling gas is reactive, such as an oxygen-containing gas, oxide nanoparticles result.

Because the top-down methods usually do not benefit much from surfactant assemblies, this topic is not further elaborated here, leaving aside the fact that surfactants can serve as grinding aids. Moreover, the physical limits of dimensions realizable by top-down methods, while creating nanoparticles or patterning them into useful devices, make them less attractive than bottom-up methods where the nanomaterial can be created by the control of the molecular organization.

III. SURFACTANTS IN NANOTECHNOLOGY APPLICATIONS

A. Nanomotors

Producing nanomotors is one of the specific goals of nanotechnology. The purpose of a motor is to obtain work. Nanotechnologists build nanomotors that convert chemical energy into work. Such a devise can involve either harnessing normal chemical energy or exploiting biochemical energy.

In normal chemical methods, work or motion is obtained by mixing two chemicals. Such self-propelling nanostructures, for example, are produced by depositing nanoscale islands of tin on a copper surface, when the islands spontaneously move around on the surface. The chemical energy released by alloying tin with copper is an exothermic process that produces the motion (Schmid et al., 2000).

The biochemical connection to nanotechnology stems from the action of biological motor proteins with nanomaterial systems. Life is made possible by the synchronization of several nanoscopic biomachines. When an entire organism moves, a cascade of nanomachines operates, as in the case of the bacterial flagella motor. For example, the power provided by the molecule adenosine triphosphate (ATP) can be materialized within the confines of a nanostructure, and effective work can be extracted as follows: The light harvesting molecule chlorophyll and the enzyme ATP synthase are incorporated in a microreactor fabricated from liposomes (spherical membranes made up of two layers of fatty lipid molecules). Chlorophyll absorbs light and initiates a series of chemical reactions enabling the enzyme to generate ATP. As the protons pass through the ATP synthase molecule to the outside of the liposome, they cause the protein to spin (Service, 2000).

Catalytic nanomotors that operate via the liquid–air interfacial tension gradient have been demonstrated for platinum/gold rods in hydrogen peroxide solution (Figure 14.17). The observed velocities of the platinum/gold nanorods are comparable to those of flagellar bacteria, which are about 2–10 body lengths per second (a typical flagellar bacterium can have a body length of >10 μ). The movement is non-Brownian with the platinum end moving forward. Out of these two metals, only platinum can decompose hydrogen peroxide and gold cannot.

The interfacial tension at the liquid–air interface, the length of the rod, its cross-sectional area, and the oxygen evolution rate were involved in the propelling motion. Experiments in an ethanol–water medium show that the velocity derived from the force on the rod axis depends linearly on the product of the oxygen evolution rate (normalized for the surface area of the rod) and the

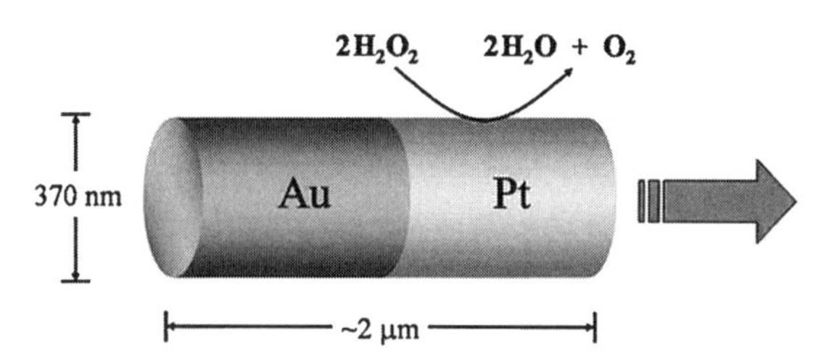

FIGURE 14.17 Platinum/gold nanorod and its dimensions. Reprinted with permission from Paxton, W. F., et al. (2006) *Angew. Chem. Int. Ed.* **45**, 5420. Copyright 2006, Wiley-VCH Verlag GmbH.

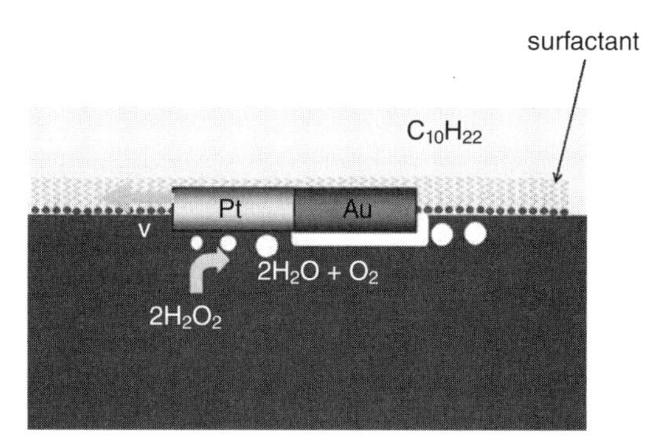

FIGURE 14.18 Oxygen evolution from hydrogen peroxide of a platinum/gold rod at the water–decane interface in the presence of surfactants. Reprinted with permission from Dhar, P., et al. (2006) *Nano Lett.* **6**, 66. Copyright 2006, American Chemical Society.

liquid–vapor interfacial tension. A similar motor action can be obtained with different classes of surfactants due to their great variation in interfacial tension(Paxton et al., 2004). Work with aqueous sodium dodecyl sulfate (SDS) solution in the concentration range of 1–10 mM and with the same rod at the decane–water interface showed that SDS forms a Gibbs monolayer at the liquid–liquid interface (Figure 14.18). The direction-dependent thermal fluctuations, resolved as perpendicular and parallel components that accompany the chemical reaction, control the collective diffusion coefficient of the body, which slows with the addition of a surfactant (SDS). Analysis of thermal orientation fluctuations and the motion of the rod generated values of surface shear viscosities of the Gibbs monolayer (Dhar et al., 2006).

A chemomechanical pulsating nanogel system responsive to pH change has been demonstrated with bromate-sulfite, pH-based oscillatory reactions within hydrogels of polyelectrolytes like acrylates, stabilized by SDS (Varga et al., 2006). Change in pH oscillates the nanogel into swollen and shrunken states, thus producing motion. Gel nanobeads that show good colloid stability over a wide range of pH, ionic strength, and temperature respond to pH stimulus, providing swelling accompanied by a 12-fold volume increase as indicated by their hydrodynamic radii. The surfactant imparts charge to the gel surface, balancing the attractive and repulsive energies of the shrunken gel at low pH values, when the surface carboxyl groups of the acrylates are completely protonated. The sudden swelling of the gel that produces more than an order of magnitude volume change as a result of the proton-activated oscillatory reaction causes work to be performed. The reactions that produce the chemical oscillations may be represented by the equations

$$BrO_3^- + 3H_2SO_3 \rightarrow 3SO_4^{2-} + Br^- + 6H^+$$

$$BrO_3^- + 6HSO_3^- \rightarrow 3S_2O_6^{2-} + Br^- + 3H_2O.$$

Ferrofluids of magnetic nanoparticles, for example, iron oxide, stabilized electrostatically by anionic surfactants, and polymers like acrylic acids or derivatives for steric stabilization, have applications as nanomotors in a dynamic magnetic field. The basic idea is to induce electricity in a coil at a desired geometric arrangement by the motion of the magnetic nanoparticles. The direction of motion of magnetic nanoparticles depends on the applied magnetic field strength: Below a critical field strength, magnetic particles move opposite to the direction of the applied field, and vice versa. This critical magnetic field is a function of the frequency, the concentration of the magnetic particles, and the fluid dynamic viscosity. The torque induced under these conditions is manifested as viscosity changes and causes the ferrofluid nanoparticles to rotate. (Zahn, 2001).

B. Other Nanodevices

Fabrication of nanodevices is achieved by an extension of techniques used by engineers engaged in microelectronics. The making of nanodevices involves surfactants at several stages. For example, surfactants like SDS help to form and stabilize metal-containing polymer latex nanoparticles in macro- and microemulsions that are used in nanolithography, a process essential in nanopatterning on microchips (Schreiber et al., 2009). Similarly, the efficiency of dip-pen nanolithography (DPN), a technique that enables the drawing or creation of nanostructures and patterns of one material over the other using scanning probe microscopy, has been improved by surfactants. For example, nonionic ethylene oxide surfactants are found to enhance the wettability of partially hydrophobic substrate surfaces. In DPN, an AFM tip (called a "nib") is adsorbed with a surfactant such as 16-mercaptohexadecanoic acid (called an "ink") and made to deposit on a substrate like gold foil (called a "paper"). The surfactant ink flows from the AFM nib and adsorbs on the gold paper, in a process analogous to writing with a pen at the substrate/water (moisture present provides an aqueous environment) interface, by capillary action to create a writing pattern called "words" (Figure 14.19). In regular black printing inks, a carbon particle-based formulation adsorbs on cellulose-based paper in an analogous manner. The importance of this process lies in the demonstrated deposition as patterns onto surfaces with alkanethiol-coated magnetic iron oxide particles (with application as magnetic storage media) and with protein and nucleic acid (potential applications in panel screening for immunoassays, proteomics, and biochips) (Piner et al., 1999; Hamley, 2003).

In another investigation, variation in the concentration of a nonionic biocompatible surfactant, polyoxyethylene (20) sorbitan monolaurate (Figure 14.20) (Chapter 1, Section IC5), in the range of 0–0.1% v/v dissolved in phosphate-buffered saline of pH = 7.2–7.4, provided a new experimental variable for the dip-pen nanowriting of maleimide-PEO$_2$-biotin (Figure 14.21) on a glass substrate functionalized with mercaptosilane. This takes advantage of the high-affinity association constant of the biotin–protein streptavidin

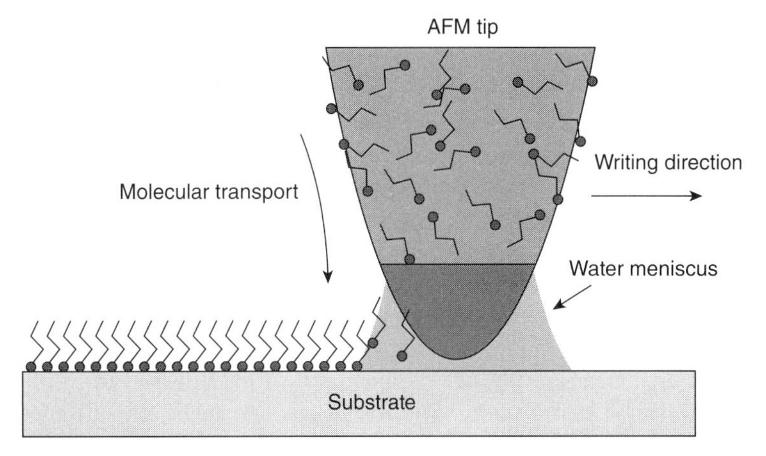

FIGURE 14.19 Depiction of the process for printing patterns on substrates with an AFM tip. Reprinted with permission from Hamley, I. W. (2003) *Angew. Chem. Int. Ed.* **42**, 1692. Copyright 2003, Wiley-VCH Verlag GmbH.

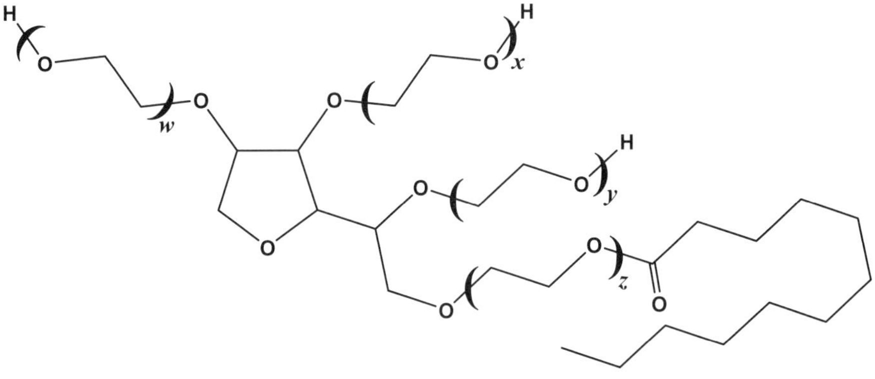

FIGURE 14.20 Structure of polyoxyethylene (20) sorbitan monolaurate ($w + x + y + z = 20$).

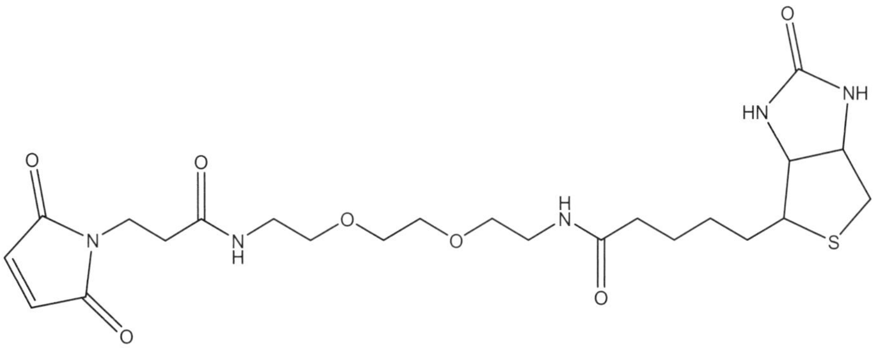

FIGURE 14.21 Structure of maleimide-PEO_2-biotin.

complex (~10^{15} L/mol). The ability to directly pattern biological molecules on the surface by the nanolithography in this case complemented well with the relative humidity, tip–substrate contact force, scan speed, and temperature (Jung et al., 2004).

Surfactant-based nanoassemblies are also candidates in tuning the refractive index of the liquid medium in immersion nanolithography, a process favored in modern computer chip manufacturing. In immersion nanolithography, patterning on the semiconductor chip is created by UV laser beam that passes through a liquid immersion medium, which helps to shift the wavelength of the laser light enhancing its photonic energy. The combination of surfactants or crown ethers and inorganic salts as components of the immersion medium has been particularly beneficial for this process. For example, the effect of cadmium chloride concentration on the degree of ionization of micellar nanoparticles of SDS and cetyl trimethyl ammonium bromide (CTAB) was found to be critical in tuning the refractive index of the nanolithographic immersion medium (Lee et al., 2005; López-Gejo et al., 2007).

C. Drug Delivery

Targeted drug delivery is an important medical process owing to its potential to mitigate the unwanted side effects of drugs on healthy tissues and organs. Nanoparticles that encapsulate the drugs selectively release the molecules on specific sites. These nanostructures are often stabilized by surfactant molecules. The nanoparticles are usually of biodegradable polymers that have high stability in biological fluids and during storage. Surfactants can further improve the pharmaceutical efficacy of the drugs by controlling the particle size, size distribution, morphology, surface chemistry, surface hydrophobicity, surface charge, drug entrapment efficiency, *in vitro* drug release, and the interaction between drug-loaded particles and cell membranes. Surfactant micelles and similar self-assemblages have been proposed and studied earlier as potential drug carriers (Kunjappu et al., 1993).

Investigations on the interaction of the drug paclitaxel (an anticancer drug with limited aqueous solubility and limited oral bioavailability—bioavailability is an index that indicates the degree to which a drug or other substance becomes available to the target tissue after administration) (Figure 14.22), encapsulated and stabilized by d-α-tocopheryl polyethylene glycol 1000 succinate (TPGS; α-tocopherol is vitamin E) (Figure 14.23) with a biomembrane composed of a dipalmitoylphosphatidylcholine (DPPC) lipid monolayer on a Langmuir film balance, indicated that TPGS exhibited a notable increase in the loading of the drug and an improvement in its release (Mu and Seow, 2006).

D. Nanostructural Architectural Control of Materials

By controlling the surfactant assembly phases, especially by manipulating the concentration of surfactants and the solvent, it is possible to achieve enhanced mechanical and catalytic properties through lyotropic liquid crystalline (LLC)

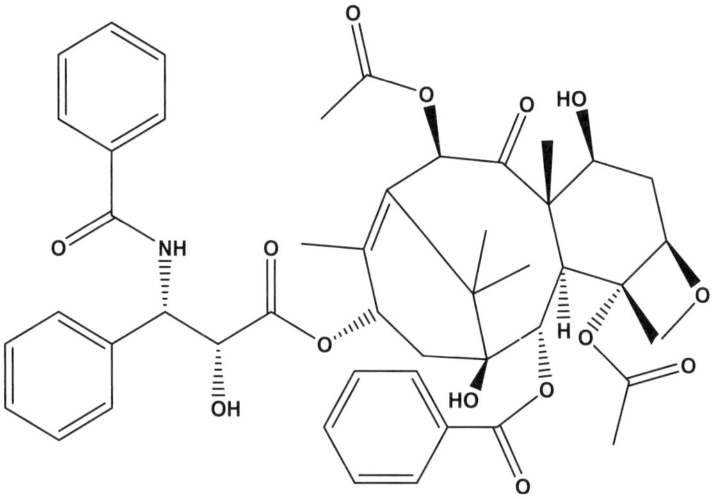

FIGURE 14.22 Molecular structure of paclitaxel (Taxol).

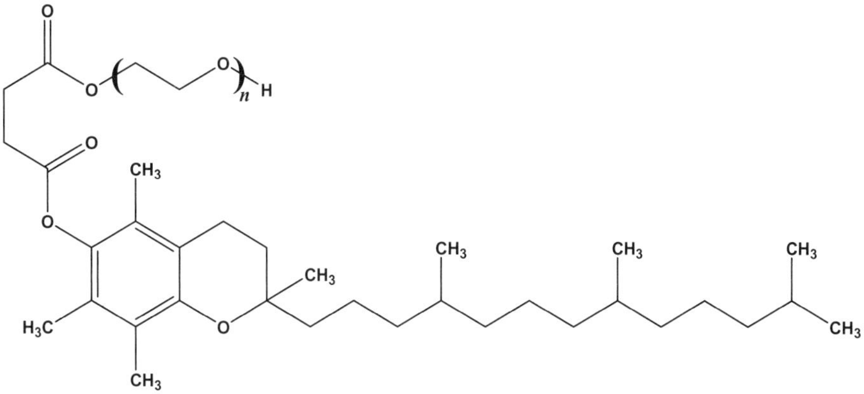

FIGURE 14.23 Molecular structure of d-α-tocopheryl polyethylene glycol 1000 succinate (TPGS).

phases (Chapter 3, Section IIC). The following scheme (Figure 14.24) depicts the formation of some of the surface-active and surfactant-like non-cross-linked polymer architectures using a variety of radical initiators ($K_2S_2O_8$, azoisobutyrylnitrile [AIBN], and UV light) (Miller et al., 1999).

Surface-active peptide molecules are selected using combinatorial phage-display libraries to select the peptides that bind to semiconductor nanosurfaces (phage display is the method used to study the interaction of proteins with biomolecules using bacteriophages; bacteriophage refers to viruses that infect bacteria; in combinatorial biology, as applied here, a large number of protein molecules are created through phage display to select the best

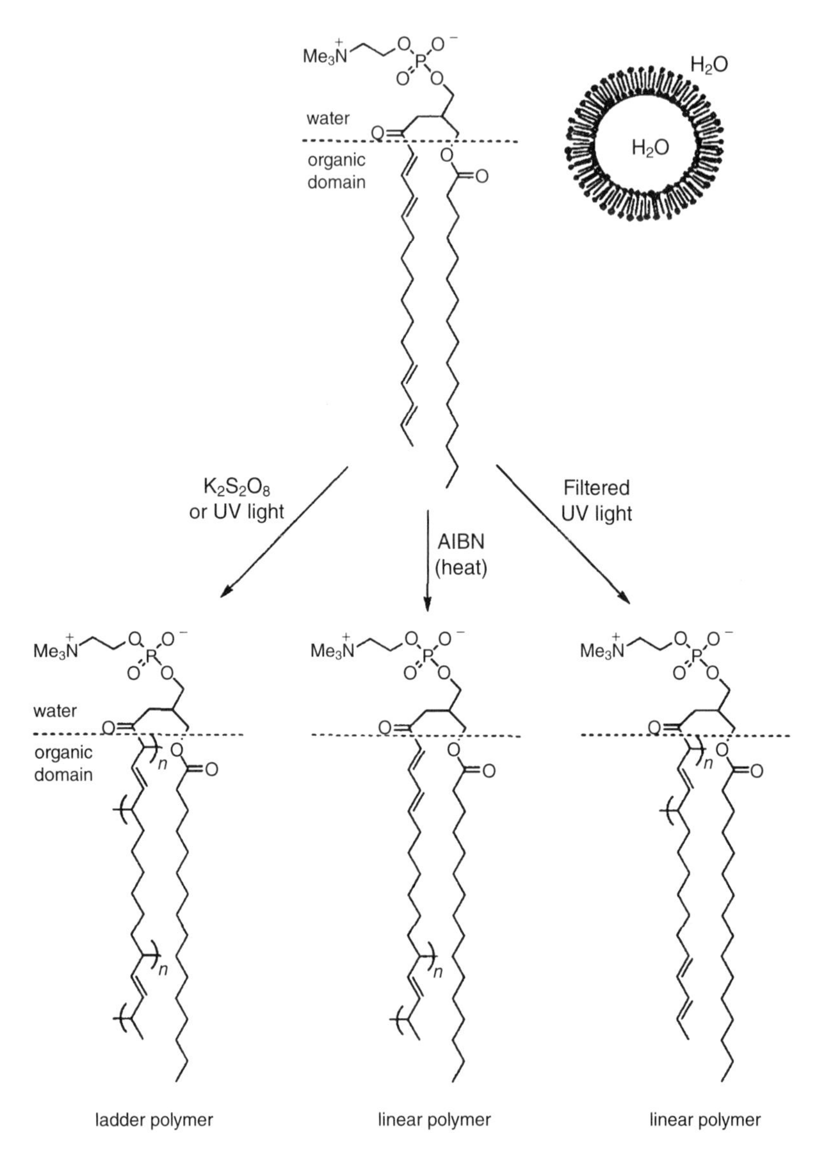

FIGURE 14.24 Reaction scheme for the synthesis of various polymers. Reprinted with permission from Miller, S. A., et al. (1999) *Curr. Opin. Colloid Interf. Sci.* **4**, 338. Copyright 1999, Elsevier.

candidate suitable for the purpose). The binding process is highly specific, depending on the crystallographic orientation and composition, and can be utilized for controlling the nucleation of the mineral phase of inorganic materials such as calcium carbonate and silica, and the assembly of crystallites and other nanoscale building blocks into complex structures needed for biological function. In these processes, the recognition capabilities and interactions inherent in biological systems are exploited. Such processes can profitably be used for assembling materials with interesting optical and electronic properties (Whaley et al., 2000).

E. Nanotubes

Single-walled nanotubes (SWNTs) are synthesized from carbon monoxide and the surfactant, lithium dodecyl sulfate, in a coagulation-based spinning process. The process produces nanotube-polyvinyl alcohol gel fibers, which can be converted to 100-m-long nanotube composite fibers roughly 50 μm in diameter (Dalton et al., 2003). Such fibers are tougher than any synthetic or natural organic fibers known so far, and exhibit twice the stiffness and strength and 20 times the toughness of steel wire of the same weight and length. The toughness is more than four times that of spider silk and 17 times greater than Kevlar fibers used in bulletproof vests.

F. Nanodetergents

For high-temperature applications, nano-based detergents are prepared by coating surfactant molecules onto a core particle such as calcium carbonate. Figure 14.25 shows such a particle composite for a benzenesulfonate detergent and calcium carbonate. A variety of amphiphilic structures are found to be useful in covering the calcium carbonate particles that include calixarenes and

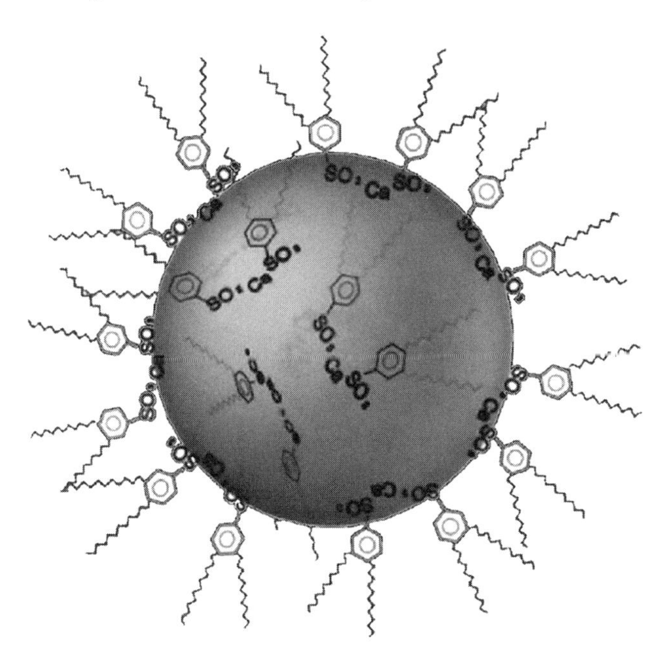

FIGURE 14.25 Cartoon representation of the adsorption of nanodetergents on calcite. Reprinted with permission from Hudson, L. K., et al. (2006) *Adv. Colloid Interf. Sci.* **123–126**, 425. Copyright 2006, Elsevier.

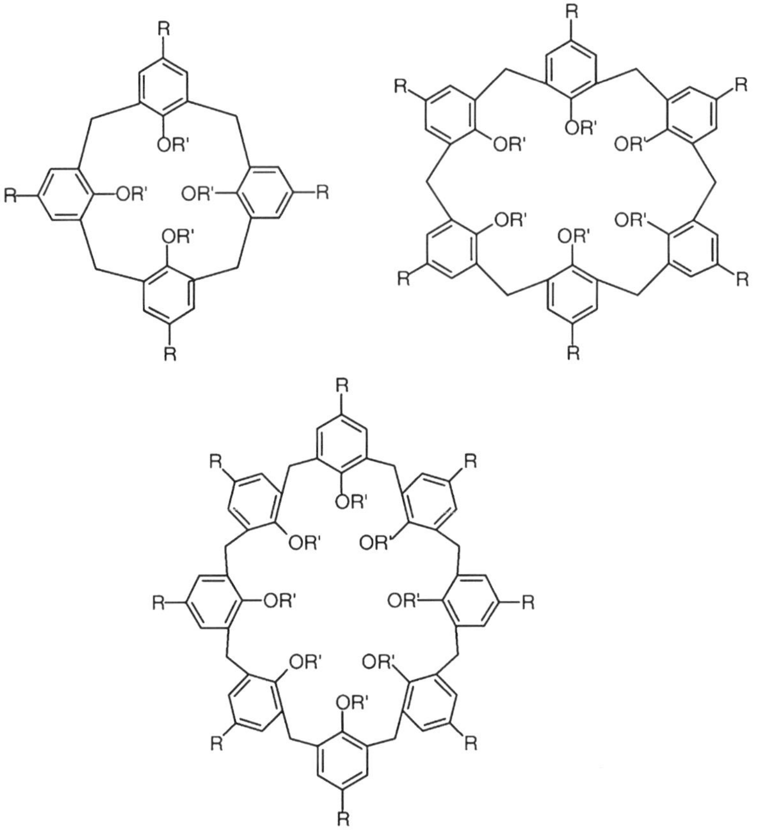

FIGURE 14.26 Structures of generic calixarene molecules. Reprinted with permission from Hudson, L. K., et al. (2006) *Adv. Colloid Interf. Sci.* **123–126**, 425. Copyright 2006, Elsevier.

phenyl derivatives with hydrophilic and hydrophobic regions (Figures 14.26 and 14.27). These detergents, in engine oil formulations, promote fuel efficiency and effective trouble-free operation of engines by enhancing (1) acid neutralization, (2) high-temperature detergency, (3) oxidation inhibition, and (4) rust prevention (Hudson et al., 2006).

G. Surfactant Nanoassemblies in the Origin of Life

Surfactants are believed to be the precursors to the nucleotide-based molecules that led to the origin of life. The lipid world theory is favored by many experts in lieu of the RNA world theory. Prebiotic molecules are proposed to have included surfactants along with amino acids, nucleic acid bases, and some simple molecules. Nano-self-assembly of surfactants looking somewhat like vesicles may have hosted the prebiotic molecules instrumental in the origin of life (Figure 14.28) (Walde, 2006). These nanoreactors could have enhanced the

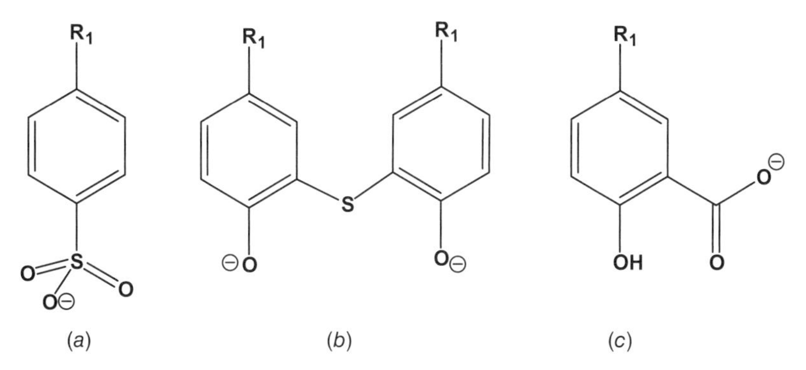

FIGURE 14.27 Molecular structures of (a) alkyl sufonate, (b) sulfurized alkyl phenol, and (c) alkyl salicylate ($R^1 = C_9–C_{16}$). Reprinted with permission from Hudson, L. K., et al. (2006) *Adv. Colloid Interf. Sci.* **123–126**, 425. Copyright 2006, Elsevier.

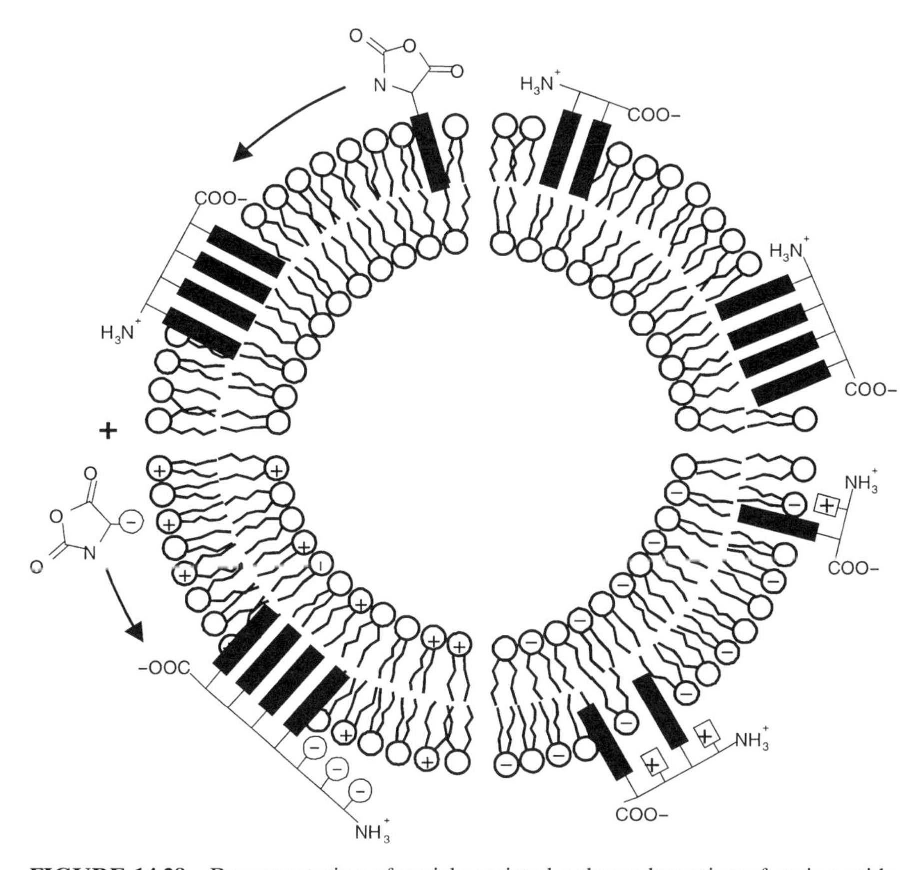

FIGURE 14.28 Representation of vesicle-assisted polycondensation of amino acids. Reprinted with permission from Walde, P. (2006) *Orig. Life Evol. Biosph.* **36**, 109. Copyright 2006, Springer.

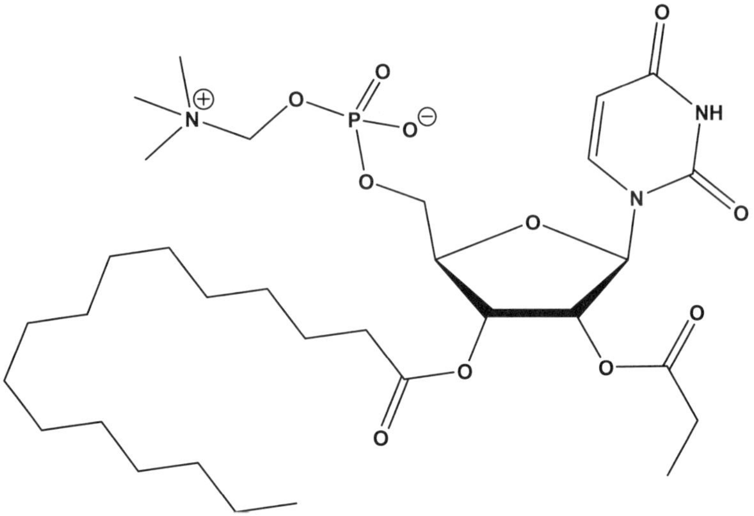

FIGURE 14.29 Surfactant structure based on a nucleotide. Reprinted with permission from Walde, P. (2006) *Orig. Life Evol. Biosph.* **36**, 109. Copyright 2006, Springer.

catalytic effects in the conversions that lead to all the effects, including locomotion, that are characteristic of life. A model nucleotide-based surfactant has a structure shown in Figure 14.29.

REFERENCES

Andres, R. P., J. D. Bielefeld, I. J. Henderson, D. B. Janes, V. R. Kolagunta, C. P. Kubiak, W. J. Mahoney, and R. G. Osifchin (1996) *Science* **273**, 1690.

Anonymous (eds.), (2004) *Nanoscience and Nanotechnology: Opportunities and Uncertainties* Document 19/04, The Royal Society and the Royal Academy of Engineering, London. http://www.nanotec.org.uk/finalReport.htm (accessed December 2011).

Bognolo, G. (2003) *Adv. Colloid Interf. Sci.* **106**, 169.

Boyd, B. J., I. Krodkiewska, C. J. Drummond, and F. Grieser (2002) *angmuir* **18**, 597.

Colvin, V. L., A. N. Goldstein, and A. P. Alivisatos (1992) *J. Am. Chem. Soc.* **114**, 5221.

Dalton, A. B., S. Collins, E. Muñoz, J. M. Razal, V. H. Ebron, J. P. Ferraris, J. N. Coleman, B. G. Kim, and R. H. Baughman (2003) *Nature* **423**, 703.

Dhar, P., T. M. Fischer, Y. Wang, T. E. Mallouk, W. F. Paxton, and A. Sen (2006) *Nano Lett.* **6**, 66.

Drexler, K. E. Engines of Creation: The Coming Era of Nanotechnology, Anchor Books Editions, 1986.

Edler, K. J. (2004) *Philos. Trans. R. Soc. Lond. A* **362**, 2635.

Hamley, I. W. (2003) *Angew. Chem. Int. Ed.* **42**, 1692.

Hirai, T., Y. Nomura, and I. Komasawa (2003) *J. Nanoparticle Res.* **5**, 61.

Hudson, L. K., J. Eastoe, and P. J. Dowding (2006) *Adv. Colloid Interf. Sci.* **123–126**, 425.

Huynh, W. U., J. J. Dittmer, and A. P. Alivisatos (2002) *Science* **295**, 2425.

Jung, H., C. K. Dalal, S. Kuntz, R. Shah, and C. P. Collier (2004) *Nano Lett.* **4**, 2171.

Khomutov, G. B., S. P. Gubin, V. V. Khanin, A. Y. Koksharov, A. Y. Obydenov, V. V. Shorokhov, E. S. Soldatov, and A. S. Trifonov (2002) *Colloids Surf.* **198–200**, 593.

Kunjappu, J. T., V. K. Kelkar, and C. Manohar (1993) *Langmuir* **9**, 352.

Lee, K., J. T. Kunjappu, S. Jockusch, N. J. Turro, T. Widerschpan, J. Zhouc, B. W. Smith, P. Zimmerman, and W. Conley Advances in Resist Technology and Processing XXII, J. L. Sturtevant (ed.), *Proceedings of SPIE*, SPIE, Bellingham, WA, Vol. 5753, 2005, p. 537.

Lianos, P. and J. K. Thomas (1986) *Chem. Phys. Lett.* **125**, 299.

López-Gejo, J., J. T. Kunjappu, N. J. Turro, and W. Conley (2007) *J. Micro/Nanolith. MEMS MOEMS* **6**, 013002.

Miller, S. A., J. H. Ding, and D. L. Gin (1999) *Curr. Opin. Colloid Interf. Sci.* **4**, 338.

Mu, L. and P. H. Seow (2006) *Colloids Surf. B: Biointerfaces* **47**, 90.

Nakanishi, T., B. Ohtani, and K. Uosaki (1998) *J. Phys. Chem. B* **102**, 1571.

Nostro, P. L., M. Ambrosi, B. W. Ninham, and P. Baglioni (2009) *J. Phys. Chem. B* **113**, 8324.

Paxton, W. F., K. C. Kistler, C. C. Olmeda, A. Sen, S. K. St. Angelo, Y. Cao, T. E. Mallouk, P. E. Lammert, and V. H. Crespi (2004) *J. Am. Chem. Soc.* **126**, 13424.

Paxton, W. F., S. Sundararajan, T. E. Mallouk, and A. Sen (2006) *Angew. Chem. Int. Ed.* **45**, 5420.

Piner, R. D., J. Zhu, F. Xu, S. Hong, and C. A. Mirkin (1999) *Science* **283**, 661.

Puntes, V. F., D. Zanchet, C. K. Erdonmez, and A. P. Alivisatos (2002) *J. Am. Chem. Soc.* **124**, 12874.

Schmid, A. K., N. C. Bartelt, and R. Q. Hwang (2000) *Science* **290**, 15617.

Schreiber, E., U. Ziener, A. Manzke, A. Plettl, P. Ziemann, and K. Landfester (2009) *Chem. Mater.* **21**, 1750.

Service, R. F. (2000) (News Section), *Science* **290**, 1528.

Shamim, N., Z. Peng, L. Hong, K. Hidajat, and M. S. Uddin (2005) *Int. J Nanosci.* **4**, 187.

Sun, S. and C. B. Murray (1999) *J. Appl. Phys.* **85**, 4325.

Varga, I., I. Szalai, R. Me′szaros, and T. Gila′nyi (2006) *J. Phys. Chem. B* **110**, 20297.

Walde, P. (2006) *Orig. Life Evol. Biosph.* **36**, 109.

Whaley, S. R., D. S. English, E. L. Hu, P. F. Barbara, and A. M. Belcher (2000) *Nature* **405**, 665.

Xiao, S., F. Liu, A. E. Rosen, J. F. Hainfeld, and N. C. J. Seeman (2002) *Nanoparticle Res.* **4**, 313.

Yan, D., Y. Zhou, and J. Hou (2004) *Science* **303**, 65.

Zahn, M. (2001) *J. Nanoparticle Res.* **3**, 73.

Zasadzinski, J. A., E. Kisak, and C. Evans (2001) *Curr. Opin. Colloid Interf. Sci.* **6**, 85.

PROBLEMS

14.1 Define the term *nanotechnology*. Give one example each for objects that lie in the range of millimeter, micrometer, nanometer, picometer, and femtometer.

14.2 What do you mean by *top-down* and *bottom-up* methods for creating nanoparticles? Exemplify.

14.3 How do surfactants find use in *dip-pen nanolithography*?

14.4 Explain how a nanoassembly of surfactants serves in *drug delivery*.

14.5 Give an example of a surfactant-based *nanodetergent*. What are its special merits?

14.6 Explain briefly the purported function of surfactants in the *origin of life*.

14.7 What is the role of surfactants in the synthesis of *nanoparticles*?

14.8 Explain how surfactants serve as templates for assembling inorganic nanostructures, drawing a parallel with the procedure used for casting molds to produce multiple replicas of statues.

15 Surfactants and Molecular Modeling

Molecular modeling is the investigation of molecular structures and properties by the collective use of computational and theoretical methods and graphical visualization techniques; that is, molecular modeling is thinking *outside the flask*. The term *molecular modeling* is not used in a narrow sense of making three-dimensional physical models of molecules, although molecular models with high precision and accuracy can be created using computer-assisted methods.

A major reason for studying surfactants by molecular modeling is to aid in the prediction of the properties and behavior of these molecules in solution and at an interface. If proper methodologies are adopted, such procedures would provide critical micelle concentrations (CMCs), micelle size, surface tension, surface excess (adsorption density), and thermodynamic parameters. The availability of such data, in turn, would enhance the efficient design of novel surfactant candidates without the need to resort to unproductive synthetic effort. The interaction of surfactants with polymeric surfactants and materials can also be investigated by these computational procedures. Moreover, the properties of an interface and of adsorbed layers may be derived for cases that are not experimentally accessible. Computer simulation methods can model the dynamic processes and produce a movie of the molecular trajectory on a real-time basis.

The determination of the energy of molecules is one of the main features of molecular modeling studies that yield molecular parameters. Thermodynamic, spectroscopic, bonding, and reaction parameters can be derived from the computed energy of the molecule. Coupling of these data with the potentials of *molecular informatics* (*cheminformatics* and *bioinformatics*) generates advanced models and interaction parameters of even complex biopolymers.

Surfactants and Interfacial Phenomena, Fourth Edition. Milton J. Rosen and Joy T. Kunjappu.

Moreover, structure–activity relationships of molecules can be quantitatively determined by the methods of *quantitative structure activity [property] relationship* (QSA[P]R) that correlate the structural features of a molecule with its chemical and biological properties.

Molecular modeling procedures used in computational chemistry using available software packages are based on two types of theories: *molecular mechanics* (MM) and *quantum mechanics* (QM).

MM, also called *force field* (FF) methods and based solely on classical physics, solves for the positions and motions of atoms in a molecule and calculates the energy of the system with fairly simple equations that can be applied even to very large molecules. Molecular modeling methods provide geometrical structures, energy levels, reaction dynamics, and spectroscopic and thermodynamic parameters of molecules (as obtained from infrared and Raman spectra). *Molecular mechanics—level 2* (MM2), *molecular mechanics—level3* (MM3), *Chemistry at HARvard Macromolecular Mechanics* (CHARMM), and *Assisted Model Building with Energy Refinement* (AMBER) are examples of programs available for performing MM calculations.

Quantum mechanical methods calculate the energy by solving the Schrödinger equation. The energy operator (Hamiltonian) and wave functions are the heart of this method. In the ab initio (from first principles) quantum mechanical method, properties of the system are computed from quantum theory and physical constants such as the mass (m) and charge (e) of the electron, the speed of light (c), and Planck's constant (h); in principle, there is no need for any other empirical parameters for the calculations. Compared to the MM method, the ab initio quantum mechanical method uses complicated equations and is extremely time-consuming to solve. As such, they are applicable only to small molecules.

In contrast, semiempirical quantum mechanical methods replace some of the complicated integrals in the ab initio procedure that are hard to compute with certain factors, such as bond parameters derived from experiments. For example, *modified neglect of differential overlap* (MNDO) and *Austin Model 1* (AM1) are based on semiempirical methods.

Another method popular in molecular modeling is the computer simulation procedures based on *molecular dynamics* and *Monte Carlo* methods.

The power of molecular modeling calculation and simulation has percolated into various chemical disciplines. Methods based on MM, QM, and their hybrids have been evolving at an alarming rate. Program packages are readily available incorporating the fine nuances of various theories and their extensions that support the study of molecular properties of small and large molecules from algorithms derived from MM and QM. MM is suitable for small and large molecules alike, but quantum mechanical methods are usually applied only to small molecules considering the computer time constraint. Computer simulation methods allied to molecular dynamics and Monte Carlo methods can be effectively used to predict the molecular properties on a time-averaged and ensemble-averaged level that even yield information on the system dynamics.

I. MOLECULAR MECHANICS METHODS

In MM (aka force field) methods, the atoms are visualized as balls and the bonds as springs connecting them, and the force and energy are related through classical laws such as Hooke's law. Here, only the nuclear positions are considered and the electrons in the system ignored.

In FF methods, the total energy of a molecule (E) is derived from the bond stretching ($E_{\text{stretching}}$), the bond bending (E_{bending}), the torsional motions ($E_{\text{torsional}}$), and the nonbonded interactions like van der Waals forces ($E_{\text{van der Waals}}$) and electrostatic force ($E_{\text{electrostatic}}$), as well as some of their combinations ($E_{\text{cross terms}}$). These contributions to the total energy of a molecule can be expressed in terms of nuclear or internal coordinates (based on bond lengths, bond angles, and torsional angles from bond rotations) by

$$E = E_{\text{stretching}} + E_{\text{bending}} + E_{\text{torsional}} + E_{\text{van der Waals}} + E_{\text{electrostatic}} + E_{\text{cross terms}}. \quad (15.1)$$

The potential energy of a diatomic molecule obtained from its rotational and vibrational motions, assuming it behaves as a *harmonic oscillator* (Hooke's law), as a function of its nuclear coordinates, is given by

$$E(R) = U(R_0) + \frac{dE}{dR}(R - R_0) + \frac{1}{2!}\frac{d^2E}{dR^2}(R - R_0)^2 + \frac{1}{3!}\frac{d^3E}{dR^3}(R - R_0)^3 + \ldots, \quad (15.2)$$

where R_0 is the equilibrium bond distance between the atoms in the molecule and $U(R_0)$ is the equilibrium energy, set arbitrarily to zero energy. By treating the derivatives in this equation as parameters that could be determined experimentally for an $R = R_0$ case, the above equation takes the form

$$E(R) = k_2(R - R_0)^2 + k_3(R - R_0)^3 + \ldots \quad (15.3)$$

The first term is set to zero and the constants are absorbed in the k terms.

The component terms for *stretch energy*, *bend energy*, *torsional energy*, *van der Waals energy*, *electrostatic energy*, and their *cross terms* are derived either from treatments similar to the above equation from consideration of bending, stretching, rotational barrier, and van der Waals forces or their possible combinations.

A. Parametrization from Experiments

The FF calculations require a number of parameters such as equilibrium bond distance, equilibrium bond angle, force constants, and torsional term, which is overwhelming for a big molecule. Some of these parameters, such as bond length, can be extracted from X-ray and neutron diffraction, and microwave and high-resolution spectroscopic experiments.

According to the concept of the *transferability principle*, the FF parameters may be *transferable* from one case to another when an MM program may fail to perform for want of parameters. The transferability principle can be applied to a variety of situations based on the type and class of molecules. The *universal force field* (UFF) program constructs the FF utilizing atomic properties such as atomic number, hybridization state, and formal oxidation state.

B. Classes of FF Methods

The FF calculations involve contributions toward energy from a number of sources, and many of them contain higher-order terms, often running to quartic terms and several types of cross terms. Such higher-level calculations in which all the contributions are considered will provide more accuracy and are referred to as *Class I* FFs. *Class I* methods such as MM 1–4, *empirical force field* (EFF), and CFF (*consistent force field*) are suitable for small- and medium-sized molecules.

Class I methods are difficult to apply to macromolecules like polymers as they need a great amount of computing time. *Class II* methods simplify the calculations by truncating the series at the quadratic term in the above equation and by ignoring the cross terms. Simplification may also be achieved by considering units of atoms like methylene group (CH_2) as a single atom (in this case, "CH_2 atom"). AMBER, CHARMM, and *GROningen MOlecular Simulation* (GROMOS package) are based on Class II methods. FF methods in tandem with quantum mechanical methods (*hybrid methods*) offer efficient solutions to many complex problems.

II. QUANTUM MECHANICAL METHODS

FF methods (MM) do not yield agreeable results in predicting spectroscopic properties because they do not treat electronic and nuclear factors for computing the molecular energy. Quantum mechanical methods consider the potential energy contribution as arising from the electrons and nuclei of atoms in molecules as a result of electron–nucleus attraction, electron–electron and nucleus–nucleus repulsions, and their kinetic energies (KEs). The theory can be simplified by neglecting factors suggesting nuclear contributions.

Broadly speaking, quantum mechanical methods are of two types:

1. Exact methods in which the theory is developed without making serious assumptions, as in ab initio procedures
2. Approximate methods like the *Hartree–Fock* method or *semiempirical methods* developed to save on computer time where certain assumptions simplify the calculations in cases where experimental data can be integrated to simplify the solution

The most common representation of QM is by

$$\hat{H}\psi = E\psi,$$

where $\hat{H}$ is the Hamiltonian operator, ψ is the wave function, and E is the energy (*eigenvalue*) of the system because $\hat{H}$ represents the total energy. The Hamiltonian function is called an energy operator because it contains factors for potential and KEs that can act on a function.

The momentum operator is given by

$$\hat{p} = -i\hbar\frac{\partial}{\partial x},$$

where

$$\hbar = \frac{h}{2\pi}$$

and h is the *Planck* constant and p is the momentum.

In its most general form, the *time-dependent* form of Schrödinger's equation for a single particle such as electrons of mass m moving in space defined by the position vector $\mathrm{r} = x\mathrm{i} + y\mathrm{j} + z\mathrm{k}$ and time t under an external field V (electrostatic potential in the field of a nucleus) is given by

$$\left[\frac{-\hbar^2}{2m}\left(\frac{\partial^2}{\partial x^2}+\frac{\partial^2}{\partial y^2}+\frac{\partial^2}{\partial z^2}\right)+V\right]\psi(r,t) = i\hbar\frac{\partial\psi}{\partial t}(r,t)$$

or

$$\hat{H}\Psi = i\hbar\frac{\partial\Psi}{\partial t}.$$

The solution of Schrödinger's equation provides energy E of the particle.

A. Application to the Electronic Problem

The Schrödinger equation is readily applicable for evaluating the energy of an atomic system. In a typical case, the Hamiltonian operator—the KE terms for electrons and nuclei and the potential energy terms for nucleus–nucleus and electron–electron repulsion as well as nucleus–electron attraction (coulombic potentials), assuming the nuclei and electrons are point charges—has to be evaluated:

$$\hat{H} = A + B + C + D + E,$$

where, A = operator for the KE of electrons;
B = operator for the KE of nuclei;
C = operator for attraction between electrons and nuclei;
D = operator for the repulsion between electrons; and
E = operator for the repulsion between nuclei.

The *Born–Oppenheimer approximation* simplifies this by assuming that the nuclei remain stationary in relation to the motion of the electrons because nuclei are much heavier than electrons. Under Born–Oppenheimer approximation conditions, the KE operator for nuclei vanishes and the nucleus–nucleus repulsion term remains a constant (the distance between nuclei does not vary). So, B becomes zero and E becomes a constant. A constant term in the operator has no effect on the operator eigenfunction. The remaining terms are collectively named as the *electronic Hamiltonian*, which describes the motion of electrons in the field of point-charged nuclei:

$$\hat{H}_{ele} = A + C + D.$$

For a one-electron system, as in a hydrogen atom, the *electron–electron repulsion* term D also vanishes, allowing the Hamiltonian to take the form

$$\hat{H} = \frac{-\hbar^2}{2m}\nabla^2 - \frac{Ze^2}{4\pi\varepsilon_0 r},$$

where r is the distance from the nucleus.

For a multielectron atom, $\hat{H}$ can be written, by the introduction of atomic units, as

$$\hat{H} = \sum_i \left(-\frac{1}{2}\nabla_i^2\right) + \sum_i V_i,$$

where V_i may be positive or negative according to whether the electron is attracted to nuclei or repelled by other electrons. Splitting these terms,

$$\hat{H} = \Sigma\left(\left(-\frac{1}{2}\nabla_i^2 - V_i\right)\right) + \frac{1}{2}\Sigma\frac{1}{r_{ij}}.$$

It is the r_{ij} term for the electron–electron interaction between the *i*th and *j*th electrons that renders the Schrödinger equation insoluble. In the mathematical sense, they are thought not be *separable*. However, numerical solutions can be obtained by Monte Carlo and other methods.

B. The Hartree Product (HP) Description

HP description for a *many-electron* system atom or molecules with N electrons assumes the wave functions of different electrons behave independent of each other so that the combined probability of their wave functions may be expressed as a *product* of their spin orbital functions. Then, the eigenfunction will be

$$\hat{H}\Psi^{HP} = E\Psi^{HP}.$$

The problem in solving the Schrödinger equation for multielectron systems lies in the difficulty in calculating the electron–electron repulsion term. A simplifying assumption to alleviate this is that electron–electron repulsion is treated in an average way so that the many-electron problem transforms into a single-electron problem. Such an approximation is called the Hartree–Fock approximation. The corresponding Hartree–Fock potential or the average field experienced by one electron depends on the spin orbitals of the other electrons. Because the Hartree–Fock formalism is a nonlinear equation, it has to be solved iteratively.

The iteration procedure involves the guessing of a spin orbital function and the calculation of the average field as seen by each electron. An iterative procedure for solving the Hartree–Fock equation is known as the *self-consistent field* (*SCF*) method.

The solution of the Hartree–Fock equation itself is difficult, and it is further simplified by assuming that each spin orbital is a linear combination of single-electron obitals:

$$\Psi = \Sigma C_i \phi_i.$$

ϕ_i, the one-electron orbitals, correspond to the atomic orbitals and is called a *basis function*. The sets of atomic orbitals for each atom are called *basis sets*. C_i is the set of coefficients that satisfies the lowest-energy wave function. The coefficients are found by minimizing the energy as

$$\frac{\partial E}{\partial C_i} = 0.$$

It is found that by making the basis set larger, the Hartree–Fock energy reaches a lower limit called the *Hartree–Fock limit*.

The simplest model used to describe a molecule is the *molecular orbital-linear combination of atomic orbitals* (*MO-LCAO*) model, which assumes that when atoms approach each other to form molecules, molecular orbitals are formed as a linear combination of atomic orbitals. The atomic orbitals are represented as an exponential function either as a *Slater orbital* or as a *Gaussian orbital*, which have different orbital *exponents*, and they have different mathematical forms. However, Gaussian-type orbitals (*GTOs*) are preferred in ab initio calculations instead of Slater-type orbitals (*STOs*) because the complex integrals in the latter are difficult to evaluate, especially when the atomic orbitals are centered on different nuclei. GTOs are further simplified into *contracted Gaussian-type orbitals* (*CGTO*).

C. Minimal and Larger Basis Sets

The minimum requirement in these treatments is to use one basis function for each atomic orbital of an atom, called the *minimal basis*, which, for example, is one for hydrogen (1S orbital only) and five for carbon (*one* 1S, *one* 2S, and *three* 2p orbitals). In a calculation scheme called *STO-3G*, each basis function is a CGTO of three Gaussian orbitals, the coefficient in the CGTO being picked to fit the corresponding STO. As the atomic orbitals deform depending on their environment, the minimal basis set cannot represent them in all the cases. Thus, we need to add extra basis functions that lead to larger basis sets. The *6-31*G* scheme utilizes larger basis sets (Pople notation for *split-valence double zeta* basis sets: *6* represents the number of primitive Gaussians comprising each core atomic orbital basis function; *3* and *1* indicate that the valence orbitals are composed of two basis functions each, the first one composed of a linear combination of 3 primitive Gaussian functions, the other composed of a linear combination of 1 primitive Gaussian function; the asterisk indicates the inclusion of *polarization functions*).

The actual application of Hartree–Fock theory to quantum mechanical calculations for evaluating molecular orbitals is done through two methods: ab initio and semiempirical methods. Ab initio (literally meaning *from the beginning*) refers to the method that relies only on input of quantities such as c, h, m, and e (speed of light, Planck's constant, mass, and charge of electron, respectively). The positions of nuclei are provided in terms of a set of internal nuclear coordinates that depend on bond length, bond angle, and torsional angle (*Z matrix*). On the other hand, semiempirical methods ignore some of the terms in the Hamiltonian and input empirical parameters for some of the integrals.

The main semiempirical methods in the order of their implementation are the following, which, in general, consider only the valence electrons of the system and treat the core electrons as a part of the nuclear core. Each advance incorporates the properties of orbitals, electrons, and nuclear core while modifying the defect in a previous method. Some of them are the following:

1. Zero-differential overlap (*ZDO*) approximation in which the overlap between pairs of different orbitals is set to zero
2. The complete neglect of differential overlap (*CNDO*)
3. The intermediate neglect of differential overlap (*INDO*)
4. The neglect of diatomic differential overlap (*NDDO*)
5. The modified INDO (*MINDO/3*)
6. The modified neglect of differential overlap (*MNDO*)
7. The *AM1*
8. Parameterization 3 of *MNDO* (*PM 3*)
9. Semi-ab initio model 1 (*SAM 1*)

D. Electron Correlation Method

Many refinements have been carried out on the ab initio method. In the Hartree–Fock formalism, which is an approximate method, only the electron correlations due to the Pauli exclusion principle are considered. Accurate treatment, therefore, needs to consider electron correlation explicitly, especially when the system is considered near its dissociation limit.

In the *configuration interaction* (*CI*) approach, the wave function of the excited states is included in the description of the electronic states, where the overall wave function is treated *as a linear combination of the ground and excited state wave functions*. The wave function is still expressed as the Hartree–Fock determinant plus many other determinants, which put electrons in different orbitals, by applying the *variational theorem* to find the best coefficients in selecting an acceptable wave function. In the *many-body perturbation treatment* by the *Møller–Plesset* theory, electron correlation appears as a perturbation to the Hartree–Fock description. This method is more accurate, but the computer time goes up. In *coupled-cluster methods*, the wave function is expressed as an exponential product, where higher-order corrections are incorporated as products of lower-order terms. An important consequence of these refinements is that the property computed turns out to be more accurate.

Because a higher-level ab initio calculation has the downside of consuming enormous computer time, it is common in actual procedures to perform the geometric optimization using a lower-level method and to carry the parameter into a higher-level calculation using the derived wave function. The mixing of two levels in one scheme is represented by separating the levels by "/," as in 6-31G*/STO-3G, which means geometry using the STO-3G basis set and the wave function using the 6-31G* basis set (the significance of these terms was discussed previously).

E. Density Functional Theory (DFT)

DFT is similar to the Hartree–Fock theory in that it treats single-electron functions, but it calculates the total electronic energy and the overall electron density distribution. Instead of calculating the wave functions of all the electrons, it determines N orbitals to form a Slater determinant. In DFT, the potential is local; in the Hartree–Fock theory, it is nonlocal. There is a direct relationship between the total electron energy and the overall electron density in an atom or molecule. Thus, electron densities can effectively replace complicated many-electron wave functions and yield accurate results. The basic idea is to minimize the energy with respect to the electron density. But DFT has the advantage that it includes electron correlation and is very similar to Hartree–Fock methodology, and the cost is only slightly more. Functionals such as *BLYP* (from the name Becke for the exchange part and Lee, Yang, and Parr for the correlation part of the formalism) and *B3LYP* (a *hybrid functional* in which Becke's exchange part is combined with the exact energy from the Hartree–Fock theory) are based on DFT. In the local spin density functional theory (*LSDFT*), the electron density is coupled with the net spin density.

III. ENERGY MINIMIZATION PROCEDURE

This is an important strategy in molecular modeling procedures. The potential energy of various conformations of a molecule is represented as a function of its coordinates, and the resulting graph is called a *potential energy surface*. The coordinates may either be the Cartesian coordinates, which is *3N* for an N atom system, or the internal coordinates, which is *3N-6* for the same system. MM methods use Cartesian coordinates, and quantum mechanical methods use the internal coordinates as input.

A typical energy surface diagram, which is represented as a contour plot or isometric plot, is characterized by a number of minima and maxima. The conformational structures with minimum energy usually correspond to the most stable state of the system. The lowest-energy minimum known as *global minima* out of a number of *low minima* is to be located. The global minima need not always correspond to the most highly populated conformation. The minimum point is located using a minimization algorithm that recognizes the *maxima* (mountain peak), *minima* (valley), and the *saddle point* (the highest point between two minima).

The minimum point of a curve represented by a function f has a first derivative with respect to an independent variable, x, as zero:

$$\left(\frac{\partial f}{\partial x}=0\right),$$

and the second derivative as positive:

$$\left(\frac{\partial^2 f}{\partial x^2} > 0\right).$$

Minimization methods work either by locating the derivative of energy or by tracing the minimum point in the energy function starting from a high energy point. In general, minimization algorithms make use of a set of initial coordinates of the system from experimental data derived from nuclear magnetic resonance (*NMR*) or *X-ray crystallography*. A purely theoretical algorithm or a hybrid of these methods can be used if experimental data are not available.

Energy minimization procedures are useful in determining the transition structures, reaction pathways, and in the identification of normal modes. They work well for systems with a small number of atoms. Energy minimization data are translated into thermodynamic and other parameters applying the principles of statistical thermodynamics that yield partition functions.

IV. COMPUTER SIMULATION METHODS

Two types of computer simulation techniques are popular: *molecular dynamics* and *Monte Carlo methods*. In the molecular dynamics method, Newtonian laws of motion are applied, assuming the particles behave as hard spheres like billiard balls. When the particles collide, the interparticle distance equals the sum of their radii. The new velocity of the particles is calculated after the collision, from which the forces are computed. The force calculations are continued at their new positions, producing a trajectory or path that describes the change in dynamic variable with time. The molecular dynamic (MD) simulation procedure is usually continued for hundreds of picoseconds in steps of femtoseconds that lead to thermodynamic parameters and conformational properties. Thus, the successive configurations generated are connected in time.

The Monte Carlo method is based on the principle of random sampling. It generates configurations randomly and checks them against certain criteria based on the Boltzmann exponential factor. A random number generated is compared with this energy to decide the viability of a configuration until the most desirable configuration is arrived at. Monte Carlo simulation does not compare the configurations in a time-dependent fashion, unlike in the molecular dynamics method, and compares the current configuration with its predecessor only.

Molecular dynamics produces *time-averaged* information and the Monte Carlo method, *ensemble-averaged* information. A property such as the internal energy U of a system may be expressed over a period T as

$$\langle U \rangle = \lim_{T \to \infty} \frac{1}{T} \int_0^T u(t)dt$$

for the time-averaged case, and

$$\langle U \rangle = \int \rho(R^n)u(R^n)dR^n$$

for the ensemble-averaged case. Here, u is a function of the microscopic positions of the n molecules, R^n.

The solvent system is usually modeled by two different methods—the *continum solvation model* (CSM) or *implicit model* and the *explicit solvation model* (ESM). In CSM, the solvent is treated as a homogeneous medium around the solute molecule and is computed by MM or QM methods. In ESM, solvent molecules are added to a single solute molecule and the system is treated by MM.

The foregoing sections were meant only to provide a basic exposure to some of the terms that will be encountered in the discussions to follow. A deeper understanding of these concepts and principles may be obtained from Szabo and Ostlund (1996), Leach (2001), Ball (2003), and Sahni (2004, 2009).

V. SURFACTANT SYSTEMS

Theoretical and computer methods are still being evolved in modeling surfactant behavior. FF, QM, and computer simulation methods are applied to model surfactant systems with different levels of success. In particular, lattice and continuum models are among the earlier attempts in modeling the surfactant systems in liquids.

Some of the instances where surfactants are modeled by various methods in solution and at an interface are described in Section VII. Before that, a few representative cases are discussed in detail.

VI. FIVE SELECTED SYSTEMS

A. Aggregation in a Liquid (i)

Smit, B., A. G. Schlijper, L. A. M. Rupert, and N. M. van Os (1990) *J. Phys. Chem.*, **94**, 6933.

An oil (decane)/water/surfactant (sodium p-alkyl benzene sulfonates [C_9, C_{10}, and C_{12}]) system is modeled in this paper by molecular dynamics. The dependence of the structure of surfactants on various thermodynamic properties is studied, for example, the effect of increasing the hydrophobic tail length on the interfacial tension.

Statistical thermodynamic treatment using lattice models or continuum mean field theories is the usual method used in investigating the relation between the chemical structure of surfactants and their thermodynamic properties.

The surfactant molecule is treated as the combination of *oil-like* and *water-like* particles (called atoms) joined by a Hookean spring (harmonic potential). The two types of particles interact with truncated Lennard–Jones potentials with energy parameter, ε_{ij}, distance parameter, σ_{ij}, and the cutoff radius, R_{ij}^c as per the following equations:

$$V_{ij} = \begin{cases} \phi_{ij}(r) - \phi_{ij}(R_{ij}^c) & r \le R_{ij}^c \\ 0 & r > R_{ij}^c \end{cases}$$

and

$$\phi_{ij}(r) = 4\varepsilon_{ij}\left[(\sigma_{ij}/r)^{12} - (\sigma_{ij}/r)^{6}\right].$$

i and j are *water-like* and *oil-like* atoms. To simplify, the Lennard–Jones parameters for various interactions, the energy parameter ε, and the distance parameter σ are assumed to be the same for both the particles. The water–water, oil–oil, and water–oil interactions are simulated by truncating the R_{ij}^c terms appropriately to reflect the attractive and repulsive interactions in these cases. Oil-like particles are successively added to the surfactant tail to increase the chain length during the simulation. The oil-like and water-like particles for simulation are arranged on a *face-centered cubic* (*fcc*) lattice at the center of a periodic box of size $7.15\sigma \times 7.15\sigma \times 21.45\sigma$ as two layers of 256 water-like particles and 512 oil-like particles at a density of $0.7\sigma^{-3}$, and at a constant temperature of $T = 1.0\varepsilon/k_B$. The total number of particles is kept constant by replacing oil and water particles by surfactants. The interfacial tension is evaluated by integrating the difference of the normal and tangential components of the pressure tensor over the interface and from the density profiles after the system is allowed to equilibrate. The values of the pressure tensor and density are calculated every 10th time step.

Experimental results and MD simulations/calculations for the same system indicated a strong dependence of the hydrophobic effect at increasing chain length in the low-surfactant concentration region. Good agreement was also observed between the experimental and theoretically evaluated surface tension values.

B. Aggregation in a Liquid (ii)

Stephenson, B. C., K. Beers, and D. Blankschtein (2006) *Langmuir*, **22**, 1500.

The micellization process is modeled by coupling the molecular-thermodynamic theory with computer simulation. The molecular-thermodynamic approach alone does not satisfactorily predict the micellization process of complex surface-active molecules in which the distinction between head group and tail piece is not well-defined. Molecular modeling strategies based on MD and Monte Carlo simulations have the potential to improve predictability with complex surfactants with dubious localized head and tail groups.

In the thermodynamic approach, many micellar parameters are calculated from the CMC of the surfactant through a relation that connects the CMC with the free energy of micellization, g^*_{mic}, where k_B and T are the Boltzmann constant and the absolute temperature, respectively:

$$\mathrm{CMC} = \exp\left(\frac{g^*_{mic}}{k_B T}\right).$$

In molecular modeling strategies, minimization of g_{mic} is performed with respect to the variables such as aggregation geometry, aggregation composition, and the aggregate core minor radius. g_{mic} is determined from g_{tr}, (the transfer free energy—computed from experimental solubility data), g_{int} (the interfacial free energy—computed from the surfactant tail/water interfacial tension data, mixing rules, and the use of an equation that approximates the dependence of the tail/water interfacial tension on curvature), g_{pack} (the packing free energy—computed from a numerical procedure in which a large number of the possible conformations of the surfactant tails within the aggregate core are generated to determine the free energy required to fix the tail moieties at the aggregate/water interface), g_{st} (steric free energy—computed from the packing of the surfactant heads at the aggregate/water interface), g_{elec} (the electrostatic free energy—computed from the electrostatic repulsions between ionic groups that may be present in the micellar aggregate), and from the degree of counterion binding through the relation

$$g_{mic} = g_{tr} + g_{int} + g_{pack} + g_{st} + g_{elec} + g_{ent} - k_B T\left(1 + \sum_j \beta_j\right) - \left(\sum_j \beta_j k_B T \ln\left(X_{c_j}\right)\right).$$

In the MD simulation, the effective head and effective tail portions of the surfactant are determined from an average position of a surfactant molecule at an oil–water interface, considering the interactions between each atom in the surfactant molecule and its environment by applying all-atom optimized performance for liquid systems (OPLS-AA) FF. To aid these simulations, depending on the nature of the surfactant and the availability of literature data, other FFs such as GROningen MAchine for Chemical Simulations (GROMACS) were used. In addition, water was treated explicitly using the simple extended point-charge model (SPC/E), and van der Waals interactions were treated using a cutoff distance of 1.2 nm, and coulombic interactions were described using particle mesh Ewald (PME) summation. The simulations were carried out under fixed bond lengths at a time step of 2 fs. The atomic charges were assigned to the surfactant molecule using either the OPLS-AA method or the CHelpG method implemented in *Gaussian 98*.

The computational approach identified the hydrated portions of the surfactant molecule in the micelle by MD simulations at an oil–water interface by modeling the micelle core/water interface. The number of contacts the surfactant segment makes with the water and oil molecules was estimated. The determination of hydrated and unhydrated segments of the surfactant molecule and molecular-thermodynamic modeling leads to (1) the free-energy change associated with forming a micellar aggregate, (2) the CMC, and (3) the optimal shape and size of the micellar aggregate. The atomic charges were determined from computer simulation and molecular-thermodynamic modeling. Both these sets of atomic charges were used in computing the micellar parameters such as CMC and aggregation number for the anionic (sodium dodecyl sulfate [SDS]), cationic (cetyltrimethylammonium bromide [CTAB]), zwitterionic (dodecylphosphocholine [DPC]), and nonionic (dodecyl poly(ethylene oxide), $C_{12}E_8$) surfactants. Such a combined modeling approach was found to hold good also in the case of the anionic surfactant, 3-hydroxy sulfonates (AOS; $n = 12–16$), and of the nonionic surfactant, decanoyl-*n*-methylglucamide (MEGA-10).

C. Liquid–liquid and Liquid–Gas Interface

Schweighofer, K. J., U. Essmann, and M. Berkowitz (1997) *J. Phys. Chem.*, **101**, 3793.

Behavior of SDS at the water–air and water/CCl_4 interfaces was studied by MD simulation. Previous experimental studies for garnering configurational information on these systems included sum frequency generation (SFG) and second harmonic generation (SHG), aside from theoretical studies using generalized van der Waals theory for nonuniform molecular fluids, and molecular modeling.

The simple point charge (SPC) water model with the bond length held constant through the SHAKE algorithm and a time integration algorithm that satisfies constraints in bond geometry in MD simulations were used. A fully flexible, nonpolarizable, five-site model was adopted for carbon tetrachloride that was in agreement with the diffusion constant, the density profile and the radial distribution functions known from polarizable models. The united atoms of 1 CH_3 group and 11 CH_2 groups were used to represent the dodecyl group, and the SO_4 group atoms were explicitly modeled.

The *in silico* (with a silicon chip, i.e., a computer) sample preparation and equilibration were performed by choosing a water–CCl_4 system consisting of 500 water and 222 carbon tetrachloride molecules in a rectangular box having X and Y dimensions of 24.834 Å and a Z dimension of 150.0 Å. SDS was inserted into the fully equilibrated water–CCl_4 system putting the sulfur atom in the center of the interfacial region and the hydrocarbon tail in the organic layer with an orientation slightly tilted relative to the interface normal. A random water molecule in the interfacial region was replaced by a sodium ion. The solvent–dodecyl sulfate interaction was simulated by turning the process on gradually over a period of 250 ps with periodic rescaling of the velocities at a constant temperature of 300 K under least perturbation of the water–CCl_4 interface, and the system was then allowed to equilibrate for another 500 ps using AMBER.

Adsorption of SDS at the water–vapor interface was simulated by replacing the CCl_4 layer with water from the fully equilibrated SDS adsorbed water–CCl_4 system. The box length remained the same as in the water–CCl_4 system, and a 500-ps equilibration followed before production runs began. For both water–CCl_4 and water–vapor simulations, production runs used a time step of 1.0 fs and data collection was done every 20.0 fs for a total time of 1.0 ns.

The Gaussian distribution of the head group (defined as the SO_3^- group plus the Na^+ counterion) and the hydrocarbon tail regions (defined as all united atom carbons from the methyl group to carbon 12) were evaluated from the liquid density profiles using the formula

$$\rho = \rho_i \exp\left(\frac{-4(Z - Z_0)^2}{\sigma^2}\right),$$

where ρ_i is the amplitude, Z_0 the center of the distribution, and σ the 1/*e*th width. The values of the *Z*-component of the average head-to-chain separations, Z_0 and σ, were calculated based on these distributions.

The amphiphilic orientation is determined from an analysis of the cosine function as defined in the equation

$$\cos(\theta) = z.\frac{\bar{R}_{1j}}{\left|\bar{R}_{1j}\right|},$$

where the *R* vectors refer to the position of the R group from the methyl group; *j* is the position of other carbon atoms 2, 3, 4, . . . ; and *z* is the unit vector in the *Z*-direction. For example, 12 subscript refers to the methyl–methylene bond vector.

Head group-water radial distribution functions were analyzed to find out the exact distribution or orientation of the individual atoms in the head group and of water.

Orientational distribution functions for water in the first shell were obtained from a plot of the cos (θ) function versus probability. Snapshots of the system during simulation substantiated the distribution and orientation of the atoms in the system.

From the values for Z_0, it was determined that the head group penetrates the water lamella by approximately 2.0 Å more in the water–vapor system than in the water–CCl_4 system. Similarly, it was observed that part of the hydrocarbon chain is solvated by the water at the vapor interface; for the water–CCl_4 system, the hydrocarbon chain is predominantly solvated by carbon tetrachloride.

The σ values could be interpreted by assuming that the hydrocarbon chain lies very close to the water surface on the average for the water–vapor system, and the chain has varying angles of tilt relative to the interface normal for the water–CCl_4 case.

The experimentally observed width, σ_{obs}, was related to l, the intrinsic width (actual width) of a fragment along the interface normal, and w, a measure of the roughness of the surface by the relation

$$\sigma_{obs}^2 = l^2 + w^2.$$

The individual probability distribution for the atoms and groups along the Z-direction for the two interfaces revealed the extent of penetration of the surfactant molecule to the interior of the interface: In the water–vapor system, some of the carbons are found to be immersed in the water. This shows that the surfactant must have a bend configuration indicating an increased probability for *gauche* defects and bending of the chain near the headgroup region. The same study revealed that the surfactant does not penetrate deeply into the water surface at the water–CCl_4 interface. Quantitative information on the bending of the surfactant molecule at the interface could be seen by examining the average position of each atom along the interface normal in the $\langle Z \rangle$ versus atom number plot from the hydrocarbon chain end.

The amphiphilic dihedral distribution showed that the C_{11}-C_{12}-O-S dihedral is almost always trans (1% gauche), in line with the argument that a gauche conformation would not have allowed the oxygen atom in the sulfate group to orient toward the waterside of the interface. The trans-gauche ratio at each carbon could also be determined for the two interfaces.

The average bulk densities obtained in this study were 0.96 g/cm^3 for water and 1.59 g/cm^3 for carbon tetrachloride. The experimental densities at 300 K are 1.0 and 1.60 g/cm^3.

The configurational behavior of the surfactant at the two interfaces are observed to be the following: At the water–vapor interface, the configuration was bent with two domains within the molecule—the head group and several methyl groups solvated in water and the rest of the molecule laying down on the water surface; at the water–carbon tetrachloride interface, the molecule was straight on the average, with an inclination of approximately 40° from the surface normal.

D. Solid–Liquid Interface

Prdip and B. Rai (2002) *Colloids and Surfaces*, **205**, 139.

The controlled design of molecular structures of surfactants with carboxylic acid functionality (fatty acid surfactants) is of industrial concern. Screening of these surfactant candidates for efficiency may be eased by exploiting the predictability of molecular modeling methodology with respect to their structure–function relationship. The candidates in this study by molecular modeling strategy (with no need for input from experimental parameters) included fatty acids for the selective flotation of calcium minerals, dicarboxylic acids as crystal growth inhibitors for calcium fluoride, carboxylic acid-type dispersants for titania-based paints, fatty acids and benzoic acid dispersants for zirconia and alumina ceramic suspensions, and phosphonic acid-based corrosion inhibitors.

The modeling was performed by the UFF method, which provided parameterisation for a wide range of atoms including calcium, titanium, barium, aluminium and zirconium atoms. In this UFF method, bond stretching was described by a harmonic term, angle bending by a three-term Fourier cosine expansion, torsion and inversion by cosine-Fourier expansion terms, the van der Waals interactions by the Lennard–Jones potential, and the electrostatic interactions by atomic monopoles (which is screened by distance-dependent coulombic term in the algorithm).

The unit cell of the mineral at a given cleavage plane (Miller index) was used to create a periodic super lattice of ~25×25 Å in the x- and y-directions. The atoms in the top layer of a given cluster and those in the bottom fixed layer immediately following the top layer were allowed to relax with respect to energy (energy minimization). The modeled unit cell lattice parameters were compared with the available reported numbers for them from the literature. For modeling the surface–surfactant interaction, the lowest energy conformation of the surfactant molecule was selected from several conformations using molecular graphics and from a consideration of the functional group that may interact with the surface. The geometry-optimized surfactant molecule was docked on the most stable surface cluster and was allowed to relax. The interaction energy was calculated from the equation $\Delta E = E_{(\text{complex})} - [E_{(\text{surfactant})} + E_{(\text{surface})}]$, where the E terms are the optimized energies for the surface–surfactant complex, the surfactant molecule and the surface cluster, respectively. A more negative ΔE value indicates a favorable interaction between the surface and the surfactant.

The effectiveness of the UFF molecular modeling strategy was validated by correlating the calculated interaction energies with the experimental values for the minerals and ceramics in the context of flotation, dispersion, and corrosion inhibition. Demonstrated in this work were the close match between the experimental and calculated values for the lattice parameters and cell dimensions of the minerals, the interaction energies of adsorption of simple molecules such as water, butane, and octane on a specific cleavage plane, and for the interaction energies for the adsorption of oleic acid and water on mineral surfaces. For example, calculated interaction energies (in kilocalorie per mole) for oleic acid and water adsorption on calcium mineral surfaces such as fluorite {111}, fluorapatite {100}, and calcite {110} planes were –52.6, –46.8, –40.2 (for oleic acid) and –23.6, –42.9, and –32.2 (for water), respectively. The approach in this work is considered to be promising because the interaction parameters for the mineral and the surfactant can be predicted by a mere knowledge of the structure of the surfactant and the surface, and no experimentally determined quantities were required to input.

E. Solid–Liquid Interface and Aggregation in a Liquid

Postmus, B. R., F. A. M. Leermakers, and M. A. C. Stuart (2008) *Langmuir*, **24**, 3960.

The adsorption at the silica–water interface and the aggregation in aqueous solution of nonionic (alkyl-ethylene oxide) (C_nE_m) surfactants were modeled by the SCF theory using pH and ionic strength as additional control parameters.

The adsorption of surfactants has been engineered to be a transport phenomenon and has been treated with basic diffusion equations involving the flux term. The theoretical model presented here envisages a single concentration gradient (1G) approach using SCF theories. The SCF theory allows for the incorporation of molecular details into the model. Several known parameters and theories were input that include short-range Flory–Huggins χ-parameters for modeling the polyethylene oxide (PEO) molecules, the small systems thermodynamics, the use of spherical lattice to mimic micelles, the Gouy–Chapman theory for modeling the electrostatic potentials, Bragg–Williams approximation for short-range interactions, and χ-parameters for all states of water and contacts of PEO. The computed parameters included translationally restricted grand potential as a function of aggregation number for various ethylene oxide (EO) chain lengths, the CMC of various surfactants as a function of the ionic strength, plotting of adsorption isotherms for the various surfactants, the relation of chain position with respect to the solid surface, plots of critical surface association concentration (CSAC) versus various aggregation parameters, and rates and kinetics of adsorption for different ionic strength and pH values. A term called "adsorption flux" was introduced as the difference between the CMC and the CSAC.

The aggregation behavior in these cases was drastically influenced at high ionic concentrations as the headgroup charge is altered. The CSAC is a function of the surfactant and the surface properties that showed remarkable dependence on both the ionic strength and the pH. The predictions include the variation of adsorption with ionic strength, the pH-dependent surface charge, and the nature of binding of the head groups onto silanol groups on the surface through H-bonding. Finally, a good match was observed between experimental and modeling data for the adsorption/desorption transitions, the various equilibria involving the surfactant, and the kinetics of adsorption. The predicted properties for the adsorption and aggregation were found to agree well with the experimental values.

Section VII lists other representative examples from the literature in chronological order in tabular form.

VII. SUMMARY OF REPRESENTATIVE MODELING STUDIES

References	Surfactant Connection	Objective
Aggregation in Liquids		
Böcker, J., J. Brickmann, and P. Bopp (1994) *J. Phys. Chem.*, **98**, 712	MD simulation study of the cationic surfactant n-decyltrimethylammonium chloride micelle in water	Molecular modeling studies help fill the gap between the predicted parameters of micellar aggregates derived from thermodynamic theories and experimental values. The focus is obtaining parameters related to the structure and shape of the micelles.
MacKerell Jr., A. D. (1995) *J. Phys. Chem.*, **99**, 1846	Micellization of the anionic surfactant SDS in aqueous solution was simulated by molecular dynamics.	The structure and dynamics of micellization cannot be fully understood due to limitations and constraints in experimental techniques. MD simulations are used to fill the gap.
Villamagna, F., M. A. Whitehead, and A. K. Chattopadyay (1995) *J. Mol. Struc. (Theochem)*, **343**, 77	Optimal solution and interfacial properties were from molecular modeling studies using known surfactant structures based on fatty acids, EO, phosphocholine, sorbitan esters, oxazoline, ethanolamine, and polyisobutylene.	Some of these surfactant structures were synthesized and their properties evaluated from experiments to test the validity of the modeling method.

Methodologies	Important Findings
The model system containing 30 surfactant molecules, 30 chloride ions, and 2166 water molecules was used in the molecular modeling simulations. The premises of the molecular modeling simulations were the following: The methylene and methyl groups are united atoms; the hydrophobic chains are flexible, with intramolecular potentials as functions of torsional angles, bond angles, and bond stretches; the tetrahedral head groups are rigid structures, whose charge distribution was obtained from ab initio calculations. The intermolecular interactions were simulated using Lennard–Jones terms and the water model from transferable intermolecular potentials with a 4 point charge (TIP4P) procedure.	The quarternary nitrogen atom carries a charge of –0.56e, and the three methyl groups and the methylene group next to the nitrogen carry a charge of 0.39e each. An analysis of the shape and the structure of the micelle indicated a slightly prolate ellipsoidal shape for the micelle from a simulation of 275 ps. The interior of the micelle was seen to be dry in contrast to monolayer–aqueous solution interfaces. Some specific information such as probability of trans bonds in a chain and bond order parameter, which may not be easily available from experiments, could be calculated from this study. The micelle surface charge is only partially neutralized by chloride counterions in the first solvation shells of the head groups. The assumed aggregation number of 30, the area per head group of 84 Å^2, and the mean radius of 14.2 Å generated from these calculations agree well with experiments.
Calculations were performed with a version of CHARMM. The water structure was simulated by the transferable intermolecular potentials with a three-point charge (TIP3P) model using periodic boundary conditions. The hydrophobic micellar core was modeled with decane: a box built on a $5 \times 5 \times 5$ lattice of 125 dodecane molecules in the all trans configuration was used. After fixing the positions of the sulfur atoms, the micelle was minimized for 100 steepest-descent (SD) steps with harmonic constraints of 5 kcal/(mol/Å) applied to all atoms, followed by 100 adopted-basis Newton–Raphson (ABNR) steps with harmonic constraints of 1 kcal/(mol/A) applied to all atoms.	The average structure and dynamic properties of the micelle were determined. The hydrocarbon interior of the micelle was shown to be less fluid than in pure alkane (dodecane), as evidenced from a decrease in the dihedral transition rates and an increased free energy barrier to dihedral rotation of the aliphatic tails as compared to pure dodecane. The ratio of trans to gauche populations was approximately the same for the micelle and dodecane. This decrease in fluidity should be taken into account when micelles are used as model systems for lipid bilayers. The sulfate head groups of the surfactant molecules in the micelle were found to interact with water while the interior of the micelle is devoid of water. Further, the terminal methyl group of the hydrocarbon chain, in some cases, may even be located at the micelle surface and exposed to solvent.
FF methods such as MM2 and molecular orbital calculations such as AM1 were performed. In each case, energy minimized conformations were selected to compute the detailed hydrated structure of the molecule in general and the head group in particular, and their van der Waals surface.	It was possible to design the overall molecular shape of a new and effective surfactant with desired features. Van der Waals surface structures of some of these surfactants were produced in this study. The optimum shape and structural features were in the 1:1:1 ratio of the volumes of the hydrocarbon tail to the compact polar core to the polar hydrogen bonding chains. The surface areas as calculated from van der Waals distances were also in the ratio 1:1:1. This study also led to new theoretically designed molecules.

(*Continued*)

References	Surfactant Connection	Objective
Palmer, B. J. and J. Liu (1996) *Langmuir*, **12**, 746	A number of surfactant structures with a head group and four, six, or eight tail sites were modeled for simulating the aggregation process.	While previous authors have used a symmetric treatment of hydrocarbon–hydrocarbon and water–water interactions, these authors have focused upon the ordering of the relative energetics of these interactions.
Derecskei, B., A. Derecskei -Kovacs, and Z. A. Schelly (1999) *Langmuir*, **15**, 1981	The aggregation behavior of bis(2-ethylhexyl) sodium sulfosuccinate in carbon tetrachloride and water was investigated	The main purpose was to improve the predictability of the solution behavior of anionic sulfosuccinate surfactants in organic and aqueous media with respect to aggregation number and aggregate size, using atomic-level molecular modeling.
Goldsipe, A. and D. Blankschtein (2005) *Langmuir*, **21**, 9850	Counterion binding on the micelles of binary mixtures of ionic and nonionic (or zwitterionic) surfactants was developed from a molecular-thermodynamic theory.	Mixtures of surfactants are common, either from deliberate addition of one surfactant to another or as isomeric or structurally close impurities, in the main surfactant. The behavior in solution and at the interface will depend upon this.

Methodologies	Important Findings
Some hypothetical surfactant molecules composed of a head group attached to tail groups through simple harmonic springs have been selected and dissolved in a Lennard–Jones solvent, and a standard MM potential involving the stretching, bending, torsional, and Lennard–Jones nonbonded interactions used in modeling. Configurations of surfactant aggregates that resemble micelles, with 100 surfactant molecules of different chain lengths in 4000 solvent molecules, were generated.	These simulations generated micellar interactions that qualitatively reproduced the energy ordering of interactions expected in aqueous solutions. Larger aggregates are found to be formed as the tail length was increased in line with experimental facts. This study describes the variation in micelle size with surfactant concentration. These simulations can throw light on the nature of packing of surfactant molecules within the micelle.
The simulations were performed by using the extensible systematic FF (ESFF) and UFFs. The UFF was applied in the conformer search, and ESFF was used for geometry optimizations and energy evaluation. The geometries of seven representative conformers bis(2-ethylhexyl) sodium sulfosuccinate were delineated and their energies were computed. They matched with experimental results. The interactions of the most probable conformer with water and with carbon tetrachloride are reported. The atomic level interactions between water (0.5-nm-thick layer) and aerosol OT (AOT) were modeled by soaking the conformer in water simulated in a box representing the real intermolecular distance.	The geometries of seven representative conformers were analyzed based on random-sampling statistics applied to the most probable conformers and were found to differ by, at most, 10 kcal/mol. The mean aggregation number for the reverse micelles in CCl_4 (14.5) determined by a truncated cone-geometry model was in good agreement with the experimental mean aggregation value of 15–17. The predicted diameter of 2.8 nm for the dry reverse micelles is comparable with the experimental apparent hydrodynamic diameter of 3.2 nm for the same surfactant/water ratio.
The thermodynamic component of the theory models the equilibrium between the surfactant monomers, the counterions, and the mixed micelles. The molecular component of the theory models the various contributions to the free-energy change, which is called the free energy of mixed micellization. Minimization of the free energy of micellization provides optimal parameters for the degree of counterion binding, the micelle composition, and the micelle shape and size. They lead to the prediction of the CMC and the average micelle aggregation number.	Good agreement with experimental results was obtained. This sheds light on the relationship between the micelle composition, counterion binding and ion condensation, and the micelle shape transition.

(*Continued*)

References	Surfactant Connection	Objective
Jdar-Reyes, A. B. and F. A. M. Leermakers (2006) *J. Phys. Chem. B*, **110**, 6300	Structural, mechanical, and thermodynamic properties of linear micelles of some EO surfactants were studied at the concentration- dependent spherical-to-rod micellar transition.	Information on the shape, size distribution, and mechanical properties of micellar solution is important in optimizing their use.
Sterpone, F., G. Marchetti, C. Pierleoni, and M. Marchi (2006) *J. Phys. Chem. B*, **110**, 11504	The hydration dynamics at the micelle/water interface of two surfactants, $C_{12}E_6$ and lauryl dimethylamine oxide (LDAO), were studied.	Hydration of biological molecules plays an important role in controlling enzymatic activity, molecular recognition, protein folding, and DNA strand stabilization. Understanding how water dynamics are affected by large changes in the hydrophilic nature of the micellar interface can be enhanced by MD simulation of the micellar surface.
Burrows, H. D., M. J. Tapia, C. L. Silva, A. A. C. C. Pais, S. M. Fonseca, J. Pina, J. S. de Melo, Y. Wang, E. F. Marques, M. Knaapila, A. P. Monkman, V. M. Garamus, S. Pradhan, and U. Scherf (2007) *J. Phys. Chem. B*, **111**, 4401	Anionic polymeric surfactant, poly{1,4-phenylene-[9,9-bis(4-phenoxy-butylsulfonate)] fluorene-2,7-diyl}copolymer (PBS-PFP), and cationic gemini surfactants α,ω-$(C_mH_{2m+1}N^+(CH_3)_2)_2(CH_2)_s(Br^-)_2$ (m-s-m; $m = 12$, $s = 2, 3, 5, 6, 10,$ and 12) were investigated.	To investigate the fluorescence emission in relation to surfactant concentration and the micellization process. The interaction of charged points, as influenced by the distance between the charges, and the effect of distance between the adjacent charges in the polymer and the spacer length in the cationic gemini.

Methodologies	Important Findings
Previous modeling studies by coarse-grained MD simulations that yielded mechanical data such as the bending modulus (persistence length) were supplemented by analyzing the partition function methods from statistical thermodynamics using SCF method (SCF-A, SCF theory for adsorption and/or association). The shape of the cylindrical micelles was analyzed for small and large lengths.	SCF treatment provided structural and thermodynamic information on the micellar system. They also yielded micelle size distribution data. Small-length rods resemble a dumbbell, and the longer ones a cylinder with two slightly swollen end caps. Longer rods showed an oscillatory behavior in the grand potential of the micelle, the wavelength of the oscillation being proportional to the surfactant tail length. Finally, the grand potential converges with the endcap energy.
The aggregation number for the two micelles were obtained from a thermodynamics model. Micelles consisting of 45 $C_{12}E_6$ monomers, 8448 water molecules, 104 LDAO monomers, and 1629 water molecules were used in the simulations, assuming a truncated octahedral geometry for the simulation cells. The interactions between the constituents of the system were modeled by an all-atom FF method and a point-charge model was used for the water molecules. The rotational relaxation and translational diffusion parameters were also involved in these calculations. The dynamics of hydration were investigated by computing the survival probability, the rotational relaxation, and the translational diffusion of the water molecules in the first hydration shell.	The hydration dynamics of these two micelles depend on the physical nature of the micellar surface in contact with water. Measurements of the residence time of water near the micelle surface and its retardation with respect to the bulk showed that the thickness of the interface is greater, and the hydrophilic character stronger, for $C_{12}E_6$ than for LDAO. The simulations indicated that the dynamics of the surface water of $C_{12}E_6$ are one to two orders of magnitude slower than that of bulk water, compared to only 18% for the LDAO system. Further, the rotational landscape experienced by water at the micellar surface shows that in the $C_{12}E_6$ micelle, the water rotation occurs in a highly anisotropic space due to confinement of water at the interface and that in LDAO, the rotational landscape is in an isotropic space. The slowdown of interfacial water relaxation near complex micelles depends on the structural properties of the interface, such as the ratio between hydrophobic/hydrophilic exposed regions and on the interface thickness and topography. Such a treatment takes into consideration both the electrostatic interactions and spatial confinement of interfacial water.
MD simulations were performed using the standard GROMACS (a package for MD simulation) FF. Topology files were generated from PROgram DRuG (PRODRG, a program for generating molecular topologies) server. The polymer and gemini surfactants were added to a box and solvated with single-point charge (SPC) model water, with their structure constrained by the SETTLE algorithm (an analytical version of the SHAKE algorithm [referred to earlier] that satisfies bond geometry constraints during MD simulations). The MD trajectory with a total length of 5 ns was generated with a time step of 2 fs.	The natural aggregation of the polymer is broken by the cationic gemini, depending on the spacer length; this, in turn, affects the fluorescence emission of the polymer compared to the polymer, again showing a strong dependence on the spacer length for the same copolymer (the charge remains the same for a given monomer). MD simulations showed the delicate balance between the hydrophobic and coulombic forces, as understood from the surface tension, aggregation, and fluorescence studies, with the shortest spacer. For the longest spacer, coulombic interactions are dominant, the driving force being hydrophobic interactions. For the intermediate space lengths, optimum balance is observed between coulombic and hydrophobic interactions.

(*Continued*)

References	Surfactant Connection	Objective
Shinoda, W., R. DeVane, and M. L. Klein (2008) *Soft Matter*, **4**, 2454	The interfacial properties of the nonionic EO surfactant, $HO(CH_2CH_2O)_5CH(CH_2O(CH_2)_{11}CH_3)_2)$, were studied in bulk aqueous solution, as well as at the air–water and oil (decane)–water interfaces.	The budding and fission of micelles (micellar dynamics) from surfactant monolayers and their repartitioning in the oil–water interface are difficult to monitor experimentally but can be simulated very effectively through molecular modeling.
Stephenson, B. C., K. A. Stafford, K. J. Beers, and D. Blankschtein (2008) *J. Phys. Chem. B*, **112**, 1641	Micellization of molecules such as SDS, octyl glucoside (OG), *n*-decyl dimethylphosphine oxide ($C_{10}PO$), *n*-decyl methyl sulfoxide ($C_{10}SO$), octylsulfinyl ethanol (C_8SE), decylsulfinyl ethanol ($C_{10}SE$), *n*-decyl methyl sulfoxide ($C_{10}SO$), and *n*-octyl methyl sulfoxide (C_8SO) was studied.	The computer simulation–free energy/molecular thermodynamics (CS-FE/MT) model for predicting micellization free energy was extended to evaluate the free-energy change associated with changing micelle composition.

Methodologies	Important Findings
The self-organization process of surfactants into micelles, and hexagonal or lamellar bulk structures, were simulated. The Coarse-grained MD simulation effectively characterizes the surfactant self-assembly process. The relative merits of the current methods for modeling surfactants in solution and at interfaces were discussed. The AA (all atomic)-MD simulation method is coupled with experimental data. Thermodynamic properties, such as surface and interfacial tension, density, and free energy for hydration and material transfer were evaluated. The segmental distribution functions from AA-MD simulations were taken into consideration by selecting a proper functional form and transferability of the CG potential.	The experimental surface pressure–area (π–A) curve for the surfactant monolayer at the air–water interface was reproduced in these simulations. Snapshots of the self-organization process were obtained to throw more light into the molecular dynamics. The study generated well-matching numbers for the surface tension of $C_{12}EO_{12}$ and an area per surfactant molecule of 72 ± 3 Å^2, compared to the experimentally determined numbers from neutron reflection at the CMC. (Neutron reflection is one of the most accurate experiments to determine the molecular area.) The surface tension was calculated as 39 mN/m, against an experimental value of 38.5 mN/m, at the CMC.
The change in the free-energy change, $\Delta\Delta G$, between the organization of a single surfactant in solution and the transformation of the micelle into a solubilizate was modeled using a hypothetical free-energy cycle reminiscent of the Born–Haber cycle, referred to as an alchemical free-energy cycle. Specifically, two methodologies, called single and dual topologies, and their combinations, were applied to morph the surfactant into the solubilizate or the cosurfactant. A coupling parameter, λ, defined the extent of the alchemical transformation. In the CS-FE/MT model, experimental CMC data or the traditional MT model was used to evaluate the free energy associated with the single surfactant micelle formation ($G_{\text{form, single}}$). An iterative approach combined the estimated value of $G_{\text{form,single,}}$ with that obtained from computer simulation for the exchange of a surfactant of one type with a cosurfactant or solubilizate of another to calculate the micelle aggregation number, the micelle bulk solution composition, and the free energy of mixed micelle formation.	Alchemical free energy methods provided very reasonable values for the free energy of micellization as a function of micelle composition. Improved computer power can predict the free-energy changes in multicomponent surfactant/cosurfactant/ solubilizate systems with great structural variations.

(*Continued*)

References	Surfactant Connection	Objective
Leclercq, L., V. Nardello-Rataj, M. Turmine, N. Azaroual and J. Aubry (2010) *Langmuir,* **26**, 1716	Aggregation of dimethyl-di-n-octylammonium chloride in aqueous solution was studied by fusing experimental and computer-aided methods.	Improvement of the predictability of surfactant behavior in solution.
Stephenson, B. C., A. Goldsipe, and D. Blankschtein (2008) *J. Phys. Chem. B*, **112**, 2357	The self-assembly process of the triterpenoid phytochemicals, asiatic acid (AA), and madecassic acid (MA), with anticancer and surface activities and structurally similar to bile salts, was investigated in aqueous solution.	Prediction of micellar properties including the shape, size, composition, and microstructure of simple surfactants are possible from the molecular-thermodynamic modeling and MD simulation methods. But these methods do not work well for structurally complex surface-active substances such as AA and MD.
Lazaridis, T., B. Mallik, and Y. Chen (2005) *J. Phys. Chem. B*, **109**, 15098	Micelle formation (CMC and aggregation number) of DPC were studied.	Computer simulations provide a means to test the assumptions of the theoretical models on the surfactant aggregation process. The micellization process studied by theoretical methods based on phase separation and mass action models can be confirmed by the computer simulation of the process.

Methodologies	Important Findings
Techniques employed were zetametry, conductimetry, dimethyl-di-n-octylammonium and chloride-selective electrodes, tensiometry, NMR spectroscopy (^{1}H and diffusion-ordered spectroscopy [DOSY]), and molecular modeling (PM3 and molecular dynamics).	By integrating the experimental and theoretical data, the aggregation process was modeled as a function of surfactant concentration. Specifically, dimers are formed in the range of 0.2–10 mM, bilayers in 10–30 mM, and vesicles in >30 mM.
The MDs simulation was performed in atomistic details: advanced FF parameters were integrated with MD simulation by a modified computer simulation/molecular-thermodynamic model (MCS-MT model).	Provides insights into the self-assembly behavior of structurally complex, unconventional surfactants in aqueous media, giving more acceptable values for their CMCs and aggregation numbers than obtained from a simple atomistic-level MD simulation method. The principal moments of the micelle radius of gyration tensor, the one-dimensional growth exhibited by AA and MA micelles as the aggregation number increases, the level of internal ordering within AA and MA micelles (quantified using two different orientational order parameters), the local environment of atoms within AA and MA in the micellar environment, and the total, hydrophilic, and hydrophobic solvent accessible surface areas of the AA and MA micelles were evaluated.
The effective energy function 1 (EEF1) solvation model for proteins was coupled with the implicit solvent model through MD simulation by using the CHARMM program and the CHARMM27 all-hydrogen lipid parameters. Snapshots of the dynamics of the micellization process revealed the aggregation phenomena—CMC and aggregation number for various concentrations—using 960 DPC molecules in the simulations.	At 20 and 100 mM DPC concentrations, the aggregation numbers were 53–56 and 90, respectively. The CMC at 20 mM DPC was 1.25 mM. These numbers obtained by MD simulation agreed well with the experimental values. The effective energy per surfactant molecule decreased initially as a function of aggregation number, which stabilized at 60 with increasing concentration. van der Waals contribution and desolvation from nonpolar groups helped micellization, while the desolvation from polar groups opposed it. The study provided translational and rotational entropy terms for free energy as 7 kcal/mol per monomer.

(*Continued*)

References	Surfactant Connection	Objective
Liquid–Liquid Interfaces		
Zhang, Y., S. E. Feller, B. R. Brooks, and R. W. Pastor (1995) *J. Chem. Phys.*, **103**, 10252	The behavior of the octane–water interface as such and after adsorbing dipalmitoylphosphatidylcholine (DPPC) lipid bilayer and monolayer was studied.	It is hard to obtain exact information on the behavior of liquid–liquid interfaces. Modeling strategy will enhance the predictability.
Laradji, M. and O. G. Mouritsen (2000) *J. Chem. Phys.*, **112**, 8621	The elastic properties of liquid–liquid interfaces were elucidated in the presence of generalized surfactant particles.	Improvement of understanding and predictability of surfactant adsorption at liquid–liquid interfaces by molecular modeling simulation.

Methodologies	Important Findings
Under constant particle conditions, this method simulates liquid–liquid interfaces using five adiabatic ensembles that include the microcanonical NVE (constant particle number, volume and energy) ensemble, and various algorithms for combinations of constant pressure, volume, surface area, and surface tension conditions. The MD simulations were performed for three interfacial systems at 293 K with 62 octanes/560 waters, 560 waters/vacuum, and 62 octanes/vacuum using CHARMM, and water was described with modified TIP3P parameters. In these simulations, octane was placed at the center of the box with dimensions of 25.6 Å in *x*- and *y*-directions and 51.1548 Å in *z* having a volume corresponding to the volume of 560 water and 62 octane molecules at bulk density, and the system was then equilibrated by adopting 400 steps of adapted-based Newton–Raphson energy minimization procedure prior to further sequences.	The simulated surface tension agreed well with experiments: for example, the octane–water interface had a surface tension of 61.5 ± 1.9 dyn/cm against an experimental value of 61.5 dyn/cm.
The modeling is based on the Lennard–Jones potential and elastic constants of the interface corresponding to the interfacial tension and the mean bending modulus. The theory also treated the tracing of the interface by a detailed consideration of the interaction potentials, real space configurations and concentration profiles, and computation of the structure factor and the elastic constants.	The interfacial tension decreased with increasing surfactant interfacial coverage and/or surfactant chain length. The bending rigidity was found to decrease with increasing surfactant interfacial coverage for small surfactant interfacial coverages but to increase as the surfactant interfacial coverage is further increased. By using a Gaussian theory of interfaces involving surfactants, it was found that the initial decrease of the bending rigidity is due to coupling of fluctuations of the surfactant orientation field to those in the interfacial height.

(Continued)

References	Surfactant Connection	Objective
Gupta, A., A. Chauhan, and D. I. Kopelevich (2008) *J. Chem. Phys.*, **128**, 234709	Mass transport in microemulsions (nonionic surfactant-covered oil–water interfaces) were investigated.	Hexadecane–water interfaces covered by monolayers of nonionic surfactants of various lengths were investigated by MD simulation methods to increase the predictability of transport phenomena across microemulsion interfaces.
Liquid–Solid Interfaces		
Aliaga, W. and P. Somasundaran (1987) *Langmuir*, **3**, 1103	Molecular orbital (MO) energies for the highest occupied molecular orbital (HOMO) and lowest unoccupied molecular orbital (LUMO) of some hydroxy oximes, members of an important class of surface-active flotation collectors, were calculated: salicylaldoxime (2) o-hydroxyacetophenone oxime; o-hydroxybutyrophenone oxime; o-hydroxybenzophenone oxime, syn isomer; o-hydroxybenzophenone oxime, anti isomer; (6) 2-hydroxy-naphthaldoxime; 2-hydroxy-5-methoxyacetophenone oxime; salicylaldazone; salicylaldehyde	Design of efficient flotation collectors

Methodologies	Important Findings
The monolayer microstructure, dynamics, and a free energy barrier to the solute transport are modeled in these simulations. Coarse-grained molecular dynamic (CGMD) simulations for flat hexadecane–water interfaces were performed to reveal the monolayer microstructure and dynamics along with the determination of the structure, size, and lifetime of pores (connected volumes inside the monolayer accessible to a spherical probe). The CGMD model (*HmTn*) approximates groups of atoms, such as several methyl or ethoxy groups, as a single united atom (bead) that simplifies the MD simulations. Coarse-grained beads were denoted as *T* for hydrophobic tail beads and *H* for hydrophilic beads. The simulation was performed for surfactant-free and surfactant-rich systems in cubic unit cells of different dimensions, depending on the value of *n*. Simulation of cell sizes for systems for various H_nT_n values were about $12 \times 12 \times 12$ nm^3. Time steps were selected to match the timescales of dynamic processes.	Interfacial tension and interfacial coverage were modeled. The interaction between two nonbonded beads was modeled from the Lennard–Jones potentials. The free energy barrier to mass transfer across the oil-water interface is related to the steric repulsion that depends on the local monolayer density, which is larger for longer surfactants. Of the pore types—voids, oil and water half-channels, and channels—there was a substantial difference in the properties of the oil and water half-channels due to the difference in the size of the oil and water molecules, which causes asymmetry of the density in the tail- and headgroup regions of the monolayer. The pore averaging (at a $t_{av} = 20$ ps) predicts that most channels connecting the bulk oil and water phases are very unstable. The solute transport mechanism proceeds by hops between smaller voids and half-channels.
UV–vis spectra, HPLC data, and the determination of proton–ligand formation constants from spectrophotometric titrations. The individual atomic electron density in conjunction with bond parameters led to HOMO and LUMO energies. The electron densities on the collector molecules were determined. Extended Hückel molecular orbital (EHMO) calculations were done with an IBM 360 with QCPE No. 344 (Quantum Chemistry Program Exchange).	MO energies were computed and correlated with the absorption energies that showed a connection with flotability. The electron density on the nitrogen atom had a strong dependence on the HOMO–LUMO gap and the flotation efficiency. The hydrophobicity and complex forming ability of these compounds through the phenolic and oxime groups ran parallel with electron densities.

(*Continued*)

References	Surfactant Connection	Objective
Herbreteau, B., C, Graff, F. Voisin, M. Lafosse, and L. Morin-Allory (1999) *Chromatographia*, **50**, 490	Perhydrogenated and perfluorinated EO surfactants were studied by chromatography and molecular modeling.	Understanding the mechanism of reverse-phase liquid chromatography (RPLC) helps in designing new column materials that improve the separation process.
Jo′dar-Reyes, A. B., J. L. Ortega-Vinuesa, A. Martı′n-Rodrı′guez, and F. A. M. Leermakers (2002) *Langmuir*, **18**, 8706	Adsorption of nonionic poly-(oxyethylene) *p*-*t*-octylphenol surfactants with the same hydrophobic chain and differing headgroup lengths on hydrophobic surfaces such as polystyrene-polystyrene-sulfonate latex beads were modeled.	Many industrial formulations use similar systems. The predictability improves the efficacy of application.
Pradip, B. Rai, T. K. Rao, S. Krishnamurthy, R. Vetrivel, J. Mielczarski, and J. M. Cases (2002) *Langmuir*, **18**, 932	The interaction of diphosphonic acid-based surfactants (alkylimino-bis-methylenediphosphonic acid [IMPA-8] and 1-hydroxy-alkylidene-1,1-diphosphonic acid [Flotol-8]) with three calcium minerals, fluorite, calcite, and fluorapatite, was studied.	The theoretical basis for choosing reagents in mineral engineering processes, such as flotation, must be rationalized for the efficient design of these processes to enhance the understanding of the molecular-recognition-based selective chemistry.

Methodologies	Important Findings
The amphiphilic behavior of the sulfur analogues of these surfactants was studied by conformation analysis in the solvent using an internal coordinate Metropolis–Monte Carlo approach, the MM2 FF, and by a continuum solvation model for water. About 6000 conformations were considered, from which structures with an energy within 10 kJ/mol of the minimum were selected.	Generic predictive QSAR relations were produced from the parameters generated. The simulations generated minimum, maximum, and mean values of molecular electrostatic potential and molecular lipophilic potential on these surfaces. Comparison between the chromatographic behavior was evaluated by comparing the correlated retention data with the calculated values.
Self-consistent field theory (SCF-A) for adsorption and association. Large numbers of conformations of a surfactant chain in solution and at interfaces require the application of statistical thermodynamic treatment approximations. The number of chains of each conformation in the equilibrium was treated by maximizing the partition function for the polymer chains. Both linear chain and branched chains were considered.	The limitations in the assumptions of the SCF model, in which homogeneous adsorbed layers were considered, fail to reproduce experimental findings. The inclusion of lateral inhomogeneities was suggested as a further improvement of the SCF model for adsorption and micellization to match the theory with the experimental findings.
Both FF (UFF) and semiempirical quantum mechanical (MNDO) methods were used in modeling the interactions. The mineral–reagent interaction energy for the adsorption of these surfactants at the (111), (110), and (100) planes of these minerals were modeled. The presence of the aqueous environment has been simulated in the FF calculations by employing an empirical dimensionless scaling factor to mediate or dampen the long-range electrostatic interactions. This scaling factor approximates the effective dielectric permittivity of the medium.	The order of the flotability of different minerals by a reagent can be accurately predicted without resorting to any experimentally determined parameters but through simple input from the structural features of the reagents and the interacting surfaces. Thus, the computer can screen many existing and novel reagents and can predict their efficiency.

(*Continued*)

References	Surfactant Connection	Objective
Cooper, T. G. and N. H. de Leeuw (2004) *Langmuir*, **20**, 3984	Adsorption of surfactant collectors like carboxylic acids, alkyl hydroxamates, hydroxyaldehydes, and alkylamines on the calcium mineral Scheelite was studied.	Selective hydrophobization of the mineral by surfactant collectors eases the removal of undesired gangue materials from the main mineral. This is especially so for calcium minerals that occur as a mixture of $CaCO_3$, CaF_2, and $CaWO_3$. Computer modeling techniques help in elucidating the modes of adsorption, the strength of interactions, and the mechanisms of adsorption of surfactants on minerals like Scheelite, [$CaWO_4$]. Design and choice of reagents are made easy through modeling.
Smith, L. A., G. B. Thomson, K. J. Roberts, D. Machin, and G. McLeod (2005) *Crystal Growth & Design*, **5**, 2164	The crystallographic properties of surfactants such as SDS and rubidium dodecyl sulfate (RDS) were computed.	Many industrial and household formulations contain high surfactant concentrations. Design of surfactant-based products from crystal structure data can help the surfactant function by manipulating the quantity, size distribution, and shape of crystallites.

Methodologies	Important Findings
The Born model of solids assumes that ionic interactions (in solids) are through long range and short range forces and the electronic polarizability of the oxygen ions in the mineral and the water molecules. Results from the previous combined DFT and interatomic potential studies showed the potential quantitative comparison of the adsorption behavior of these different collectors on Scheelite surfaces. The hydrated mineral surface was modeled with one monolayer of water molecules. Further addition of water molecules was done on energetically favored locations on the adsorbed surface after they had been identified. Supercells of 105.3–194.6 Å^2 size were selected as surface simulation cells to avoid computational artifacts at the periodic boundary conditions parallel to the surface.	The geometric and structural parameters for the unit cell dimension and for the solid were computed and found to be in good agreement with experimental data. For example, Ca–O distances at 2.43 and 2.46 Å and W–O bond lengths at 1.69 Å as well as the O–W–O angles at 106.6° and 115.4° all agreed well with experimental values. The adsorption and solvent exchange energy were predicted and were found to agree with experiments. Results obtained by the replacement of preadsorbed water by methanoic acid, hydroxymethanamide, hydroxyethanal molecules, and methylamine showed that the solvent exchange was exothermic.
The known molecular FF parameters for these surfactant crystallites were optimized and refined and validated against the known bond lengths, unit cell parameters, and lattice energies. The total lattice energy of the crystal was divided into the attachment energy, E_{att}, the energy released on the addition of a growth layer (slice) of thickness d_{hkl} to the surface (*hkl*), and (E_{sl}), the energy released on the formation of this slice. Attachment energy is related to the growth rate of the crystal, the faces with the lowest attachment energy having the slowest growth rate and being morphologically important. E_{att} and E_{sl} are computed by the lattice summation of the partition of the intermolecular bonds. The partial charges on the atoms in SDS and RDS were determined using the modified neglect of differential overlap (MNDO) approximation. The enthalpies of sublimation and the intermolecular bonding parameters were used in describing these systems.	The lattice energies computed with these systems showed that the experimental sublimation enthalpies are in good agreement with the calculated values (–173.13, 145.50 kcal/mol; –176.40, 155.76 kcal/mol for SDS and RDS, respectively). Simulation studies revealed a platelike morphology for both of these materials.

GENERAL REFERENCES

Ball, D. W. *Physical Chemsitry*, Thomson (Brooks/Cole), California, 2003.

Leach, A. R. *Molecular Modeling (Principles and Applications)*, 2nd ed., Prentice Hall, England, 2001.

Sahni, V. *Quantal Density Functional Theory*, (Vols. 1 and 2), Springer, New York, 2004 & 2009.

Szabo, A. and N. S. Ostlund *Modern Quantum Chemistry (Introduction to Advanced Electronic Structure Theory)*, Dover Publications, New York, 1996.

PROBLEMS

15.1 How do the molecular modeling methods complement/supplement the experimental investigations on surfactant and similar systems? Exemplify.

15.2 How does the quantum mechanical method for molecular modeling differ from the molecular mechanics methodology?

15.3 *STO–3G* is a well-known quantum mechanics-based calculation used in molecular modeling investigations: Explain the significance of each alphabet and numeral in this contraction and check against your understanding of the theory.

15.4 How is the Hartree–Fock formalism related to the density functional theory?

15.5 How are the molecular dynamic and Monte Carlo methods related to the energy minimization procedure adopted in molecular modeling studies?

15.6 Read any two papers from Section VII (other than the ones elaborated in Section VI) or from the current literature on molecular modeling studies of surfactant systems and prepare a detailed account of their content, carefully linking the theory applied in those papers with the materials you picked up from this chapter.

Answers to Selected Problems

CHAPTER 1

1.1 $RCOO^{-}Me^{+}$; $ROSO_2O^{-}Me^{+}$

1.2 $RCOOCH_2CHOHCH_2OH$

1.3 $RN(CH_3)_3^{+}X$

1.4 $C_{12}H_{25}C_6H_5SO_3^{-}Na^{+}$

1.5 $H(OC_2H_4)_x[OCH(CH_3)CH_2]_y(OC_2H_4)_zOH$

1.6 $RN^{+}(CH_3)_2(CH_2)_xSO_3^{-}$

1.7 $H(OC_2H_4)_x[OCH(CH_3)CH_2]_y(OC_2H_4)_zOH$

1.8 $RCOOCH_2CH_2SO_3^{-}Na^{+}$

1.10 $RN^{+}(CH_2C_6H_5)(CH_3)_2Cl^{-}$

1.13 (ii) With increase in EO groups, water solubility increases; workable at low temperatures; effect of salinity is minimal. The alkyl chains will also modify the properties of EO surfactant for a given EO length as in (i). (iii) With increasing fluorination, the ionization of carboxylates increases and hence less affected by acids and bases or hard water; low surface tension values that increases with extent fluorination. Fluorinated surfactants are surface active in organic solvents also; they have better chemical and thermal stability than hydrocarbon analogs; adsorption on hydrophilic and hydrophobic surfaces increase.

1.15 Synthesis of linear alkane benzene sulfonates from the abundant inexpensive hydrocarbon, benzene, is achieved by two well developed synthetic methods in organic chemistry—Friedel–Crafts and sulfonation reactions—to yield a reasonably pure and effective anionic surfactant.

Surfactants and Interfacial Phenomena, Fourth Edition. Milton J. Rosen and Joy T. Kunjappu.

CHAPTER 2

2.1 **(a)** 4.54×10^{-10}mol cm^{-2}
(b) 36.6 $Å^2$
(c) –47.4 $kJmol^{-1}$

2.2 **d < c < a < b < e**

2.4 5.8 Å

2.5 60. $Å^2$

2.6 $X_1 = 0.40$

2.7 **(a)** 1.1×10^{-5} M.
(b) Soak glass beaker in cationic surfactant solution overnight. Replace solution with fresh one.

2.8 A number very close to $\Gamma_m = 4.4 \times 10^{-10}$ mol/cm^2.

CHAPTER 3

3.1 Cylindrical; $V_H/l_c.a_o = 0.44$

3.4 –29.5 $kJmol^{-1}$

3.7 **e > c > a ≅ d > b**

CHAPTER 4

4.4 Micellar catalysis of acid hydrolysis of the ester linkage may occur.

CHAPTER 5

5.2 **a, d, e** cause increase; **b, c** cause decrease.

5.4 **(a)** 34.6 mN/m
(b) 31.3 mN/m

5.5 ~1.3 layers

CHAPTER 6

6.1 **(a)** Weak interaction; $\phi = 0.62$
(b) 7 dyne/cm

6.2 Since $\gamma_C = \gamma_{LA}$ of the wetting liquid and $\cos\theta = 1$, then $\gamma_{SL} = 0$ (Equation 6.3).

6.3 Increase of 17.6 ergs/cm^2

6.4 On cellulosic substrates, the POE nonionics may initially be adsorbed via hydrogen bonding of the POE chain to the surface, making the latter less wettable momentarily.

6.5 **c > b > e > a > d**

6.6 **(a)** Hydrocarbon-chain surfactants and hydrocarbon substrates (e.g., paraffin, polyethylene)

(b) Siloxane-chain surfactants and hydrocarbon substrates

CHAPTER 7

7.1 See Equation 7.3. Γ and C_s reach maximum values near the CMC.

7.5 1.6×10^{-4} s

7.7 **(a)** POE nonionic

(b) Cloud point

7.8 For antifoaming efficiency: Low-foaming situations cannot be always obtained by manipulating the surfactant structure alone. Antifoaming agents then supplement the system to prevent further growth of the foam. They act (i) by removing the surfactant film from the bubble surface. Silica particles dispersed in silicon oil desorbs the surfactant from the bubble interface; (ii) by replacing the surfactant film by another type of film. Tertiary acetylenic glycols is effective in this respect; (iii) by promoting drainage in the foam lamellae–tributyl phosphate and tertiary alkyl ammonium ions (though not typical surfactants), which reduces surface viscosity by interacting with surface films.

CHAPTER 8

8.5 61% $C_{12}H_{25}(OC_2H_4)_2$; 39% $C_{12}H_{25}(OC_2H_4)_8OH$, using the relationship 20 $M_H/(M_H + M_L)$

8.6 See Equations 8.10 and 8.11

8.7 **(a)** Temperature change

(b) Electrolyte addition to the solution

CHAPTER 9

9.1 Increase the dielectric constant of the dispersing liquid; increase the surface potential of the solid; increase the thickness of the electrical double layer.

9.2 **(a)** Reduction of the effective Hamaker constant

(b) Production of steric barrier.

9.3 It produces a greater reduction of $1/\kappa$.

9.4 Answer: B is correct. **A** is too soluble in heptane to adsorb efficiently; $C_{12}H_{25}N^{+}(CH_3)_3$ would not adsorb; the ligninsulfonate group would not produce an effective steric barrier.

9.5 It is a positively charged solid, probably with polyvalent cation sites, as evidenced by precipitation with POE nonionic.

9.6 **(a)** and **(b)** flocculation

(c) no effect

(d) flocculation, followed by redispersion.

9.8 The typical values of Hamaker constant are in the range of 10^{-18} to 10^{-20} J. The magnitude of Hamaker constant indicates the extent of van der Waals contribution toward the stability of a lyophobic colloid as depicted in the DLVO theory, the other factor not represented by the Hamaker constant being the repulsion between double layers that is affected considerably by the ionic strength.

CHAPTER 10

10.1 Substrate and soil may both acquire a positive charge that prevents adsorption of the cationic onto them via its hydrophilic head.

10.2 Neutralization of negative charge of textile surface by cationic surfactant enhances adsorption onto it of anionic surfactant.

10.3 **(a)** Spreading coefficient will decrease because γ_{SB} is increased.

(b) **1** Cationic surfactant on negatively charged textile surface

2 POE nonionic on hydroxylated (e.g., cellulose) surface

10.4 Most of the natural surfaces are negative and a cationic surfactant interacts with negative proteinaceous materials electrostatically. Removal of intersegmental repulsion on the backbone of the biopolymer modifies the conformation of the protein, and it loses its biological activity by a phenomenon that is called denaturing. Hydrophobic interactions between the surfactant tail and the hydrophobic regions of the protein are also possible.

10.8 Builders are an additive in detergent formulations that annuls the deleterious effect of polyvalent hardness ions. Polyphosphates, silicates, and organic polymers function as builders, each one having a different mechanistic role in enhancing the detergent property. The major paths of their action are sequestration, precipitation, ion-exchange, dispersion, and pH stabilization.

CHAPTER 11

11.1 **(a)** From their pC_{20} values, $|\ln C_A/C_B = 1.39$. Thus, synergism of this type should exist in the mixture.

(b) $\alpha^* = 0.25$ (for surfactant A); $C_{12,\min} = 2.1 \times 10^{-4}$ molar (for $\pi = 20$ dyn/cm).

11.2 **(a)** $\beta^M = +0.7$

(b) No. $\left|\ln C_1^M / C_2^M\right| = 1.13$

11.3 $\beta^\sigma = +1.4$; $\beta^\sigma - \beta^M$ consequently $= +0.7$. Since $\left|\ln C_1^{0,\mathrm{CMC}} / C_2^{0,\mathrm{CMC}} \cdot C_2^M / C_1^M\right| = 0.35$, the system will exhibit antagonism in surface tension reduction effectiveness.

11.4 **(b)** = least negative β^σ value; **(d)**; **(a)**; **(c)** = most negative β^σ value.

CHAPTER 12

12.1 Increased distortion of solvent by double (or triple) the number of hydrophobic groups in the molecule

12.2 Greater difficulty of accommodating two (or three) hydrophobic groups in the interior of a spherical or cylindrical micelle than at a planar surface

12.3 Larger molecular size of the gemini, hence a smaller diffusion rate.

12.4 Closer packing of the hydrophobic groups of the gemini at the relevant interface

12.5 Geminis can adsorb with one hydrophilic group facing the solid and the second facing the aqueous phase, producing a hydrophilic surface. The first layer of a conventional surfactant adsorbs to produce a hydrophobic surface.

12.6 As the length of the hydrophobic methylene spacer groups increase, methylene linkages loop into the micellar interior (for > 10 CH_2 groups) and the C_{20}, CMC, and the head group area decrease. On the other hand, if the spacer group is rigid, the head group area increases and an increase in these parameters will be observed. In the case of hydrophilic spacers only a monotonic increase in these properties is observed as the number of hydrophilic spacer linkage increases.

CHAPTER 13

13.1 Cell membranes are made up of amphiphilic lipid molecules such as glycerophospholipids and sphingolipids that form bilayered structures. Various subcellular organelles such as nucleus, mitochondria, and chloroplasts may be considered as materials solubilized within the surfactant assemblies mixed with other substances like proteins.

13.4 The electrostatic interactions between the charged groups in the protein and the ionic surfactant affect the repulsions in certain segments of the protein, and thus its shape. This accelerates the denaturization of proteins of the cell membrane, a process preceding cell lysis.

13.5 The 3-D structures of proteins are generally studied by X-ray diffraction that requires crystalline proteins. Protein–surfactant complex helps the surfactant–protein complex remain in the solubilized state, while making reasonable contact with the crystal.

CHAPTER 14

14.1 *Nanotechnology* is the study, design, characterization, production, and application of structures, devices, and systems by controlling shape and size at the nanometer level. millimeter = human hair; micrometer = bacterium; nanometer = DNA helix; picometer = wavelength of X-rays; and femtometer = size of a proton.

14.3 In DPN, a surfactant like 16-mercaptohexadecanoic acid is adsorbed on an atomic force microscopy tip for writing on a substrate like gold foil to create nanopatterns. Nonionic EO surfactants enhance the wettability of partially hydrophobic substrate surfaces.

14.7 Surfactants primarily control the stability of nanodispersions of particles aside from altering the surface properties of nanoparticles by adsorbing onto them that causes changes in their electrical, hydrophobic, and related properties.

CHAPTER 15

15.1 *In silico* computer-based molecular modeling methods have the advantage in many cases, where the system is not readily accessible to experimentation. The dynamics of aggregation and adsorption phenomena that may not be discernible even by fast reaction kinetics techniques can be predicted with a reasonable level of success with molecular modeling strategies. These methods often help bypass unfruitful synthetic effort in the development of new surfactants with predictable structure–activity relationship.

15.4 Hartree–Fock description uses assumptions so that quantum mechanical methods for a *many-electron* system atom or molecules can be simplified into a single electron problem, by treating the electron–electron repulsion in an average way. Density functional theory, which is similar to *Hartree–Fock* theory in that it treats single electron functions, determines the total electronic energy and the overall electron density distribution. In DFT, the potential is local and in HF theory it is nonlocal; also, the direct relationship between the total electron energy and the overall electron density in an atom or molecule substitutes many-electron wavefunctions with electron densities.

15.5 The final aim of the modeling methods is to find out the molecular state with minimum energy from which the system properties are derived. Molecular dynamics and Monte Carlo methods determine the energy states by different means—molecular dynamics from the velocity and position of the particles after collisions and Monte Carlo method from the configuration by random sampling.

INDEX

Surfactants and Interfacial Phenomena, Fourth Edition. Milton J. Rosen and Joy T. Kunjappu.